LABORATORY OUTLINES IN BIOLOGY VI

LABORATORY OUTLINES IN BIOLOGY VI

Peter Abramoff
Robert G. Thomson
Marquette University

W. H. Freeman and Company
New York

Library of Congress Cataloging-in-Publication Data

Abramoff, Peter, 1927–
Laboratory outlines in biology—VI / Peter Abramoff, Robert G.
Thomson. — [6th ed.]
p. cm.
Includes bibliographical references.
ISBN 0-7167-2633-5 (soft cover)
1. Biology—Laboratory manuals. 2. Biology—Experiments.
I. Thomson, Robert G. II. Title.
QH317.A277 1994
574′.078—dc20 94-32653
CIP

Printed in the United States of America

1 2 3 4 5 6 7 8 9 VB 9 9 8 7 6 5 4

Contents

Preface

Tell me and I will forget.
Show me and I might remember.
Involve me and I will understand.

Chinese proverb

These words still convey our philosophy that laboratory studies are paramount in developing students' ability to think critically and to increase their appreciation of the methods by which biologists obtain and analyze information. In laboratory studies, students become personally and intensely involved in the knowledge they acquire.

We have, in our sixth edition of *Laboratory Outlines in Biology*, retained many of the popular laboratory studies and features of previous editions. In addition, we have organized this manual into units, each unit consisting of several exercises contributing to a common theme. Each unit is introduced by a brief summary of its contents. New to this edition, at the request of numerous users, are studies in human genetics, physiology and histology of the nervous system, and a section on environmental biology.

As in previous editions, all laboratory exercises have been thoroughly and critically reviewed. With few exceptions, each exercise has been edited, revised, and rewritten to reflect the excellent critiques and comments of selected reviewers and the many others who have used our manuals over the years. All new material reflects contemporary information found in current editions of the most widely used introductory textbooks. The taxonomic studies follow the Whittaker five-kingdom system of classification, with modifications reflecting contemporary schemes.

All flow diagrams, figures, tables, charts, and artwork have been carefully reviewed, examined, and updated for clarity and labeling. We feel confident that these changes will provide the student and the instructor with an exceptionally usable laboratory manual designed to (1) introduce the complexity and diversity of living organisms and their relationships; (2) provide a solid foundation for those students who elect to study science; and (3) convey something of the meaning, scope, and excitement of biology.

We have made every effort to keep expenditures for supplies and equipment reasonable and within the budgets of most colleges and universities. The accompanying *Instructor's Handbook* has also been revised and updated and is designed to save time for the instructor and his laboratory assistants. The handbook contains detailed lists of materials and equipment, sources of supplies, instructions for the preparation of reagents, and suggestions for the culture and maintenance of the various living organisms used in these studies.

Our special thanks go to the following reviewers whose detailed and thoughtful comments proved especially helpful: Claire Oswald, College of St. Mary; and Edith Robbins, Borough of Manhattan Community College.

We also appreciate comments offered to us by the numerous users of our manual, both students and faculty. In our desire to improve this sixth edition, we invite and welcome all comments from those who use it.

October 1994

Peter Abramoff
Robert G. Thomson

LABORATORY OUTLINES IN BIOLOGY VI

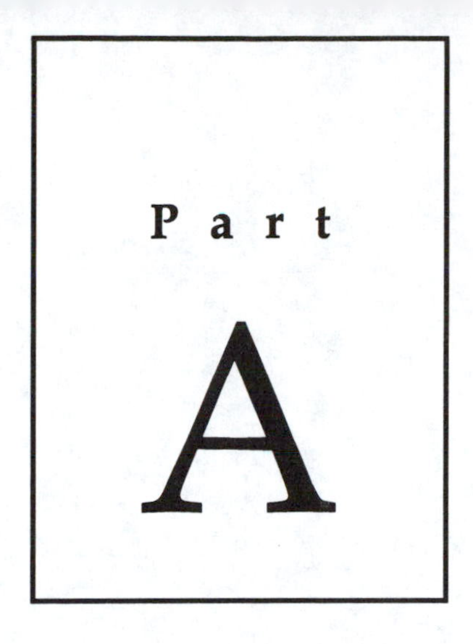

The Cell

The cell theory, which affirms that the cell is the basic structural, functional, and developmental unit of life, has long been recognized as one of the fundamental truths of biology. As Carl Swanson, a noted cytologist, has said, ". . . we understand life itself only to the extent that we understand the structure and function of cells." In 1665 the Englishman Robert Hooke described the microscopic structure of thin slices of cork and used the word "cell" to describe the tiny structures he viewed. Here is how Hooke described what he saw:

> "I took a good clear piece of Cork, and with a Pen-knife sharpened as keen as a Razor, I cut . . . an exceeding thin piece of it, and placing it on a black object Plate, because it was itself a white body, and casting the light on it with a deep plano-convex Glass, I could exceeding plainly perceive it to be all perforated and porous, much like a Honey-comb, but that the pores of it were not regular; yet it was not unlike a Honey-comb in these particulars. First, in that it had a very little solid substance, in comparison of the empty cavity that was contained between. Next, in that these pores, or cells, were not very deep, but consisted of a great many little Boxes, separated out of one continued long pore."

However, it was not until the early 1800s that it became commonly accepted that all plants and animals were composed of cells. This cell theory, which was first set forth by the German biologists Matthias Schleiden and Theodor Schwann, has served as a "base line" for all subsequent investigations into the nature of the cell.

Just as life is diverse in form, so are the forms and functions of the various cells that make up living organisms. A single cell, such as the amoeba, *Euglena*, or paramecium, can be a free-living organism capable of carrying on an independent existence. Some cells, such as those in *Volvox*, function as part of a loosely organized colony of cells that move from place to place. Other cells are immovably fixed as part of the tissues of higher plants and animals and depend on the closely integrated activities of other cells for their existence. Cells also vary greatly in size. A cell such as the pleuropneumonia microbe is 1/250,000ths of an inch long. On the other hand, the yolk of an ostrich egg is the size of a small orange. Some cells, such as red blood cells, transport oxygen and carbon dioxide. Other cells have different specialties. Whatever its form or function, the cell is recognized today as the basic unit of living organisms and contains all those properties and processes that we collectively call life.

Biologically Important Molecules: Proteins, Carbohydrates, Lipids, and Nucleic Acids

Exercise

1

The bulk of the "dry matter" of cells consists of carbon, oxygen, nitrogen, and hydrogen organized into small units called **monomers,** which in turn are combined to form large molecules called **polymers.** The polymers consist of four types: proteins, carbohydrates, lipids (fats), and nucleic acids.

A. PROTEINS

Protein molecules are large and complex macromolecules constructed of smaller nitrogen-containing molecules known as **amino acids.** The amino acids are linked to each other by **peptide bonds,** bonds formed between the carboxyl of one amino acid and amino group of another amino acid. The structures resulting from the formation of peptide bonds are called **dipeptides, tripeptides,** or **polypeptides,** depending on the number of amino acids in a chain. Polypeptides range in molecular weight from about 5000 (insulin) to 40 million (tobacco mosaic virus protein). Protein molecules exhibit an unlimited variety of sizes, configurations, and physical properties for various reasons:

- The large number of amino acids in a protein molecule
- The almost infinite number of combinations that can be formed from the various amino acids
- The reactivity of the side groups of particular amino acids

Humans are composed of thousands, possibly hundreds of thousands, of different proteins. Each protein has a special function, and each is specifically suited to its function by its unique chemical structure.

Proteins have no competitors in the diversity of roles they play in biological systems. They make up a significant portion of the structure of cells and form a large part of most cellular membranes and organelles (e.g., mitochondria, chloroplasts, ribosomes, spindle fibers, microtubules, and chromosomes). In addition to their structural role, proteins act as biological catalysts (enzymes), regulate cellular and tissue functions (hormones), and form a major line of defense against foreign organisms (antibodies).

1. Qualitative Tests to Detect Proteins

Chemical reactions involving the free terminal amino (—NH_2) and carboxyl (—COOH) groups in protein molecules are used for detecting the types of proteins in complex mixtures of molecules. In this part of the exercise, you will become familiar with several tests used to indicate the presence of protein molecules and will learn how to determine the amino acid composition of casein, a milk protein.

a. Ninhydrin Reaction

Ninhydrin is a powerful oxidizing agent that removes the amino groups of amino acids. The reaction liberates ammonia, carbon dioxide, and a reduced form of ninhydrin. The ammonia then reacts with ninhydrin and the reduced ninhydrin to form a purple color. The appearance of the purple color is a positive test for protein.

Add 3.0 ml of distilled water to one test tube and 3.0 ml of a 0.1% egg albumin solution (or other protein provided by your instructor) to a second tube. Add solid sodium acetate to each test tube (scoop-type spatula loaded to a depth of 1 inch). Add eight drops of ninhydrin to each tube. Heat the tubes for 3 minutes in a boiling water bath, then cool them. Record your observations in Table 1-1.

b. Sakaguchi Test

Alkaline solutions of proteins that contain the amino acid arginine react with *a*-naphthol and sodium hypobromite to produce an intense red color. This color disappears rapidly unless stabilized by the addition of urea. Thus, the Sakaguchi test is a useful tool for the detection of proteins that contain the amino acid arginine.

Pipet 3.0 ml of distilled water into test tube 1, 3.0 ml of a 1% albumin solution (or other protein provided by your instructor) into test tube 2, and 3.0 ml of a 0.1% arginine solution into test tube 3. Add 1.0 ml of 10 *N* sodium hydroxide to each tube, followed by 1.0 ml of a 0.02% *a*-naphthol solution. Add two drops of sodium hypobromite to tube 1, followed immediately (within 10 seconds) by 1.0 ml of a 40% urea solution. Repeat this procedure on tubes 2 and 3. Record your observations in Table 1-1.

c. Pauly Test

When the amino acids tyrosine and/or histidine are present in a protein hydrolysate (product of enzymatic hydrolysis of protein), they react in alkaline solution with sulfanilic acid to give an intense red color. No other amino acids react with this reagent. Thus, this test is useful to confirm the presence of histidine and/or tyrosine in a protein molecule.

TABLE 1-1 Qualitative chemical reactions of amino acids and proteins

Test	Tube	Reagents Tested	Observations
Ninhydrin	1	Distilled H_2O	
	2	0.1% egg albumin (or other protein)	
Sakaguchi	1	Distilled H_2O	
	2	1.0% egg albumin (or other protein)	
	3	0.1% arginine	
Pauly	1	Distilled H_2O	
	2	10 mg/ml tyrosine	
	3	10 mg/ml glycine	
	4	10 mg/ml histidine	
	5	Hydrolysate of casein	

Pipet 2.0 ml of distilled water into tube 1, 2.0 ml of tyrosine (10 mg/ml) into tube 2, 2.0 ml of glycine (10 mg/ml) into tube 3, 2.0 ml of histidine (10 mg/ml) into tube 4, and 2.0 ml of a hydrolysate of casein (a milk protein) into tube 5. Add 1.0 ml of sulfanilic acid reagent and 1 ml of 5% sodium nitrite to each tube. Mix and let stand for 30 minutes. Add 3.0 ml of 20% sodium carbonate to each tube and mix. Record your observations in Table 1-1.

2. Quantitative Chemical Determination of Protein

Biuret, a simple molecule prepared from urea, contains what can be regarded as two peptide bonds and, thus, is structurally similar to a simple tripeptide. When reacted with copper sulfate, an intense purple color forms as a result of the reaction between the copper ions and the peptide bonds. Proteins give a particularly strong **biuret reaction** because they contain large numbers of peptide bonds. The biuret reaction can be used to determine concentrations of proteins in substances because most proteins contain approximately the same number of peptide bonds per gram.

In this experiment, you will determine the unknown concentration of two protein solutions by colorimetrically measuring the intensity of the color produced in the biuret reaction compared with the color produced by a known concentration of the protein bovine serum albumin (BSA). Refer to Appendix A for information on the use of the Milton Roy Spectronic® colorimeter and to Appendix B for the principles of spectrophotometry.

1. Prepare a set of five test tubes, each containing 5 ml of increasing concentrations of BSA (Table 1-2). Also prepare two test tubes containing 5 ml each of unknown concentrations of the same or two different protein solutions.

2. Add 2.5 ml of biuret reagent to each of the tubes, and mix thoroughly by rotating the tubes between the palms of your hands. The color develops fully in 30 minutes and is stable for at least one hour. While waiting for the color to develop, standardize, or "blank," the colorimeter using tube 1, which contains 5 ml of distilled water and 2.5 ml of biuret reagent. Set the instrument at a wavelength of 540 nanometers nm. After blanking the instrument, determine the percent transmittance (%T) for tubes 2–5. Record your readings in Table 1-2. Convert the %T into absorbance (A) for tubes 2–5 using Table 1-3, and then plot your data in Fig. 1-1 to obtain a **standard curve** (concentration) for BSA. Using this curve, determine the concentrations of the two unknown protein solutions.

3. Chromatographic Separation of Amino Acids

Several procedures are available for the isolation and purification of proteins and amino acids. One of the simplest techniques is **chromatography.** (See Appendix C for a more detailed discussion of chromatography.)

In this study, you will use thin-layer chromatography to separate and identify the amino acids in a hydrolysate of casein, a milk protein. If possible,

TABLE 1-2 Protocol for quantitative determination of protein

Tube	Protein (BSA)		Biuret reagent (ml)	%T	A
	Volume (ml)	Concentration (μg/ml)			
1 (blank)	5	0 (H_2O)	2.5	100	0.0
2	5	250	2.5		
3	5	500	2.5		
4	5	1000	2.5		
5	5	2000	2.5		
Unknown 1	5		2.5		
Unknown 2	5		2.5		

TABLE 1-3 Conversion of percent transmittance (%*T*) into absorbance (*A*)

%*T*	Absorbance (*A*)				%*T*	Absorbance (*A*)			
	(.00)	1 (.25)	2 (.50)	3 (.75)		(.00)	1 (.25)	2 (.50)	3 (.75)
1	2.000	1.903	1.824	1.757	51	.2924	.2903	.2882	.2861
2	1.699	1.648	1.602	1.561	52	.2840	.2819	.2798	.2777
3	1.523	1.488	1.456	1.426	53	.2756	.2736	.2716	.2696
4	1.398	1.372	1.347	1.323	54	.2676	.2656	.2636	.2616
5	1.301	1.280	1.260	1.240	55	.2596	.2577	.2557	.2537
6	1.222	1.204	1.187	1.171	56	.2518	.2499	.2480	.2460
7	1.155	1.140	1.126	1.112	57	.2441	.2422	.2403	.2384
8	1.097	1.083	1.071	1.059	58	.2366	.2347	.2328	.2310
9	1.046	1.034	1.022	1.011	59	.2291	.2273	.2255	.2236
10	1.000	.989	.979	.969	60	.2218	.2200	.2182	.2164
11	.959	.949	.939	.930	61	.2147	.2129	.2111	.2093
12	.921	.912	.903	.894	62	.2076	.2059	.2041	.2024
13	.886	.878	.870	.862	63	.2007	.1990	.1973	.1956
14	.854	.846	.838	.831	64	.1939	.1922	.1905	.1888
15	.824	.817	.810	.803	65	.1871	.1855	.1838	.1821
16	.796	.789	.782	.776	66	.1805	.1788	.1772	.1756
17	.770	.763	.757	.751	67	.1739	.1723	.1707	.1691
18	.745	.739	.733	.727	68	.1675	.1659	.1643	.1627
19	.721	.716	.710	.704	69	.1612	.1596	.1580	.1565
20	.699	.694	.688	.683	70	.1549	.1534	.1518	.1503
21	.678	.673	.668	.663	71	.1487	.1472	.1457	.1442
22	.658	.653	.648	.643	72	.1427	.1412	.1397	.1382
23	.638	.634	.629	.624	73	.1367	.1352	.1337	.1322
24	.620	.615	.611	.606	74	.1308	.1293	.1278	.1264
25	.602	.598	.594	.589	75	.1249	.1235	.1221	.1206
26	.585	.581	.577	.573	76	.1192	.1177	.1163	.1149
27	.569	.565	.561	.557	77	.1135	.1121	.1107	.1083
28	.553	.549	.545	.542	78	.1079	.1065	.1051	.1037
29	.538	.534	.530	.527	79	.1024	.1010	.0996	.0982
30	.532	.520	.516	.512	80	.0969	.0955	.0942	.0928
31	.509	.505	.502	.498	81	.0915	.0901	.0888	.0875
32	.495	.491	.488	.485	82	.0862	.0848	.0835	.0822
33	.482	.478	.475	.472	83	.0809	.0796	.0783	.0770
34	.469	.465	.462	.459	84	.0757	.0744	.0731	.0719
35	.456	.453	.450	.447	85	.0706	.0693	.0680	.0667
36	.444	.441	.438	.435	86	.0655	.0642	.0630	.0617
37	.432	.429	.426	.423	87	.0605	.0593	.0580	.0568
38	.420	.417	.414	.412	88	.0555	.0543	.0531	.0518
39	.409	.406	.403	.401	89	.0505	.0494	.0482	.0470
40	.398	.395	.392	.390	90	.0458	.0446	.0434	.0422
41	.387	.385	.382	.380	91	.0410	.0398	.0386	.0374
42	.377	.374	.372	.369	92	.0362	.0351	.0339	.0327
43	.367	.364	.362	.359	93	.0315	.0304	.0292	.0281
44	.357	.354	.352	.349	94	.0269	.0257	.0246	.0235
45	.347	.344	.342	.340	95	.0223	.0212	.0200	.0188
46	.337	.335	.332	.330	96	.0177	.0166	.0155	.1044
47	.328	.325	.323	.321	97	.0132	.0121	.0110	.0099
48	.319	.317	.314	.312	98	.0088	.0077	.0066	.0055
49	.310	.308	.305	.303	99	.0044	.0033	.0022	.0011
50	.301	.299	.297	.295	100	.0000	.0000	.0000	.0000

Note: Intermediate values can be arrived at by using the .25, .50, and .75 columns. For example, if %*T* equals 85, the absorbance equals .0706; if %*T* equals 85.75, the absorbance equals .0667.

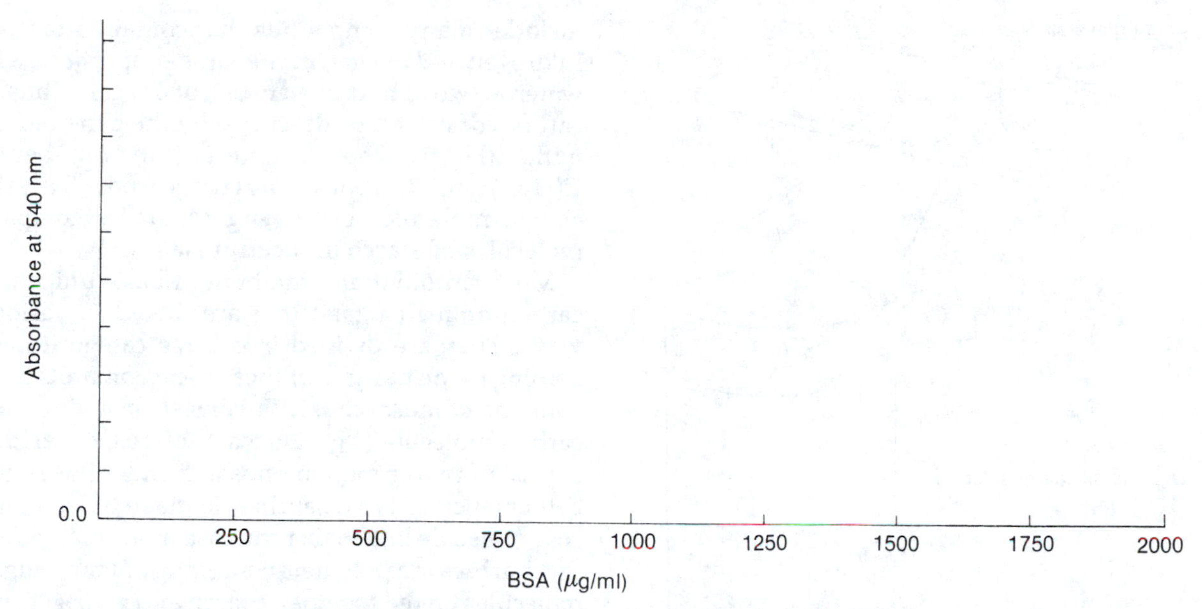

FIGURE 1-1 Bovine serum albumin (BSA) standard curve

precoated silica-gel chromatographic plates should be purchased for this exercise. Otherwise, the following procedure can be used to coat your own slides with silica gel.

1. Holding a clean glass slide by the edges, immerse it as far as possible into a jar containing silica gel. Swirl the slide several times. Stop, then carefully lift the slide *straight up* (Fig. 1-2).

2. Allow the slide to air-dry. (*Note:* The white silica-gel coat is very fragile. Do not damage the surface.)

3. Select the side of the slide with the smoothest surface. Then remove the silica gel from the other side by wiping with a paper towel. (*Note:* Avoid excessive handling of the slide because your hands may contaminate it with amino acids contained in the oils on the surface of the skin. Touch it only at the edges.)

4. Lay the slide down with the coated side up. "Spot" the coated surface at two points approximately 12 mm apart and 6 mm from the bottom. Add a small drop of casein hydrolysate to spot 1 using a capillary tube. To spot 2, add one of the known amino acids (aspartic acid, glutamic acid, methionine, proline, tyrosine, histidine, alanine, or lysine) supplied by your instructor. Each of the other students in the class will be given one of the other known amino acids.

5. Allow the spots to dry. Then carefully place the slide in a chromatographic jar containing solvent and cover. When the **solvent front** (i.e., the leading edge of the solvent) has moved to about 6–12 mm from the top edge of the silica gel (30–45 minutes), remove the slide from the jar. Allow the slide to dry.

6. In a hood or other well-ventilated area, cautiously spray the surface of the slide with ninhydrin.

Caution: *Do not inhale the fumes or get any spray in your eyes. Use eye protection and a nose/mouth mask.*

Allow the slide to dry, then heat the slide as directed by your instructor for 2 or 3 minutes.

The amino acids in the hydrolysate will react with the ninhydrin and appear as colored spots on the slide. The ninhydrin test yields purple colors with most amino acids and a yellow color with the amino acid proline. Record the colors of the spots on your chromatogram in Table 1-4.

Estimate the center of each amino acid spot and measure the distance it has traveled up the slide from its point of application. This distance, divided by the total distance traveled by the solvent from the origin line is known as the **R_f value.** Because two substances with the same R_f are probably identical, this value can be used to identify amino acids separated from a mixture.

Record your observations and those of students who were given other known amino acids in Table 1-4.

B. CARBOHYDRATES

The term **carbohydrate** means "hydrate of carbon." This name is used because a carbohydrate

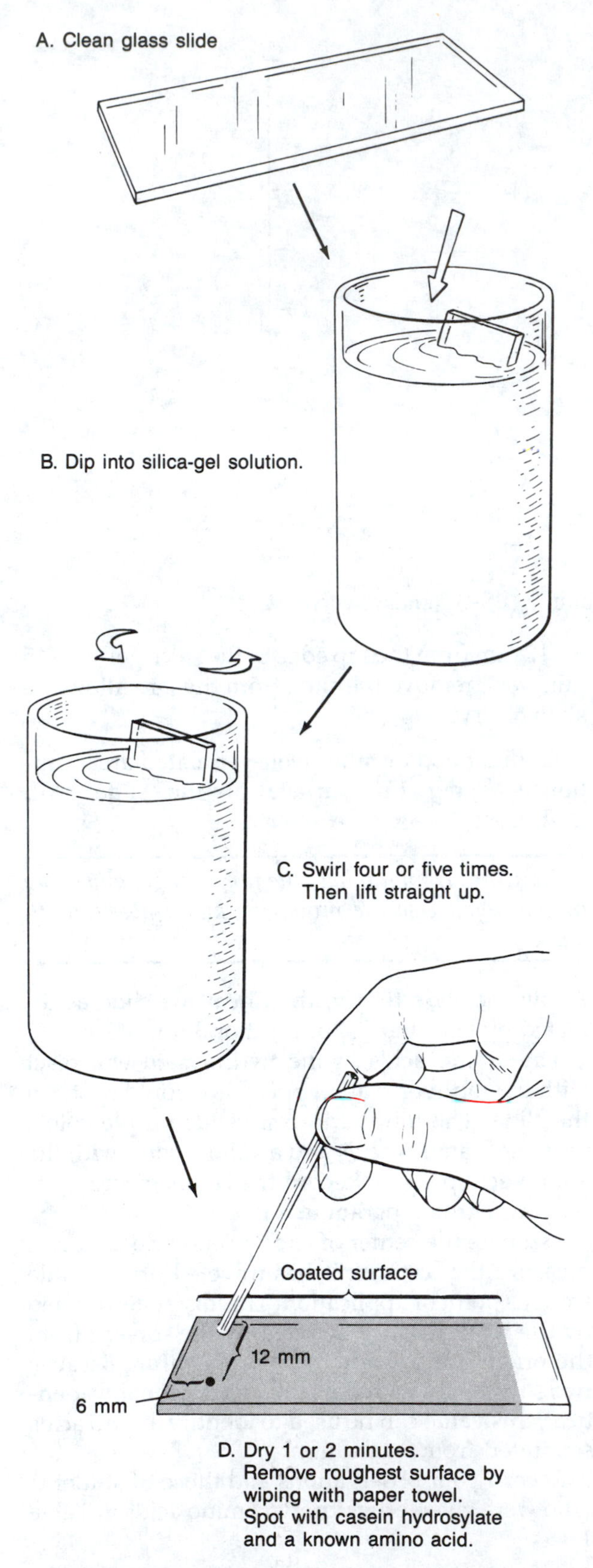

FIGURE 1-2 Preparation of plate for thin-layer chromatography

includes many compounds that contain atoms of hydrogen and oxygen in the same proportion as in water—two of hydrogen to one of oxygen. Thus, a carbohydrate can be described by the general formula $C(H_2O)_n$, where *n* represents the number of $C(H_2O)$ units. Carbohydrates range from relatively simple molecules called **sugars** to the complex molecules of **starch** and **cellulose.**

Most carbohydrates are built of basic units of 6 carbon atoms (sugars) that are linked in various ways. They are divided into three categories according to the number of these 6-carbon units they contain: **monosaccharides** consist of a single 6-carbon molecule (e.g., glucose); **oligosaccharides** consist of two or more monosaccharides linked together (sucrose is a disaccharide made up of a glucose molecule linked to a fructose molecule); **polysaccharides** are polymers consisting of many sugar molecules linked together. Starch and glycogen are long polysaccharides with chains of glucose molecules that serve as storage forms of carbohydrates; starch is usually produced in plants and glycogen in animals.

Carbohydrates can be identified by color reactions with specific reagents. These tests can determine the approximate amount as well as the kind of carbohydrate in a substance by measuring the variations in color obtained with different concentrations of the test reagent. Most of these procedures require heating the carbohydrate and reagent together in a hot-water bath. It is advisable to carry out a procedure on all the carbohydrates in the same water bath at the same time so that a direct comparison can be made of the relative response of all the carbohydrates in a test.

In this study, you will become familiar with two of the more common tests for detecting the presence of specific types of carbohydrates. You will also be asked to identify the carbohydrate composition of an unknown solution containing one or more carbohydrates. Only the carbohydrates you test will be found in the unknown. *The unknown should always be run simultaneously with the known carbohydrates.*

1. Tests for Reducing Sugars

A reducing sugar is one with a free or a potentially free aldehyde group

$$(-C\!\overset{\nearrow\!\!\!\!\!\!=O}{\searrow})$$

or ketone group

$$(\,\rangle C{=}O)$$

TABLE 1-4 Chromatography of amino acids

Amino acid	Color of spot	Distance solvent moved	Distance spot moved	R_f value
Aspartic acid				
Glutamic acid				
Methionine				
Proline				
Tyrosine				
Histidine				
Alanine				
Lysine				

In a solution of sufficiently high pH, these sugars can reduce weak oxidizing agents such as cupric, silver, and ferricyanide ions. For example, Cu^{2+} ions react with glucose to form a colored precipitate of cuprous oxide. The color of the precipitate will range from green to reddish brown, depending on the quantity of the reducing sugar present.

$$\text{glucose} + Cu(OH)_2 \xrightarrow{\text{heat}} \underset{\left(\substack{\text{colored}\\\text{precipitate}}\right)}{Cu_2O} + H_2O + \text{oxidized glucose}$$

a. Benedict's Test

Benedict's reagent contains sodium bicarbonate, sodium citrate, and copper sulfate. When combined with a reducing sugar (such as glucose or fructose) and heated, the divalent copper ion (Cu^{2+}) of copper sulfate ($CuSO_4$) is reduced to the monovalent copper ion (Cu^+) of cuprous oxide (Cu_2O), which forms a precipitate.

Place 5.0 ml of Benedict's reagent in a test tube, and add eight drops of a 1% solution of the sugar to be tested. Heat the contents for 2 minutes over a Bunsen burner (do not boil) and then allow to cool to room temperature.

Caution: *Do not point the open end of the tube toward yourself, or anybody else.*

Alternatively, place the test tube in a boiling water bath for 3 minutes and allow to cool to room temperature. When several sugars are tested at one time, the latter method is preferred.

Record the color and amount of precipitate formed for each of the sugars tested in Table 1-5. Use (+) to indicate a small amount of precipitate, (++) for a moderate amount, and (+++) for a large amount.

b. Barfoed's Test

Barfoed's test is used to distinguish between monosaccharides and oligosaccharides. The reagent is similar to Benedict's reagent except that it is slightly acidic, having a pH of about 4.5. At this pH, oligosaccharides, when heated for 2 minutes, *will not* reduce the Cu^{2+} to Cu_2O, whereas monosaccharides *will* reduce the Cu^{2+}. Heating of oligosaccharides beyond 2 minutes, however, may lead to some reduction because of the formation of monosaccharides by hydrolysis. Therefore, all the sugars must be treated in exactly the same way and the exact time of appearance of a precipitate must be noted.

For each sugar to be tested, put 5 ml of Barfoed's reagent in a test tube and add 0.5 ml of a 1% solution of the sugar. Mix the solutions well. Place the test tubes in a boiling water bath for 2 minutes. A positive test for monosaccharides is the appearance of a red precipitate of cuprous oxide within 1–2 minutes. Record the results in Table 1-5, noting the time of appearance of the precipitate.

TABLE 1-5 Qualitative sugar tests

Sugars	Observations	
	Benedict's test	Barfoed's test
Glucose		
Fructose		
Galactose		
Mannose		
Xylose		
Lactose		
Maltose		
Sucrose		
Starch		
Unknown		

2. Unknown Carbohydrates

Using Benedict's and Barfoed's tests, determine whether the unknown you have been given to analyze contains a monosaccharide or an oligosaccharide. Confirm your results with your instructor.

C. LIPIDS

Lipids are a diverse group of fatty or oily substances that are classified together because they are insoluble in water and soluble in the so-called fat solvents (e.g., ether, acetone, and carbon tetrachloride). The simplest lipids, or **triglycerides** (butter, coconut oil, and animal and plants fats), are composed of carbon, hydrogen, and oxygen and, on hydrolysis, yield glycerol and three fatty acids (Fig. 1-3). They have a higher proportion of carbon–hydrogen bonds than do carbohydrates and consequently release a larger amount of energy on oxidation than do other organic substances. Fats, for example, release about twice the calories as equal amounts of carbohydrates. **Phospholipids,** triglycerides in which one of the fatty acids is replaced by a negatively charged phosphate group, are a major component of most cellular membranes, which control the movements of nonmembranous lipids and lipid-soluble materials into and out of cells.

In this exercise, you will separate a lipid extract into its component fatty acids by using thin-layer chromatography. If possible, use precoated silica-gel chromatographic plates for this exercise. Otherwise, use the procedure that was given in part A.3 of this exercise to prepare your own slides.

Lay the slide down with the coated side up. Spot the hydrolyzed lipid extract on the coated surface about 6 mm from the bottom and allow the spot to dry. Then carefully place the slide in a chromatographic jar containing a lipid solvent and cover the jar. When the solvent front has moved to about 6–12 mm from the top edge of the gel (30–45 minutes), remove the slide from the jar. Allow the slide to dry 4–5 minutes.

Then place the slide in a jar containing iodine crystals, cover it, and leave it until brownish spots appear on the surface of the silica gel.

Only fatty acids migrate in this solvent. Other components in the extract either do not absorb onto the silica gel or remain at the origin. For example, triglycerides will be found at the edge of the solvent front and phospholipids will remain at the origin. Locate these components on your slide.

D. NUCLEIC ACIDS

Nucleic acids are so named because they were found in the nuclei of fish sperm by Miescher in 1874. There are two types of nucleic acids in all

H_2O Hydrolysis

H—C—OH + $C_{17}H_{35}COOH$
H—C—OH + $C_{17}H_{35}COOH$
H—C—OH + $C_{17}H_{35}COOH$

Lipid — Glycerol + 3 fatty acids

FIGURE 1-3 Hydrolysis of a lipid to glycerol and fatty acids

cells: **deoxyribonucleic acid (DNA)** and **ribonucleic acid (RNA).** DNA is composed of (1) the purine nitrogen bases adenine and guanine, (2) the pyrimidine nitrogen bases cytosine and thymine, (3) the pentose sugar deoxyribose, and (4) phosphoric acid. RNA is composed of essentially the same structural units except that the pyrimidine uracil is present instead of thymine and the sugar ribose is present instead of deoxyribose. DNA is found predominantly in the nucleus; RNA is most abundant in the cytoplasm but is also found in the nucleoli of the nucleus.

The presence of ribose in RNA and deoxyribose in DNA can be used to identify and differentiate these nucleic acids. The colorimetric measurement of the green color that is obtained when ribose reacts with Bial's orcinol reagent can be used as a quantitative assay of ribose and thereby of the RNA from which it has been hydrolyzed. Deoxyribose can be quantitatively determined by measuring the blue color that is formed when it reacts with Dische diphenylamine reagent.

The orcinol and diphenylamine reagents can be applied directly to RNA and DNA solutions because the strong acids in these reagents hydrolyze the nucleic acids to sugars, bases, and phosphoric acid. The sugars react with the appropriate reagent to yield color. The sugars attached to pyrimidine bases do not react under these conditions because the bond linking the sugar to the pyrimidine base is resistant to hydrolysis. Because purine and pyrimidine bases are present in a ratio of approximately 1 : 1 in nucleic acids, about half the total sugar in a sample is measured under these conditions.

In this part of the exercise, you will prepare extracts of DNA and RNA from bovine spleen tissue. Bial's and Dische reactions will be used to measure the amount of RNA and DNA in the extracts by colorimetric comparison with known amounts of these nucleic acids.

1. Extraction of DNA

Because the amount of DNA in most cells is rather small, it is important to select a tissue or organ that contains cells with a relatively high nucleus-to-cytoplasm ratio (i.e., large nuclei surrounded by a relatively small amount of cytoplasm). Lymphocytes are ideal sources for DNA extraction. Therefore, lymphoid tissues such as spleen and thymus are routinely used since large numbers of lymphocytes are found within these organs.

In this experiment you will use bovine (cow) spleen or thymus that was obtained from a slaughter house. (*Note:* This entire extraction should be completed in one laboratory period. If this is impossible, the first six steps should be completed, after which the extract can be put into a flask, labeled with your name, and frozen until you can complete steps 7–10.)

1. Your instructor will give you cubes of frozen bovine spleen or thymus tissue. Carefully weigh out 15 g of the tissue and return the excess to your instructor.

2. Pour 150 ml of a cold (4°C) citrate buffer solution (pH 7.2–7.4) into a chilled Waring or similar blender. Start the blender and add the frozen cubes of tissue one at a time, blending until each cube has been thoroughly homogenized. This is an excellent procedure for breaking the cell and nuclear membranes to release their cytoplasmic and nuclear contents. Blend for 30–60 seconds after adding the last cube.

The citrate buffer is used to inhibit the activity of intracellular DNA-hydrolyzing enzymes (DNases), which are released from lysosomes during homogenization. These enzymes require magnesium ions (Mg^{2+}) for their activity. Because citrate has a strong affinity for Mg^{2+}, it binds these ions and prevents the DNAases from inactivating the DNA in the course of extraction. Carrying out the extraction procedure at 0–4°C also retards the activity of these enzymes.

3. Pour the homogenate into a centrifuge tube and centrifuge for 15 minutes at 4000 × *G*. The homogenate is centrifuged to separate unbroken cells, debris, and **deoxyribonucleoprotein (DNP),** which contains nucleoproteins (protamines, histones) attached to the DNA molecule. DNP is insoluble in the citrate buffer while RNA is soluble under these conditions and will be found in the supernatant fluid.

4. Carefully decant the supernatant, measure its volume, and freeze it for later use in the extraction of RNA (part D.3).

5. After decanting the supernatant, wash the sediment, or pellet, by pouring enough cold citrate buffer into the centrifuge tube to fill it about half full. Stopper the tube and shake it for several minutes to break up the pellet. When the pellet is thoroughly dispersed, recentrifuge for 15 minutes at 4000 × *G*.

6. Decant this second supernatant and discard it. Pour cold 2.6 M sodium chloride (NaCl) (about 15% concentration) into the centrifuge tube until it is about half full. Break up the pellet with a glass rod, stopper the tube, and shake it vigorously for

several minutes to dissolve the DNP, which is soluble in 2.6 M NaCl. The NaCl also dissociates the nucleoproteins (protamines and histones) from the DNP to yield free DNA. The nucleoproteins form a fine precipitate that can be removed at high centrifuge speeds, leaving the DNA dissolved in the supernatant.

To dissolve all of the DNA in the sample, pour the suspension into the blender and homogenize it for 1 minute. If the suspension is quite thick, add a little more of the 2.6 M NaCl. If any lumps remain, homogenize for another 30 seconds. If necessary, the contents of your centrifuge tube can be poured into a 250-ml beaker and the suspension mixed on a magnetic stirrer for 10 minutes.

(*Note:* At this point, the procedure can be stopped and the extract frozen until the next lab. If you have sufficient time left (about 1 hour), continue with step 7.)

7. Transfer the DNP preparation into a centrifuge tube and centrifuge at 20,000 × *G* for 20 minutes to precipitate the protein (thaw first if frozen).

8. Pour this supernatant, containing the dissolved DNA, into a 50-ml beaker. The DNA can now be precipitated by *slowly* adding (down the inside wall of the tube) approximately two volumes of 95% ethyl alcohol (ethanol) so that it forms a layer over the supernatant. A mass of white fibrous material will form at the interface of the DNA solution and the ethanol. Stir the precipitate with a glass rod to force the alcohol into the supernatant. As you do this, the DNA precipitate will become wound around the glass rod. Continue this procedure until you can no longer see precipitate being added to the rod.

9. Transfer the glass rod with the precipitated DNA to a 250-ml flask. Add 200 ml of distilled water, stopper the flask, and shake. In a few minutes, the DNA should dissolve away from the glass rod to form a viscous, colorless solution. If necessary, use a magnetic stirrer.

10. Transfer 5 ml of this DNA solution to a test tube and give the rest to your instructor to save. Using the 5-ml sample, you can measure the concentration of DNA in your preparation by means of the Dische diphenylamine reaction.

2. DNA Detection by the Dische Diphenylamine Reaction

The presence of deoxyribose can be used to characterize DNA and to differentiate this nucleic acid from RNA, which contains ribose. Furthermore, the concentration of DNA can be quantitatively determined by measuring the intensity of the blue color that is formed when deoxyribose reacts with Dische diphenylamine reagent.

1. Prepare a test tube rack containing six test tubes numbered 1–6. Tubes 1–5 will be used to prepare a standard DNA concentration curve; tube 5 will also serve as a blank control. Tube 6 will contain a DNA solution of unknown concentration.

2. Dissolve 5 mg of commercially prepared DNA in 5 ml of distilled water. This will be the stock DNA solution containing 1 mg of DNA per milliliter.

3. Pipet 2 ml of the stock DNA solution into tubes 1 and 2. Pipet 2 ml of distilled water into tubes 2, 3, 4, and 5. Mix the contents of tube 2 thoroughly and then transfer 2 ml to tube 3. Thoroughly mix the contents of tube 3 and transfer 2 ml to tube 4. Mix tube 4 well and then discard 2 ml. Pipet 2 ml of the unknown DNA solution into tube 6. Each of the tubes should now contain 2 ml of the solutions listed in Table 1-6.

4. Pipet 4 ml of Dische diphenylamine reagent into each of the six test tubes and mix thoroughly. Place all tubes in a boiling water bath for 10 minutes. While the tubes are being heated, prepare an ice bath by placing crushed ice in a 500-ml beaker and adding water until the beaker is about two-thirds full. After heating the six tubes, transfer them to the ice bath and agitate them gently for 5 minutes to cool the contents rapidly.

5. Turn on the colorimeter and allow 5 minutes for the instrument to warm up. (See Appendix A for instructions on the use of the Milton Roy Spectronic 20® colorimeter and Appendix B for a discussion of spectrophotometry.) Check to make sure the sample holder is empty. Adjust the dial so that it reads 0% transmittance (%*T*) at a wavelength of 500 nm. Place the blank tube (tube 5) in the tube holder, close the cover, and adjust the light control until the dial reads 100% transmittance. Remove the tube, then determine the percent transmittance of tubes 1, 2, 3, and 4. Convert these readings into absorbance (*A*), using Table 1-3. Record these data in Table 1-6.

6. Prepare a standard DNA curve (Fig. 1-4) by plotting the absorbance of tubes 1, 2, 3, and 4 against the known concentrations of DNA in each tube. Using this standard curve, determine and record the concentration of DNA in your unknown preparation.

TABLE 1-6 Protocol for quantitative determination of DNA

Tube	Contents and concentration	%*T*	*A*
1	DNA (1 mg/ml)		
2	DNA (0.5 mg/ml)		
3	DNA (0.25 mg/ml)		
4	DNA (0.125 mg/ml)		
5	Distilled water (0, blank)		
6	Unknown DNA preparation		

3. Extraction of RNA from Bovine Spleen

There are several methods for extracting RNA from tissues. The technique used in this experiment is not as sophisticated as some other procedures but is adequate to prepare RNA for our purpose. The RNA preparation is somewhat impure and the yield is lower than that obtained by other methods.

1. Thaw the frozen supernatant obtained in step 4 of part D.1. Add an equal volume of cold (4°C) 30% trichloroacetic acid (TCA), stir gently, and allow to stand for 5 minutes.

Caution: *TCA is a strong acid. Use extreme care in handling it. Wear protective glasses.*

2. Centrifuge the solution at 2000 × *G* for 5 minutes to collect the precipitate that forms. Pour off and discard the supernatant. Add enough cold (4°C) acetone to half fill the tube and stir to resuspend the pellet. Recentrifuge at 2000 × *G* for 5 minutes, discard the supernatant, and again add cold acetone. Recentrifuge, discard the supernatant, and this time add room-temperature acetone. Recentrifuge, then discard the supernatant and retain the precipitate.

3. Place the precipitate in a small beaker and allow to air-dry until a fine powder. The powder is a mixture of RNA and proteins. Suspend the powder in a test tube that contains approximately 10 ml of 10% sodium chloride.

4. Cover the test tube with a loose cap and place it in a boiling water bath for 40 minutes. If the volume of liquid in the test tube decreases while heating (due to evaporation), add distilled water to restore the original 10-ml volume.

5. Cool the contents of the test tube to room temperature and then centrifuge the suspension at 2000 × *G* for 10 minutes. Collect the supernatant, which contains dissolved RNA, and discard the precipitate, which contains the proteins.

6. Add 2 volumes of absolute ethyl alcohol to the supernatant and place the tube in an ice bath

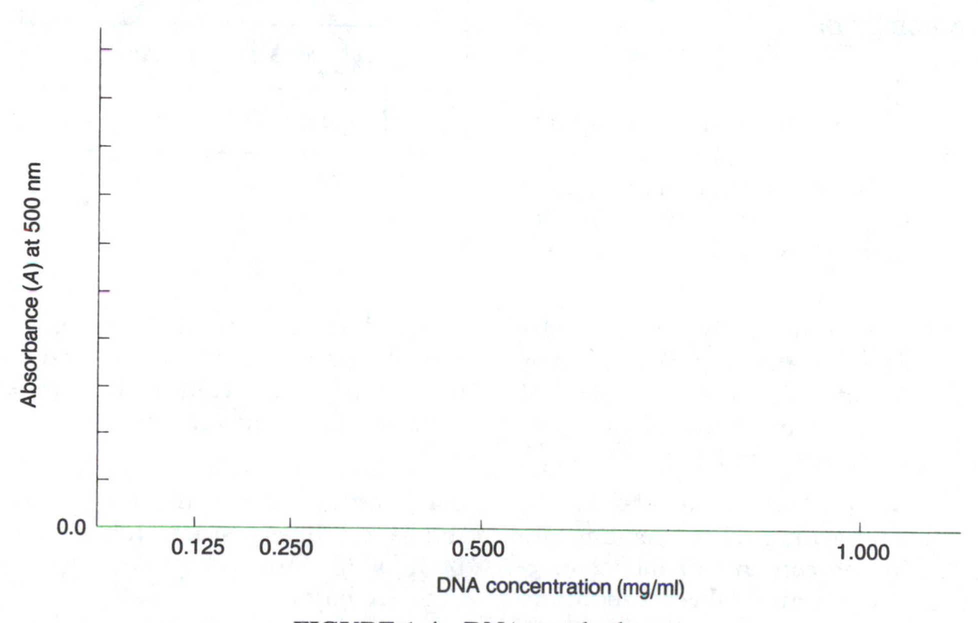

FIGURE 1-4 DNA standard curve

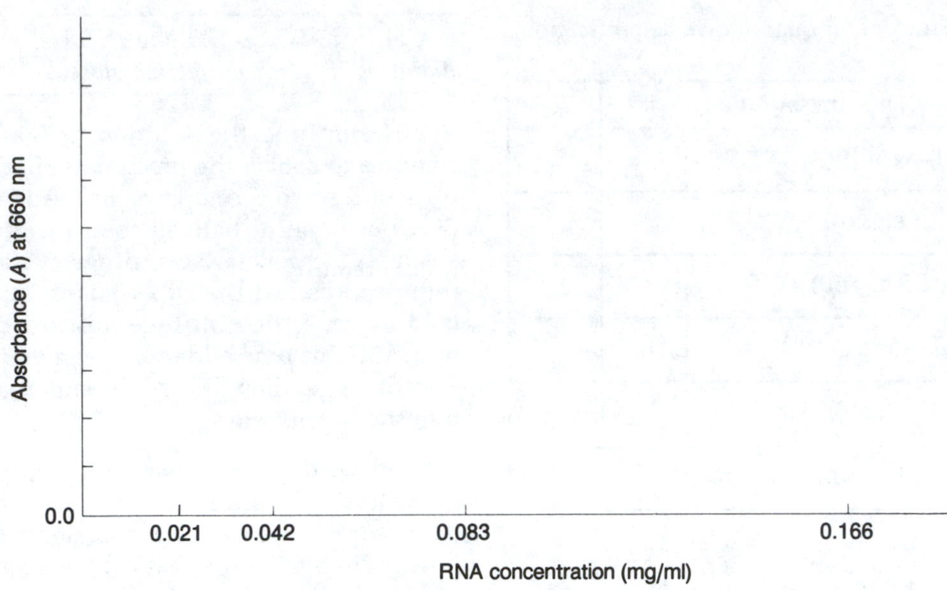

FIGURE 1-5 RNA standard curve

for 5 minutes. Colle :t u:e precipitate containing the RNA by centrifuging at 3000 × *G* for 10 minutes. Discard the supernatant. Wash the precipitate by adding acetone and stirring for several minutes. Centrifuge at 2000 × *G* for 10 minutes and discard the supernatant.

7. Place the precipitate in a beaker and allow to air-dry to obtain your final RNA preparation. This material can be frozen for use at a later time.

4. RNA Detection by the Orcinol Reaction

1. Prepare a test tube rack containing six test tubes numbered 1–6. Tubes 1–4 will be used to prepare a standard RNA concentration curve, tube 5 will serve as a blank, and tube 6 will contain the extracted RNA solution of unknown concentration.

2. Dissolve 1 mg of commercially prepared yeast RNA in 6 ml of distilled water. The dissolution of RNA can be aided by adding several drops of 0.1 N HCl. This will be the standard RNA solution containing 0.166 mg of RNA per milliliter.

3. Pipet 3 ml of distilled water into tubes 2, 3, 4, and 5. Pipet 3 ml of the standard RNA solution into tubes 1 and 2. Mix the contents of tube 2 thoroughly and then pipet 3 ml of this solution into tube 3. Mix the contents thoroughly and transfer 3 ml to tube 4. Mix the contents and discard 3 ml of solution from tube 4. Into tube 6, pipet 3 ml of the extracted RNA preparation.

4. Pipet 6 ml of the acid–orcinol reagent and 0.4 ml of the alcohol–orcinol reagent into each of the six test tubes. Place all tubes in a boiling-water bath for 20 minutes, then cool by immersing them in an ice bath.

TABLE 1-7 Protocol for quantitative determination of RNA

Tube	Contents and concentration	%*T*	*A*
1	RNA (0.166 mg/ml)		
2	RNA (0.083 mg/ml)		
3	RNA (0.042 mg/ml)		
4	RNA (0.021 mg/ml)		
5	Distilled water (0, blank)		
6	Unknown RNA preparation		

5. Determine the absorbance (A) for each of the tubes at 660 nm following the procedure used for the DNA measurements. Record these data in Table 1-7. Prepare a standard RNA curve (Fig. 1-5) and from this determine and record the concentration of RNA in the spleen extraction.

REFERENCES

Alberts, B., et al. 1989. *Molecular Biology of the Cell.* 2d ed. Garland.

Stryer, L. 1995. *Biochemistry.* 4th ed. W. H. Freeman and Company.

Swanson, C. P., and P. L. Webster. 1985. *The Cell.* 5th ed. Prentice-Hall.

Exercise

2

Light Microscopy

The light microscope can increase our ability to see detail by 1000 times, so that we can see objects as small as 0.1 **micrometer** (μm), or 100 **nanometers** (nm), in diameter. The transmission electron microscope extends this capability to objects as small as 0.5 nm in diameter, 1/200,000th the size of objects that are visible to the naked eye. Without microscopes, our understanding of the structures and functions of cells and tissues would be severely limited.

Revealing the structure of small objects, however, is not so much a function of the microscope's ability to magnify as of its ability to distinguish detail. The ability to distinguish detail is called **resolving power** (RP) and depends on the wavelength (λ) of light used and on the **numerical aperture** (NA), a characteristic of microscopes that determines how much light enters the lens. In its simplest form, resolving power, or **resolution,** can be expressed by the formula

$$\mathrm{RP} = \frac{\lambda}{2 \times \mathrm{NA}}$$

Under normal viewing conditions, resolution is increased by decreasing the wavelength of the light source. For example, if you use a green filter that permits light with a wavelength of 500 nm to pass through a microscope lens having a numerical aperture of 1, then the resolving power is 500 nm/ 2 × 1 or 250 nm. This means that you can see two objects (e.g., mitochondria or bacteria) that are 250 nm or father apart as distinct objects; if they are closer together than 250 nm, they appear fuzzy or as one object.

If you use blue light (or a blue filter that provides light at a wavelength of 400 nm) and a lens having an NA of 1, the resolving power would be equal to 400 nm/2 × 1 or 200 nm. The two objects observed under these conditions could be 50 nm closer together than those seen with a green filter and still be seen as separate objects.

Knowing the significance of the wavelength of light in the ability to distinguish detail, you can appreciate the role of electron microscopes that use electrons (λ = 0.005 nm) as a source of light and microscopes that use ultraviolet light (λ = 280 mm) in elucidating the structure and function relationships of cells and organelles (subcellular structures).

The microscope is a major tool of the biologist. Without the microscope, the cell theory would not have been developed and we would lack most of our present knowledge of objects too small to be seen with the unaided eye. This exercise, therefore, has been designed to familiarize you with the use and care of this indispensible instrument.

A. PARTS OF A COMPOUND MICROSCOPE

Referring to Fig. 2-1, locate the following features of the microscope available in your laboratory; it should have all or most of the features described.

1. Ocular Lens

The **oculars** are the lenses you look through. If there is only one ocular, you are using a **monocular** microscope; if there are two, it is a **binocular** microscope. The oculars of many binocular microscopes can be adjusted to accommodate the distance between the eyes of different observers **(interpupillary adjustment).** One of the oculars may also have a knurled knob that can be rotated so that you can move the ocular in or out to compensate for any focusing disparity between your eyes. Your instructor will demonstrate how these adjustments are made.

Oculars on different microscopes may have dif-

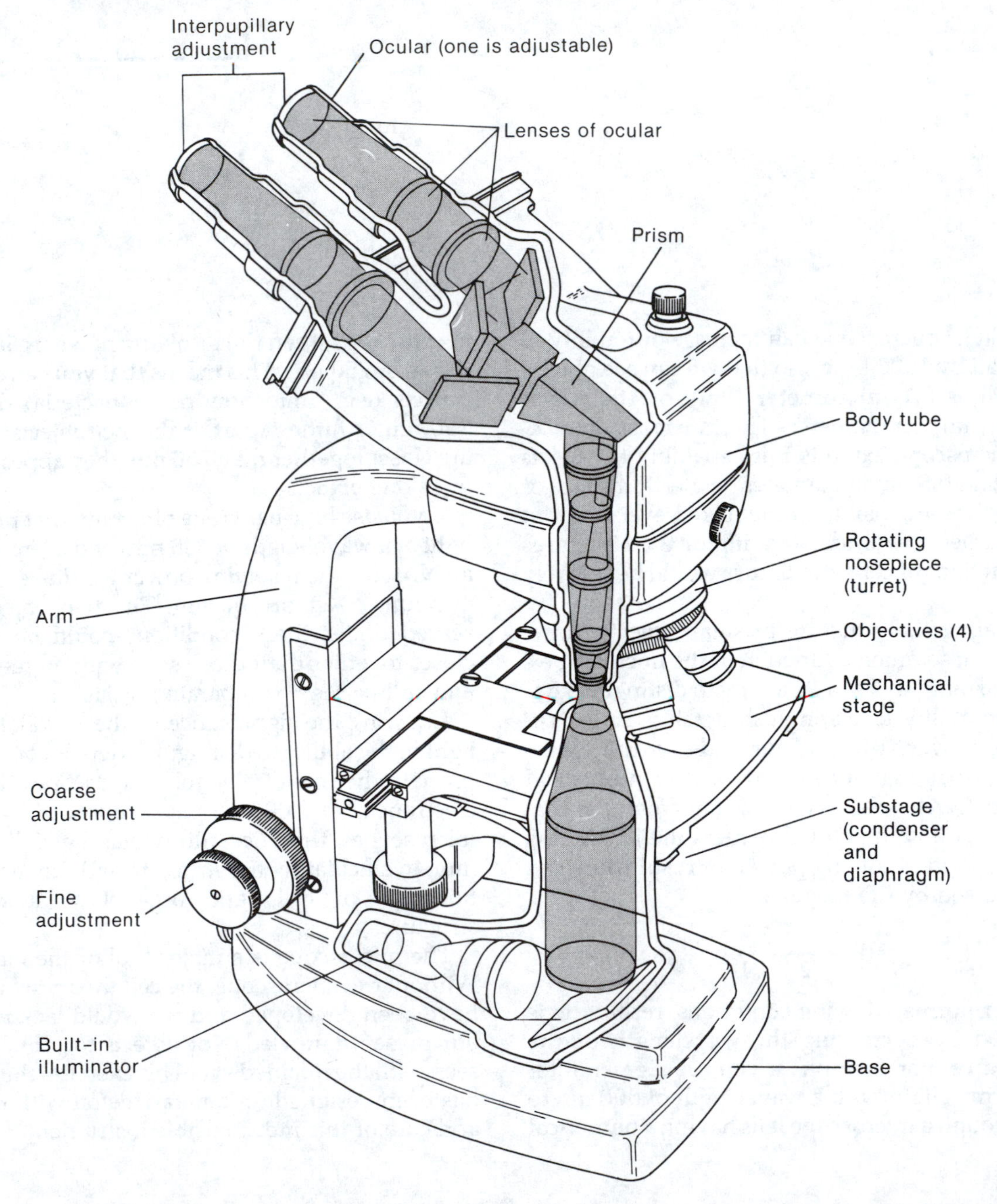

FIGURE 2-1 Binocular microscope sectioned to show pathway of light from illuminator through various lenses and prisms

ferent magnifications (i.e., 5×, 10×). You may have to remove the ocular from the tube it is contained in to determine its magnification. What is the magnification stamped on the housing of the oculars on your microscope?

___

2. Objective Lens

Attached to a rotating nosepiece, or turret, at the base of the body tube is a group of three or four **objectives.** Rotate the nosepiece and listen for the click as each objective comes into position.

The objective lenses focus the light that comes through the specimen, up the body tube, and through the oculars. Each objective has numbers stamped on it. One of these numbers identifies the magnification of the objective (e.g., 43×). Objective lenses are usually named according to their magnifying power, as follows:

scanning power—4×

low power—10×

high power—43×

oil immersion—93×

What are the magnifications of each of the objectives on your microscope?

___

The **total magnification** is calculated by multiplying the magnifications of the ocular and objective lenses on the microscope being used. In Table 2-1, calculate the total magnification for each ocular/objective combination on your microscope.

A second set of numbers on the objective, usually a decimal, represents the numerical aperture for the lens; the abbreviation NA may precede the number. In Table 2-2, list the magnification and numerical aperture for each objective on your microscope.

3. Body Tube

Light travels from the objectives through a series of magnifying lenses in the body tube to the ocular. In some microscopes, the body tube is straight. In others, the oculars are held at an angle, as in Fig. 2-1, and the body tube contains a prism that bends the light rays so that they will pass through the oculars.

4. Stage

The surface or platform on which you place the microscope slide is the **stage.** Note the opening **(stage aperture)** in the center of the stage. On some microscopes, the stage is stationary and has clips to hold the slide in place. On other microscopes, the stage is movable and is called a **mechanical stage.** Movement is controlled by two knobs located on the top, side, or bottom of the stage. Note the **horizontal** and **vertical scales** on the mechanical stage. Suggest a function for these scales.

___

___

___

TABLE 2-1 Calculation of total magnification for various ocular/objective combinations

Ocular	×	Objective	=	Total magnification

TABLE 2-2 Numerical aperture and magnification for various objectives

Magnification of objective	Numerical aperture (NA)

How are slides held in position on a mechanical stage?

5. Substage

The area under the stage, called the substage, may contain a diaphragm, a condenser, or both.

a. Diaphragm

The diaphragm regulates the amount of light passing from the light source through the specimen and through the lens system of the microscope. By properly adjusting the diaphragm, you can provide for better contrast between the surrounding medium and the specimen, thus greatly improving the image of the specimen. The diaphragm may be either annular or iris.

- An **annular diaphragm** consists of a circular plate with holes of different diameters. You can rotate the plate to position the various holes in the light path, thereby regulating the amount of light that passes from the light source through the specimen.
- An **iris diaphragm** consists of a circle of overlapping thin metal plates. The lever that projects from the side of the iris diaphragm opens and closes the plates, thereby regulating the amount of light that enters the microscope.

What type of diaphragm does your microscope have?

b. Condenser

The condenser contains a series of lenses that focus light onto the specimen. It is moved up and down by a knob at its side or by a lever projecting from its housing. By properly adjusting the condenser, you can greatly improve the clarity of the specimen image.

A filter holder may be attached to the bottom of the condenser. It usually contains a blue filter. Why would you use a blue filter instead of a green or red filter when making microscopic observations?

6. Light Source

A microscope may have an attached mirror or a built-in illuminator. The mirror is usually concave on one side and flat on the other. The flat side of the mirror is used with the scanning and low-power objectives and the concave mirror with higher power objectives. The light source for the mirror is usually a lamp. Natural light may be used but is not preferred because its intensity is too variable.

The illuminator of most microscopes is built into the base of the microscope and controlled by an on/off switch. You can control the amount of light entering the specimen by adjusting the diaphragm. You can also control the light intensity by adjusting the voltage of a transformer attached to the illuminator. Use low- or medium-voltage settings for most microscopic observations. You will need a higher setting when using the oil-immersion lens. Why?

7. Focusing

You focus a microscope by using the coarse and fine adjustment knobs that raise or lower the body tube or the stage, depending on the type of microscope you are using. With the low-power objective in position about ¼ inch above the stage, rotate the coarse adjustment knob one half turn clockwise while watching the movement of the low-power objective. Do the same with the fine adjustment knob. Based on these observations, why should you not use the coarse adjustment knob for focusing when the high-power objective or oil-immersion objective is in its normal operating position?

____

____

____

8. Eyeglasses and Microscope Usage

If you are nearsighted or farsighted, you need not wear glasses for microscopic observations. The adjustments made in focusing the microscope compensate for these eye problems. Wear glasses if you have astigmatism (a defect in the eye's refractive surface), since this problem is not corrected by the lenses of the microscope. In either case, when using a monocular microscope keep both eyes open to prevent eyestrain.

B. PROPER USE OF MICROSCOPES

Before using a microscope, thoroughly clean the ocular and objective lenses with lens paper. Use a circular cleaning motion to avoid scratching the lenses. When using the microscope, keep your eyelashes from touching the ocular lens. Oil from the lashes adheres to the ocular lenses and smears them. After using salt solutions or other harsh chemicals to prepare wet mounts, thoroughly clean the oculars and objectives, stage, and microscope slides to prevent damage to the microscope.

Despite its sturdy appearance, a microscope is a delicate precision instrument. It should be handled carefully and with common sense. The following suggestions will help you avoid some common mishaps.

- Avoid dropping the microscope, banging it against a laboratory bench, or having the oculars fall out.
  a. Carry the microscope upright using both hands.
  b. Keep the microscope away from the edge of the bench, particularly when not in use.
  c. Move power cords out of the way so that you cannot trip on them and pull the microscope or transformer down.
- Avoid breaking a coverslip and/or microscope slide while focusing.
  a. First locate the specimen using the low-power objective, and then switch to the higher power objectives.
  b. Never focus the high-power objective with the coarse adjustment knob, and never use the high-power objective when examining thick specimens or whole mounts of specimens.
- Avoid mechanical difficulties with various parts of a microscope.
  a. Never force microscope parts to work.
  b. When changing the bulb in the illuminator, never force it; it might shatter in your fingers.
  c. Never try to dismantle the microscope.

C. USING A COMPOUND MICROSCOPE

1 Focusing

1. Clean the oculars and objectives using lens paper.

2. Cut out a letter *e* from a newspaper or other printed page. Clean a microscope slide and prepare a wet mount of the letter, using the procedure described in Fig. 2-2. Put the scanning (4×) or low-power (10×) objective in position, and then place the slide on the stage in its normal viewing position.

3. Turn on the illuminator and open the diaphragm fully. If there is a condenser, position it as high as it will go, so that the top lens of the condenser unit is level with the top of the stage.

4. Center the specimen over the stage aperture.

5. Position the objective as close to the slide as possible without touching it; then, while looking through the oculars, use the coarse adjustment knob to back off slowly until the specimen comes into focus.

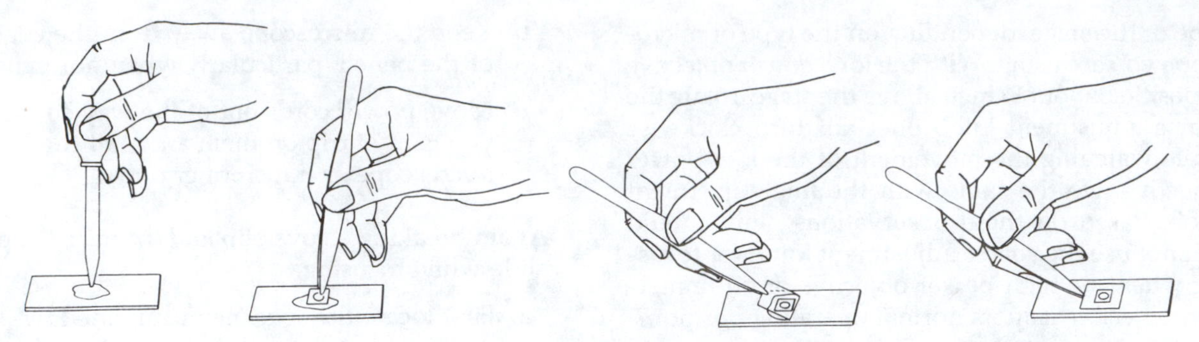

A. Add a drop of water to a slide.

B. Place the specimen in the water.

C. Place the edge of a coverslip on the slide so that it touches the edge of the water.

D. Slowly lower the coverslip to prevent forming and trapping air bubbles.

FIGURE 2-2 Preparing a wet mount slide

6. Use the diaphragm (and/or adjust the transformer voltage) to readjust the light intensity as necessary, and again center the specimen by moving the slide.

7. If you are using the scanning (4×) lens, switch from the scanning lens to the low-power objective (10×). Make certain the objective clicks into position. You can sharpen the focus by small adjustments of the fine adjustment knob.

Caution: *If the letter* e *is not in focus after changing objectives, you may have to use the coarse adjustment knob and then the fine adjustment knob to bring it back into focus. But remember, do not do this with the high-power or oil-immersion objectives in position. Ask your instructor for help if you have difficulty focusing the microscope.*

Recenter the specimen, adjust the diaphragm, and adjust the position of the condenser to increase the contrast of the specimen.

8. Switch to the high-power objective (43×) and adjust the focus using the fine adjustment knob.

These procedures are usually used when examining a wet mount or a commercially prepared microscope slide. Always use clean microscope slides. Always proceed from the lowest power to the highest power objectives, making minor focus and light corrections as necessary. Learn to fine-tune your microscope.

2. The Microscopic Image

The image you see in the microscope is affected by several factors: the orientation of the image, the total magnification, the size and brightness of the field of view, the plane of focus, the depth of focus, and the contrast of the specimen.

a. Orientation of the Image

Hold the slide of the letter *e* so that the letter is in a normal reading position. Then place it on the stage in that position and examine it with the low-power objective. What difference is there, if any, in the way the image is oriented when viewed through the oculars compared with looking at it directly with your eyes?

____

While looking through the microscope, attempt to make the image move to the right. In which direction did you have to move the slide?

____

Try to move the image upward in the field of view (away from you). Which way did you have to move the slide?

____

In what direction do you have to move the letter to make the image move to the right and then up?

____

When you want to point out something of interest to someone, you can describe its approximate location by referring to the field of view as a clock. Thus, you could tell them to "look at three o'clock," or "look just off-center toward nine o'clock," and so forth. Alternatively, in some microscopes a thin black line appears to cut across the field of view. This is a **pointer.** You can move an object under observation to the end of the pointer.

b. Brightness of the Field of View and Working Distance

Examine the specimen on the slide, starting with the lowest power objective and progressing to the highest power objective. Describe any changes in the brightness of the field of view when you change objectives.

____

____

Note that when the object on the slide is in focus for each objective, the **working distance,** the distance between the slide and the objective lens, decreases as the objective magnification increases. Of what use can this observation be when setting up the microscope to view a specimen?

____

____

____

c. Depth of Focus

Like the human eye, the lenses of a microscope provide a limited **depth of focus.** This means that only part of the object is in sharp focus; areas above and below that part will be slightly out of focus or not in focus at all.

To visualize three-dimensional form and the concept of depth of focus, place a white thread across a red thread on a microscope slide. Add a drop of water and a coverslip. Using the scanning objective (4×), focus on the point at which the threads intersect and determine the depth of focus at this magnification. For example, are both of the threads in sharp focus, or are they in sharp focus only where they intersect?

____

____

Change to the low-power objective (10×). Describe any changes in the depth of focus.

____

____

Switch to the high-power objective (43×) and describe any changes in the depth of focus.

____

____

At higher magnifications, it is difficult but not impossible to determine three-dimensional form. You can do this by building a series of **optical sections** in your mind as you focus through the specimen (Fig. 2-3).

Try to determine the three-dimensional structure of your preparation at high power by making and visualizing a series of optical sections as demonstrated in Fig. 2-3. Begin by focusing on the surface of the top thread and work through to the lower surface of the bottom thread.

d. Contrast

Even with sufficient magnification and resolution, you can visualize an object under a microscope only if there is sufficient contrast between the object and its surroundings or between various parts of the object.

You can improve image contrast by regulating the opening of the diaphragm. This deflects the light rays from the edge of the diaphragm and causes them to enter the specimen at an angle. Such scattering makes the specimen look darker.

In addition, cells and subcellular structures may contain natural pigments (e.g., chlorophyll in chloroplasts and hemoglobin in red blood cells) that provide contrast and make these structures visible.

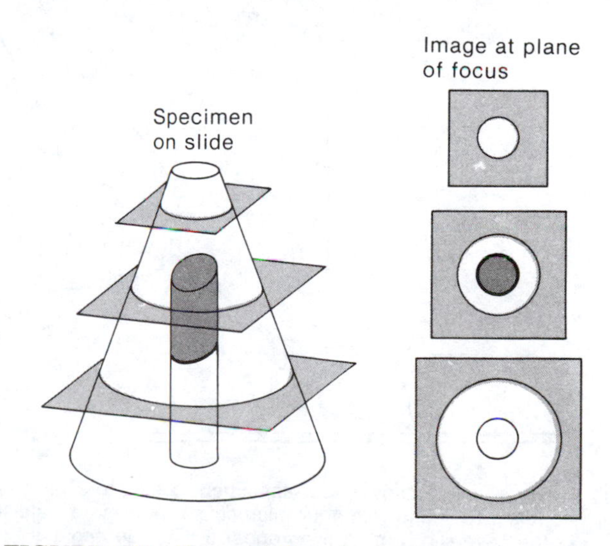

FIGURE 2-3 Determining three-dimensional image through optical sectioning

However, many cells and parts of cells are translucent. One way you can improve contrast in such cells is to use dyes or stains that bind to or are taken up by various subcellular structures, which then absorb enough light to provide contrast.

Following the instructions in Fig. 2-4A–C, examine epithelial cells obtained from the inner lining of your cheek.

Caution: *Do not give your toothpick to any other student. Discard the toothpick in the designated container as soon as you have prepared your slide.*

First try to determine something of their structure by adjusting the diaphragm and the condenser. Next add a drop of methylene blue stain to the edge of the coverslip and draw it under as shown in Fig.

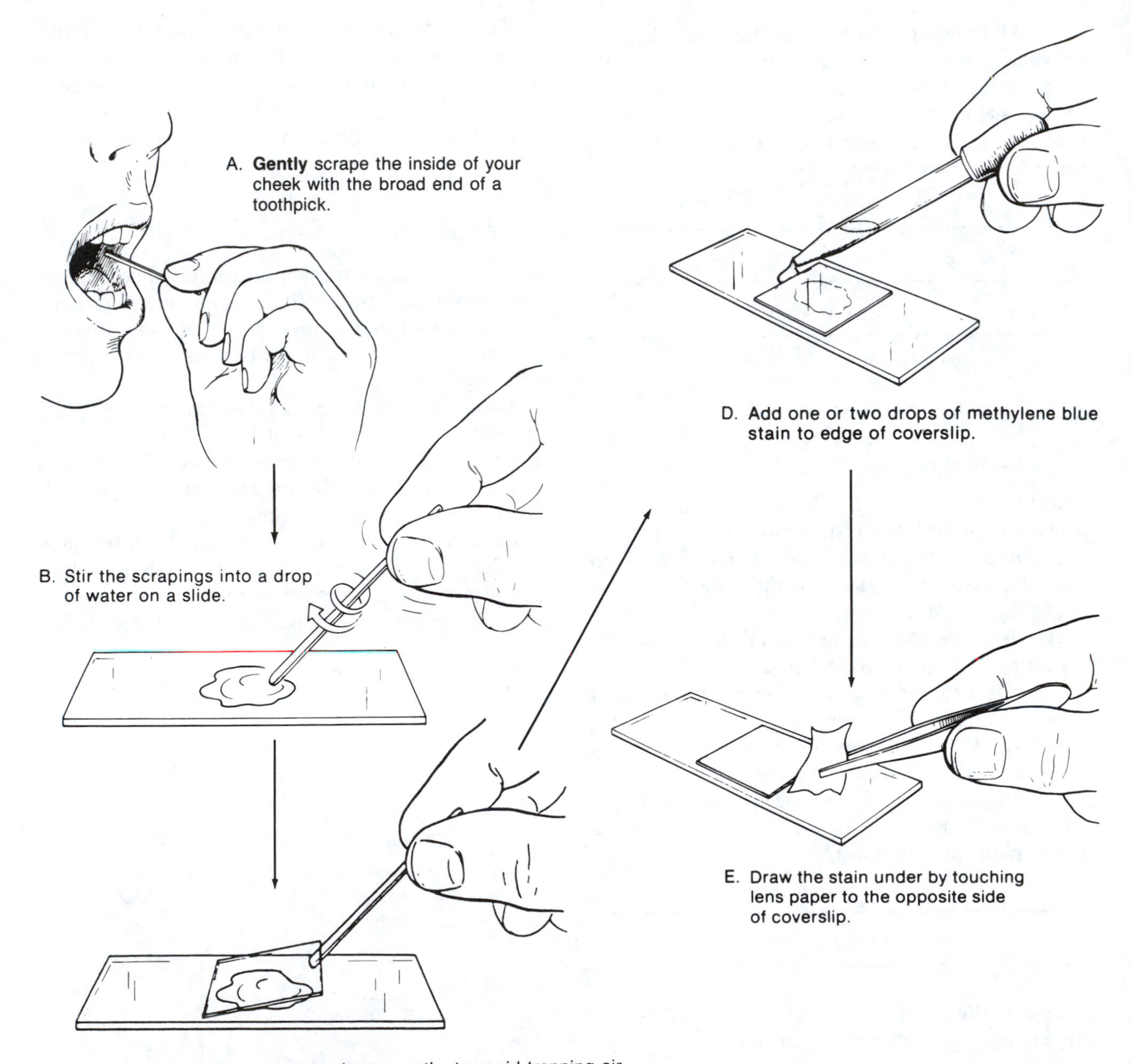

FIGURE 2-4 Preparing a wet mount slide and staining cells to improve image contrast

2-4D, E. Describe any changes in contrast or visibility of the structures in the cells.

____

____

____

____

e. Measurement of Microscopic Specimens

If you know the diameter of the field at each magnification, you can estimate the size of object you are examining. To determine the diameter of the field, place a transparent millimeter rule on the stage, focus on the rule, and measure the diameter of the field for the scanning and low-power objectives.

It is very difficult to measure the high-power and oil-immersion fields, but you can get a good approximation from the following formula:

$$\frac{D_H}{D_L} = \frac{X_L}{X_H} \quad \text{or} \quad D_H = \frac{D_L \times X_L}{X_H}$$

where

D_L is diameter of the low-power field

D_H is diameter of the high-power field

X_L is magnification of the low-power objective lens

X_H is magnification of the high-power objective lens

Insert the appropriate values in the formula and calculate the approximate diameters of the high-power and oil-immersion objective fields of your microscope.

____

____

A more precise method of measurement involves the use of an **ocular micrometer,** a small glass disc on which uniformly spaced lines of *unknown* distances are etched (Fig. 2-5A). The ocular micrometer is inserted into the ocular of the microscope and then calibrated against a **stage micrometer,** which has uniformly spaced lines of *known* distances (Fig. 2-5B). The stage micrometer with the ocular micrometer in place appears as shown in Fig. 2-5C. To calibrate the ocular micrometer, use the following procedure.

1. Turn the ocular in the body tube until the lines of the ocular micrometer are parallel with those of the stage micrometer. Match the lines at the left edges of the two micrometers by moving the stage micrometer (Fig. 2-5D).

2. Calculate the actual distance in millimeters (mm) between the lines of the ocular micrometer by observing how many spaces of the stage micrometer are included within a given number of spaces on the ocular micrometer. Because the smallest space on the stage micrometer equals 0.01 mm, you can calibrate the ocular micrometer using the following:

a. 10 spaces on the ocular micrometer equals X spaces on the stage micrometer.

b. Since the smallest space on the stage micrometer equals 0.01 mm, 10 spaces on ocular micrometer = X spaces on stage micrometer × 0.01 mm

c. Therefore,

$$1 \text{ space on ocular micrometer} = \frac{X \text{ spaces on stage micrometer} \times 0.01 \text{ mm}}{10}$$

d. Since 1 mm equals 1000 μm,

$$1 \text{ space on ocular micrometer} = \frac{X \times 10\ \mu\text{m}}{10}$$

e. Example: If 10 spaces on the ocular micrometer equal 6 spaces on the stage micrometer, then

$$1 \text{ ocular space (in mm)} = \frac{6 \times 0.001 \text{ mm}}{1} = 0.006 \text{ mm}$$

or

$$1 \text{ ocular space (in } \mu\text{m)} = \frac{6 \times 1\ \mu\text{m}}{1} = 6\ \mu\text{m}$$

Note: The numerical value obtained holds only for the specific ocular/objective lens combination that is used. Each time the objective or ocular lens is changed, the ocular micrometer must be recalibrated.

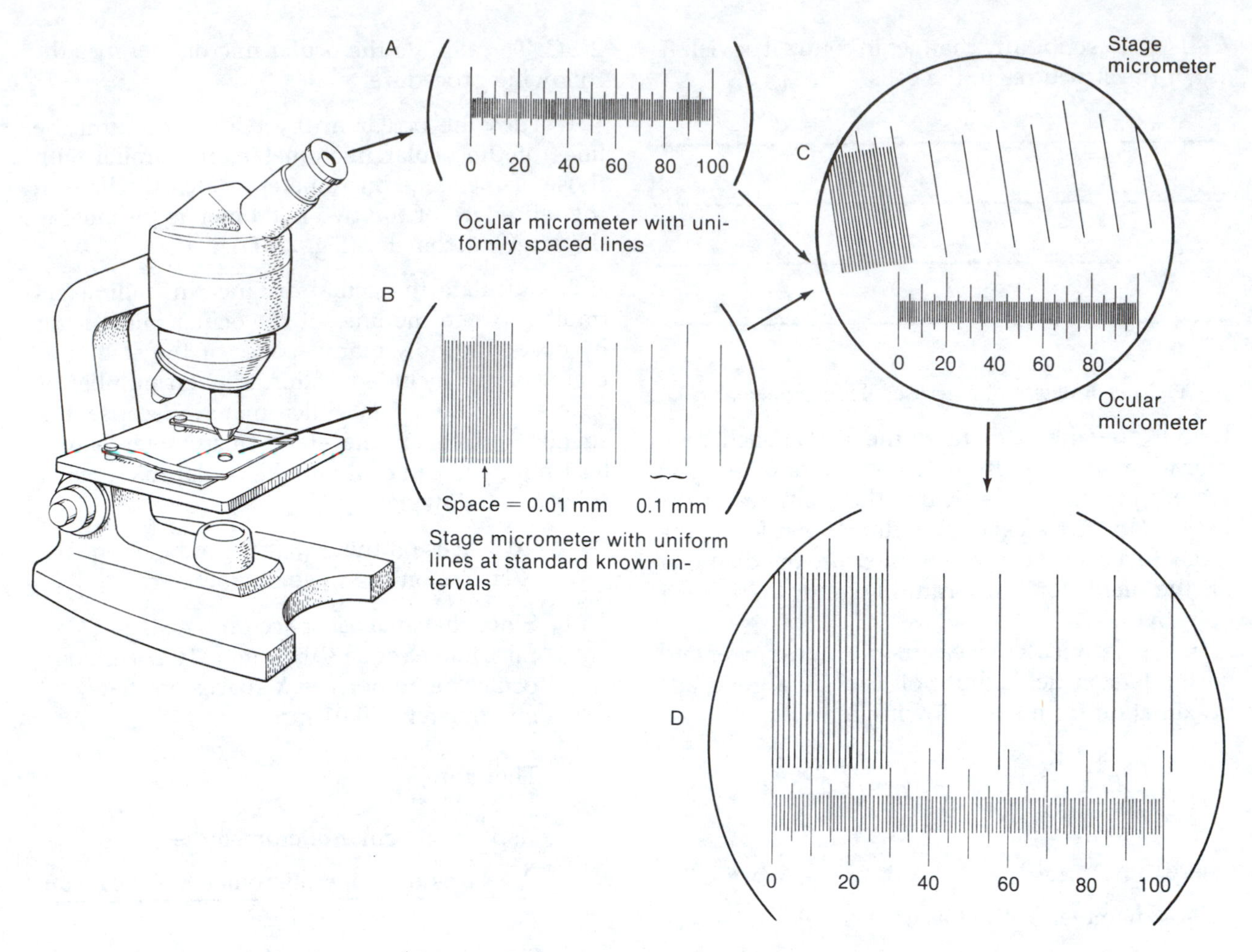

FIGURE 2-5 Using an ocular and stage micrometer to determine the size of microscopic objects

D. USE AND CARE OF THE STEREOSCOPIC (DISSECTING) MICROSCOPE

The stereoscopic microscope shown in Fig. 2-6 has two advantages over the compound microscope: (1) it enables you to examine objects that are too large or too thick to be seen with the higher magnifications of the compound microscope, and (2) it gives you a three-dimensional view of the specimen.

In this microscope, the light source may be *reflected* from an illuminator above the specimen or *transmitted* through the specimen from a mirror below the stage. The choice of light source depends on the nature of the specimen. Use reflected light for opaque objects and transmitted light for transparent objects.

Using a stereoscopic microscope, examine your finger or some other opaque object. Adjust the oculars for interpupillary distance and focus as you did with the compound microscope (part A.1). Change the magnification by using the magnification knob on the top of the body tube. On some stereoscopic microscopes, the magnification is varied by switching ocular lenses, as with the compound microscope. How does the movement of the image compare with that of the compound microscope (part C.2)?

____

____

How do you adjust the brightness of the field?

____

____

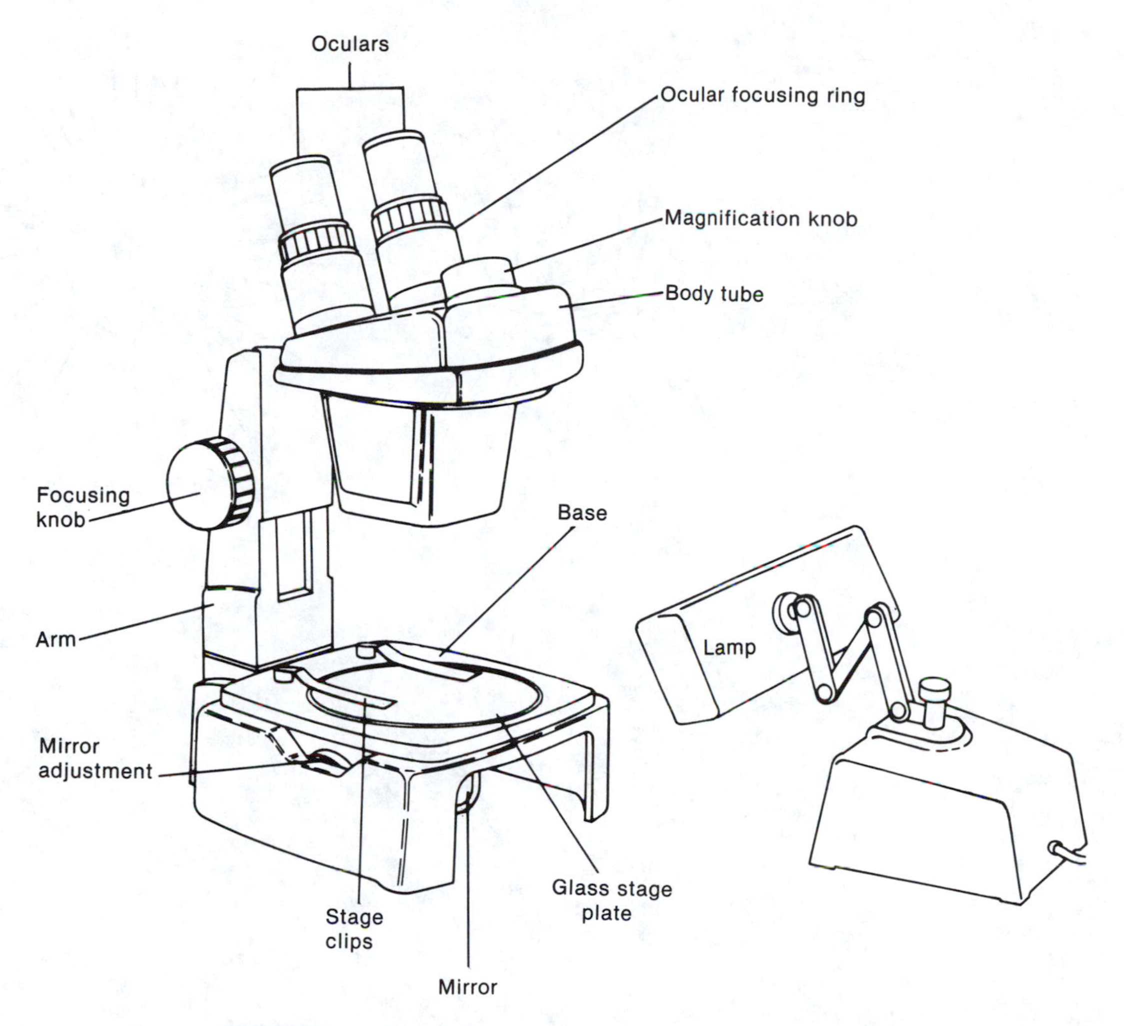

FIGURE 2-6 Parts of a stereoscopic (dissecting) microscope

Examine the previously prepared slide of the crossed threads. First use light reflected from the mirror and then use light transmitted from a lamp. Describe any advantage of one type of lighting over the other.

E. EXAMINATION OF POND WATER

In your laboratory work with the compound microscope, many of your observations will be of living organisms or tissues or parts of organisms you will want to keep alive. To allow them to dry out greatly distorts them, to say nothing of the effect of death on a study of their movements. To observe living material, prepare a wet mount of a drop of pond water as shown in Fig. 2-2.

Excess water under the coverslip can be soaked up by carefully touching a piece of paper toweling to the edge of the coverslip. If the preparation begins to dry out while under observation, add one drop of water at the edge of the coverslip.

Under low power and with reduced light, survey the drop of pond water. Identify as many of the organisms as you can. Figs. 2-7, 2-8, 2-9, and 2-10 should help you identify what you see. Carefully study the differences in structure of the organisms and their methods of movement.

Prepare additional wet mounts with samples

(*Text continued on page 32*)

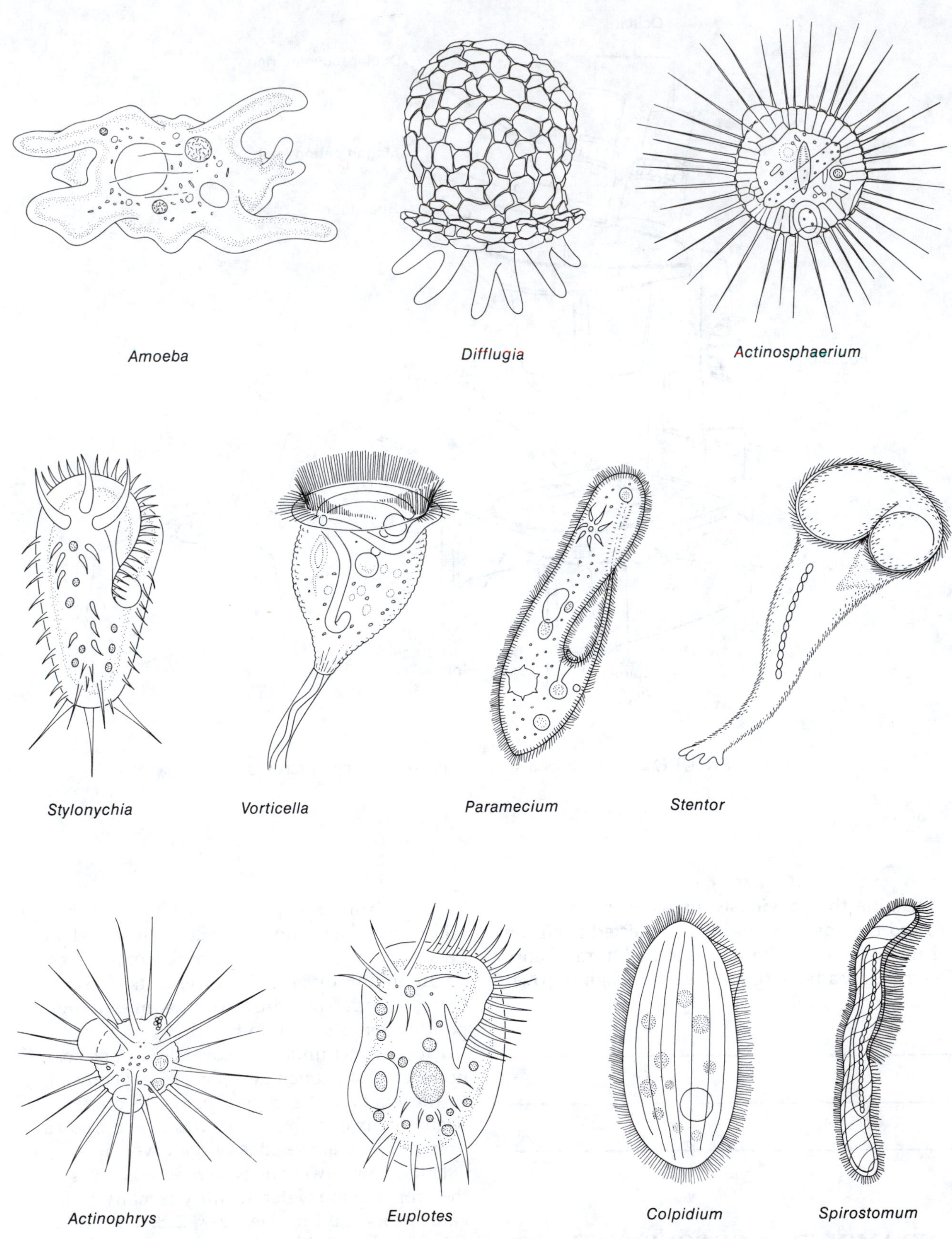

FIGURE 2-7 Protozoans commonly found in pond water

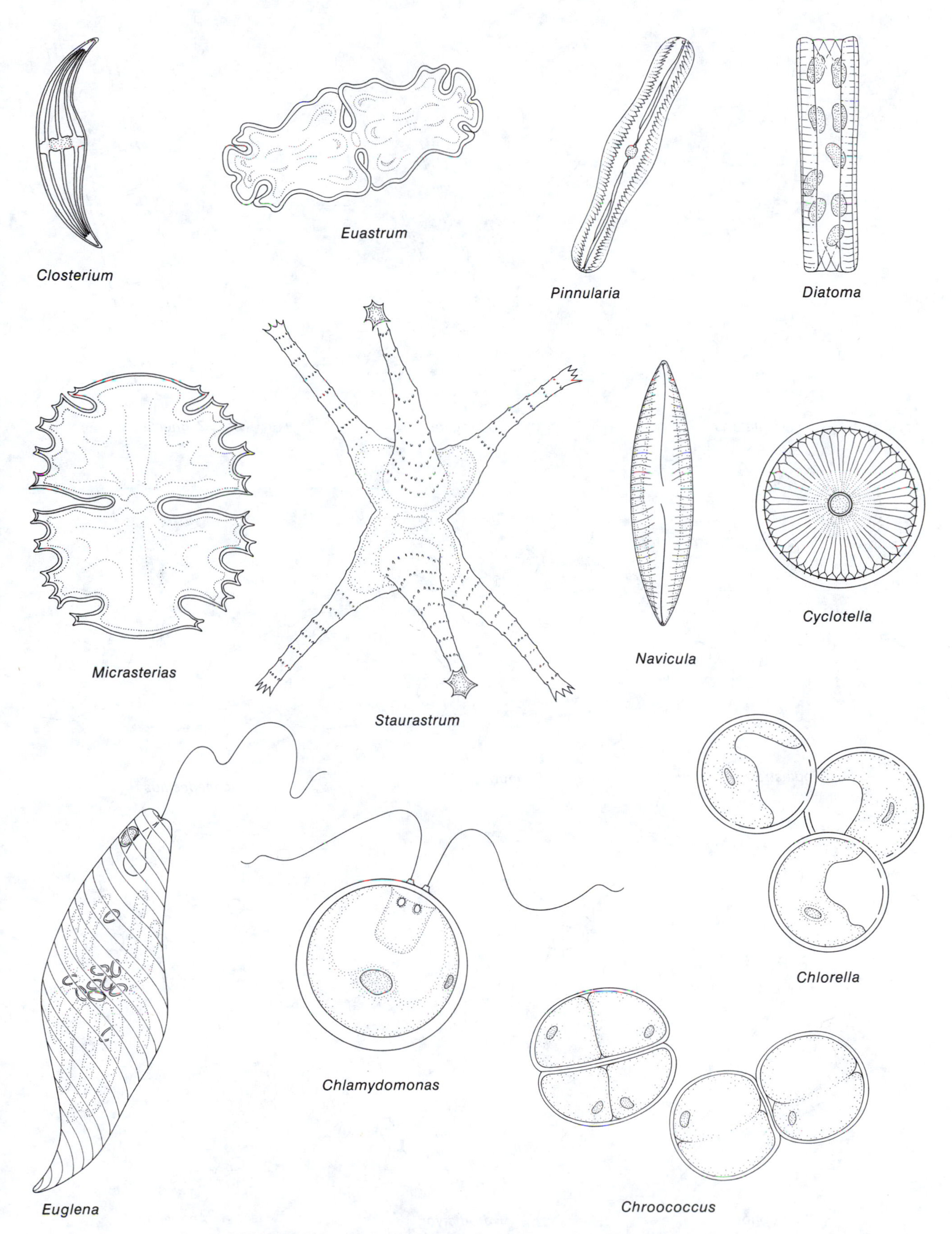

FIGURE 2-8 Unicellular algae commonly found in pond water

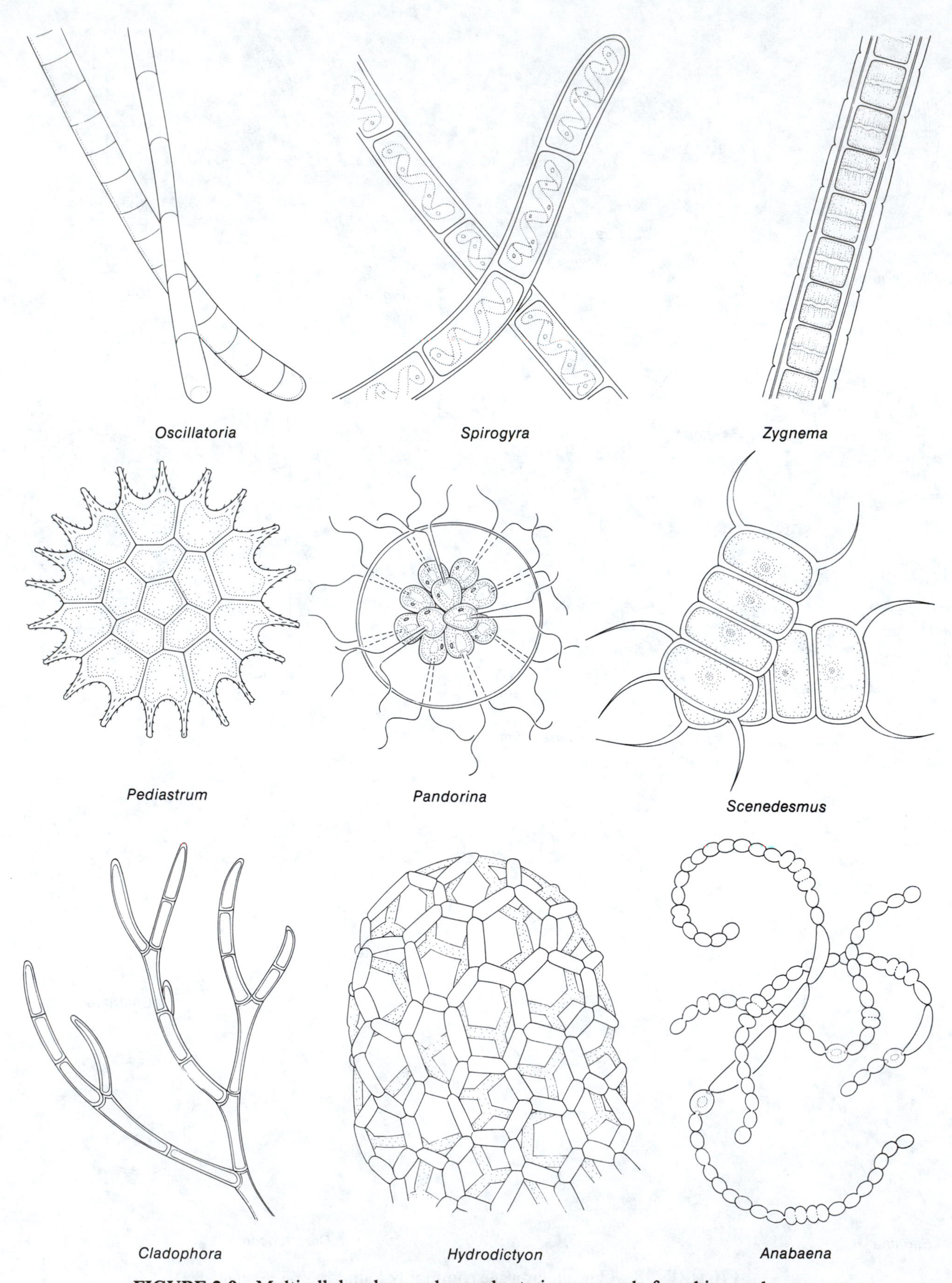

FIGURE 2-9 Multicellular algae and cyanobacteria commonly found in pond water

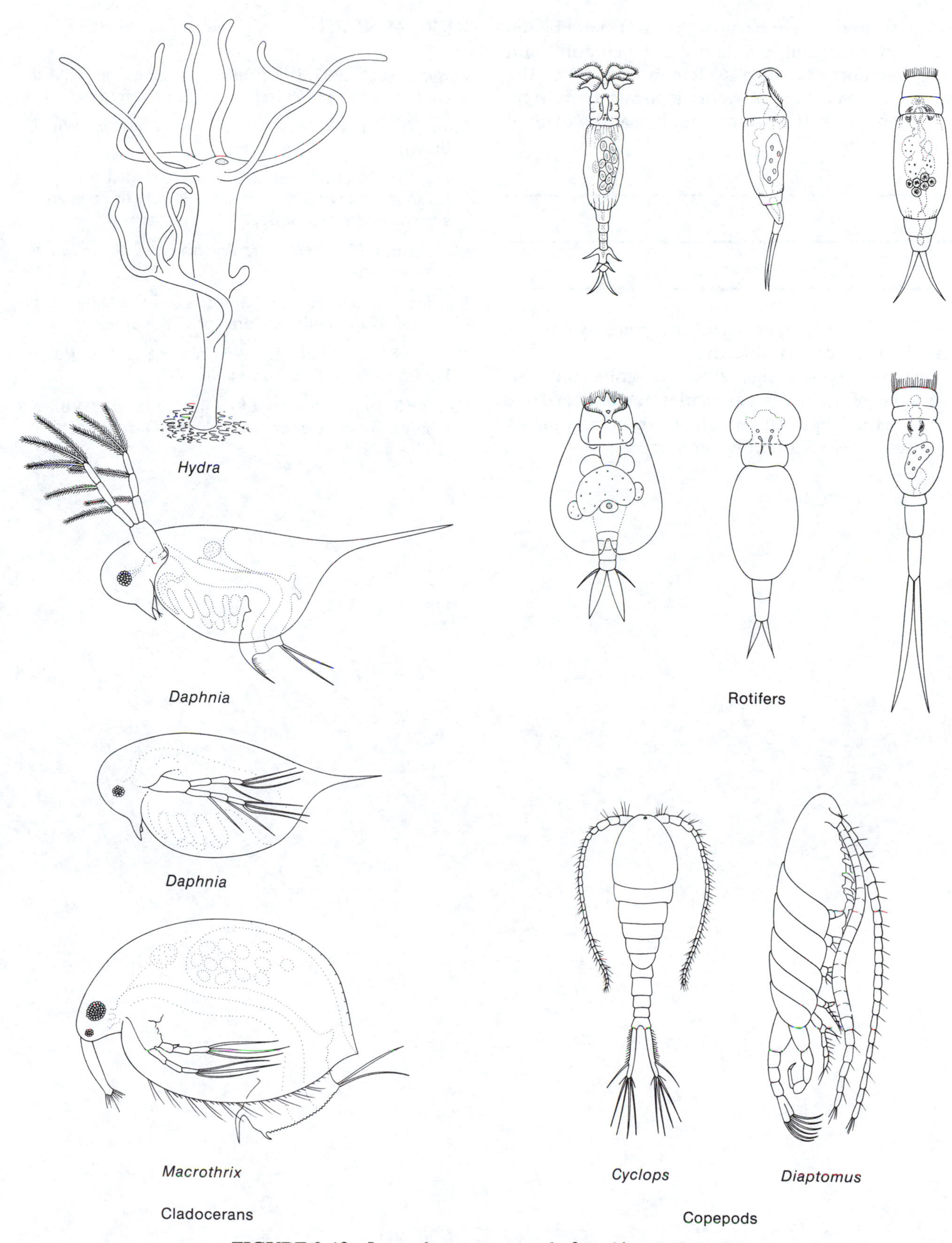

FIGURE 2-10 Invertebrates commonly found in pond water

taken from different parts of the jar. Do not be too hasty in discarding a slide because you don't find any microorganisms; a systematic survey of the preparation is often necessary to locate them. Why do the organisms often accumulate at the edge of the coverslip?

___

___

___

To identify the smaller organisms, you may have to use the high-power objective.

Using the measuring methods described (i.e., the diameter of the field or an ocular micrometer), determine the length and width of various organisms observed in your pond water sample.

REFERENCES

Eddy, S., et al. 1982. *Taxonomic Keys to the Common Animals of the North Central States.* 4th ed. Burgess.

John, T. L. 1949. *How to Know the Protozoa.* Wm C Brown.

Lee, J. J., S. H. Hunter, and E. C. Bovee (eds.). 1985. *An Illustrated Guide to the Protozoa.* Society of Protozoologists, Lawrence, Kansas.

Needham, G. H. 1977. *The Practical Use of the Microscope.* Thomas.

Needham, J., and P. Needham. 1941. *Guide to the Study of Fresh-Water Biology.* Comstock.

Palmer, M. C. 1959. *Algae in Water Supplies.* Public Health Service, Publication No. 657.

Taft, C. E. 1961. A Revised Key for the Field Identification of Some Genera of Algae. *Turtox News* 39 (4):98–103.

Exercise

3

Cell Structure and Function

Just as there is diversity of form in organisms, so there is diversity in the form and function of cells that make up organisms. Single cells, such as *Amoeba* and *Paramecium*, can be free-living organisms able to exist independently. Some cells live in a loosely organized colony of similar cells that move from place to place. Others are immovably fixed as part of the tissues of higher plants and animals and depend on closely integrated activities with other cells.

Cells vary in size. Many bacteria are roughly 1 micrometer (μm) long (1 millionth of a meter), while the yolk of an ostrich egg, also a single cell, is the size of a small orange. Cells have special functions, such as the transport of oxygen and carbon dioxide by red blood cells. Whatever its form and function, the cell is recognized as the basic unit of living matter, containing all those properties and processes that are collectively called *life*.

Contemporary biologists recognize two basic types of cells. **Eukaryotic** (Greek *karyon* means "kernel" or "nucleus," *eu* means "good" or "true") cells have a well-defined nucleus, which is separated by a double-layered membrane from the rest of the cell in which the organelles are found. Examples include protozoa and the constituent cells of fungi, plants, and animals. **Prokaryotic** cells, as exemplified by the kingdom Monera (bacteria and cyanobacteria), lack a nuclear membrane and membrane-bound cytoplasmic organelles. The cyanobacteria (formerly the blue-green algae) have a well-developed membranous photosynthetic apparatus that is similar to the components of the chloroplasts of higher plant cells.

The differences between prokaryotic and eukaryotic cells are striking, but they have several characteristics in common. Both are surrounded by a plasma membrane that is very similar in structure although functionally very different. The plasma membrane of prokaryotic cells is the site of energy-yielding reactions that take place in the membranes of mitochondria in eukaryotic cells. Both types have similar enzymes, deoxyribonucleic acid (DNA) as the genetic material, and ribosomes that function in protein synthesis.

Eukaryotic cells are considered more advanced (they evolved later than prokaryotic cells and have a more complex organization), but their many similar characteristics suggest a common ancestor in the evolutionary past.

In this exercise, you will examine examples of prokaryotic and eukaryotic cells and become familiar with the diversity in these organismic groups.

A. PROKARYOTIC CELLS

1. Bacteria

Examine Fig. 3-1A, an electron micrograph of *Azotobacter vinelandii*, a bacterium commonly found in garden soils. In this micrograph, the cell has nearly divided: each cell is about 1 × 1.5 μm in size. The diagram of this bacterium (Fig. 3-1B) shows the cell wall, plasma membrane, ribosomes, and the nucleoid (meaning "resembling a nucleus") region, which contains the DNA but is not membrane-bound.

Because of the small size of prokaryotic cells, you cannot see the structural details with the light microscope that you can with the electron microscope. However, you can observe some of their features if they are stained. To do this, place a drop of the stain crystal violet on a clean slide. Then, using an inoculating loop, transfer a loopful of a broth culture of *Bacillus subtilis* to the drop of stain on the slide. Mix the bacteria in the drop, then add a coverslip. Examine the preparation microscopically. What cellular structures and organelles can you identify that are also evident in the electron micrograph of *Azotobacter*?

____

____

____

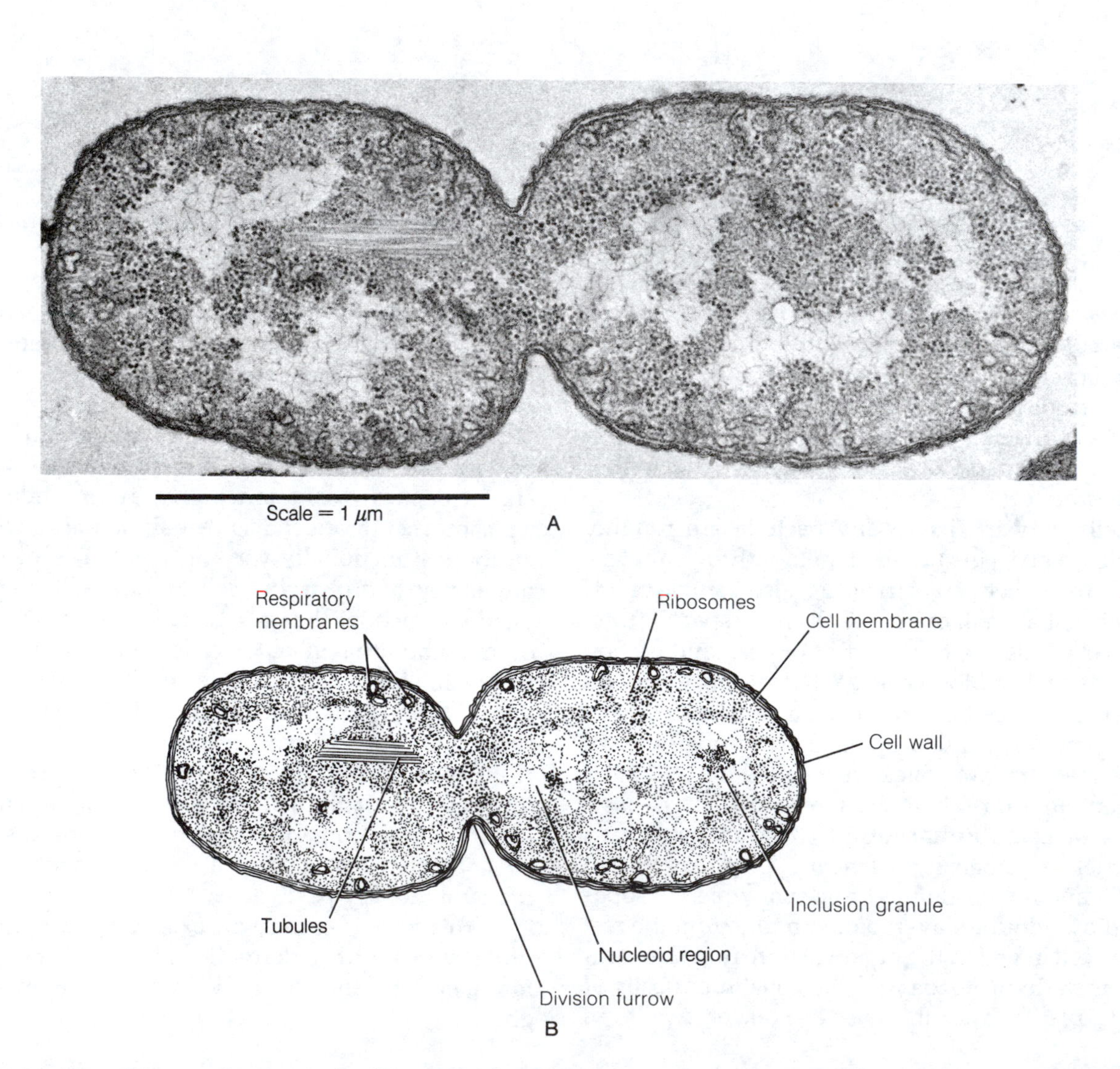

FIGURE 3-1 (A) *Azotobacter vinelandii*, a bacterium found in garden soils. In this transmission electron micrograph, division into two cells is nearly complete. (Micrograph courtesy of W. J. Brill.) (B) Diagram based on the electron micrograph details the subcellular structure of this bacterium. (Diagram by I. Atema.) (From *Five Kingdoms*, 2d ed., by Margulis and Schwartz. W. H. Freeman and Company. © 1987.)

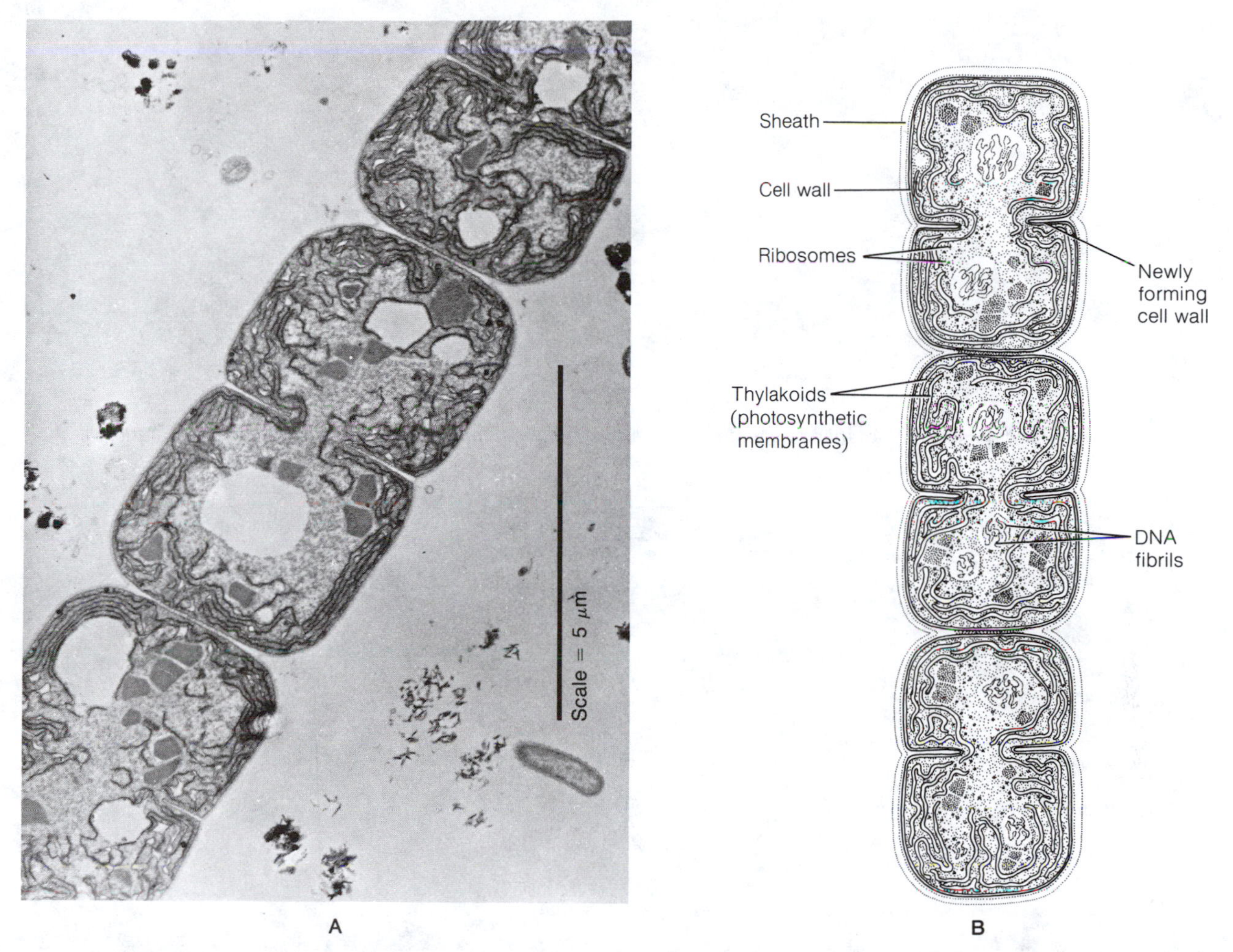

FIGURE 3-2 (A) Transmission electron micrograph of *Anabaena*, a filamentous cyanobacterium that grows in freshwater ponds and lakes. Within the gelatinous sheath, the cells divide by forming crosswalls. (Micrograph courtesy of N. J. Lang.) (B) Diagram of *Anabaena* details the cellular ultrastructure of this organism. (Diagram courtesy of R. Golder.) (From *Five Kingdoms*, 2d ed., by Margulis and Schwartz. W. H. Freeman and Company © 1987.)

2. Cyanobacteria

The internal organization of cyanobacteria is similar to that of bacteria. A major difference is that cyanobacteria contain membranous structures called **thylakoids,** in which the photosynthetic pigments are embedded (Fig. 3-2).

Cyanobacteria are among the common inhabitants of surface waters. This poses some problems because, as human population and industrial demands increase, villages and cities are turning from groundwater to surface waters, such as lakes, streams, and reservoirs, for their sources of water. Groundwaters are essentially free from contaminating organisms, but surface waters contain many organisms that make the water unpalatable. Such organisms, especially some of the cyanobacteria, affect the odor and taste of water; clog filters; grow in pipes, cooling towers, and on reservoir walls; form mats or blooms on the surface of water; and produce toxic materials. Furthermore, the methods by which some companies dispose of waste aggravate the situation. Materials such as sewage and organic wastes from paper mills, fish-processing factories, slaughterhouses, and milk plants greatly increase the growth of algae and other organisms.

To become familiar with the diversity of these organisms, examine a commercially prepared slide that contains several different species of cyanobacteria. Alternatively, collect surface-water samples from various sources, such as lakes, streams, ponds, reservoirs (including the walls), swimming pool walls and filters, and water treatment plants. Compare the organisms you observe with Fig. 3-3

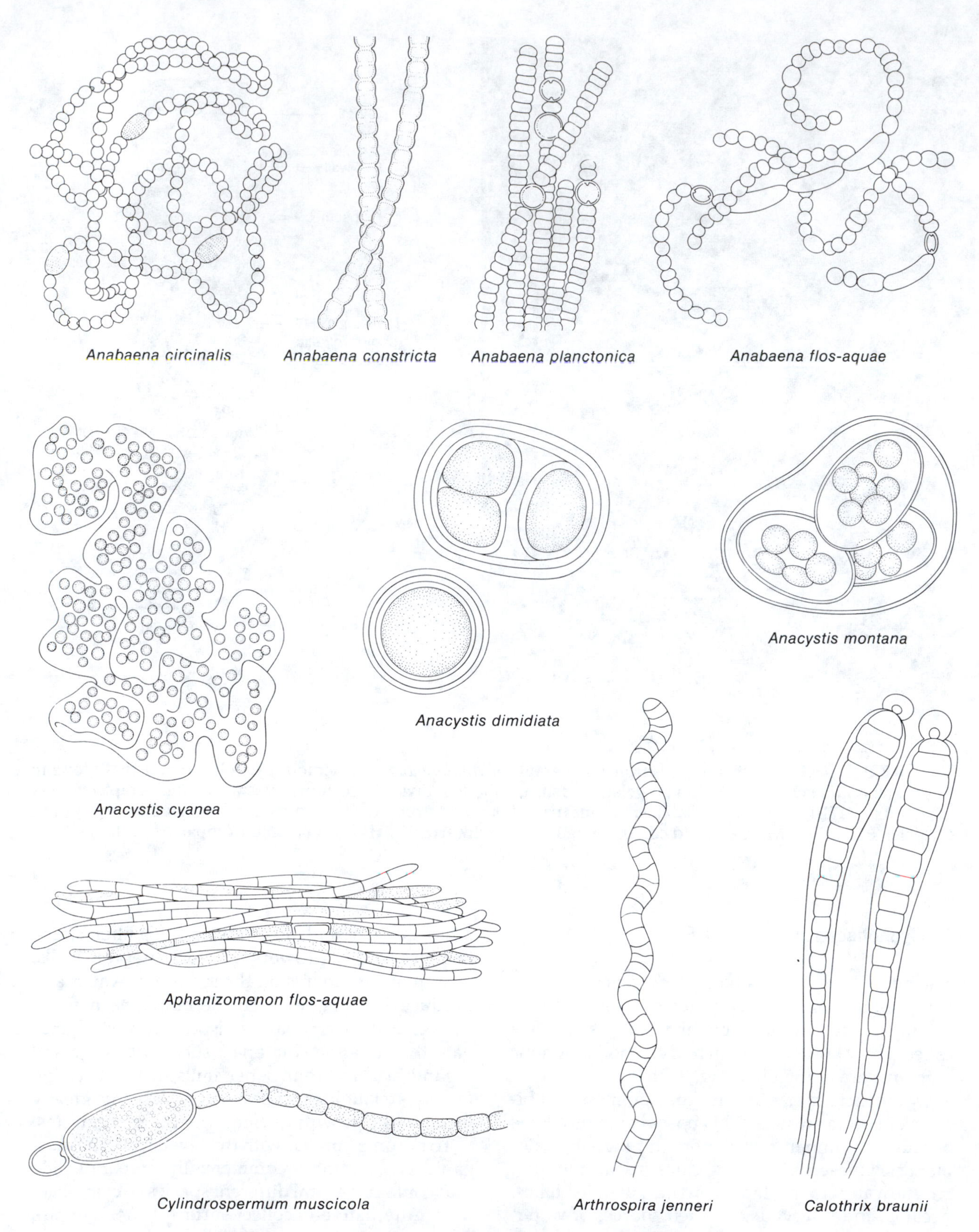

FIGURE 3-3 Cyanobacteria that contaminate surface water supplies

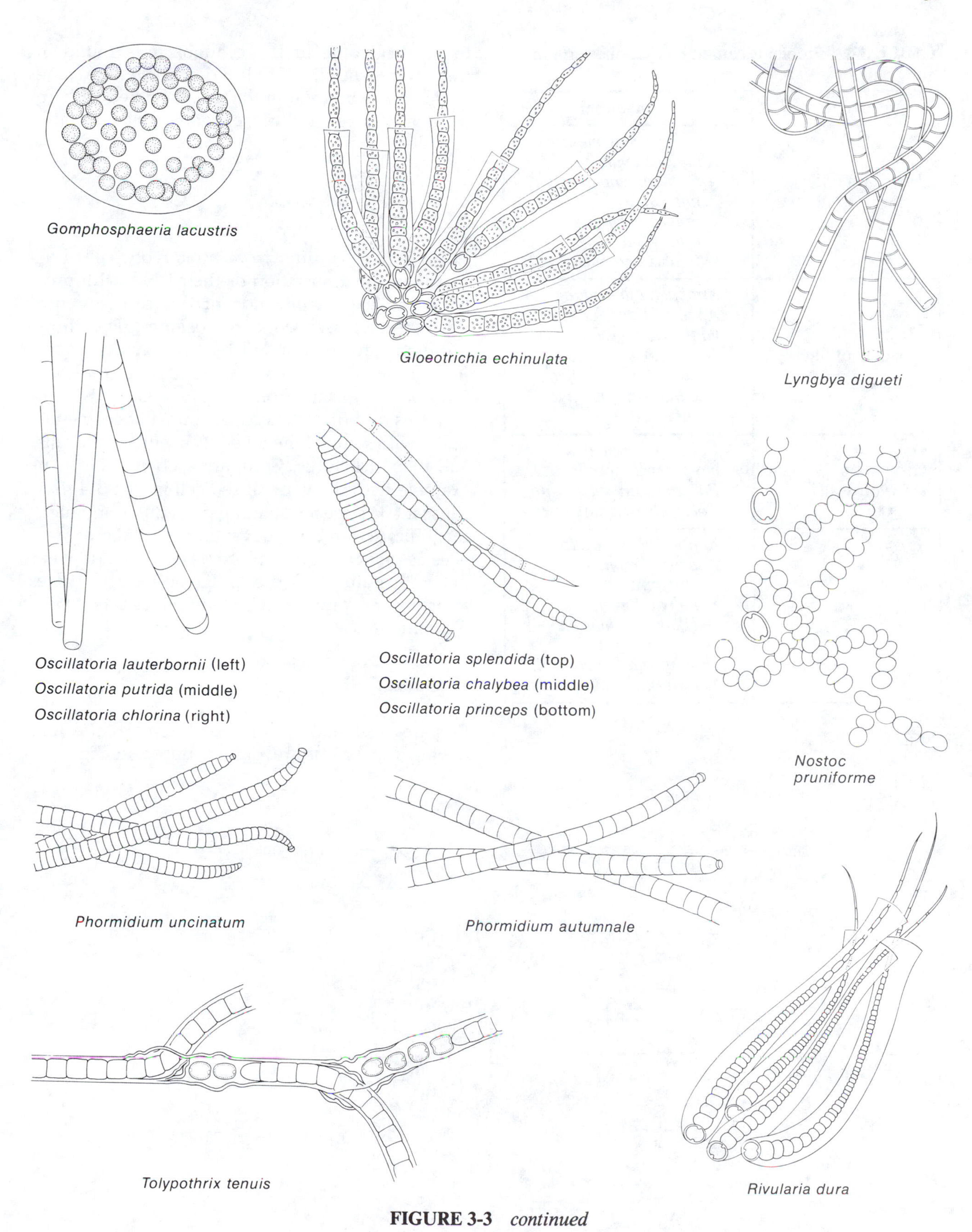

FIGURE 3-3 *continued*

TABLE 3-1 Problems caused by cyanobacteria in water supplies

Problem	Organism
Taste and odor	*Anabaena circinalis* *Anacystis cyanea* *Aphanizomenon flos-aquae* *Cylindrospermum muscicola* *Gomphosphaeria lacustris*
Clogging of filters	*Anabaena flos-aquae* *Anacystis dimidiata* *Gloeotrichia echinulata* *Oscillatoria princeps* *Oscillatoria chalybea* *Oscillatoria splendida* *Rivularia dura*
Growth on reservoir wall	*Calothrix braunii* *Nostoc pruniforme* *Phormidium uncinatum* *Tolypothrix tenuis*
Polluted water	*Anabaena constricta* *Anacystis montana* *Arthrospira jenneri* *Lyngbya digueti* *Oscillatoria chlorina* *Oscillatoria putrida* *Oscillatoria lauterbornii* *Phormidium autumnale*

to identify some of the cyanobacteria found in these water supplies. Table 3-1 characterizes some of the more common problems associated with an overabundance of these cyanobacteria in water supplies.

B. EUKARYOTIC CELLS

Eukaryotic cells differ from prokaryotic cells primarily in the association of their DNA with proteins and the organization of this complex into large structures called **chromosomes.** The chromosomes are surrounded by the **nuclear envelope,** a double membrane that separates the contents of the nucleus from the cytoplasm.

In this part of the exercise, you will become familiar with some of the diversity in eukaryotic cells and their structures. Compare each cell that you examine with the generalized cell in Fig. 3-4. Because it takes special staining procedures and high magnification to bring out such subcellular organelles as mitochondria, chloroplasts, the endoplasmic reticulum, and the Golgi apparatus, these structures may not be evident in the cells that you examine.

1. Onion Cells

Prepare a wet mount of onion epidermal tissue, following the procedure demonstrated in Fig. 3-5

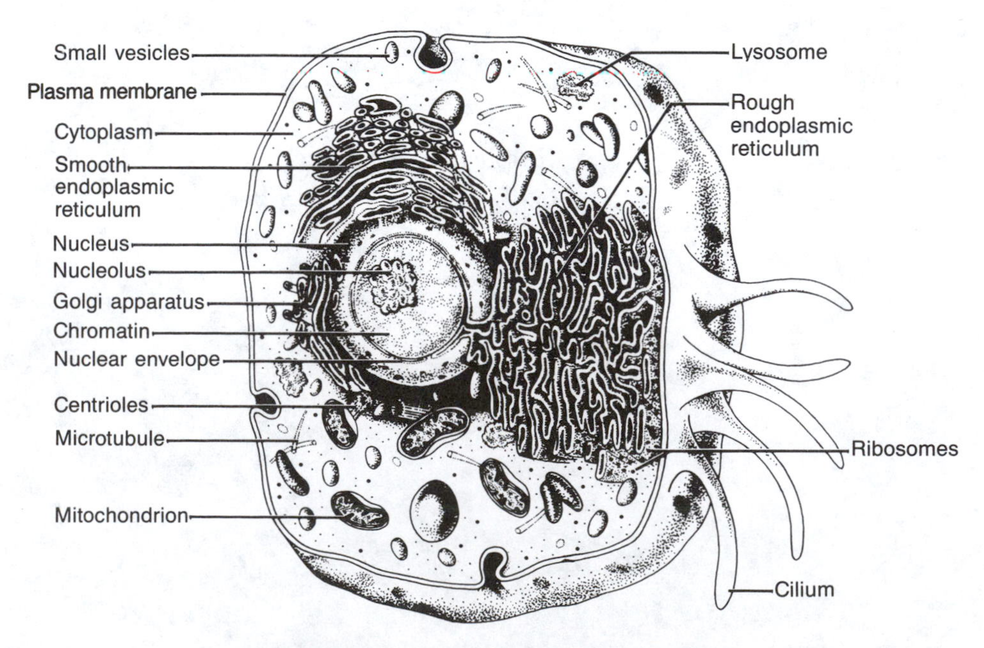

FIGURE 3-4 Some intracellular components

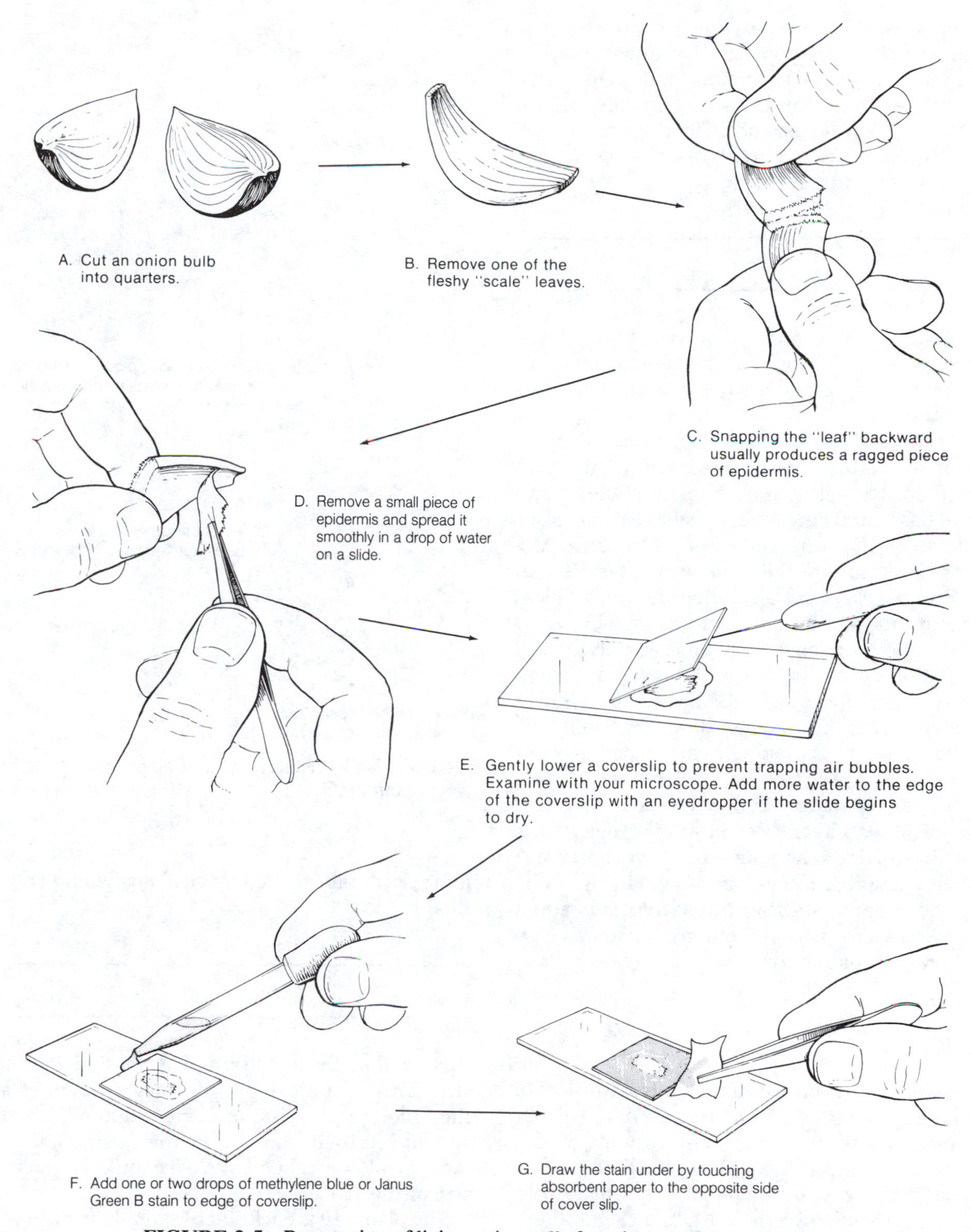

FIGURE 3-5 Preparation of living onion cells for microscopic examination

A–E. Examine this tissue with the low-power objective.

The "lines" that form the network between the cells are nonliving cell walls composed chiefly of cellulose. The cell wall surrounds the **plasma membrane,** which encloses the **cytoplasm.** The central part of many plant cells (which is difficult to observe in living cells) is taken up by a **vacuole** that is filled with water and salts.

Next examine the cells under high power. Locate

the **nucleus,** which appears as a dense structure in the translucent cytoplasm. Note that in some cells, the nucleus looks circular and seems to be lying in the central part of the cell. In other cells, it seems to be compressed and pushed against the cell wall. Explain this apparent discrepancy in the shape and position of the nucleus.

___

___

The central vacuole, nucleus, and cell wall are separated from the cytoplasm by membranes, but the membranes are difficult to observe in this preparation.

Mitochondria, organelles that are involved in cellular respiration, can be observed in the onion cell. To do this, select another piece of onion epidermal tissue and cut it with a razor so that it is approximately 1 × 3 millimeters (mm) in size. Mix three to five drops of the stain Janus Green B with one drop of 5% sucrose solution on a clean glass slide and mount the tissue. Add a coverslip.

If your preparation is truly one cell thick, it will look like a nearly transparent brick wall when viewed under low power. Using the high-power objective, locate the mitochondria, which will look like very small rods or spheres at the periphery of the cell. They should be blue in color when you first view your preparation. If they are not, add a few drops of Janus Green B stain at one edge of the coverslip and draw the stain through your preparation with absorbent paper, as shown in Fig. 3-5F, G. Stained mitochondria lose their color after about 5 minutes as a result of the action of an enzyme located on their membranes.

2. *Elodea* Cells

In this study, you will examine cells from the leaf of an aquatic plant called *Elodea* (Fig. 3-6). These cells are green because they have **chloroplasts,** which contain a pigment called **chlorophyll. Photosynthesis** is the process by which this pigment absorbs light energy and converts it into the chemical energy of organic molecules.

Place a young leaf from the tip of the plant in a drop of water on a slide and add a coverslip. Examine the preparation with the low-power objective. Locate the nucleus, cytoplasm, and cell wall.

Examine a group of cells near the center of the leaf. Carefully switch to high power. Bring the cells into focus by using the fine adjustment and determine the number of cell layers. How would you do this?

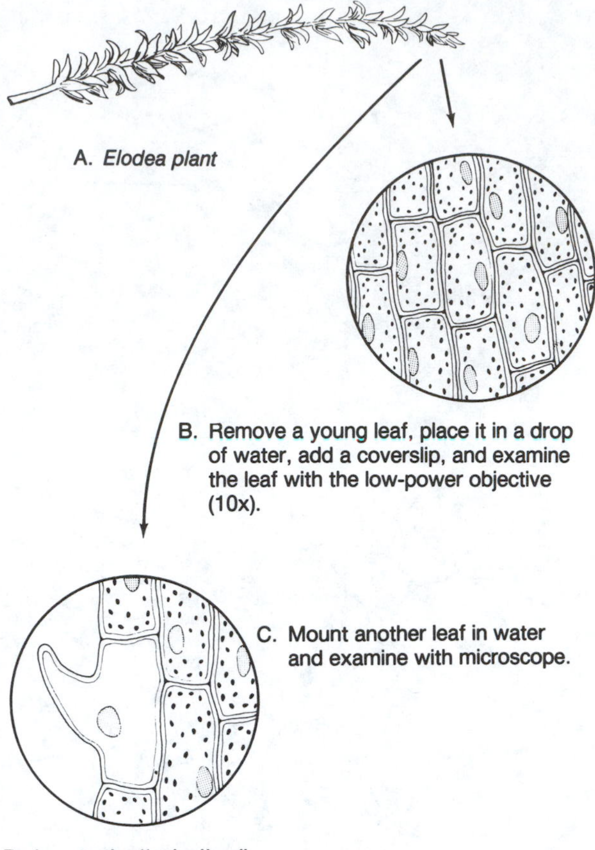

FIGURE 3-6 Preparation of *Elodea* cells for microscopic examination

___

___

Note that the chlorophyll is located in small structures in the cytoplasm. These structures are the chloroplasts. Examine the chloroplasts; you should see them moving in the cytoplasm. This movement is called **cyclosis** or **cytoplasmic streaming.**

The plant cell is enclosed by a nonliving cell wall and a plasma membrane that is difficult to observe because it is pushed tightly against the inside of the cell wall by the pressure of the cytoplasmic fluid. This membrane can be observed by placing the cell in a hypertonic saline solution (i.e., a solution that is more concentrated that the cytoplasm). Under these conditions, water moves out of the cell and the membrane shrinks away from the cell wall. If you adjust the diaphragm for more contrast, you should easily see the plasma membrane.

Select another young *Elodea* leaf, mount it in a drop of water, and add a coverslip. Examine the preparation with the low-power objective. Along the edges of the leaf, locate "spine" cells (Fig. 3-6D). Switch to high power and study this cell.

Add one or two drops of a hypertonic saline solution to one edge of the coverslip and draw it under the coverslip with a piece of absorbent paper. Repeat this step twice more to be sure that the original water has been replaced by the saline. Examine the spine cell closely. Describe your observations and account for them.

___

___

___

3. Human Epidermal Cells

Following the procedures outlined in Fig. 3-7, examine cells obtained from the epidermal lining of your inner cheek.

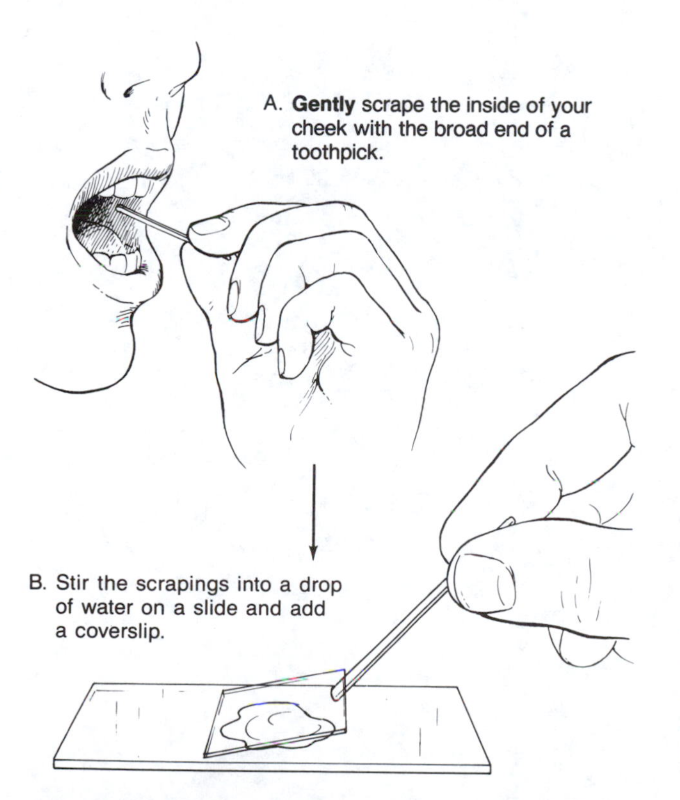

FIGURE 3-7 Preparation of human epidermal cells for microscopic examination

---

Caution: *Do not give your toothpick to any other student. Discard the toothpick in the designated container as soon as you have prepared your slide.*

Locate the cells under high power. What do these epidermal cells have in common with plant cells?

___

___

How are they different?

___

___

The edges of many of the epidermal cells may be folded over. What does this indicate about the thickness of these cells?

___

___

Add a drop of methylene blue to the edge of the coverslip and draw it under as demonstrated in Fig. 3-5F, G. What structure in the cell has been stained by this dye?

___

C. CYTOPLASMIC STREAMING (CYCLOSIS)

The phenomenon of cyclosis is easily observed in the plasmodium of the slime mold *Physarum polycephalum.* The **plasmodium** is a multinucleate mass of protoplasm that lacks a cell wall. This organism is easily propagated from its dormant stage, which is a hard, crusty structure called a **sclerotium,** by placing pieces of the sclerotium on a moist substrate containing nutrients (e.g., agar). In a short time, the organism begins to grow out over the surface. Channels of streaming cytoplasm become visible, usually after 72 hours of growth.

Examine a plasmodium of *Physarum* growing on agar in a petri dish. Leave the cover on while you examine the organism carefully using a dissecting microscope. What seems unusual about cytoplasmic streaming in *Physarum*?

____

____

____

____

____

____

REFERENCES

Alberts, B., et al. 1983. *Molecular Biology of the Cell.* Garland.

Cairns, J. 1966. The Bacterial Chromosome. *Scientific American* 214(1):36–44 (Offprint 1030). *Scientific American* Offprints are available from W. H. Freeman and Company, 41 Madison Avenue, New York, NY 10010, and 20 Beaumont Street, Oxford OX1 2NQ, England. Please order by number.

Holtzman, E., and A. B. Novikoff. 1984. *Cells and Organelles.* 3d ed. Saunders.

Karp, G. 1984. *Cell Biology.* 2nd ed. McGraw-Hill.

Lodish, H. F., et al. 1995. *Molecular Cell Biology.* 3d ed. W. H. Freeman.

Prescott, D. M. 1988. *Cells: Principles of Molecular Structure and Function.* Jones and Bartlett.

Staehelin, L. A., and B. E. Hull. 1978. Junctions Between Living Cells. *Scientific American* 238(5):140–152 (Offprint 1388).

Swanson, C. P., and P. L. Webster. 1985. *The Cell.* 5th ed. Prentice-Hall.

Exercise

4

Subcellular Structure and Function

A basic problem in microscopy is the limited resolving power (resolution) of light microscopes. For example, subcellular structures that are smaller or closer together than 0.1 micrometer (μm) cannot be clearly seen through a light microscope. Cell membranes and other organelles are indeed smaller or closer together than this.

Microscopes that utilize electrons rather than light for illumination have been in use for some time now. Theoretically, the resolution of such instruments could be at least 100,000 times greater than that of light microscopes because electrons have a wavelength of 0.005 nanometer (nm) compared with the 500-nm average wavelength of visible light. Thus, the development of the **electron microscope** significantly enhanced our understanding of the subcellular structure of cells.

Exercise 3 introduced you to the structure of plant and animal cells as viewed with the light microscope, and you examined a number of different types of cells to become acquainted with their diversity of structure and function. In this exercise, you will become familiar with **transmission** and **scanning electron microscopes** and with the structure and function of selected subcellular components. You also will learn how to determine the dimensions of objects seen with the electron microscope.

A. SUBCELLULAR ORGANIZATION

Microscopic analysis has shown that eukaryotic cells contain numerous specialized structures that carry out a variety of activities. These activities include acquisition and assimilation of nutrients, elimination of wastes, synthesis of new cellular materials, movement, and reproduction. All cells have an internal structure that includes specialized organelles that carry out these various functions. It is important to understand that the cell is not a random assortment of parts but a highly structured and integrated entity (Fig. 4-1).

Not long ago, cells were visualized as fluid-filled bags containing enzymes and other dissolved molecules, along with the nucleus, some mitochondria, and a few other organelles that could be made visible by using special staining techniques. With the development of modern microscopic procedures, many more structures have been identified, each specialized to carry out specific functions. In addition to this variety of organelles, electron microscopy has revealed previously unknown interconnections among filamentous structures in the cytoplasm. These structures form the **cytoskeleton,** which maintains cell shape, enables cell movements, and anchors the various organelles.

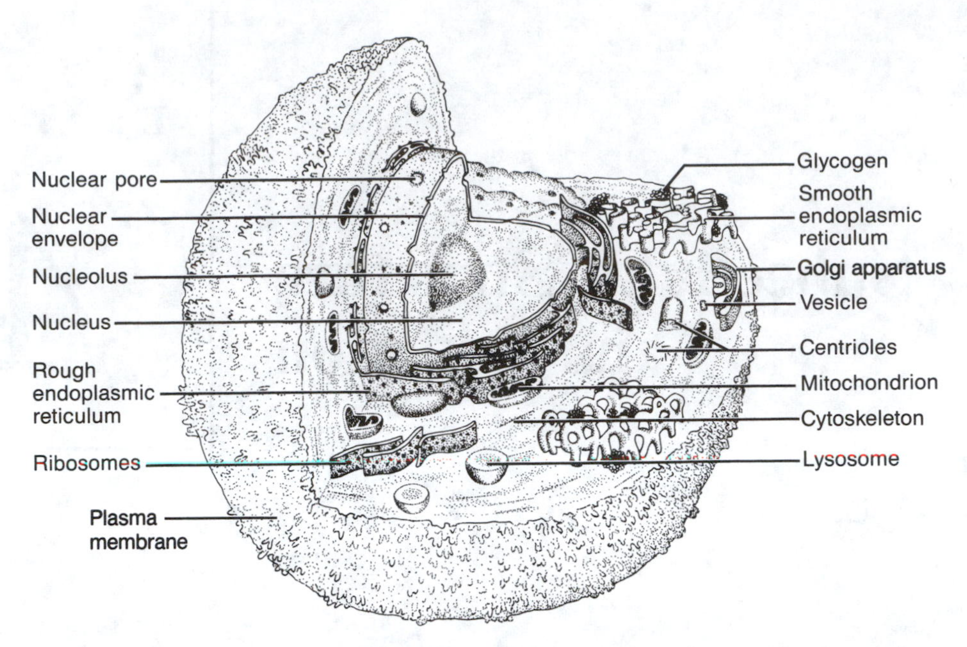

FIGURE 4-1 Composite cell as interpreted from electron microscopic observations

1. The Plasma Membrane

Cells exist as separate and distinct entities because they are surrounded by a **plasma membrane** that regulates the movement of materials into and out of the cell. The plasma (or cell) membrane, the structure of which cannot be seen through the light microscope, is approximately 6–9 nm thick and appears as a thin double line through the transmission electron microscope (Fig. 4-2A). The **fluid mosaic model** of membrane structure (Fig. 4-2B) shows that the framework of animal cell membranes is a lipid bilayer consisting of cholesterol and phospholipid molecules positioned so that their hydrophobic ends are pointing inward. Embedded in or attached to this lipid bilayer are proteins with various functions, which can move within the layer because it is in a fluid state. Short carbohydrate chains, which are attached to some of the protein and phospholipid molecules, protrude out to the external environment and serve as molecular recognition sites for the attachment of hormones, antibodies, and other regulatory molecules. Other proteins are attached to the side of the membrane facing the cytoplasm. All plasma membranes have this structure, though there are differences in the types of lipids, proteins, and carbohydrates they contain. These differences are important because the variations in membrane molecules confer different properties to membranes, correlated with the distinctive functions of the various types of cells and their organelles.

Prokaryotic and plant cells have essentially the same membranous structure as eukaryotic cells, except that most prokaryotes do not have cholesterol in the lipid bilayer. Rather, a different steroid is present whose function in the membrane is similar to that of cholesterol.

2. Nucleus

The nucleus is a large, usually spherical structure bounded by two lipoprotein membranes that together constitute the nuclear envelope (Figs. 4-1 and 4-3). The membranes are fused together at frequent intervals to create pores through which materials pass between the nucleus and the cytoplasm.

Every nucleus (except those in the gametes) contains a copy of the organism's entire complement of the genetic information (deoxyribonucleic acid—DNA) that determines the organism's development and governs its various functions. It does this by influencing the activities of each cell to ensure that it synthesizes the various complex molecules needed by that cell and, in some cases, other cells.

3. Endoplasmic Reticulum and Ribosomes

Throughout the cytoplasm is a network of interconnecting, flattened sacs, tubes, and channels that form the **endoplasmic reticulum** (ER) (Fig. 4-4). The ER has the same membrane structure as the plasma membranes and is frequently continu-

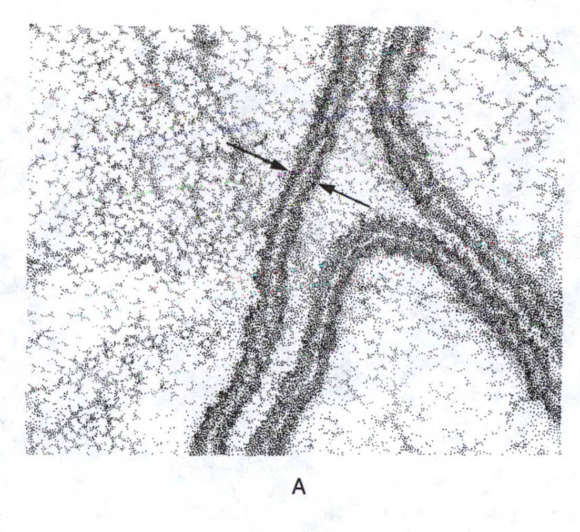

A

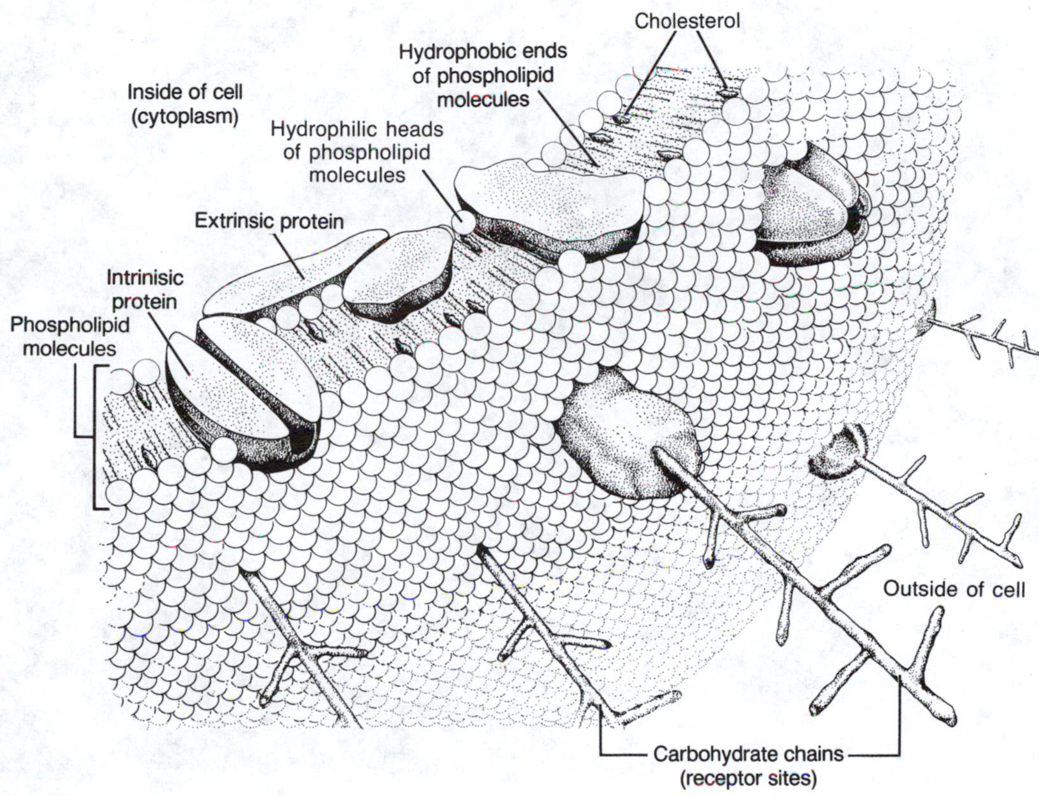

B

FIGURE 4-2 (A) Transmission electron micrograph of plasma membrane (between arrows) consisting of two dark layers, the hydrophilic heads of lipid molecules, and a light layer, the hydrophobic tails of lipid molecules. (B) Fluid mosaic model of the plasma membrane showing orientations of the hydrophilic and hydrophobic ends of lipid molecules, intrinsic proteins (those that penetrate the bilipid layer), extrinsic proteins (those attached to surface of the lipid layer), and carbohydrate chains (that serve as receptor sites for hormones and antibodies).

ous with the plasma and nuclear membranes. It can be made up of rough or smooth membranes. Those cells that are active in protein synthesis contain rough ER, which is made "rough" by the presence of structures called **ribosomes.** Ribosomes are composed of protein and ribonucleic acid (RNA) (Fig. 4-4). Smooth ER lacks the large accumulations of ribosomes.

It is thought that proteins, synthesized by the ribosomes are released into the channels of the endoplasmic reticulum, stored, and then transported to other parts of the cell or to the outside.

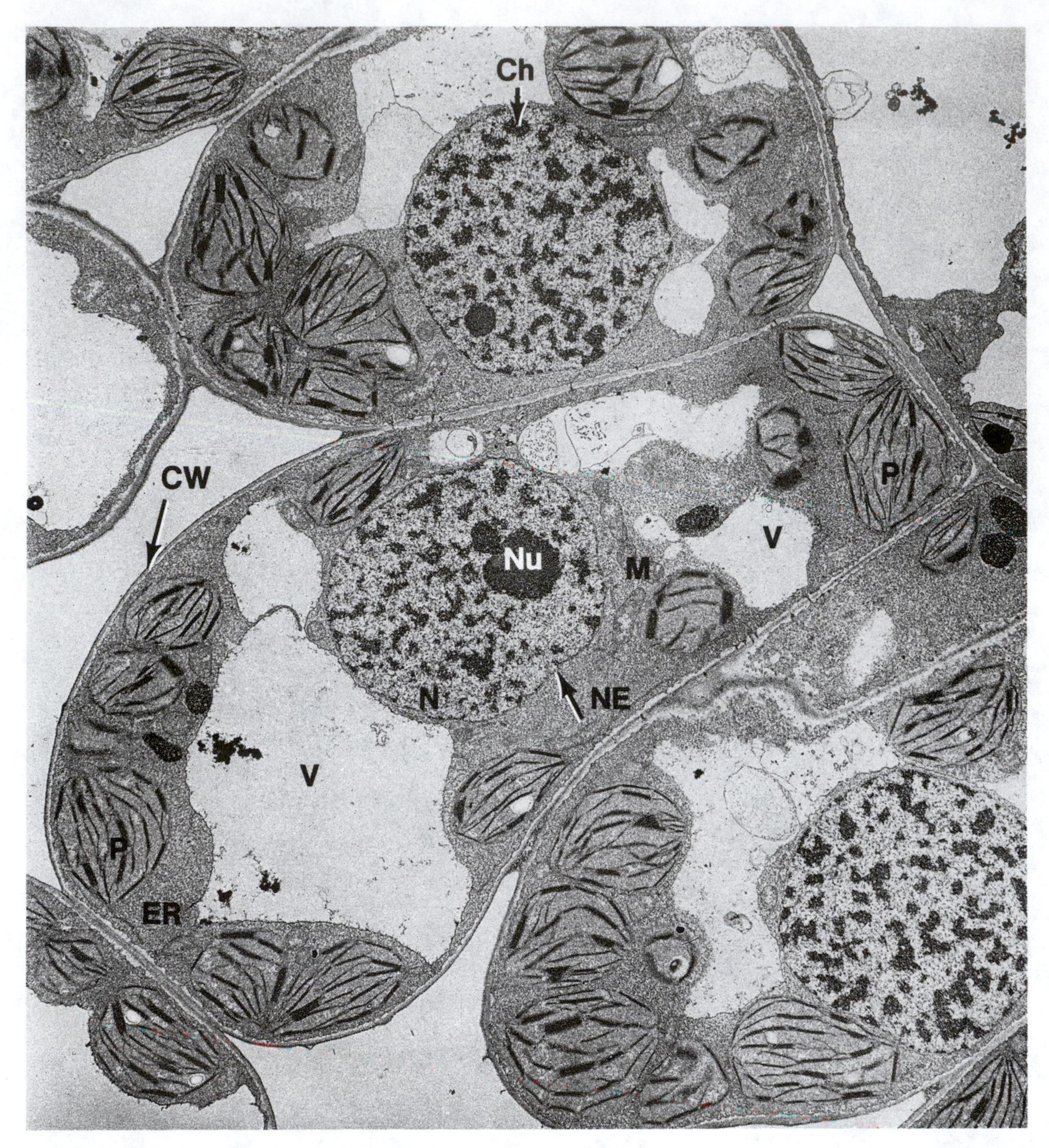

FIGURE 4-3 Structure of a higher plant cell: N, nucleus; Nu, nucleolus; NE, nuclear envelope; Ch, chromatin; CW, cell wall; V, vacuole; P, plastids; M, mitochondrion; ER, endoplasmic reticulum.

4. Mitochondria

Mitochondria are subcellular structures that participate in cell respiration. Fine structure (i.e., electron microscopic) analysis of these organelles shows them to consist of two membranes: an outer membrane and an inner membrane that invaginates (folds in upon itself) to the interior to form **istae** (Fig. 4-5). The more active the mitochon- the more cristae it is likely to have. The space between the membranes is filled with a fluid that contains some of the enzymes of the Krebs cycle. The inner and outer membranes have thousands of small particles on their surfaces. Those on the inner membrane are attached by small stalks. These particles contain many of the enzymes and electron carrier molecules involved in cellular respiration.

The stalkless particles on the outer surface of the outer membrane carry out the various reactions that provide electrons to the interior of the mitochondrion. The stalked particles on the surface of the inner membrane are involved in the transfer of electrons along a chain of electron transport molecules and ultimately synthesize adenosine triphosphate (ATP) through a process called **oxidative phosphorylation.**

5. Golgi Apparatus

The **Golgi apparatus** is another system of membranes found in both plant cells and animal cells. As shown in Fig. 4-6, the apparatus consists of parallel flattened sacs, the **cisternae.** At the margins,

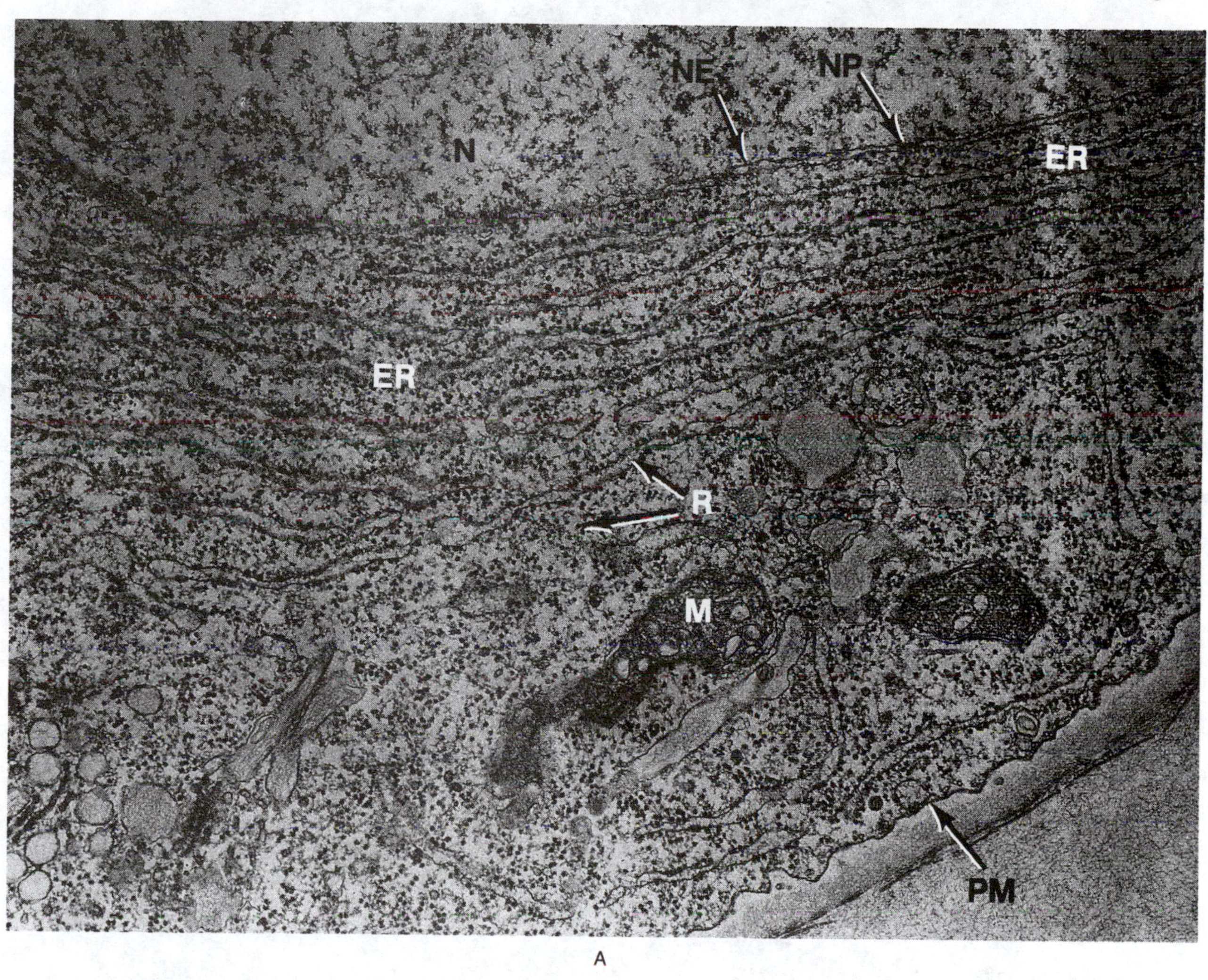

A

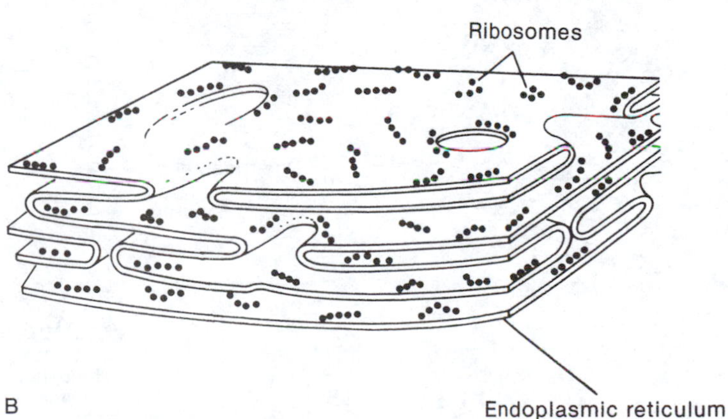

B

FIGURE 4-4 (A) Transmission electron micrograph showing subcellular structure of cells: N, nucleus; NP, nuclear pore; NE, nuclear envelope; ER, endoplasmic reticulum; M, mitochondrion; R, ribosome; PM, plasma membrane. (B) Structural organization of the endoplasmic reticulum.

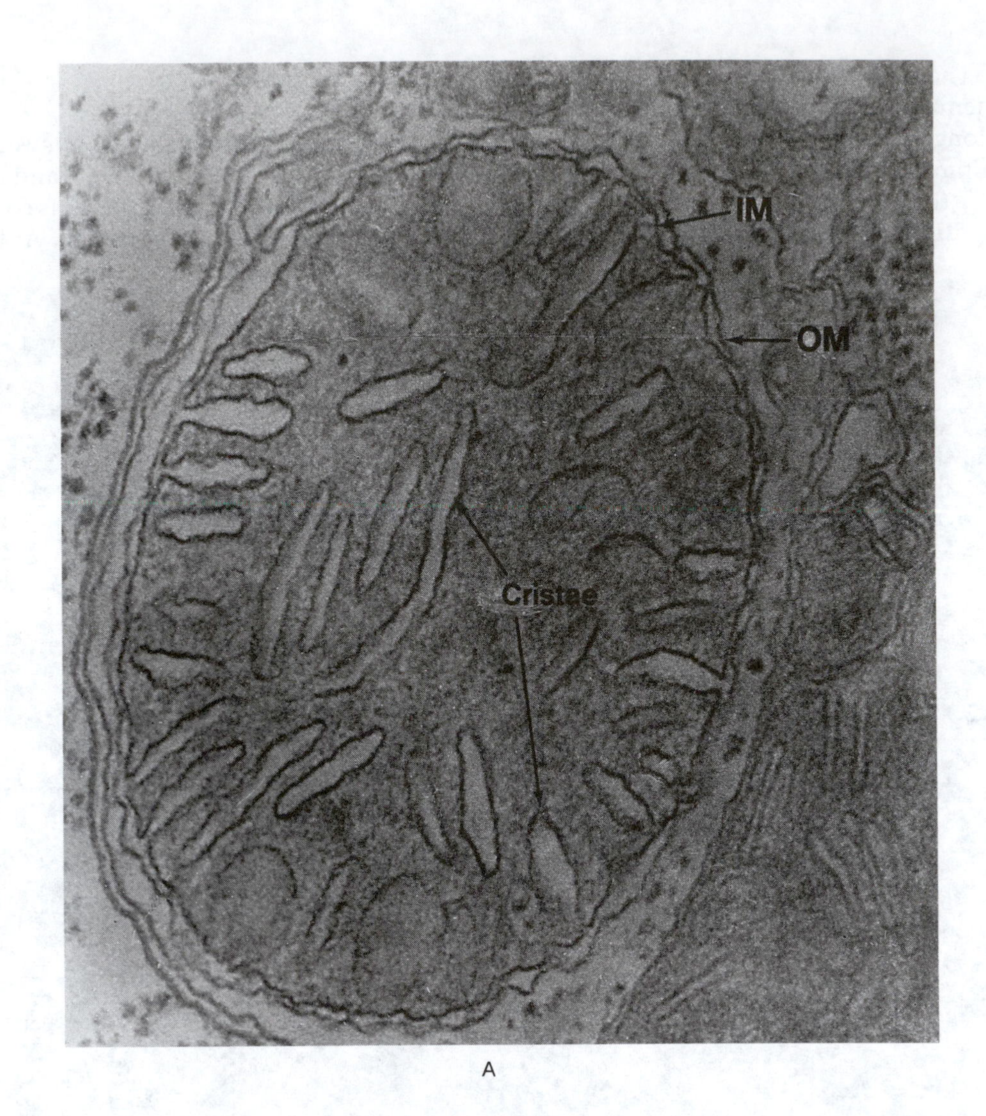

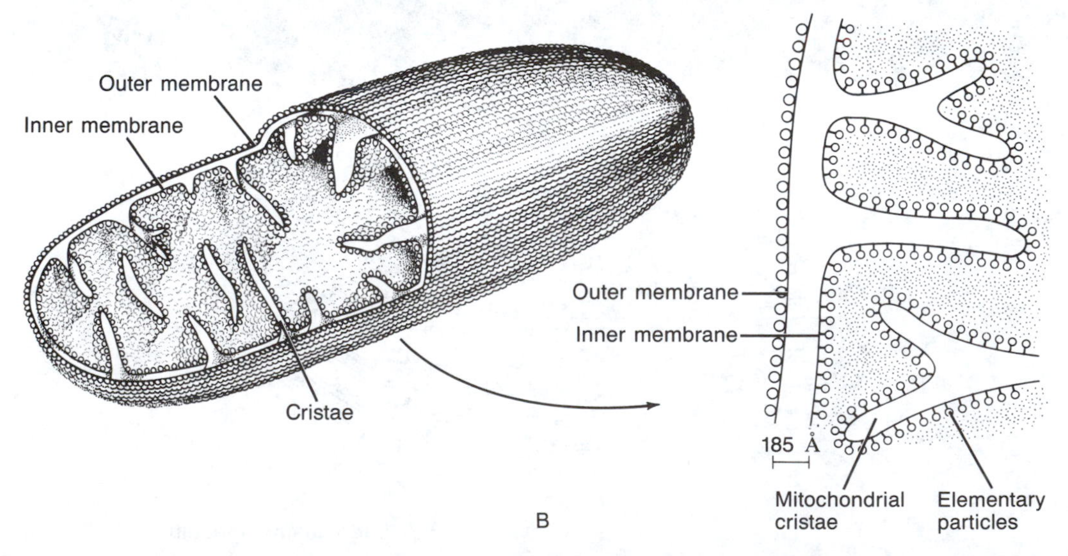

FIGURE 4-5 (A) Fine structure of a mitochondrion: IM, inner membrane; OM, outer membrane; cristae. (B) Structural organization of mitochondria showing arrangement of elementary particles, stalked on inner membrane surface and stalkless on outer membrane surface.

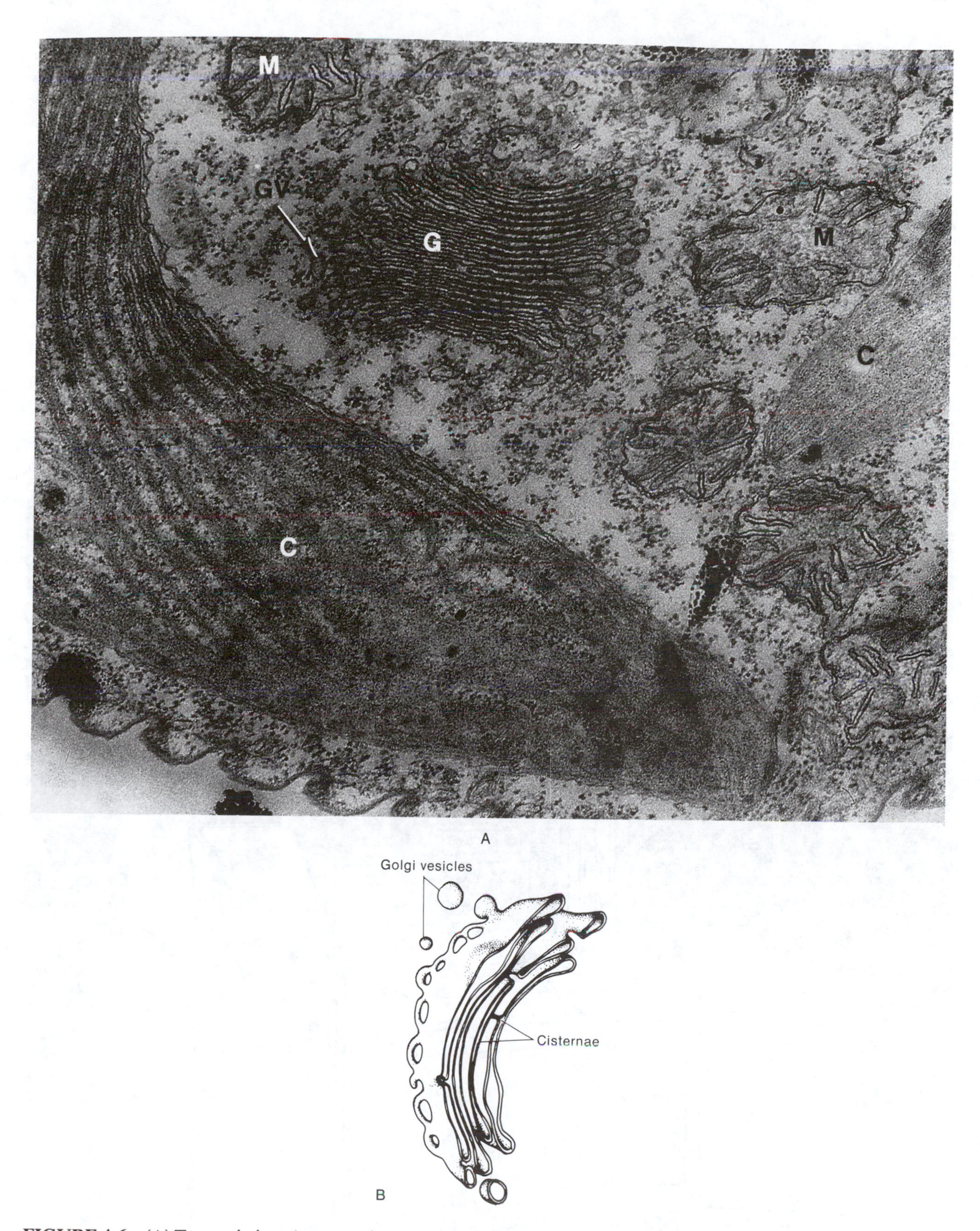

FIGURE 4-6 (A) Transmission electron micrograph of the alga *Euglena* showing G, Golgi apparatus; GV, Golgi vesicle; M, mitochondrion; C, chloroplast. (B) Structural organization of the Golgi apparatus.

the cisternae form spherical vesicles by a "pinching off" process. The Golgi apparatus also receives vesicles formed by the endoplasmic reticulum, modifies the membranes of these vesicles, and further processes and distributes their contents to other parts of the cell, particularly the cell surface. The Golgi apparatus is the packaging and distribution center of the cell. Indeed, the final assembly of the various proteins and carbohydrates associated with cell membranes and organelles takes place in these structures.

B. TRANSMISSION ELECTRON MICROSCOPE

A transmission electron microscope (TEM) is shown in Fig. 4-7. The microscope consists of a tall, central column and various electronic equipment including a beam detector and cathode ray viewing screen, vacuum pumps, and electrical and plumbing components.

The TEM is essentially a vertical television tube; the electron gun (or source) is at the top and the viewing screen is at the bottom. The electrons emerge from the hot filament of the electron gun, and high voltages accelerate them down the tube to the viewing screen. The column must be evacuated of air so that the electrons can pass without interference. Magnets on the column act like the adjustment knobs on the light microscope to focus the streaming electrons onto the specimen.

When the electrons strike the screen, it glows. If a properly prepared specimen is placed in a holder in the column, an image of the specimen appears on the screen as dark and light areas. The dark areas

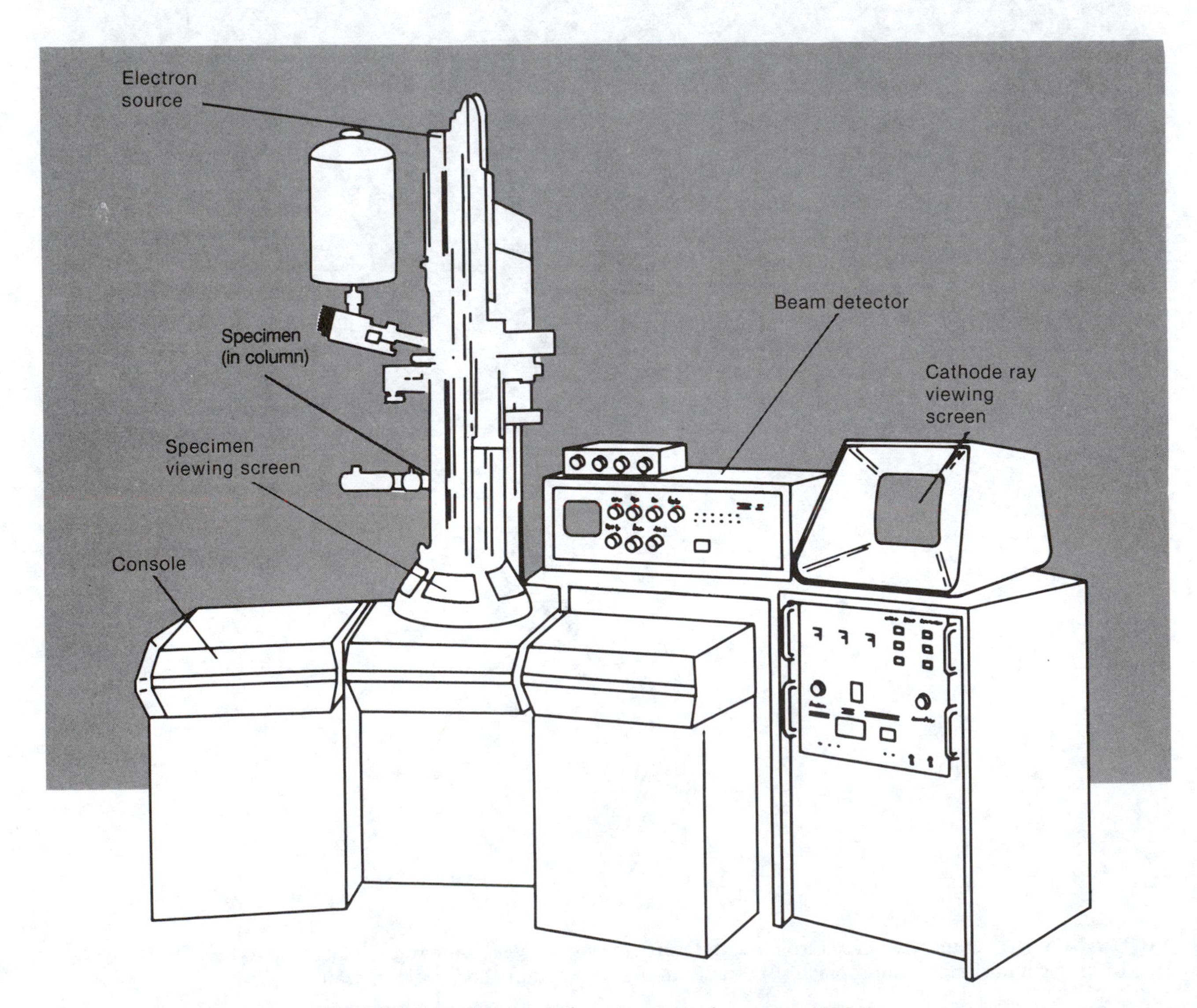

FIGURE 4-7 Transmission electron microscope (TEM).

correspond to electron-opaque regions of the specimen because electrons are deflected and do not pass through these regions. The light areas represent the electron-transparent areas of the specimen. Normally, the images that form are observed by photographing them. This is done by moving the screen out of the way and allowing the electrons to strike a photographic film below. This technique produces images like those shown in Figs. 4-8 and 4-9. Keep in mind, however, that TEM photographs can be taken only of very thin sections of material. Thus a structure that is present in the cell may not appear in the photograph because it happened not to be in that particular slice. This is somewhat similar to slicing hard salami: not all slices contain a peppercorn. Remember this concept when using electron micrographs to analyze cells for the presence of various organelles.

C. SCANNING ELECTRON MICROSCOPE

If we consider the transmission electron microscope as being analogous to the compound light microscope, then the scanning electron microscope (SEM) is the counterpart of the dissecting microscope. In preparing specimens for the SEM, whole,

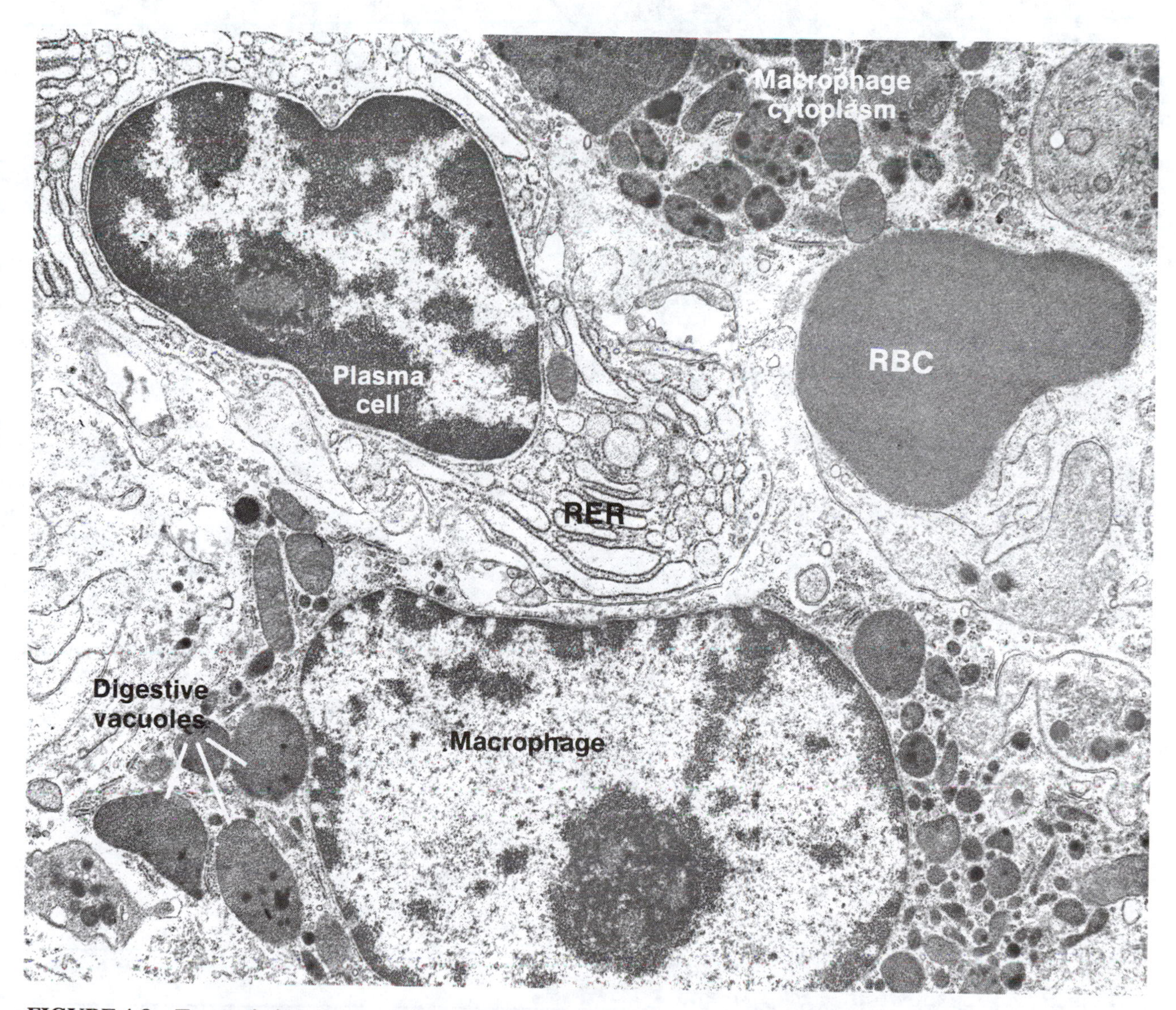

FIGURE 4-8 Transmission electron micrograph showing a macrophage, a plasma cell with rough endoplasmic reticulum (RER), and a red blood cell (RBC). Digestive vacuoles in the macrophage function in hydrolysis of material taken into cell by phagocytosis (×11,390). (From *Tissues and Organs: A Text-Atlas of Scanning Electron Microscopy* by R. G. Kessel and R. H. Kardon. W. H. Freeman and Company © 1979.)

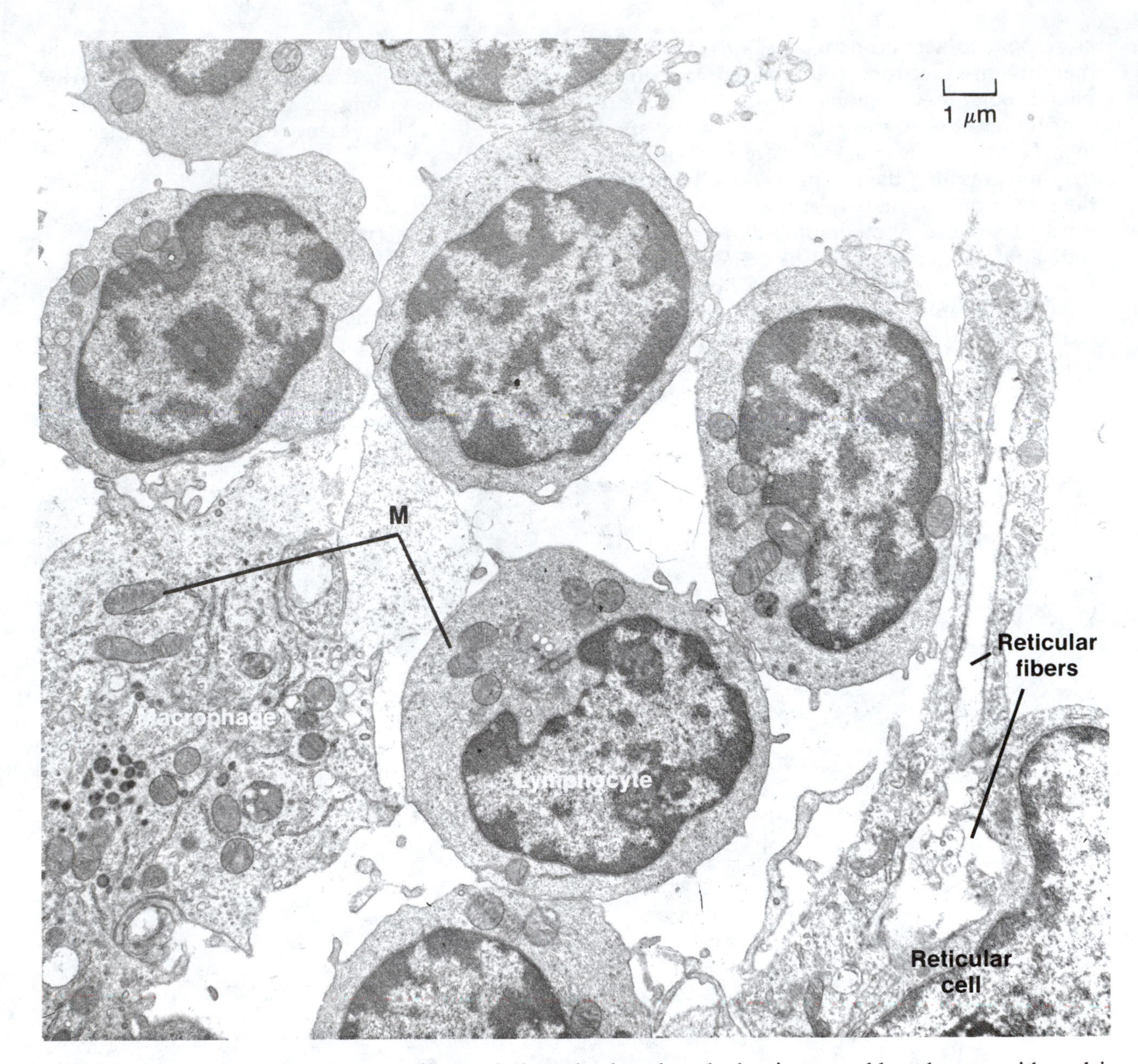

FIGURE 4-9 Transmission electron micrograph through a lymph node showing several lymphocytes with nuclei (N), macrophages with mitochondria (M), and a reticular cell with reticular fibers (×8545). (From *Tissues and Organs: A Text-Atlas of Scanning Electron Microscopy* by R. G. Kessel and R. H. Kardon. W. H. Freeman and Company. Copyright © 1979.)

not sectioned, materials are used because the image is formed by reflected rather than transmitted electrons. As shown in Figs. 4-10 and 4-11, SEM images are three-dimensional. What advantages are provided by the SEM over the TEM?

____

____

____

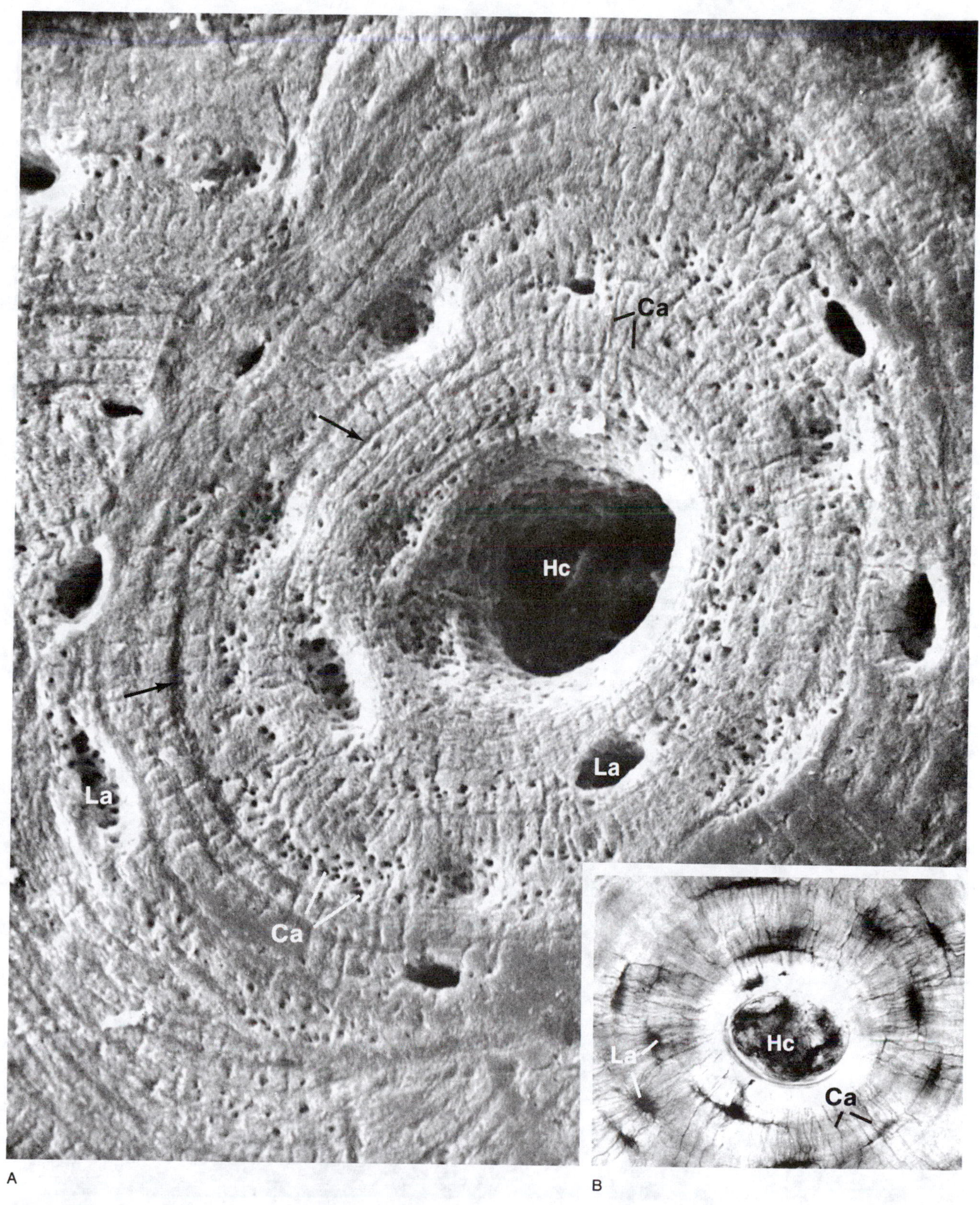

FIGURE 4-10 (A) Scanning electron micrograph of compact bone (×1345). (B) Light microscope photograph of a similar specimen. La, lacunae or spaces in which bone cells, called osteocytes, are found; Ca, canaliculi (small channels between lacunae); Hc, Haversian canal containing blood vessels. (From *Tissues and Organs: A Text-Atlas of Scanning Electron Microscopy* by R. G. Kessel and R. H. Kardon. W. H. Freeman and Company. Copyright © 1979.)

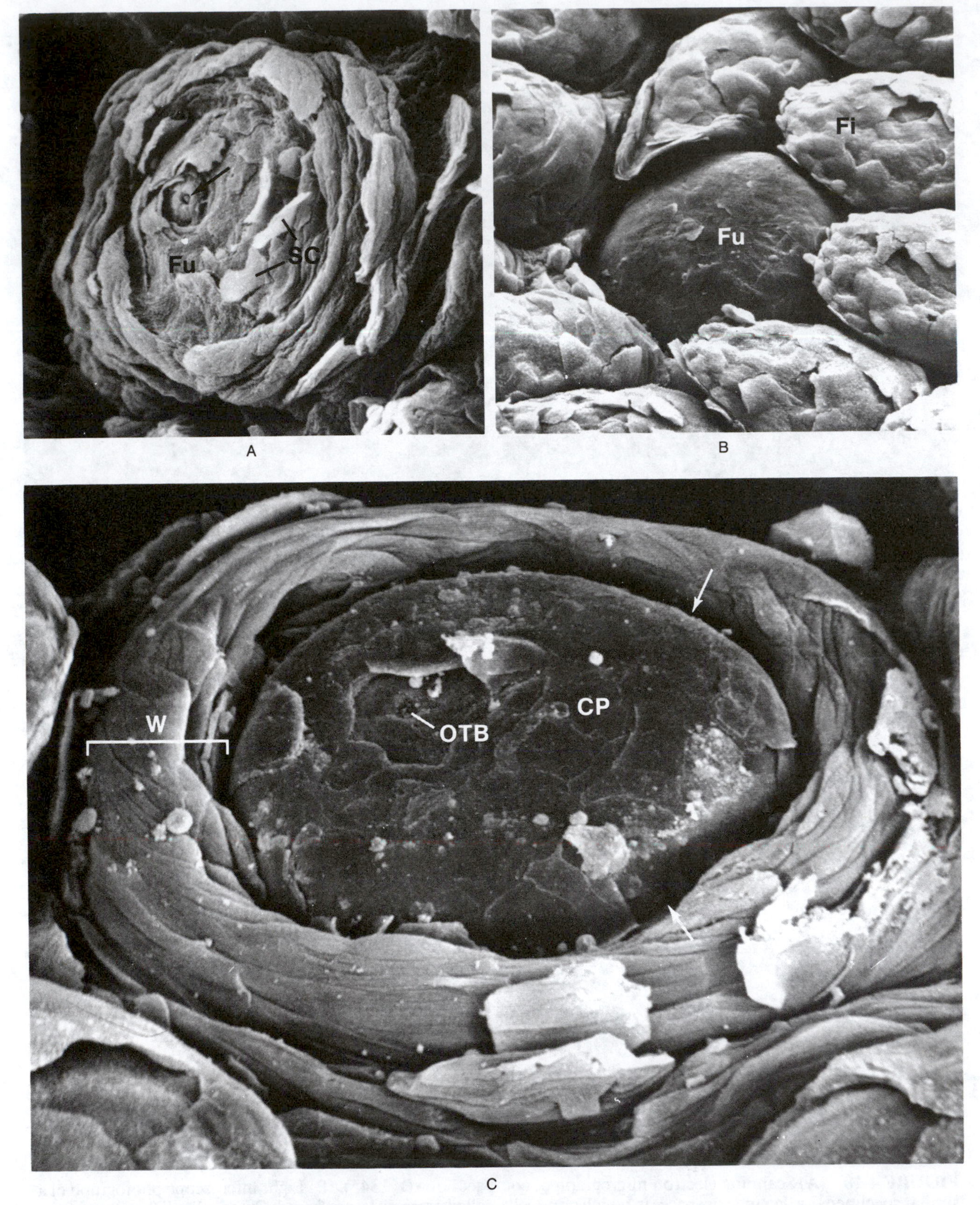

FIGURE 4-11 Scanning electron micrographs of papillae, found on the surface of the tongue, that have taste buds associated with them. (A) and (B) are fungiform (Fu) and filliform (Fi) papillae; (C) is a circumvallate papilla (CP); OTB, opening of pore to taste bud; SC, squamous cells. (Magnifications: (A) ×615; (B) ×245; (C) ×910.) (From *Tissues and Organs: A Text-Atlas of Scanning Electron Microscopy* by R. G. Kessel and R. H. Kardon. W. H. Freeman and Company © 1979.)

D. INTERPRETATION AND MEASUREMENT IN ELECTRON MICROGRAPHS

1. Using Magnification of Micrograph

1. Examine the micrographs in Figs. 4-8 and 4-9. How many and what types of cells are visible?

____

____

____

Look for different kinds of structures in the cells. Find nuclei, mitochondria, endoplasmic reticulum, Golgi apparatus, and plasma membranes. If you cannot find a particular structure, can you assume that it is not present in the cell? Explain.

____

____

____

2. Note that Fig. 4-8 has a magnification of ×11,390 and Fig. 4-9, ×8545. In Fig. 4-8, measure the longest dimension of the macrophage in millimeters (mm) ______ and then convert this figure to micrometers (μm) ______. To determine the actual size of this cell, use the following formula:

$$\text{actual size} = \frac{\text{measured size of cell } (\mu\text{m})}{\text{magnification of micrograph}}$$

What is the longest dimension of macrophage in micrometers?

____

If the macrophage in this micrograph were cut into sections 40 nm thick, how many sections would you get if you cut through the entire cell?

____

What advantages are there to examining more than one section of a cell or tissue?

____

____

____

3. Measure the following cells and cellular organelles and determine their sizes:

a. From Fig. 4-8:
size of a plasma cell (longest dimension)

______ μm

size of red blood cell (longest dimension)

______ μm

thickness of a single endoplasmic reticulum

______ μm

b. From Fig. 4-9:

length of mitochondrion ______ μm

Why do you think some mitochondria in these micrographs look round while others are elongated?

____

____

size of lymphocyte nucleus ______ μm

width of the plasma membrane ______ μm

c. From Fig. 4-10:
size of a bone lacuna (longest dimension)

______ μm

width of Haversian canal ______ μm

size of canaliculus (length) ______ μm

d. From Fig. 4-11:
sizes of fungiform papillae

______ μm; ______ μm

size of circumvallate papillae ______ μm

2. Using Indicator Scale on Micrograph

Some electron micrographs have a scale indicating the length of 1 μm. You can use this indicator scale to measure sizes of subcellular structures; alternatively, you can do the following:

1. Measure the size of this scale in millimeters and determine the decimal factor obtained from calculating the number of micrometers equal to 1 millimeter. Thus,

$$\text{decimal factor} = \frac{1\ \mu\text{m}}{\text{length of indicator scale in mm}}$$

2. Then measure any object in the micrograph in millimeters and multiply this number by the decimal factor obtained in (1). This will give you the size, in micrometers, of the object measured. Example:

a. Length of an indicator scale = 10 mm

$$\text{decimal factor} = \frac{1\ \mu\text{m}}{10} = 0.10$$

b. Length of mitochondrion (in millimeters) as measured in micrograph = 14 mm

$$14\ \text{mm} \times 0.10 = 1.4\ \mu\text{m (length of mitochondrion)}$$

Using the indicator scale in Fig. 4-9, measure the size of some of the cells, nuclei, and mitochondria in the micrograph.

REFERENCES

Alberts, B., et al. 1989. *Molecular Biology of the Cell.* 2d ed. Garland.

Capaldi, R. A. 1974. A Dynamic Model of Cell Membranes. *Scientific American* 230(3):26–33 (Offprint 1292). *Scientific American* Offprints are available from W. H. Freeman and Company, 41 Madison Avenue, New York, NY 10010, and 20 Beaumont Street, Oxford OX1 2NQ, England. Please order by number.

DeDuve, C. 1984. *A Guided Tour of the Living Cell.* Scientific American Library. W. H. Freeman and Company.

Everhart, T. E., and T. L. Hayes. 1972. The Scanning Electron Microscope. *Scientific American* 226:54–67.

Fawcett, D. W. 1981. *An Atlas of Fine Structure: The Cell.* 2d ed. Saunders.

Flegler, S., et al. 1993. *Scanning and Transmission Electron Microscopy.* Oxford.

Karp, G. 1984. *Cell Biology.* 2d ed. McGraw-Hill.

Kessel, R. G., and R. H. Kardon. 1979. *Tissues and Organs: A Text-Atlas of Scanning Electron Microscopy.* W. H. Freeman and Company.

Marx, J. L. 1985. A Potpourri of Membrane Receptors. *Science,* 230:649–651.

Staehelin, L. A., and B. E. Hull. 1978. Junctions Between Living Cells. *Scientific American* 238(5):140–152 (Offprint 1388).

Swanson, C. P., and P. L. Webster. 1985. *The Cell.* 5th ed. Prentice Hall.

Exercise

5

Cellular Reproduction

Mitosis and meiosis are the two mechanisms by which the nuclei of cells divide. In mitosis, the two daughter nuclei produced are identical to each other and to the parental nucleus in chromosome number and genetic makeup. Meiosis, in contrast, results in four daughter nuclei each containing one-half the chromosome number of the parental cell and different genetic composition. Meiosis is an important part of the sexual life cycle in which the daughter nuclei produced are found in cells that differentiate as male and female gametes: the sperm and eggs.

A. THE CELL CYCLE

The remarkable diversity of form and function that cells assume is even more remarkable when you consider that a multicellular organism begins life as a single fertilized cell, the zygote. The complex series of events that encompasses the life span of an actively dividing cell is termed the **cell cycle,** which consists of two phases; the **M** (for **mitosis) phase,** during which the nucleus and cell are actively dividing, and **interphase** (Fig. 5-1).

For the convenience of discussion, the mitotic (M) phase is divided into four distinct stages: **prophase, metaphase, anaphase,** and **telophase.** During interphase the replication (duplication) and synthesis of deoxyribonucleic acid (DNA) and the synthesis of the ribonucleic acid (RNA) and most proteins that are essential for mitosis occur. Note in Fig. 5-1 that the replication of DNA and the synthesis of histones (proteins associated with the DNA molecule), occur only during a period of interphase called the **S** (for **synthesis) phase.** The doubling of DNA during the S phase provides a full complement of DNA for the daughter cells that will result from the next mitotic division. During interphase, there are also two phases called **G** (for **gap) phases.**

The G1 gap preceding DNA replication is the period between the end of mitosis and the beginning of the S phase of the next division. It is during G1 that a cell may follow a path that leads to differentiation, rather than continue the cell cycle. This possibility is indicated in Fig. 5-1 by the arrow labeled "Differentiation." During the G2 gap, structures directly involved with mitosis, such as the spindle fibers, are assembled.

Although the mitotic process occupies only about 10% of the total time taken by the cell cycle, it is important to distinguish among the several parts of mitosis. Thus, nuclear division involving the organization of DNA strands into identifiable chromosomes and the separation of the chromosomes is called **karyokinesis.** Division of the cell body (i.e., cytoplasm and its organelles) is termed **cyto-**

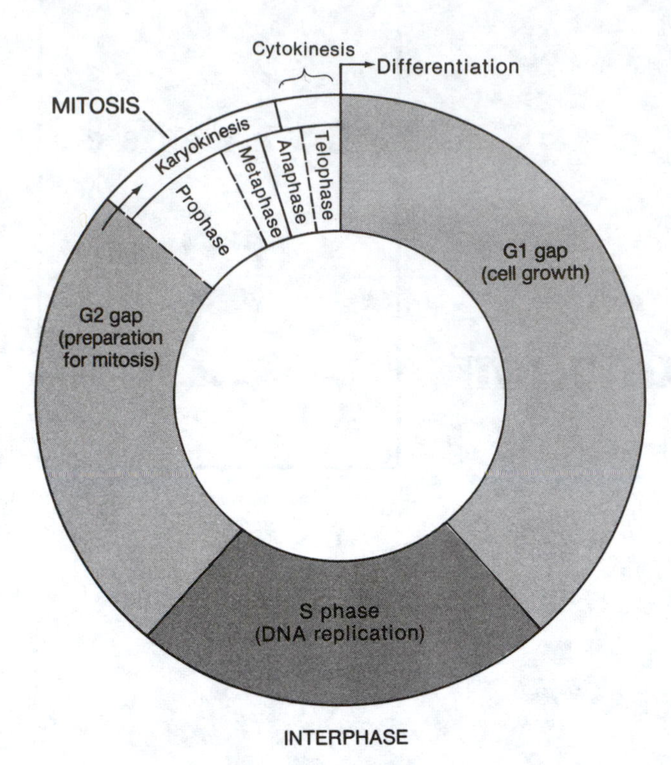

FIGURE 5-1 Stages of the cell cycle

kinesis. Cytokinesis and karyokinesis need not occur simultaneously.

Although the cell cycle is essentially the same in all organisms, just as plant and animal cells differ to some extent structurally there are some differences between them in this process. It is the objective of this part of Exercise 5 to examine the essential steps in the cell cycle and to characterize the similarities and differences in this process between plant and animal cells. Completion of Table 5-1 will help you summarize these similarities and differences.

1. The Cell Cycle in Plant Cells

The onion root tip is one of the most widely used materials for the study of the cell cycle because it is readily available, preparation of the dividing cells is easy, and the chromosomes are large and few in number—hence, easier to study than the cells of many other organisms. Since root tips are regions of active cell division, chances are good that in a specimen of such tissues one can find every stage of this process.

Resist the temptation to think of the various stages of the cell cycle as a series of individual stages in which one stage is discretely followed by another; the process is more like a movie. However, it is easier to study the process as a series of individual stages (like frames of a movie) that show the important characteristics of each stage.

Obtain a slide of onion root tips, and note a series of dark streaks on it (Fig. 5-2A, B, C). Each streak is a very thin longitudinal section through an onion root tip.

Place the slide on the stage of your microscope and locate one of the sections under low power. Because each section is very thin, not all will be equally good for study. After preliminary examination under low power, change to high power, being careful not to break the slide. Keep in mind the sequence in which the stages occur, but do not try to find them in sequence. Thus, if you happen to find an anaphase first, study it before proceeding to another stage. Because cells remain in interphase and prophase longer than in the other stages, chances are that most of the cells will be in interphase; many will be in prophase, and only a few will be in metaphase, anaphase, and telophase.

a. Interphase

The interphase cell, so named because early biologists thought it was in a resting phase, is actively undergoing respiration and the synthesis of DNA, RNA, and protein in preparation for mitosis (Fig. 5-2D).

TABLE 5-1 Similarities and differences between the cell cycle in plant and animal cells

Phase	Plant cells	Animal cells
Interphase		
Prophase		
Metaphase		
Anaphase		
Telophase		

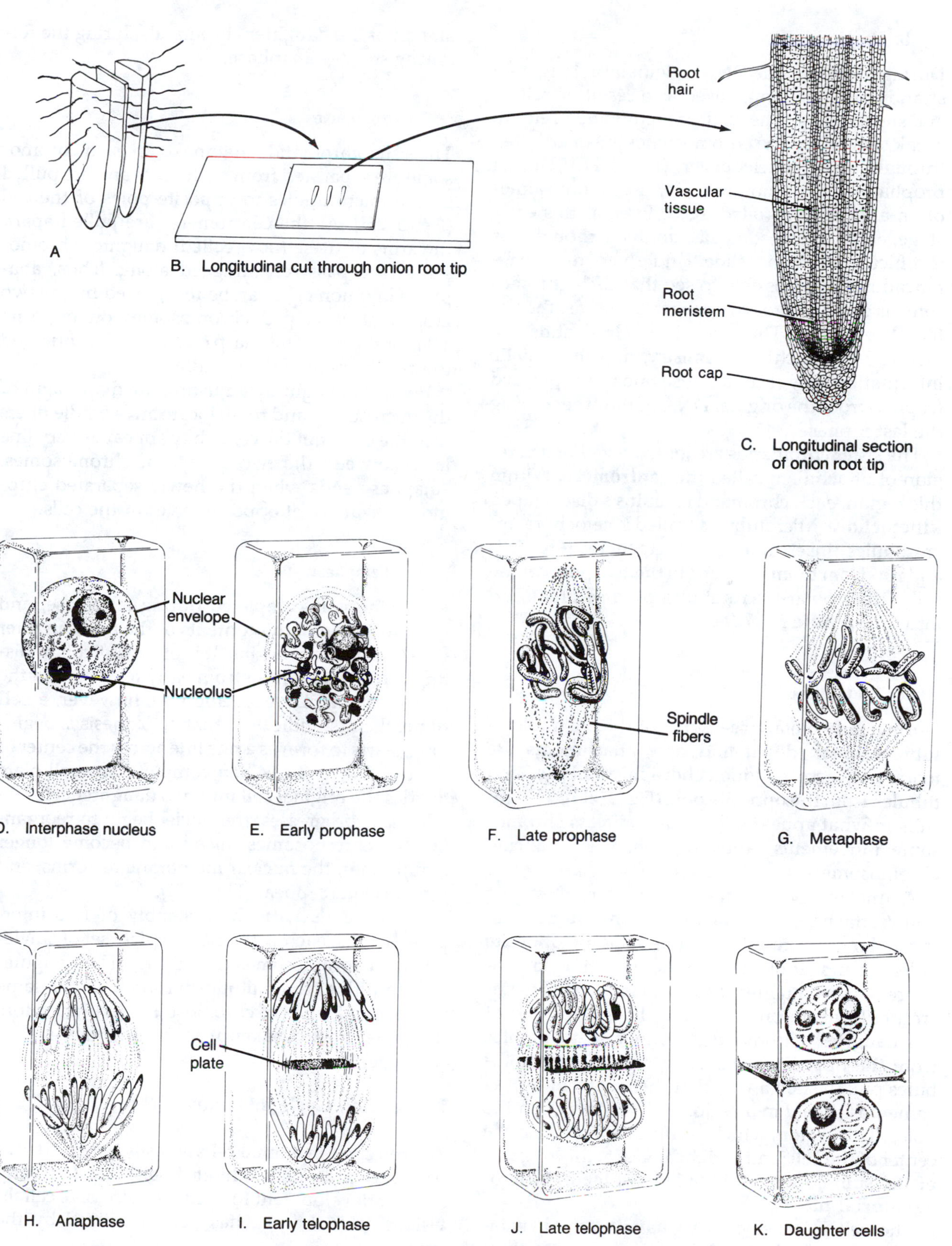

FIGURE 5-2 Diagrammatic representation of the cell cycle stages in the onion root tip

b. Prophase

During prophase, the DNA, originally in long, thin strands, becomes condensed as a result of coiling and supercoiling. The nuclear membrane begins to break down, and the chromosomes are distributed throughout the nucleoplasm (Fig. 5-2E). During prophase in the onion root tip, the chromosomes often appear as a coiled mass. Even at this early stage, each chromosome has doubled, though this is difficult to see on a slide. Under very high magnifications, it is possible to see that each chromosome is composed of two separate strands, the sister **chromatids.** The two sister chromatids are identical in structure, chemistry, and the genetic information they carry because one was replicated (copied) from the original DNA of the other during the last S phase.

The sister chromatids are joined together at a region of attachment called the **centromere.** Within this region, each chromatid contains a disc-shaped **kinetochore.** Microtubules (called kinetochore microtubules) insert into the kinetochores (Fig. 5-3) and run from them outward to the two poles of the cell. Other polar microtubules become organized into the **spindle fibers.**

c. Metaphase

During early metaphase, some of the polar microtubules break down and new attachments are made between the kinetochore microtubules and tubules from the opposite pole (Fig. 5-2G). This results in what appear to be rather aimless chromosome movements, aptly described as "dancing chromosomes."

As metaphase progresses, a random breaking and reattachment of kinetochore microtubules to the polar microtubules of the same or opposite poles occurs until (again randomly) the kinetochore of one daughter chromatid is attached to microtubules from one pole and the kinetochore of the daughter chromatid is connected to tubules from the opposite pole. Then the polar microtubules pull in such a way that the kinetochores become positioned in a region halfway between the poles. This region, which occupies a plane near the center of the cell (and at right angles to the long axis of the spindle fibers), is called the **metaphase** or **equatorial plate.**

The cell is considered to have reached metaphase when the kinetochores of all chromosomes have arrived at this equatorial plate region. At this time the centromeres divide in preparation for separation of the daughter chromatids during the following stage—anaphase.

d. Anaphase

The sister chromatids that make up each chromosome are separated from each other and are pulled by the microtubules to opposite poles of the cell (Fig. 5-2H). As the centromeres are pulled apart, the arms of these (now called) daughter chromosomes are passively dragged along. Thus, anaphase in onion cells can be recognized by the two groups of V-shaped chromosomes on opposite sides of the cell. The sharp end of the V is oriented toward the pole of the spindle.

Reduce the light by adjusting the diaphragm of the microscope, and try to locate any spindle fibers near the center of the cell. They appear as very fine lines between the two groups of chromosomes. Anaphase ends when the newly separated chromosomes arrive at opposite poles of the cells.

e. Telophase

Karyokinesis is completed during telophase, and reorganization of the contents of the two daughter cells (cytokinesis) begins. It is often difficult to distinguish late anaphase from early telophase in the cells of plants. During telophase, however, a **cell plate,** the first indication that cytokinesis is beginning, starts to form as a fine line across the center of the cell (Fig. 5-2I). When complete, the cell plate divides the original cell into two daughter cells. As telophase progresses, the nuclei begin to reorganize: the chromosomes uncoil and become longer and thinner, the nuclear membrane re-forms, and the nucleoli reappear (Fig. 5-2J).

Mitosis ends with the assembly of two interphase nuclei, each with one complete set of single-stranded chromosomes (Fig. 5-2K). The daughter cells resulting from mitotic division have the same number and kinds of chromosomes (and therefore the same genetic makeup) as the original cell.

2. The Cell Cycle in Animal Cells

The cell cycle in animal cells is easily observed on a prepared slide of a whitefish blastula (an early stage of development formed by successive cell divisions after the egg has been fertilized by the sperm).

Obtain a slide of whitefish blastula cells stained to show various stages of the cell cycle. As with

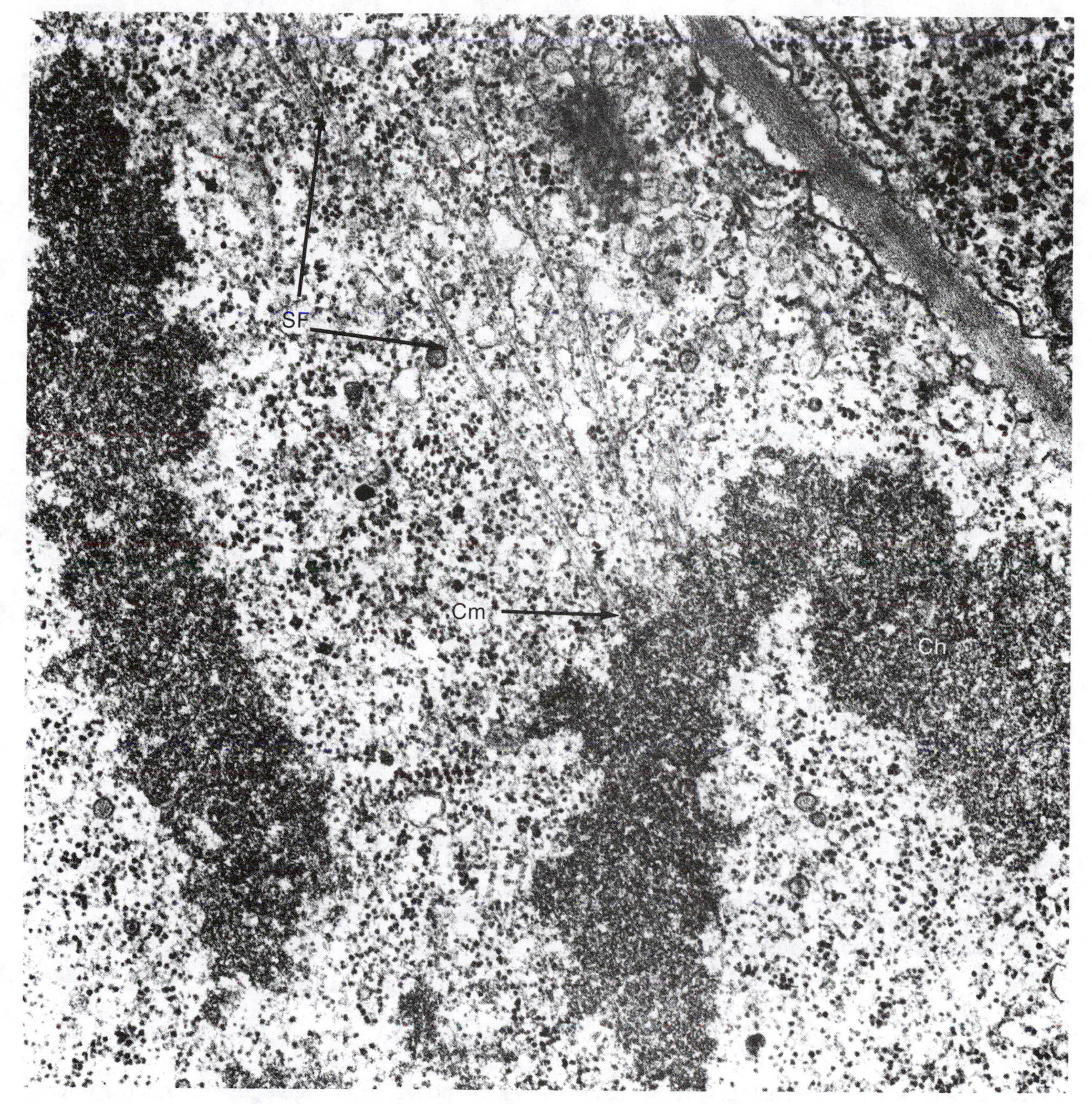

FIGURE 5-3 Spindle fibers (SF) attached to a chromosome (Ch) at the centromere (Cm). Note the tubular structure of the spindle fibers.

your study of the onion slides, locate the various stages of animal cell mitosis.

a. Interphase

Interphase cells are characterized by a distinct nucleus and nucleolus bounded by a nuclear membrane (Fig. 5-4A). Lying next to the nuclear envelope is a pair of cytoplasmic organelles called **centrosomes,** which contain the centrioles.

b. Prophase

During prophase (Fig. 5-4A), in contrast to plant cells, the **centrioles,** found within the centrosome, begin to move apart, as if repelled by each other,

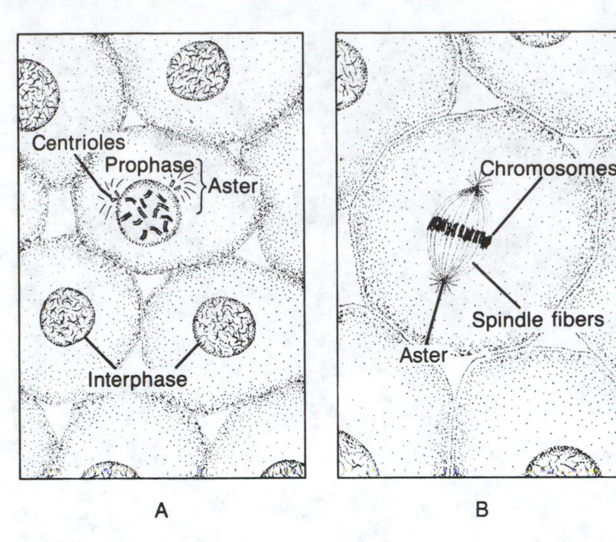

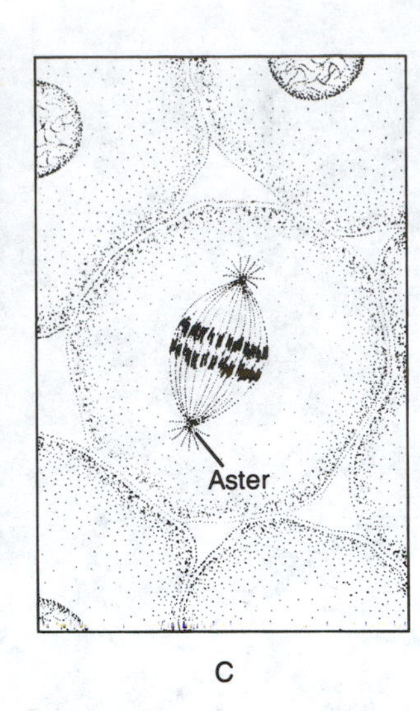

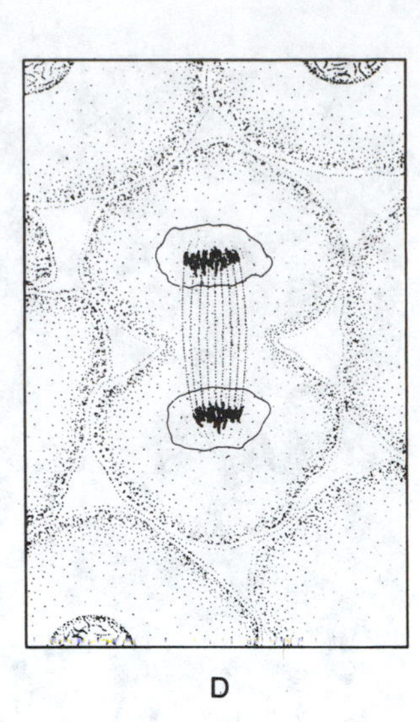

FIGURE 5-4 Mitotic stages in the whitefish blastula. (A) Interphase and prophase. (B) Metaphase. (C) Anaphase. (D) Telophase.

and migrate around the nucleus toward opposite poles of the cell. Microtubules radiate from each pair of centrioles like spokes on a wheel, forming a configuration known as an **aster.** When the nuclear envelope disintegrates, the region between the centrioles becomes visible. This relatively transparent region is called the spindle. The microtubules of the spindle are arranged so as to form the spindle fibers.

c. Metaphase

During metaphase, the chromosomes move toward the central region of the spindle to form the **metaphase (or equatorial) plate** (Fig. 5-4B). The chromosomes are maneuvered into position by the spindle fibers that are attached to the kinetochore of each chromosome.

d. Anaphase

Anaphase begins when the pairs of chromatids are pulled apart and become daughter chromosomes, after which they are pulled toward the poles of the cell (Fig. 5-4C). When the chromosomes reach the poles, telophase begins.

e. Telophase

During telophase, the spindle disappears, two daughter nuclei are organized, the nucleoli appear, and the nuclear membranes are formed by fusion of parts of the endoplasmic reticulum.

In late telophase, the cytoplasm becomes deeply furrowed, or pinched in between the two nuclei, and cytokinesis takes place. This results in two daughter cells having equivalent nuclear contents and equal amounts of cytoplasm (Fig. 5-4D).

B. MEIOSIS

In meiosis, immature or primordial germ cells undergo a reduction from the diploid number to the haploid number of chromosomes and become mature gametes. As a result, as you will see, meiosis maintains the chromosome number constant and provides genetic variability because of crossing over and the subsequent exchange of genes between chromosomes.

1. Meiosis in the Lily

Meiosis will be studied as it occurs in the development of mature pollen grains of flowering plants. These pollen grains give rise to male gametes, which fuse with an egg to produce a zygote.

As you examine a series of slides in the meiotic sequence, refer to Fig. 5-5 for help in locating the stages.

a. Meiosis I

Examine a lily flower and locate the **anthers,** or pollen sacs, which contain numerous pollen mother cells (Fig. 5-5A, B). These cells undergo meiosis and produce mature pollen grains. Next, with your microscope, examine slides of a cross section through a young lily anther, and locate the pollen mother cells (Fig. 5-5C, D). The nuclei of

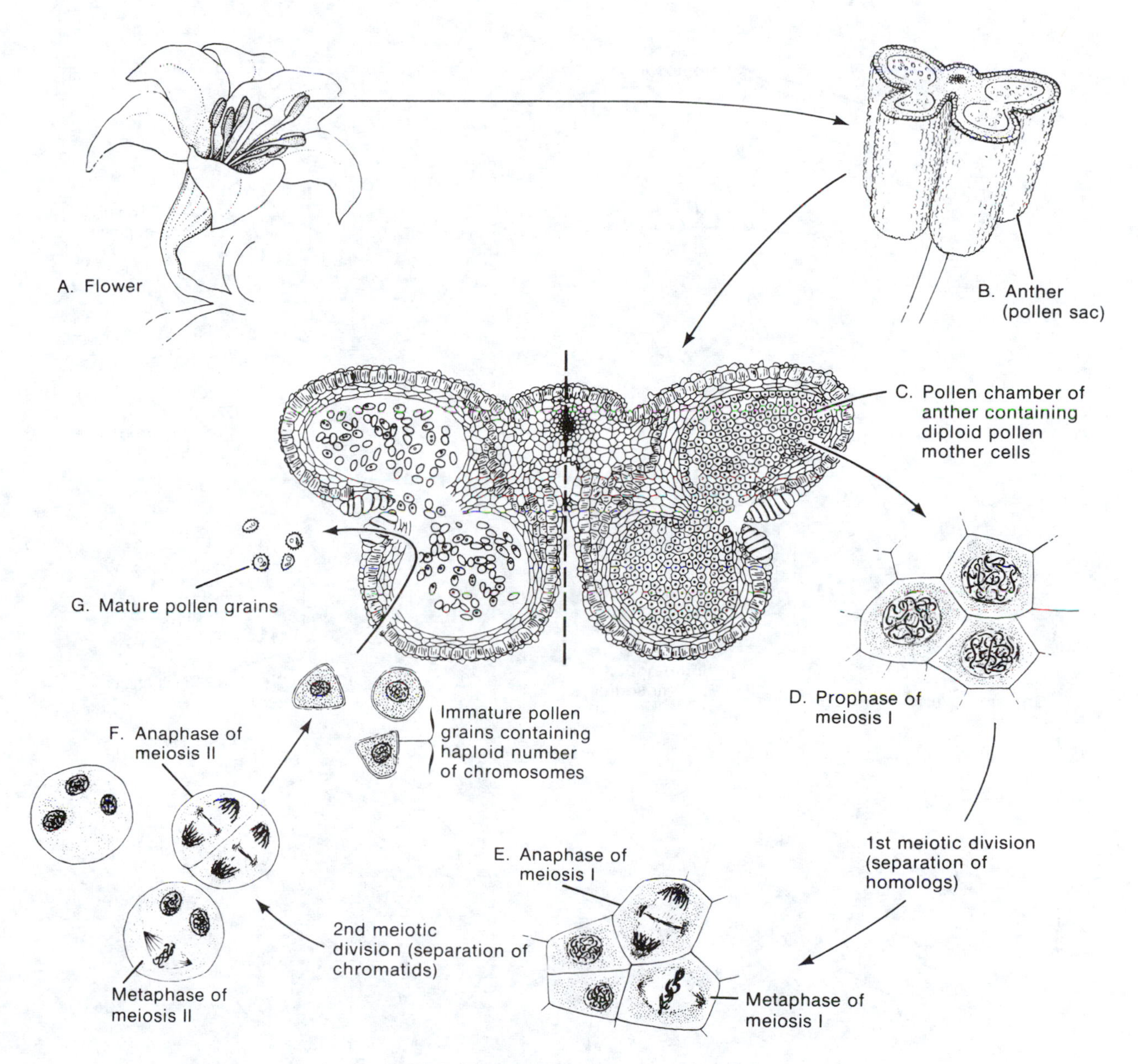

FIGURE 5-5 Meiosis in the lily anther

these cells contain the diploid number of chromosomes. Many of these pollen mother cells are in prophase of the first meiotic division. During this phase, **homologous chromosomes,** each composed of two chromatids, synapse (or lie adjacent to one another), form **tetrads** (the association of the four chromatids of the homologous chromosomes), and exchange genetic components by a physical mechanism called **crossing over** (Fig. 5-6). What is the significance of this process?

____

____

The homologous chromosomes then separate during anaphase, with one chromosome of each pair moving to each pole of the cell. Because this is a separation of pairs of chromosomes, not of chromatids, the chromosome content of the cells at the end of meiosis I has been reduced from the diploid to the haploid condition. Examine slides of a lily anther showing separation of homologous chromosomes (Fig. 5-5E). The diploid number in the lily is 24. What is the chromosome number of the cells formed following the first meiotic division?

____

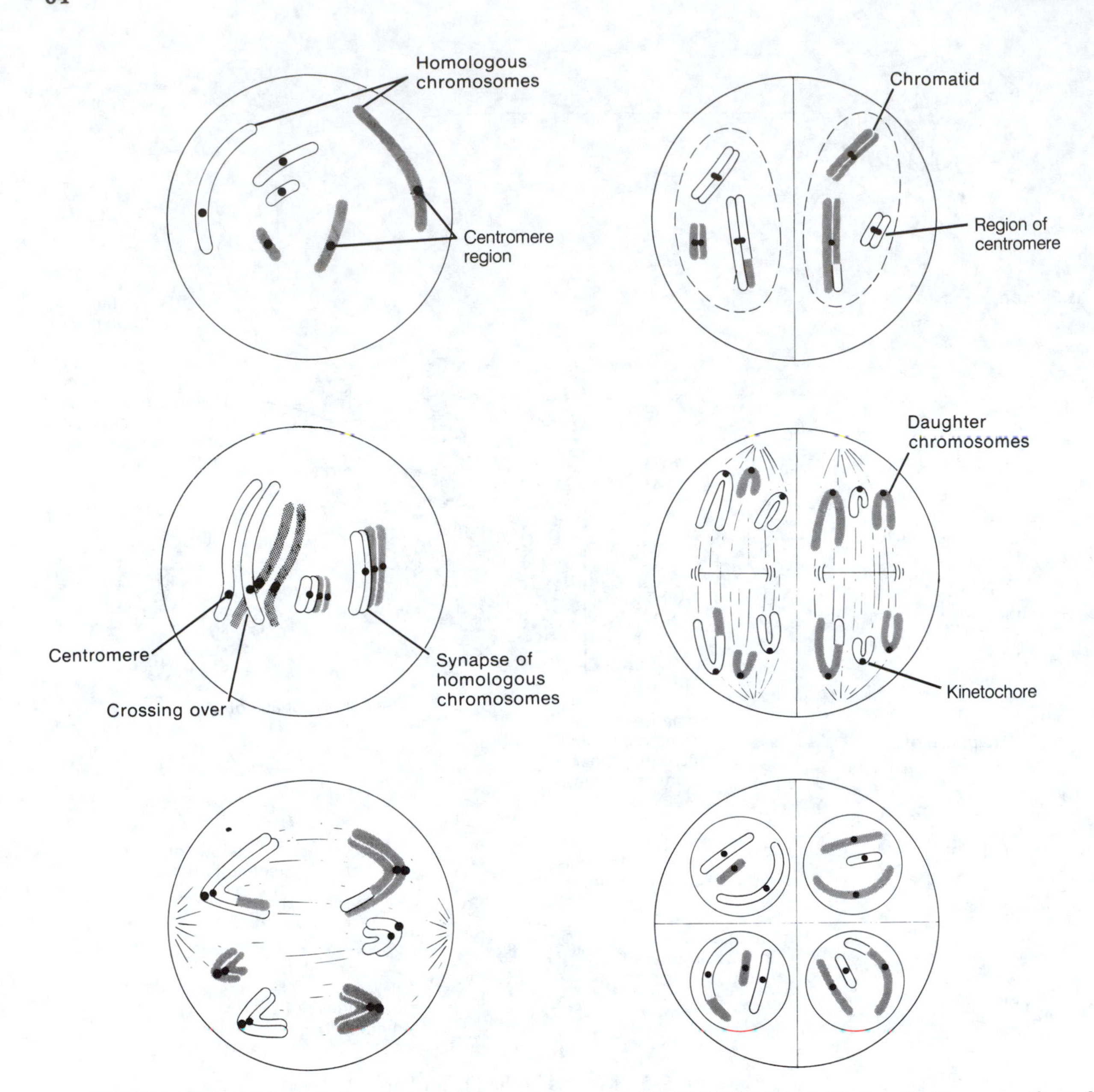

FIGURE 5-6 Meiosis I, showing crossing over

FIGURE 5-7 Meiosis II, showing the separation of the daughter chromosomes

b. Meiosis II

Examine slides of anthers in which the cells resulting from the first meiotic division are in metaphase and/or anaphase of meiosis II (Fig. 5-5F). In these cells, the chromatids that make up each chromosome separate and migrate to the poles (Fig. 5-7). As in mitosis, the chromatids, when separated from each other, are called **daughter chromosomes.** At the poles, the daughter chromosomes become enclosed in a nuclear membrane.

Cytokinesis follows the division of the nucleus. How many pollen grains are formed as a result of the two meiotic divisions of the pollen mother cells?

____

What is the chromosome complement of each pollen grain (i.e., *N* or 2*N*)?

____

Examine lily anthers showing mature pollen grains (Fig. 5-5G).

2. Meiosis in *Ascaris*

In male animals, the formation of gametes **(gametogenesis)** occurs in the testes; in females, gametogenesis occurs in the ovaries or oviducts. In this part of the exercise, you will study the meiotic events in the formation of the egg **(oogenesis)** as they occur in the parasitic roundworm *Ascaris*. Because the diploid number of chromosomes in *Ascaris* is only four, it is ideal for the study of this process.

The reproductive organ in a female *Ascaris* consists of a pair of long, coiled tubes that are regionally divided into the ovary, oviduct, and uterus (Fig. 5-8A). The "eggs," which are produced in the ovaries, pass into the oviducts, where they are fertilized by sperm.

Examine slides containing the oviduct and uterus of *Ascaris*. Locate the oviduct, which has a large number of triangular sperm interspersed among numerous "eggs" (Fig. 5-8B). The eggs at this stage are still diploid, because oogenesis does not begin until after the egg has been penetrated by a sperm. The term *egg* at this stage of development, therefore, is not quite accurate. A more correct term would be **primary oocyte,** a cell that will undergo meiosis and become the mature egg. On the slide, some of them may have been penetrated by sperm.

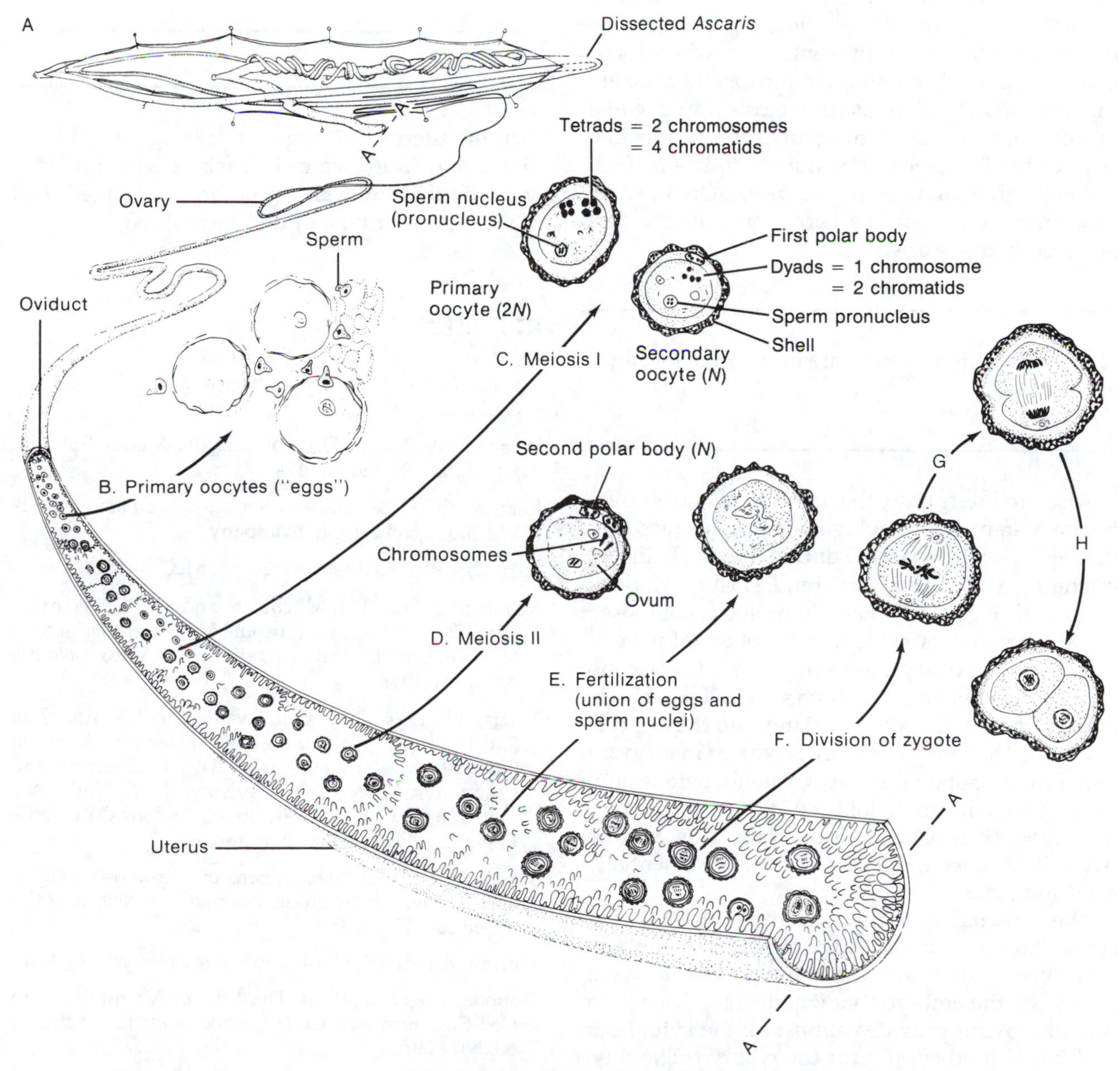

FIGURE 5-8 Meiosis in *Ascaris*

How many chromosomes are in the primary oocyte of *Ascaris*?

Shortly after fertilization, the homologous chromosomes of the primary oocyte duplicate to form chromatids and then, during prophase of meiosis I, come to lie adjacent to each other (synapse), forming tetrads (Fig. 5-8C). At this time, crossing over may occur, which results in an exchange of genetic components of homologous chromosomes (Fig. 5-6). Locate primary oocytes in which synapsis has taken place.

At anaphase of meiosis I, the members of homologous pairs of chromosomes, each **homolog** still consisting of a pair of chromatids, move to opposite poles (Fig. 5-8C). Since the centromeres do not separate, the sister chromatids remain together and are called **dyads.** At telophase of meiosis I, an unequal division of the cytoplasm occurs. This results in a large cell called a **secondary oocyte** and a small cell called the **first polar body** (Fig. 5-8C). On the slide, how many chromosomes are found in the first polar body of *Ascaris*?

How many chromosomes are found in the secondary oocyte?

After an interphase, the second meiotic division begins when the homologous chromosomes (homologs), which separated during meiosis I, line up on the equator of a division spindle in metaphase of meiosis II (Fig. 5-8D). Each homolog is composed of two chromatids, which, at anaphase of meiosis II, separate and migrate to the poles. The second meiotic division produces both a second polar body and a large cell that differentiates into the egg cell, or ovum. The first polar body produced in meiosis I may or may not go through a second meiotic division. Thus, when a diploid cell in the *Ascaris* ovary undergoes complete meiosis, only one mature ovum is produced; the polar bodies are essentially nonfunctional.

The unequal cytokinesis during oogenesis ensures that an unusually large supply of cytoplasm and stored food is allotted to the nonmotile ovum for use by the embryo that will develop from it. In fact, the ovum provides almost all the cytoplasm and initial food supply for the embryo. The tiny, highly motile sperm cell contributes only its genetic material.

During maturation of the egg, the sperm nucleus **(pronucleus)** has been lying inactive in the cytoplasm of the egg. Following the second meiotic division, the egg and the sperm pronucleus unite **(fertilization)** and form a single cell (zygote). On the slide, locate eggs that show separate sperm and egg nuclei and eggs in which these nuclei have fused (Fig. 5-8E). How many chromosomes does the zygote contain?

If you did not know the actual number of chromosomes in the zygote, how would you describe the chromosome content of this cell?

In the uterus, the zygote nucleus and cell soon divide and form two cells, each of which divides again until a multicellular embryo is formed (Fig. 5-8F–H). What type of division is this?

REFERENCES

Alberts, B., et al. *Molecular Biology of the Cell.* 2d ed. Garland.

Inoue, S. 1981. Cell Division and the Mitotic Spindle. *J. Cell Biol.* 91:132s–147s.

John, B., and K. Lewis. 1980. *Somatic Cell Division.* Carolina Biological Supply Company.

Karp, G. 1984. *Cell Biology.* 2d ed. McGraw-Hill.

Koshland, D. E. , T. J. Mitchison, and M. W. Kirschner. 1988. Polewards Chromosome Movement Driven by Microtubule Depolymerization in Vitro. *Nature* 331:499–504.

Mazia, D. 1974. The Cell Cycle. *Scientific American* 230(1):54–64 (Offprint 1288). *Scientific American* Offprints are available from W. H. Freeman and Company, 41 Madison Avenue, New York, NY 10010, and 20 Beaumont Street, Oxford OX1 2NG, England. Please order by number.

Moens, P. B. 1973. Mechanisms of Chromosome Synapsis at Meiotic Prophase. *International Review of Cytology* 35:117–134.

Murray, A., and T. Hunt. 1993. *The Cell Cycle.* Oxford.

Sloboda, Roger D. 1980. The Role of Microtubules in Cell Structure and Cell Division. *American Scientist* 68:290–298.

Smith-Klein, C., and V. Kish. 1988. *Principles of Cell Biology.* Harper and Row.

Exercise 6

Movement of Materials Through Plasma Membranes

To perform its functions, the body must maintain a condition in which its internal environment remains within relatively constant limits **(homeostasis).** One of the mechanisms by which homeostasis is achieved involves the plasma membrane's regulation of the movement of materials into and out of the cell. Because not all substances penetrate the plasma membrane equally well, the membrane is said to be **differentially permeable.**

The external and internal environments of a cell are aqueous solutions of dissolved inorganic and organic molecules and ions. Movement of these molecules and ions in the solutions and through the plasma membranes is by **diffusion.** This is a physical process in which the kinetic energy of molecules and ions causes them to move from regions in which their concentration is high to regions in which their concentration is lower, until they become distributed throughout the available space. Thus, a gas set free in a room eventually becomes uniformly distributed throughout the room. And when a crystal of salt (NaCl) dissolves in a glass of water, the sodium and chlorine ions of which it is composed become uniformly distributed throughout the water. This type of diffusion, which results from the random motions of the solute and solvent molecules and requires no added energy, is called **passive diffusion.**

Active transport is a type of diffusion in which dissolved particles (solutes) move *against* a concentration gradient; it does require an energy input. For example, human erythrocytes have almost 30 times more potassium than does the surrounding blood plasma.

A special case of diffusion that occurs in biological systems is **osmosis.** Simply defined, osmosis is the diffusion of water molecules through a differentially permeable membrane from a region in which they are more highly concentrated to a region in which their concentration is lower. This phenomenon is diagrammed in Fig. 6-1.

Although diffusion and osmosis result from the kinetic activity of molecules or ions, they are affected by a number of other factors, such as temperature, the molecular weight of the diffusing substance, and the lipid solubility of the solute. In this study, you will become familiar with diffusion and osmosis and examine some of the factors regulating these processes.

A. OBSERVING DIFFUSION

In this part of the exercise, you will observe various types of diffusion and study the effects of several factors on the rate at which particles diffuse.

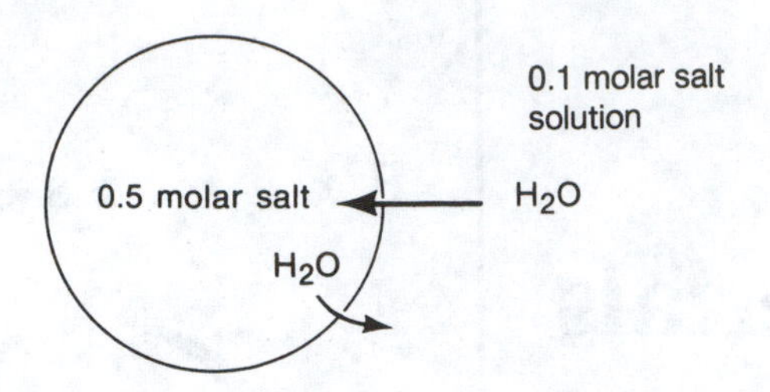

A. More water enters cell than leaves.

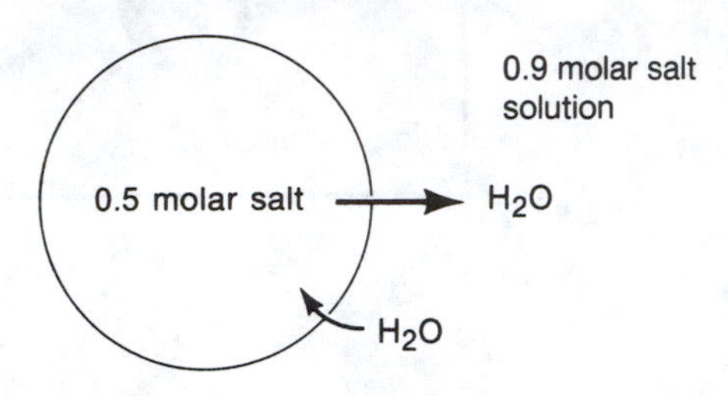

B. More water leaves cell than enters.

FIGURE 6-1 Osmosis

1. Diffusion of a Gas in a Gas

A striking demonstration of diffusion is the movement of gases through air. An apparatus for demonstrating such diffusion is shown in Fig. 6-2.

Saturate a piece of absorbent cotton on one side with ammonium hydroxide (NH_4OH) and another piece of cotton with hydrochloric acid (HCl).

Caution: *These are corrosive chemicals, so rubber (vinyl) gloves should be used to handle them. Open only one reagent bottle at a time.*

Simultaneously place the pieces of cotton in opposite ends of a glass tube, as shown in Fig. 6-2. Why simultaneously?

____

____

Ammonium hydroxide and hydrochloric acid react to form ammonium chloride (NH_4Cl), a cloudy, white precipitate, and water. The equation for this reaction is

$$NH_4OH + HCl \rightarrow NH_4Cl + H_2O$$

The molecular weights of the ammonium (NH_4^+) and chlorine (Cl^-) ions are 18 and 35.5, respectively. At which end of the tube would you predict the precipitate to form?

____

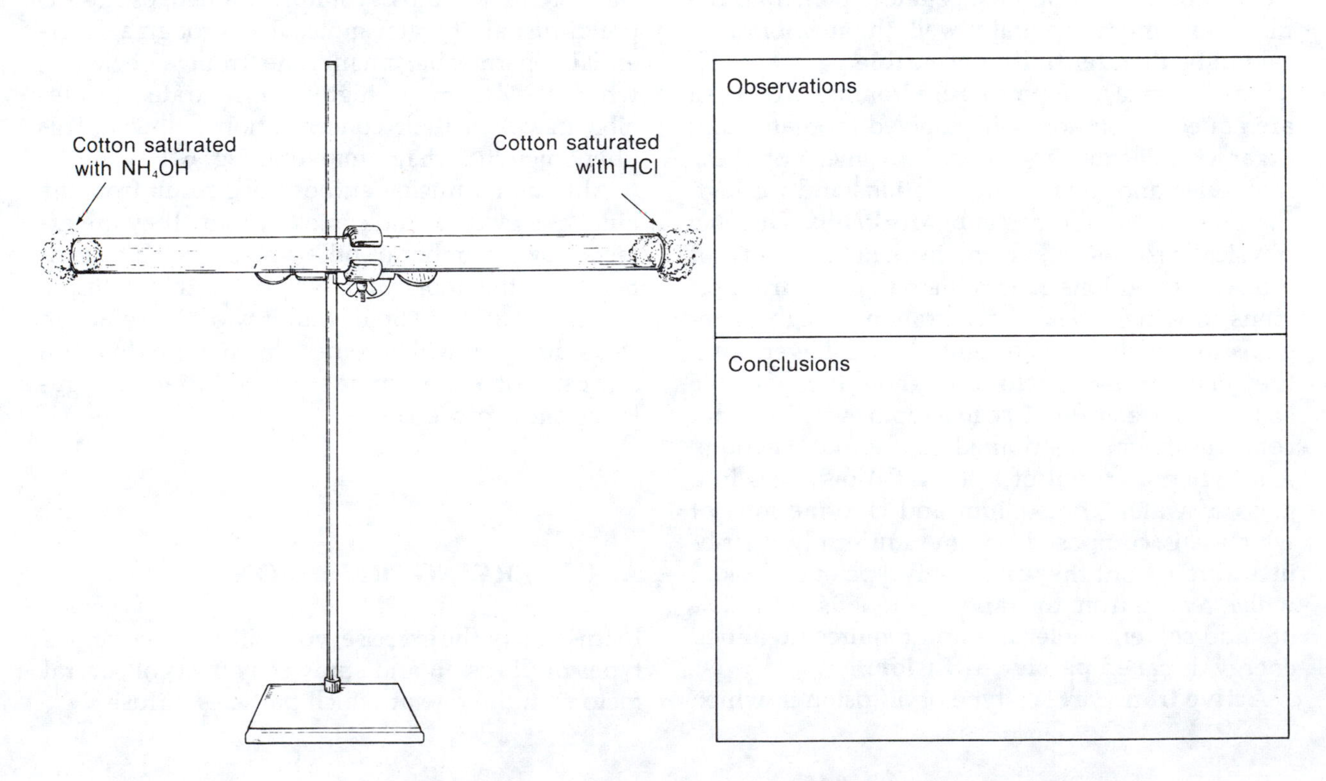

FIGURE 6-2 Apparatus for studying gas diffusion

Describe what happens when these two gases meet in the tube.

___

___

___

What, if any, relationship is there between the molecular weights of these gases and their rates of diffusion?

___

___

___

___

Caution: *Dispose of the saturated cotton plugs by putting them in separate beakers of water.*

2. Diffusion of a Liquid in a Solid

In this exercise, you will use a procedure, called double immunodiffusion or the **Ouchterlony technique,** to study the effect of molecular weight on diffusion. This procedure is a modification of a simple but useful and informative technique used to characterize the complex relationships between antigens (substances that induce an immune response) and antibodies (specific protein molecules that are produced by the body in response to exposure to antigens). The technique has widespread clinical application, though it is being replaced by more sensitive procedures.

Basically, the procedure consists of pouring agar, a gel obtained from certain seaweeds, into a petri dish, allowing it to solidify, and then punching circular wells close to one another in the gel. The liquid substances that are to be studied are then added to the wells and allowed to diffuse outward until they meet, react, and form a line of precipitate.

Your instructor will give you a disposable petri dish containing agar (Fig. 6-3A). Using a No. 5 cork borer, punch four holes in the agar as shown in Fig. 6-3B, C. *Note:* By placing your petri dish over Fig. 6-3C, you can use this as a template to accurately position your holes. Fill each of the holes uniformly with a small amount of 1 N solutions of sodium chloride (NaCl), potassium bromide (KBr), potassium ferricyanide ($K_3Fe[CN]_6$), and silver nitrate ($AgNO_3$). The approximate molecular weights of each of the migrating anions that form when these molecules are placed in solution are chloride (Cl^-), 35; bromide (Br^-), 80; ferricyanide ($Fe(CN)_6^-$), 212; nitrate (NO_3^-), 62. Periodically examine the petri dish and record your observations in Fig. 6-4.

From this study, what can you conclude about the relationship between the rate of diffusion and molecular weight?

___

___

___

B. DIALYSIS

Essentially, **dialysis** is diffusion through a differentially permeable membrane that separates small molecules or ions from large molecules or ions. The principle of dialysis is used in artificial kidney machines, which pass a patient's blood through a tube of **dialyzing membrane.** This artificial membrane takes the place of damaged, defective, or missing kidneys. As the blood moves through the membranous tube, small-particle waste products (urea, sulfate) move, by diffusion, from the blood into a solution surrounding the membrane. The purified blood is then returned to the body.

In this study you will become familiar with the concept of dialysis by removing chloride ions from a solution of starch and sodium chloride. To detect the presence of these compounds in solution, you must be familiar with simple tests for identifying chloride ions and starch.

1. Tests for Chloride Ions and Starch

1. Fill six numbered test tubes with 5 ml of one solution or the other or water, as shown in Fig. 6-5.

2. To tubes 1, 3, and 5, add three drops of silver nitrate ($AgNO_3$).

3. To tubes 2, 4, and 6, add three drops of iodine solution.

4. Mix the contents of each tube by swirling, and record your observations in Table 6-1.

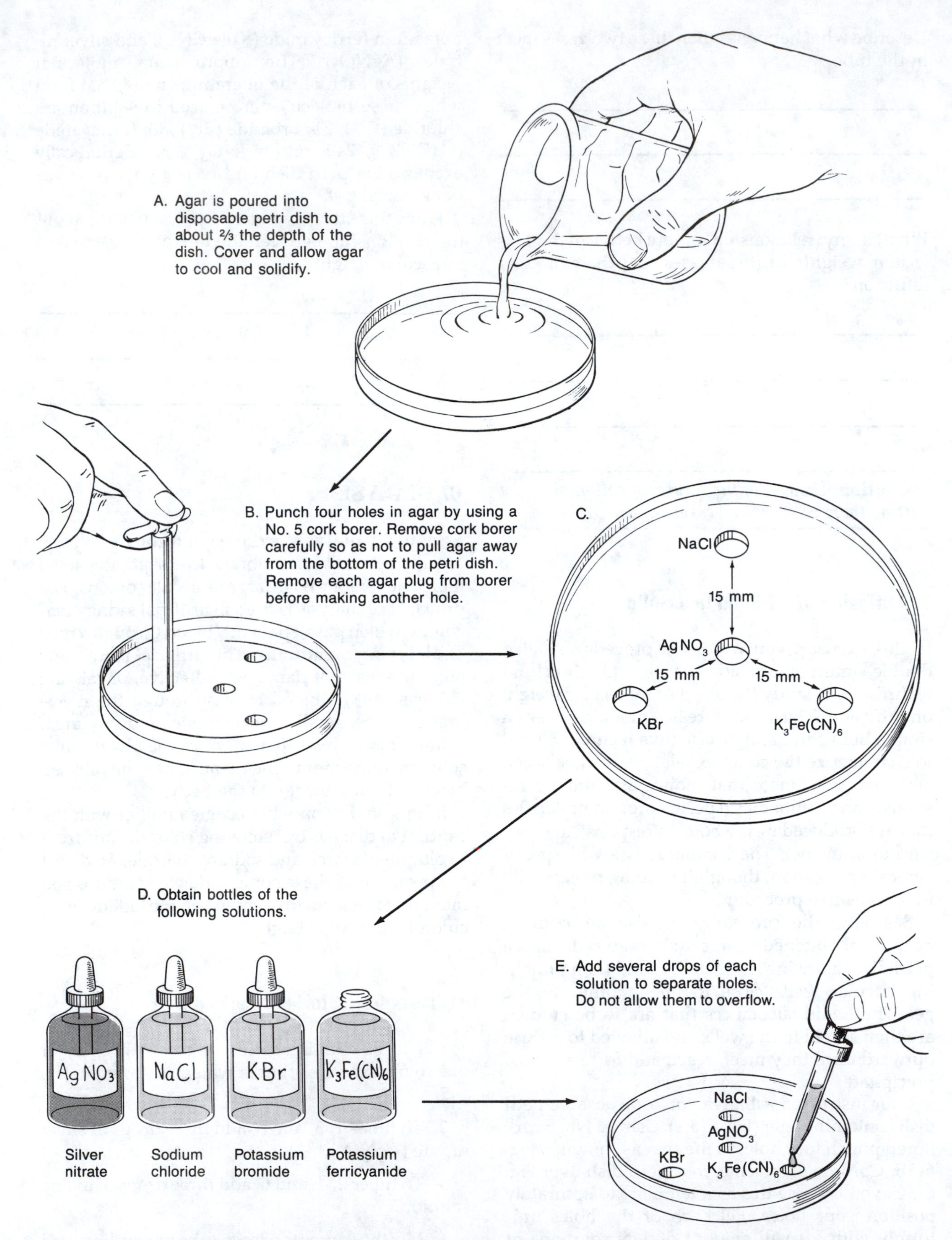

FIGURE 6-3 Procedure for determining the effect of molecular weight on diffusion

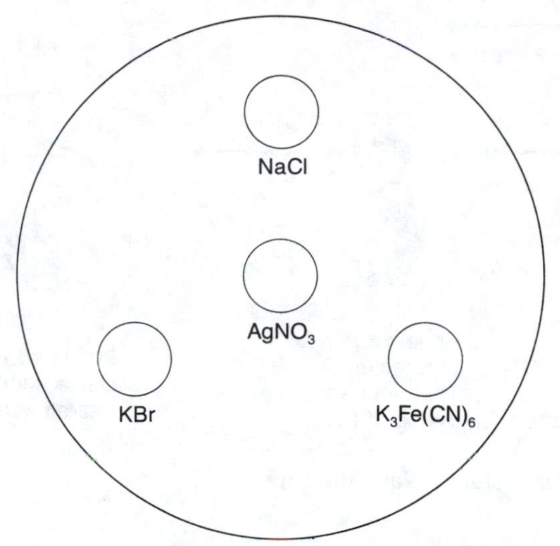

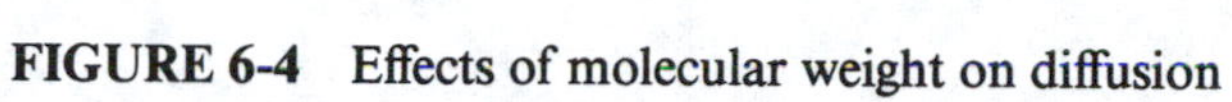

FIGURE 6-4 Effects of molecular weight on diffusion

2. Dialysis of a Starch/Sodium Chloride Mixture

Following the procedure in Fig. 6-6, fill a dialyzing membrane with 15 ml of a solution containing starch and sodium chloride. Seal the dialysis bag as

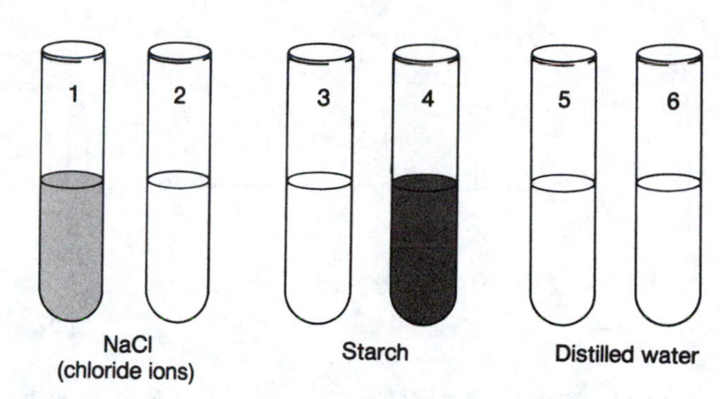

FIGURE 6-5 Tests for chloride ions and starch molecules

shown and place in a 250-ml beaker of distilled water. Why *not* use tap water?

____

____

____

____

TABLE 6-1 Test for chloride ions and starch

Test solution	Reagent/Observations	
	Silver nitrate	Iodine
Sodium chloride (chloride ions)		
Starch		
Distilled water		
Conclusions:		

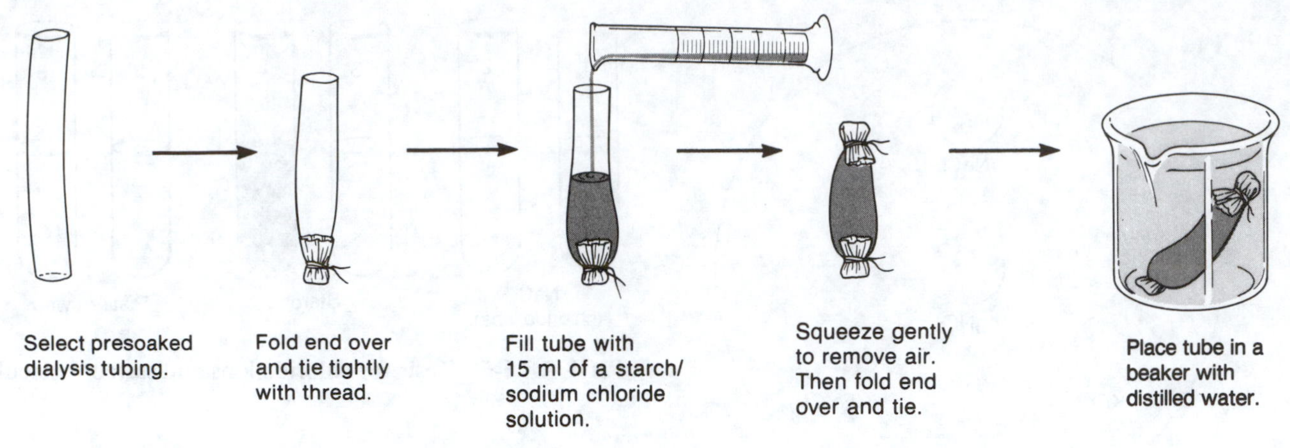

FIGURE 6-6 Procedure for dialysis of starch/NaCl mixture

Describe how you would determine whether NaCl or starch had diffused through the membrane into the distilled water?

___

___

___

Carry out the tests you have described and explain the results you obtain.

___

___

C. OSMOSIS

1. Hemolysis

The cell membrane of **erythrocytes** (red blood cells) is freely permeable to water molecules but relatively impermeable to salts. Thus, if red blood cells are placed in an **isotonic** saline solution (i.e., one that has the same salt concentration as found in the plasma and cytoplasm of the red blood cell—equivalent to 0.85% NaCl), the cell will retain its normal shape and size. Why?

___

___

___

If red blood cells are placed in a **hypotonic** saline solution (i.e, one that has a lower salt concentration than does plasma or cytoplasm), water enters the cells more rapidly than it leaves. As a consequence, the red blood cells swell and ultimately burst, releasing hemoglobin. This phenomenon is called **hemolysis.** Red blood cells placed in a **hypertonic** saline solution (i.e., one that has a higher salt concentration than does plasma or cytoplasm) shrink and appear to have a bumpy, irregular outline. These cells are said to be **crenated.**

To demonstrate the changes in red blood cells under these conditions, carry out the following procedure.

1. Put a small drop of 0.85% NaCl on a clean glass slide.

2. Add a small drop of sheep red blood cells to the saline on the slide. Add a coverslip.

3. Examine the red blood cells with the high-power (43×) lens of your microscope. Observe a region where the cells are not too dense. Note the sizes and shapes of these normal cells and draw a few of them in Fig. 6-7.

4. Add two or three drops of 5% NaCl (hypertonic saline solution) to one edge of the coverslip. Continue to observe the blood cells and watch the changes that occur as the more concentrated saline solution reaches them. Record your observations in Fig. 6-7.

5. Put a drop of distilled water and a drop of sheep red blood cells on a second slide. Add a coverslip and observe the cells in this hypotonic solution for several minutes. Record any changes that occur in Fig. 6-7.

Isotonic Hypotonic Hypertonic

FIGURE 6-7 Appearance of red blood cells in isotonic, hypotonic, and hypertonic saline solutions

Knowledge of the changes that may occur in the **tonicity** (solute concentration) of plasma and/or tissue fluids has practical applications. For example, one of several symptoms of diabetes mellitus (sugar diabetes) is extreme thirst caused by the decreased production of insulin by the endocrine tissue of the pancreas. What brings about this feeling of thirst? (Refer to any physiology textbook for the answer to this question.)

____

____

____

____

2. Effect of Solute Concentration on the Rate of Osmosis

The rate at which osmosis occurs (i.e., the rate at which water molecules move into or out of cells) is a function of the tonicity of the cytoplasm of the cell or of the extracellular fluid. In this part of the exercise, you will use an artificial membrane to measure the effect on osmosis of varying the tonicity of the fluid inside this differentially permeable membrane.

1. Obtain five dialysis bags, which will function as artificial, differentially permeable membranes. Fold over the end of each bag and tie it with a thread (Fig. 6-8).

2. Fill the bags as follows.

Bag 1: 15 ml of tap water

Bag 2: 15 ml of 20% sucrose solution

Bag 3: 15 ml of 40% sucrose solution

Bag 4: 15 ml of 60% sucrose solution

Bag 5: 15 ml of tap water

3. As each bag is filled, remove the air by gently squeezing the bottom end of the bag to bring the liquid to the top of it. Press the sides of the bag together so that air does not reenter. Fold over the end of the bag about 5 cm and tie the end securely with a thread. Wipe each bag dry and weigh it to the nearest 0.5 g. Record the weights of the bags in Table 6-2 at zero time.

4. Place bags 1, 2, 3, and 4 in separate beakers of water, and place bag 5 in a beaker of 60% sucrose solution.

5. At 10-minute intervals (i.e., after 10, 20, 30, 40, and 50 minutes), remove the bags from the breakers, carefully wipe off all water, and again weigh each bag separately. Record the data in Table 6-2. Plot the changes in weight (Δ wt) of each bag against time in Fig. 6-9. What relationship (if any) is there between the concentration of sucrose and the rate of osmosis? How do you account for the differences observed?

____

____

____

____

____

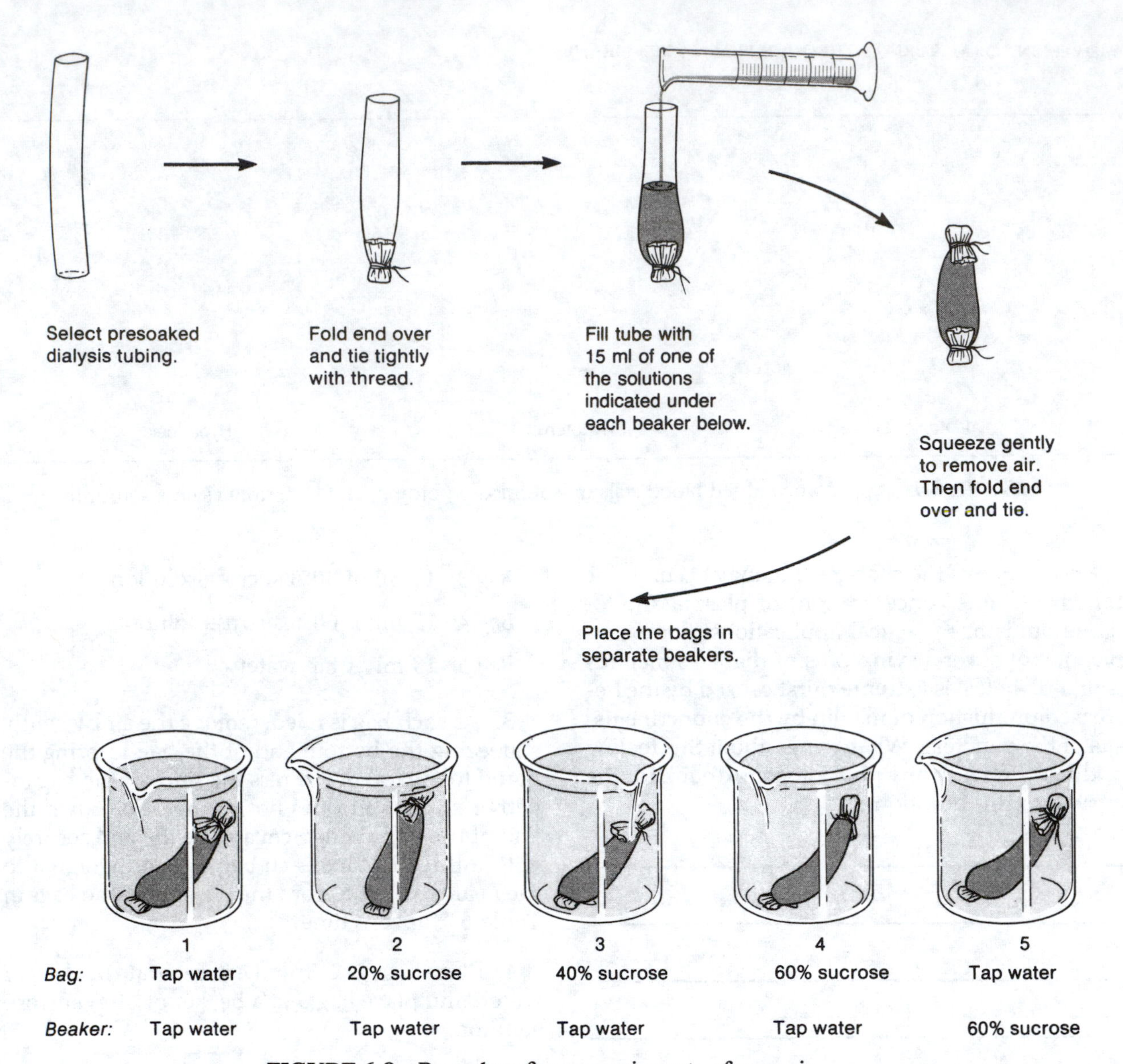

FIGURE 6-8 Procedure for measuring rate of osmosis

TABLE 6-2 Osmosis data

Time (minutes)	Bag 1		Bag 2		Bag 3		Bag 4		Bag 5	
	Weight	Δ wt[a]	Weight	Δ wt	Weight	Δ wt	Weight	Δ wt	Weight	Δ wt
0		0		0		0		0		0
10										
20										
30										
40										
50										

[a]Change in weight (Δ wt) may be recorded as differences (+) or (−) between each reading or as differences for each reading from 0 time.

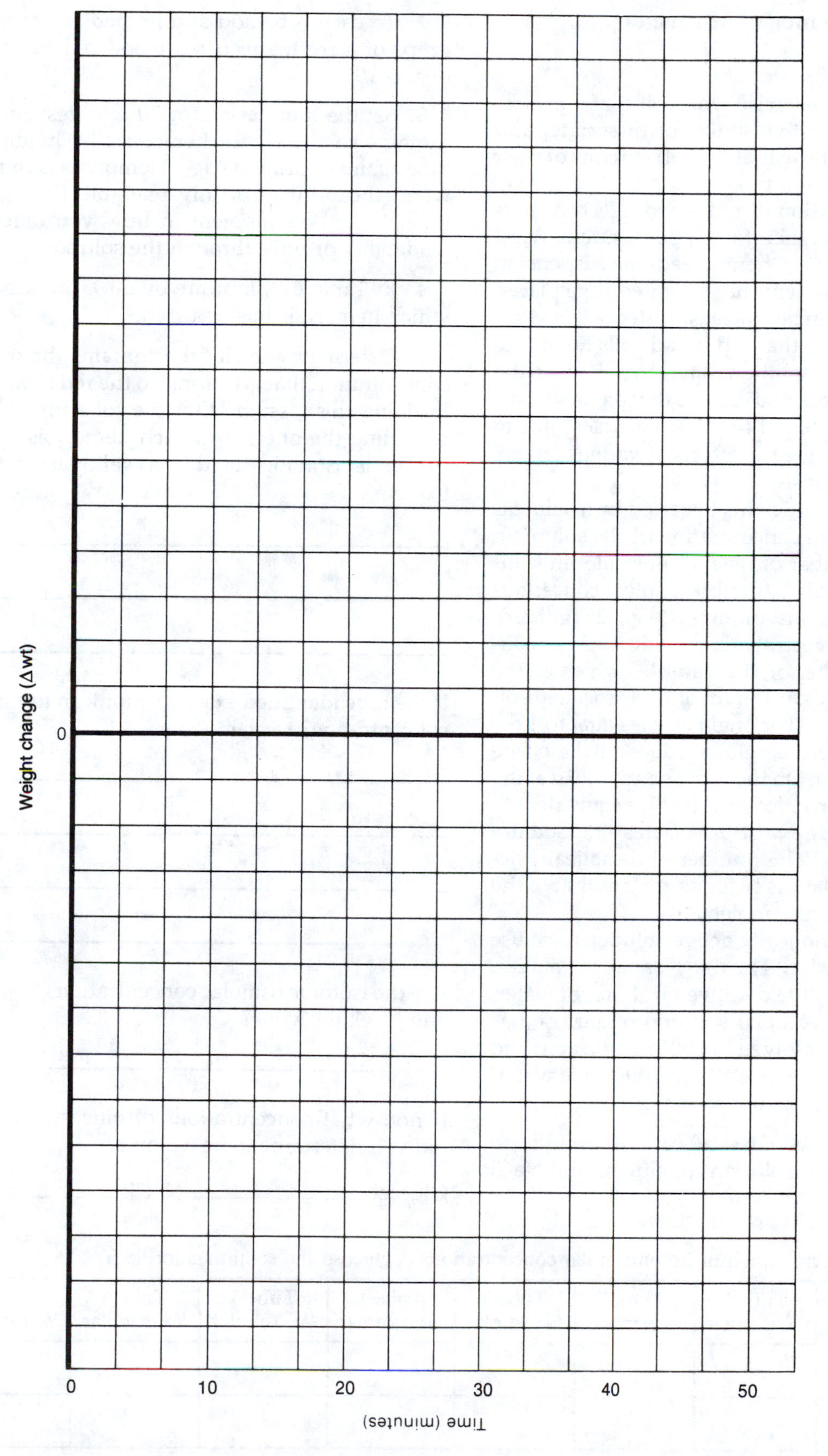

FIGURE 6-9 Osmosis data

3. Effect of Ionization of Molecules on Osmosis

In this part of the exercise, you will examine electrolytes (molecules that ionize or dissociate) and nonelectrolytes (those that do not) in terms of their osmotic effects on red blood cells.

A dilute suspension of red blood cells transmits very little light and therefore appears turbid. But if the red blood cells are hemolyzed, the suspension becomes so transparent that a printed page placed behind the tube can be read easily. Recall from Part C.1 of this exercise that red blood cells hemolyze when exposed to a solution that is hypotonic to the cells. For this reason, you can use the rate at which a suspension of red blood cells becomes clear due to hemolysis to examine the effects of various factors on the rate of osmosis.

Because the osmotic effect exerted by a solute is proportional to the concentration of the solute (in terms of the number of ions or molecules in solution), it is meaningless to express solute concentration merely in terms of mass (e.g., 25 g/liter). Instead, for nonelectrolytes we will express concentration in terms of the **osmole,** which is the number of particles in 1 gram of undissociated solute. For example, 180 g of glucose is equal to 1 osmole of glucose because glucose does not dissociate (ionize) in solution. On the other hand, 58.5 g of the electrolyte sodium chloride (NaCl) is equivalent to 2 osmoles because NaCl dissociates into sodium and chloride ions. The number of osmotically active particles in NaCl is twice as great as that of the undissociated glucose molecules.

In defining osmotically active solutions, we use the term **osmolarity.** Thus, a 1 **osmolar** solution has 1 osmole of solute dissolved in 1 liter of water. For example, a 1 osmolar solution of glucose has 180 g of glucose dissolved in 1000 ml of water, and a 0.1 osmolar solution has 18 g in 1000 ml of water.

Carry out the following procedure.

1. Set up two rows of test tubes corresponding to the various osmolar solutions of glucose and NaCl shown in Table 6-3.

2. To each tube add and immediately mix two drops of a fresh sheep red blood cell suspension (Fig. 6-10).

3. Set the tubes aside for 30 minutes, and then examine each tube for hemolysis by holding the tube against a printed page. Hemolysis is complete when the print is plainly readable through the tube. *Note:* Be consistent in how you determine readability of print through the solution.

4. In Table 6-3, indicate by an X the tube(s) in which hemolysis has occurred.

5. Determine which tube contains the osmolar concentration that is isotonic to the red blood cells. In doing this, assume that the tube immediately preceding the one(s) in which hemolysis had occurred is isotonic. Is this a valid assumption? Explain.

____

____

____

Is the tube identified exactly isotonic in terms of its concentration? Explain.

____

____

____

____

Are the isotonic osmolar concentrations of glucose and NaCl the same?

____

If not, what concentrations of glucose and NaCl did you determine to be isotonic?

Glucose: ____________ NaCl: ____________

TABLE 6-3 Determination of isotonic molar concentrations of glucose and sodium chloride

Hemolysis solutions[a]	Tube 1 ¼ osmolar	Tube 2 ⅙ osmolar	Tube 3 ⅛ osmolar	Tube 4 1/10 osmolar	Tube 5 1/12 osmolar	Tube 6 1/14 osmolar	Tube 7 1/16 osmolar
Glucose							
Sodium chloride							

[a]Indicate the occurrence of hemolysis by X.

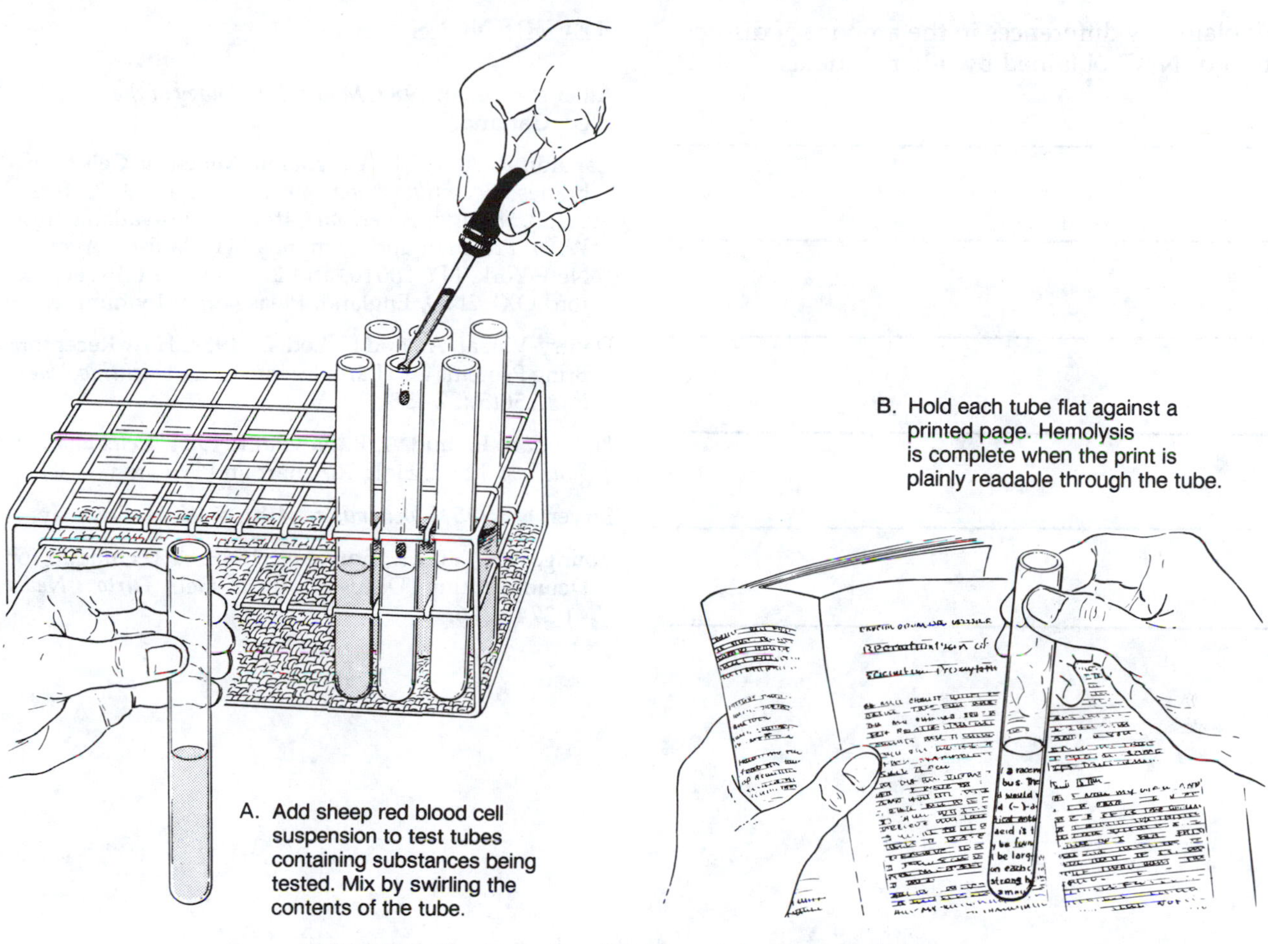

FIGURE 6-10 Procedure for determining hemolysis time

Explain any difference in isotonic osmolar concentration between glucose and NaCl.

____

____

____

____

____

____

____

With the data you have obtained, you can calculate the degree of dissociation of NaCl using the formula

$$a = \frac{i}{1 + (k - 1)}$$

where

$$i = \frac{\text{isotonic osmolar concentration of glucose}}{\text{isotonic osmolar concentration of NaCl}}$$

k = number of ions from each molecule of NaCl

a = degree of dissociation (to obtain percent dissociation, multiply by 100)

What is the percent dissociation of NaCl in your study?

____

What is the percent dissociation of NaCl obtained by several other students?

____

Explain any differences in the amount of dissociation of NaCl obtained by other students.

____

____

____

____

____

____

REFERENCES

Alberts, B., et al. 1989. *Molecular Biology of the Cell.* 2nd ed. Garland.

Capaldi, R. A. 1974. A Dynamic Model of Cell Membranes. *Scientific American* 230(3):26–33 (Offprint 1292). *Scientific American* Offprints are available from W. H. Freeman and Company, 41 Madison Avenue, New York, NY 10010, and 20 Beaumont Street, Oxford OX1 2NQ, England. Please order by number.

Dautry-Varsat, A., and H. Lodish. 1984. How Receptors Bring Proteins and Particles into Cells. *Scientific American* 250(5):52–58.

Holtzman, E., and A. B. Novikoff. 1984. *Cells and Organelles.* 3d ed. Holt, Rinehart and Winston.

Stryer, L. 1995. *Biochemistry.* 4th ed. W. H. Freeman.

Young, J. H., and L. Brubaker. 1963. A Technique for Demonstrating Diffusion in a Gel. *Turtox News* 41:274–276.

Exercise 7

Enzymes

Many of the chemical reactions that are characteristic of cellular activity can be made to proceed rapidly in a test tube but only at temperatures and pressures that are incompatible with life. In a living organism, these complex metabolic reactions proceed at relatively low body temperatures and precise rates because of the presence of organic catalysts called **enzymes.**

Enzymes are powerful catalysts, and many are specific to the reactions they affect. Enzymes do not alter the direction of a chemical reaction but accelerate it equally in either direction. For example, the enzyme that hastens hydrolysis of a compound is also capable of speeding up the dehydration synthesis of the same substance. Enzymes are needed only in minute amounts because they are not used up in the reactions they catalyze.

Enzyme activity is usually expressed as the rate of the reaction that the enzyme catalyzes. The rate of reaction is defined as the amount of substrate (the material with which the enzyme reacts) transformed, or the amount of product formed, per unit of time. The study of the rates of enzyme-catalyzed reactions is called **enzyme kinetics.**

In this exercise, you will examine some of the factors that affect enzyme-catalyzed reactions by studying two enzymes—amylase and phosphorylase—that are involved in the hydrolysis and the synthesis of starch, respectively. Starch is a polysaccharide composed of glucose molecules linked together, as shown in Fig. 7-1. Starches are stored by many plants (in roots) and by animals (in the liver as glycogen) for later use as an energy source. Starch is broken down into its component sugar units by enzymatic hydrolysis. This sugar (glucose) is important because (1) it is a key compound in cellular metabolism (recall that it is synthesized from CO_2 and H_2O by photosynthesizing plants) and (2) it contains energy (in the chemical bonds holding the molecule together) that can be released to do the work of the cell.

A. STARCH HYDROLYSIS BY AMYLASES

Starch is hydrolyzed by **amylases,** enzymes that cleave the starch molecule into smaller and smaller subunits until maltose, a reducing sugar, is obtained. This sequence is shown in Fig. 7-2.

Maltose is enzymatically converted into glucose by the enzyme maltase, which can enter the glycolytic and Krebs cycles, where it is broken down into carbon dioxide, water, and energy in the form of adenosine triphosphate (ATP). This process, called **cellular respiration,** will be considered in Exercise 8.

In the experiments that follow, you will measure

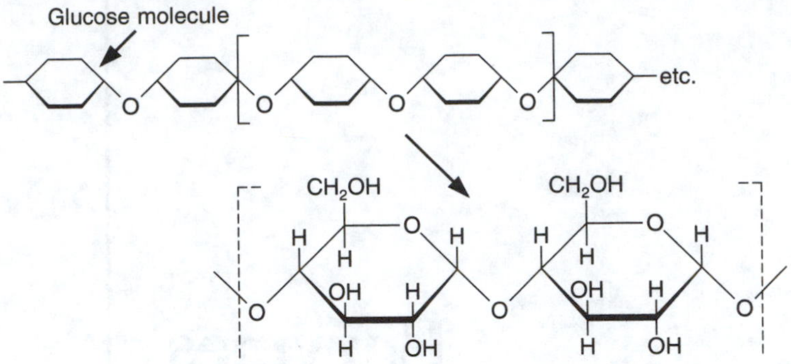

FIGURE 7-1 Structure of a starch molecule

colorimetrically the rate of hydrolysis of starch by amylase under various conditions of temperature, pH, and enzyme and substrate concentrations. You will test for the presence of starch by adding several drops of iodine to the sample. If starch is present, the solution turns a deep blue-black.

As hydrolysis proceeds, the amount of starch in the sample gradually decreases. This decrease is reflected in the color of the sample after you add the iodine. When you test the sample at various intervals after adding the enzyme, you will observe a graded series of colors from deep blue (starch present) to red (partial hydrolysis) to the color of iodine (complete hydrolysis). Thus, you can *qualitatively* measure starch hydrolysis by observing the color changes. You will *quantitatively* measure the rate of hydrolysis by determining the amount of light the sample absorbs when placed in a colorimeter (refer to Appendix A for use of the Milton Roy Spectronic 20® colorimeter and Appendix B for the principles of spectrophotometry).

Before beginning the experiment, carry out the following preliminary procedures to obtain the enzyme salivary amylase and to standardize the colorimeter.

Caution: *Each student should use only his or her own saliva for these studies. In addition, each student*

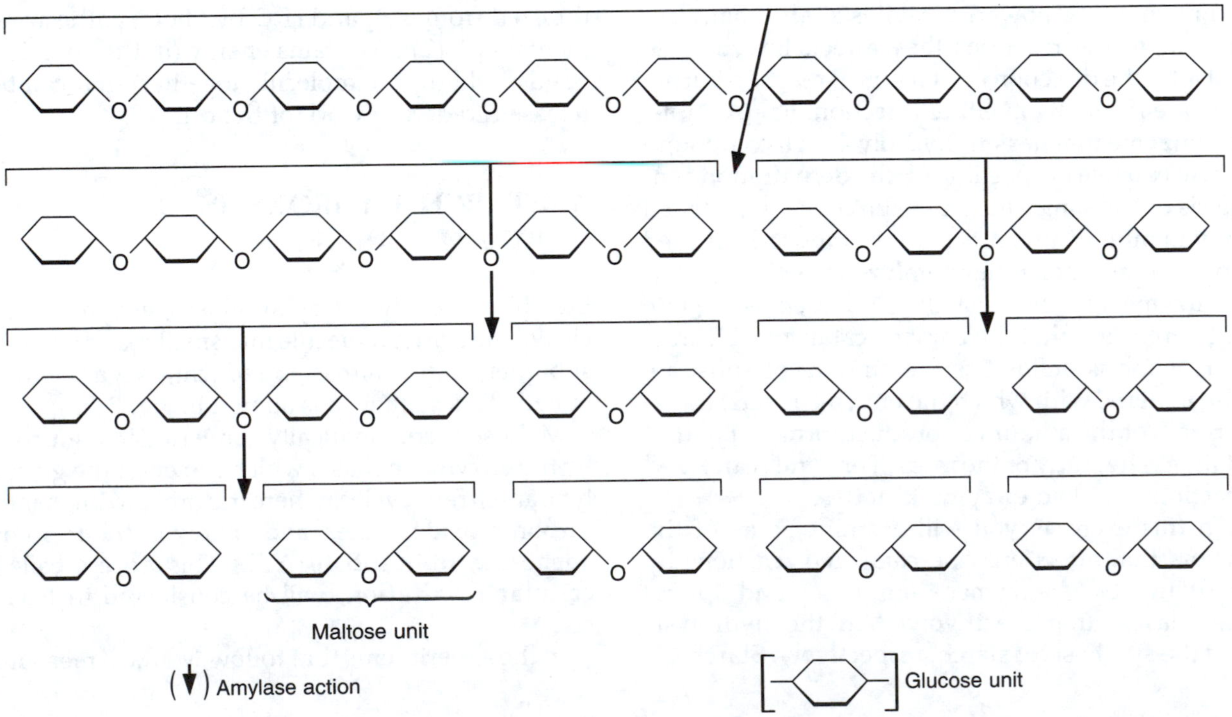

FIGURE 7-2 Action of amylase on starch

should be the only *one who carries out pipetting or other transfer and mixing procedures involving his or her saliva sample. Students should use disposable pipets, test tubes, and colorimeter tubes and discard them in the special biological disposal bags provided by the instructor.*

1. Collect 10 ml of saliva in a clean test tube. (You can stimulate the flow of saliva by chewing a small piece of paraffin.) Filter the saliva through a double layer of cheesecloth into a small beaker. This is the *stock enzyme solution.*

2. Turn on the colorimeter and allow the instrument to warm up for 5 minutes. Set the meter needle to record 0% transmittance at a wavelength of 560 nanometers (nm) with no test tube in the holder.

3. To blank the instrument, prepare an iodine control by adding three drops of iodine to 3 ml of water in a colorimeter tube. Place the iodine control in the colorimeter, close the cover, and adjust the light control until the needle records 100% transmittance. Why is this step required?

___

1. Effect of Substrate Concentration on Activity of Amylase

In this part of the exercise, you will determine the effect that altering the amount of substrate available (starch) has on the activity of the enzyme.

1. To each of four 125-ml Erlenmeyer flasks, numbered 1–4, add 50 ml of distilled water.

2. Add 50 ml of starch solution to flask 1. This flask now has a 1 : 2 dilution of starch.

3. Make additional starch dilutions of 1 : 4, 1 : 8, and 1 : 16 as follows.

   a. Remove 50 ml of the 1 : 2 starch dilution from flask 1 and pour it into flask 2; mix thoroughly to make a 1 : 4 dilution.

   b. Remove 50 ml from flask 2 and pour it into flask 3; mix thoroughly to make a 1 : 8 dilution.

   c. Remove 50 ml from flask 3 and pour it into flask 4; mix thoroughly to make a 1 : 16 solution. Remove and discard 50 ml from flask 4 into the bottle provided by your instructor.

4. Add 0.2 ml of saliva (your stock enzyme solution) to flask 1, mix thoroughly, and note the time.

5. After 2 minutes, pipet 3 ml of the mixture from flask 1 into a colorimeter tube.

6. Add three drops of iodine to the tube. Mix the contents of the tube thoroughly and place the tube in the colorimeter. (*Note:* Make certain the instrument has been blanked with the iodine control (preliminary procedure, step 3) before taking your reading of the first tube.)

7. Determine the percent transmittance and record this value in Table 7-1.

8. Repeat steps 5–7, using a new sample, every 2 minutes until either the contents of the flask are used up or a reading of 90% transmittance is reached. (*Note:* For these experiments, a reading of 90% *T* or higher indicates that hydrolysis is complete.)

9. Repeat steps 4–7 for flasks 2, 3, and 4, sampling one flask at a time.

10. Plot the data from Table 7-1 in Fig. 7-3. Interpret the results in terms of the amount of substrate available to a constant amount of enzyme.

___

___

___

___

2. Effect of Enzyme Concentration on the Activity of Amylase

1. To each of four 125-ml Erlenmeyer flasks, numbered 1–4, add 20 ml of distilled water.

2. Add 5 ml of the stock enzyme solution to flask 1 to make a 1 : 5 enzyme dilution.

3. Prepare enzyme dilutions of 1 : 25, 1 : 125, and 1 : 625 as follows.

   a. Remove 5 ml of the 1 : 5 dilution of enzyme from flask 1 and add it to flask 2. Mix thoroughly to make a 1 : 25 enzyme dilution.

   b. Remove 5 ml of the 1 : 25 enzyme dilution from flask 2 and add it to flask 3 to make a 1 : 125 enzyme dilution.

   c. Remove 5 ml of the 1 : 125 enzyme dilution

TABLE 7-1 Effect of substrate concentration on the activity of salivary amylase (percent transmittance)

Time (min)	Substrate dilution			
	Flask 1 (1:2)	Flask 2 (1:4)	Flask 3 (1:8)	Flask 4 (1:16)
Conclusion:				

from flask 3 and add it to flask 4 to make a 1:625 enzyme dilution. Remove and discard 5 ml of solution from this flask into the bottle provided by your instructor.

4. Add 25 ml of starch solution to flask 1, mix thoroughly, and note the time.

5. Two minutes after adding the starch, pipet 3 ml of the mixture from flask 1 and transfer it to a colorimeter tube.

6. Immediately test for starch by adding three drops of iodine.

7. Mix the contents of the tube thoroughly; determine the percent transmittance and record this value in Table 7-2. (*Note:* Do not forget to blank the instrument before using the iodine control.)

8. Repeat steps 5–7, using a new sample every 2 minutes as described in Part A.1.

9. Repeat steps 4–8 for flasks 2, 3, and 4, doing only one flask at a time.

10. Plot the data in Fig. 7-4.

3. Effect of pH on Activity of Amylase

1. To each of four 125-ml Erlenmeyer flasks, numbered 1–4, add 25 ml of buffer solution at pH 5, 6, 7, and 9, respectively.

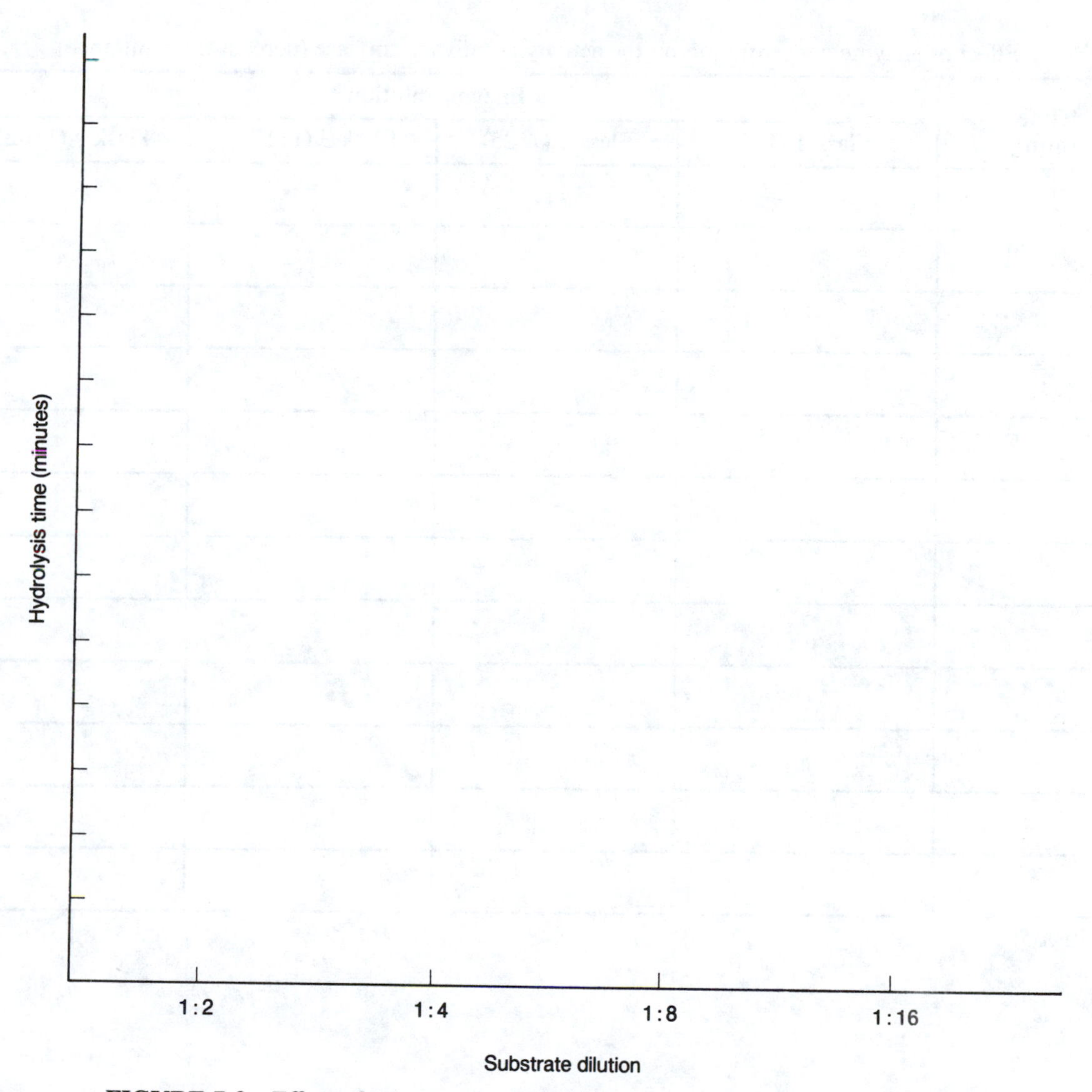

FIGURE 7-3 Effect of substrate concentration on the activity of salivary amylase

2. Add 0.5 ml of the solution having the optimum enzyme concentration (determined in Part A.2) to each flask.

3. To flask 1 add 25 ml of the solution having the optimum substrate concentration (determined in Part A.1). Mix thoroughly and note the time.

4. After 2 minutes, pipet 3 ml of the starch–buffer–enzyme mixture from flask 1 to a colorimeter tube or cuvette.

5. Immediately test for starch by adding three drops of iodine.

6. Mix the contents of the tube thoroughly; determine the percent transmittance and record this value in Table 7-3.

7. Repeat steps 4–6, using a new sample every 2 minutes as described in Part A.1.

8. Repeat steps 3–7 for flasks 2, 3, and 4, doing only one flask at a time.

9. Plot the data in Fig. 7-5.

What is the optimum pH for the activity of salivary amylase?

____

TABLE 7-2 Effect of enzyme concentration on the activity of salivary amylase (percent transmittance)

Time (min)	Enzyme dilution			
	Flask 1 (1:5)	Flask 2 (1:25)	Flask 3 (1:125)	Flask 4 (1:625)
Conclusion:				

Are your results consistent with your knowledge of where this enzyme functions in the body? Explain.

4. Effect of Temperature on Activity of Amylase

1. To each of five 125-ml Erlenmeyer flasks, numbered 1–5, add 50 ml of the solution having optimum starch concentration (determined in Part A.1).

2. Immerse one flask in each of five large beakers of water, which you have adjusted to temperatures of 5, 15, 30, 45, and 70°C by adding ice water or hot water as needed. These temperatures should be maintained throughout the experiment and should not vary more than ±3°C.

3. To flask 1 add 1 ml of the solution having the optimum enzyme concentration (determined in Part A.2). Mix thoroughly and note the time.

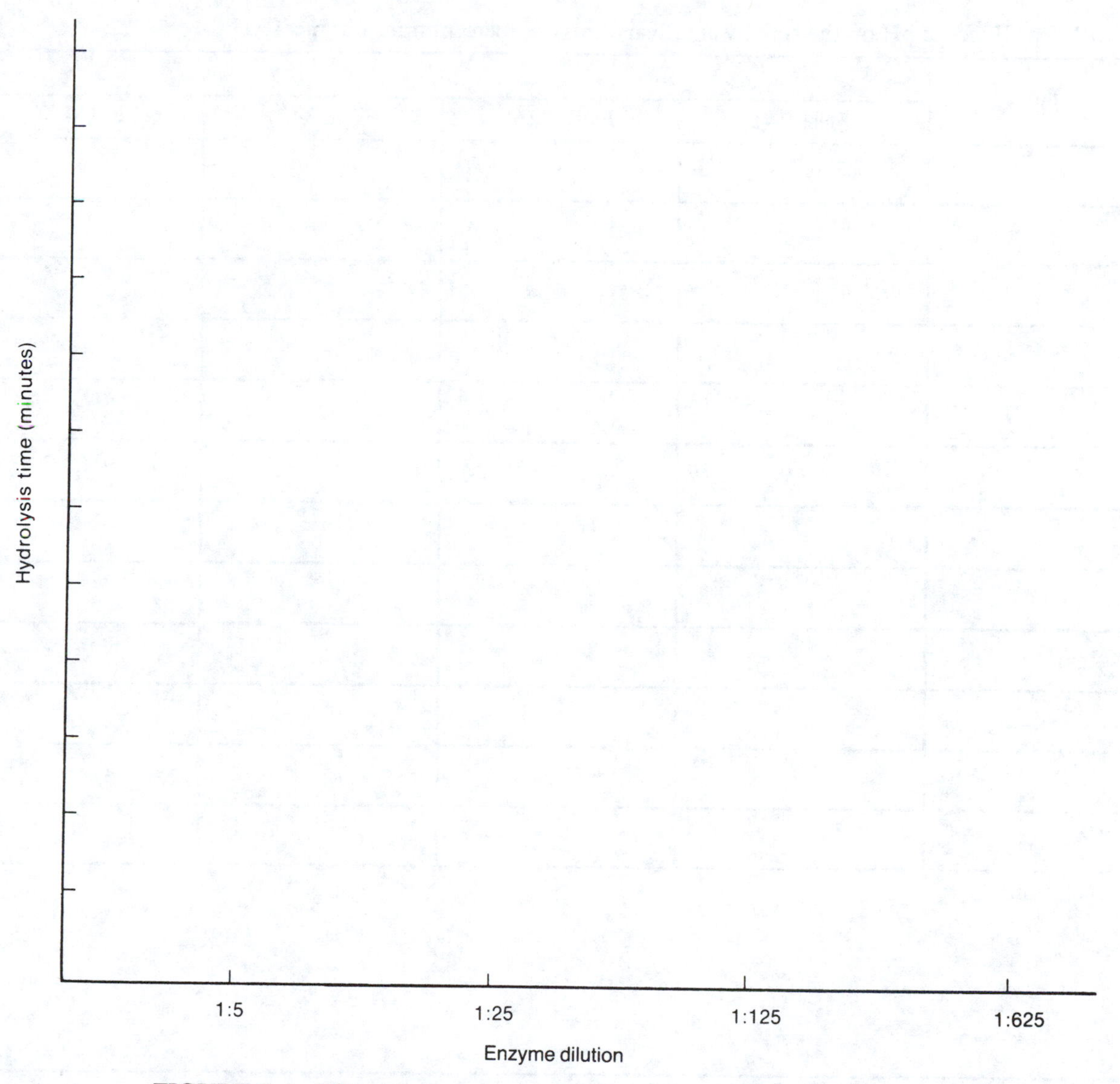

FIGURE 7-4 Effect of enzyme concentration on the activity of salivary amylase

4. After 2 minutes, pipet 3 ml of the mixture from flask 1 into a colorimeter tube.

5. Immediately test for starch by adding three drops of iodine.

6. Mix the tube thoroughly; determine the percent transmittance and record this value in Table 7-4.

7. Repeat steps 4–6, using a new sample every 2 minutes as described in Part A.1.

8. Repeat steps 3–8 for flasks 2, 3, 4, and 5, doing one flask at a time.

9. Plot the data in Fig. 7-6.

What is the effect of temperature on the activity of salivary amylase?

____

____

B. ENZYMATIC SYNTHESIS OF STARCH BY PHOSPHORYLASES

In Part A of this exercise, you observed the effect of the hydrolysis of starch by a class of enzymes called amylases. Enzymes called **phosphorylases**

TABLE 7-3 Effect of pH on the activity of salivary amylase (percent transmittance)

Time (min)	pH			
	Flask 1 (5)	Flask 2 (6)	Flask 3 (7)	Flask 4 (9)
Conclusion:				

can also hydrolyze starch. In fact, they degrade starch more rapidly than any known amylase.

The activities of these two classes of enzymes differ in a number of ways besides their rates of hydrolysis. For example, amylases cleave starch into maltose units and require a second enzyme, called **maltase,** to complete the conversion to glucose. Phosphorylases cleave starch into glucose–phosphate units, which are further hydrolyzed to glucose and phosphoric acid by the activity of the enzyme **phosphatase.** The most important difference between amylase and phosphorylase lies in the reversibility of the phosphorylase reaction, which goes in either direction, whereas the amylase reaction is almost irreversible (Fig. 7-7).

In the following procedure (Fig. 7-8), you will isolate phosphorylase from fresh potatoes and then synthesize starch using this enzyme. You can monitor the formation of starch by periodically adding iodine to samples of the reaction mixture. As the starch forms, the color, after addition of iodine, progresses from that of the iodine solution (no starch present) to blue-black (indicating the presence of starch).

1. Cut a peeled potato into quarters. Place the pieces into a blender, add 65 ml of water, and homogenize until a slurry is formed. Pour the slurry through four layers of cheesecloth into a 125-ml beaker to remove any remaining pieces of potato.

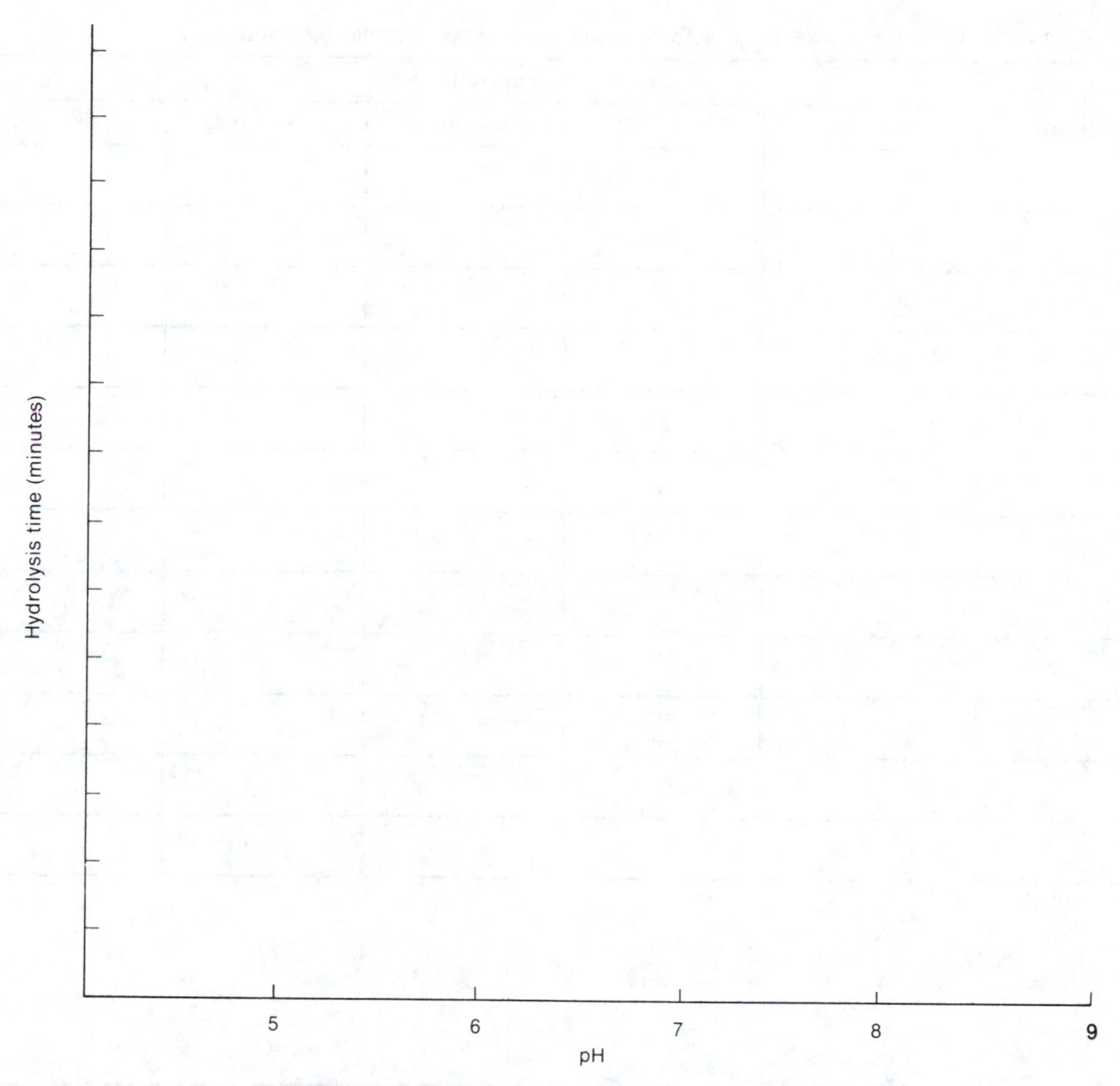

FIGURE 7-5 Effect of pH on the activity of salivary amylase

2. Place a piece of Whatman #1 filter paper on a Buchner funnel and wet the paper. Attach rubber tubing from the flask to the water aspirator and turn on the water faucet.

3. Pour the potato filtrate into the funnel. After filtration is complete, first remove the tubing from the aspirator and then turn off the faucet.

4. Pour the filtrate into a beaker and place the beaker in a 50°C water bath. Amylases are denatured at 50°C. Why is it necessary to destroy these enzymes in your preparation?

____

____

5. After 5 minutes, pour 25 ml of the heated filtrate into a beaker and slowly add 5 g of ammonium sulfate [$(NH_4)_2SO_4$] crystals. Stir until the crystals are dissolved and a brown precipitate of amylase forms.

6. With a wax pencil, mark a conical centrifuge tube 10 cm from the bottom and pour the amylase filtrate into the tube up to this mark.

7. Place two large beakers on the platforms of a double-beam balance, and add water to the lightest of the two beakers until the beakers balance. Place your filled conical centrifuge tube in one of the beakers; have another student place his tube in the other beaker. Add water to the centrifuge tubes as necessary to balance them. Place the balanced centrifuge tubes directly opposite one another in the

TABLE 7-4 Effect of temperature on the activity of salivary amylase (percent transmittance)

Time (min)	Temperature (°C)				
	Flask 1 (5)	Flask 2 (15)	Flask 3 (30)	Flask 4 (45)	Flask 5 (70)
Conclusion:					

head of a clinical centrifuge. Centrifuge these preparations for 5 minutes at full speed.

8. Remove the centrifuge tubes and pour the supernatant liquid into a graduated cylinder. The pellet may not be packed tightly, so it is best to pour the supernatant with a quick motion. Discard the pellet by washing it out of the centrifuge tube.

9. Add water to the supernatant to bring the volume up to 25 ml. Add 4 g of ammonium sulfate crystals to the supernatant, stirring until all crystals dissolve and another brown precipitate forms. *This is the precipitate you want to keep because it is mostly phosphorylase.*

10. Pour the mixture into a conical centrifuge tube and repeat the balancing procedures in step 7. Centrifuge at full speed for 5 minutes. Remove the centrifuge tube after the centrifuge has stopped and discard the supernatant.

11. Add 1 ml of the phosphate buffer solution to the pellet and stir. Then add 0.5 ml of glucose-1-phosphate solution to the resuspended pellet in the centrifuge tube and swirl to mix. Note the time: ________________

12. Obtain a spot plate and place one drop of iodine solution in one of the wells.

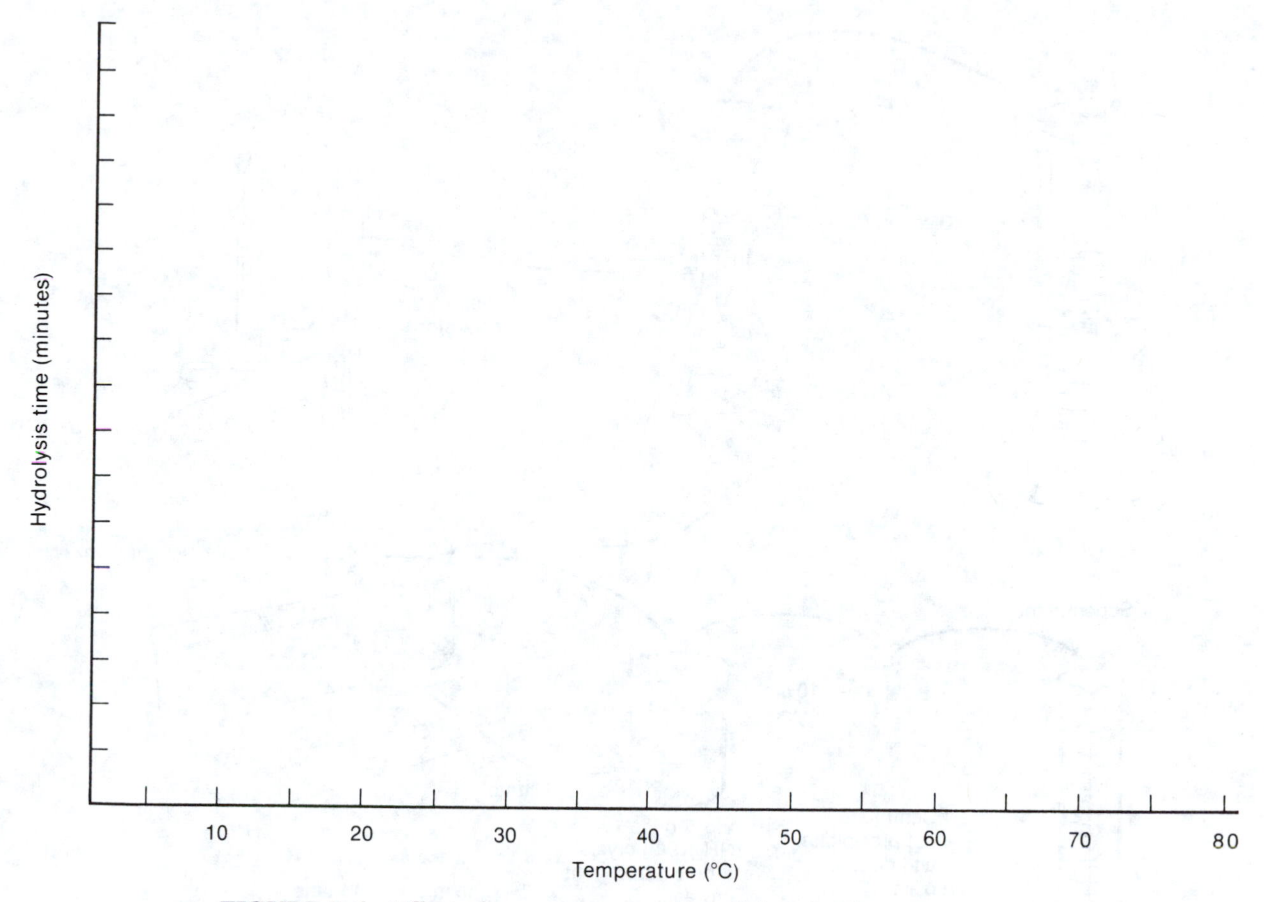

FIGURE 7-6 Effect of temperature on the activity of salivary amylase

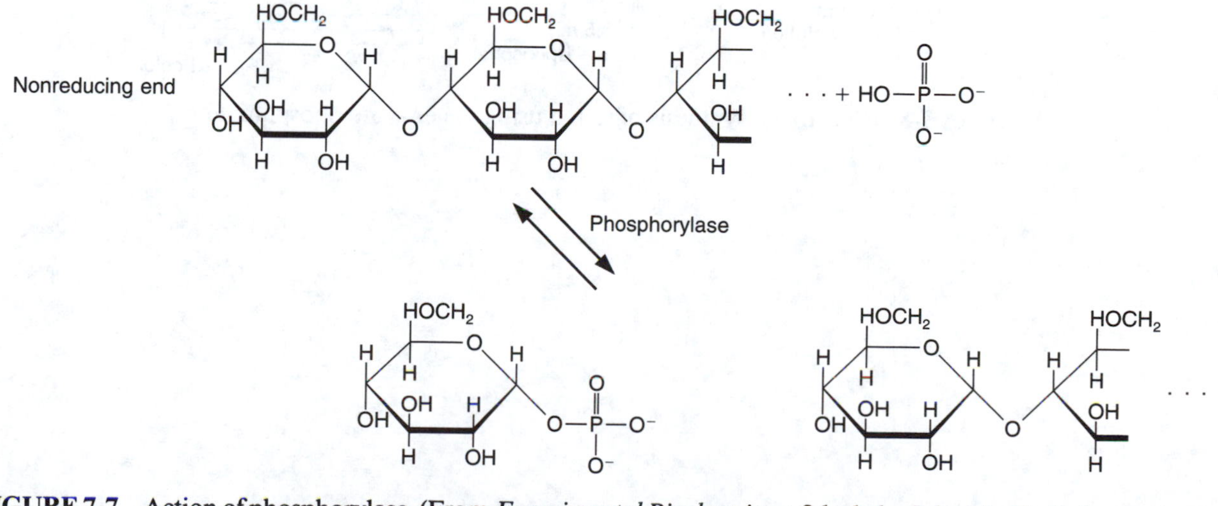

Nonreducing end

Phosphorylase

FIGURE 7-7 Action of phosphorylase. (From *Experimental Biochemistry*, 2d ed., by John M. Clark, Jr., and Robert L. Switzer. W.H. Freeman and Company © 1977.)

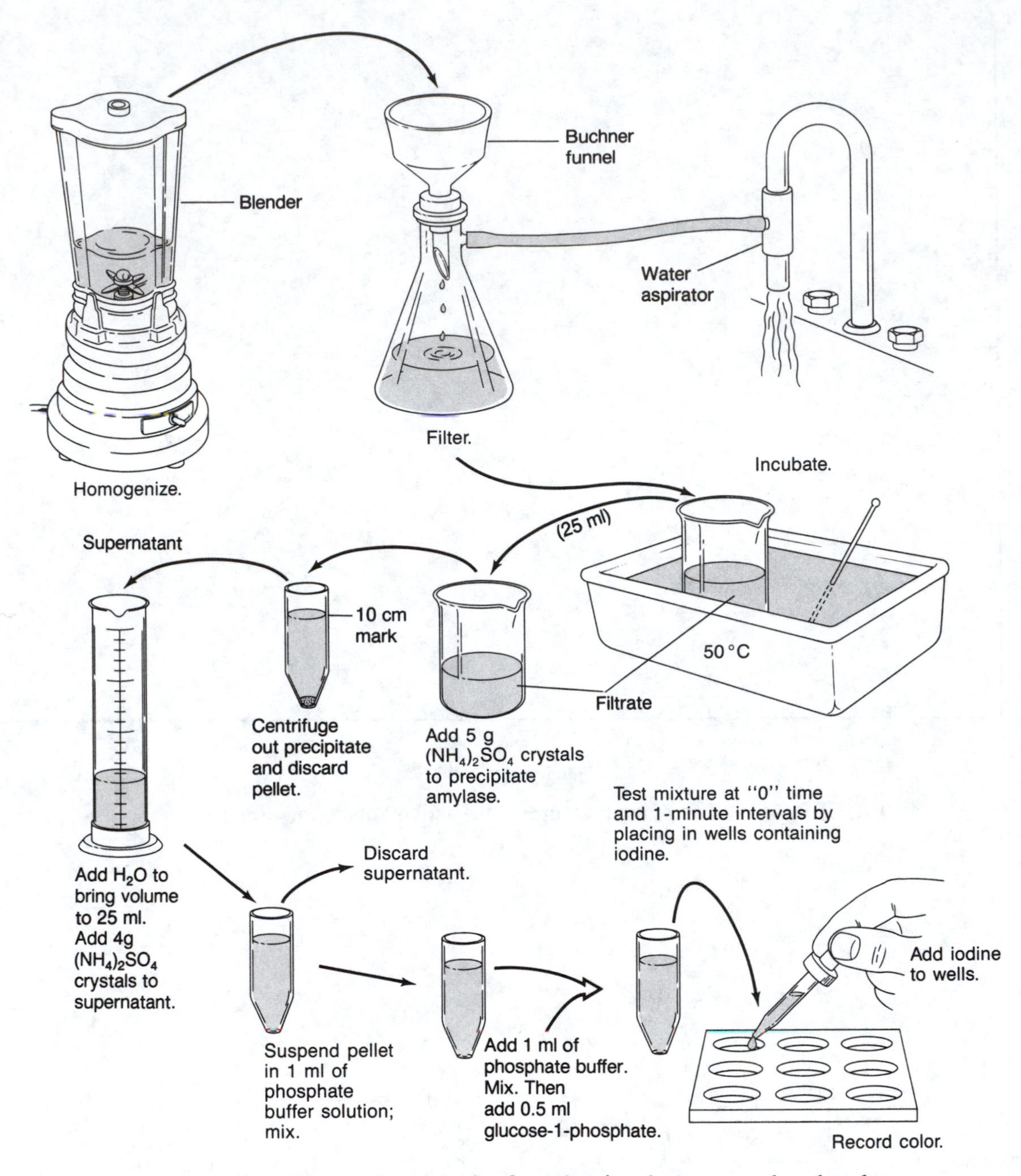

FIGURE 7-8 Enzymatic synthesis of starch using the enzyme phosphorylase

13. Remove some of the mixture in the centrifuge tube and add it to the first well containing the drop of iodine. Then add an additional three drops of iodine to the well with a Pasteur pipet. Record the color changes in Table 7-5.

14. Place another drop of iodine in the next well on the spot plate and repeat step 13. Record the color changes in Table 7-5.

15. Continue to make 1-minute readings for a total of 6 minutes. How long did it take for starch to begin to be synthesized in measurable quantities?

____

____

Indicate three ways to speed up this reaction.

____

____

____

____

____

TABLE 7-5 Enzymatic synthesis of starch by phosphorylase

Time (min)	Color intensity
0	
1	
2	
3	
4	
5	
6	

REFERENCES

Alberts, B., et al. 1989. *Molecular Biology of the Cell.* 2d ed. Garland.

Fersht, A. 1984. *Enzyme Structure and Mechanism.* W. H. Freeman.

Lehninger, A. L. 1982.*Principles of Biochemistry.* Worth.

Lodish, H. F., et al. 1995. *Molecular Cell Biology.* 3d ed. W. H. Freeman.

Stryer, L. 1988. *Biochemistry.* 3d ed. W. H. Freeman.

Exercise

8

Cellular Respiration

The energy that plants and animals need for living is obtained from the chemical bonds in nutrient molecules. The energy in these chemical bonds is solar energy that was transformed to chemical energy by the photosynthesis of green plants. The conversion of the chemical bond energy into a usable form such as the high-energy phosphate bond of **adenosine triphosphate (ATP)** and the final use of energy by an organism are processes that are collectively referred to as **respiration.** Respiration can be classified broadly as **aerobic** (with oxygen) or **anaerobic** (without oxygen). Anaerobic respiration evolved first and is still an important part of many of the metabolic activities of plants and animals. Aerobic respiration is a biochemical innovation (in the evolutionary sense) and is more efficient in terms of energy recovery from nutrient molecules. Fig. 8-1 summarizes the many steps in the anaerobic and aerobic respiration involving the simple 6-carbon sugar molecule glucose.

Besides yielding much more energy than does anaerobic respiration, aerobic respiration completely breaks down glucose to carbon dioxide and water, whereas anaerobic respiration generally leads to end products such as organic acids and alcohols, which may be toxic.

In aerobic respiration, the energy of the glucose molecule is released in three stages: **glycolysis,** the **Krebs cycle,** and the **electron transport chain** (Fig. 8-2). During **glycolysis,** the 6-carbon sugar molecule is converted by a complex series of enzyme-catalyzed reactions into two 3-carbon molecules of pyruvic acid and ATP. In the presence of oxygen, pyruvic acid enters a series of enzyme-catalyzed reactions called the **Krebs cycle.** During the Krebs cycle, hydrogen atoms, removed from Krebs cycle compounds, are split into protons (positively charged) and high-energy electrons (negatively charged). The electrons are passed on to an **electron transport chain** that consists of a series of molecules that are alternately oxidized (lose electrons) and reduced (gain electrons). During these oxidation-reduction reactions, some of the energy of the electrons is incorporated into the high-energy phosphate bonds of ATP. Finally, at the end of the chain, the protons (H in the figure) combine with the now low-energy electrons and oxygen to form water.

That respiration is taking place in an organism can be demonstrated by measuring the energy given off in the form of heat, the amount of glucose used, the amount of oxygen consumed, or the amount of carbon dioxide released. In this exercise, you will study respiration by determining indirectly the amount of oxygen consumed.

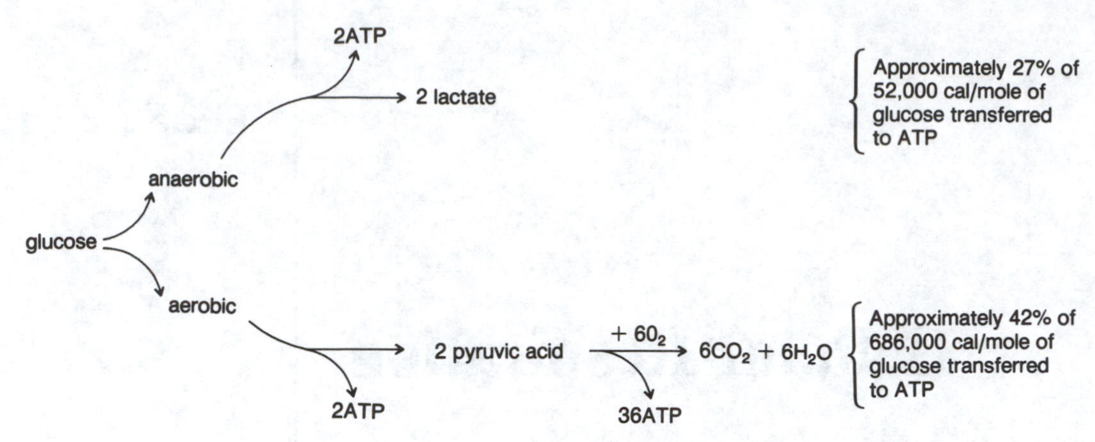

FIGURE 8-1 Anaerobic and aerobic respiration of glucose

A. MEASURING THE EFFECTS OF RESPIRATION

Many methods used in the study of respiration depend on measuring changes in the volume or pressure of CO_2 or O_2. Any change in the volume or pressure of one of these gases in a closed system in which an organism is respiring represents the net difference between oxygen consumption (which, by itself, would decrease pressure and volume in the closed container) and carbon dioxide production (which, by itself, would increase pressure and volume). Thus, if the carbon dioxide that is produced by an organism is absorbed in some way, changes can be attributed to oxygen consumption.

Oxygen Consumption by Germinating Peas

A simple **respirometer** (used to detect changes in gas pressure and volume) is shown in Fig. 8-3. This equipment consists of two vessels that can be closed to the outside. Respiring material is placed in one of the containers, along with potassium hydroxide (KOH), which absorbs carbon dioxide. Because gas volume is affected by such factors as atmospheric pressure and temperature, the second container (identical except for the living material) is employed as a **compensation chamber.** Be sure to consider changes in the volume of gas in the compensation chamber when evaluating changes in the respiration chamber.

In this part of the exercise, you will determine the respiratory rate of germinating peas in terms of oxygen intake.

1. Following the diagram in Fig. 8-3A, fill the respiration test tube half full with germinating peas.

2. Place a loose wad of cotton over the peas. Place about 12 mm of KOH pellets over the cotton. The cotton separates the KOH from the living seeds; it should not be packed tightly. KOH will remove the CO_2 as fast as it is given off by the respiring peas. Why is it necessary to remove the CO_2 from the tube?

____

____

____

3. Prepare the compensation tube in the same way but use glass beads in place of peas. Why use the inert material?

____

____

4. Insert a rubber stopper, with attached capillary tubing, firmly into each tube (Fig. 8-3B).

5. Place the tubes in a vertical position by clamping them to a ring stand (Fig. 8-3C). Using an eyedropper, add enough dye to the end of each capillary tube so that about 12 mm of the dye is drawn into the tube (Fig. 8-3D).

6. After allowing 2–3 minutes for the gas pressures to reach equilibrium, note the position of the

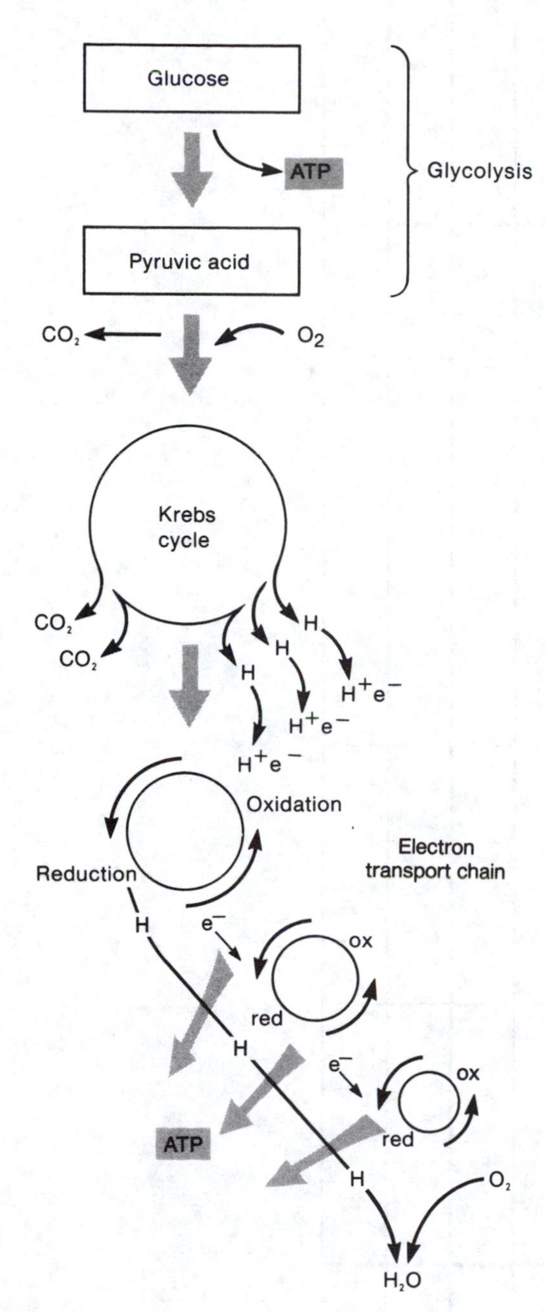

FIGURE 8-2 Energy, in the form of ATP, produced during respiration

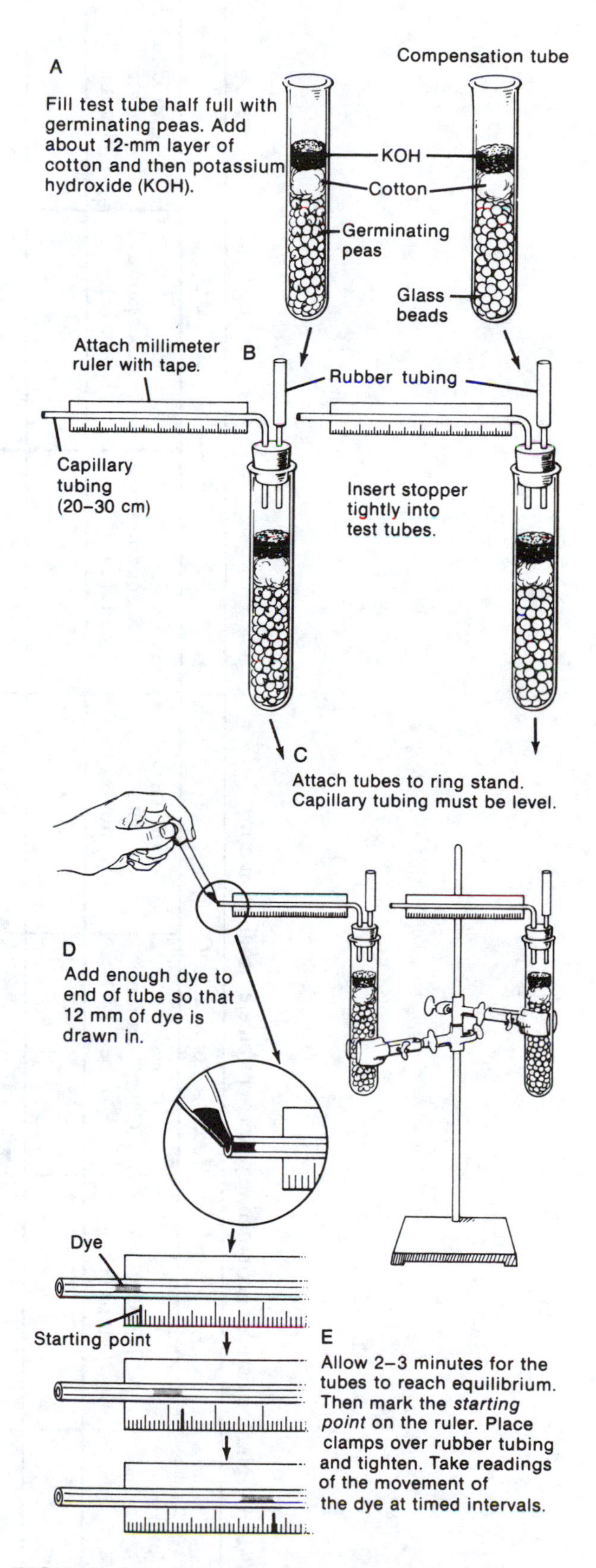

FIGURE 8-3 Procedure for measuring oxygen consumption in germinating peas

inner end of the dye column on the millimeter scale (Fig. 8-3E). Record this reading in Table 8-1. Then attach pinch clamps to the rubber tubing on both test tubes. (*Note:* Because the respirometer is very sensitive to volume changes due to heat, keep it away from heat sources such as lamps and hot plates or outside windows facing the sun.)

7. Take readings of the location of the column at 1-minute intervals for the next 5 minutes. Record

TABLE 8-1 Respiration data for germinating peas (respirometer readings in mm)

Time (min)	Room temp: ______°C			______°C		
	Respiration tube (1)	Compensation tube (2)	Corrected data (1 minus 2)	Respiration tube (1)	Compensation tube (2)	Corrected data (1 minus 2)

the data in Table 8-1. (*Note:* If the movement of the dye is fairly rapid, take readings at shorter intervals—20 or 30 seconds—or the column may reach the bent portion of the tube before you have enough data.) The dye column can be returned to the outer end of the tube by opening the pinch clamp and tilting the capillary tube. Why does the dye move toward the respiration chamber and not away?

____

____

Under what circumstances might the dye move away from the chamber?

____

____

____

8. Repeat the procedure to determine the effects of temperature on respiration and record the data in Table 8-1. Students should select various temperatures to reflect a variety of temperature effects.

9. Plot the data from Table 8-1 in Figs. 8-4 and 8-5. Identify each line on the graph with the appropriate label.

B. MEASURING BIOLOGICAL OXIDATION

As described at the beginning of this exercise, oxidation–reduction reactions are important in releasing the energy in glucose and incorporating this energy into ATP molecules. Iron-containing compounds called **cytochromes** take part in these reactions, which are carried out in the mitochondria. In these reactions, oxidation takes place in the presence of oxygen. An example of an oxidation–reduction reaction is shown in Reaction 1, in which A is the hydrogen donor and B is the hydrogen acceptor. Thus, every oxidation is accompanied by a simultaneous reduction. The energy required for the removal of hydrogens in oxidation reactions is supplied by the accompanying reduction. For this

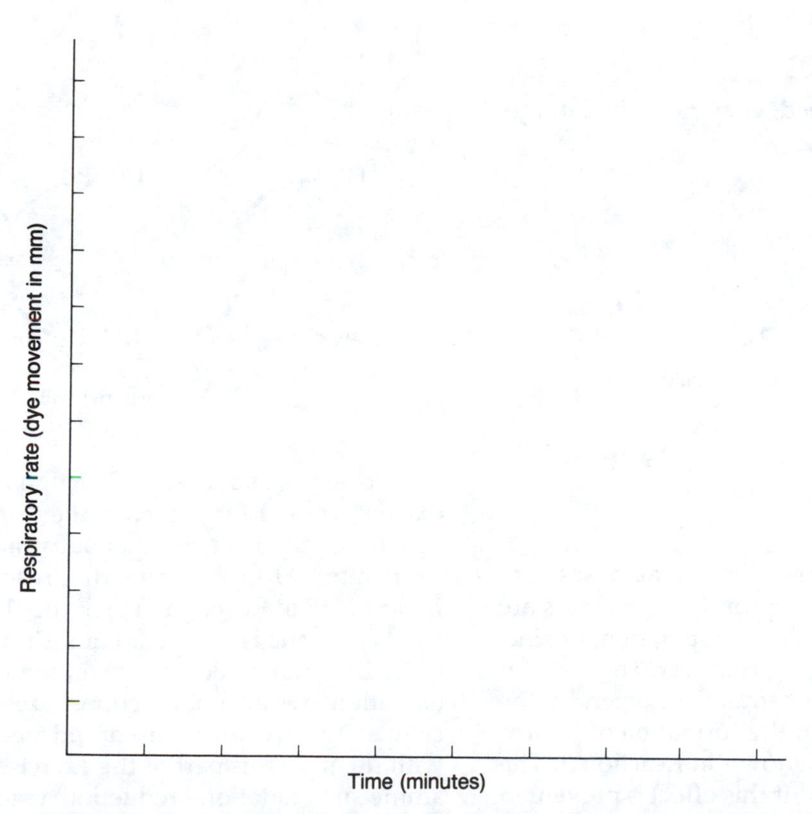

FIGURE 8-4 Respiratory rate of germinating peas at room temperature

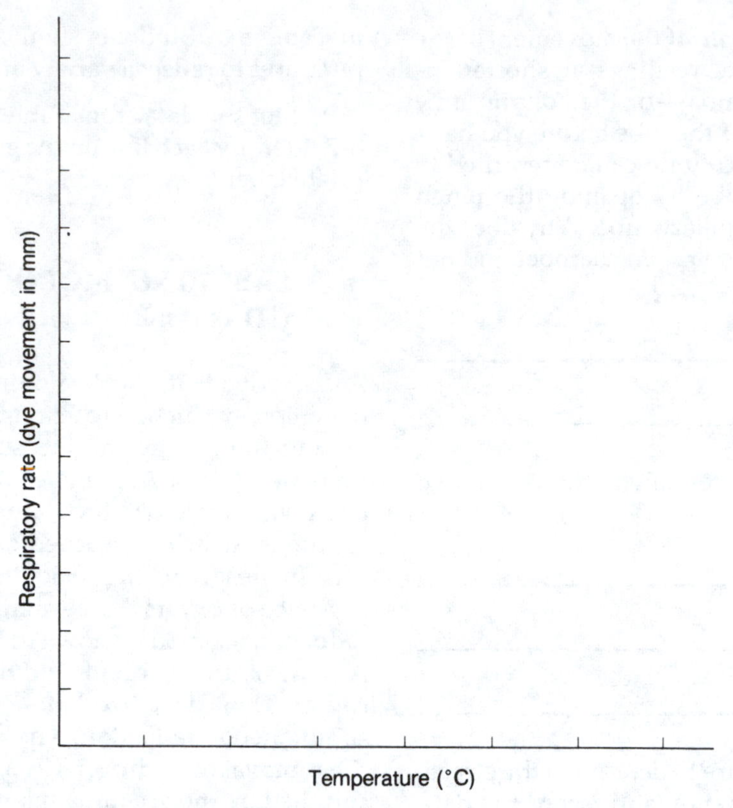

FIGURE 8-5 Effect of temperature on respiratory rate

type of reaction to proceed, enzymes called **dehydrogenases** are required.

Reaction 1

A · H_2 B

dehydrogenases

A B · H_2

Specific dehydrogenases, called **oxidases,** use oxygen as a hydrogen acceptor. Most oxidases are composed of a metal, such as copper, iron, or zinc, and a riboflavin-containing complex. The transfer of hydrogen from the substrate to oxygen by the oxidase usually results in the formation of hydrogen peroxide (H_2O_2), as shown in Reaction 2. The H_2O_2 is toxic to tissues, but this effect is prevented by two enzymes: **peroxidase** and **catalase.**

Reaction 2

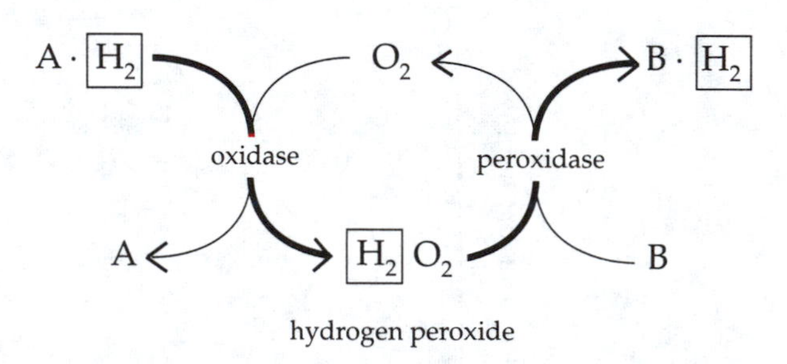

Peroxidase, as shown in Reaction 2, removes the oxygen from H_2O_2, which is then prepared to accept hydrogen from another substrate molecule to form more H_2O_2. During this reaction, another molecule (B in Reaction 2) picks up the H_2 released from H_2O_2 and is reduced. Thus, the toxic potential of hydrogen peroxide has been removed. Because oxidation–reduction reactions are important in cellular respiration, you should become familiar with them. In this part of the exercise, you will examine an oxidation–reduction reaction that involves peroxidase.

Guaiacol is oxidized to a colored product by H_2O_2 in the presence of peroxidase (Reaction 3). The reaction can be followed colorimetrically by measuring the amount of light absorbed by the guaiacol. (Refer to Appendix A for information on the use of the Milton Roy Spectronic 20® colorimeter and Appendix B for the principles of spectrophotometry.)

Reaction 3

$$\text{guaiacol (reduced)} + H_2O_2 \xrightarrow{\text{peroxidase}} \text{guaiacol (oxidized colored complex)} + 2H_2O$$

guaiacol (reduced) (ring with OCH_3 and OH) guaiacol (oxidized colored complex) (ring with two O)

1. Peel a turnip, cut it into small pieces, place them in a blender, and blend for 30 seconds. Or, place the pieces in a mortar with quartz sand, and grind to pulp with a pestle (Fig. 8-6).

2. Filter the homogenate through a double layer of cheesecloth into a beaker. Squeeze out as much of the juice as you can. Then dilute 1 ml of the juice with 200 ml of distilled water.

3. Using a wavelength of 500 nm, blank the colorimeter with a tube containing 0.01 ml of guaiacol, 0.2 ml of 0.9% H_2O_2, and 9.7 ml of water.

4. Combine 0.01 ml of guaiacol, 0.2 ml of 0.9% H_2O_2, and 9.7 ml of distilled water in a test tube. Put 1.0 ml of the diluted turnip extract and 4 ml of water into a second colorimeter tube. Pour the

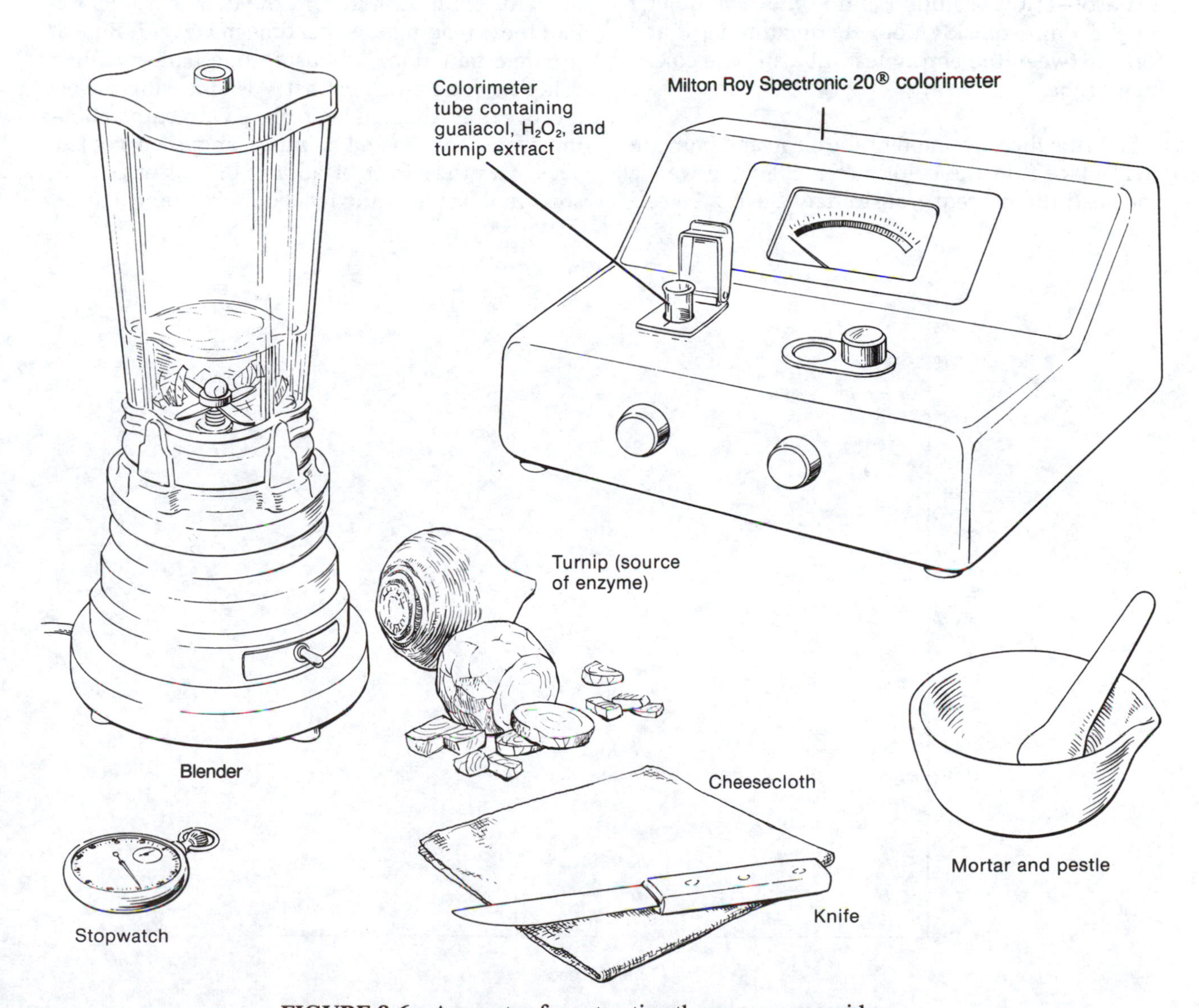

FIGURE 8-6 Apparatus for extracting the enzyme peroxidase

TABLE 8-2 Peroxidase activity in turnip tissue (percent transmittance)

Time (sec)	1 ml turnip extract	0.5 ml turnip extract	2.0 ml turnip extract	1 ml boiled turnip extract	1 ml extract +0.5 ml NaF
20					
40					
60					
80					
100					
120					

guaiacol–H_2O_2 solution into the colorimeter tube. To mix, quickly pour the mixture back and forth between the empty test tube and the colorimeter tube.

5. Wipe the colorimeter tube clean and immediately place it in the colorimeter. Start the watch, and read the percent transmittance every 20 seconds for 2 minutes. Record the data in Table 8-2. Plot these readings against time in Fig. 8-7. Repeat the determination, first using one-half and then using twice as much extract. Also measure the effect of extract that has been placed in a bath of boiling water for several minutes and then cooled. Then test the effect of adding 0.5 ml of 0.01 *M* sodium fluoride to the reaction mixture.

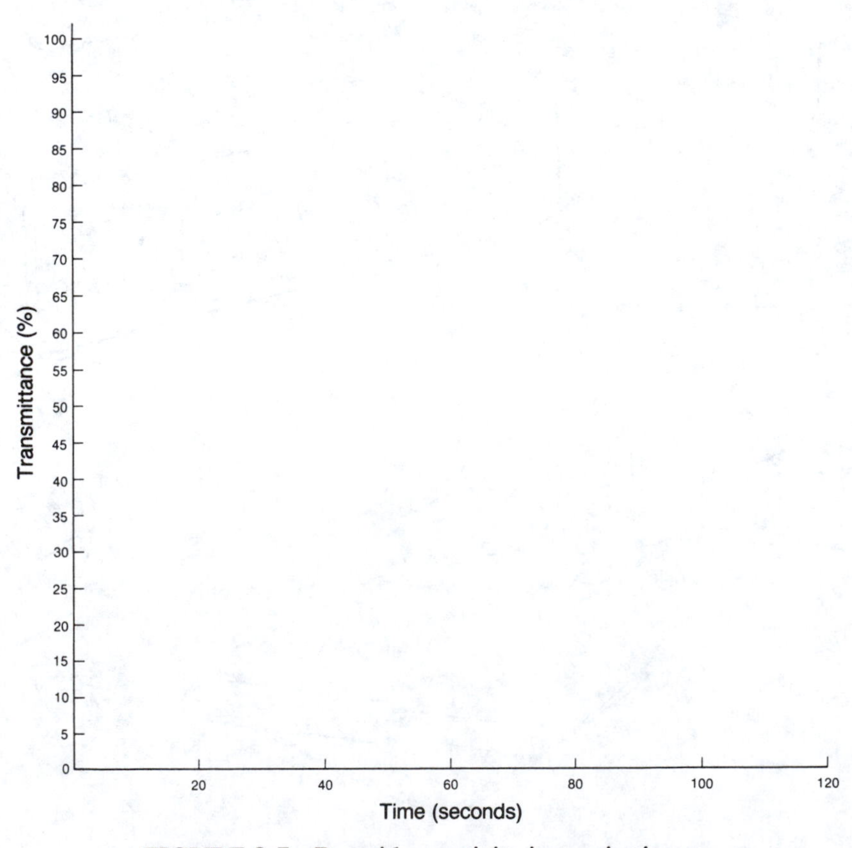

FIGURE 8-7 Peroxidase activity in turnip tissue

Caution: *Poison!*

From these data, what can you conclude about the activity of peroxidase in turnip tissue?

____

____

Fill in the blanks (heavy lines) in Reaction 4. Indicate hydrogen donor, hydrogen acceptor, the reaction catalyzed by peroxidase, and the reactions catalyzed by dehydrogenases.

Reaction 4

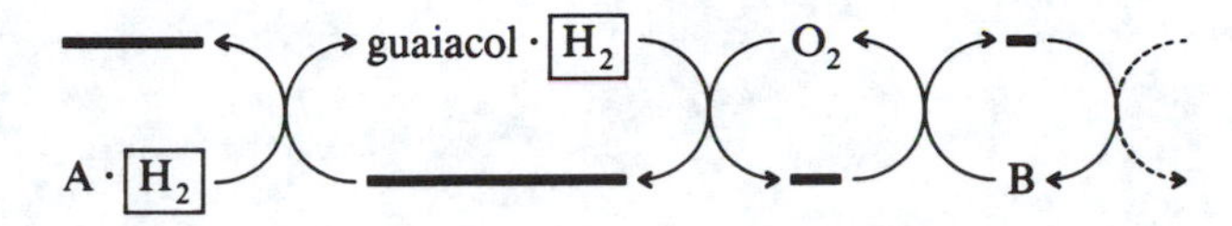

In oxidation–reduction reactions, what component of the hydrogen atom is actually being shunted through the various carrier molecules?

____

REFERENCES

Alberts, B., et al. 1989. *Molecular Biology of the Cell.* 2d ed. Garland.

Hinkle, P. C., and R. E. McCarty. 1978. How Cells Make ATP. *Scientific American* 238:104–123 (Offprint 1383). *Scientific American* Offprints are available from W. H. Freeman and Company, 41 Madison Avenue, New York, NY 10010, and 20 Beaumont Street, Oxford OX1 2NQ, England. Please order by number.

Karp, G. 1984. *Cell Biology.* 2d ed. McGraw-Hill.

Lodish, H. F., et al. 1995. *Molecular Cell Biology.* 3d ed. W. H. Freeman.

Prescott, D. M. 1988. *Principles of Molecular Structure and Function.* Jones and Bartlett.

Stryer, L. 1995. *Biochemistry.* 4th ed. W. H. Freeman.

Exercise

9

Photosynthesis

The living world exists almost entirely on the energy captured by the photosynthetic machinery of green plants and numerous other photosynthetic organisms. That is, the ultimate source of the energy expended by living organisms is the converted energy of sunlight that is trapped in newly synthesized organic molecules during photosynthesis. From the products of photosynthesis and from a small number of inorganic compounds which are available in the environment, living organisms synthesize the numerous complex molecules that make up their cellular structure or are essential to their existence in other ways.

Classically, the reaction in plant photosynthesis is

$$\underset{\text{carbon dioxide}}{6CO_2} + \underset{\text{water}}{12H_2O} \xrightarrow[\text{chlorophyll}]{\text{light}} \underset{\text{sugar}}{C_6H_{12}O_6} + \underset{\text{oxygen}}{6O_2} + \underset{\text{water}}{6H_2O}$$

This equation suggests that carbohydrate synthesis is the central feature in this process. Photosynthesis, however, is not a single-step reaction, as you might think from this equation. It is a complex process involving the interactions of many compounds. The large number of reactions can be divided into two groups: (1) the light, or photochemical, reactions in which light is required; and (2) the so-called dark, or biosynthetic, reactions that do not require light.

In the **photochemical (light) reactions,** radiant energy is used for two purposes:

- Light energy splits water molecules into oxygen and hydrogen. The hydrogen is then transferred to $NADP^+$ (nicotinamide adenine dinucleotide phosphate) to form NADPH, which in turn transfers hydrogen to other molecules.
- The light energy absorbed by chlorophyll is converted into chemical energy, which is stored in the molecule adenosine triphosphate (ATP). This conversion occurs in the chloroplast and consists of the transport of electrons from "excited" chlorophyll molecules through a series of acceptor molecules (including the cytochromes) that constitutes an electron transport system. This light-dependent generation of ATP is called **photophosphorylation** to differentiate it from **oxidative phosphorylation,** which occurs in mitochondria.

Thus, the light reactions result in the formation of NADPH and ATP and in the release of oxygen. In the **biosynthetic (dark) reactions,** NADPH and ATP are used to reduce CO_2 to carbohydrate.

In this exercise, you will examine the roles of light, carbon dioxide, and chloroplast pigments in the photosynthetic process.

A. ROLE OF LIGHT

1. Necessity of Light for Photosynthesis

It has been shown experimentally that much of the sugar produced in the leaves of a plant is rapidly condensed into starch. Although starch is not a direct product of photosynthesis (it is a condensation product of the glucose produced by photosynthesis), we can use its synthesis as indirect evidence of photosynthetic activity.

Your instructor will supply geranium plants that have been kept in the dark for 48 hours and plants that have been in the light for 48 hours. Test a leaf from a "dark" plant for the presence of starch using the following procedure (Fig. 9-1A–E).

1. Boil the leaf in a water bath for several minutes; then remove the pigment by putting the leaf in hot alcohol.

Caution: *Heat the alcohol in a separate beaker, on a hot plate. Do not heat it over a bunsen burner.*

---

2. Transfer the leaf to a petri dish containing iodine. If starch is present, the leaf will turn a deep blue-black.

3. Similarly, test for the presence of starch in the plants that were continuously exposed to light for 48 hours.

From your observations, what can you conclude about the necessity of light for photosynthesis?

____

____

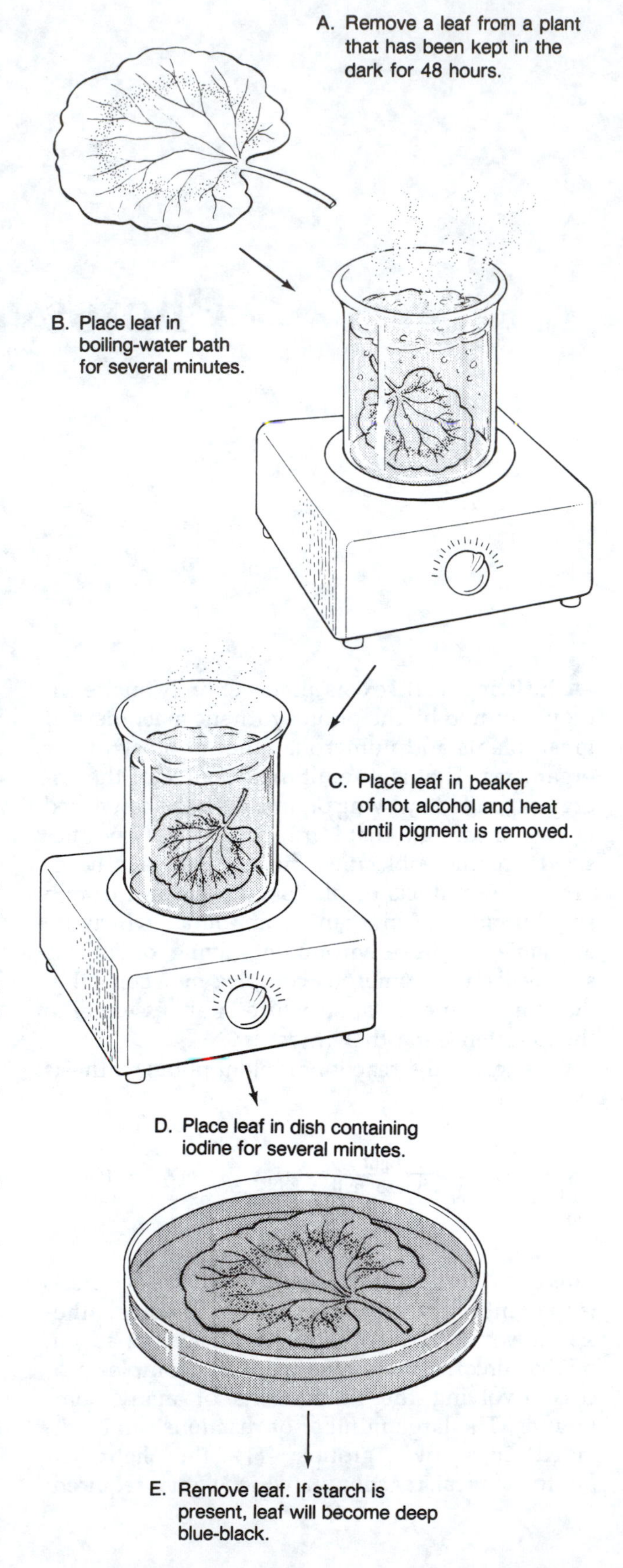

FIGURE 9-1 Procedure for determining the necessity of light for photosynthesis

2. Effect of Light Intensity on the Rate of Photosynthesis

Because oxygen is a byproduct of photosynthesis, oxygen liberation can be used in designing an experiment to measure the effect of variations in light intensity on photosynthesis. In this study, light intensity is varied by placing an *Elodea* sprig at varying distances from a constant light source.

You can use either of the following methods. Method 1 is a semiquantitative procedure in which the amount of oxygen produced is equated with the

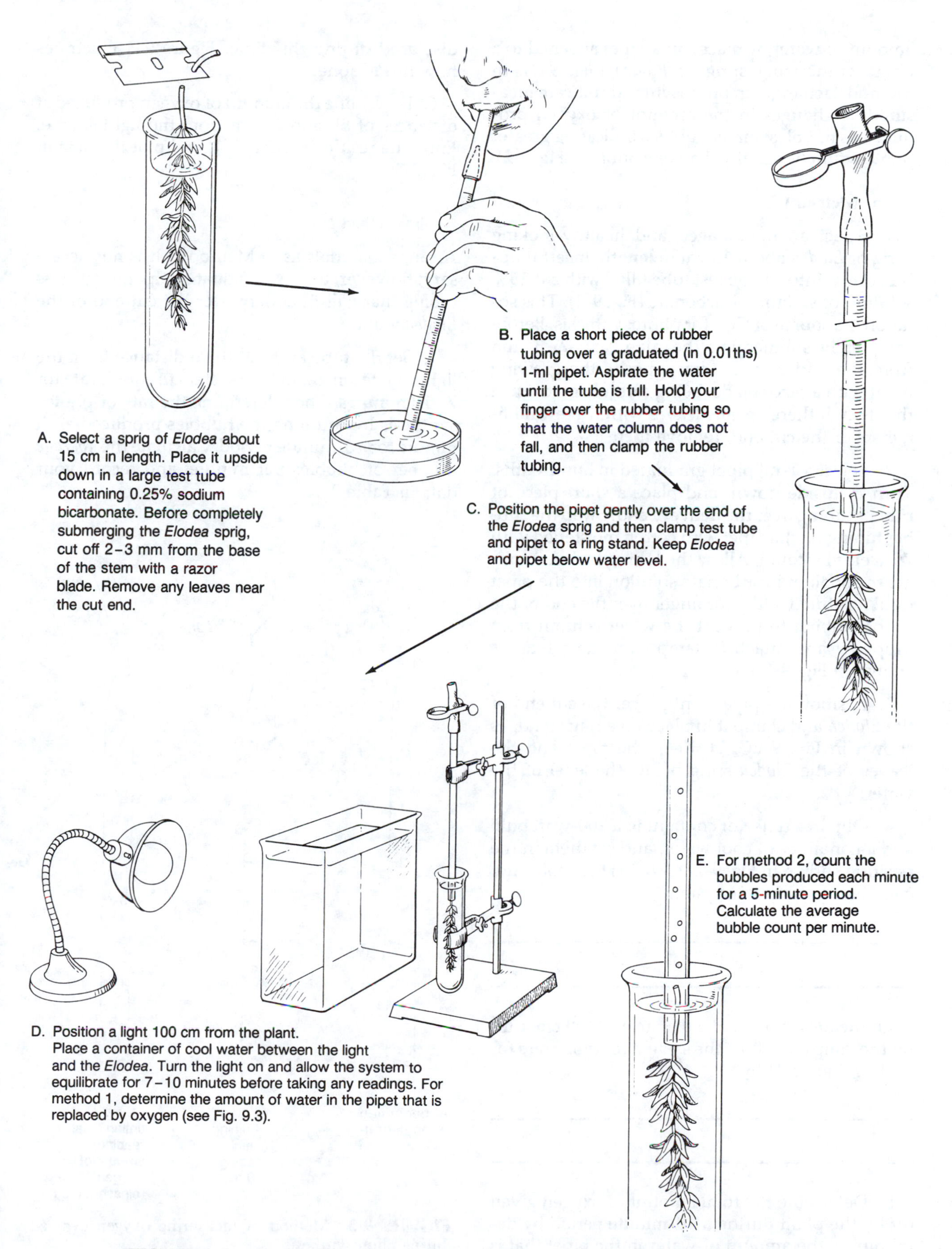

FIGURE 9-2 Procedures for determining the effect of light intensity on photosynthesis

amount of water displaced in a pipet attached to a photosynthesizing sprig of *Elodea* (Fig. 9-2). In Method 2, changes in photosynthetic rate are measured as changes in the amount of oxygen produced as bubbles, many bubbles indicating greater photosynthetic activity than few bubbles (Fig. 9-2).

a. Method 1

1. Select an undamaged and healthy-looking sprig of *Elodea* about 15 cm in length. Insert it upside down into a large test tube filled with a 0.25% solution of sodium bicarbonate (Fig. 9-2). This solution is a source of CO_2 for photosynthesis. Before completely submerging the plant, cut 2–3 mm from the end of the stem opposite the growing point with a razor blade, being careful not to crush the stem. If there are any leaves within a few millimeters of the cut end, remove them.

2. Select a 1-ml pipet graduated in hundredths. Turn it upside down, and place a short piece of rubber tubing over the delivery end. Swab the rubber tubing on the pipet with cotton moistened with 70% ethyl alcohol. Allow the tubing to dry. Aspirate the sodium bicarbonate solution into the pipet until it is full. Hold your finger over the end of the rubber tubing to prevent the water column from dropping and attach a clamp over the tubing as shown in Fig. 9-2B, C.

3. Position the pipet gently over the cut end of the *Elodea* and clamp it in place on a ring stand, as shown in Fig. 9-2C, D. Keep the pipet and the leaves of the *Elodea* sprig below the level of the water.

4. Obtain a reflector containing a 200-watt bulb and a container of cool water, and set them in the position shown in Fig. 9-2D. Why is the cool water container used in this system?

With **the *Elodea*** plant at a distance of 100 cm, turn on the lamp and allow the system to equilibrate for 7–10 minutes. Why?

5. Determine the total amount of oxygen given off by the plant during a 10-minute period by determining the amount of water in the pipet that is displaced during this time. Figure 9-3 illustrates how this is done.

6. Determine the amount of oxygen produced at distances of 50 and 10 cm from the light source. Enter the results in Table 9-1, and plot the data in Fig. 9-4.

b. Method 2

Arrange materials as in Method 1. It is not necessary, however, to use a graduated pipet. Any glass tubing that will fit closely over the cut end of the *Elodea* will do.

1. Set the tube at the 100-cm distance from the light source, and allow the system to equilibrate for 7–10 minutes. Then determine the rate of photosynthesis by counting the bubbles produced each minute for a 5-minute period. Calculate the *average* number of bubbles per minute, and record your data in Table 9-1.

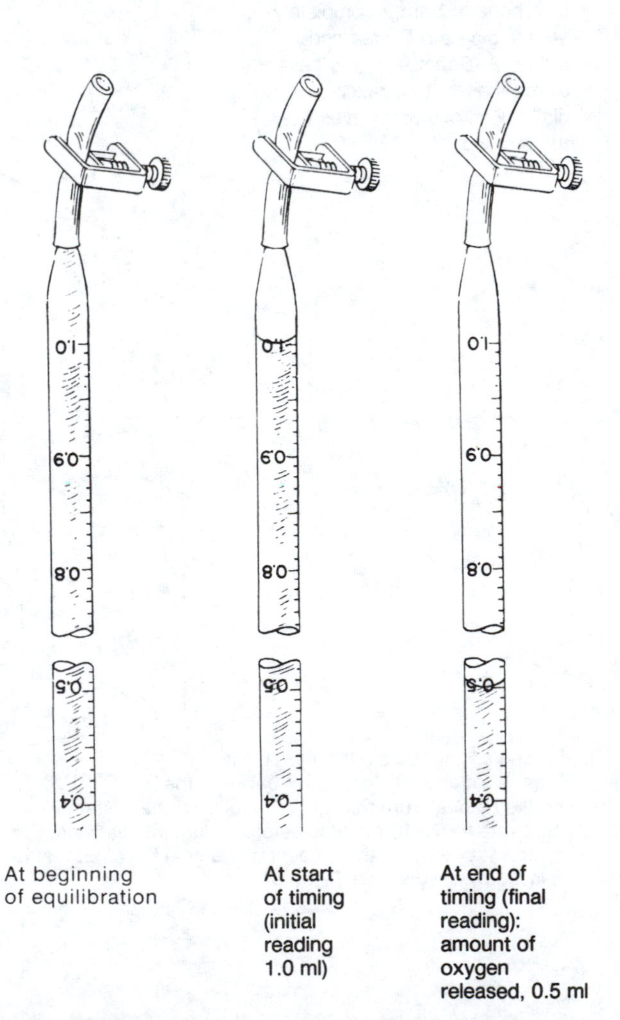

FIGURE 9-3 Method of measuring oxygen evolved during photosynthesis

TABLE 9-1 Effects of light intensity on photosynthetic activity

Distance from light (cm)	Volume of oxygen (ml)	Average bubble count
100		
50		
10		

2. Move the tube to the 50-cm distance, allow the system to equilibrate for 7–10 minutes, and calculate the average number of bubbles. Repeat this procedure for the 10-cm distance. Plot the data in Fig. 9-4.

As light intensity increases, does the rate of photosynthesis (as measured in oxygen production) increase with it? If not, what does this suggest?

____

____

____

____

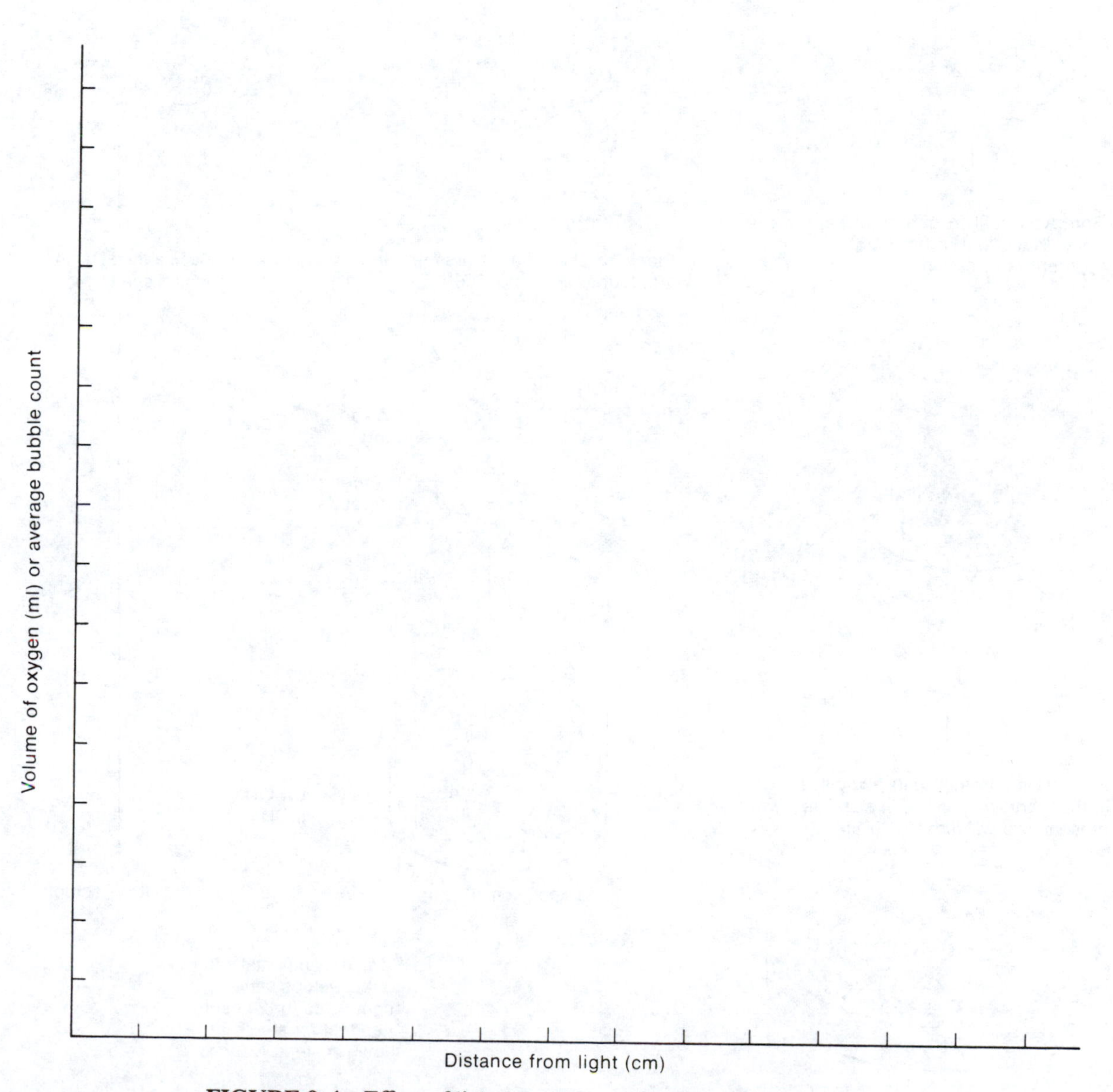

FIGURE 9-4 Effect of light intensity on photosynthetic rate

B. ROLE OF CARBON DIOXIDE

1. Necessity of CO_2 for Photosynthesis

To determine the necessity of CO_2 in photosynthesis, the apparatus shown in Fig. 9-5C will be used.

1. Remove a leaf from a geranium plant that has been in the dark for 24 hours. Test the leaf for the presence of starch by immersing it in hot alcohol until it loses its green color, then place it in a petri dish containing iodine (Fig. 9-5A, B).

Caution: *Heat the alcohol in a separate beaker, on a hot plate. Do not heat it over a bunsen burner.*

Return the plant to the dark while performing the starch test. If the test results in a strong positive starch reaction, select another plant and test a leaf for starch. Repeat this process until you find a plant that gives a negative or very weak starch reaction. Why is this step necessary?

____

____

____

2. Select another leaf from the plant that gives the negative starch reaction and place the leaf in

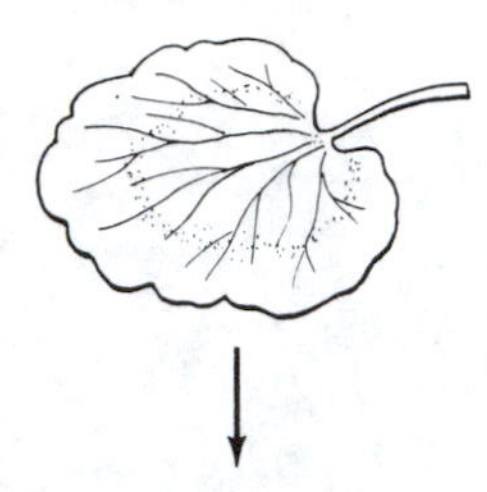

A. Remove a leaf from a plant kept in the dark. Place leaf in hot alcohol until pigment is removed.

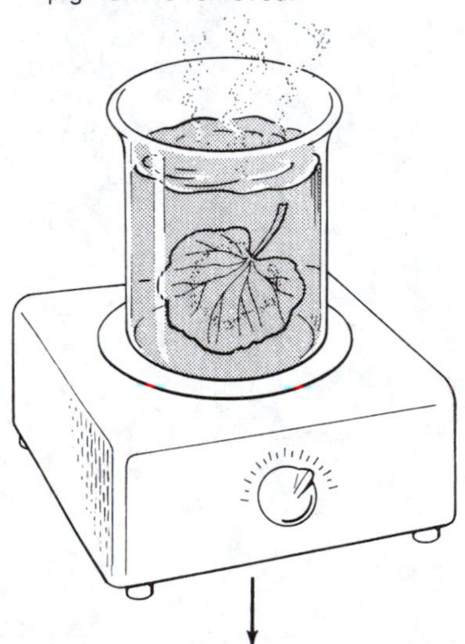

B. Remove leaf from alcohol and place in dish containing iodine. If starch is present, leaf will turn blue-black.

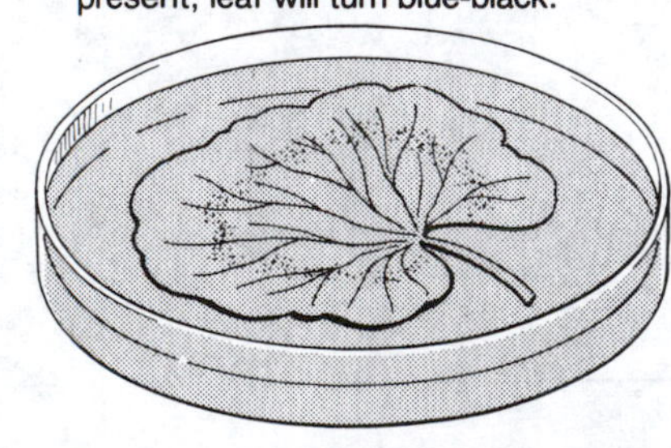

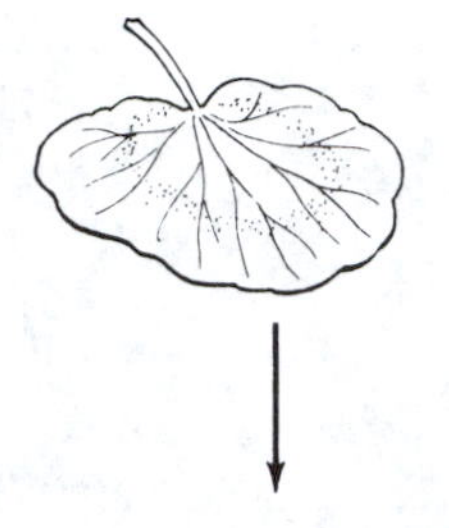

C. Place another leaf from same plant in atmosphere lacking CO_2.

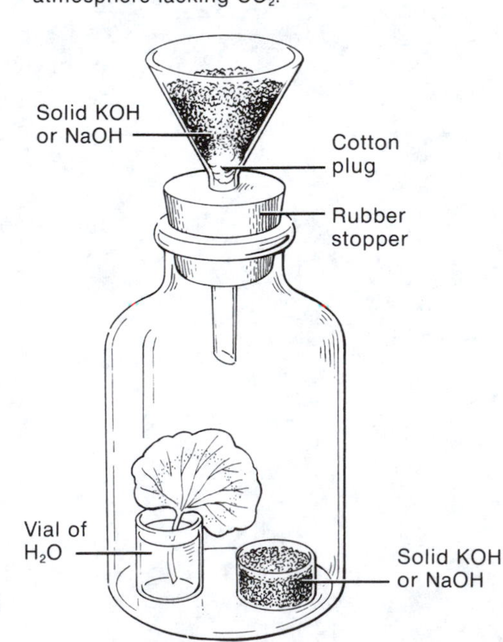

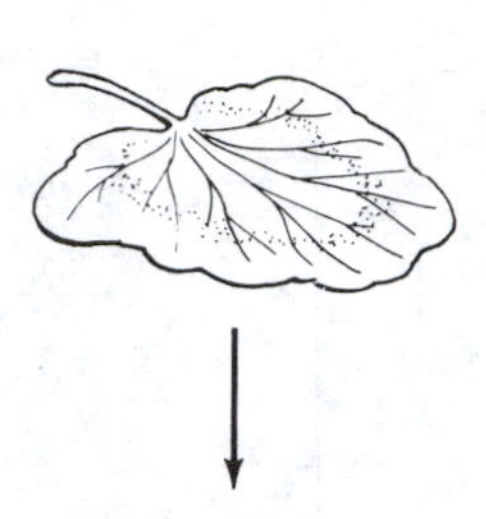

D. Place a third leaf in control conditions.

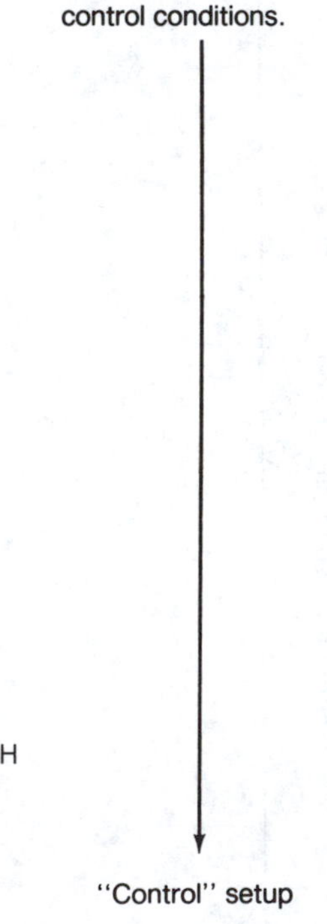

E. Place experimental and control setups under bright lights for 24 hours. Then test for starch as shown in steps A and B.

FIGURE 9-5 Procedure for determining the necessity of CO_2 for photosynthesis

the experimental setup shown in Fig. 9-5C. Place a 200-watt shielded lamp near the setup but not close enough to heat the jar. What is the control for this experiment?

Run the control simultaneously with the experimental setup. This experiment and the control should run for about 24 hours, after which you should remove the leaves and test them for photosynthetic activity in terms of starch production.

After setting up your experiment, examine the demonstration setup arranged by your instructor (Fig. 9-6). Why is there a beaker of $Ba(OH)_2$ in the bell jar?

What control would be needed for this experiment?

To save time, your instructor has tested the leaves of the experimental and control plants in the demonstration for photosynthetic activity. Under which condition is the starch test negative?

If the results of your experiment do not agree with those of your instructor, suggest reasons for this difference.

2. Uptake of CO_2 by Aquatic Plants

You can demonstrate that CO_2 is used during photosynthesis by placing a healthy *Elodea* plant in a test tube containing a chemical indicator that changes color with the presence or absence of CO_2. Phenol red is red in an alkaline solution and turns yellow in an acid solution. Using this information, devise and run an adequately controlled experiment showing (1) that CO_2 is taken up by *Elodea* and (2) that light affects the plant's ability to take up CO_2.

C. ROLE OF CHLOROPLAST PIGMENTS

To maintain life, a constant source of energy must be available. For cells this energy is either radiant (light) energy from the sun or potential energy stored in chemical bonds. Light energy, which is a type of electromagnetic radiation (Fig. 9-7), must first be transformed into chemical (bond) energy before living cells can use it. This transformation takes place in green plant cells. Because only absorbed light can transfer its energy, the colored components of plant cells must be absorbing visi-

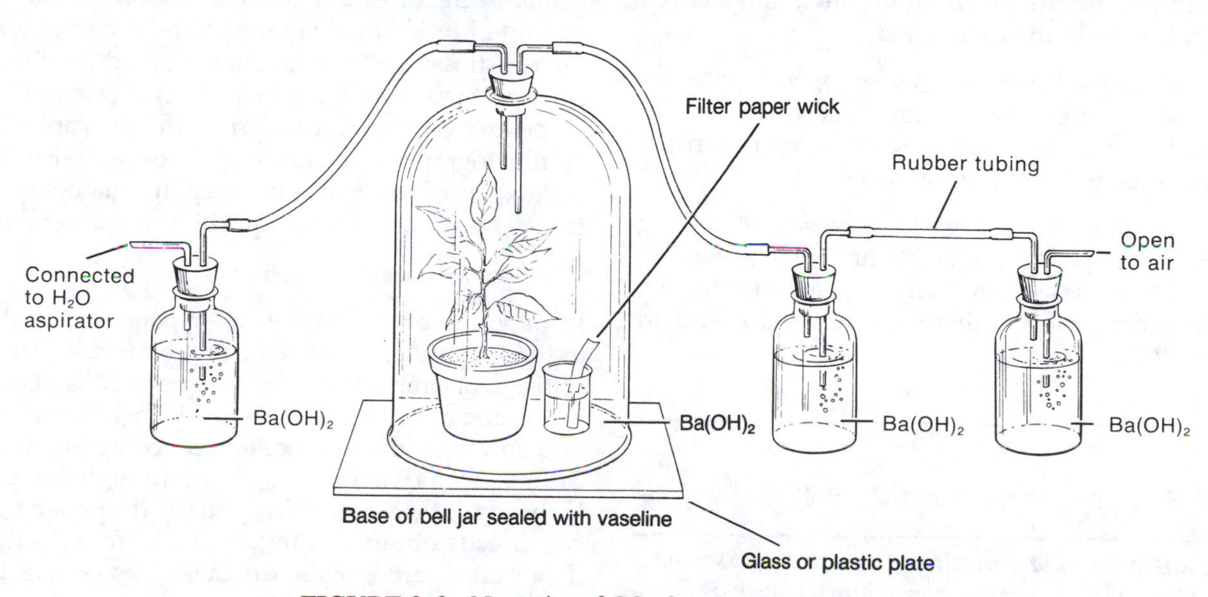

FIGURE 9-6 Necessity of CO_2 for photosynthesis

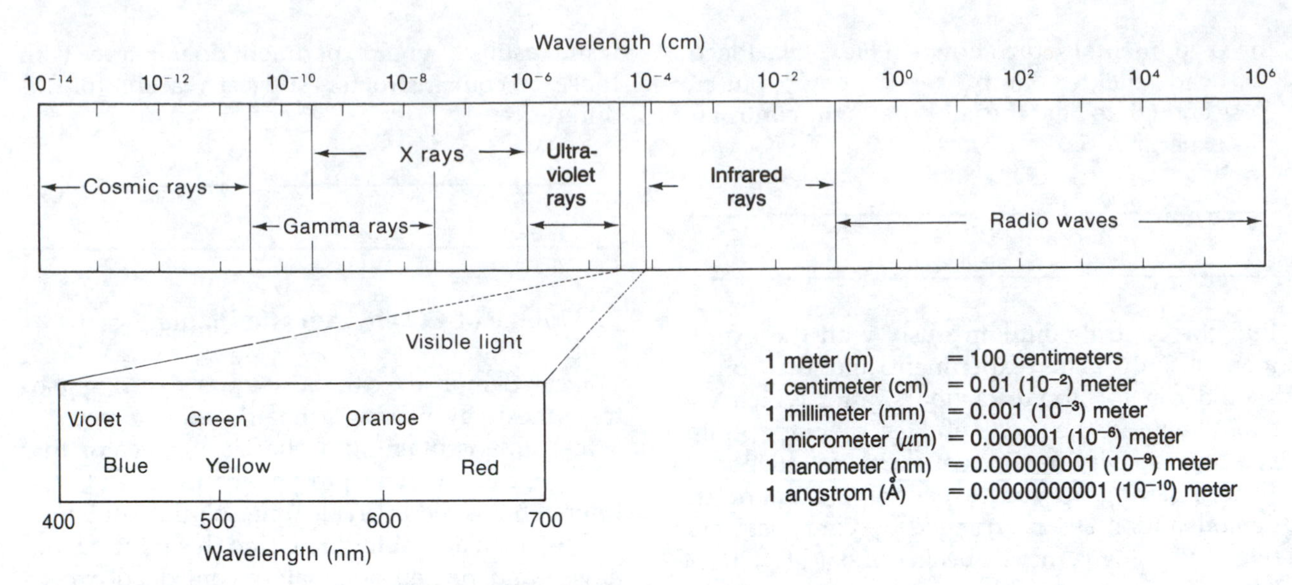

FIGURE 9-7 Electromagnetic spectrum

ble light. Substances that have the ability to absorb light selectively are called **pigments.**

In this part of the exercise, you will determine experimentally the necessity for chlorophyll in photosynthesis, the nature of the green color of plants, and the absorption spectrum of a chloroplast pigment solution.

1. Necessity of Chlorophyll for Photosynthesis

1. Select a leaf from a variegated *Coleus* and a leaf from a silver-leafed geranium plant. In Row 1 of Table 9-2, draw an outline of each leaf, showing the distribution of the pigments. The obvious pigments will be the green chlorophyll and the red anthocyanin in the *Coleus* leaf.

2. Place the leaves in a beaker of cold water for several minutes. Then remove the leaves and record your observations (by drawings or written comments) in Row 2 of Table 9-2.

3. Transfer the leaves to a beaker of boiling water for several minutes. Remove the leaves and record any changes in Table 9-2. Account for the differences observed between Rows 2 and 3 in Table 9-2.

____

4. Place the leaves in hot alcohol.

Caution: *Heat the alcohol in a separate beaker, on a hot plate. Do not heat it over a bunsen burner.*

After several minutes, the leaves will become whitish. At this point, transfer them to a petri dish containing iodine. Swirl the dish gently. Outline the distribution of starch in each leaf in Row 4 of Table 9-2. How do these experiments demonstrate the necessity for chlorophyll in photosynthesis?

____

____

2. Isolation and Characterization of Chloroplast Pigments

Complex mixtures of chemical substances can be separated by chromatography. The separation of the constituents of the mixture is based on differences in their solubilities in various solvents. (See Appendix C for a discussion of the principles of chromatography.) In this study, you will use the techniques of paper or thin-layer chromatography to analyze the pigment composition of chlorophyll.

a. Paper Chromatography

In paper chromatography, filter paper is usually used to separate components of mixtures. A streak of the substance to be chromatographed is placed at one end of the paper. This end is then immersed in a solvent, which separates the components of the mixture as it migrates upward through the spot to the top of the paper. After you dry the paper, you can directly observe materials that have separated if they are colored, or you can make them visible by using various spray reagents.

TABLE 9-2 Role of chlorophyll in photosynthesis

Row	Treatment	Observations	
		Coleus	Silver-leaved geranium
1	None		
2	Cold H_2O for several minutes		
3	Boiling H_2O for several minutes		
4	Hot alcohol for several minutes. Place in dish with iodine.		

1. Prepare a chlorophyll extract by grinding two or three fresh (not frozen) spinach leaves in 5 ml of acetone (Fig. 9-8A). Adding a small quantity of quartz sand will make the grinding easier.

2. Using a small paint brush, apply a narrow strip of chlorophyll extract to the filter paper (Fig. 9-8B). Dry the paper thoroughly by blowing on it or waving it in the air. Apply the extract five or six more times. Let it dry thoroughly after each application.

3. Place the strip in a test tube containing benzene-petroleum ether, as shown in Fig. 9-8C. Examine the chromatogram for the next several minutes. How long does it take for the solvent to reach the top of the paper?

____

Describe any separation that occurs.

____

____

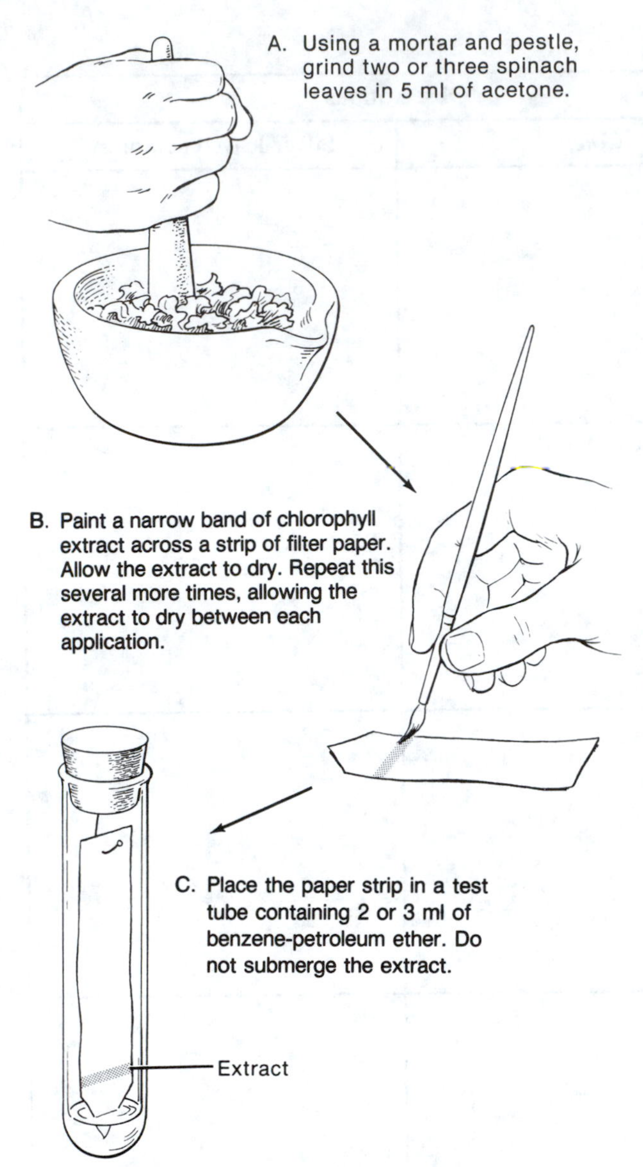

FIGURE 9-8 Procedure for separating chlorophyll pigments using paper chromatography

b. Thin-Layer Chromatography

Alternatively, you may use commercially prepared sheets, consisting of silica gel on acetate backing, to ensure uniformity of adsorbent thickness. These sheets have the added advantage that you can cut them into any desired shape and size.

An advantage of thin-layer chromatography is the speed at which separation occurs. Paper chromatography can require up to 24 hours to separate a complex chemical mixture, but the same separation by thin-layer methods can be done in an hour.

1. Place about 0.5 cm of solvent (isooctane–acetone–diethyl ether, 2 : 1 : 1) into a chromatographic jar. Cover the jar to allow the interior to become saturated with the fumes of the solvent (Fig. 9-9).

2. Using a capillary hematocrit tube, apply several drops of the previously prepared chloroplast extract in a spot about 2 cm from the bottom of a strip of thin-layer silica gel. Try to make the spot about 3 or 4 mm in diameter. Dry the gel thoroughly between the application of each drop.

3. Insert the chromatogram into the jar and cover. Allow the test to run until solvent reaches within 2 cm of the top of the sheet.

Approximately how long does it take to separate the pigments using the thin-layer method compared with the paper chromatographic method?

____

3. Absorption Spectra of Chloroplast Pigments

The wavelengths of the visible spectrum that are absorbed by the chloroplast pigments can be determined by using a **spectroscope** (Fig. 9-10A). In this instrument, a highly polished plate with closely spaced, etched lines disperses visible light into its component spectral bands (Fig. 9-10B) and then projects them onto a scale. The disappearance from the spectrum of various colors (wavelengths) as the light passes through a pigment solution indicates that those wavelengths were absorbed by the pigments. A graph of the amount of light a substance absorbs versus the wavelength of the light is called an **absorption spectrum.** The absorption spectrum for a hypothetical substance is shown in Fig. 9-11. What wavelengths are strongly absorbed by this hypothetical substance?

____

What wavelengths are weakly absorbed?

____

In this study, you will determine the absorption spectrum of a chloroplast extract using two different methods.

a. Method 1: Spectroscopic Determination

Pipet a sample of the chloroplast extract available in the laboratory into a small test tube. Your instructor will help you use the spectroscope to determine the absorption spectrum for the extract. In

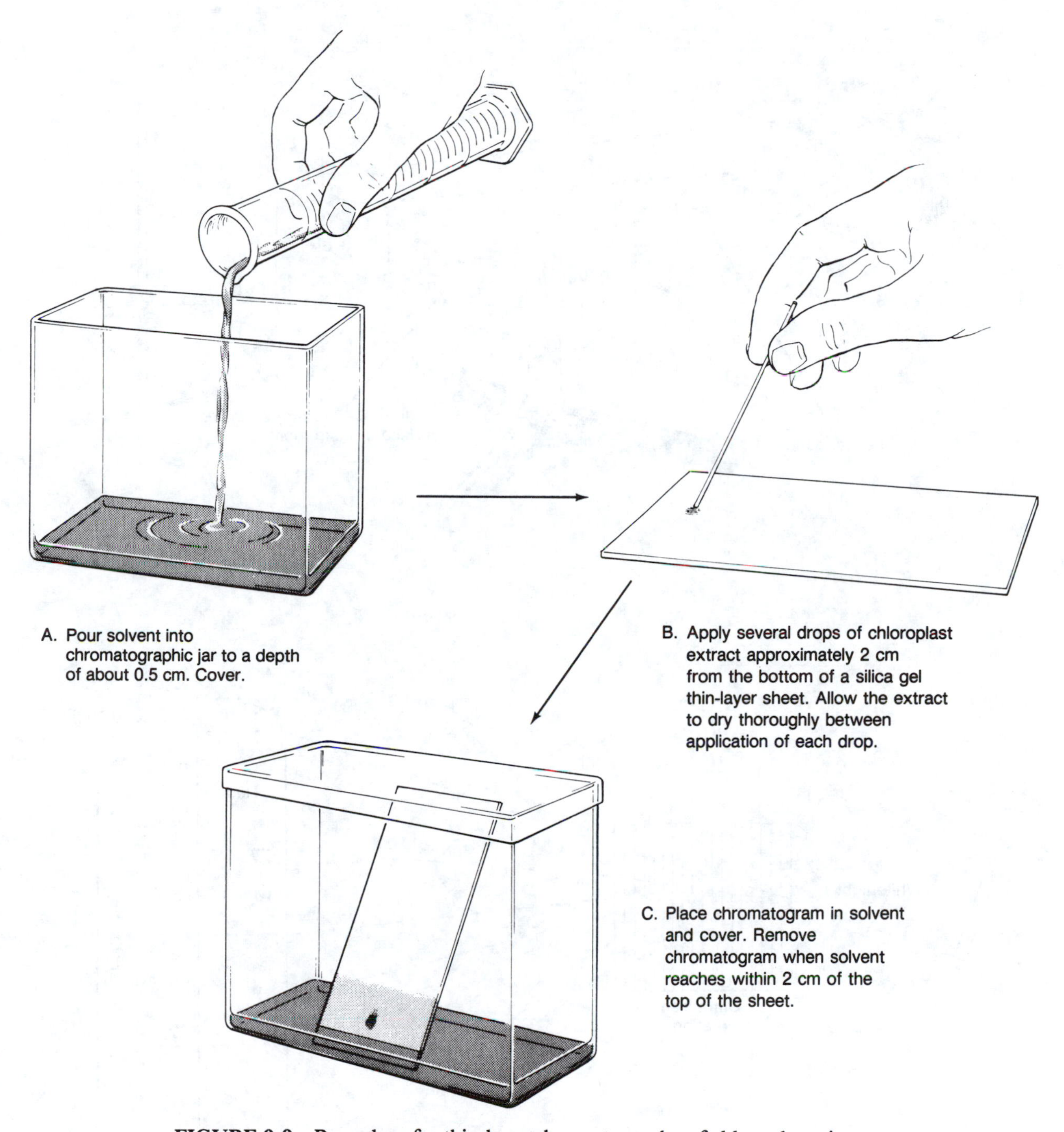

FIGURE 9-9 Procedure for thin-layer chromatography of chloroplast pigments

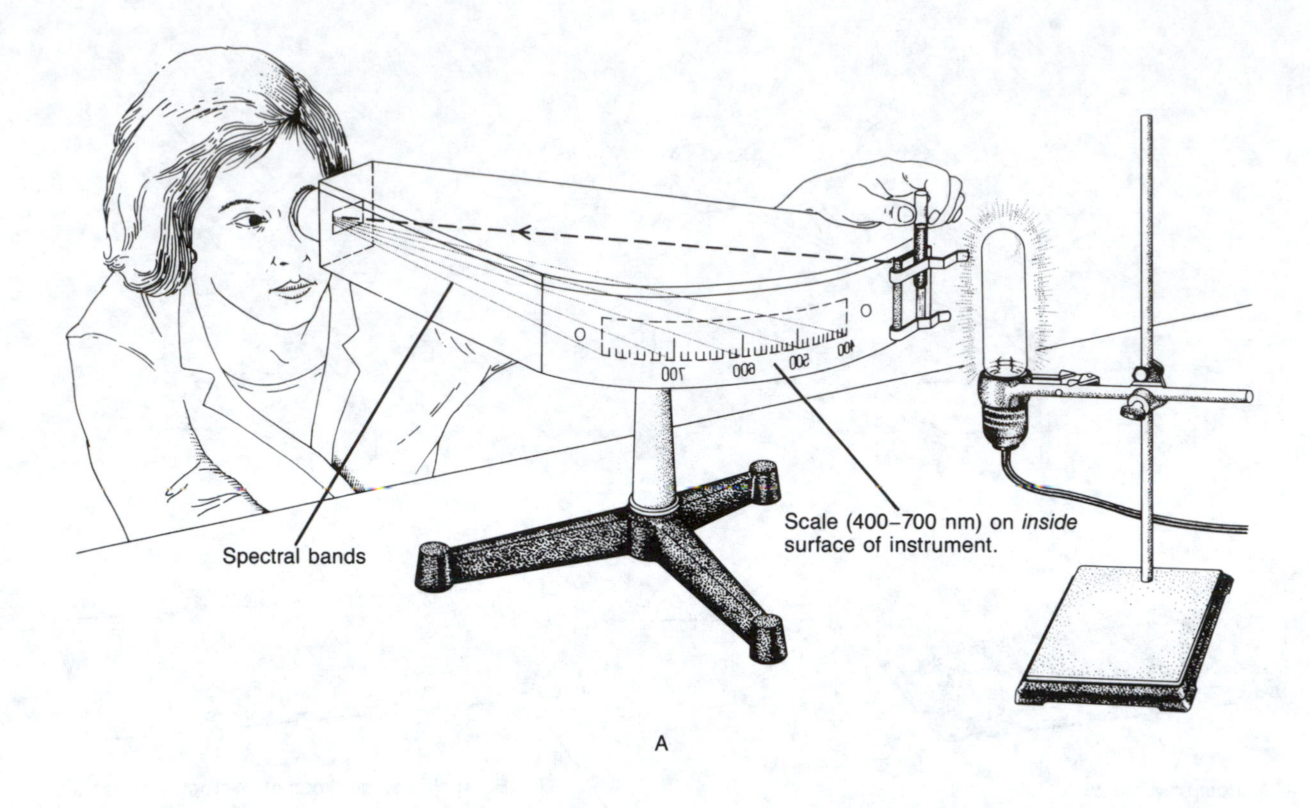

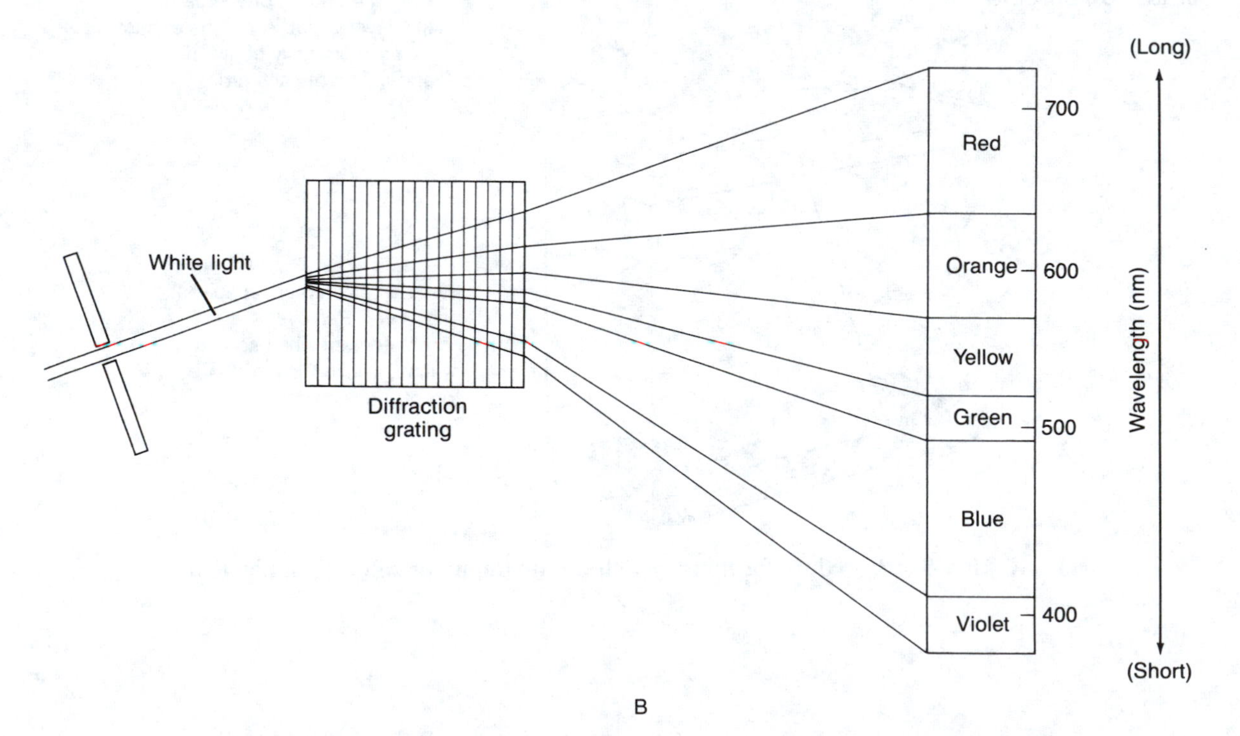

FIGURE 9-10 (A) Determination of the absorption spectrum of chloroplast pigments by a spectroscope. (B) Dispersion of white light by a diffraction grating.

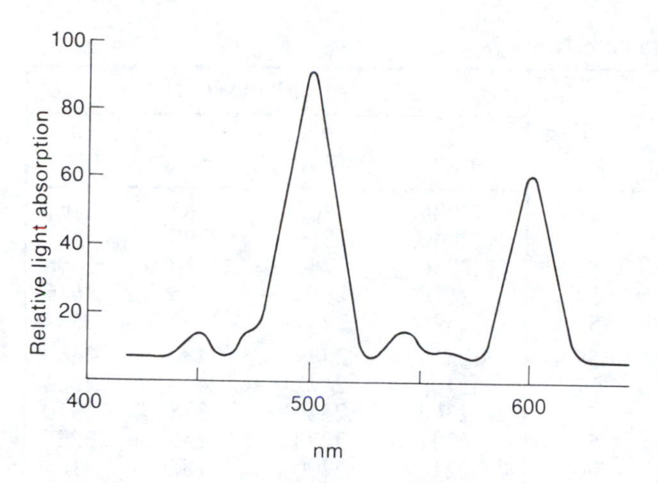

FIGURE 9-11 Simple absorption spectrum

the particular instrument used here, two spectra are projected onto a scale (400–700 nm) on the back inside surface of the instrument. The upper reference spectrum shows the various colors (wavelengths) of light. The lower sample spectrum results from passage of the light through your sample. In Table 9-3, indicate the wavelengths of light that are absorbed by the chloroplast extract.

b. Method 2: Spectrophotometric Determination

In this method, a Milton Roy Spectronic 20® colorimeter is used to determine the absorption spectrum of the chloroplast pigments more accurately. (Refer to Appendices A and B for a description of the theory and mechanics of using this instrument.)

1. Beginning at a wavelength of 400 nm, blank the instrument with the acetone–ethanol solvent used to extract the chloroplast pigments. Why is this solvent used to blank the spectrophotometer?

____

____

____

2. Place the tube containing the extract into the sample holder and determine the percent transmittance (%*T*).

3. Remove the sample and reset the wavelength control to 425 nm. Reblank the instrument to 0% and 100% transmittance, and then determine the percent transmittance of the sample.

4. Repeat step 3 at 25-nm intervals. It will be necessary to insert an accessory red filter and red-sensitive phototube for determinations above 625 nm.

5. Convert the percent transmittance (%*T*) into absorbance (*A*) using Table 9-4, and then plot the data in Fig. 9-12. At what wavelength(s) does the chlorophyll extract absorb maximally?

____

____

Why do leaves of those plants in which chlorophyll is the predominant pigment appear green?

____

____

____

Because the chloroplast pigments consist of both chlorophylls and carotenoids, you cannot tell from the absorption spectrum which pigments are absorbing which wavelengths. How could you determine this?

____

____

____

TABLE 9-3 Wavelengths of light absorbed by chlorophyll extract

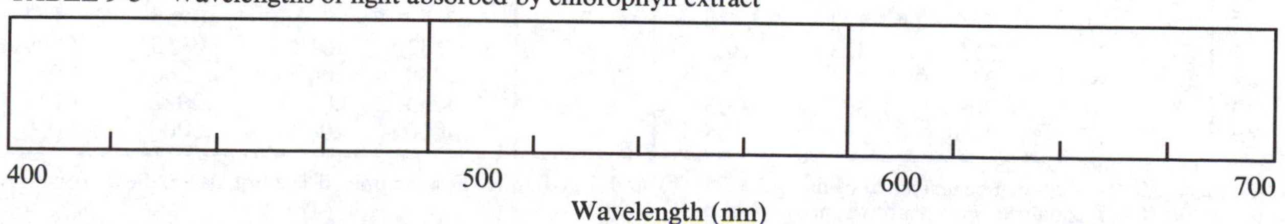

TABLE 9-4 Conversion of percent transmittance (%T) to absorbance (A)

%T	Absorbance (A)				%T	Absorbance (A)			
	1 (.00)	2 (.25)	3 (.50)	4 (.75)		1 (.00)	2 (.25)	3 (.50)	4 (.75)
1	2.000	1.903	1.824	1.757	51	.2924	.2903	.2882	.2861
2	1.699	1.648	1.602	1.561	52	.2840	.2819	.2798	.2777
3	1.523	1.488	1.456	1.426	53	.2756	.2736	.2716	.2696
4	1.398	1.372	1.347	1.323	54	.2676	.2656	.2636	.2616
5	1.301	1.280	1.260	1.240	55	.2596	.2577	.2557	.2537
6	1.222	1.204	1.187	1.171	56	.2518	.2499	.2480	.2460
7	1.155	1.140	1.126	1.112	57	.2441	.2422	.2403	.2384
8	1.097	1.083	1.071	1.059	58	.2366	.2347	.2328	.2310
9	1.046	1.034	1.022	1.011	59	.2291	.2273	.2255	.2236
10	1.000	.989	.979	.969	60	.2218	.2200	.2182	.2164
11	.959	.949	.939	.930	61	.2147	.2129	.2111	.2093
12	.921	.912	.903	.894	62	.2076	.2059	.2041	.2024
13	.886	.878	.870	.862	63	.2007	.1990	.1973	.1956
14	.854	.846	.838	.831	64	.1939	.1922	.1905	.1888
15	.824	.817	.810	.803	65	.1871	.1855	.1838	.1821
16	.796	.789	.782	.776	66	.1805	.1788	.1772	.1756
17	.770	.763	.757	.751	67	.1739	.1723	.1707	.1691
18	.745	.739	.733	.727	68	.1675	.1659	.1643	.1627
19	.721	.716	.710	.704	69	.1612	.1596	.1580	.1565
20	.699	.694	.688	.683	70	.1549	.1534	.1518	.1503
21	.678	.673	.668	.663	71	.1487	.1472	.1457	.1442
22	.658	.653	.648	.643	72	.1427	.1412	.1397	.1382
23	.638	.634	.629	.624	73	.1367	.1352	.1337	.1322
24	.620	.615	.611	.606	74	.1308	.1293	.1278	.1264
25	.602	.598	.594	.589	75	.1249	.1235	.1221	.1206
26	.585	.581	.577	.573	76	.1192	.1177	.1163	.1149
27	.569	.565	.561	.557	77	.1135	.1121	.1107	.1093
28	.553	.549	.545	.542	78	.1079	.1065	.1051	.1037
29	.538	.534	.530	.527	79	.1024	.1010	.0996	.0982
30	.532	.520	.516	.512	80	.0969	.0955	.0942	.0928
31	.509	.505	.502	.498	81	.0915	.0901	.0888	.0875
32	.495	.491	.488	.485	82	.0862	.0848	.0835	.0822
33	.482	.478	.475	.472	83	.0809	.0796	.0783	.0770
34	.469	.465	.462	.459	84	.0757	.0744	.0731	.0718
35	.456	.453	.450	.447	85	.0706	.0693	.0680	.0667
36	.444	.441	.438	.435	86	.0655	.0642	.0630	.0617
37	.432	.429	.426	.423	87	.0605	.0593	.0580	.0568
38	.420	.417	.414	.412	88	.0555	.0543	.0531	.0518
39	.409	.406	.403	.401	89	.0505	.0494	.0482	.0470
40	.398	.395	.392	.390	90	.0458	.0446	.0434	.0422
41	.387	.385	.382	.380	91	.0410	.0398	.0386	.0374
42	.377	.374	.372	.369	92	.0362	.0351	.0339	.0327
43	.367	.364	.362	.359	93	.0315	.0304	.0292	.0281
44	.357	.354	.352	.349	94	.0269	.0257	.0246	.0235
45	.347	.344	.342	.340	95	.0223	.0212	.0200	.0188
46	.337	.335	.332	.330	96	.0177	.0166	.0155	.0144
47	.328	.325	.323	.321	97	.0132	.0121	.0110	.0099
48	.319	.317	.314	.312	98	.0088	.0077	.0066	.0055
49	.310	.308	.305	.303	99	.0044	.0033	.0022	.0011
50	.301	.299	.297	.295	100	.0000	.0000	.0000	.0000

Note: Intermediate values can be arrived at by using the .25, .50, and .75 columns. For example, if % T equals 85, the absorbance equals .0706; if % T equals 85.75, the absorbance equals .0667.

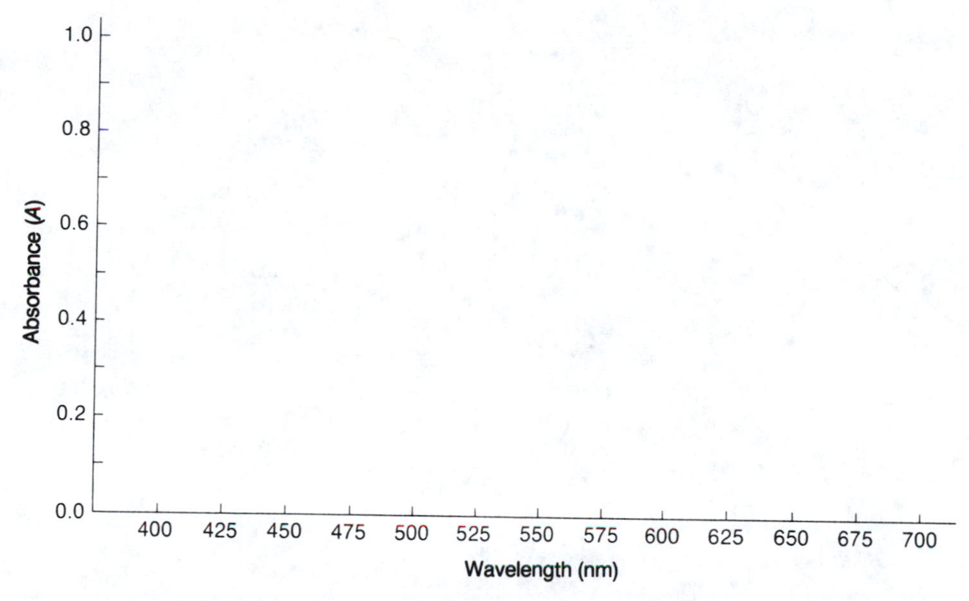

FIGURE 9-12 Absorption spectrum of chloroplast extract

REFERENCES

Alberts, B., et al. 1989. *Molecular Biology of the Cell.* 2d ed. Garland.

Lodish, H. F., et al. 1995. *Molecular Cell Biology.* 3d ed. W. H. Freeman.

Stryer, L. 1995. *Biochemistry.* 4th ed. W. H. Freeman.

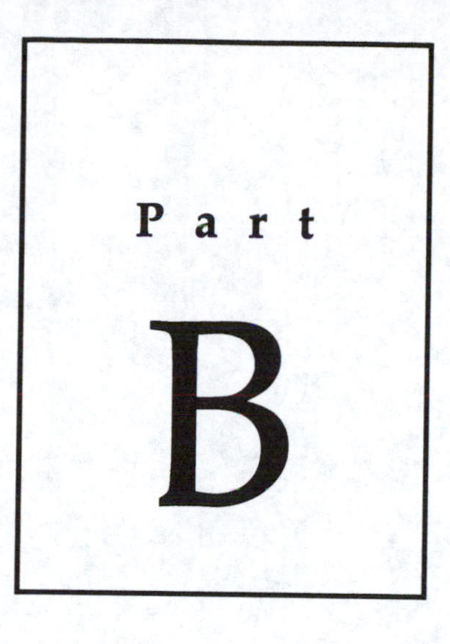

Part B

Genetics

The science of genetics has passed through several phases during the past 100 years. Each period was initiated by some important event.

The first period, beginning with the rediscovery of Mendel's papers in 1900, demonstrated the universal application of the laws of heredity. That is, the transmission of heritable characteristics is brought about by the same mechanism that regulates Mendelian segregation.

The second period, beginning around 1910, saw the introduction of a new "tool" to be used in genetic research—the fruit fly, *Drosophila melanogaster*. From this period came the experimental evidence that genes are located in linear order on the chromosomes.

Further studies on the structure of chromosomes and the discovery that radiation (such as X rays) causes mutations marked the beginning of the third period of activity. These studies were further developed during succeeding years.

The determination by James Watson and Francis Crick of the molecular structure of deoxyribonucleic acid (DNA) opened the era of molecular genetics. It is now known that genetic information, carried within the structure of the DNA molecule, consists of a "code." The role that the genetic code plays in determining the structure of proteins is regarded as one of the greatest discoveries of all time.

Exercise

10

Mendelian Genetics

Modern genetics, and indeed much of contemporary evolutionary theory, is founded on experimental evidence established by Gregor Mendel. His studies of the inherited characteristics of the sweet pea led him to the discovery of the laws of segregation and independent assortment.

In this exercise, you will become familiar with the laws of segregation and independent assortment through studies involving monohybrid and dihybrid crosses. A variety of organisms can be used to demonstrate Mendel's laws. You will study the fruit fly *Drosophila* and maize (corn) because they are two of the eukaryotic organisms for which substantial genetic knowledge has accumulated. Before beginning these studies, read Appendix D for a discussion of Mendel's laws of inheritance.

A. GENETIC STUDIES USING *DROSOPHILA*

The common fruit fly, *Drosophila melanogaster*, is one of the more widely used organisms for genetics studies. It is easily bred, and its generation time is only 9 or 10 days at room temperature (25°C). Because *Drosophila* is small, its cultures occupy little space and it is therefore a convenient and inexpensive organism with which to work. *Drosophila* is well understood genetically; **wild-type** (normal) and mutant strains of *Drosophila* can be obtained easily. An enormous number of spontaneous mutations have been found in this fly, and many others have been induced by radiation, thus making it ideal for the investigation of genetic crosses.

1. Examination of Wild-Type *Drosophila*

In this part of the exercise, you will become familiar with the characteristics of the wild type and some of the common mutants of *Drosophila*. Then you will be given several flies whose mutant traits you are to identify.

Obtain a vial of wild-type fruit flies from your instructor and etherize them according to the procedure outlined in Fig. 10-1. When all the flies have stopped moving, turn them onto a white card and carefully examine them with a stereoscopic microscope or hand lens. Become familiar with male and female characteristics (Fig. 10-2), and record your observations in Table 10-1.

From your instructor, obtain a numbered vial containing a mixture of the following mutant flies.

Vestigial: Wing mutant characterized by reduced, withered wings

White: Mutant whose eyes look white

Bar: Eye mutant in which the number of facets of

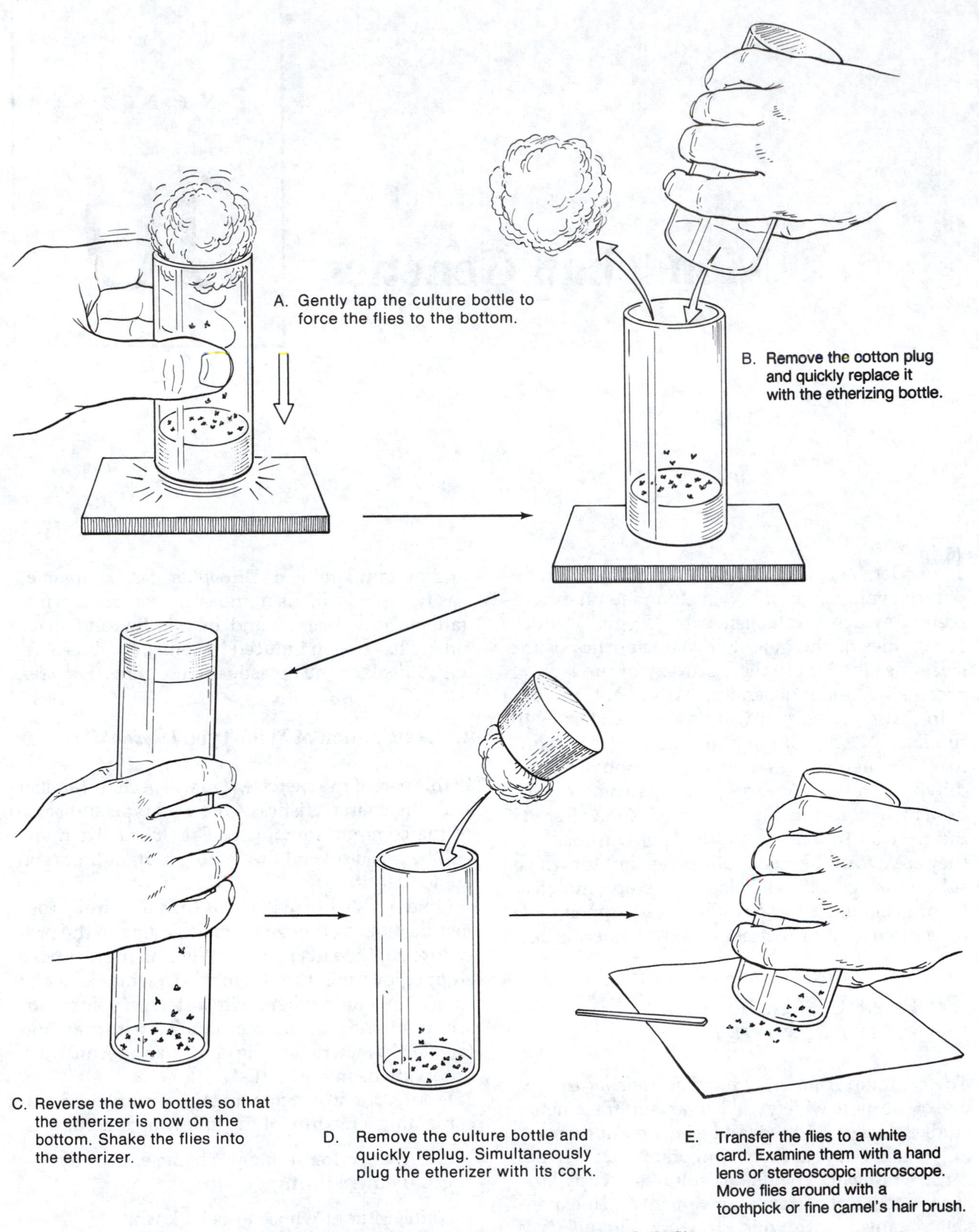

FIGURE 10-1 Procedure for etherizing fruit flies

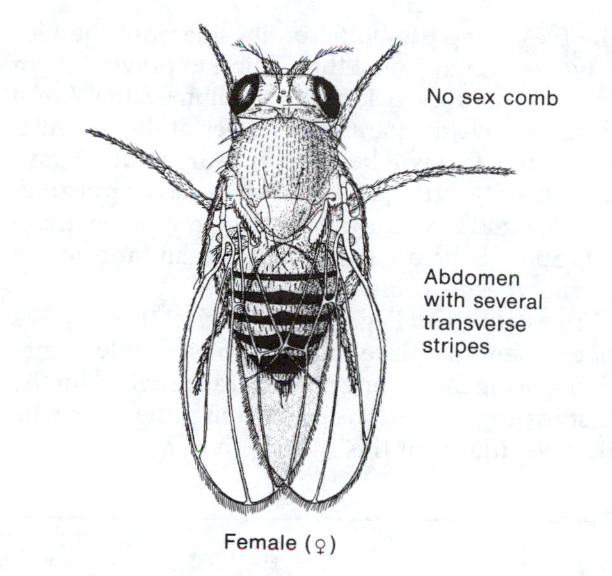

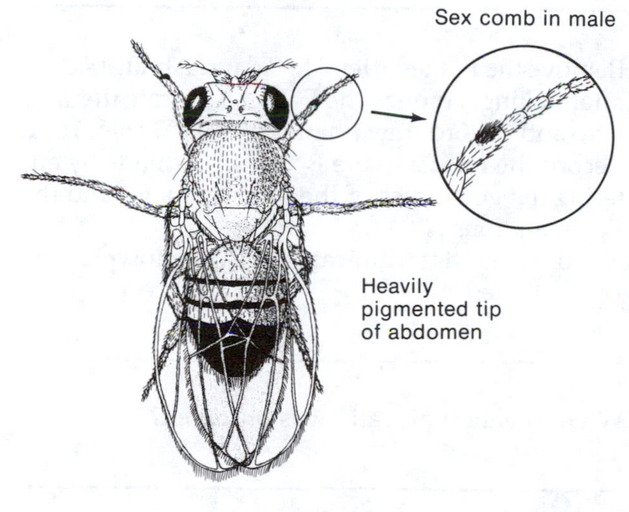

FIGURE 10-2 The adult fruit fly, *Drosophila melanogaster*

the eye is reduced, resulting in the appearance of a "bar" down the middle of the eye

Black: Mutant whose body is black

"Unknown"

Etherize the flies and determine the nature of the mutant flies in your vial. Record your observations in Table 10-2.

2. Experimental Genetic Crosses

To make these genetic crosses, you will be supplied with pedigreed stocks, maintained in the laboratory, that carry mutant genes. Flies in these cultures

TABLE 10-1 Comparison of male and female fruit flies

Characteristic	Observation	
	Male (♂)	Female (♀)
Which is larger in overall size?		
What is the difference in banding on the abdomen?		
What is the shape of the tip of the abdomen?		
Are sex combs present or absent?		

tend to breed true as long as they mate among themselves. These studies will use flies with easily recognizable traits.

Breeding stocks are conveniently designated by symbols indicating the particular mutant gene(s) carried. Flies that exhibit traits that can be considered standard or normal are designated as wild type.

- A plus sign (+) indicates the wild-type gene (allele).
- A lower-case letter indicates that the mutant allele is recessive to the wild-type allele. The

TABLE 10-2 "Unknown" mutant flies

Mutation	Sex

symbol *e*, for example, represents the recessive mutant allele for ebony body color, and e^+ represents the dominant allele for the wild-type gray-brown body.

- Homozygous ebony flies are represented as *ee*, and homozygous wild-type flies are e^+e^+.

When making a cross between two varieties of fruit flies, you should consider only those characteristics in which the parent flies (i.e., the parental, or P_1, generation) differ. For example, in crossing flies having vestigial wings and those having sepia eyes, only wing shape and eye color should be observed carefully.

B. MENDEL'S LAW OF SEGREGATION

According to Mendel's **law of segregation,** genes do not blend but behave as independent units. They pass intact from one generation to the next, in which they may or may not produce visible traits depending on which characteristics are dominant. Furthermore, genes segregate at random, thus producing predictable ratios of traits in the offspring.

1. In *Drosophila*

You will now attempt several crosses (matings) to demonstrate Mendel's law of segregation—either crosses given in Table 10-3 or others suggested by your instructor. If time is limited, your instructor may have prepared the necessary matings for you. If you are to make the necessary crosses, your instructor will supply you with appropriate cultures. As you make the crosses, complete the information indicated in Table 10-4.

Isolate an etherized male of one variety and an etherized virgin female of the other, using the procedures described in Appendix D. While holding a culture bottle on its side, place these flies in the bottle. Be sure to add some dry yeast granules or yeast suspension to the medium *before* introducing the flies. Keep the bottle on its side until the flies have recovered from etherization to prevent them from becoming stuck in the medium. After 7 or 8 days, remove the parent flies. Flies of the first filial generation (F_1) will begin to appear about 10 days after mating. After several F_1 flies have appeared, etherize and examine them under low-power magnification. Separate the flies by sex and record the phenotypes in Table 10-4.

To establish an F_2 generation, select three F_1 flies of each sex and place them in a fresh bottle of medium, using the mating procedure described for the first mating. It is not necessary that the F_1 female flies be virgins for this mating. Why?

____

____

Remove the F_1 flies after 7 to 8 days. About 14 days after mating, etherize the F_2 flies, separate them by sex, and record the phenotypes in Table 10-4. Record the results of the F_2 cross you made by entering the genotypes of the parent, the F_1, and the F_2 generations.

From these data, indicate the F_2 phenotypic ratio you obtained.

____

What F_2 genotypic ratio was observed?

____

Which trait is dominant in your crosses?

____

TABLE 10-3 Suggested matings

Parent (P_1) generation		
Female (♀)	×	Male (♂)
Sepia eye color *(se)*	×	Wild-type (red eye color)
Dumpy wings *(dp)*	×	Wild-type (long wings)
Vestigial wings *(vg)*	×	Wild-type (long wings)
Ebony body color *(e)*	×	Wild-type (gray-brown body color)

2. Chi-Square Analysis of *Drosophila* Data

In general, when you carry out crosses involving one or more pairs of alleles, the results are fairly predictable. For example, if you carry out the following hypothetical crosses, you would expect the typical 3 : 1 monohybrid phenotypic ratio in the F_2 generation:

P_1: *AA* × *aa*

F_1: *Aa*

F_1: *Aa* × *Aa*

F_2: *AA* *Aa* *aa*

1 : 2 : 1 genotypic ratio

3 : 1 phenotypic ratio

TABLE 10-4 Monohybrid crosses in *Drosophila*

______________ Student name

______________ P_1 female × ______________ P_1 male

Date P_1 mated ______________ Date P_1 removed ______________

Phenotype of F_1 females ______________ Phenotype of F_1 males ______________

______________ F_1 female × ______________ F_1 male ______________ Date F_1 mated

Date F_1 removed ______________ Date F_2 examined ______________

F_2 males		F_2 females	
F_2 phenotypes	Number	F_2 phenotypes	Number
Total =		Total =	

	Male	Female
Genotype of P_1		
Genotype of F_1		
Genotypes of F_2		

The same prediction could be made for a dihybrid cross in which the 9:3:3:1 phenotypic ratio would be expected.

These are the **expected ratios.** In natural populations, the expected ratios are rarely identical to the **experimental** or **observed ratios** you get when counting the individuals expressing the characteristic in which you are interested.

The purpose of the **chi-square test** is to determine whether the experimentally obtained data (i.e., the observed data) are a satisfactory approximation to the expected data for a given ratio. That is, this test determines whether any deviations from the expected values are due to something other than chance.

Using the numerical data used to determine the F_2 ratios in Table 10-4, carry out a chi-square analysis of the data to determine if the experimental (observed) ratios fit the expected ratios and thus support the 3:1 monohybrid phenotypic ratio. Record the results in Table 10-5. (Read Appendix E for a description of how to do a chi-square analysis.) Are your observed ratios small enough to be within the limits expected by chance alone?

If not, what factors, other than chance, could account for the larger deviations of the observed ratios from the ratios you expected to obtain?

---

3. In Maize (Corn)

You will be given F_1 corn seeds that, when planted, will grow into F_2 plants, some of which will appear normal and others of which will exhibit an obvious deficiency. Your instructor will give you planting and watering instructions. It takes about 7–10 days for the seedlings to emerge from the soil. When they are 50–75 mm tall, count the numbers of each of the two types of seedlings in your own tray and in those of the rest of the class. You may use whatever symbols you feel are appropriate to distinguish one allele from the other. Set up appropriate tables, properly labeled, to tabulate your data.

What is the deficiency expressed by one of the alleles?

What is the F_2 ratio obtained?

Is the ratio you obtained in your tray the same as or different from that obtained when you counted the plants of the entire class? Explain any differences.

TABLE 10-5 Data for chi-square analysis of monohybrid cross in *Drosophila*

Phenotype	Genotype	Observed (O)	Expected (E)	$(O-E)$	$(O-E)^2$	$(O-E)^2/E$
Totals						
Conclusion:						

What phenotypes are expressed in the F_2 generation?

What genotypes are expressed in the F_2 generation?

What would be the phenotype and genotype of the F_1 generation?

What would be the phenotypes and genotypes for the original parents (i.e., the P_1 generation)?

Would it be possible to have had sexually mature P_1's for each of the traits expressed in the F_2 generation? Explain.

Analyze your results using the chi-square test (see Appendix E). Are your experimental values consistent with what you would expect to obtain? If not, suggest why the deviation from the expected value could have occurred.

C. MENDEL'S LAW OF INDEPENDENT ASSORTMENT

Mendel's **law of independent assortment** was derived from his studies involving crosses of two gene pairs. He demonstrated that whether you are dealing with one, two, three, or more alleles, each acts independently of the others and is not changed in transmission from one generation to the next. However, we now know that the law is valid only if the genes involved are located on different chromosomes, as happened to be the case in all of Mendel's studies. Thus, Mendel always obtained the 9:3:3:1 dihybrid ratio. Subsequent studies have shown other ratios for dihybrid crosses in which the genes are linked, that is, located on the same chromosomes. Give examples of other dihybrid ratios (i.e., 9:7) and the circumstances under which they would be obtained (i.e., linkage).

1. In *Drosophila*

The gene for ebony body color (*e*) is located on the third chromosome and the gene for vestigial wings (*vg*) is located on the second chromosome. Reciprocal crosses should be made from these stocks as follows:

vestigial ♂ *(vgvg; e^+e^+)* × ebony ♀ *(vg^+vg^+; ee)*

and

vestigial ♀ *(vgvg; e^+e^+)* × ebony ♂ *(vg^+vg^+; ee)*

Note: Make certain only virgin females are used in these crosses. Why?

After about 8 days, remove the parental stock and discard them. When the F_1 flies emerge, count them and determine the characteristic(s) that they express. Record the data you obtain in Table 10-6 or 10-7. What do you predict the F_1 phenotype and genotype to be?

Mate the F_1 males and F_1 females in new bottles. Again, after 8 days, remove and discard all F_1 flies. When the F_2 generation flies have emerged, count

TABLE 10-6 Dihybrid crosses in *Drosophila*

__________ Student name

__________ P_1 female × __________ P_1 male

Date P_1 mated __________ Date P_1 removed __________

Phenotype of F_1 females __________ Phenotype of F_1 males __________

__________ F_1 female × __________ F_1 male

__________ Date F_1 mated

Date F_1 removed __________ Date F_2 examined __________

F_2 males		F_2 females	
F_2 phenotypes	Number	F_2 phenotypes	Number
Total =		Total =	

	Male	Female
Genotype of P_1	__________	__________
Genotype of F_1	__________	__________
Genotypes of F_2	__________	__________
	__________	__________
	__________	__________
	__________	__________
	__________	__________

TABLE 10-7 Dihybrid crosses in *Drosophila* (reciprocal cross)

____________ Student name	____________ P_1 female ×	____________ P_1 male
Date P_1 mated ____________	Date P_1 removed ____________	
Phenotype of F_1 females ____________	Phenotype of F_1 males ____________	
____________ F_1 female ×	____________ F_1 male	____________ Date F_1 mated
Date F_1 removed ____________	Date F_2 examined ____________	

F_2 males		F_2 females	
F_2 phenotypes	Number	F_2 phenotypes	Number
Total =		Total =	

	Male	Female
Genotype of P_1	____________	____________
Genotype of F_1	____________	____________
Genotypes of F_2	____________	____________
	____________	____________
	____________	____________
	____________	____________
	____________	____________

them and determine the numbers and ratios for the characteristics being studied. Record this information in Table 10-6 or 10-7. Also, fill in Table 10-8 and then carry out a chi-square analysis to determine if your experimental values are consistent with what you would theoretically expect to obtain. (See Appendix E for a discussion of chi-square analysis.)

2. In Maize (Corn)

Your instructor will provide you with F_2 maize seeds removed from an ear of corn similar to the one shown in Fig. 10-3.

Determine the four phenotypes being expressed. Count the number of kernels represented by each phenotype and record them. Select the symbols you feel would represent the phenotypes being expressed.

Allele	Phenotype
______	______
______	______
______	______
______	______

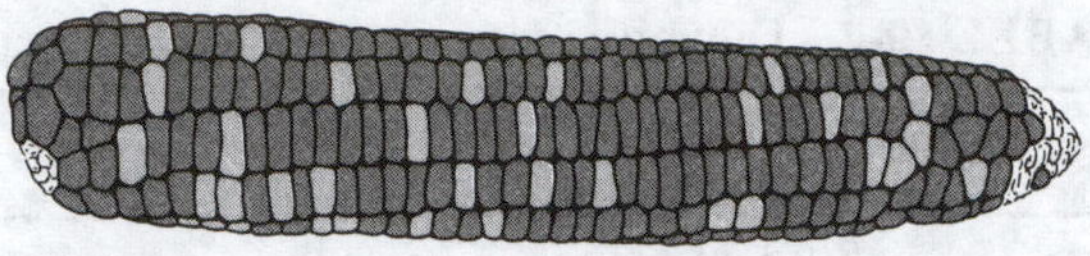

FIGURE 10-3 Example of a dihybrid cross in maize. Typical 9:3:3:1 ratio expressed for genes located on different chromosomes.

Which characteristics are recessive?

____

Which characteristics are dominant?

____

Give the phenotype and genotype of the F_1 generation.

__________ __________

Give the phenotype and genotype of the original P_1, homozygous parents.

__________ __________

Complete Table 10-9 and carry out a chi-square analysis to determine if the experimental values are consistent with what you would expect to obtain in this dihybrid cross.

TABLE 10-8 Data for chi-square analysis of dihybrid cross in *Drosophila*

Phenotype	Genotype	Observed (O)	Expected (E)	$(O-E)$	$(O-E)^2$	$(O-E)^2/E$
Totals						
Conclusion:						

TABLE 10-9 Data for chi-square analysis of dihybrid cross in maize

Phenotype	Genotype	Observed (O)	Expected (E)	$(O-E)$	$(O-E)^2$	$(O-E)^2/E$
Totals						
Conclusion:						

REFERENCES

Crow, J. F. 1979. Genes that Violate Mendel's Rules. *Scientific American* 240:134–146 (Offprint 1418). *Scientific American* Offprints are available from W. H. Freeman and Company, 41 Madison Avenue, New York, NY 10010, and 20 Beaumont Street, Oxford OX1 2NQ, England. Please order by number.

Demerc, M., and B. P. Kaufmann. 1969. *Drosophila Guide: Introduction to the Genetics and Cytology of* Drosophila melanogaster. 8th ed. Carnegie Institution.

Flagg, R. O. 1971. *Drosophila Manual.* Carolina Biological Supply Company.

Griffiths. A. J. F., et al. 1993. *An Introduction to Genetic Analysis.* 5th ed. W. H. Freeman.

Singer, S. 1985. *Human Genetics.* 2d ed. W. H. Freeman.

Strickberger, M. W. 1985. *Genetics.* 3d ed. Macmillan.

Exercise 11

Chromosomal Basis of Heredity

One of the most striking attributes of living organisms is their ability to transmit hereditary characteristics from one generation to another. The physical basis involved in this transmission, however, was not understood until the early part of the twentieth century when the chromosomal basis of heredity was established. Walter S. Sutton (1903), studying sex determination in insects, suspected that **chromosomes** were the carriers of the hereditary "factors" transmitted from parents to offspring. He emphasized the importance of the fact that, in the diploid condition, each cell contains two morphologically similar sets of chromosomes and that during the formation of gametes by meiosis, each haploid gamete receives only one chromosome of each homologous pair. He further explained that there are far more hereditary factors than there are chromosomes, therefore requiring that a number of different hereditary factors **(genes)** must be associated with a single chromosome

Though Sutton did not prove the chromosomal basis of heredity, one of his professors, Edmund B. Wilson (1905), demonstrated the role of chromosomes in heredity through studies on the nature of sex determination in insects. He showed that although the cells of female insects contain a pair of **X chromosomes,** the cells of males contain only one. Furthermore, in some species (including humans), male gametes contain a chromosome not found in females, called the **Y chromosome;** while every egg contains one X chromosome, only half the sperm contain this chromosome while the remainder carry the Y chromosome. Thus, fertilization of an egg by an X-bearing sperm leads to an XX zygote, which develops into a female, while fertilization by a sperm carrying a Y chromosome gives rise to male offspring. Subsequent research has firmly established the same chromosome basis of sex determination in most sexually reproducing organisms.

A. HUMAN SEX CHROMATIN

In 1959, Murray L. Barr observed a morphological distinction between the nuclei of male and female cells that were not undergoing mitosis. He observed that a small body was found adjacent to the nucleolus at a higher frequency in the cells of female cats than in cells of male cats. It has since been established that many other mammals (including humans) exhibit the same sexual differentiation, or **dimorphism.** Because staining techniques have demonstrated the chemical properties of this body to be similar to that of the chromosomes, it is now commonly referred to as the **sex chromatin** or **Barr body.**

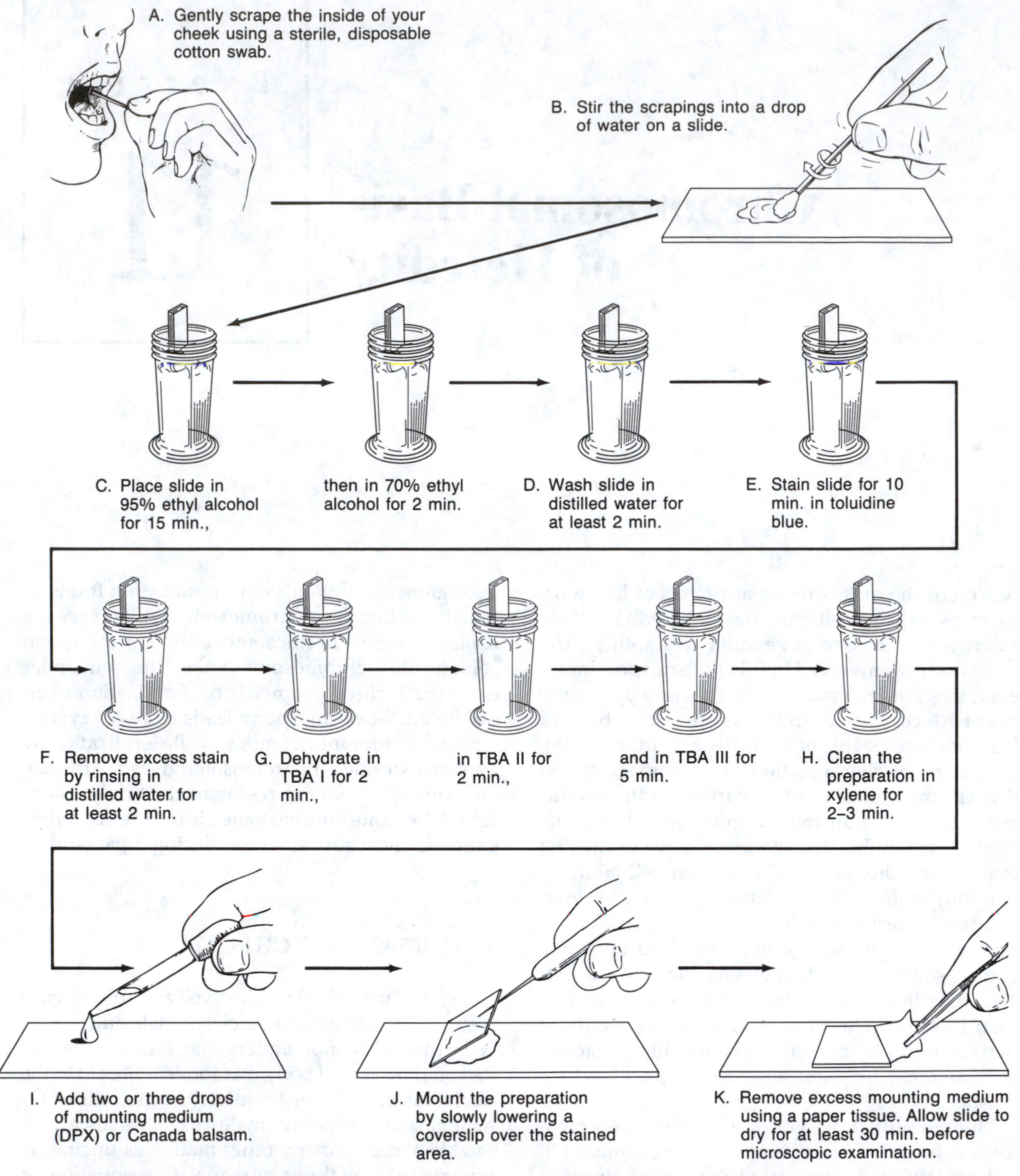

FIGURE 11-1 Procedure for demonstrating the presence of Barr bodies in cheek epithelial cells

The Barr body represents one of the two X chromosomes of female cells. When a cell is in interphase of mitosis, the substance of one X chromosome is in an attenuated or uncoiled form throughout the nucleus and hence is not visible. The other X chromosome remains coiled so that it stains deeply and is therefore identifiable as the Barr body. In human female cells the Barr bodies are attached to the inside of the nuclear membrane. Nuclei of male cells lack Barr bodies. Here then is a means of distinguishing male from female cells even when chromosomes are not visible.

Because of the discovery of the Barr body, it is now possible to identify the sex of an unborn child at an early age. Using the technique of amniocentesis, a small quantity of amniotic fluid is removed during the early months of pregnancy. Cells in the fluid are then examined for the presence of Barr bodies to determine the sex of the child.

In this study you will prepare a buccal smear of cells from the inside of your mouth and, using the procedure outlined in Figure 11-1, determine the presence (or absence) of the Barr body in these cells.

Locate the cells under the low power of the compound microscope. Add a drop of immersion oil to the coverslip directly below the lower-power objective. Carefully rotate the oil immersion objective into position and examine the preparation. The nucleus should be stained a pale blue. The Barr body will appear as a blue-black body adjacent to the nuclear membrane. Observe 50 cells and count the number having a Barr body. Do not count cells that appear abnormal (shriveled, shrunken, folded, broken) or are stacked on top of one another. Do not be upset if you are a male and find Barr bodies in your cells. About 2% of all male cells normally contain the sex chromatin. Record your results and those of the class in Table 11-1.

B. CHROMOSOME MORPHOLOGY

1. Salivary Gland Chromosomes in *Drosophila*

The giant chromosomes of the salivary gland cells of *Drosophila* and other flies afford an excellent opportunity to study chromosomal morphology. Although these salivary gland cells are highly specialized, they are like most other cells in terms of their major nucleotide components. They are unlike many other cells, however, in that the first gland primordium is formed in the embryo and the cells that make up the gland do not divide. The larval salivary gland thus contains a constant number of cells from before the egg hatches until the gland degenerates during formation of the puparium. The gland grows entirely through cell enlargement; as the cell enlarges, the nucleus and its chromosomes also enlarge.

The chromosomes attain their maximum size just before pupation. The salivary gland degenerates (undergoes histolysis) in *Drosophila* during formation of the puparium. An important physiological function of these glands is the secretion of "silk," which is used to spin a cocoon or to attach the puparium to a substratum. Salivary glands also secrete digestive enzymes during the active ingestion phase of larval growth and are well developed in some fly larvae that do not appear to secrete silk. These large, multistranded, or **polytene,** chromosomes are not confined to the salivary gland cells but are also present in cells of the gut epithelium, in the Malpighian tubules of the larva, and in the foot pads of adult flies.

In this part of the exercise, you will study the morphology of salivary gland chromosomes from the larva of the fruit fly *Drosophila melanogaster.*

1. Examine a culture and locate wormlike larvae. Select one of the larger, slower-moving ones,

TABLE 11-1 Percent of Barr bodies in cheek cells

Sex	Your cells	Class average	Class range
Male			
Female			

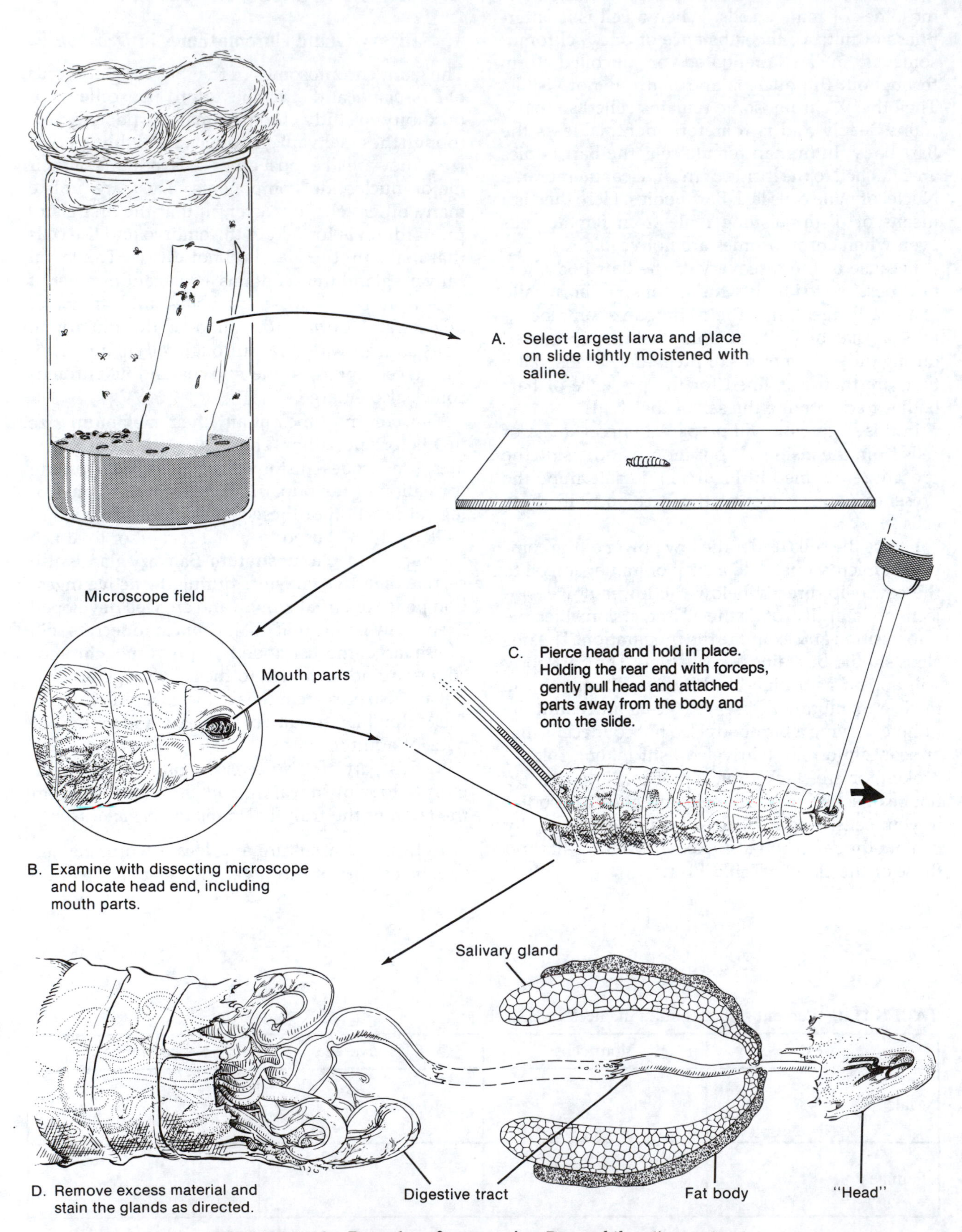

FIGURE 11-2 Procedure for removing *Drosophila* salivary glands

preferably one that is crawling up the side of the culture container. Gently remove it with forceps and place it on a glass slide lightly moistened with saline solution. Examine it with the stereoscopic microscope. Note that it has a blunt rear end and a pointed head end that contains black mouth parts. To obtain the salivary glands, the head end must be dissected away from the rest of the body (Fig. 11-2).

2. With a finely pointed needle, pierce the head as close to the anterior end as possible. The larva wriggles, so you will probably have to make several attempts before you are successful.

3. After you have secured the head, hold the rear end with a pair of finely pointed forceps. Then, with one smooth, quick motion at the anterior end, stretch the larva until the mouth parts are torn off and pulled onto the moist slide. All this can be done while observing the operation with the stereoscopic microscope.

4. The salivary glands, when pulled out, will probably be accompanied by part of the digestive tract and fat bodies. Add a drop of saline to the slide. When you are sure that you have the salivary glands, remove the digestive tube and any other extraneous parts with a razor blade or scalpel. Discard these parts, leaving only the salivary glands on the slide.

5. The salivary glands are now ready to be stained. Remove the excess saline solution from the slide by soaking it up with a small piece of paper toweling or filter paper. Try not to touch the glands with the paper because they might adhere to it. Cover the glands with a drop of aceto-orcein stain.

6. After staining for 5 minutes, carefully place a plastic coverslip over the glands. Then place a small piece of paper toweling over the coverslip and press hard with your thumb to squash the glands. Examine the preparation with your compound microscope. If properly squashed, the cells of the gland will be separated and you should be able to see nuclei in most of the cells.

7. If the preparation has been adequately stained, you should also be able to see banding along the length of the chromosomes. The bands may be better observed if the chromosomes are re-

FIGURE 11-3 Salivary gland chromosomes of *Sciara coprophila*, showing distinct longitudinal differentiation. (Photomicrograph courtesy of Dr. Ellen M. Rasch, East Tennessee State University.)

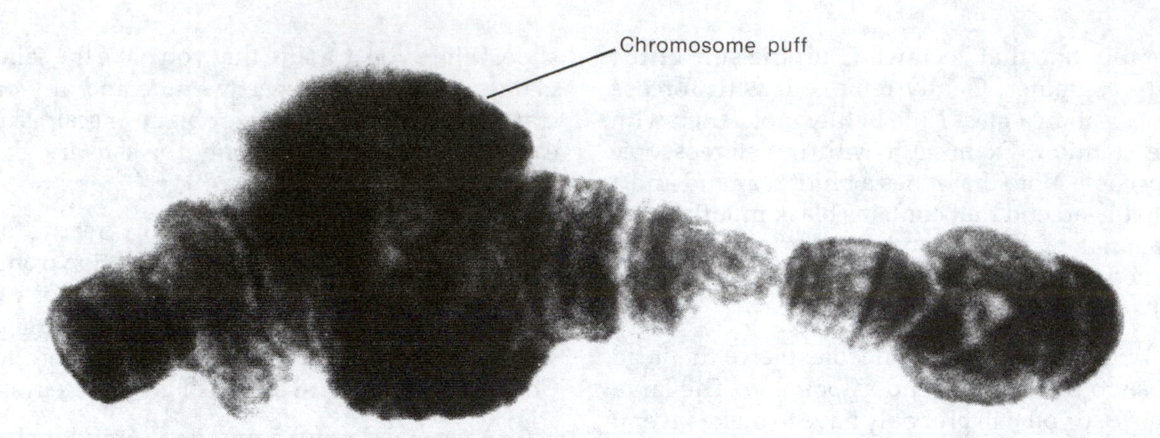

FIGURE 11-4 Giant salivary gland chromosomes of *Chironomus*. (Photomicrograph by Claus Pelling, Max Planck Institute for Biology.)

leased from the nuclei. To do this, firmly tap the coverslip with a pencil eraser several times. Examination under the high-power objective of the microscope should show the separated chromosomes and the bands sharply stained. Compare your preparation with that of the *Sciara* chromosomes shown in Fig. 11-3.

2. Chromosome "Puffs"

In your studies of mitosis and meiosis, the chromosomes appeared as heavily stained bodies that apparently lacked distinct morphology. However, when you examined the giant chromosomes in the salivary glands of *Drosophila* larvae, you could see structural detail that is not apparent in smaller chromosomes.

Figure 11-4 is a photograph of the giant salivary gland chromosomes of the midge fly, *Chironomus*. (If they are available, supplement your studies with specially prepared slides of these chromosomes.) Note the darkly stained bands separated from each other by lighter areas. These bands are areas of DNA, which show up when these chromosomes are stained by Feulgen's reagent, a dye that stains DNA.

Note the chromosome "puff" in Fig. 11-4. The pattern of puffing changes during the development of the organism. Experiments suggest that the puff regions are sites of special gene activity.

It has been shown that the hormone involved in insect molting can also induce puffing. This hormone, called **ecdysone,** appears to act on specific genes, which then generate a message that is transferred to the cytoplasm, where it initiates the synthesis of a protein.

REFERENCES

Alberts, B., et al. 1989. *Molecular Biology of the Cell.* 2d ed. Garland.

Demerc, M., and B. P. Kaufmann. 1969. *Drosophila Guide: Introduction to the Genetics and Cytology of* Drosophila melanogaster. 8th ed. Carnegie Institution.

Hadorn, E., and H. K. Mitchell. 1951. Properties of Mutants of *Drosophila melanogaster* and Changes During Development as Revealed by Paper Chromatography. Proc. Natl. Acad. Sci. USA 37:650–665.

Mange, A. P., and E. J. Mange. 1990. *Genetics: Human Aspects.* 2d ed. Sinauer Associates.

Watson, J. D., et al. 1989. *Molecular Biology of the Gene.* 4th ed. Vols I and II. Benjamin.

Weiss, R. 1989. Genetic Testing Possible Before Conception. *Sci. News* 136(21):326.

Exercise 12

Human Genetics

The specific characteristics of an individual, whether plant or animal, are established at fertilization following the union of the male and female chromosome complements carried by the gametes. These chromosomes carry the genes that determine the various characteristics expressed by the organism. In this exercise, you will examine the inheritance of certain morphological traits (e.g., tongue rolling, hair whorl, hair color) and physiological traits (e.g., PTC and sodium benzoate taste responses), as well as the human blood groups.

A. INHERITANCE OF MORPHOLOGICAL CHARACTERISTICS

1. Tongue Rolling

Many people can turn the sides of their tongues so that near the tip the sides nearly touch on top (Fig. 12-1). When everyone in the class has tried to do this, record the results in Table 12-1. Also record the data of other class sections and determine the percentages of "rollers" and "nonrollers." Percentages will not tell you whether the ability to roll the tongue is inherited or, if it is inherited, whether a dominant or recessive gene is involved. To learn this, determine how many members of your family have this trait and record your findings in Fig. 12-1. Write + in the circle or square to indicate a roller and − for a nonroller.

To help you determine if this is an inherited trait, use T to represent the dominant characteristic (rolling) and t for the recessive characteristic (nonrolling). Remember, if the trait inherited is recessive, both alleles must be recessive (i.e., tt). If, however, it is inherited as a dominant characteristic, then the alleles may be present in the homozygous or heterozygous condition (i.e., TT or Tt, respectively). Examining siblings, parents, and grandparents will help determine whether you are homozygous or heterozygous for the trait. Will it be possible to determine the specific genotype (i.e., TT, Tt or tt) for each family member? Explain.

____

____

____

____

To help you to decide on the method of inheritance, it may be useful to examine the inheritance of hair color in human beings as illustrated in Fig. 12-2. For example, it is possible for two parents

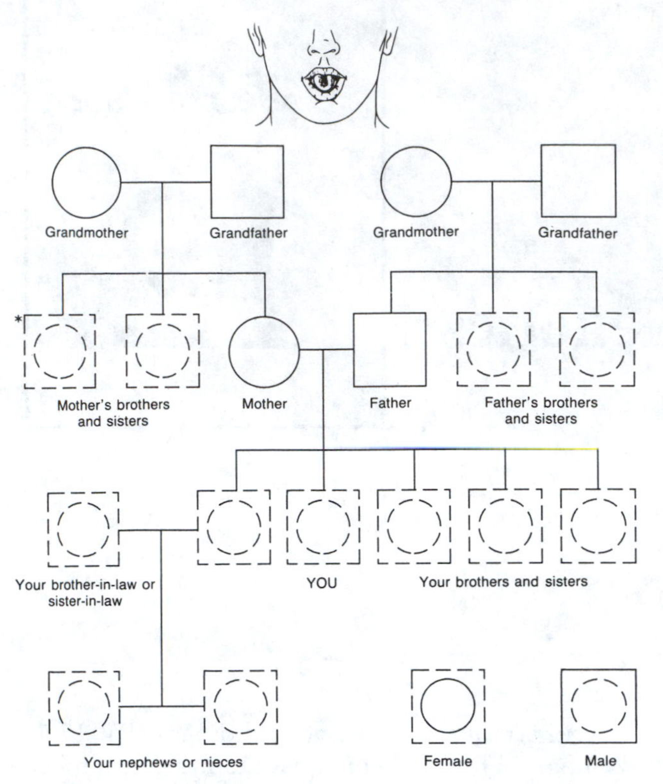

FIGURE 12-1 Inheritance of tongue rolling

who do not have red hair to have a child who does have red hair. However, it is not possible for two parents with red hair to have a child who does not have red hair.

Based on the data in Fig. 12-2, is red hair determined by a dominant or recessive gene? Explain.

What genotypes would you expect the parents to have to obtain the data in square A?

The data in square B?

Explain.

On the basis of the information you have collected and using the insights gained for the transmission of hair color, is the ability to roll the tongue inherited as a dominant or recessive gene?

---

2. Hair Whorl ("Cowlick")

Near the top, rear part of the head locate a whorl of hair which may rotate in either a clockwise or counterclockwise direction. Have your laboratory partner determine the direction of your hair whorl and those of other classmates. Record these data in Table 12-2. Also record the data of other class sections. On the basis of the information obtained, which direction of hair whorl appears to be dominant?

---

3. Darwin's Ear Point

The presence of a conspicuous point on the outer rim of the ear is inherited as a dominant character (Fig. 12-3A). This dominant gene is rather rare in the human population and shows variability in the way it is expressed. For example, some individuals have a Darwin's point on only one ear. Furthermore, some individuals who manifest the recessive gene transmit the dominant gene.

Have your laboratory partner examine your ears for this characteristic. What percent of the students in your class have this characteristic?

---

4. Widow's Peak

Individuals whose hairline dips down in the middle of the forehead (Fig. 12-3B) are said to have a "widow's peak." Examine your hairline and those of your classmates. Does this characteristic appear to be dominant or recessive?

---

5. Tongue Folding

The capacity to fold the tongue backward, without pressing it against the upper teeth (Fig. 12-3C), is very rare, occurring in the human population at a frequency of less than once per thousand individ-

TABLE 12-1 Analysis of the ability to roll the tongue

Class section number	Number of students in class	Number of tongue rollers	Number of nonrollers	% rollers	% nonrollers
1					
2					
3					
4					
5					
6					
7					
8					
9					
10					
11					
12					
13					
14					
15					

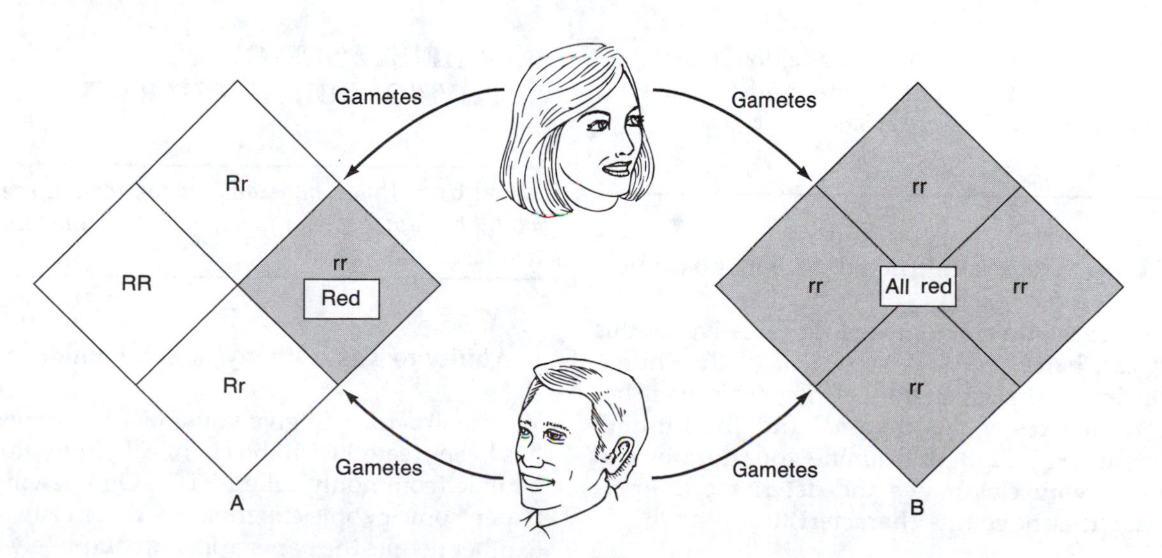

FIGURE 12-2 Inheritance of human hair color

TABLE 12-2 Analysis of human hair whorl

Class section number	Number of students in class	Number with clockwise rotation	Number with counterclockwise rotation	% clockwise rotation	% counterclockwise rotation
1					
2					
3					
4					
5					
6					
7					
8					
9					
10					
11					
12					
13					
14					
15					

uals. This capacity is inherited as a dominant characteristic. What percent of the students in your class possess this characteristic?

____

6. Hyperextension of the Distal Thumb Joint

Individuals who are homozygous recessive for this trait can bend the distal segment of the thumb backward so that an angle of 60 degrees is made between the axes of the proximal and distal thumb segments (Fig. 12-3D). Examine your thumbs and those of your classmates and determine the percentage that have this characteristic.

____

B. INHERITANCE OF A PHYSIOLOGICAL CHARACTERISTIC

Caution: *If you have any known food allergies, it would be advisable not to participate in the following two taste tests.*

1. Ability to Taste Phenylthiocarbamide (PTC)

Your instructor will give you a piece of paper that has been treated with the chemical phenylthiocarbamide (commonly called PTC). On chewing this paper, some people experience a bitter taste, while to other people the paper appears completely tasteless. Furthermore, for those who can taste PTC the

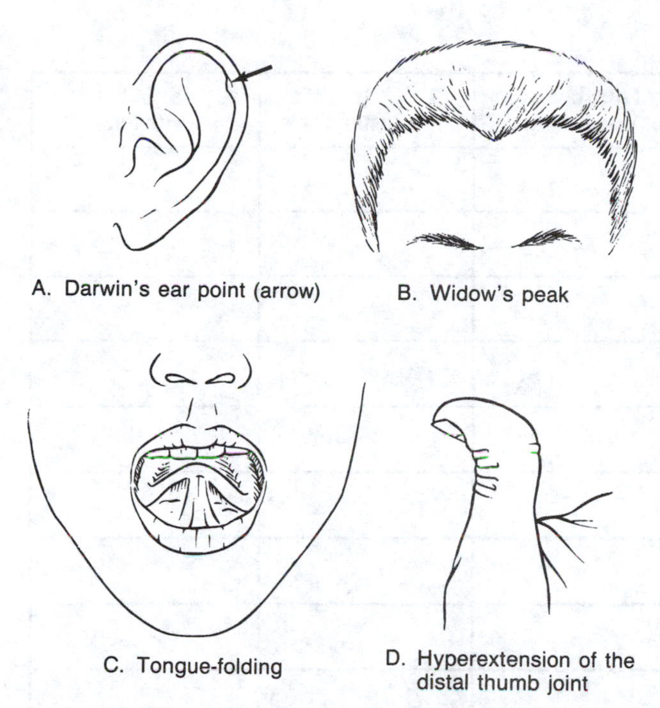

FIGURE 12-3 Inheritance of human morphological characteristics

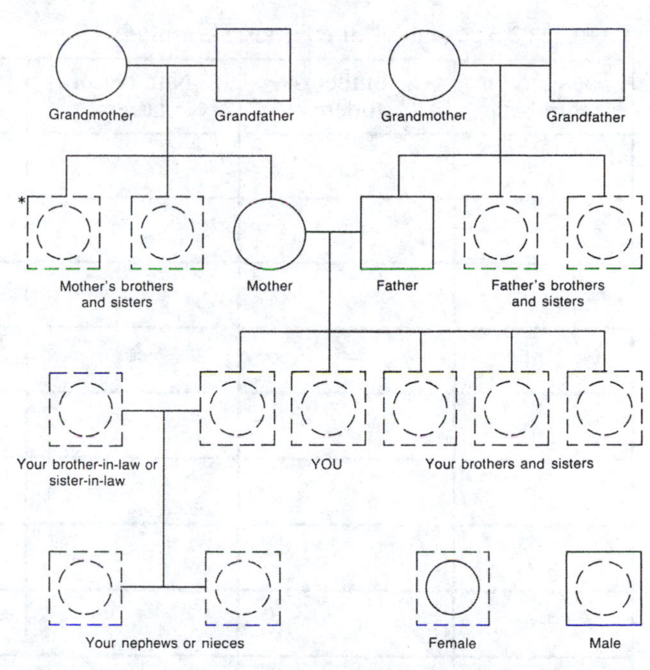

FIGURE 12-4 Inheritance of the ability to taste PTC

taste varies from bitter to sweet. Are you a "taster" or "nontaster"?

___

When everyone in the class has tried tasting PTC, record the results of your section and those of other class sections in Table 12-3.

In order to determine the nature of inheritance of PTC tasting, your instructor will give you several strips of PTC paper to take home. Gather as much family information as you can in order to complete Fig. 12-4. Use + to indicate a taster and − for a nontaster. Using T to represent the dominant characteristic and t the recessive characteristic, indicate the genotypes (TT, Tt, or tt) of each individual tested in Fig. 12-4.

On the basis of the information you collected, is the ability to taste PTC transmitted as a dominant or recessive gene?

___

2. Ability to Taste Sodium Benzoate

Sodium benzoate is another substance that can be tasted by some people but is tasteless to others. This compound, at a concentration of 0.1%, is sometimes used as a food preservative. Since there is some question about whether it produces detrimental effects on health, regulations regarding the use of benzoates vary widely. Some states prohibit its use while others place severe restrictions on its use; still others have liberal regulations allowing the use of benzoates as food preservatives.

Your instructor will give you a piece of paper that has been treated with sodium benzoate. Chew the paper and record the kind of taste you experience (salty, sweet, sour, bitter, tasteless).

___

What percent of the students in your class are tasters? nontasters?

___ ___

Of those who can taste sodium benzoate, which of the kinds of taste are most common? least common?

___ ___

C. INHERITANCE OF HUMAN BLOOD GROUPS

The presence of different blood groups in the human population was reported for the first time

TABLE 12-3 Analysis of PTC taste sensitivity

Class section number	Number of students	Number of tasters	Number of nontasters	% tasters	% nontasters
1					
2					
3					
4					
5					
6					
7					
8					
9					
10					
11					
12					
13					
14					
15					

in 1900 by Dr. Karl Landsteiner. His work led to the establishment of the presence of four basic blood types, A, B, AB, and O. These are determined by the presence or absence of molecules called **antigens** on the surface of red blood cells and other molecules called **antibodies** in the liquid portion of the blood (plasma). The factors that determine to which group a person's blood belongs are inherited. Furthermore, it is the presence or absence of two different antigens called **A** and **B** that determines blood type. A person's blood cells may contain either the A or the B antigen, both A and B, or neither A nor B. These possibilities are outlined in Table 12-4.

Blood may also contain antibodies that cause the clumping **(agglutination)** of red blood cells that are foreign to the blood. The distribution of these two antibodies (anti-A and anti-B) in the four blood types is shown in Table 12-4. Table 12-5 shows their distribution in several ethnic groups in the United States. It should be noted that no individual's blood contains antibodies in the plasma that will agglutinate his/her own blood cells. Thus, blood transfusions from one person to another can safely be made only when the blood groups are compatible. This means that no one should receive blood that contains red blood cells with antigens different from those of their own red blood cells. If foreign red blood cells are transferred, they react with antibodies present in the recipient's blood and cause the agglutination of the foreign red blood cells.

TABLE 12-4 Human ABO blood groups

Blood group	Antigen on red blood cell surface	Antibody in blood plasma
A	A	anti-B
B	B	anti-A
AB	A and B	none
O	none	anti-A and anti-B

TABLE 12-5 Incidence of human blood groups in the United States (%)

Group	Caucasians	Blacks	Chinese	Native Americans
O	45	48	36	23
A	41	27	28	76
B	10	21	23	0
AB	4	4	13	1

The antigens determining the four blood groups are the result of the expression of three genes, O, A, and B; A and B are dominant to O. The genotype AA cannot, by chemical analysis, be distinguished from AO, and genotype BB cannot be distinguished from BO. Thus, both genotypes are classified as phenotypes A and B, respectively, and although six genotypes occur (OO, AO, AA, BO, BB, AB), only four phenotypes (A, B, AB, O) can be recognized.

Using the phenotypes of the parents listed in Table 12-6, list all possible genotypes of the parents and of the children and all possible phenotypes of the children.

D. SOME OTHER INHERITED CHARACTERISTICS

For the following traits, try to determine your phenotype and possible genotypes. Recall that if you have the dominant characteristic, you may be either homozygous or heterozygous dominant. Since you will not be examining other family members, you will not know if you carry the recessive allele of the gene. In this case, use a – to indicate the "unknown" second allele (i.e., XX vs. X–). On the other hand, if you have a recessive characteristic you will have both recessive alleles. Record your phenotype and possible genotypes for the following traits in Table 12-7.

1. Interlocking Fingers

When people interlock their fingers, some generally place their left thumb on top of the right one (dominant characteristic; allele F). Others place the right thumb over the left (recessive allele f).

2. Dimpled Chin

A cleft in the chin is a dominant characteristic (allele for dimpling D). Absence of the cleft is recessive (allele d).

3. Pigmented Iris

If you are recessive for this characteristic (genotype pp), no pigment is present in the front of your eyes and a blue layer at the back of the iris shows through. Thus, you have blue eyes. If you have at least one dominant allele (P–), pigment is in the eye and masks the blue to varying degrees depending on other genes that regulate the amount of this masking pigment. Thus you have brown, violet, green, hazel or another eye color, depending on the amount and density of this pigment.

4. Mid-Digit Hair

Some people have hair on the middle (second) joint on one or more of their fingers. Complete absence

TABLE 12-6 Inheritance of human blood types

Phenotypes of parents		Genotypes of parents		Genotypes of children	Phenotypes of children
Mother	Father	Mother	Father		
A	O				
A	B				
B	O				
AB	A				
AB	B				
AB	O				
O	O				

TABLE 12-7 Other inherited characteristics

Characteristic	Phenotype	Possible genotypes
Interlocking fingers		
Dimpled chin		
Pigmented iris		
Mid-digit hair		
Bent little finger		

of hair is due to the recessive allele m; the dominant characteristic is due to the allele M. Since the hair can be very fine in nature, you may need to use a magnifying glass or hand lens to determine the presence or absence of these hairs.

5. Bent Little Finger

The dominant allele B results in the last joint of the little finger bending inward toward the fourth finger. The recessive allele b results in a straight little finger. To determine whether you have this characteristic, lay your hands flat on the table and relax your muscles.

REFERENCES

Cummings, M. 1991. *Human Heredity: Principles and Issues.* West.

Delisi, C. 1988. The Human Genome Project. *American Scientist*, Vol. 76:488–493.

Edlin, G. 1988. *Genetic Principles: Human and Social Consequences.* Jones and Bartlett.

Griffiths, A. J. F., et al. 1993. *An Introduction to Genetic Analysis.* 5th ed. W. H. Freeman.

Mange, A. P., and E. J. Mange. 1990. *Genetics: Human Aspects.* 2d ed. Sinauer Associates.

Strickberger, M. N. 1986. *Genetics.* 3d ed. Macmillan.

Exercise

13

Expression of Gene Activity

The present era of molecular genetics and recombinant DNA technology was initiated by the formulation of the Watson–Crick model of the molecular structure of deoxyribonucleic acid **(DNA).** DNA is unusual in three respects.

- It is a very large molecule, having uniformity of size, rigidity, and shape. Despite this uniformity, however, the huge number of possible permutations of its internal structure gives it the complexity required for carrying information.
- It can make remarkably exact copies of itself; that is, it can **replicate** itself.
- The information in its chemical structure is transmitted from the nucleus to the cytoplasm (in eukaryotes) or to other parts of the cell (in prokaryotes), where it is used to synthesize proteins that govern the behavior of the cell.

A. CHROMATOGRAPHIC SEPARATION OF *DROSOPHILA* EYE PIGMENTS

In 1941, George W. Beadle and Edward L. Tatum presented experimental evidence supporting the concept that genes control the chemical activities of the cell by controlling the production of proteins called **enzymes.** Enzymes catalyze the numerous chemical reactions that take place in cells and that are ultimately expressed in the morphology, physiology, biochemistry, and behavior of an adult organism.

Eye pigmentation in the fruit fly *Drosophila melanogaster* is genetically controlled. The normal red eye color of this insect is correlated with the presence of a characteristic series of substances called **pteridines** (Greek *pteron,* "wing"). This term was chosen because the first substances in this class of compounds were extracted from butterfly wings. These compounds are easily separated by chromatography and, if viewed under ultraviolet light, produce distinctive fluorescent patterns in wild-type *Drosophila* (Fig. 13-1). Pteridines are present in many invertebrates, in certain pigment cells of amphibians and fishes, and in plants (in which they may participate in photosynthesis). The pathway of pteridine synthesis is not well understood, but certain steps have been identified.

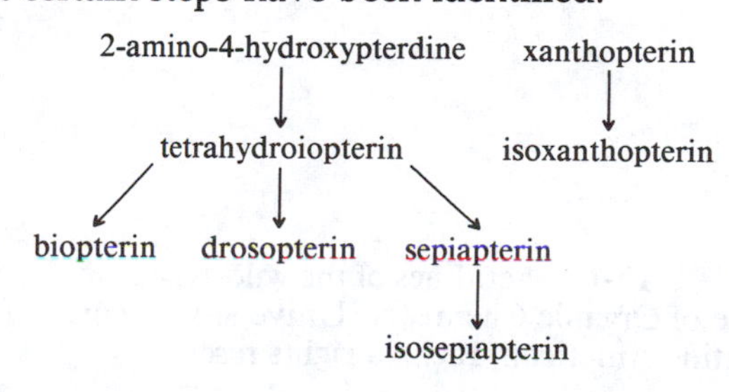

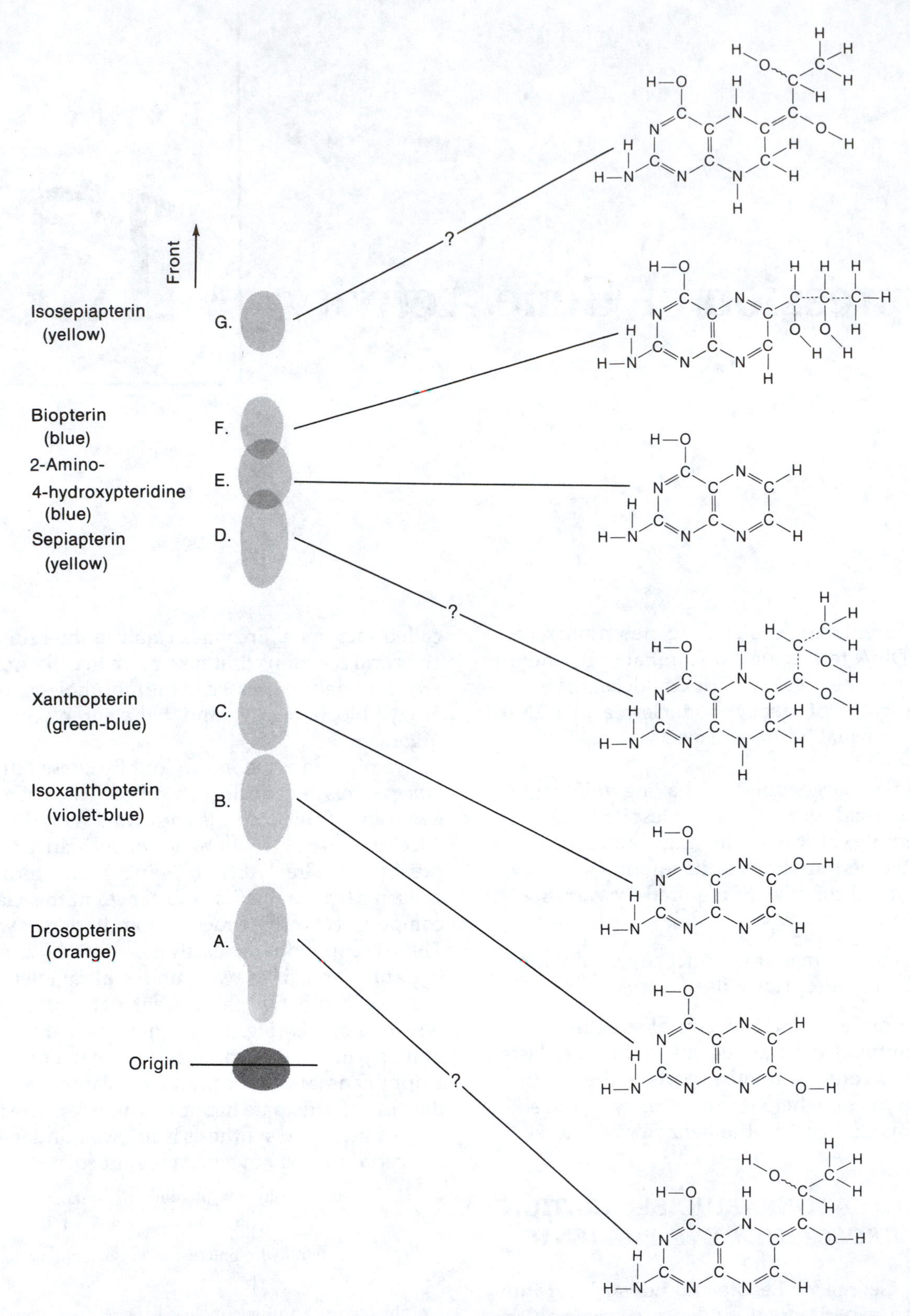

FIGURE 13-1 Pteridines of the wild-type fruit fly. The chemical structures were worked out by Max Viscontini, Institute of Organic Chemistry, University of Zurich. (From *Fractionating the Fruit Fly*, by Ernst Hadorn. © 1962 by Scientific American, Inc. All rights reserved.)

The pteridine patterns of flies with mutant eye colors are distinctly different from that of the wild-type flies. In these flies certain normal wild-type pteridines may be completely missing whereas others may be present in abnormally large quantities.

Two mutants that have been independently isolated have dull reddish-brown eyes instead of the bright red eyes of normal flies. The two genes producing this trait reside in two locations in the *Drosophila* genome. One of them, termed *maroonlike* (*mal*), is located on the first (or sex) chromosome (Chromosome I), whereas the other, termed *rosy* (*ry*), is located on the third chromosome (Chromosome III). Enzymatic analysis has revealed that both mutant types are deficient in xanthine dehydrogenase, the enzyme that converts 2-amino-4-hydroxypteridine into isoxanthopterin. As a result of the enzymatic deficiency, mutant flies accumulate 2-amino-4-hydroxypteridine; the wild-type flies have little or none of this substrate but do contain considerable quantities of the product isoxanthopterin. This example supports the generalization that genes produce their effects through the action of enzymes.

In this part of the exercise, you will chromatographically analyze the pteridines in the normal (wild-type) fruit fly and compare the wild type with several eye-color mutants to establish that the eye-color mutations are accompanied by differences in pteridine patterns. (See Appendix C for a discussion of chromatography.)

1. On a 20-by-20-cm silica-gel thin-layer chromatographic plate, draw a pencil line parallel to and about 25 mm from one edge (Fig. 13-2). Handle the silica-gel plate as little as possible and only by the edges, because fingerprints interfere with separation.

2. Obtain three wild-type flies, three each of the eye-color mutants rosy (*ry*) and maroonlike (*mal*), and three each of one or more of the following mutants:

Sepia (*se*)	Cinnabar (*cn*)
Brown (*bw*)	Vermilion (*v*)
Plum (*pm*)	Eosin (w^e)
Scarlet (*st*)	Apricot (w^a)
	White (*w*)

Because sex differences exist with respect to the pteridines, the following analyses should be made using flies of the same sex. Choose adult males or females.

The wild-type fruit fly has dark red eyes, a tannish, bristle-covered body, and a long, straight pair of wings reaching beyond the tip of the abdomen (Fig. 13-3). The adult male has a somewhat smaller body than the adult female, a rounded, heavily pigmented abdomen, and a "sex comb" (a tuft of bristles) on the forelegs. The abdomen of the female is somewhat pointed, is traversed by several dark stripes, and may have a terminal tuft of short bristles.

3. Etherize the three wild-type flies and place them in a vial containing about 0.25 ml of chromatographic solvent. Crush the flies with a glass rod to dissolve the pigments. Using a capillary tube, add several drops of the fly extract to the silica gel, as shown in Fig. 13-2. Allow the spot to dry between applications. Be careful not to disturb the silica-gel coating any more than necessary. Label for identification. Repeat this procedure for each of the eye-color mutants. What is one control that should be used?

___

Add this control to the thin-layer chromatographic sheet.

4. Allow the spots to dry for several minutes. Then place the silica-gel sheet in the solvent in the chromatographic jar (Fig. 13-2). Because the pteridines are light sensitive, chromatograms should be developed in a dark room (or cover the jar with foil).

5. Allow the chromatogram to develop until the solvent front reaches to within 3.5 cm of the top of the sheet.

6. Remove the sheet from the tank, mark the solvent front with a pencil, and air-dry for several minutes. Examine the sheet using ultraviolet (UV) light of long wavelengths (360 nm).

Caution: *Do not look directly into the UV lamp. Wear goggles to protect your eyes from UV reflection from the laboratory table.*

Note the fluorescent colors of the various pteridines (Fig. 13-1). Outline each spot with a pencil. The pteridines are listed in Table 13-1 in the order in which they should separate on the chromatogram, with the pteridine listed last being found at the bottom of the chromatogram, nearest the origin. Indicate which pteridines are present in the flies you examined by checking the appropriate boxes in Table 13-1.

Calculate the R_f values for each of the pteridines isolated on your chromatogram and record them in

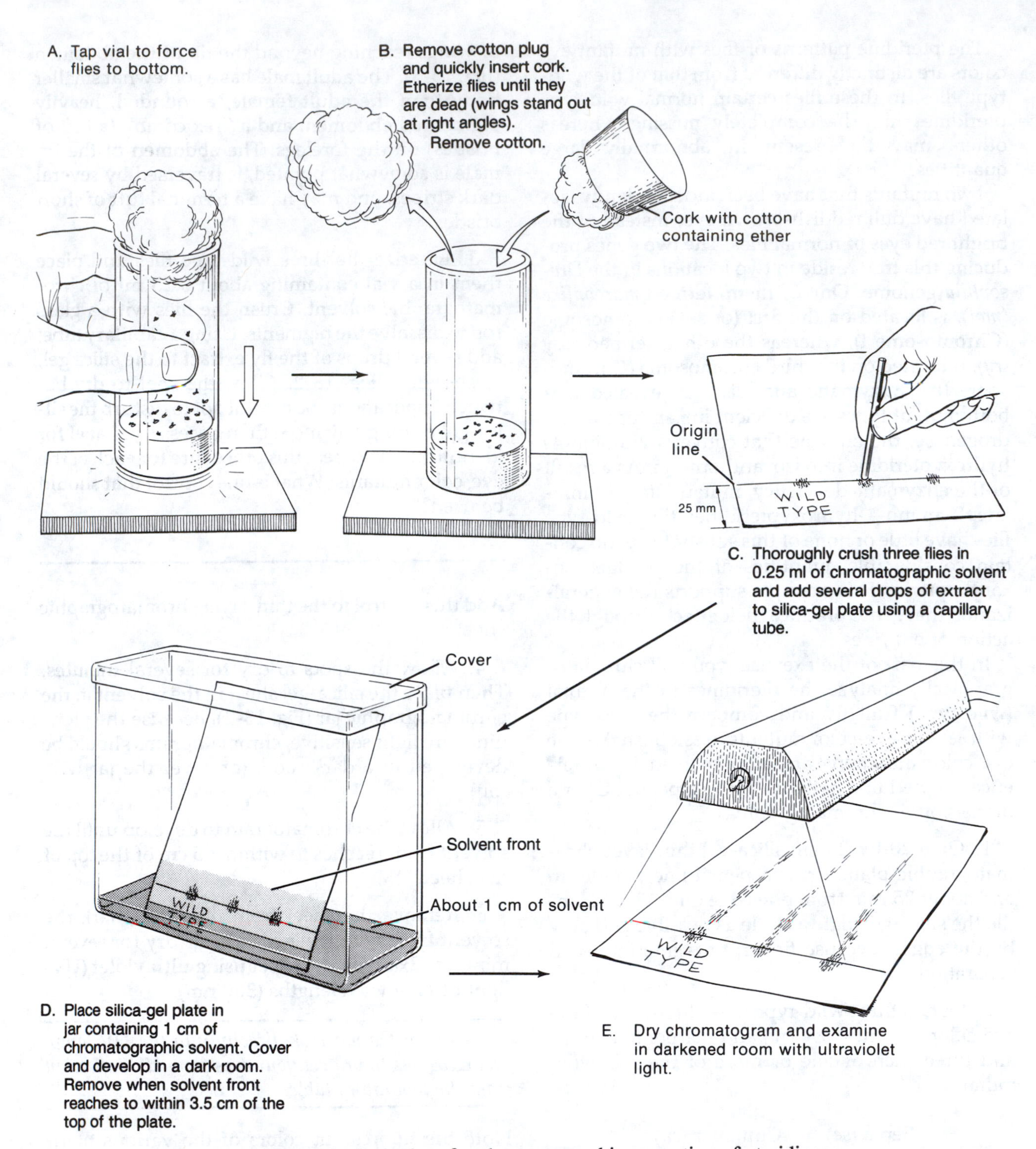

FIGURE 13-2 Procedure for chromatographic separation of pteridines

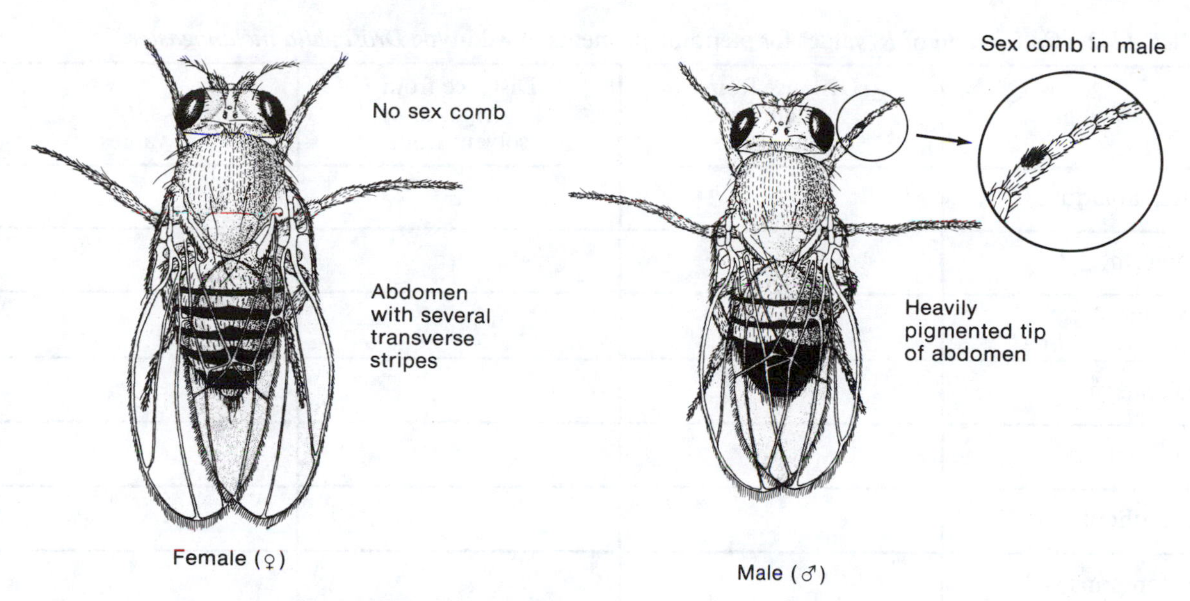

FIGURE 13-3 Adult fruit fly *Drosophila melanogaster*

TABLE 13-1 Distribution of pteridines in the wild-type *Drosophila* and in eye-color mutants

Pteridine (color)	Wild type	Mutants				
		Rosy	Maroonlike	___	___	___
	Sex: ___	Sex: ___	Sex: ___	Sex: ___	Sex: ___	Sex: ___
Isosepiapterin (yellow)						
Biopterin (blue)						
2-Amino-4-hydroxypteridine (blue)						
Sepiapterin (yellow)						
Xanthopterin (green-blue)						
Isoxanthopterin (violet-blue)						
Drosopterins (orange)						

TABLE 13-2 Calculation of R_f values for pteridine pigments of wild-type *Drosophila melanogaster*

Pigment	Distance from origin to center of spot	Distance from origin to solvent front	R_f values
Isosepiapterin			
Biopterin			
2-Amino-4 hydroxypteridine			
Sepiapterin			
Xanthopterin			
Isoxanthopterin			
Drosopterins			

Table 13-2. Use the procedure outlined in Appendix C. Do chromatograms of sepia-eyed flies show any pteridines present in greater amount than in the wild type? ________. If so, what pigments?

____

What pteridines did you or your classmates observe in male and female rosy and maroonlike eye-color mutant flies?

____

____

____

____

Explain any differences you observed.

____

____

____

____

Discuss the eye-color mutations in terms of the genetic control of enzyme synthesis and activity.

____

____

____

____

If time permits, consider using this technique to answer the following questions.

- Do adult male and female flies have the same pteridine patterns?
- Do the patterns of pteridine synthesis change during the development of the fly from egg to adult?
- Are there pigment differences in body-color mutants similar to those in eye-color mutants?

B. INDUCTION OF MUTATION BY ULTRAVIOLET LIGHT

The capacity for mutation is inherent in the genetic material of all living organisms and viruses. Although mutations occur spontaneously, the frequency with which they occur can be increased by a variety of agents called **mutagens.** One of the more commonly used mutagenic agents is ultraviolet light, which produces its mutagenic effects (at least in part) by causing chromosomes to break, which results in the loss or rearrangement of their parts.

In this experiment ultraviolet radiation will be used to study the mutation rate of the antibiotic-producing mold *Penicillium.* This study will be limited to such easily detected mutations as the shape and pigmentation of the colony and the effects on growth rate. (See Fig. 13-4.)

1. Working with a partner, obtain two petri plates containing nutrient agar. Label one "Control" and the other "UV," and add your names, the date, and your laboratory section number.

2. Add 1 ml of a *Penicillium* spore suspension to each of the plates as follows.

   a. Gently shake the flask containing the spores to suspend the contents uniformly.

   b. Pipet 1 ml (20 drops) of the spore suspension from the flask and place it on the surface of the agar in the control dish. Then, holding the agar plate at eye level, tilt the plate, allowing the suspension to run to the edge.

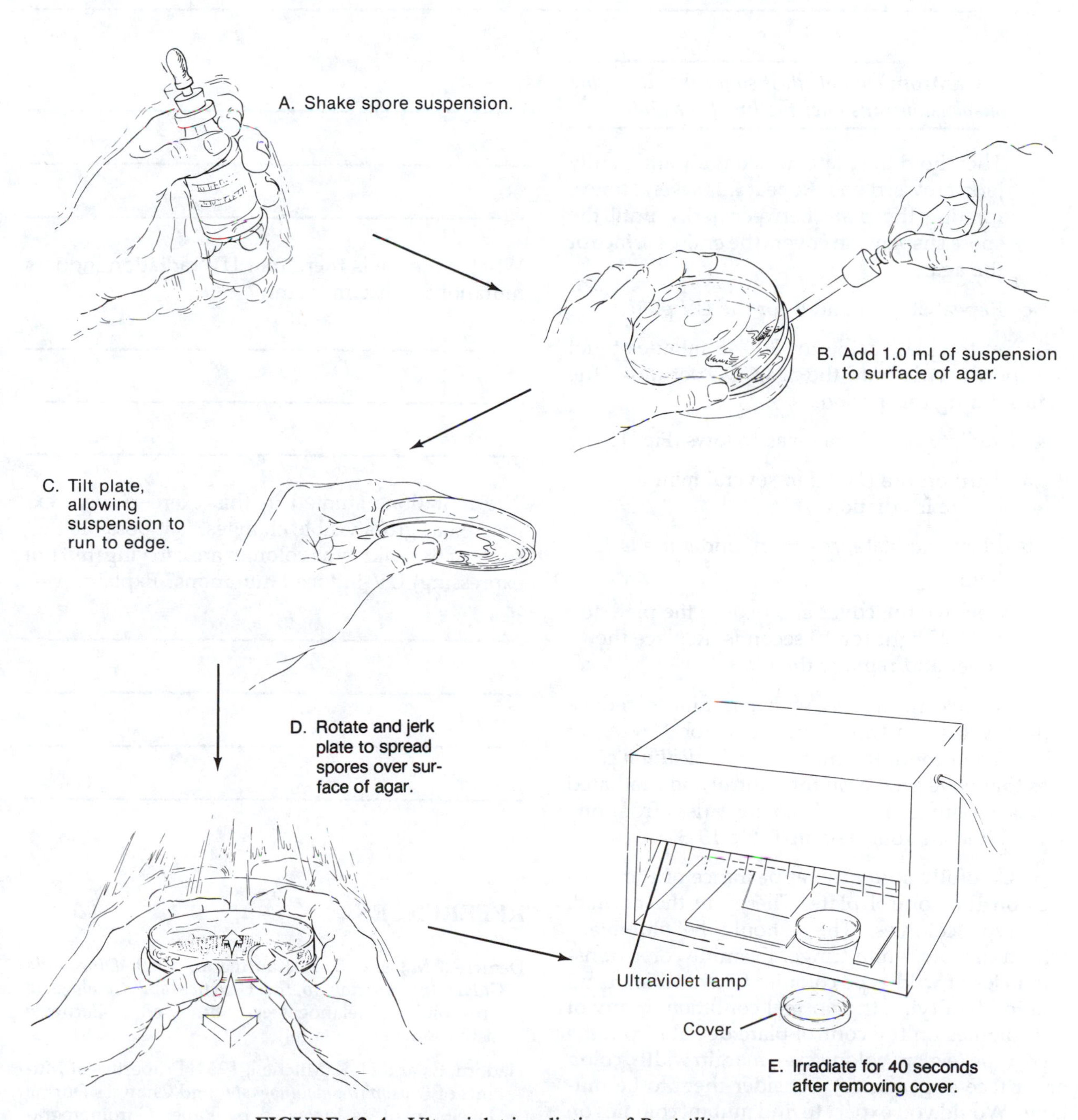

FIGURE 13-4 Ultraviolet irradiation of *Penicillium* spores

TABLE 13-3 Effects of ultraviolet radiation on *Penicillium*

Penicillium	Class data		Team data	
	Control	UV	Control	UV
Total number of colonies				
Number of mutant colonies				
Percentage of mutant colonies				
Percentage of spores surviving irradiation				

Caution: *Do not tilt it so far that the spore suspension runs over the lip of the plate.*

Then hold the plate horizontally and gently jerk it toward you. Repeat this several times, rotating the plate between jerks, until the spore suspension covers the entire surface of the agar.

c. Repeat steps a and b for the UV plate.

3. Set the plate aside for 30–45 minutes to let the spores settle onto the agar. Do not move the plates during this period.

4. Irradiate the UV plate as follows (Fig. 13-4).

a. Turn on the UV lamp several minutes before irradiation.

b. Place the plate, cover on, under the UV lamp.

c. Remove the cover and expose the plate to the UV light for 40 seconds. Replace the cover and remove the plate.

5. Encircle the plates with parafilm to reduce moisture loss and incubate them for 1 week at 28°C. Then count the number of *Penicillium* colonies that have grown on the control and irradiated plates. (Assume that each colony arises from one spore.) Record your data in Table 13-3.

6. Carefully study the appearance of the colonies on the control plate. These are the normal, wild-type colonies. They should be blue-black with a narrow white fringe. Examine color transparencies of wild-type colonies (if available) to aid you in identifying the normal condition. If any of the colonies on the control plate deviate from the wild-type (e.g., in colony size, margin width, color, or surface appearance), consider them to be mutants. Would you expect to find mutant colonies on the control plate? Explain.

____

____

____

What evidence is there that UV radiation induces mutations indiscriminately?

____

____

____

The mutations studied in this exercise were expressed as gross visible changes. Is it possible that some of the wild-type colonies are carrying (but not expressing) UV-induced mutations? Explain.

____

____

____

____

REFERENCES

Demerec, M., and B. P. Kaufmann. 1969. *Drosophila Guide: Introduction to the Genetics and Cytology of* Drosophila melanogaster. 8th ed. Carnegie Institution.

Hadorn, E., and H. K. Mitchell. 1951. Properties of Mutants of *Drosophila melanogaster* and Changes During Development as Revealed by Paper Chromatography. *Proc. Nat. Acad. Sci. USA* 37:650–665.

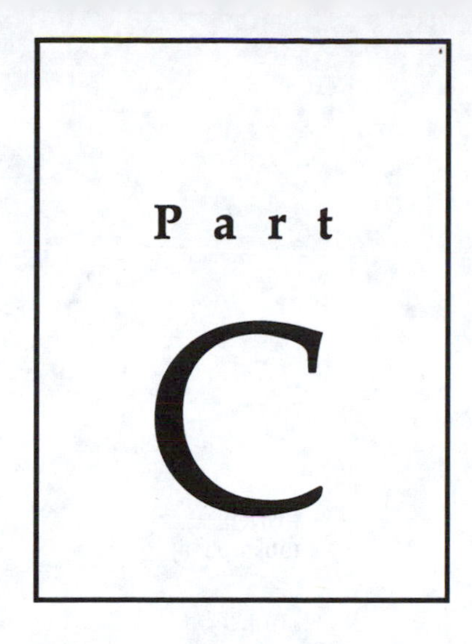

Diversity of Life: Monera, Protista, Fungi

There are some 5 million distinct kinds of living organisms on earth. Of these, about 1.5 million have been named and classified. To more efficiently discuss and study these organisms, they are grouped into categories, each of which is given a name. For example, each organism is grouped into a genus (each of which consists of one or more species). Genera are grouped into families, families into orders, orders into classes, classes into phyla (or divisions, in plants), and phyla into kingdoms.

In this manual, we use the Whittaker five-kingdom system of classification (shown on the next page) consisting of the Monera (all **prokaryotes)** and four kingdoms of **eukaryotes.** The eukaryotic organisms consist of the Protista, a catchall group containing mostly unicellular organisms, and three kingdoms consisting mostly of multicellular forms (plants, animals, and fungi).

REFERENCE

Whittaker, R. H. 1969. New Concepts of Kingdoms of Organisms. *Science* 163:150–160.

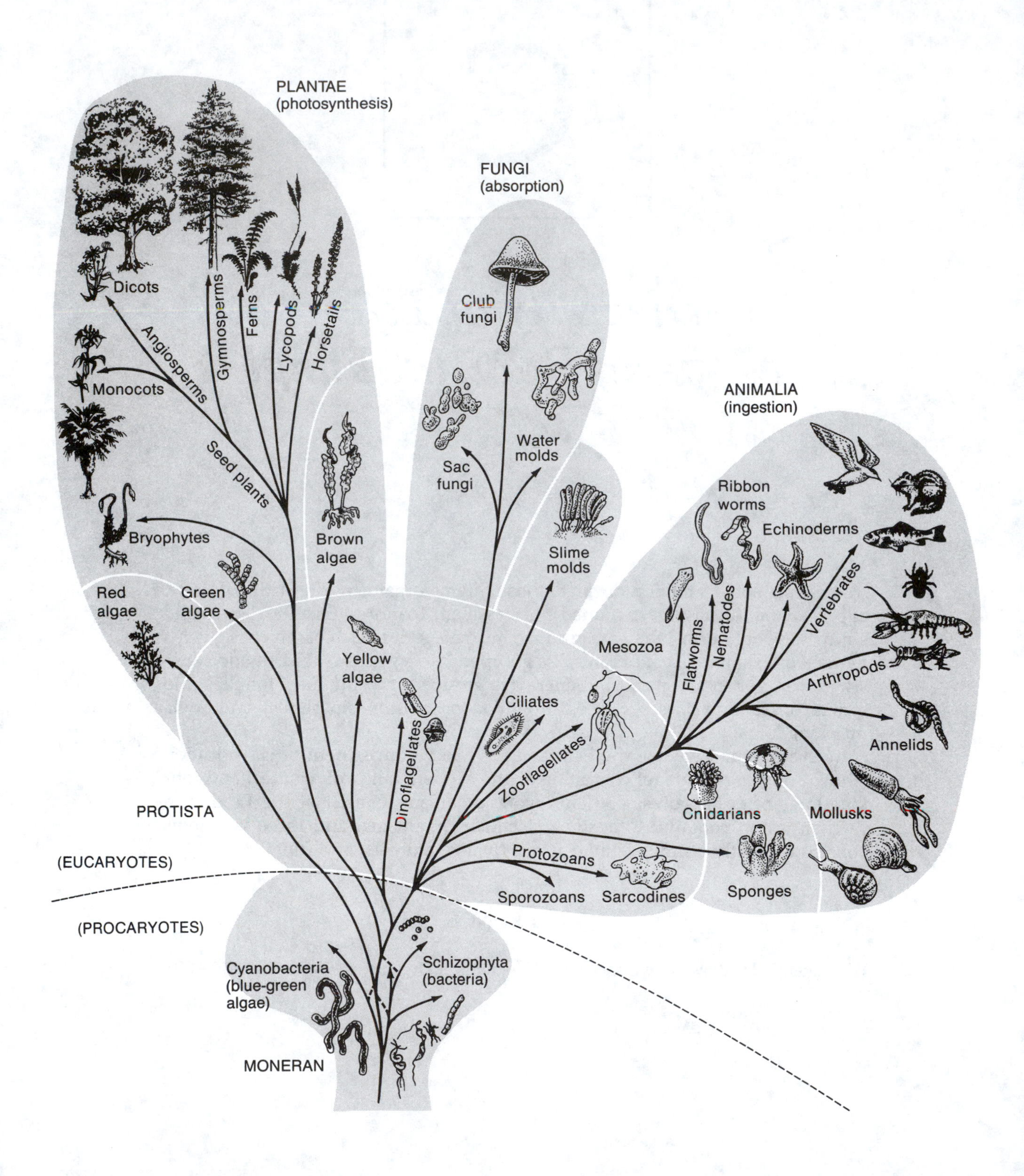
PLANTAE
(photosynthesis)
FUNGI
(absorption)
ANIMALIA
(ingestion)
Dicots
Gymnosperms
Ferns
Lycopods
Horsetails
Angiosperms
Monocots
Seed plants
Bryophytes
Red algae
Green algae
Brown algae
Club fungi
Water molds
Sac fungi
Slime molds
Ribbon worms
Echinoderms
Vertebrates
Mesozoa
Flatworms
Nematodes
Arthropods
Annelids
Yellow algae
Dinoflagellates
Ciliates
Zooflagellates
Cnidarians
Mollusks
PROTISTA
(EUCARYOTES)
Protozoans
Sporozoans
Sarcodines
Sponges
(PROCARYOTES)
Cyanobacteria
(blue-green
algae)
Schizophyta
(bacteria)
MONERAN

Exercise 14

Kingdom Monera

Members of the kingdom Monera, which includes all prokaryotes, are unicellular, but sometimes aggregate into filaments or other loose collections of cells. Prokaryotic cells differ from eukaryotic cells in four important ways. First, their DNA is not organized within a membrane-bounded nucleus, and it is not associated with histone protein to form chromatin as in eukaryotes. Second, the DNA in prokaryotic cells is circular, whereas in eukaryotes it is organized into distinct chromosomes. Third, prokaryotes lack membrane-bounded organelles such as mitochondria, plastids, lysosomes, Golgi complexes, and endoplasmic reticulum, and the plasma membranes of many prokaryotes have folds and convolutions extending into the cytoplasm, which increase the surface area. Fourth, the cell walls of most prokaryotes contain peptidoglycan, a molecule that is unique to the Monera.

In this exercise, you will study the following two divisions of Monera:

- Division Schizophyta (bacteria) are unicellular, usually reproduce asexually by cell division, and their nutrition is usually **heterotrophic** (all their carbohydrates, fats, and amino acids must exist ready made in their diet) though some are autotrophic (photosynthetic).
- Division Cyanobacteria (formerly called the blue-green algae) are unicellular or colonial, have chlorophyll but no plastids, reproduce by fission, and their nutrition is usually **autotrophic** (they can synthesize their carbohydrates, fats, and amino acids).

A. DIVISION SCHIZOPHYTA (BACTERIA)

As a group, bacteria are the most ancient of all organisms; their fossils have been found in rocks that have been dated as far back as 3.5 billion years. This is far earlier than fossil eukaryotes, which are believed to be about 800 million years old. Despite their minute size and apparently simple structure, bacteria have adapted to a variety of environments. In fact, they are more widely distributed in nature than any other group of organisms and can survive in environments that support no other form of life. Some bacteria are **obligate anaerobes;** that is, they can survive and multiply only in the absence of free oxygen. Other bacteria, the **facultative anaerobes,** can survive without oxygen but grow more vigorously when it is available in their environment.

Bacteria are especially abundant in the soil. Some species play an important role in the nitrogen cycle, in which nitrogen gas in the atmosphere is converted into nitrogenous salts that plants can use for growth.

Bacteria are responsible for the decay of dead organic matter, which releases carbon dioxide that can be reused in photosynthesis. Although most bacteria are heterotrophs, some are autotrophs. One group of bacteria **(photoautotrophs)** obtains energy for synthesis by converting light energy into chemical energy in a process that is similar to photosynthesis. Photosynthesis in bacteria, however, differs from that in eukaryotic organisms in that hydrogen sulfide (H_2S) instead of water (H_2O) is used as the reducing agent for carbon dioxide. Thus, photosynthetic bacteria evolve sulfur rather than oxygen.

A second group of autotrophic bacteria **(chemoautotrophs)** synthesizes organic materials from carbon dioxide, ammonia, or nitrate by using energy from the oxidation of inorganic substances. For example, one species of bacterium in the soil oxidizes ammonia to nitrate and, in the process, generates useful energy. Chemoautotrophic bacteria are magnificent "factories" for making protoplasm.

Other bacteria are used in the commercial production of bakery goods, alcohol, vinegar, chemicals, enzymes, antibiotics, and many other products. These microorganisms are also responsible for food spoilage, food poisoning, and many animal diseases, including tuberculosis, scarlet fever, pneumonia, and diphtheria.

Today, some bacteria are used in the new field of recombinant biotechnology. For example, the human gene for insulin has been inserted into a bacterial chromosome where it is then activated to produce human insulin. This not only results in production of insulin at lower cost but greatly reduces the possibility of an allergic reaction to insulin that is isolated from the pancreas of other animals. Bacteria thus directly or indirectly affect the survival of humanity.

Most bacterial species exist as single-celled forms, but some form colonies or filaments of loosely joined cells. There are three basic shapes of bacteria (Fig. 14-1): spherical (coccus), rod-shaped (bacillus), and spiral (spirillum). For many years, bacteria were thought to reproduce only by an asexual process called *fission*, during which the cell pinches in two. It is now known that bacterial cells also exchange genetic material through sexual reproduction. Research has revealed that sexual reproduction is actually widespread among bacteria.

Note: Before beginning this study, become familiar with the aseptic procedures described in Appendix G. These are essential when working with microorganisms to prevent contamination of you, your partners or the environment.

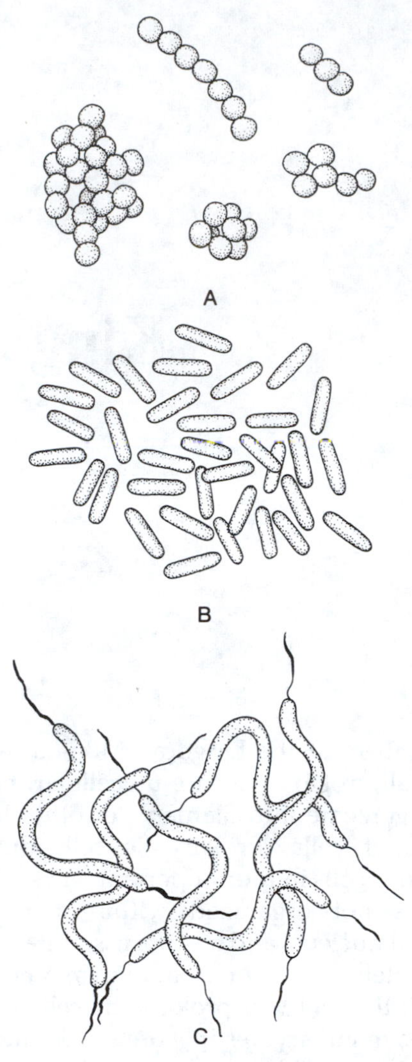

FIGURE 14-1 Shapes of bacteria. (A) Spherical (coccus). (B) Rod-shaped (bacillus). (C) Spiral (spirillum).

1. Bacterial Staining

Living bacteria are almost colorless and do not show enough contrast with the water in which they are suspended to be seen clearly under the microscope. For this reason, they are usually stained to increase their contrast with their surroundings and make them more visible.

Stains are dyes that are usually salts, which consist of negatively and positively changed ions. One of the ions is colored and is termed the **chromophore.** For example, methylene blue chloride dissociates as follows:

$$\text{methylene blue chloride} \rightarrow \text{methylene blue}^+ + \text{chloride}^-$$

The chromophore of this simple stain is the positively charged methylene blue ion. Thus, methylene blue is a **basic dye** or **stain,** which reacts with negatively charged components of cells such as nucleic acids and some polysaccharides. Basic stains are sometimes referred to as **nuclear stains** because they stain nuclei in those cells that have them. Stains in which the chromophore is the negative ion are **acidic stains.** They react with positively charged constituents of cells (i.e., many proteins), thus staining only cytoplasm. They are therefore referred to as **cytoplasmic stains.**

Although there are numerous and varied techniques for staining bacteria, you will use those that involve simple stains, differential stains and one that can determine the nature of materials that are stored in the bacterial cell.

a. Preparation of Slides

Clean, grease-free slides are required to obtain good preparations. Two methods for cleaning slides follow; use either one.

- *With alcohol:* Wipe both surfaces of a slide with alcohol. Allow the alcohol to dry and then, using a forceps to hold the slide, pass the slide through the flame of an alcohol lamp or bunsen burner to flame off residual alcohol. Allow the hot slide to cool. *Note:* Always hold a clean slide by its edges. Grease or oil from your fingers causes water to form into drops, which interfere with the even spreading of bacteria on the slide and can interfere with the adherence of bacteria to the slide in the heat fixing process.
- *With cleanser:* Moisten the tip of your finger and then rub it over a bar of cleanser such as Lava, or Fels Naptha. Rub the paste that forms over both surfaces of the slide. Allow the paste to air-dry, then remove it with a clean paper towel.

b. Fixing Bacteria to Slides

Before you can stain bacteria they must be smeared on the slide and fixed—that is, made to stick to the surface. Otherwise, the cells will wash away during the staining procedure. You will use the method for fixing bacteria shown in Fig. 14-2 on bacteria from a broth (liquid) culture, but the same procedure can be used for bacterial cultures grown on agar.

Obtain cultures from your instructor and fix smears of bacteria to three or four slides as shown in Fig. 14-2. Label the slides with the names of the bacteria used.

c. Positive (Basic) Staining

For this procedure, you will use methylene blue to stain a bacterial preparation.

1. Place the slides you fixed earlier onto a wire screen or other support in your staining tray (Fig. 14-3A).

2. Apply methylene blue stain a drop at a time until the smear is fully covered.

3. Stain for 60 seconds. (*Note:* If you notice that the stain is drying on the slide, add more stain drop by drop. *Do not allow the stain to dry!*)

4. Gently and *briefly* wash the stain from the slide with distilled water (Fig. 14-3B).

5. Remove excess water from the slide by touching one corner to a paper towel or other absorbent paper, then blot the slide dry between pieces of the paper towel (Fig. 14-3C).

6. Examine the preparations with the oil-immersion objective lens. (*Note:* Your instructor will demonstrate the use of this lens.) It will be evident if you have made your smear too thick (that is, transferred too many bacteria to the slide or not spread them out enough). If you prepared the smear properly, you will be able to observe numerous individual bacterial cells.

d. Negative (Acidic) Staining

An acidic dye, such as nigrosin (or sodium eosinate, acid fuchsin, or Congo red) is used in this procedure. The chromophore in these dyes is the negatively charged ion. Because most bacterial cells carry a net negative charge, the stain is not incorporated into the cells but forms a deposit around them. Thus the cells appear as clear, colorless objects against a dark background. Under these conditions, it is relatively easy to observe the cell size and shape.

1. Transfer a loopful of *Escherichia coli* (*E. coli*), or other bacteria provided by your instructor to a clean slide. *Do not smear!*

2. Place a drop of nigrosin solution next to the bacterial suspension and mix them together, but do not smear them (Fig. 14-4A).

3. Following the procedure shown in Fig. 14-4B, spread the suspension smoothly over the slide.

4. Air-dry the smear; *do not heat fix.*

5. Examine the slide microscopically. If the smear was properly made, you should see a graduated thickness of the smear from one end of the

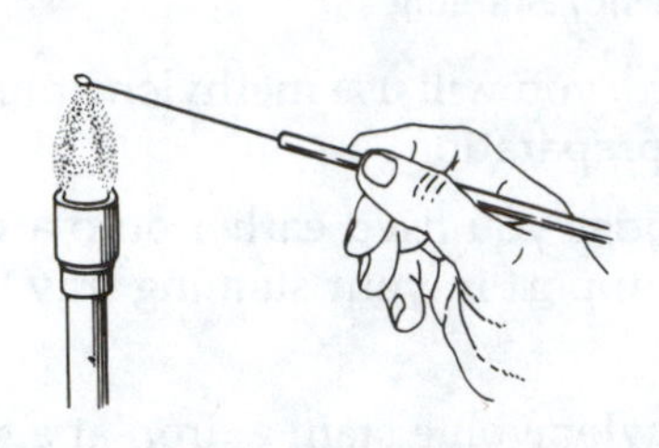

A. Insert inoculating loop into upper cone of flame. Heat entire wire to redness.

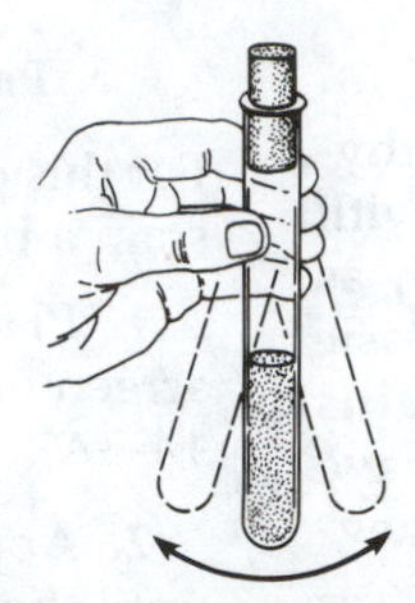

B. While holding the loop, pick up culture tube with free hand. Shake tube gently from side to side.

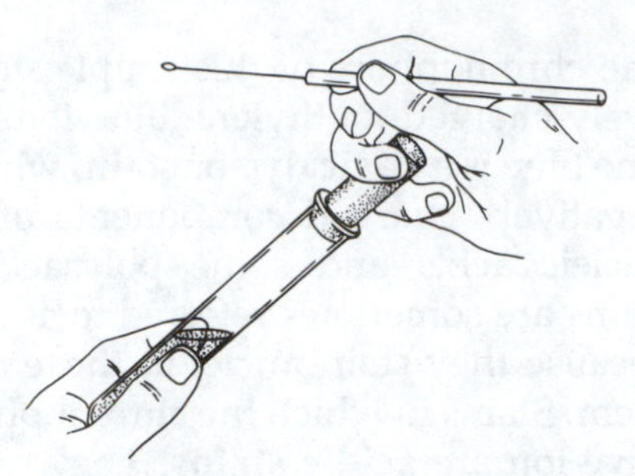

C. Remove cap or plug from tube with free fingers of the hand holding the loop. Hold cap in your fingers. Do not put cap down.

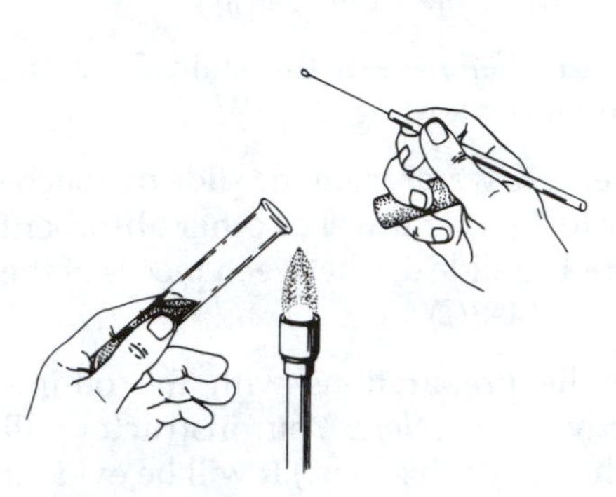

D. Quickly pass top of tube through flame.

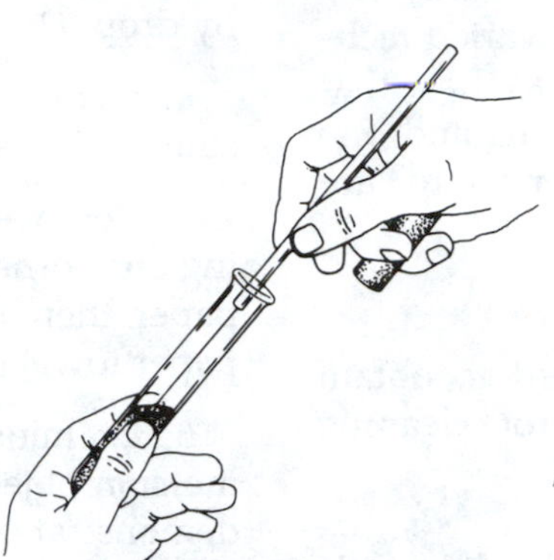

E. Insert sterilized loop into culture and remove small amount of culture. *Note:* Because the loop is hot, you may hear a sizzling noise. Do not be disturbed by this.

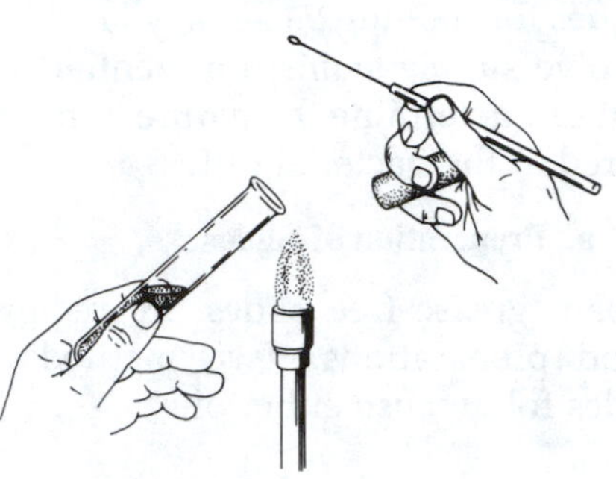

F. Reflame the tube and replace the cap or plug.

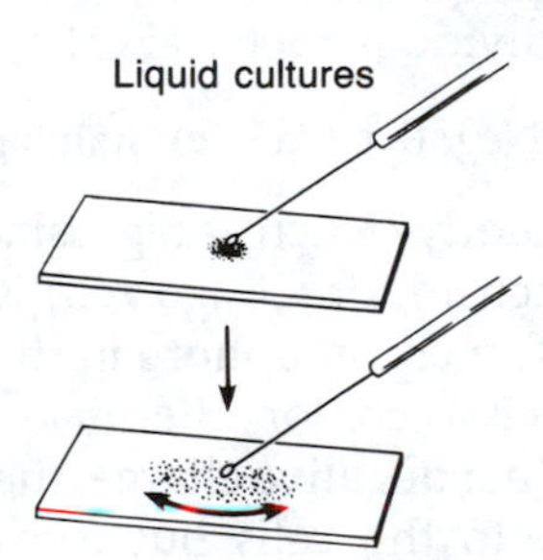

G. If you are transferring from a liquid culture, place the loopful of culture on the slide and spread it over a small area. Allow this smear to air-dry.

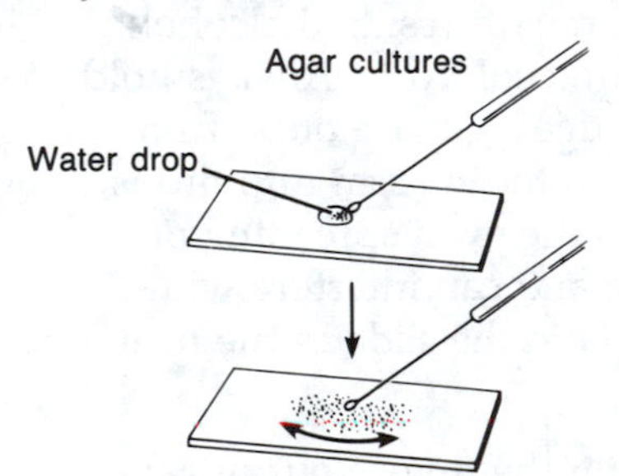

H. If you are using an agar culture, place a drop of water on a slide and mix a small amount of culture into it. Spread the culture over a small area and air-dry.

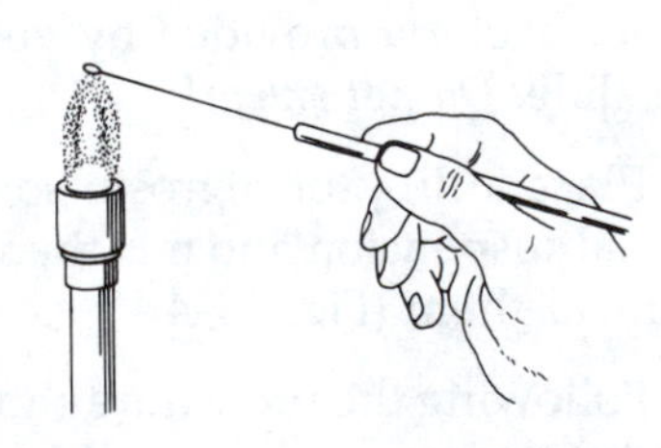

I. Reflame the inoculating loop to redness and put away or down on table.

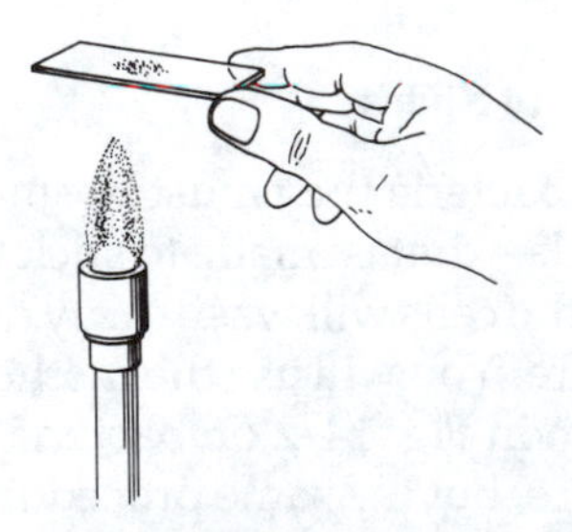

J. Pass the slide (smear uppermost) through the flame several times quickly. Do not overheat.

FIGURE 14-2 Preparation of bacterial smears

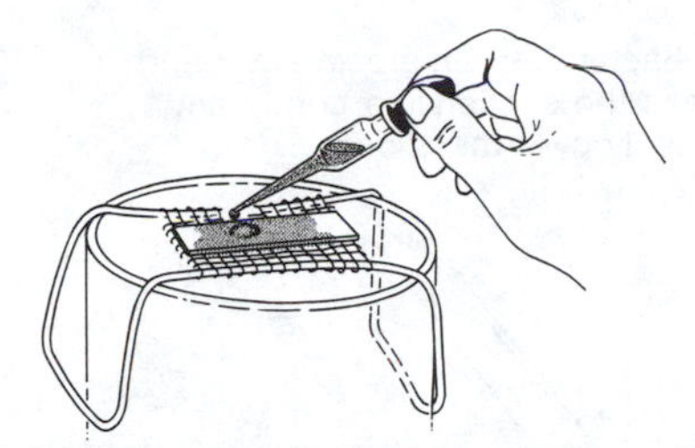

A. Add the methylene blue stain dropwise until the entire smear is covered. Stain for 60 seconds.

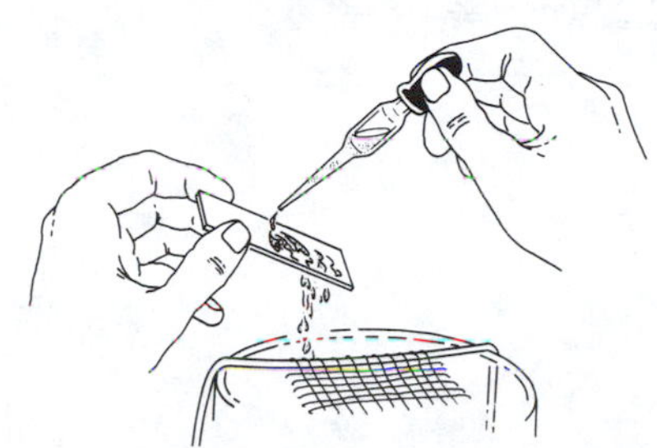

B. Wash off the stain.

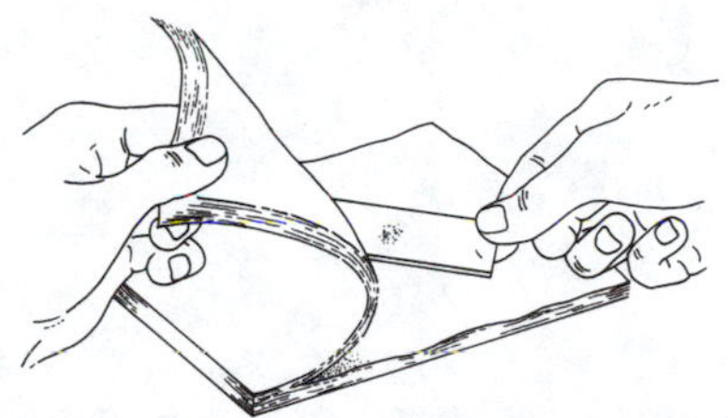

C. Remove excess water by touching one end of the slide to absorbent paper, and then dry the slide between pieces of the paper.

FIGURE 14-3 Procedure for staining bacterial smears

slide to the other. It will probably be too thick at one end, so you will not see any detail (a--a in Fig. 14-4E). At the other extreme, it will be too thin, so there will be no contrast. However, at some point, the smear will be sufficiently thin (or thick) to provide good contrast and individual cells will be outlined against the darker background provided by the stain (b--b in Fig. 14-4E).

e. The Gram Stain

This staining procedure is one of the most widely used in microbiology. It is a **differential stain,** which means that it can chemically distinguish between two species of bacteria that may be morphologically indistinguishable. The **gram stain reactions** enable us to classify bacteria into two major groups. Those that retain the crystal violet stain throughout the procedure and appear blue to violet are **gram-positive.** Those that lose crystal violet after rinsing with alcohol and appear pink to red when subsequently counterstained with safranine are **gram-negative.** This difference in the reaction to the gram stain is due in part to differences in the structure and composition of the cell walls of gram-positive and gram-negative bacterial cells.

The gram-stain procedure requires four solutions:

Crystal violet, a basic dye

Iodine, which acts as a **mordant**—that is, a substance that increases the affinity of cells for the crystal violet dye

Alcohol (or acetone), which removes the dye from the cell **(decolorization)**

Safranine, a basic dye of a different color that is used to **counterstain** the cell if the crystal violet stain is removed by the decolorizing solution

1. Prepare smears of *E. coli* and *Bacillus subtilis* (*B. subtilis*) on a clean microscope slide as shown in Fig. 14-5A. Make certain that the smears are separate from each other.

2. Air-dry and heat fix the smears.

3. Place the slide on wire mesh or a staining rack and flood it with crystal violet for 30 seconds (Fig. 14-5B). *Do not allow the stain to dry on the slide. Add extra stain if necessary.*

4. Pour off the excess stain and *gently* wash the slide with tap water (Fig. 14-5C).

5. Flood the slide with iodine solution and leave on for 1 minute (Fig. 14-5D).

6. Wash off the iodine solution with tap water and blot the slide *gently* between pieces of paper toweling. (Fig. 14-5E)

7. Holding the slide at an angle, apply 95% alcohol a drop at a time to all smears until the violet color no longer appears in the runoff (Fig. 14-5F). *Do not continue to rinse with alcohol when no more stain is seen in the rinse.*

8. Quickly rinse off the alcohol with tap water and blot the slide dry.

9. Counterstain the smears by flooding the slide with safranine for 30 seconds.

10. Gently wash off the stain with tap water. Drain off the excess water, blot the slide between paper toweling and then allow the slide to air-dry.

11. Examine the center smear with oil immersion. This will enable you to contrast gram-negative and gram-positive bacteria. Then examine the

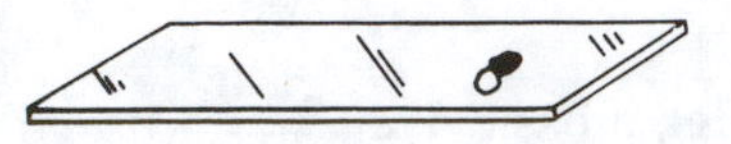

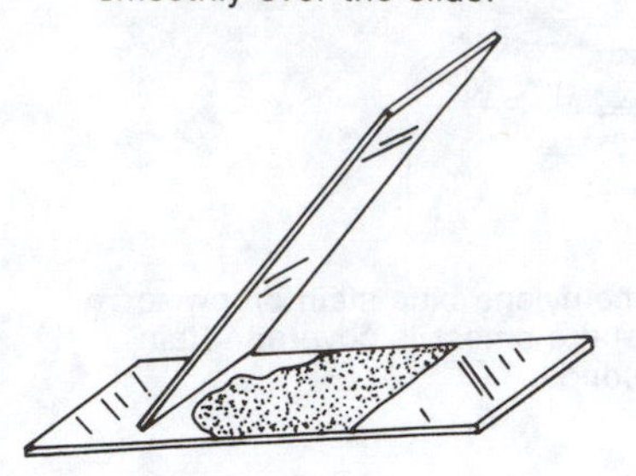

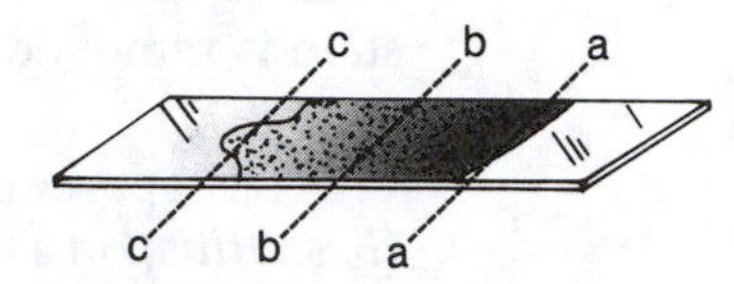

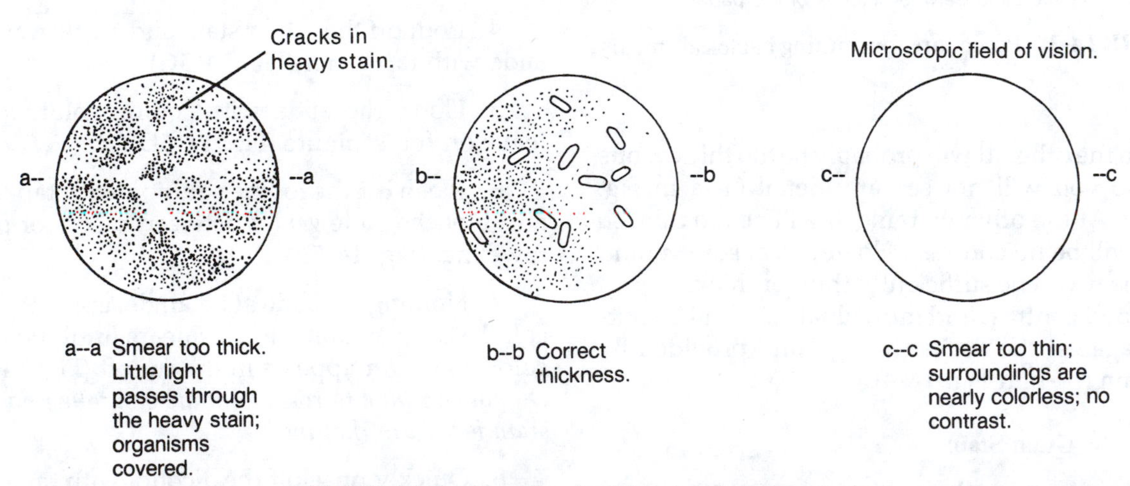

FIGURE 14-4 Negative (acidic) staining of bacteria

other two smears. Is *E. coli* gram-negative or gram-positive?

___

Is *B. subtilis* gram-negative or gram-positive?

___

12. If available, examine other bacterial species and determine which are gram-negative and which are gram-positive.

2. Control of Bacterial Growth

In 1929, the British biologist Alexander Fleming found that a petri dish containing a bacterial cul-

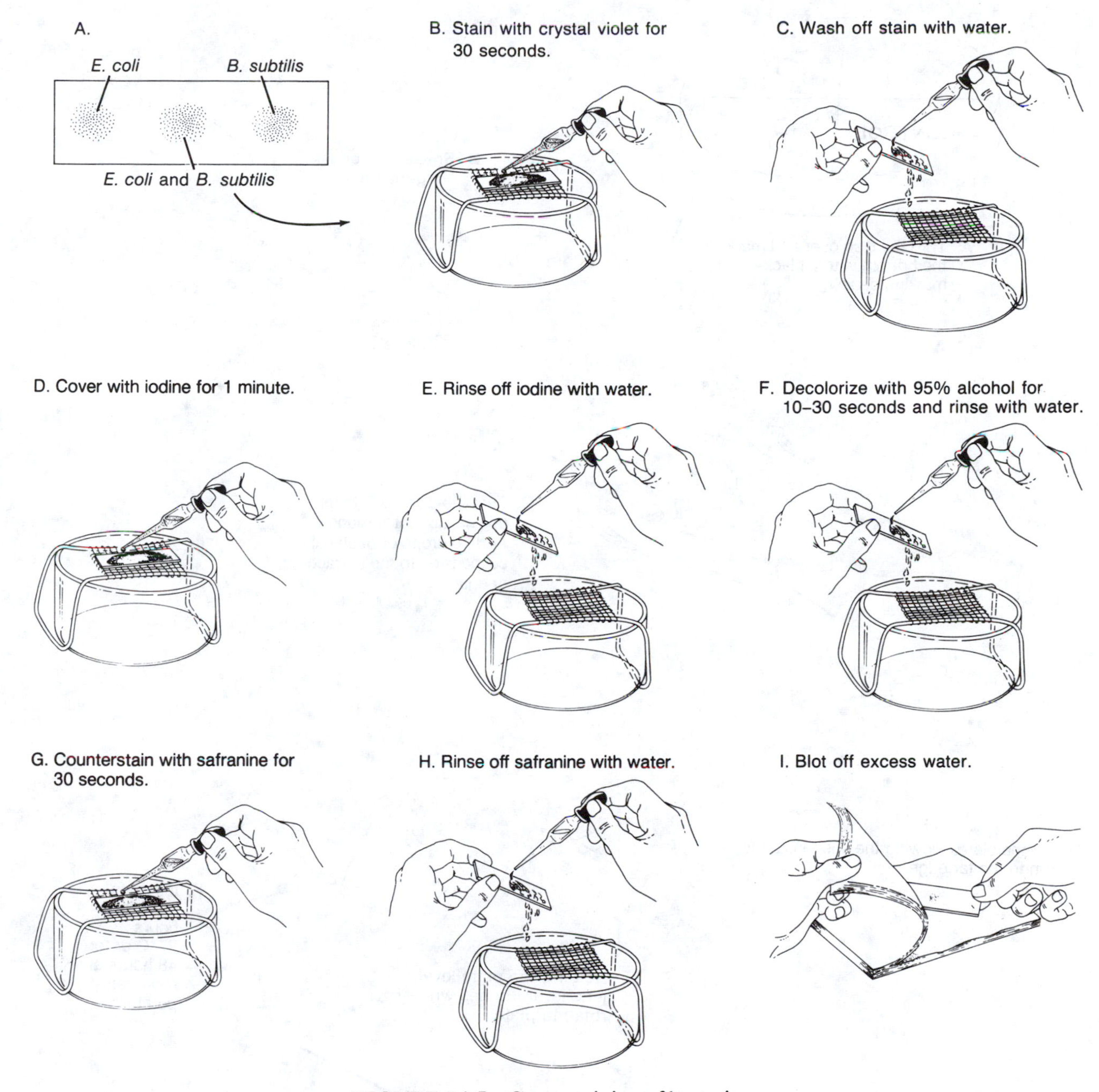

FIGURE 14-5 Gram staining of bacteria

ture had become contaminated by a mold called *Penicillium.* Noticing that the growth of the bacteria around the mold colony was inhibited, he thought that the mold produced a diffusable chemical agent capable of inhibiting bacterial growth. Fleming later isolated this antibacterial chemical and called it *penicillin.* Chemicals that have the ability to retard the growth of certain other organisms are called **antibiotics.** Among the large variety of such antibiotics are streptomycin, chloromycetin, neomycin, novobiocin, erythromycin, kanamycin, penicillin, and tetracycline. The use of antibiotics has significantly reduced the occurrence and virulence (the ability to cause disease) of many bacteria.

In this part of the exercise, you will examine the effect of various antibiotics on the growth of one of three common species of bacteria.

1. Obtain a petri dish containing nutrient agar, a mixture of chemicals that is optimal for the growth of the bacterium you are to study. Divide the plate

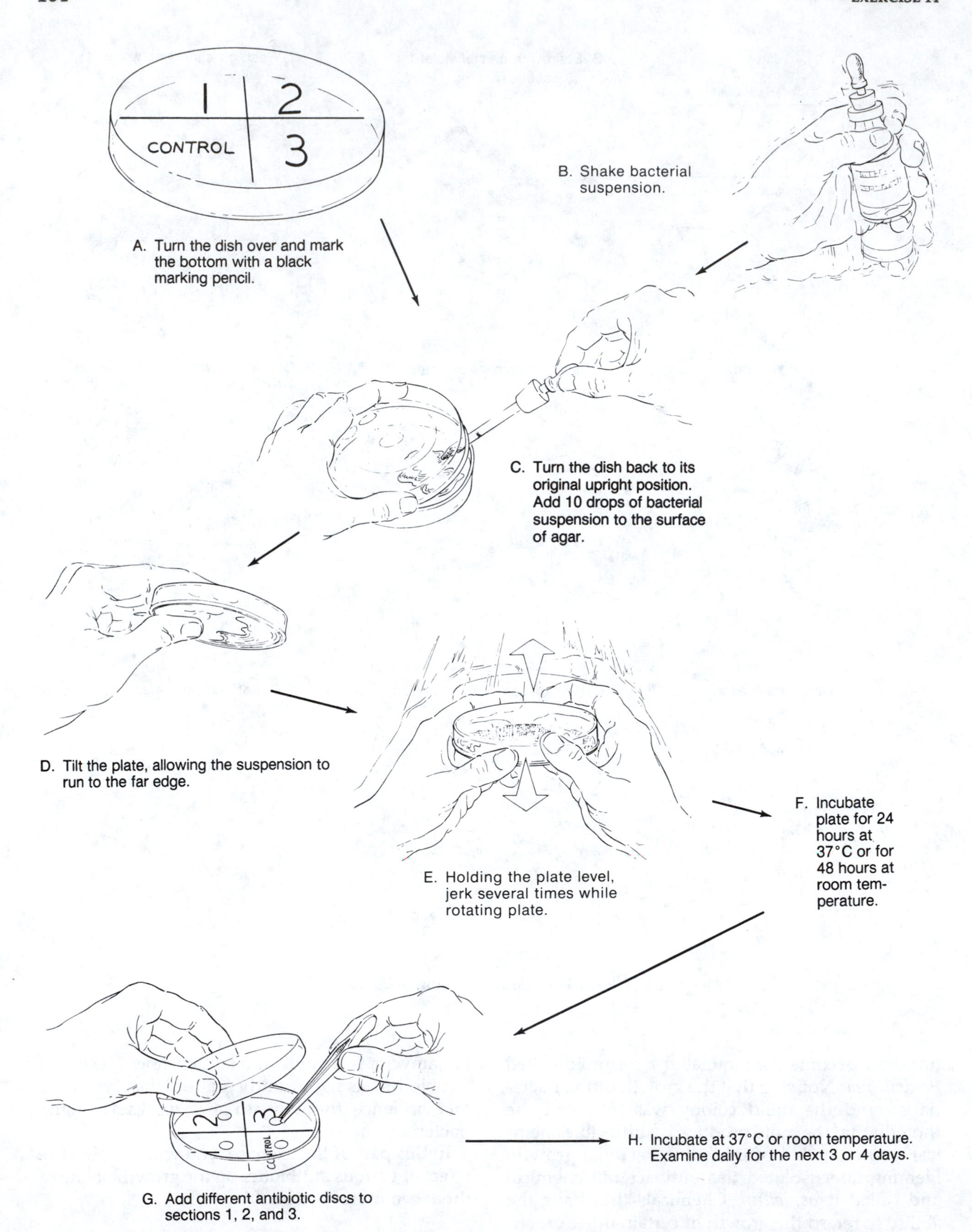

FIGURE 14-6 Procedure for determining the effect of antibiotics on bacterial growth

into four sectors labeled 1, 2, 3, and Control by marking the bottom of the dish with a black marking pencil (Fig. 14-6A).

2. Lift the cover of the petri dish slightly and add ten drops of the bacterial suspension (provided by your instructor) to the plate (Fig. 14-6B, C).

One of the following three bacterial cultures will be provided by your instructor for use in this study:

Escherichia coli (*E. coli*)

Bacillus subtilis (*B. subtilis*)

Serratia marcescens (*S. marcescens*)

Your instructor may provide alternative or additional cultures.

3. Hold the plate and tilt it to let the suspension run to the far edge (Fig. 14-6D). Then, with the plate level, jerk it toward you. Repeat this several times, rotating the plate a little between jerks (Fig. 14-6E) until the bacterial suspension covers the entire surface of the agar.

4. Incubate the plate for 24 hours at 37°C (or 48 hours at room temperature) before proceeding with the next step. Why might incubation for a period of time be desirable before adding the antibiotics?

____

____

____

____

5. After incubation, partially lift the cover and, using flamed (but cooled) forceps, place a disc with a different antibiotic in each of the numbered sectors of the plate (Fig. 14-6G). The following commercially available antibiotic discs can be used in this study:

Chloromycetin	Novobiocin
Erythromycin	Penicillin
Kanamycin	Streptomycin
Neomycin	Tetracycline

What should be placed in the Control section of the plate?

____

6. Incubate the plate at 37°C or at room temperature as before. Examine the plate daily for the next 3 or 4 days. Record your observations in Fig. 14-7 and Table 14-1 using the following symbols:

R (for resistant) if there is no clear zone of inhibition around the colony and growth goes up to the disc

HS (highly sensitive) if there is a distinct zone of no growth around the disc (no matter what size the zone is)

S (sensitive) if there is an inhibited zone, but some colonies appear to have grown back into the zone

Some bacteria have an enzyme called *penicillinase.* Would the growth of such bacteria be inhibited by penicillin? Explain.

____

____

____

____

____

____

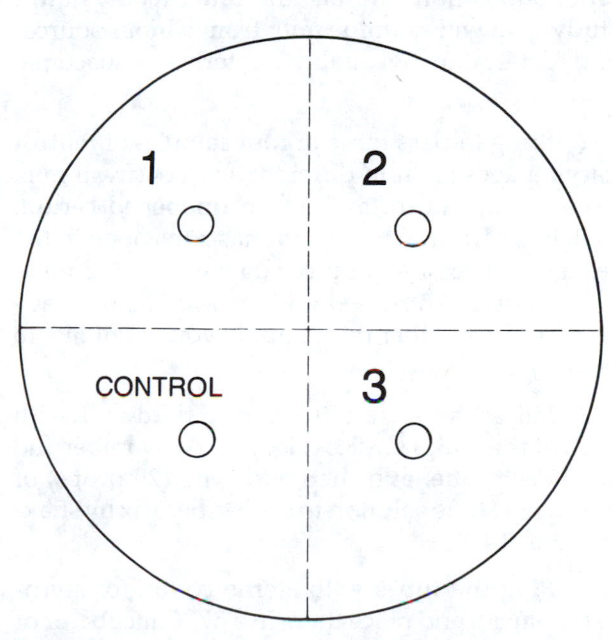

FIGURE 14-7 Effect of antibiotics on bacterial growth. Use R for resistant, HS for highly sensitive, and S for sensitive.

TABLE 14-1 Bacterial sensitivity to antibiotics. Use R for resistant, HS for highly sensitive, and S for sensitive.

Antibiotic	Bacterium			
	Escherichia coli	*Bacillus subtilis*	*Serratia marcescens*	Other
Chloromycetin				
Erythromycin				
Kanamycin				
Neomycin				
Novobiocin				
Penicillin				
Streptomycin				
Tetracycline				

3. Bacteria in Milk

Because milk is an excellent medium for the growth of bacteria, great care must be used in its processing. Harmful bacteria in milk are killed by **pasteurization,** a process that uses low heat to reduce bacterial populations in milk and other foods. In this study, you will examine milk from various sources and determine its quality in terms of bacterial growth.

1. Bring to class three or four samples of milk of varying ages or from different sources (fresh milk from a dairy farm, milk in an unopened carton, milk in a carton or bottle that has been open in the refrigerator for 1–3 or more days, out-dated milk, powdered milk, canned milk, raw cow's or goat's milk, and any other milk sample you might like to test).

2. Fill separate test tubes one-third full with each of the milk samples (Fig. 14-8A). Number and label each tube, and then add 1 ml (20 drops) of methylene blue solution to each tube. Mix by shaking (Fig. 14-8B).

3. Plug the tubes with sterile cotton (or styrofoam plugs), and place them in a 37°C incubator or water bath at 37°C (Fig. 14-8C). Record the time in Table 14-2.

When bacteria are actively growing in milk, they consume oxygen. You can detect the reduction in oxygen content by using the fact that methylene blue loses color as the oxygen content of the milk diminishes. If the number of bacteria in the milk is high, the mixture of methylene blue and milk will lose its color rapidly. If the bacterial population is low, more time will be required for decolorization.

For the purposes of this study, the quality of the milk with respect to the number of bacteria present can be rated by the time it takes to decolorize the methylene-blue/milk solution. Periodically examine the milk samples. In Table 14-2, record the time it takes to decolorize each sample. Rate each sample according to the description given in Table 14-3. Which sample rated the highest in bacterial contamination?

____

____

Which rated the lowest?

____

From the results, what appears to lead to the contamination of milk?

____

____

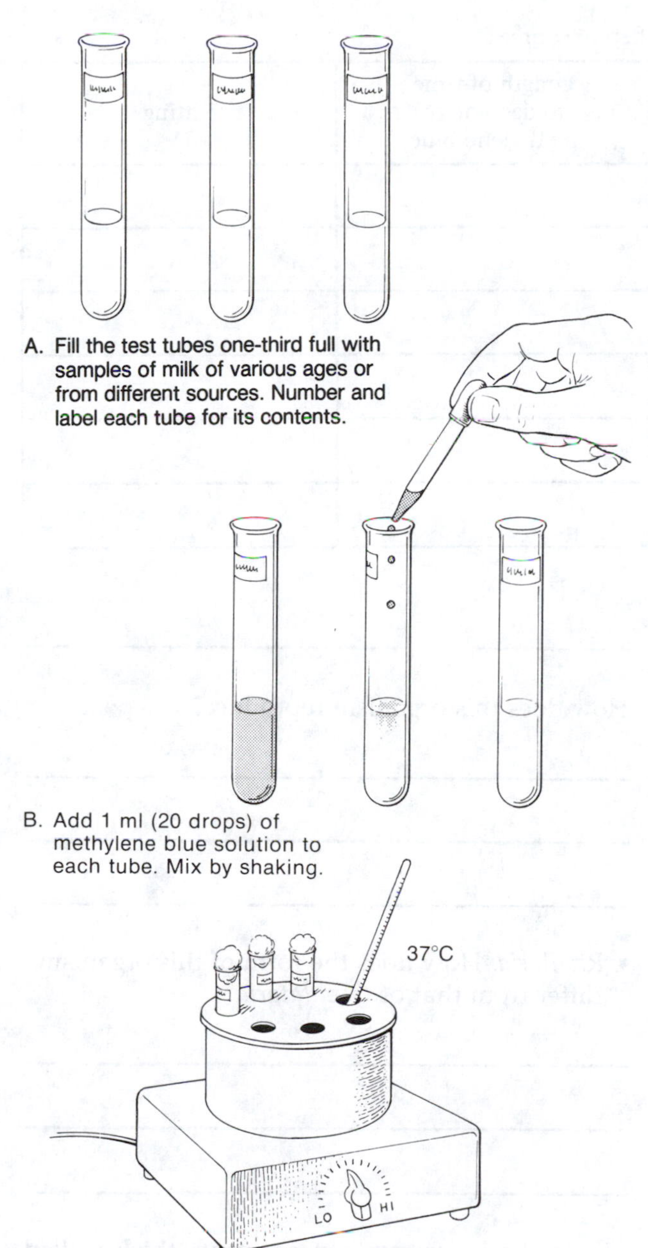

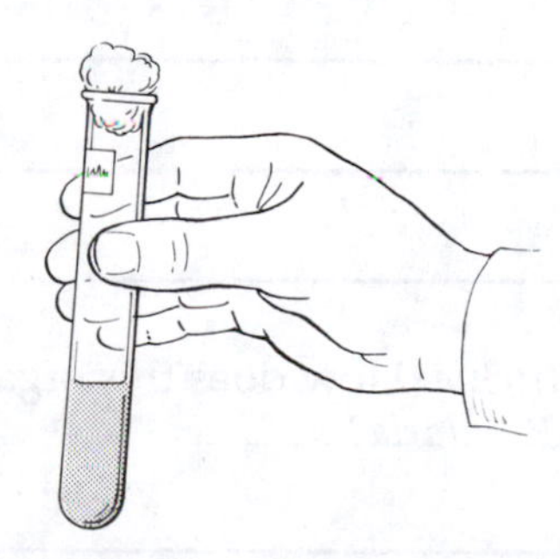

D. Examine each tube periodically. Rate the quality of the milk according to its bacterial population as described in Table 14-3.

FIGURE 14-8 Procedure for estimating bacterial contamination of milk

What appears to reduce contamination?

____

____

B. DIVISION CYANOBACTERIA

The group of organisms known as **cyanobacteria,** formerly called the blue-green algae, contains unicellular and multicellular forms. The unicellular form is considered to be more primitive. In addition, cyanobacteria—though considered to be prokaryotic organisms because they lack true nuclei—use water in photosynthesis as the reducing agent for carbon dioxide and, therefore, evolve oxygen as a byproduct. In this respect, they are biochemically similar to photosynthetic eukaryotic organisms such as algae and plants.

Cyanobacteria contain chlorophyll and accessory pigments known as **phycobilins:** phycocyanin, a blue pigment that is always present, and phycoerythrin, a red pigment that is often present.

In this study, you will become familiar with the cellular morphology and range of complexity of cyanobacteria.

1. Unicellular Forms

Prepare wet mounts of *Chroococcus* and *Gloeocapsa*, two common cyanobacteria. To see the characteristic gelatinous sheath that surrounds the cells of these organisms, add a drop of India ink to the slide. The sheath will stand out against the dark background. Although these are unicellular forms, they frequently form small clusters of cells. Do clustered cells share a common gelatinous sheath?

____

____

Would you consider these clusters to represent multicellular organisms? Explain.

____

____

____

TABLE 14-2 Rating milk samples of different ages or different sources

Contents of tube	Time methylene blue added	Length of time to decolorize methylene blue	Rating
1			
2			
3			
4			

2. Multicellular (Colonial) Forms

The shapes of colonies of cyanobacteria are largely determined by the planes of cell division, which may result in filaments, flat sheets one cell thick, or balls of cells. Examine the following cyanobacteria and classify each as to body shape (e.g., filamentous) and as to the plane of cell division (single, double, or irregular) that resulted in the shape of the colony.

- *Merismopedia.* Locate the gelatinous sheath. Cells in the colony frequently are seen to be dividing.
- *Oscillatoria.* What evidence is there of cellular differentiation in this colony?

Describe any motility you observe.

TABLE 14-3 Decolorization times for milk samples

Time to decolorize methylene blue	Rating
Less than 20 minutes	Highly contaminated
20 minutes to 2 hours	Poor
2–5½ hours	Fair
5½–8 hours	Good
More than 8 hours	Excellent

How does this organism reproduce?

- *Rivularia.* How does the form of this organism differ from that of *Oscillatoria*?

Locate the heterocyst, an enlarged, thick-walled cell. What is its function?

- *Gloeotrichia.* How does this organism differ from *Rivularia*?

- *Anabaena.* Crush the cells of the water fern *Azolla* to release the cyanobacterium, which exists symbiotically inside of this plant. What function might this organism carry out in the cells of *Azolla*?

____

____

____

REFERENCES

Bold, H. C., and C. L. Hundell. 1987. *The Plant Kingdom.* 5th ed. Prentice Hall.

Brock, T. and M. T. Madigan. 1988. *Biology of Microorganisms.* 5th ed. Prentice Hall.

Delevoryas, T. 1977. *Plant Classification.* 2d ed. Holt, Rinehart and Winston.

Pelczar, M. J., E. C. S. Chan, and N. R. Krieg. 1986. *Microbiology.* 5th ed. McGraw-Hill.

Stanier, R. Y., et al. 1986. *The Microbial World.* 5th ed. Prentice Hall.

Walsby, A. E. 1977. The Gas Vacuoles of Blue-Green Algae. *Scientific American* 237 (2):90–97 (Offprint 1367). *Scientific American* offprints are available from W. H. Freeman and Company, 41 Madison Avenue, New York, NY 10010, and 20 Beaumont Street, Oxford OX1 2NQ, England. Please order by number.

Exercise

15

Kingdom Protista I: Algae and Slime Molds

All protists are eukaryotic organisms and all evolved from the Monera. Although they are often called simple or primitive, they possess a tremendous degree of specialization at the cellular level. This has enabled them to adapt to a wide range of habitats. The protists include a number of parasitic heterotrophs, photosynthetic autotrophs (photoautotrophs), and a few species that are both photosynthetic and heterotrophic. **Autotrophs** (Greek *autos,* "self;" *trophe,* "nourishment") are organisms that are able to synthesize all needed organic molecules from simple inorganic substances and sunlight by means of photosynthesis. **Heterotrophs** (Greek *heteros,* "other;" *trophe,* "nourishment") obtain their nourishment from organic materials produced by other organisms. These organisms are found in brackish, fresh, and marine waters and exhibit a variety of symbiotic and parasitic relationships.

This kingdom has a number of divisions, but only the following algae and slime molds will be studied in this exercise:

- Division Euglenophyta (**euglenoids**): Mostly autotrophic plantlike organisms; some are heterotrophs. Food is stored as **paramylon** (a polysaccharide) and fat.
- Division Chrysophyta (diatoms, golden-brown and yellow-green algae): Unicellular algae with plastids containing gold-yellow or yellow-green pigments; most are unicellular but some are filamentous. Food is stored as a carbohydrate called **chrysolaminarin.**
- Division Chlorophyta (green algae): Unicellular, colonial, or multicellular organisms having chlorophylls *a* and *b* and various carotenoids. Food is stored as starch. Motile cells have flagella.
- Division Phaeophyta (brown algae): Multicellular marine organisms having chlorophylls *a* and *c* and the pigment fucoxanthin. Food is stored as a carbohydrate called **laminarin.** Motile cells have two flagella, one anterior, the other trailing. Multicellular forms, unlike green algae, show differentiation of the body structure. Some have specialized conducting cells.
- Division Rhodophyta (red algae): Mostly marine organisms having chlorophyll *a* and *d* and the phycobilins phycoerythrin and phycocyanin. Food is stored as a carbohydrate called **floridean starch.** No motile cells in life cycle. Lack specialized conducting cells found in brown algae.
- Division Myxomycota (plasmodial slime molds): Heterotrophic amoeboid organisms, most of which lack cell walls but form sporangia at some stage in their life cycles.

A. ALGAE

More than 20,000 species of algae are grouped in several divisions. Members of these divisions (with few exceptions) are photosynthetic. They have a simple body structure that may consist of a single cell, filaments of cells, or plates of cells or a structure somewhat comparable to that of some land plants. They do not have the complex organization of tissues found in the vascular plants.

Algae differ from each other in the type of flagella (if they produce motile cells) and in several biochemical characteristics. Although all algae contain chlorophyll, a variety of carotenoids is also distributed in the various groups. Indeed, the names of the divisions of algae are derived from the various pigments that mask the green color imparted by the chlorophylls. Great diversity also exists in the type of food storage products in each of the divisions.

Reproduction is accomplished asexually by fragmentation of the parent body or by production of spores that develop into new individuals. In addition, new individuals may arise sexually as a result of the union of two gametes. The zygote that is formed as a result of this union may develop directly into another alga or may produce spores.

1. Division Euglenophyta (Euglenoids)

The euglenoids, during their long course of evolution, have acquired chloroplasts with biochemical properties similar to those of the green algae. The division takes its name from one of its members, *Euglena*, which is a photoautotroph. *Euglena* is a green, unicellular organism commonly found in the surface scum of standing or very slowly moving waters.

Place a drop of culture of living *Euglena* on a slide. Add a drop of 10% methyl cellulose, which will slow the rapid movements of the organism. Add a coverslip and examine the slide microscopically. Observe the method of locomotion of an active specimen. The organism moves by means of a long **flagellum,** which pulls the organism through the water. Does the beating of the flagellum move from the base toward the tip or in the opposite direction?

Euglena also exhibits a wormlike movement during which a series of contractions pass along the body. Because this type of movement is unique to *Euglena*, it has been termed **euglenoid movement.**

Using high power and with the help of Fig. 15-1A, locate the following morphological features of *Euglena*. You can supplement your observations by studying a commercially prepared slide of this organism. Identify the **pellicle,** a thin, elastic membrane that may be striated because of spiral thickenings; the **ectoplasm,** peripheral cytoplasm; the **endoplasm,** denser, internal cytoplasm; the **chloroplasts,** organelles containing the chlorophylls; and the **nucleus,** which is centrally located. More easily seen in prepared slides is a dense staining body in the nucleus, the **nucleolus.** The **cytostome** (cell "mouth") is a funnel-shaped depression near the anterior end that leads to the **cytopharynx** (cell "gullet"), which is enlarged at the base to form the **reservoir.** Adjacent to the reservoir is a water-expulsion vesicle (**contractile vacuole**), which periodically collects excess cell water and discharges it into the reservoir and then out through the cytopharynx. A reddish-orange photoreceptor, located near the anterior end of the ani-

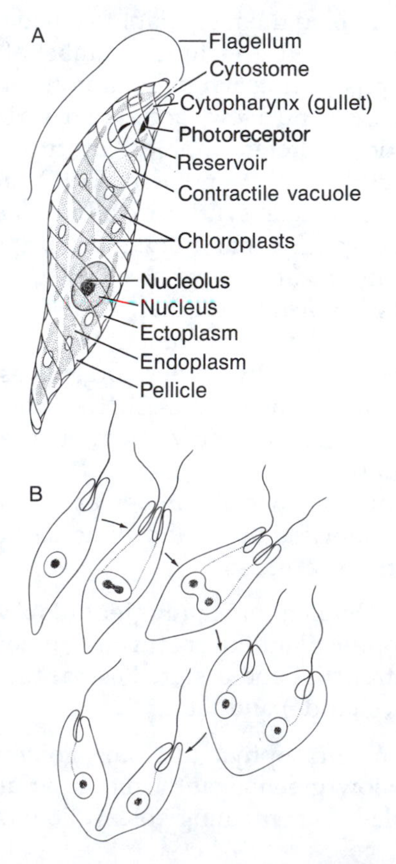

FIGURE 15-1 *Euglena.* (A) Morphology. (B) Binary fission.

mal and adjacent to the cytopharynx, is a light-sensitive organelle. What function might it serve?

When you have finished making your observations of the living organism, add a drop of iodine solution to the edge of the coverslip and let it be drawn under by capillary action. The iodine will stain the flagellum so that it is seen more easily.

Most *Euglena* are autotrophic. Some species of *Euglena*, however, are able to grow and reproduce without any visible chloroplasts or chlorophyll. They exhibit heterotrophic nutrition.

Reproduction in *Euglena* is by **binary fission** (Fig. 15-1B). The nucleus divides by **mitotic division,** and then the anterior organelles (e.g., reservoir, flagella, and photoreceptor) are duplicated. This is followed by a longitudinal splitting of the protist beginning at the anterior end. If they are available, examine demonstration slides showing this type of reproduction.

2. Division Chrysophyta (Diatoms, Golden-Brown and Yellow-Green Algae)

This division includes two major classes: the diatoms, with almost 10,000 species, the golden-brown algae, with about 1500 species, and a minor class, the yellow-green algae, with about 600 species. The chrysophytes are characterized by **plastids,** which have gold-yellow or yellow-green pigments; rigid cell walls, many of which are impregnated with silica; and food storage in the form of oils and a carbohydrate called chrysolaminarin. Fresh water that contains large numbers of chrysophytes may have an unpleasant oily taste, as do the fish caught in such waters.

Diatoms are found in the soil and in fresh and salt water throughout the world. They are a major component of the plankton (small marine organisms that live in large numbers in the upper levels of the oceans) and are an important source of food for marine animals.

Place a drop of a culture of living diatoms on a slide, add a coverslip, and examine microscopically. If living diatoms are unavailable, examine a prepared slide containing a mixture of various diatoms. The most striking feature of diatoms is their beautiful, ornamented cell walls, which are composed of pectin impregnated with silica. Observe the fine markings of the cell wall, which give each species of diatom its distinctive morphologic features. These markings are pores that connect the living protoplasm within the shell to the outside environment.

The walls of diatoms consist of two overlapping portions, called **valves,** that fit together like the top and bottom of a petri dish. The valves are either pennate or centric. **Pennate diatoms** are boat-shaped or rod-shaped and have bilaterally symmetrical patterns of markings. In the center of the valve of most pennate diatoms is a groove or **raphe. Centric diatoms** are circular, oval, or elliptical and have radial symmetry. Examine your slide preparation and identify pennate and centric diatoms.

Diatomaceous earth (diatomite) is found as vast deposits of siliceous shells of diatoms that were deposited over millions of years in former ocean bottoms. It is used as a fine abrasive material in silver polish, some toothbrushes, and for filtering and insulating materials.

3. Division Chlorophyta (Green Algae)

The green algae exhibit the greatest diversity among protists, both in form and in reproductive patterns. Of the approximately 9000 species, most are aquatic, though other habitats include the surface of snow, hot springs, trunks of trees, and (as symbionts) lichens, protozoa, and hydra.

a. The Volvocine Line of Evolution in Green Algae

Two evolutionary trends are evident among the members of this division.

- Members range in complexity of the thallus (structure of "body") from a single cell to a multicellular colonial type. Unicellular forms are considered the most primitive. Colonial forms probably originated as a unicellular alga that achieved complexity through cell divisions in different planes.
- The types of gametes and/or reproductive organs range from **isogamy** (sexual reproduction in which the gametes are morphologically alike), which is considered the most primitive, to **oogamy** (sexual reproduction in which one of the gametes, usually the larger, is not motile), which is considered the most advanced.

In this study, you will examine *Chlamydomonas,* which can be considered the primitive form;

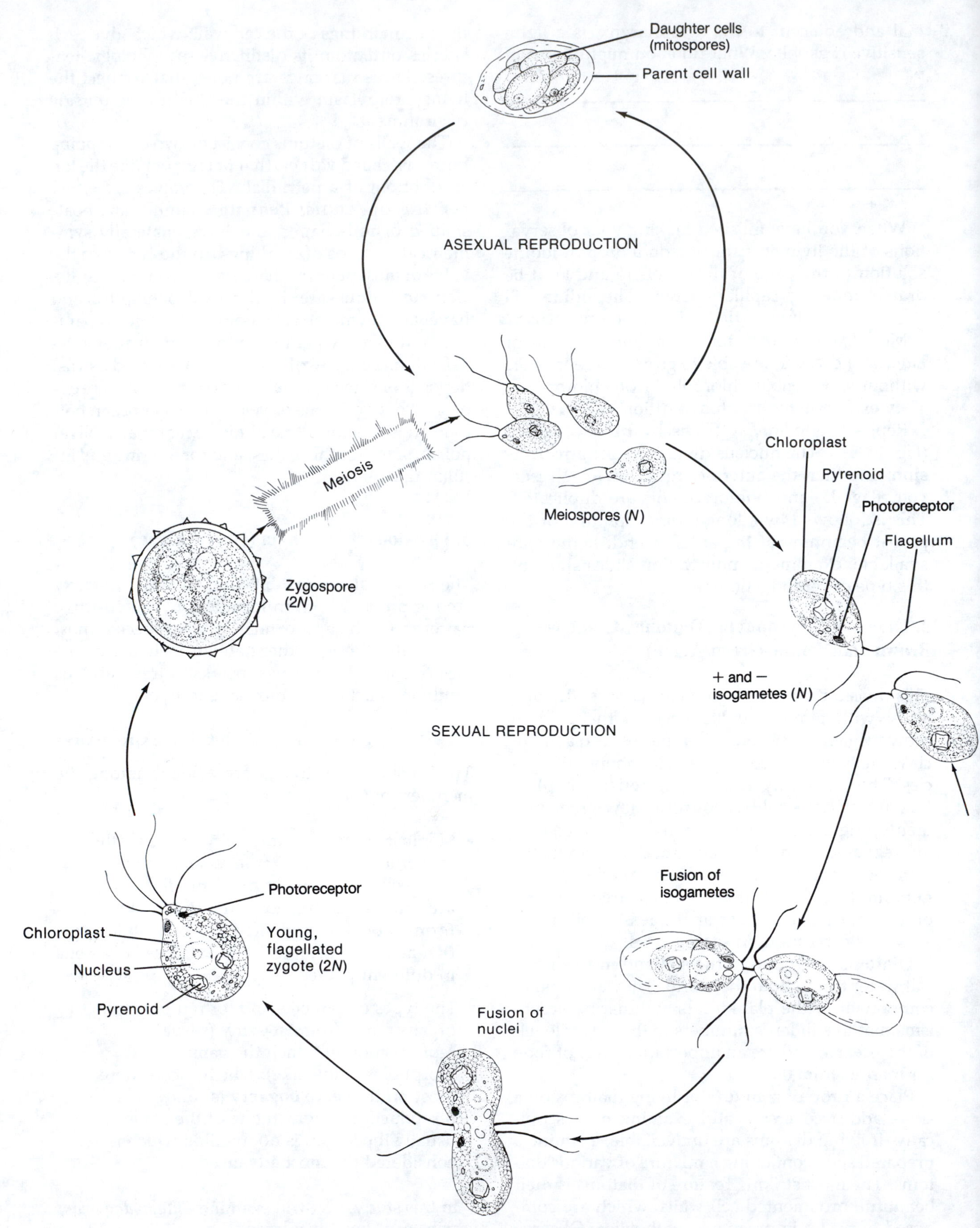

FIGURE 15-2 Life cycle of *Chlamydomonas*

Gonium and *Pandorina*, progressively complex forms; and *Volvox*, the peak of evolutionary complexity. Observe the differences as you examine the organisms.

Chlamydomonas. *Chlamydomonas* is a single-celled, motile alga found in damp soil, lakes, and ditches. It is typically egg-shaped and has a large cup-shaped chloroplast containing a proteinaceous body, the **pyrenoid,** which functions in starch formation. A photoreceptor is located inside the chloroplast. The nucleus is difficult to observe in a living cell.

Prepare a slide of living *Chlamydomonas* and examine it microscopically (Fig. 15-2). Add a drop of methyl cellulose to slow the movement of this alga. Locate the conspicuous chloroplast and the pyrenoid. The flagella, located at the anterior end of the cell, enables the organism to move through water. The flagella can be observed more easily by closing the iris diaphram to reduce the light and provide more contrast.

At the start of **asexual reproduction,** the flagella are retracted and movement ceases. Mitosis takes place during this quiescent period and results in the formation of daughter **protoplasts,** which may divide a second and third time (Fig. 15-2). A cell wall is formed around each daughter protoplast, forming temporary colonies that soon rupture and release individual daughter cells called **mitospores.** Locate nonmotile, vegetative colonies on your slide that show two, four, or eight daughter cells.

Most green algae form haploid gametes in **sexual reproduction.** In *Chlamydomonas*, however, two vegetative cells can function as gametes, one as the male and the other as the female. These gametes are usually identical in size and appearance, though in some species the female gamete may be slightly larger. Gametes that are morphologically indistinguishable are called **isogametes.**

Sexual reproduction in *Chlamydomonas* can be demonstrated with mating strains, which we characterize as plus (+) and minus (−). Place a drop of each mating type next to each other on a slide. Do not mix. Then, while observing them with the stereoscopic microscope, mix the drops together. Note the peculiar clumping phenomenon that precedes union of the gametes. Using a compound microscope, locate cells that have paired. At which end of the cell has union occurred?

As a result of sexual union, a diploid ($2N$) zygote is formed that secretes a thick, spiny wall to form a **zygospore,** which then enters a period of dormancy. Examine the preparation closely and observe the zygospores. Under favorable conditions, the zygote nucleus undergoes meiotic division to form four haploid (N) nuclei. The protoplast then divides and forms four uninucleate **meiospores,** which escape from the old zygospore, develop flagella, and swim away.

Gonium. *Gonium* is a colonial organism made up of *Chlamydomonas*-like cells that become loosely arranged in a flat, platelike colony (Fig. 15-3A). The cells are held together by a gelatinous matrix.

Gonium represents a stage in evolution in which the plant body has become larger by the assembly

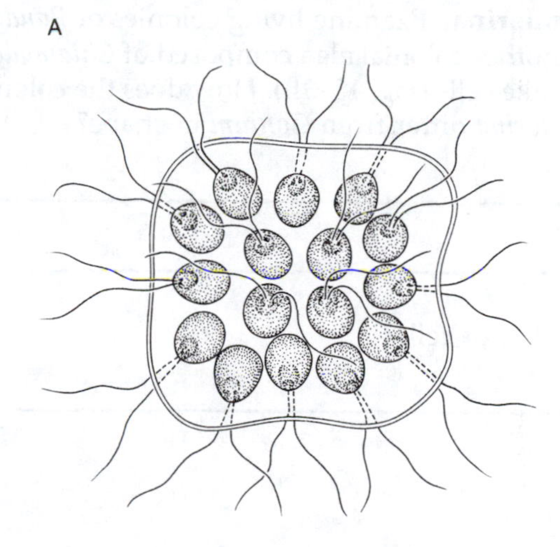

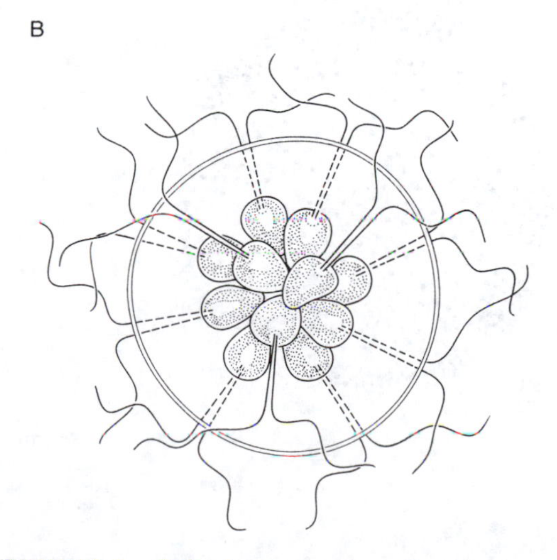

FIGURE 15-3 Colonial green algae. (A) *Gonium.* (B) *Pandorina.*

of a few cells that exhibit simple coordination. The cells in this colony swim in unison, so the entire plate of cells moves as a unit.

Examine living colonies of *Gonium*. Is the number of cells in each colony constant?

____

If not, describe the variability in cell number found in different colonies.

____

____

____

____

Pandorina. Examine living colonies of *Pandorina*, another colonial alga composed of *Chlamydomonas*-like cells (Fig. 15-3B). How does the colony of *Pandorina* differ from *Gonium* in shape?

____

____

In number of cells?

____

In *Pandorina*, as well as in *Gonium*, each cell of the colony, when mature, gives rise asexually to a new colony within the gelatinous envelope of the parent colony.

In the sexual reproduction of these algae, isogametes fuse into a zygote that undergoes meiosis to form meiospores. Each meiospore is capable of developing into a new colony.

Volvox. *Volvox* is a colonial green alga in which the thallus is a hollow, spherical colony of 500 to 50,000 cells bound together in a common matrix (Fig. 15-4A). Individually, the cells exhibit many of the features seen in *Chlamydomonas*; they have a photoreceptor, flagella, and a large chloroplast. In *Volvox*, however, specialized cells function in reproduction.

Asexual reproduction is accomplished by the enlargement and subsequent division of specialized cells in the colony. In early development, these cells form a flat plate, which turns into a sphere with a small pore at the posterior end. The new colony increases in size, and the sphere evaginates through the pore and turns itself inside out to form a daughter colony, which is released when the parent colony disintegrates.

Examine a sample of living *Volvox* with a stereoscopic microscope. Describe the motility of *Volvox*.

____

____

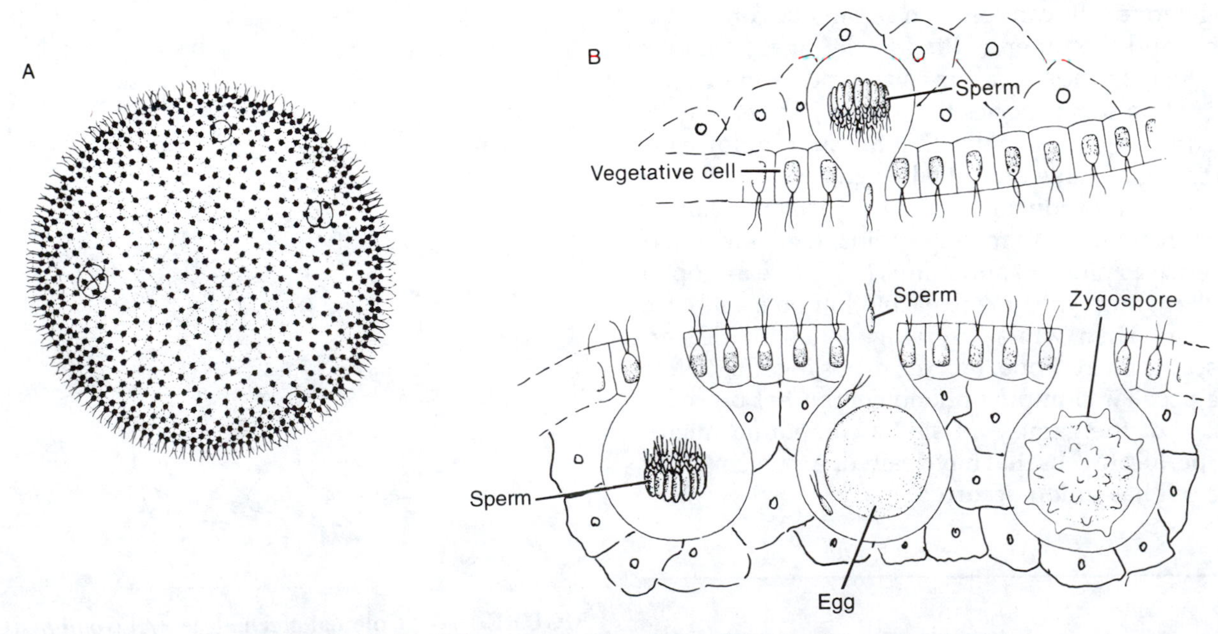

FIGURE 15-4 *Volvox.* (A) Adult colony. (B) Sexual reproduction.

Describe the shape of the colony.

___

___

Examine living material and prepared slides of *Volvox* for daughter colonies.

All the green algae studied thus far produce isogametes. *Volvox* produces **oogametes,** which are morphologically differentiated into sperm and eggs (Fig. 15-4B). These are developed from cells that become differentiated as the colony grows. In forming an egg, one differentiated cell increases greatly in size, takes on a rounded form, and becomes filled with food materials, especially lipids. The male gametes are formed from other cells that give rise to flat bundles of flagellated sperm. When mature, the eggs are fertilized by the sperm and then develop heavy spiny walls and become **zygospores.** Germination of the zygospore occurs in the spring and results in a new colony. Examine prepared slides of *Volvox* and locate sperm, eggs, and zygospores.

b. Filamentous Green Algae

Spirogyra. *Spirogyra* is a floating green alga found in small freshwater pools in the spring. It is frequently referred to as "pond scum." (Fig. 15-5.)

Prepare a fresh mount of *Spirogyra* and examine it microscopically. Is any branching evident?

___

___

Why do you think this alga is called *Spirogyra*?

___

___

Locate several small pyrenoids within the chloroplast. The nucleus, suspended in the center of the cell by the cytoskeleton, is difficult to observe unless stained. Apply a drop of methylene blue to the edge of the coverglass. After a few minutes, reexamine the cell and locate the nucleus, which will appear as a bluish body in the central part of the cell.

In sexual reproduction, filaments of opposite mating types come to lie adjacent to one another, and small projections appear in opposing cells of each filament (Fig. 15-5). The projections increase in length and eventually contact each other. At the point of junction, the cell walls dissolve, forming a **conjugation tube.** The protoplasts of the conjugating cells become isogametes. One gamete functions as the male and migrates through the conjugation tube to unite with the nonmotile female gamete. Fusion of the gametes produces a zygospore, which is released when the filament disintegrates. Before germination, the nucleus of the zygospore undergoes meiosis, and three of the four haploid nuclei that are formed disintegrate. On germination of the zygospore, a short protuberance is formed, which contains the fourth nucleus.

Mitosis, followed by cell division, results in a filament of cells that is similar to the parental filament. Examine the living material for various stages in gametic union. If your preparation does not show any stages of conjugation or zygospore formation, examine prepared slides showing this process.

No means of asexual reproduction is known to occur in *Spirogyra* other than fragmentation of the filaments.

Ulothrix. *Ulothrix*, like *Spirogyra*, consists of a simple, unbranched filament of cells. Unlike *Spirogyra*, it is not free-floating but has a basal **holdfast** cell that attaches it to rocks or other objects in fresh water and, in some species, salt water (Fig. 15-6).

Examine a prepared slide or living specimens of *Ulothrix.* Note that all cells, with the exception of the holdfast, are alike. The holdfast cell may not be present in your specimen. Why?

___

___

___

The chloroplast is C-shaped and may have one or more pyrenoids. Is a photoreceptor present?

___

If not, why would you expect this organism to have such a structure?

___

___

___

With the exception of the holdfast, any cell can

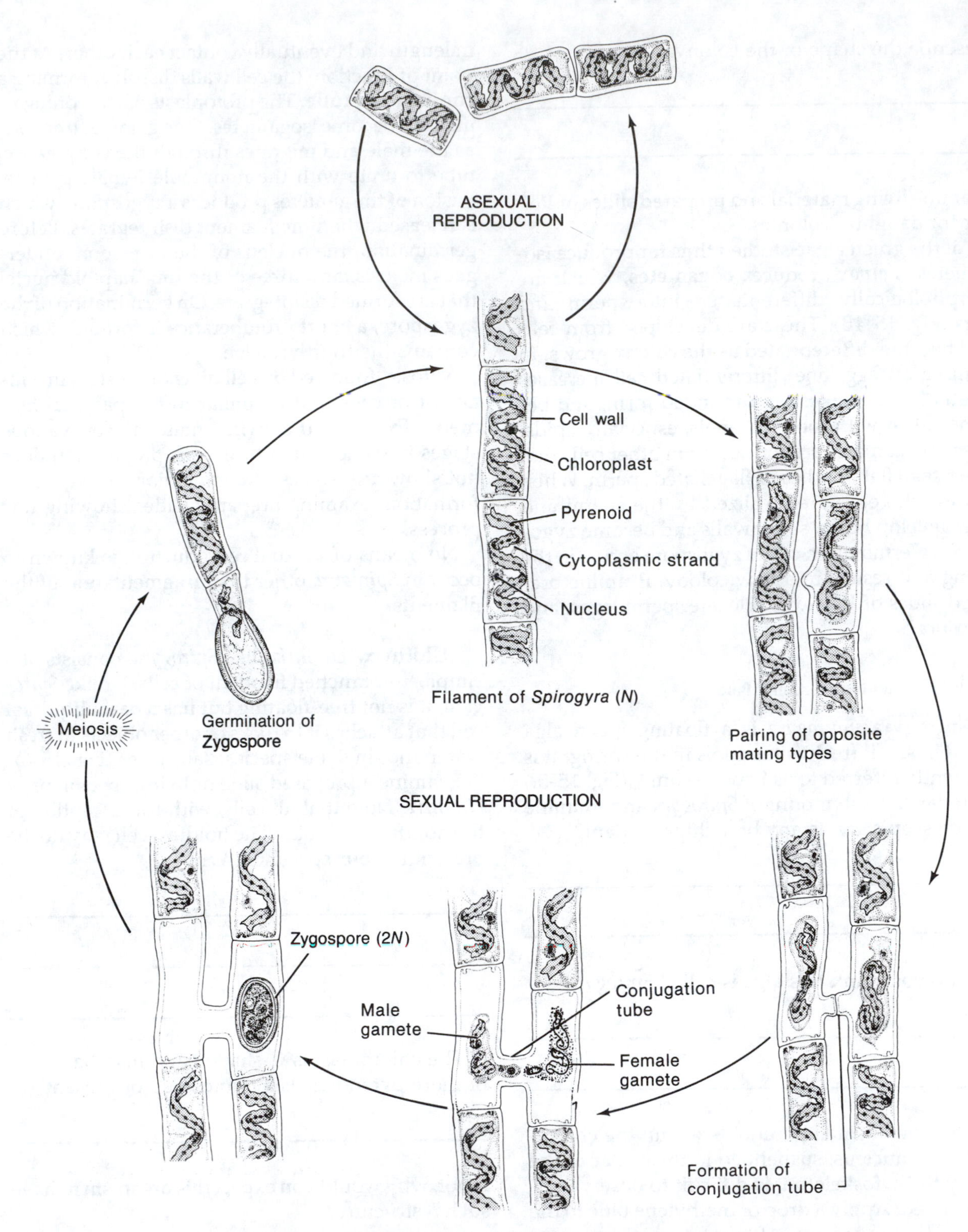

FIGURE 15-5 Life cycle of *Spirogyra*

reproduce the plant body by asexual or sexual means. In asexual reproduction, the protoplast of the parent cell undergoes mitosis and produces four to eight daughter cells that, when released from the parent cell (the **zoosporangium**), become flagellated **mitospores** (Fig. 15-6). After swimming around for a short time, a mitospore loses its flagella, settles to the bottom, and through a series of mitotic divisions gives rise to a new filament.

In sexual reproduction, the parent cell produces

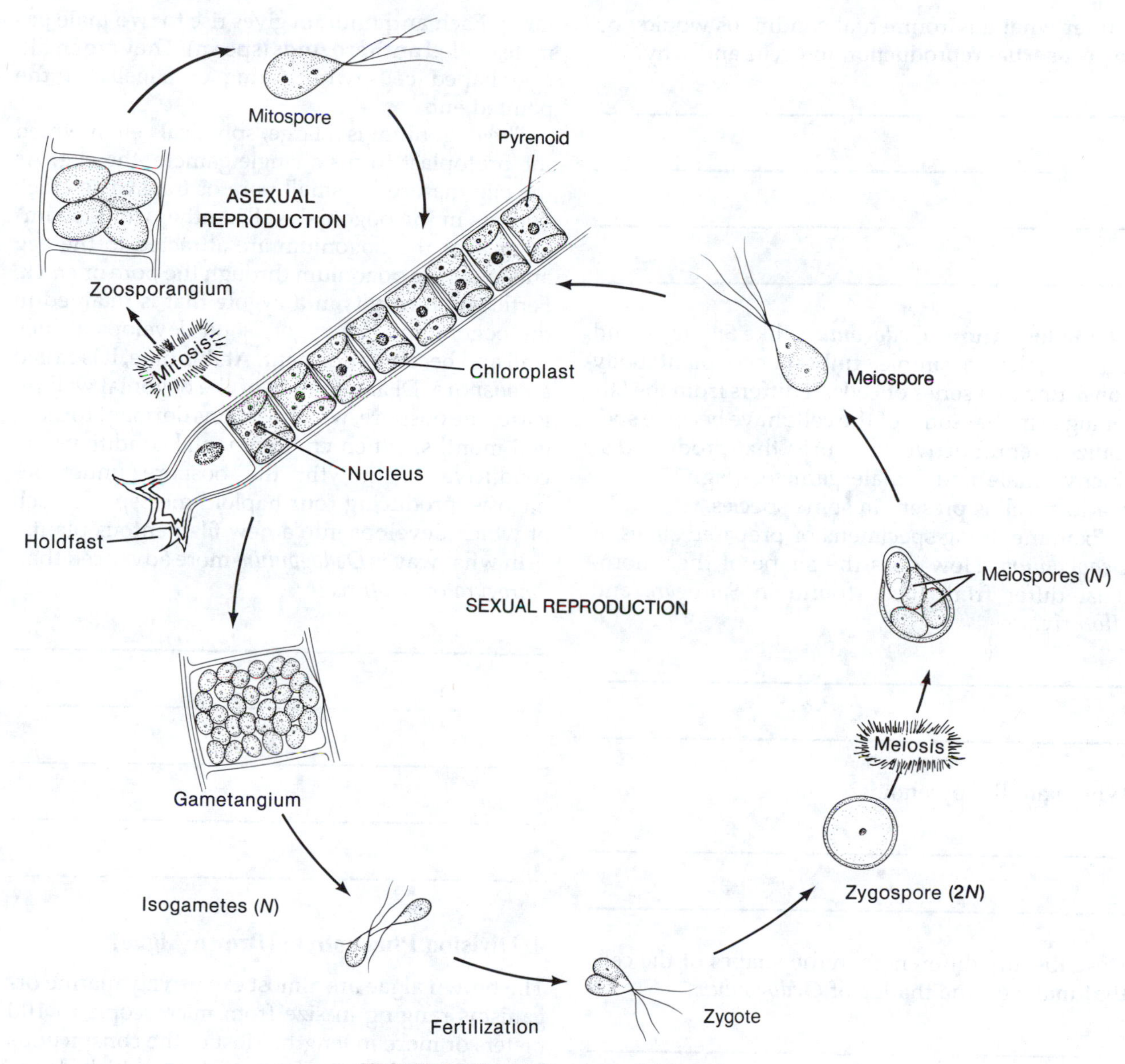

FIGURE 15-6 Life cycle of *Ulothrix*

32–64 isogametes to form a **gametangium.** These differ from mitospores in being smaller and having two rather than four flagella. Gametes from different filaments fuse to form a zygote that enters a dormant period, at which time it is called a zygospore. Under favorable conditions, this resting spore undergoes meiosis, producing four haploid meiospores, each of which develops into a new filament of cells.

Is the plant body in *Ulothrix* haploid or diploid? Explain.

____

____

____

Which type of reproduction (asexual or sexual) is primarily responsible for increasing the population? Explain.

____

____

____

____

Under what environmental conditions would you expect sexual reproduction to occur and why?

Oedogonium. *Oedogonium*, like *Spirogyra* and *Ulothrix*, has a simple, unbranched plant body consisting of a series of cells. It differs from the latter algae in that some of the cells have become specialized reproductive structures that produce distinctive male and female gametes (Fig. 15-7). A holdfast cell is present in some species.

Examine living specimens or prepared slides of *Oedogonium*. How does the shape of the chloroplast differ from those found in *Spirogyra* and *Ulothrix*?

Where are the pyrenoids?

Describe any differences in the shapes of the cells that make up the thallus of *Oedogonium*.

Asexual reproduction can take place in two ways. The plant body can simply fragment, with each fragment increasing in size by cell division, or the protoplast of any vegetative cell (i.e., nonreproductive cell or holdfast cell) can become a mitospore that, when released, actively swims for a while and then produces a new filament of cells (Fig. 15-7).

To study the sexual phase, examine a prepared slide or living material showing **antheridia** and **oogonia,** the specialized reproductive cells. The antheridia are shortened, disc-shaped cells in contrast to the elongated vegetative cells of the filament. Each antheridium gives rise to two male gametes called **antherozoids** (sperm). They are small, egg-shaped cells with a ring of flagella at the pointed end.

The oogonium is a large, spherical cell in which the protoplast forms a single gamete, the egg. As the egg matures, a small pore or transverse crack appears in the oogonial wall. Antherozoids swimming near the oogonium are attracted to the egg and enter the oogonium through the pore or crack. Fertilization results in a zygote that is retained in the oogonium. The zygote soon develops a thick wall and becomes dormant. At this time, it is called an **oospore.** Disintegration of the oogonial wall releases the oospore, which remains dormant for several months. When environmental conditions are conducive to growth, the oospore undergoes meiosis, producing four haploid meiospores each of which develops into a new filamentous plant.

In what way is *Oedogonium* more advanced than *Spirogyra* or *Ulothrix*?

4. Division Phaeophyta (Brown Algae)

The brown algae are almost exclusively marine organisms ranging in size from microscopic to 100 meters or more in length. Most of the conspicuous seaweeds of the temperate regions, which dominate rocky shores, are brown algae. The brown algae provide us with several economically useful products, such as alginic acid, which is used as a stabilizer to give ice cream a smooth texture. It is also used in the manufacture of fire-resistant paints, cosmetics, and many pharmaceuticals.

In many parts of the world, the brown algae are an important food source. In Oriental countries, *Laminaria* (Fig. 15-8A) is grown on ropes suspended between bamboo poles driven into the sea bottom along the coasts. A food product prepared from these algae, called kimbri, is a dietary staple in Japan.

Macrocystis, the giant Pacific kelp, is being cultivated on an experimental basis along the California coast to determine its potential as a source of methane fuel.

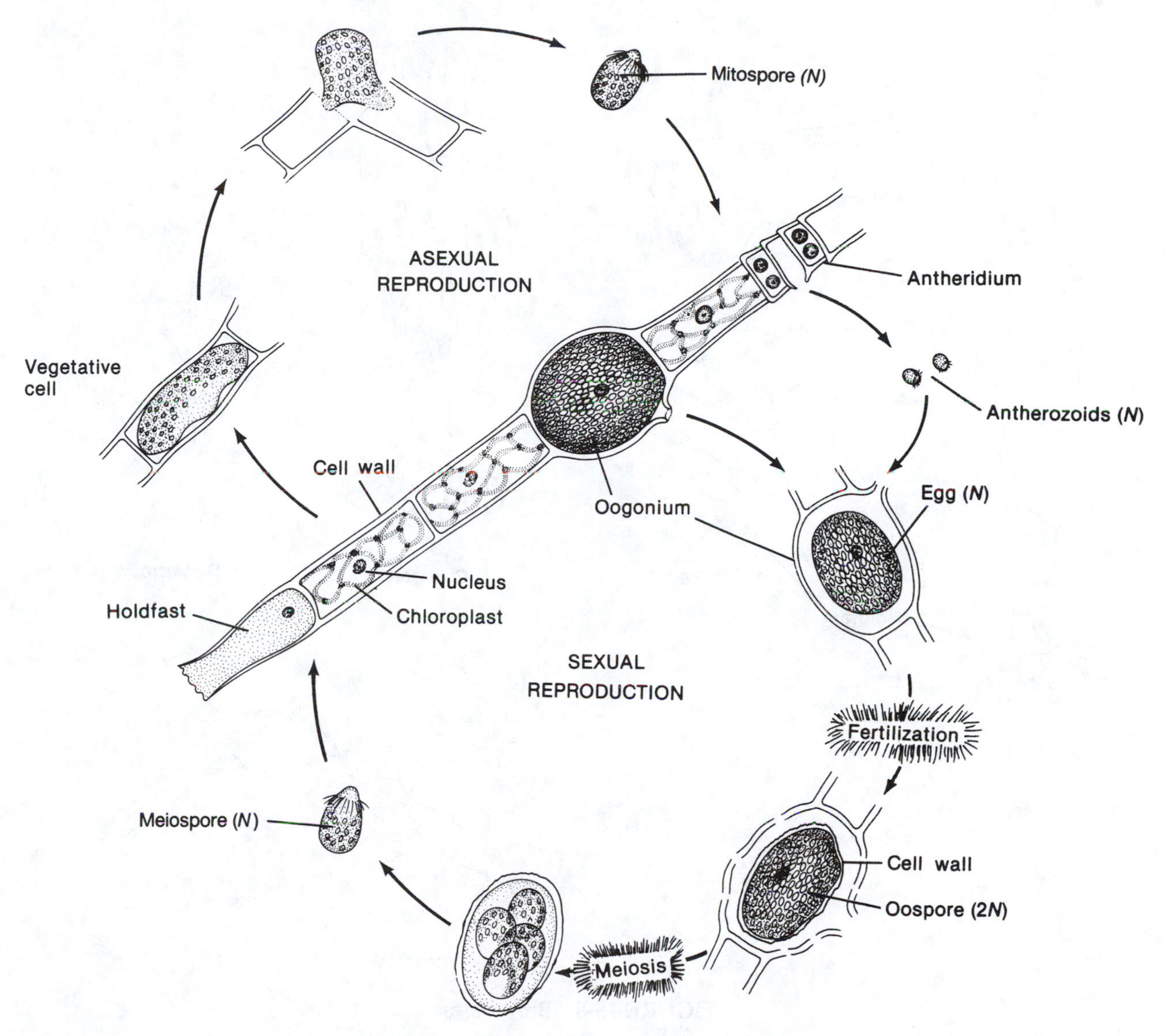

FIGURE 15-7 Life cycle of *Oedogonium*

Examine specimens of brown algae, noting the variation in size and complexity of the thallus. Many of the kelps (Fig. 15-8B, C, D) are externally differentiated into parts that look like roots (the **holdfast**), stems (the **stipe**), and leaves (the **blade**). Also observe that many of the specimens have air-filled bladders. How are such floats advantageous to the organism?

____

____

____

If available, examine *Sargassum*, a floating brown alga found in tropical waters (Fig. 15-8E). The thalli of this alga occur in large numbers and may extend over thousands of acres of the ocean's surface. The Sargasso Sea in the mid-Atlantic takes its name from this alga.

5. Division Rhodophyta (Red Algae)

Like the brown algae, red algae are primarily marine organisms. Although occasionally found in cooler regions, they are most abundant in tropical

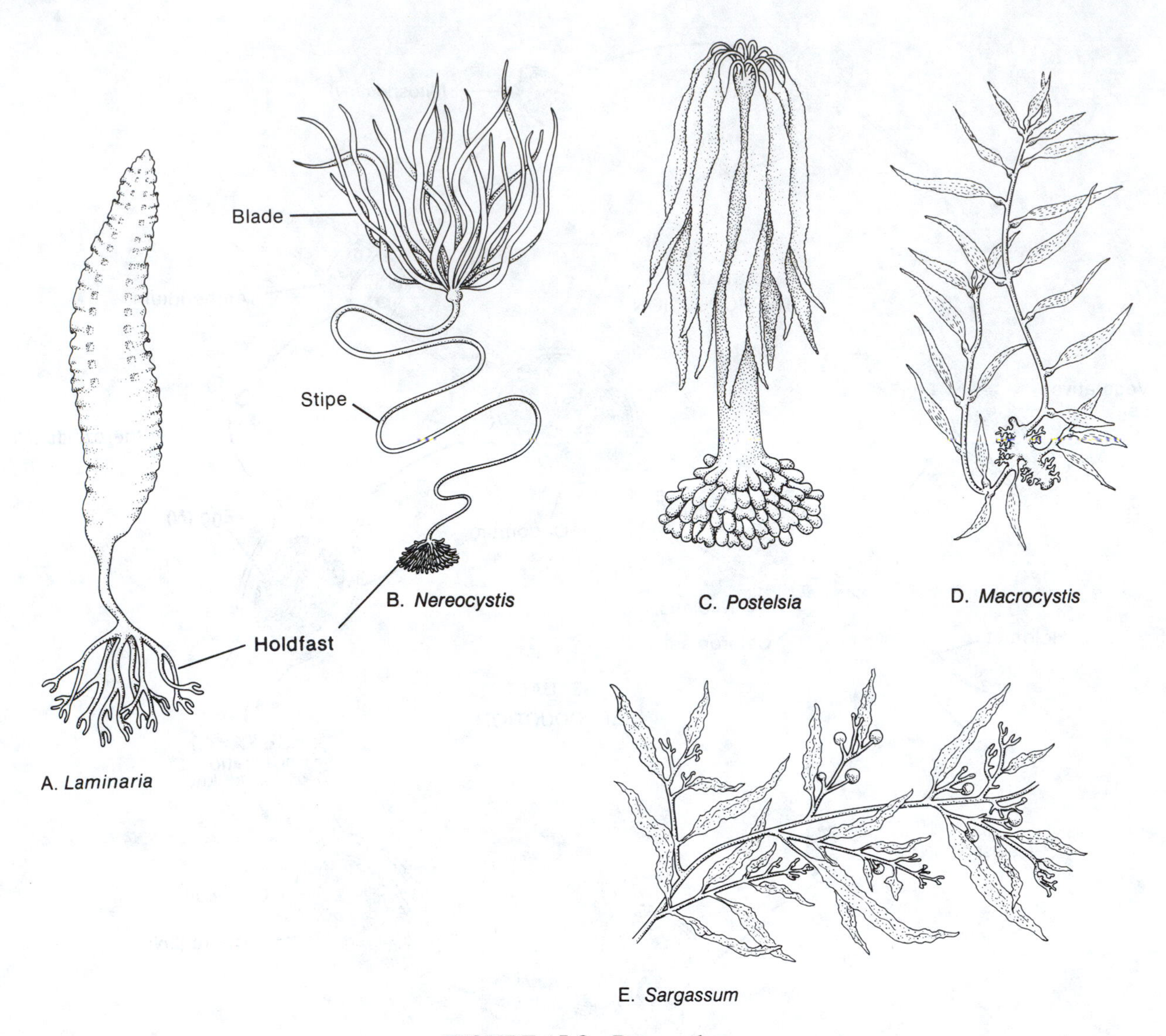

FIGURE 15-8 Brown algae

and warm waters. Unlike brown algae, the red algae always grow attached to solid objects and usually below the tide level. In warm waters, they are found at greater depths than any other group of algae, typically 100 – 200 meters below the surface. The phycobilin pigments, phycoerythrin and phycocyanin, mask the color of chlorophyll *a* and give these algae their distinctive red color. These pigments can absorb the wavelengths of light that penetrate into deep water.

The thallus of red algae is similar to that of brown algae in that it is differentiated into a holdfast, stipe, and blade. In some red algae, known as **corallines,** the thalli are heavily impregnated with limestone and are considered to be as important to the formation of coral reefs and atolls as are the coral animals.

Red algae provide several useful products. For example, a gelatinous cell-wall material provides agar, which is used extensively in laboratories for the culture of bacteria and fungi and some higher plants. Carrageenan, a colloid, is used as an emulsifying agent in dairy products such as chocolate milk, in which it prevents the chocolate from settling out.

The red alga *Porphyra* is used as food in several Oriental dishes. This is the alga that is used to wrap the rice and bits of raw fish for sushi.

Examine various specimens of red algae on demonstration in the lab.

B. PLASMODIAL SLIME MOLDS (DIVISION MYXOMYCOTA)

Slime molds are classified into two groups based on the form of the vegetative feeding phase of their life cycles. Because the vegetative phase of the **cellular slime molds** consists of masses of single amoeboid cells, they are thought to be more closely related to the amoebas than to any other group and, therefore, have been traditionally classified as **protozoa.** The acellular, or **plasmodial slime molds,** have a vegetative phase consisting of masses of protoplasm (plasmodia) of indefinite size and shape. Both types of slime mold live predominantly on decaying plant material, primarily microorganisms (especially bacteria). In this part of the exercise, you will observe the growth and production of fruiting structures of a plasmodial slime mold.

Obtain a small piece of filter paper that contains slime mold **sclerotia** (a dry, resting phase of the organism) and place it on the surface of the agar in a petri dish (Fig. 15-9). Sprinkle a few oatmeal flakes over the sclerotium, and moisten the oatmeal with two or three drops of water.

Replace the cover on the dish and set it in a dark place. After 24 hours, examine the plates for growth of the slime mold. Record your observations by drawing the growing organism in Fig. 15-10.

After growth has begun, examine the slime mold with a hand lens, a stereoscopic microscope, or the low power of the compound microscope. Describe the pattern of cytoplasmic movement observed in the plasmodium of the mold.

____

____

____

____

Puncture a branch of the slime mold with a needle and watch it for a few minutes.

After the dish becomes covered with the plasmodium of the slime mold, partially remove the cover. The slime mold will begin to dry out and will initiate the formation of fruiting bodies. Examine the culture during the next few days and describe the shape of the fruiting body that is formed.

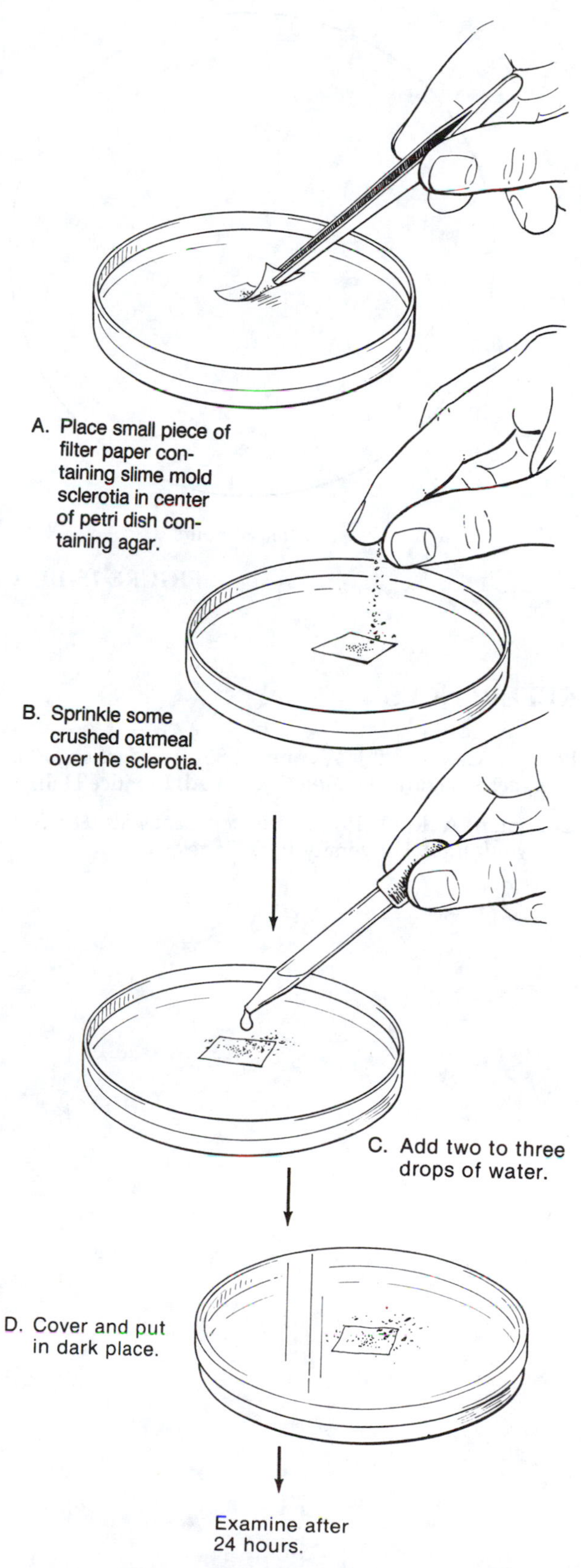

FIGURE 15-9 Procedure for growing slime mold

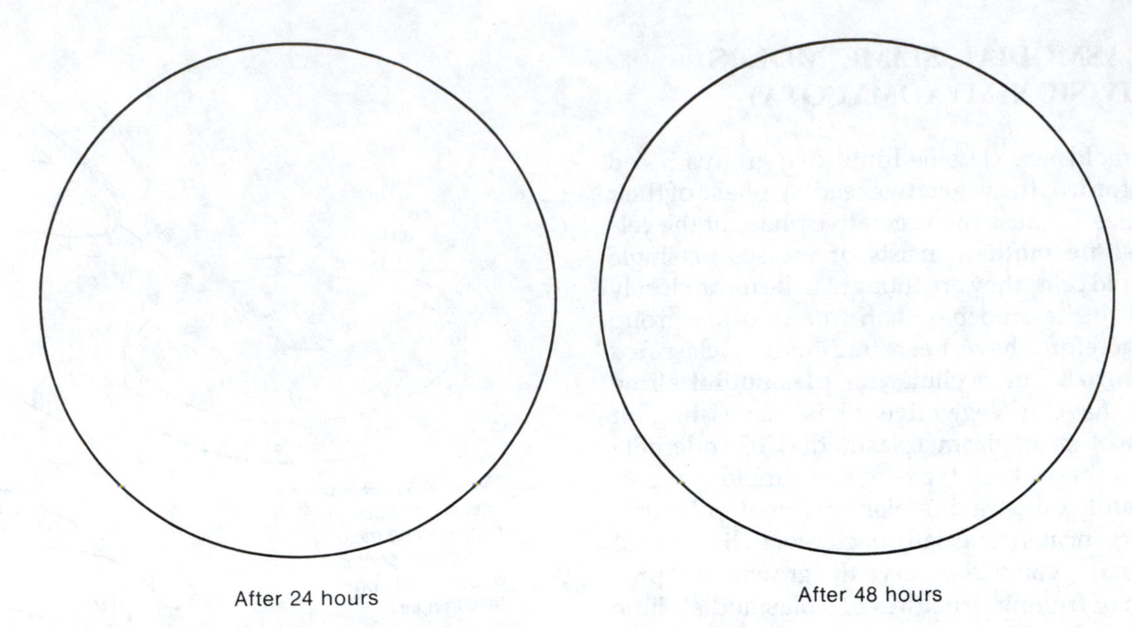

FIGURE 15-10 Growth of slime molds

REFERENCES

Bold, H. C., and M. J. Wynne. 1985. *Introduction to the Algae: Structure and Function.* 2d ed. Prentice Hall.

Chapman, A. R. O. 1979. *Biology of Seaweeds: Levels of Organization.* University Park Series.

Prescott, G. W. 1978. *How to Know the Freshwater Algae.* 3d ed. Brown.

Raven, P. H., R. F. Evert, and S. E. Eichhorn. 1986. *Biology of Plants.* 4th ed. Worth.

Exercise

16

Kingdom Protista II: Protozoa

Protozoa are heterotrophic, one-celled organisms that exhibit a remarkable degree of subcellular organization. Instead of organs and tissues, they have functionally equivalent subcellular structures called **organelles.**

Protozoans are found in a great variety of habitats. Most are free-living and inhabit fresh and marine waters. A number of protozoans also inhabit the bodies of other organisms in relationships described as **commensalistic** (in which one organism benefits and the other is neither harmed nor benefited), **mutualistic** (in which both benefit), or **parasitic** (in which one benefits and the other is harmed). Of the five phyla of unicellular heterotrophs that are commonly referred to as protozoa, you will study the following four in this exercise:

- Phylum Mastigophora (flagellates): Move by means of one or more long, whiplike **flagella.**
- Phylum Ciliophora (ciliates): Move by cilia. Typically have macro- and micronuclei and unusual reproductive patterns.
- Phylum Sarcodina (amoebas and related forms): Lack cilia or flagella and move (and also feed) by irregular cytoplasmic extensions called **pseudopodia.** No definite body shape. Some have elaborate shells.
- Phylum Sporozoa (sporozoans): Parasitic organisms that live part of their life cycle in cells of other organisms. Most have complex reproductive cycles involving asexual and sexual reproduction.

A. PHYLUM MASTIGOPHORA (FLAGELLATES)

The members of this phylum move by means of flagella. A few of the flagellates are free-living organisms in fresh or salt water, but most live in the bodies of higher plants or animals.

1. *Trichonympha*

Trichonympha (Fig. 16-1A) inhabits the intestines of termites. Although termites ingest bits of wood, they are unable to digest cellulose, the chief constituent of wood. *Trichonympha* forms **pseudopodia** that engulf the wood fragments ingested by the termite. The cell walls of the wood are digested into soluble carbohydrates that can then be used by the termite. Neither the termite nor *Trichonympha* can live without the other.

Examine slides of this organism. Note the small fragments of wood in the cytoplasm and the large number of flagella that cover the upper part of the organism.

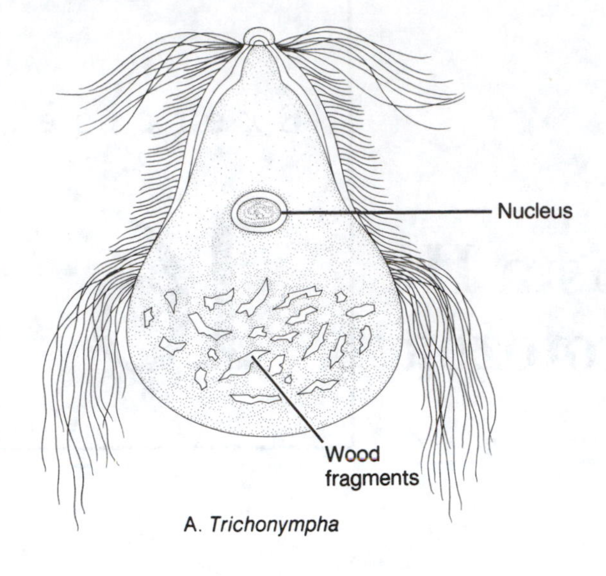

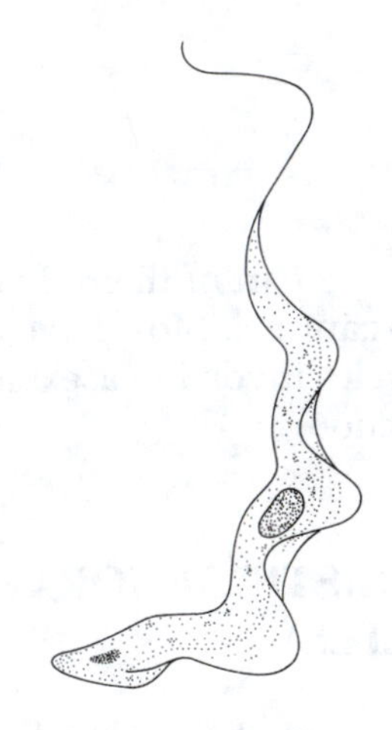

FIGURE 16-1 Flagellates

2. *Trypanosoma*

Trypanosoma (Fig. 16-1B) inhabits the blood of vertebrates and is transmitted from one host to another by bloodsucking insects. For example, the **trypanosomes** that cause human sleeping sickness in Africa are transmitted by the bloodsucking tsetse fly. Examine stained slides of human blood smears that contain these parasitic flagellates.

B. PHYLUM CILIOPHORA (CILIATES)

Compared with other protozoans, the **ciliates** are relatively large and complex. They are also distinguished from other protozoans by having **cilia,** two types of nuclei (a smaller **micronucleus** that functions in reproduction and a larger **macronucleus** that controls cell metabolism and growth), and a type of reproduction involving **conjugation** between two animals and the exchange of genetic material. Most ciliates are free-living and are commonly found in fresh and salt water. A few are parasitic to human beings. These organisms also play an important role in the aquatic food chain, serving as food for small multicellular animals, which, in turn, are eaten by larger animals.

1. *Paramecium caudatum*

a. Morphology

Paramecium caudatum is often selected as a typical ciliate to study because it is easy to obtain and, owing to its size, is easy to observe. This ciliate is frequently found in pond water that contains large amounts of decaying vegetation. Because it can be easily grown in large numbers under laboratory conditions, it is used extensively in studies of nutrition, cancer, behavior, genetics, and ecology.

Place a small drop of *Paramecium* culture on a clean slide and add a small drop of methyl cellulose to slow its movements. Add a coverslip and examine microscopically with the low-power objective. You may have to adjust the iris diaphragm to increase contrast.

Paramecium caudatum resembles a twisted slipper. With the aid of Fig. 16-2 and commercially prepared slides, locate the following structures on the living organism.

The **oral groove** begins near the anterior end, runs diagonally toward the posterior end, and leads to the **cytostome** (mouth) and then to the **cytopharynx** (gullet). Food is carried into the oral groove by beating cilia, which can be seen by using the high-power objective and adjusting the iris diaphragm to provide more contrast.

A **contractile vacuole** is located at each end of the organism. Do these vacuoles move or are they stationary?

____

Do they contract at the same time or alternately?

____

Locate the canals that radiate outward from each vacuole. Suggest a function for these structures.

____

____

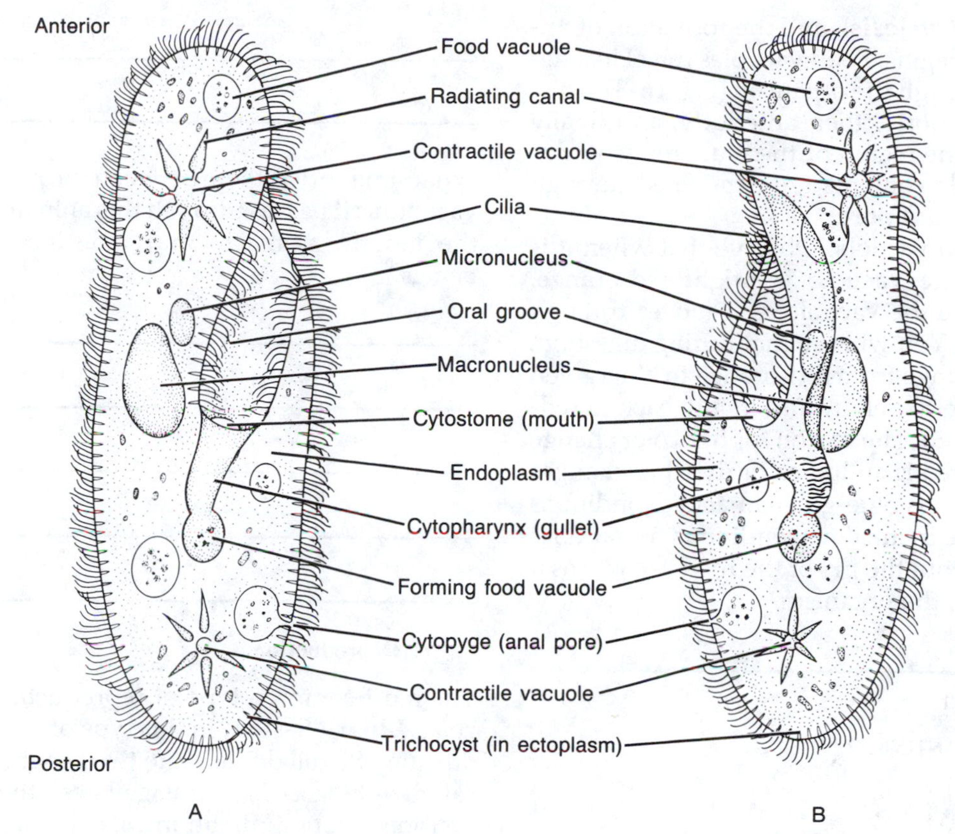

FIGURE 16-2 *Paramecium.* (A) Ventral view. (B) Lateral view.

What is the function of contractile vacuoles?

___

___

___

The macronucleus can be seen as a relatively large, clear area in the **endoplasm,** the central part of the cell. A micronucleus is located adjacent to the macronucleus. These nuclei are difficult to observe in the living animal but can be stained. Add a drop of acetocarmine or methyl green stain to one side of the coverslip and draw it under by touching a piece of paper toweling to the other side. (Supplement these observations with commercially prepared slides of *Paramecium* stained to show the nuclei.)

The **trichocysts,** located in the **ectoplasm,** are carrot-shaped structures that contain coiled, barbed filaments that can be ejected and are believed to aid the organism in capturing and holding the smaller organisms on which it feeds. Add a small drop of iodine or acetic acid at the edge of the coverslip. Allow it to run underneath and observe the discharge of the trichocysts.

b. Nutrition and Feeding

Most protozoa, including *Paramecium*, are **holozoic;** that is, they ingest solid food such as other protozoa, bacteria, or decaying organic material **(detritus)** from the water. These food particles must be digested before they can be used for growth, repair, or reproduction.

Place a drop of *Paramecium* culture on a slide and add a *small* drop of yeast stained with Congo red. (*Note:* The resulting color should be pink, not red.) Add a drop of methyl cellulose and a coverslip, and locate a *Paramecium* under high magnification. Observe the vortex of water produced by the cilia in the vicinity of the oral groove that carries the stained yeast into the oral groove and then to the cytostome and cytopharynx. A **food vacuole** (a membrane-bounded sac containing water and suspended food particles) is formed at the base of the cytopharynx. As soon as the food vacuole is formed, it is carried away by a rotary movement of the cytoplasm **(cyclosis)**, and the formation of another vacuole begins. Food vacuoles travel in a defined route through the organism (Fig. 16-3). They first pass posteriorly, then anteriorly, and finally posteriorly to the region of the oral groove where the undigestible contents are eliminated through the **cytopyge (anal pore).**

Locate a food vacuole. Observe that when it is first formed, the vacuole is bright red-orange. Closely observe the vacuole as it moves through the protozoan. You will see that during digestion, its contents change from red-orange to blue-green to yellow-green to yellow and finally back to red-orange (Fig. 16-3). The reason for this color change is that Congo red is an indicator dye that changes color with pH: it is blue-green in acidic conditions and red-orange in alkaline conditions. What does this indicate about the pH of the food vacuole as it moves through the organism?

____

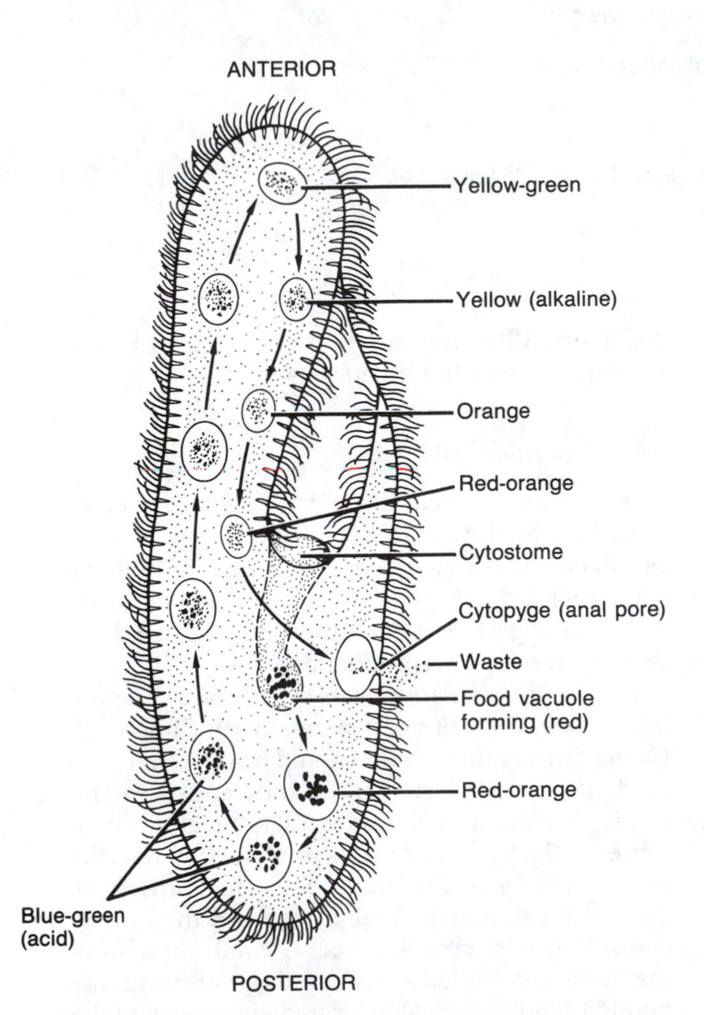

FIGURE 16-3 Digestion in *Paramecium.* Arrows indicate pathway of movement of food vacuoles.

____

____

What similarity is there between the pH of the food vacuole as it passes through this animal and that of the mouth, stomach, and intestines in human beings?

____

____

____

____

____

c. Reproduction

The most common type of reproduction in protozoa is **binary fission.** In this type of asexual reproduction, the cell divides into two genetically identical daughter cells. In flagellates, the plane of division is longitudinal; in ciliates, the parent cell divides transversely. Examine demonstration slides of *Paramecium* showing the various stages of binary fission (Fig. 16-4).

Occasionally, paramecia reproduce sexually by conjugation (Fig. 16-5), in which micronuclei are exchanged. You can observe conjugation in *Paramecium* by carrying out the following procedure. For this study, you will use mating strains of *Paramecium bursaria,* which is symbiotic with a green alga. Supplement your observations by examining commercially prepared slides showing various stages of conjugation.

Put a *small* drop of one of the two mating strains into the depression of a deep-well slide. While observing the paramecia with a stereoscopic microscope, add a drop of the second mating strain. You should observe, almost immediately, the **agglutination** (or clumping) of opposite mating strains, which brings the cells together for the transfer of nuclear material. Place the slide in a covered petri dish containing moist filter paper to prevent dessication of the culture. Examine periodically. Conjugating paramecia can be seen for up to 48 hours, after which few or no conjugants are found.

2. Other Ciliates

Obtain samples of the following ciliates and examine them microscopically.

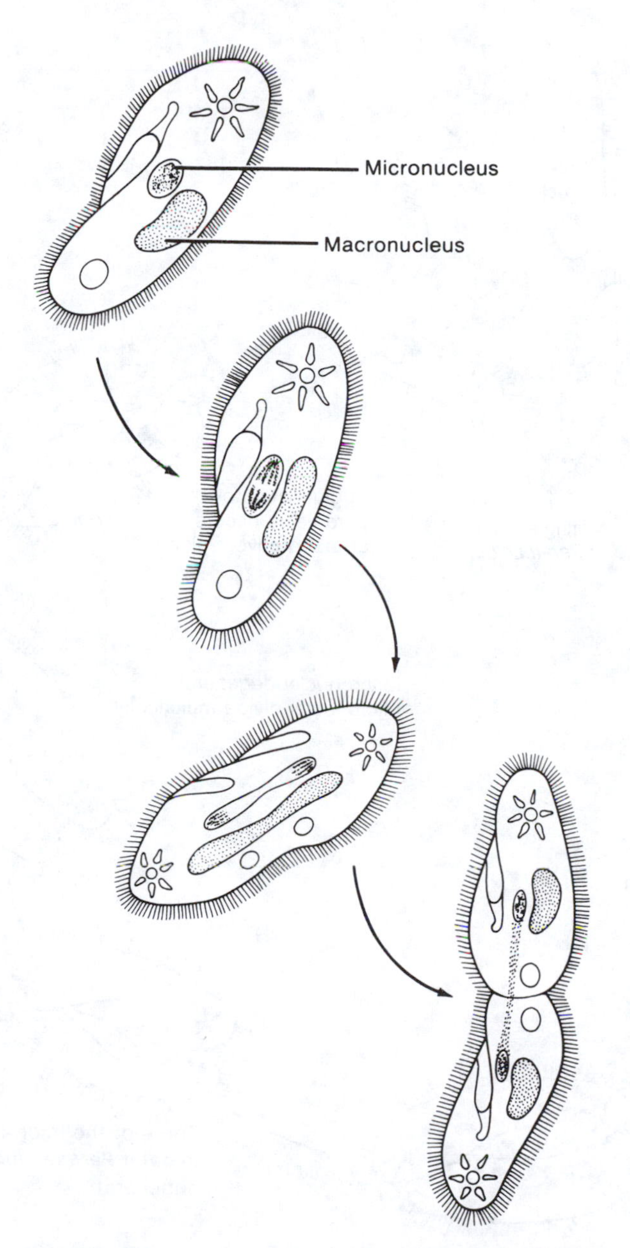

FIGURE 16-4 Asexual reproduction (binary fission) in *Paramecium*. The micronucleus divides by means of a mitotic spindle that is confined within the nuclear envelope. The macronucleus is pulled apart.

a. *Stentor*

This organism is trumpet-shaped, is blue when living, and has a macronucleus shaped like a string of beads (Fig. 16-6A). Food is brought into the cytostome of the organism through the activity of cilia.

b. *Vorticella*

This freshwater ciliate resembles an inverted bell attached to a stalk that anchors it to submerged vegetation and stones (Fig. 16-6B). Locate an organism in which the stalk is straight and extended. While watching the animal, gently tap the slide and describe what occurs.

____

____

____

c. *Balantidium coli*

Balantidium (Fig. 16-6C) is the only ciliate parasite of human beings. It burrows into the wall of the colon and causes ulcers. It is **pathogenic** (disease producing) and may cause symptoms similar to amoebic dysentery. The most common source of infection is uncooked pork.

Examine prepared slides of *Balantidium coli* and note that the whole body is covered with cilia that are arranged in rows. The macronucleus is slightly curved and is associated with a very small micronucleus. Food particles are swept into the cytostome by currents set up by the beating cilia. There are two contractile vacuoles, and food vacuoles circulate in the cytoplasm. Like other ciliates, *Balantidium* divides by transverse fission.

d. *Tetrahymena*

This small ciliate (Fig. 16-6D), which is relatively easy to grow in pure culture, has been widely used in physiological and genetic research. Because its mitotic division can be synchronized by appropriate heat shocks, this protozoan is a useful tool in studying the details of mitosis. Examine living specimens or prepared slides of *Tetrahymena*.

C. PHYLUM SARCODINA

1. Amoeba

Amoebas (Greek *amoibe*, "change") are protozoa that are common in freshwater ponds and streams. Microscopically they look like gray, irregular masses that continually change shape due to the extension and withdrawal of fingerlike protuberances called pseudopodia (Fig. 16-7).

Using a clean dropper pipet, obtain a sample of amoebas from the bottom of the culture dish where they aggregate. Place a few drops of the culture in a depression slide (alternatively, place a few drops on a clean glass slide and add a few pieces of debris from the culture and a few grains of sand or some

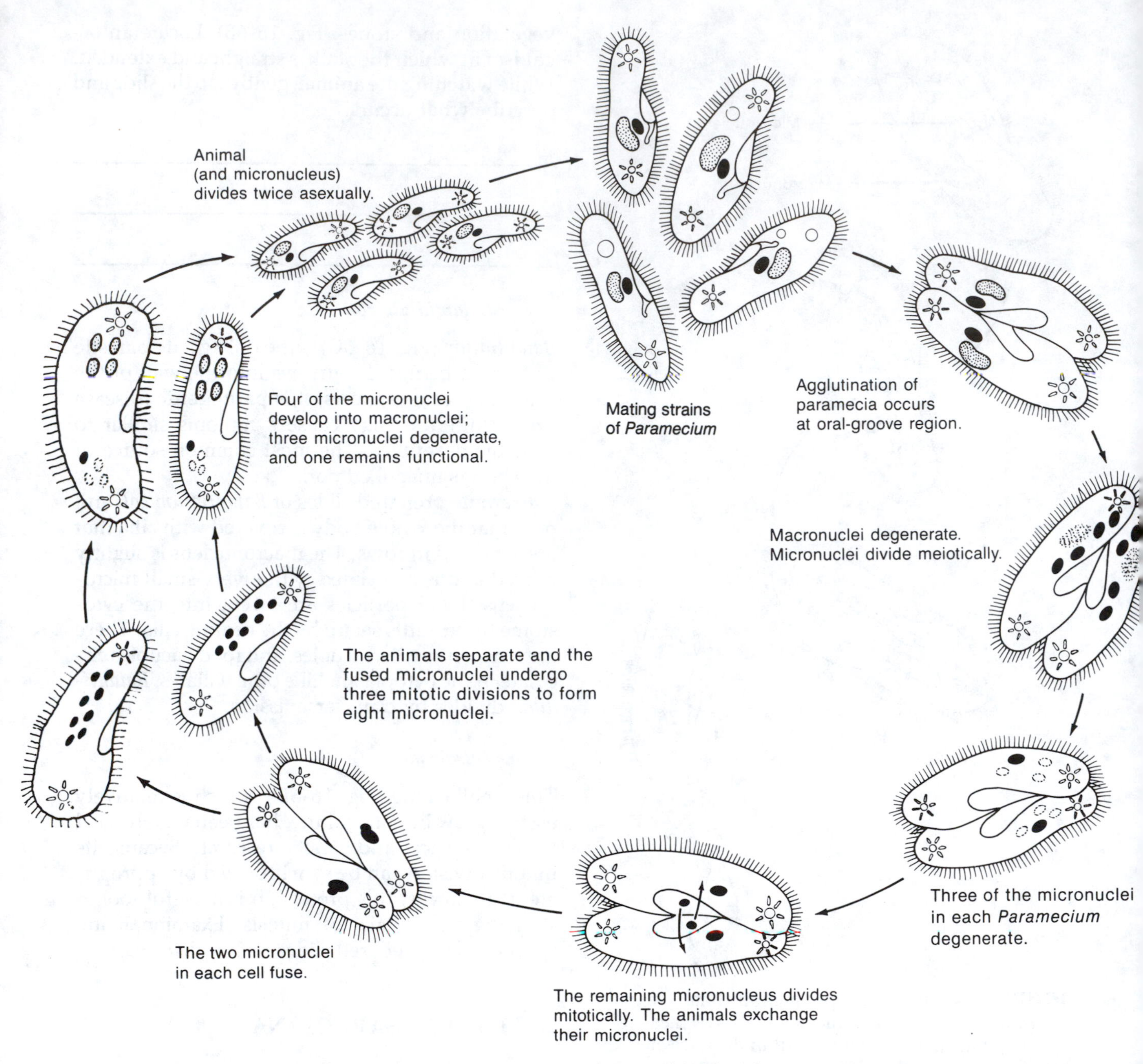

FIGURE 16-5 Sexual reproduction (conjugation) in *Paramecium*

small pieces of broken coverslips to prevent the organism from being crushed when you place a coverslip over the preparation). To visualize the three-dimensional form of an amoeba, first examine the culture using a stereoscopic microscope. Adjust the spot lamp to give transmitted light first and then reflected light. You may have to adjust the intensity of the light.

Study the preparation using the low power (10 ×) lens of your compound microscope. You will have to reduce the amount of light by closing the iris diaphragm because the amoebas are nearly transparent and thus almost invisible under bright light.

With the help of Fig. 16-7 and commercially prepared slides, locate the following structures: the

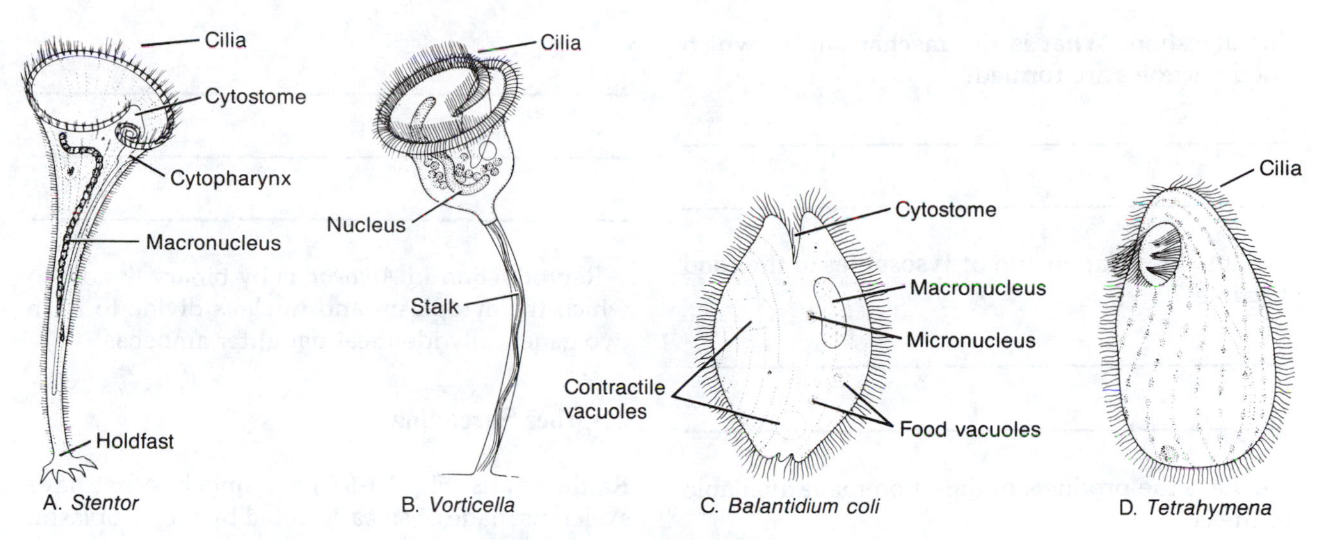

FIGURE 16-6 Other ciliates

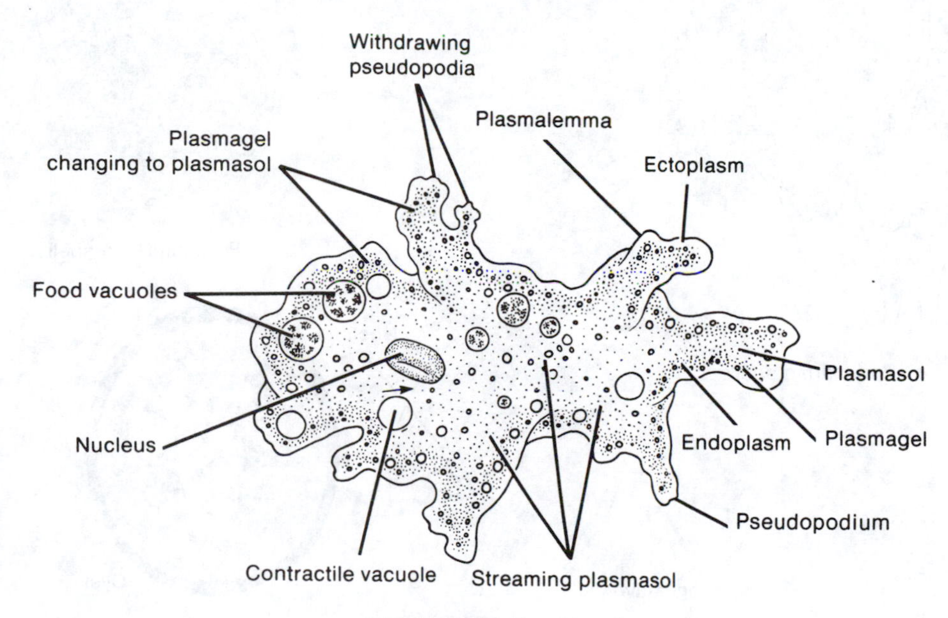

FIGURE 16-7 *Amoeba*

pseudopodia; the **endoplasm,** the inner granular material that makes up most of the cytoplasm; the **ectoplasm,** a thin layer of cytoplasm surrounding the endoplasm; the **cell membrane (plasmalemma);** the **plasmagel,** the gel-like outer layer of the endoplasm; the **plasmasol,** the fluid, central region of the endoplasm (streaming of the cytoplasm should be evident in this area); the nucleus, a somewhat transparent structure that is not fixed in position and sometimes appears wrinkled or folded; and the **contractile vacuoles,** clear spherical vacuoles found in the endoplasm that collect water from the cell and discharge it to the outside. These vacuoles function to maintain water balance in the cell. Would you expect a marine (saltwater) form of amoeba to have contractile vacuoles? Explain.

____

____

Food vacuoles contain ingested food and enzymes

for digestion. What is the mechanism by which food vacuoles are formed?

____

____

What is the relationship of **lysosomes** to the food vacuoles?

____

____

How are the products of digestion made available to the cell?

____

____

____

Reproduction in *Amoeba* is by binary fission, in which the cytoplasm and nucleus divide to form two genetically identical daughter amoebas.

2. Other Sarcodina

Radiolarians (Fig. 16-8A) are amoebas that have skeletons made of silica secreted by the cytoplasm,

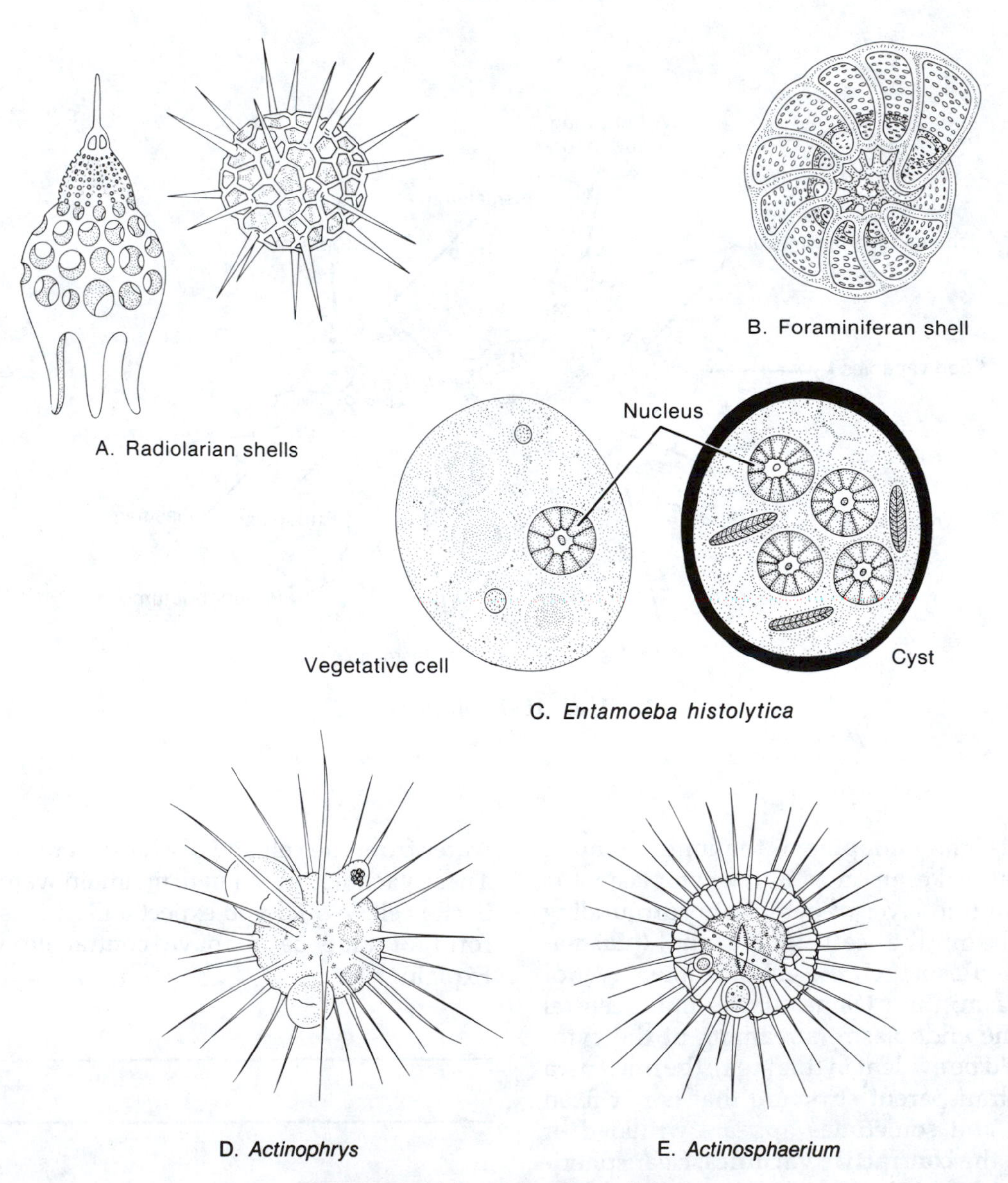

FIGURE 16-8 Other Sarcodina

usually taking the form of an intricate latticework, through which extend stiff, radiating spines. When these protozoans die, they drop to the ocean floor and eventually become compressed to form siliceous rock. Examine prepared slides of radiolarian skeletons.

Foraminiferans (Fig. 16-8B) are a large group of marine amoebas that secrete shells, resembling those of snails, that are made of calcium carbonate. Long, delicate feeding pseudopodia extend out of minute pores that perforate the surface of the shells. Dead foraminifera sink to the ocean floor, where their shells form a gray mud that is gradually transformed into chalk. Examine prepared slides of a variety of foraminiferan shells.

Entamoeba histolytica (Fig. 16-8C) is the organism that causes amoebic dysentery. Examine prepared slides of both the active (vegetative) stages and the cysts (resistant, infective stages) of the organism. In its vegetative stage, the organism has a single nucleus; the cysts have four nuclei, as well as a heavy wall surrounding the entire cell.

Actinophrys and *Actinosphaerium* (Fig. 16-8D, E) are members of a group of spherical amoebas with delicate pseudopodia that are supported by axial rods. Most species of this group of protozoa live in fresh water. Examine slides of *Actinosphaerium* and *Actinophrys.* Note in particular the delicate pseudopodia, which have suggested the name "sun animalcules" for these organisms.

D. PHYLUM SPOROZOA

All members of this phylum are parasites and can be found infecting nearly every major group of the animal kingdom, ranging from simple invertebrates to human beings. In this study, you will become familiar with *Plasmodium vivax,* a sporozoan that causes one form of malaria in human beings.

Malaria is one of the most devastating diseases of human beings with respect to sickness, mortality (death), and economic loss.

There are four kinds of malaria. **Benign tertian malaria,** caused by *Plasmodium vivax,* is characterized by attacks of fever every 48 hours. *Plasmodium ovale* also causes fever every 48 hours. **Quartan malaria,** which usually produces fever every 72 hours, is caused by *Plasmodium malariae.* **Pernicious malaria,** in which the fever is virtually continuous, is produced by *Plasmodium falciparum;* this form has the highest mortality rate.

Because all four malarias have similar life cycles, *Plasmodium vivax* will be the representative sporozoan for this study. Refer to Fig. 16-9 and commercially prepared slides of various stages of the life cycle.

Malaria is transmitted by the bite of female mosquitoes; male mosquitoes cannot infect because they lack the mouth parts for piercing the skin and sucking blood. Many animals can be infected with malaria that is transmitted by a variety of mosquitoes; however, the malarial parasite is transmitted to human beings only by female mosquitoes of the genus *Anopheles.*

When a mosquito's mouth parts enter the skin, saliva containing anticoagulants enters the wound. If this mosquito carries malaria, **sporozoites,** the form of the parasite in its infective stage, enter the bloodstream (Fig. 16-9A) but do not invade the **erythrocytes** (red blood cells). Instead, they penetrate cells of the liver, where they grow and multiply to form **merozoites.** On release from the liver cells, the merozoites enter the bloodstream and penetrate the erythrocytes, where they become ring-shaped and then irregular in shape. At this stage, the parasite is called a **trophozoite** (Fig. 16-9B). The trophozoite undergoes maturation (the time required depends on the species) and then divides, by a type of fission called **schizogony,** to form more merozoites (Fig. 16-9B). At regular intervals, depending on the species of *Plasmodium,* all infected red blood cells burst, releasing merozoites. Toxic substances that are released along with the merozoites account for the chills and fever that typify malarial attacks. Each merozoite in turn penetrates another erythrocyte and becomes a trophozoite. This cycle is repeated a number of times, with ever-increasing numbers of erythrocytes being affected.

Eventually, some merozoites, instead of becoming trophozoites, develop into male and female **gametocytes** (Fig. 16-9C). As long as the gametocytes remain in the human host, they are not significant. However, if they are sucked up by an *Anopheles* mosquito, they pass into the insect's stomach and become active. One female gametocyte develops into one egg; one male gametocyte gives rise to several sperm by a process termed **exflagellation.** Union of the egg and sperm produces a zygote called an **ookinete,** which migrates through the stomach epithelium and becomes embedded in the wall of the stomach. The nucleus undergoes successive mitotic divisions and produces large numbers of sporozoites within a structure now called an **oocyst.** When the oocyst ruptures, the sporozoites enter the body cavity of the mosquito and then migrate to the salivary glands. The next time a female mosquito bites a host, sporozoites are injected into the human body and the life cycle is repeated.

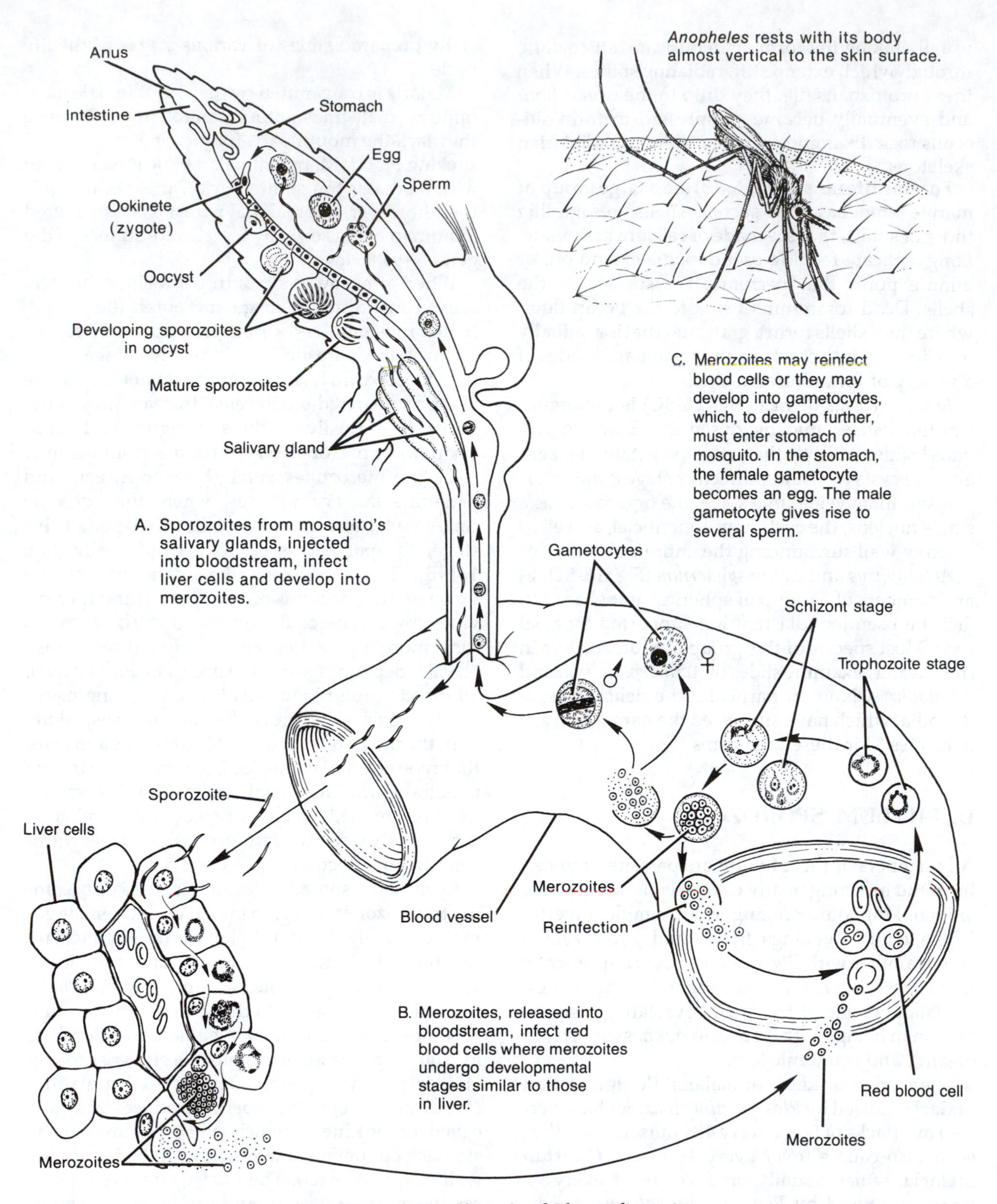

FIGURE 16-9 Life cycle of *Plasmodium vivax*

REFERENCES

Corliss, J. O. 1984. The Kingdom Protista and Its 45 Phyla. *BioSystems* 17:87–126.

Lee, J. J., S. H. Hunter and E. C. Bouvee, eds. 1985. *An Illustrated Guide to the Protozoa.* Soc. of Protozoologists. Lawrence, Kansas.

Sherman, I. W., and V. G. Sherman. 1976. *The Invertebrates: Function and Form: A Laboratory Guide.* 2d ed. Macmillan.

Villee, C. A., W. F. Walker, and R. D. Barnes. 1984. *General Zoology.* 6th ed. Saunders.

Exercise 17

Kingdom Fungi

The fungi and some bacteria are decomposers and, as such, they are as necessary to the biosphere as are the food producers. Their metabolism releases carbon dioxide into the atmosphere and nitrogenous materials into the soil and surface waters where they can be used by green plants, which are then used by animals.

Fungi are primarily terrestrial (land) organisms. Some are unicellular, but most are filamentous and may be organized into highly structured shapes, such as mushrooms. All fungi are **heterotrophic** and obtain their food as **saprobes** (i.e., they live on nonliving organic matter) or as **parasites** (i.e., they feed on living organic matter). Fungi do not ingest their food but absorb it. They secrete enzymes that break down the food outside the fungus. Partially digested molecules are then transported through the fungal membrane. All fungi have cell walls and most of them produce some type of spore.

This kingdom has three major divisions.

- Division Zygomycota (zygomycetes): Mostly saprophytic and terrestrial; some are parasitic on plants and insects. Sexual reproduction results in the formation of the primary characteristics of this division—a thick-walled structure called a **zygospore.** Hyphae are nonseptate (lack cross walls).
- Division Ascomycota (ascomycetes): Largest division, which includes yeasts, powdery mildews, molds, morels, and truffles. Many ascomycetes are pathogenic. Sexual reproduction often results in the formation of structures called **ascocarps.**
- Division Basidiomycota (basidiomycetes): Terrestrial fungi whose most familiar members are the mushrooms. Sexual basidiospores are produced on specialized branches called **basidia.**

In addition to these three major divisions of fungi, two other groups are often considered when discussing the fungi: the fungi imperfecti (deuteromycetes) and the lichens. The **fungi imperfecti** have many characteristics of the ascomycetes, but, in most cases, sexual reproduction has not been observed. Many are parasitic and pathogenic to plants and animals, including humans, causing infections of the skin and mucous membranes (ringworm and athlete's foot). Some fungi imperfecti are used in the production of certain cheeses (roquefort and camembert) and antibiotics, including penicillin.

The **lichens** are a large and diverse group of mainly ascomycete fungi that exist in a mutualistic relationship with a green alga or cyanobacterum. Occasionally, a basidiomycete or deuteromycete may be involved. The body of a lichen is quite distinctive in appearance. Three major forms are evident: crustose, foliose, and fruticose. Reproduction frequently involves the formation of ascospores by the fungal symbiont or the production of soredia (small fragments consisting of at least one algal cell surrounded by fungal hyphae).

In this exercise, you will study representative examples of the zygomycetes, ascomycetes, basidiomycetes, and lichens.

A. DIVISION ZYGOMYCOTA (ZYGOMYCETES OR BREAD MOLDS)

One of the more common members of this division is *Rhizopus stolonifer*, black bread mold, which grows as cottonlike masses on bread, fruit, and other organic material that is high in carbohydrate. Infection begins when a haploid spore germinates and grows into masses of filamentous hyphae that differentiate into **rhizoids** (these anchor the fungus to its substrate, secrete digestive enzymes, and absorb partially digested organic materials) and **sporangiophores** (aerial hyphae that produce **sporangia** containing spores at their tips). As sporangia mature they become black, thus giving the fungus its common name.

1. Asexual Reproduction

Examine black bread mold in a petri dish. The mold is growing not on bread but on agar containing various organic compounds needed for growth. If you wish, bring some bread from home (remember that most mass-produced breads contain mold inhibitors) and place a piece in a petri dish. Moisten it with several drops of water, and then place a small piece of the mold from the agar on the bread. Cover the dish and set it aside for 1–2 days.

Leave the cover on the petri dish to prevent spores from being released into the air, while examining the mold with the dissecting microscope. Note the whitish mass of filaments growing over the surface of the agar (Fig. 17-1A). Each filament is called a **hypha** (plural **hyphae).** The total mass of hyphae is called a **mycelium.** Some hyphae grow upward and form small, black, globelike structures called sporangia (Fig. 17-1C). Inside the sporangia are cells called **spores,** which are released when the sporangia open. What is the function of the spores?

____

____

Other specialized hyphae penetrate the agar (or bread). Turn the dish over, and focus downward through the agar to locate small, rootlike hyphae called **rhizoids** growing into the agar (Fig. 17-1C). What is the function of the rhizoids?

____

____

Place a small piece of the mold in a drop of water on a slide (Fig. 17-1B). Add a coverslip and examine the mold with the microscope. Locate sporangia, spores, hyphae, and rhizoids.

2. Sexual Reproduction

Sexual reproduction occurs when the hyphae of two mating strains meet and fuse, attracted by hormones that diffuse from the hyphae. The two strains are designated + and − because they show no morphological differences by which to designate them as male or female.

Your instructor has inoculated an agar plate with a + strain and a − strain. The growth of both strains has brought them into contact. At the points of contact, structures, called **gametangia,** containing + and − nuclei are formed (Fig. 17-2A). Fusion of the gametangia results in the formation of a line of black, thick-walled **zygospores** (Fig. 17-2B, C), each containing several diploid nuclei. Zygospores can remain dormant for several months. Meiosis usually occurs just before the zygospore germinates. All haploid nuclei except one degenerate. An aerial hypha develops with a sporangium at its tip. Haploid spores, produced in the sporangium, can, on release, germinate and begin a new cycle. Keeping the dish closed, located the zygospores with the dissecting microscope.

B. DIVISION ASCOMYCOTA (SAC FUNGI)

When reproducing sexually, most ascomycetes produce **asci (meiocytes)** in ascocarps. Each ascus typically produces eight **ascospores (meiospores).**

1. Yeasts

Mount some living yeast cells on a slide and examine them microscopically. Look for nuclei and small glistening food granules (Fig. 17-3A). Note that some of the yeast cells exhibit small, rounded projections called **buds** (Fig. 17-3B, C, D). What type of reproduction do these buds represent?

____

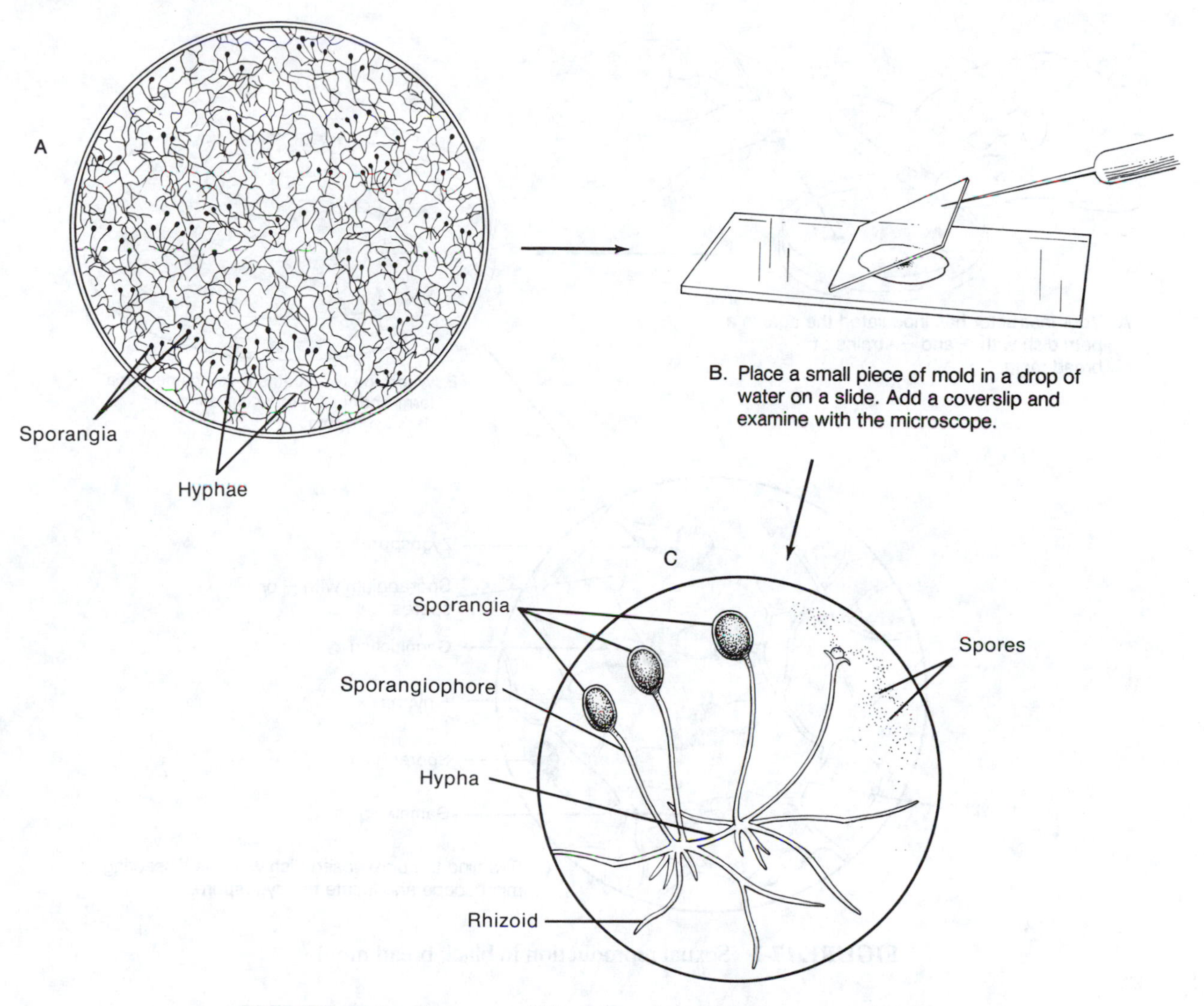

FIGURE 17-1 Structure of black bread mold (*Rhizopus stolonifer*)

If none of the yeast cells are budding, examine a demonstration slide that shows this process.

Yeasts are classified as ascomycetes because at some period in their life cycle an ascus is formed. Examine a demonstration slide that shows the ascus of yeast and its contents. List several ways in which yeasts are economically important to human beings.

____

____

____

____

2. Powdery Mildew (Microsphaera)

Examine lilac plant leaves infected with this fungus. Why is this organism called *powdery* mildew?

____

____

____

____

The hyphae that make up the mycelium that grows over the surface of the leaf occasionally pen-

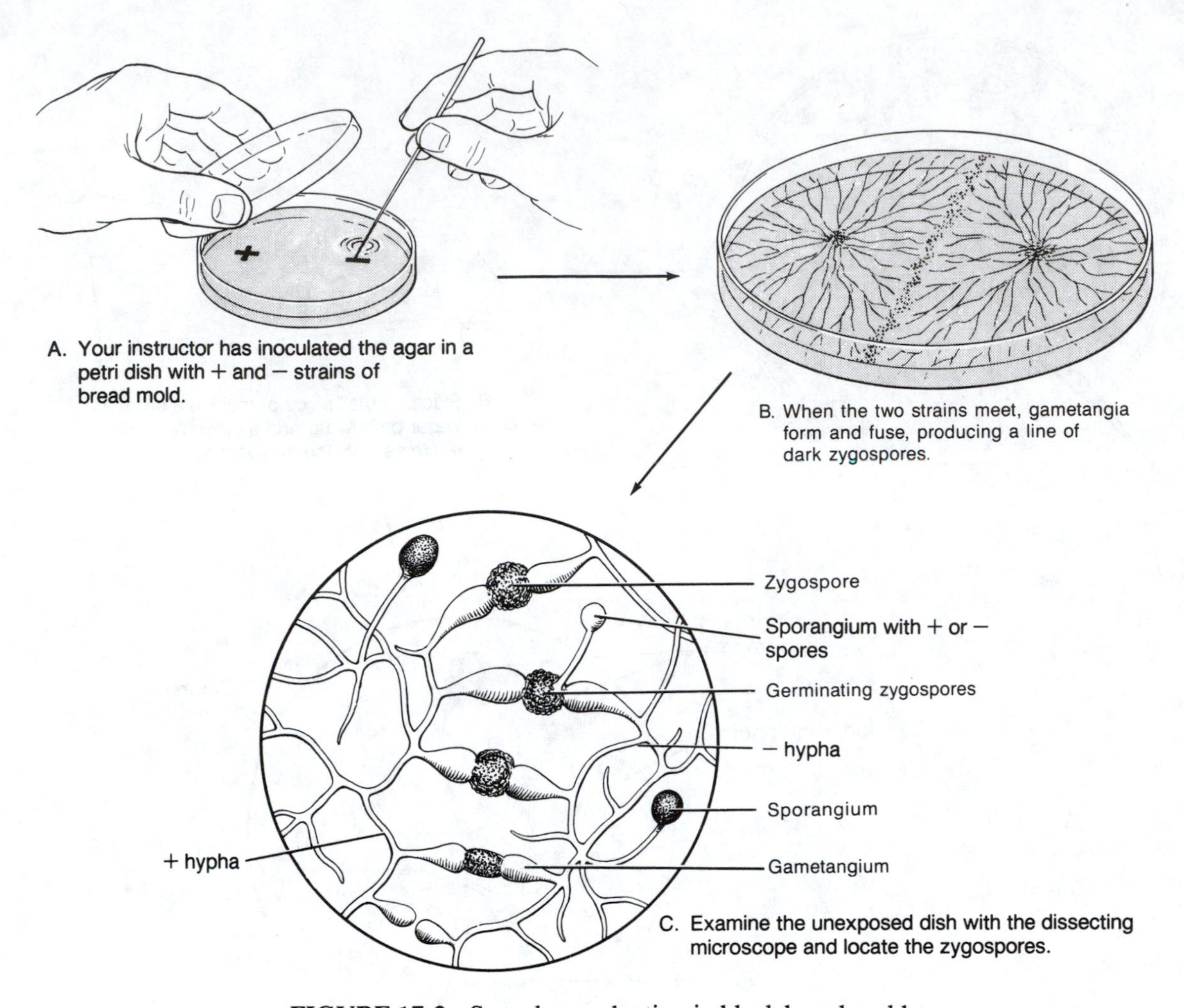

FIGURE 17-2 Sexual reproduction in black bread mold

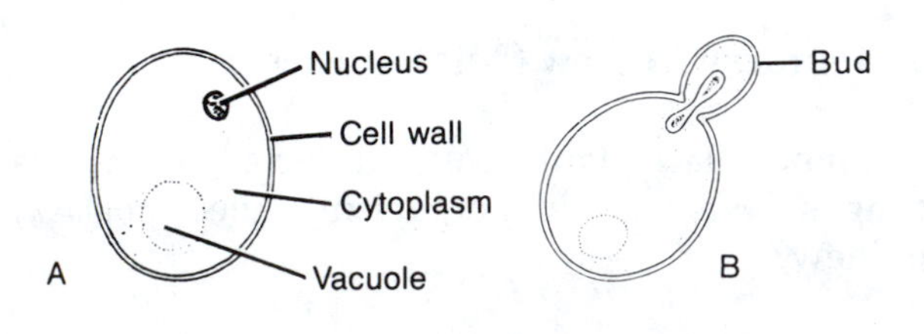

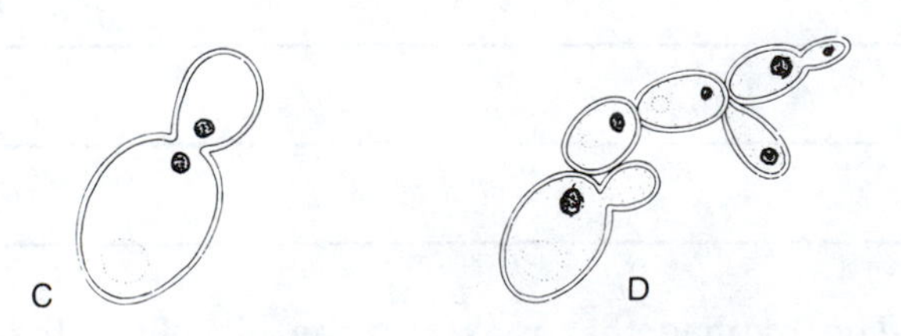

FIGURE 17-3 Budding in yeast

etrate the epidermal cells (Fig. 17-4). These penetrating hyphae are called **haustoria** (singular **haustorium).** What is their function?

___

Examine demonstration slides that show haustorial penetration of the epidermal cells of the host plant. What kind of nutrition does this fungus exhibit?

___

During late spring, large numbers of asexual spores **(conidia)** are produced at the ends of specialized aerial hyphae and disseminated by the wind, spreading the infection to the same or other lilac plants (Fig. 17-4, top). Examine slides of powdery mildew and locate conidia.

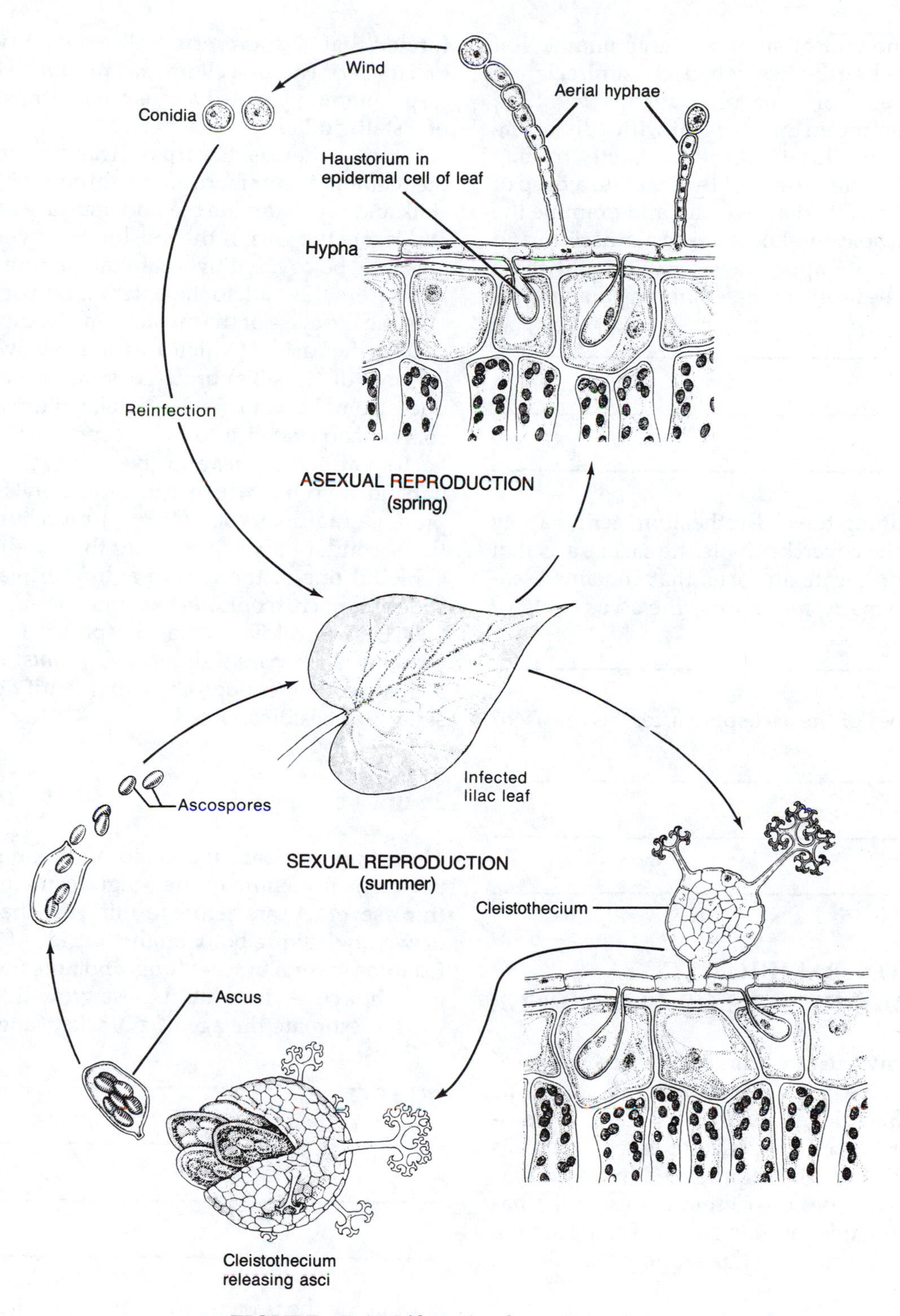

FIGURE 17-4 Life cycle of powdery mildew

Toward the end of summer, large numbers of spherical fruiting bodies called **cleistothecia** are formed as a result of sexual reproduction (Fig. 17-4, bottom). Examine an infected leaf with a dissecting microscope and locate some of these fruiting bodies. Scrape the surface of the leaf into a drop of water on a slide. Add a coverslip and examine the slide microscopically. Locate a cleistothecium and note its elaborate appendages. How might these appendages be useful in disseminating the fungus?

____

____

____

While examining the cleistothecium, gently apply pressure to the coverslip. Note the saclike asci that are extruded. Locate an ascus that contains ascospores. How many spores does the ascus contain?

____

What becomes of the ascospore after it is released?

____

____

____

C. DIVISION BASIDIOMYCOTA (BASIDIOMYCETES OR CLUB FUNGI)

The basidiomycetes are a large and varied group of fungi that have both saprophytic and parasitic members. **Basidiospores** (meiospores) are produced by structures composed of one to four cells called **basidia,** which are produced in large numbers in several types of fruiting bodies called **basidiocarps.** Included in this group of fungi are the toadstools, mushrooms, smuts, and rusts.

1. Mushrooms

Mushrooms are characterized by plates or gills on the undersurface of the basidiocarp. Prior to the formation of this elaborate fruiting body, the primary mycelia, which contain haploid nuclei, grow beneath the surface of their substrate (e.g., the ground or a rotting log). When haploid mycelia of different strains unite, a secondary mycelium is formed that is **dikaryotic** (cells contain two haploid nuclei). This mycelium may produce a basidiocarp, commonly called a mushroom, that consists of a stalk and cap.

Examine the basidiocarp of *Agaricus campestris*, the common commercial mushroom. Note the stalk and cap. Examine the undersurface of the cap and locate the gills. If the mushroom is young, the gills may be covered by a thin membrane that extends from the stalk to the outer margin of the cap.

In the life cycle of the mushroom, the diploid nucleus in the basidia (which are found all over all the surfaces of the gills) undergoes meiosis and produces four haploid nuclei, which ultimately become incorporated into basidiospores (Fig. 17-5). Each spore germinates and gives rise to the primary haploid mycelia, which can ultimately fuse and produce the dikaryotic ($N + N$) mycelium. From this secondary mycelium arises the basidiocarp.

Mount one of the gills in a drop of water on a slide and microscopically examine the edges of the gills. Locate basidia and basidiospores. If available, examine a prepared slide of *Coprinus,* showing a cross section through the cap. Identify gills, basidia, and basidiospores.

2. Bracket Fungi

These fungi are parasitic or saprophytic on various trees. The mycelium of the fungus can grow in the trunk several years before forming the characteristic woody fruiting body on the outside of the tree. Examine several bracket fungi and note the several growth layers. How could these growth layers be used to estimate the age of a bracket fungus?

____

____

____

____

Examine the undersurface of one of the fruiting bodies with a dissecting microscope. How does the undersurface of a bracket fungus differ from a mushroom?

____

____

Examine a prepared slide showing a cross section

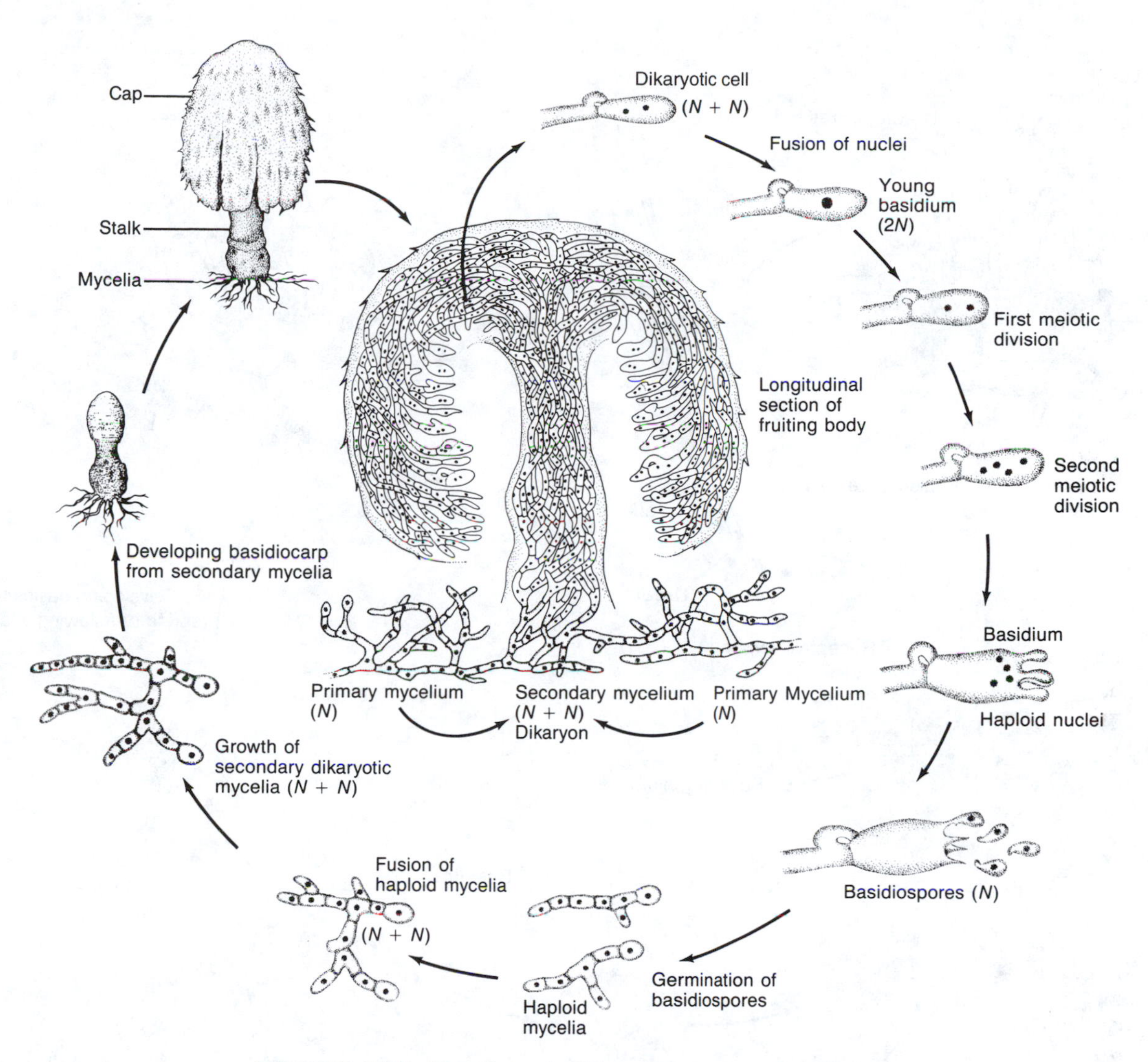

FIGURE 17-5 Details in the life cycle of a mushroom

of the fruiting body of a bracket fungus. What do the circular openings in the cross section represent?

Locate basidia and basidiospores.

3. Rusts

The rusts are a group of parasitic basidiomycetes that infect many species of seed plants as well as some of the ferns. All species of rust produce at least two distinct types of spores; some have three, four, or five types.

The most widely known rust, which causes the greatest economic loss, is *Puccinia graminis*, wheat rust. Infection by this parasite affects the wheat plant in several ways. First, many of the host's cells are killed because the fungus uses the content of the cells for its own growth. Second, the fungus robs the host of food. Third, due to the killing of cells that contain chloroplasts, photosynthetic processes are greatly reduced. Thus, as a result of infection by *Puccinia*, the wheat plant becomes pale green and stunted and the wheat ripens prematurely, with its small, shrunken, kernels having very limited food reserves.

a. Phases of the Rust in the Wheat Plant

Urediospores. Examine stems and leaves of wheat that show reddish patches (hence the name *rust*) on the surface. These masses (called **uredia;** singular **uredium)** contain large numbers of red-orange, binucleate urediospores (Fig. 17-6A).

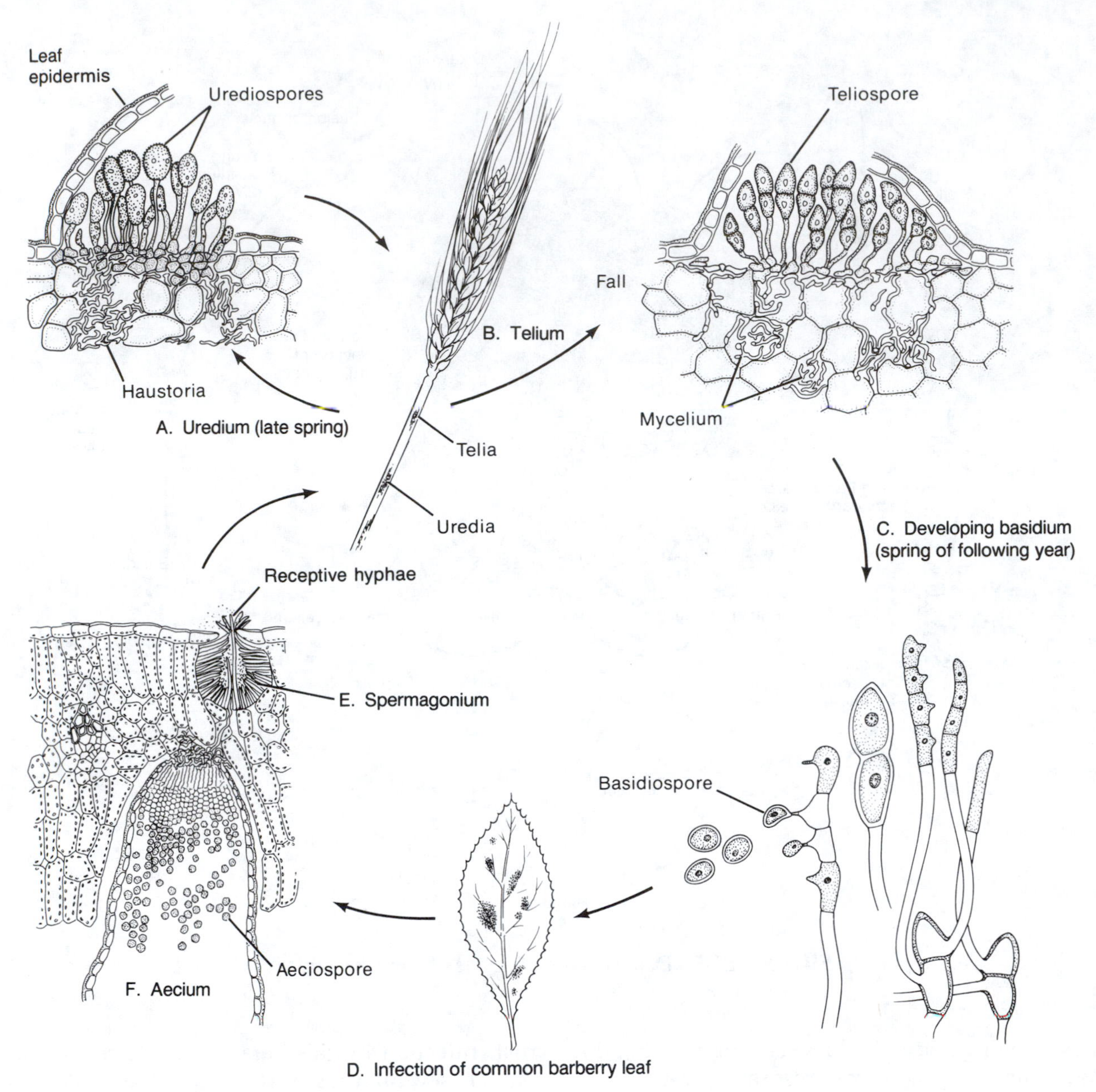

FIGURE 17-6 Life cycle of wheat rust

These, the first of two types of spores, appear in late spring and are produced throughout most of the growing season until the plant is mature. Urediospores are transported by various means (but primarily by wind) to other wheat plants, where the spores germinate and send out hyphal branches that enter leaf stoma and penetrate the intercellular spaces of the leaf. The mycelium becomes highly branched. Haustoria penetrate host cells and absorb food. After a period of growth, the hyphal mass forms a layer of cells just under the epidermis and then develops into the stalked urediospores, which are produced in such abundance that the epidermis ruptures.

With a razor blade, scrape a small amount of the uredia into a drop of water on a slide and add a coverslip. Locate and examine some urediospores. Next examine a slide of a cross section through a uredium. Locate urediospores, mycelia, and haustoria.

Teliospores. As the wheat plant approaches maturity, the mycelium, which until this time was producing urediospores, now begins to produce teliospores (Fig. 17-6B), located in masses called **telia.** These spores have thicker walls than urediospores and do not usually germinate until the following spring, after the wheat matures.

Scrape a small amount of material from a wheat stem or leaf containing telia into a drop of water on a slide. Add a coverslip and examine microscopically. Locate and examine some teliospores. How many cells make up each teliospore?

Next examine a slide of a longitudinal section through a telium. Locate immature teliospores, which when first formed are binucleate, and mature uninucleate spores, in which the two early nuclei are fused.

b. Phases of the Rust in the Common Barberry Plant (Alternate Host)

In the spring of the following year, the nuclei of the cells making up the teliospore undergo meiosis; each cell then contains four haploid nuclei. Each cell then forms a short hypha that develops into a four-celled basidium. The nuclei migrate into the basidium, each cell of which then contains a haploid nucleus. Each cell of the basidium forms a small projection, into which the nucleus will pass. The swollen end of this projection becomes a basidiospore.

The wheat rust requires an alternate host to complete its life cycle. Basidiospores cannot reinfect wheat. They can infect only the alternate host, the common barberry (Fig. 17-6C, D). When it lands on a fruit, branch, or leaf of the barberry, the basidiospore germinates and forms a hyphal branch that penetrates the tissue. In the tissue, the mycelium proliferates extensively. Ultimately, two types of structures are formed: **spermagonia** (singular **spermagonium)** and **aecia** (singular **aecium)** (see Fig. 17-6E, F). These are best observed by examining longitudinal sections through them.

Spermagonium. The spermagonium is a flask-shaped structure just below the epidermis. It contains numerous hairlike hyphae, at the ends of which are sporelike cells called **spermatia** (singular **spermatium).**

Spermatia are carried by insects to **receptive hyphae** protruding from other spermagonia where the nucleus from the spermatium enters the receptive hypha. This initiates the development of another type of mycelium that ultimately produces the cuplike aecia that protrude from the lower surface of the barberry leaf (Fig. 17-6E).

Aecium. Masses of **aeciospores** are formed within the aecia (Fig. 17–6F). If these spores are carried by air currents to the leaves or stems of wheat, they germinate and form an intercellular mycelium that then forms urediospores to complete the life cycle.

Examine infected leaves of barberry and locate spermagonia and aecia. Then examine slides showing longitudinal sections through these structures. Locate spermatia, receptive hyphae, and aeciospores.

How would you control the spread of a fungus, such as wheat rust, whose infective spores can be carried by air currents to infect plants thousands of miles away?

D. FUNGI IMPERFECTI

Much spoilage of food, leather, and cloth is caused by the fungus *Penicillium.* Examine a living culture of *Penicillium.* Why is it sometimes referred to as a blue-green mold?

Examine prepared slides showing the special hyphal branches that produce asexual spores or **conidia** (Fig. 17-7). How does *Penicillium* differ from *Rhizopus* in the way in which spores are produced?

How is it similar to *Microsphaera*?

One species, *Penicillium notatum,* produces a potent antibiotic, penicillin. It is one of the more effective antibiotics for combating bacterial infections.

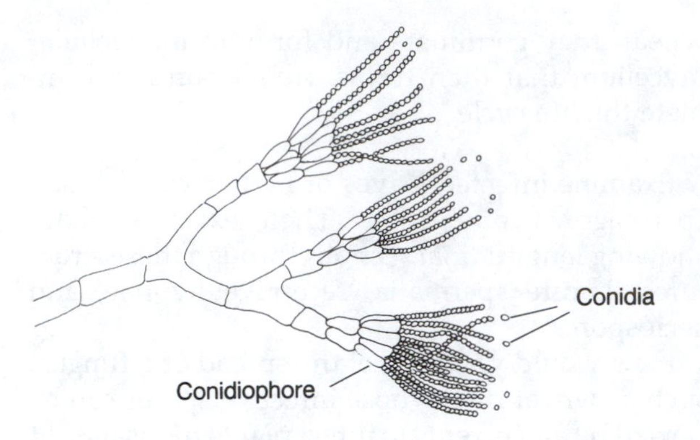

FIGURE 17-7 Conidia of *Penicillium* on branched conidiophore

E. LICHENS

Lichens are called **pioneer plants** because they often grow on bare rock and are the first "plants" to cover burnt-out regions. Lichens, however, are not plants. They are a symbiotic partnership between a fungus (usually an ascomycete) and a cyanobacterium (blue-green alga). Because of their ability to obtain nutrition from the photosynthetic partner, lichens have invaded some of the harshest environments in the world. They occur in arid desert regions and the arctic and grow on bare soil, tree trunks, rocks, and under water.

1. Vegetative Features

The plant body, or thallus, of lichens is distinctive in appearance, and lichens are classified largely according to their form of growth. The following are descriptions of three main types of plant body or thallus formed.

- **Crustose** Grows closely appressed to its substrate, with only the upper surface visible
- **Foliose** More loosely attached to its substrate, with both upper and lower surfaces visible
- **Fruticose** Attached at one point to its substrate and grows erect or pendant

Examine various lichens with the naked eye and with a dissecting microscope. Using the descriptions given, determine which type of plant body or thallus each lichen has. Record your observations in Table 17-1.

TABLE 17-1 Classification of vegetative growth of lichen

Name	Type of plant body (thallus)

2. Microscopic Anatomy

With a sharp razor blade, cut a thin cross section through a foliose or crustose lichen. Alternatively, use commercially prepared slides. Examine microscopically and, referring to Fig. 17-8, locate the following parts.

Upper cortex: Protective, dense aggregation of fungal hyphae

Algal layer: Layer of algal cells and loosely interwoven masses of thin-walled fungal hyphae

Medulla: Somewhat thick layer of loosely interwoven, colorless hyphae. This layer comprises about two-thirds of the plant body or thallus and is thought to serve as a storage region.

Lower cortex: Layer thinner than the upper cortex, with projections called **rhizines** that attach the lichen to its substrate

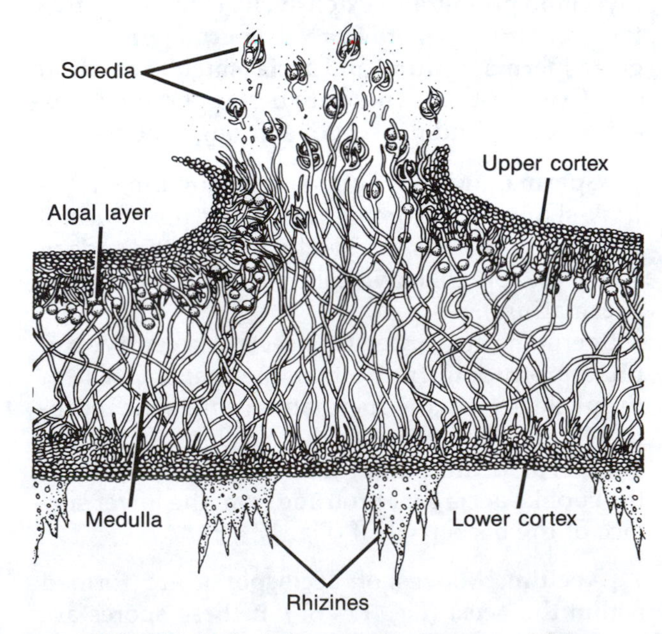

FIGURE 17-8 Microscopic structure of lichen

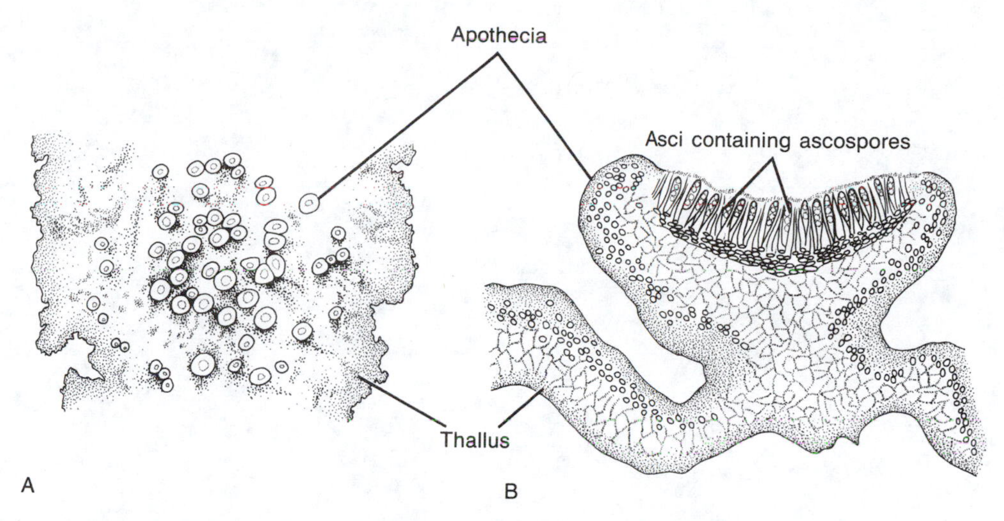

FIGURE 17-9 (A) Apothecia on surface of lichen. (B) Cross section through apothecium showing asci containing ascospores.

3. Asexual Reproduction

Examine various specimens with a dissecting microscope and look for small, granular masses on the surface. These are **soredia,** which contain one or more algal cells surrounded by the fungal hyphae. Soredia are dispersed by wind or rain, and each one is capable of growing into a new lichen thallus.

Scrape a few soredia into a drop of potassium hydroxide (or other wetting agent) and examine with the compound microscope. Supplement your observations with commercially prepared slides of a lichen thallus (Fig. 17-8).

4. Sexual Reproduction

Select a lichen that has small, cup-shaped structures on its surface, called **apothecia** (Fig. 17-9A). Cut a thin longitudinal section through an apothecium and examine microscopically (or examine a prepared slide). Locate elongated asci (Fig. 17-9B). How many spores are found in each ascus?

____

How are they arranged?

____

Which of the lichen symbionts is reproducing sexually? Explain.

____

____

____

____

____

____

____

REFERENCES

Ahmadjian, V. and S. Paracer. 1986. *Symbiosis: An Introduction to Biological Associations.* University Press of New England.

Alexopoulos, C. J., and C. W. Mims. 1979. *Introductory Mycology.* 3d ed. Wiley.

Bold, H. C., and J. LaClaire, eds. 1984. *The Plant Kingdom.* 5th ed. Prentice-Hall.

Margolis, L., and K. V. Schwartz. 1987. *Five Kingdoms: An Illustrated Guide to the Phyla on Earth.* 2d ed. W. H. Freeman.

Richardson, D. H. S. 1975. *The Vanishing Lichens: Their History, Biology and Importance.* Hafner.

Smith, A. H. 1980. *The Mushroom Hunter's Field Guide.* University of Michigan Press.

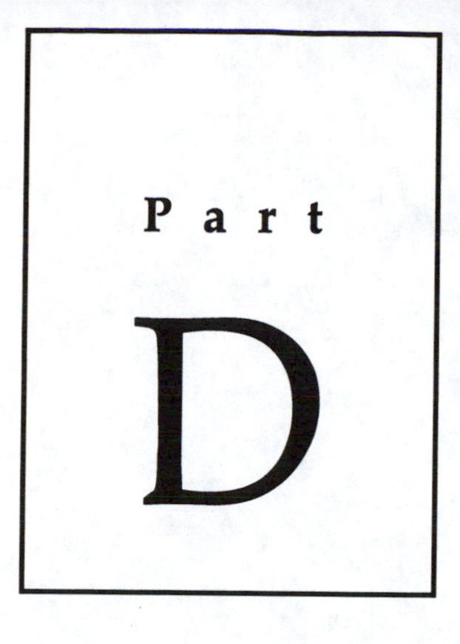

Diversity of Life: Kingdom Plantae

Only a few kinds of living organisms—some bacteria, algae, and plants—have the ability to capture light energy and, by the process of photosynthesis, use it to synthesize organic compounds. Directly, or indirectly, all living creatures rely on these photosynthetic organisms for their own energy requirements.

Plants are basically terrestrial organisms that have evolved from the green algae. During the course of their evolution they have acquired a multitude of specialized features that adapt them to living on land. Not long after transition to land, plants evolved into at least two major groups. One group is the bryophytes, which includes the present-day mosses and liverworts; the second group includes the vascular plants. A primary difference between the groups is that vascular plants, as implied by their name, have a well-developed vascular system that transports water, minerals, sugars, and other regulatory substances through the plant body. Bryophytes, on the other hand, are rather simply constructed and relatively small. Lacking a vascular system, bryophytes are typically restricted to moist environments and generally grow close to the ground in contrast to the many vascular plants that achieve great heights.

In these studies you will become familiar with examples of these groups and examine their structure, reproduction, and some factors that regulate their growth and development.

Exercise 18

Kingdom Plantae: Division Bryophyta

Bryophytes, which consists of mosses, liverworts, and hornworts, are inconspicuous plants that grow in moist habitats. Many bryophytes, particularly some of the mosses, can withstand prolonged periods of drought, and for this reason they are successful in regions where temperatures are below freezing for much of the year. However, bryophytes do require water during their reproductive periods because their sperm must swim through water to fertilize the egg. In addition, because they lack vascular tissue, most forms rely on the surrounding water to transport fluids and salts necessary for growth.

Bryophytes, like all plants, exhibit an **alternation of generations** in which a diploid phase, or **sporophyte,** alternates with a haploid phase called a **gametophyte.** In bryophytes, the predominant phase is the gametophyte plant, which produces gametes by mitosis. When the gametes fuse, they form a diploid zygote. The zygote develops into the sporophyte, in which meiosis eventually occurs and haploid spores are formed. These spores germinate and grow into a gametophyte.

In bryophytes, the gametophyte is typically the larger of the two phases and is photosynthetic. The sporophyte is almost always smaller than the gametophyte and always obtains its food from the gametophyte. Indeed, in some bryophytes, the sporophyte is almost completely enclosed in the gametophyte.

The three classes of bryophytes have the following characteristics.

- Class Hepaticae (liverworts): The gametophytes are green and photosynthetic, dorsoventrally flattened, and bilaterally symmetrical; reproductive structures are multicellular. Water is necessary for fertilization.
- Class Musci (mosses): The gametophyte is erect and radially symmetrical and has stemlike and leaflike structures. Reproductive structures are multicellular and require the presence of water for fertilization. The mosses lack true vascular tissues.
- Class Anthocerotae (hornworts): The gametophyte is a multilobed thallus, similar to the liverworts, and grows closely appressed to rocks and soil. Gametophytes may contain nitrogen-fixing cyanobacteria. The reproductive structures are multicellular.

In this exercise, you will study representatives of the liverworts and mosses, with special emphasis on their structures, habitat, and reproductive processes.

A. CLASS HEPATICAE (LIVERWORTS)

1. The Gametophyte

The liverworts were given their odd name because the lobing of the gametophyte plant in some species resembles the lobes of the liver. The term *wort* means "plant." Examine living gametophyte plants of *Marchantia*, and note the Y-shaped **(dichotomous)** branching of the plant body **(thallus).** Mitotic divisions in the cells located in the notch of the Y produces the peculiar growth pattern seen here.

1. Remove a small piece of the gametophyte thallus and examine the ventral (bottom) surface with a dissecting microscope. Locate slender hairlike structures, called **rhizoids.** Suggest a function of the rhizoids.

___

___

2. Examine the dorsal surface of the gametophyte thallus. Observe the minute "pores" in the center of small, diamond-shaped areas. Because the gametophyte of *Marchantia* is photosynthetic, what role might be ascribed to these openings?

___

___

___

3. Examine a prepared slide of a cross section of *Marchantia* (Fig. 18-1). Locate the upper epidermis and the pores that were seen in the living plant. Below the epidermis, find a series of small air chambers that are partitioned by branching filaments arising from the floor of the chambers. In living material, the cells of these filaments contain chloroplasts. Suggest a function for the air chambers.

___

___

___

The tissue underlying the air chambers is several cells thick. Many of the cells contain colorless plastids called **leucoplasts.** A few large cells contain **mucilage,** a gelatinous substance that absorbs water. What is the advantage of having this substance in the cells?

___

___

On the lower epidermis, locate the rhizoids.

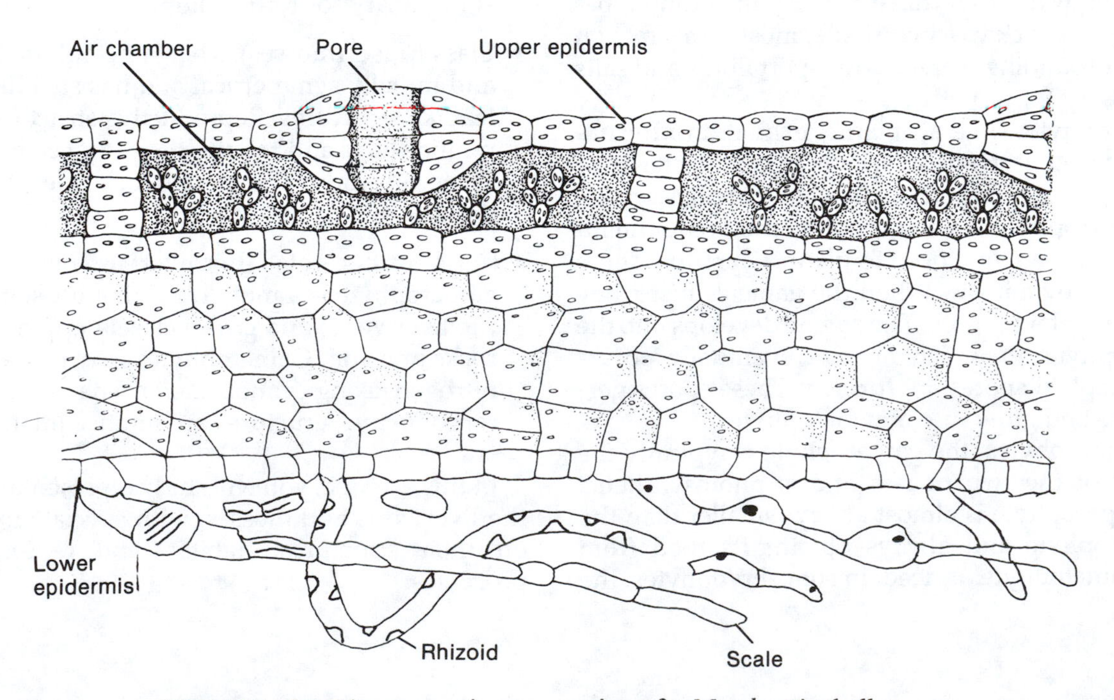

FIGURE 18-1 Diagrammatic cross section of a *Marchantia* thallus

2. Asexual Reproduction

The gametophyte of *Marchantia* reproduces asexually by means of small plantlets called **gemmae.** These are produced by the mitotic division of cells located in the bottom of gemma cups, which are found on the dorsal surface of the plant (Fig. 18-2). Examine gemma cups using a stereoscopic micro-

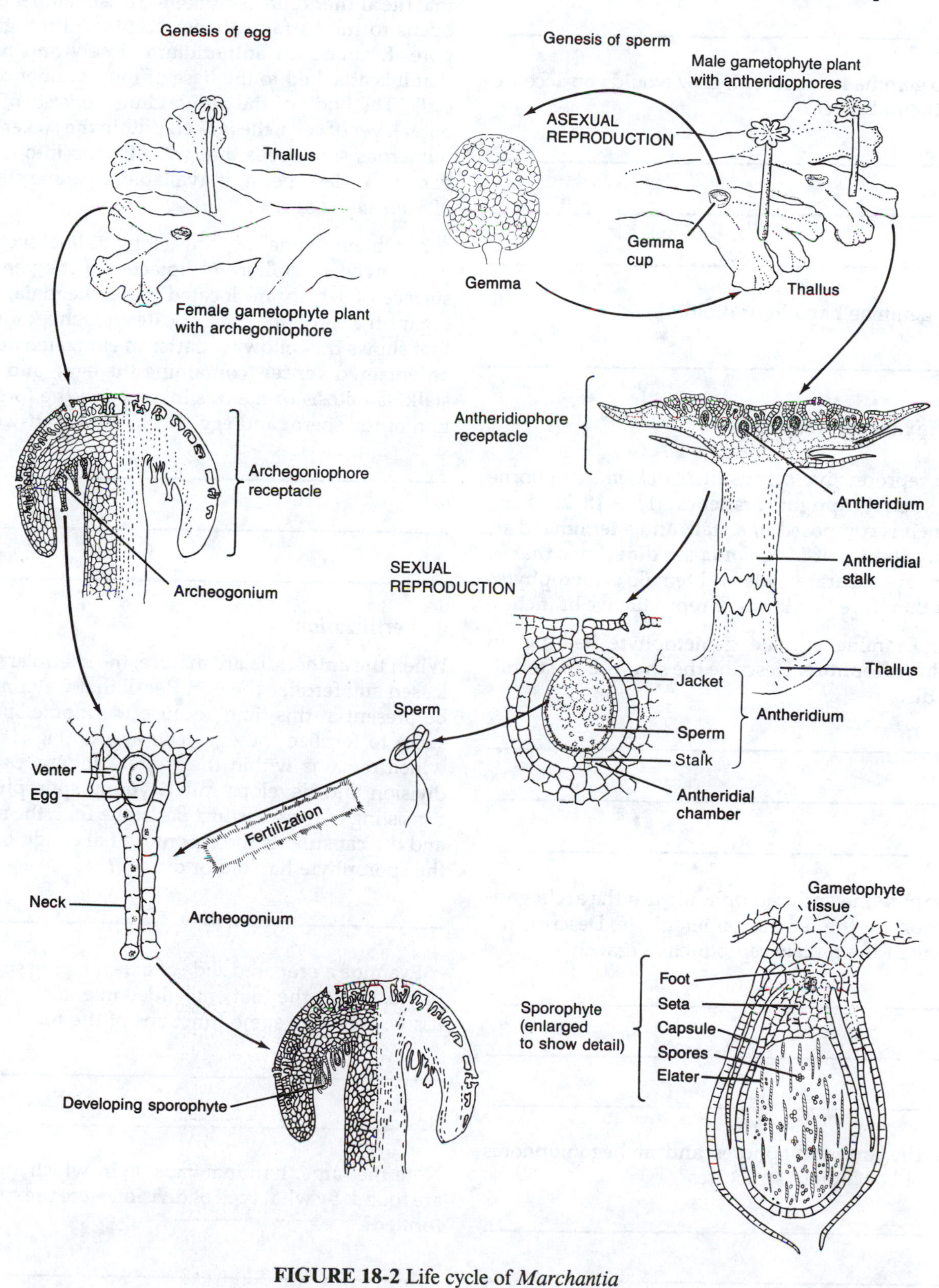

FIGURE 18-2 Life cycle of *Marchantia*

scope. Describe the appearance of any gemmae present in the cups.

If no gemmae are present, how would you account for their absence?

Are gemmae haploid or diploid?

3. Sexual Reproduction

The reproductive organs of *Marchantia* are borne on special upright branches (Fig. 18-2). Each branch is composed of a stalk and a terminal disc. Many species of *Marchantia* are **dioecious;** that is, there are separate male and female gametophytes and therefore two kinds of reproductive branches.

1. Examine a male gametophyte bearing an **antheridiophore.** Describe the shape of the terminal disc.

Compare the antheridiophore with the **archegoniophore** of the female gametophyte. Describe the shape of the female reproductive branch.

Are the antheridiophores and archegoniophores haploid or diploid structures?

2. Examine slides showing longitudinal sections of the antheridiophore receptacle. Locate the antheridia, sex organs that produce sperm, just below the upper surface of the receptacle (Fig. 18-2). Note that the antheridium is contained in a chamber that opens to the surface of the receptacle through a pore. Examine an antheridium closely and note that it is attached to the base of the chamber by a stalk. The body of the antheridium consists of an outer layer of cells, the **jacket.** Within the jacket are numerous small cells that will develop into male gametes called sperm. If available, examine slides of *Marchantia* sperm.

3. Obtain a slide showing a longitudinal section of an archegoniophore receptacle, on the ventral surface of which are located the archegonia, sex organs that produce eggs. Locate an **archegonium** that shows the following parts: an elongated neck, an enlarged **venter** (containing the egg), and the stalk. Is mitosis or meiosis involved in the formation of the sperm and egg? Explain your answer.

4. Fertilization

When the antheridia are mature, the sperm are released and fertilize the egg. Recall that water must be present at this time because the motile sperm swim to fertilize the egg. The zygote (Fig. 18-2), which remains within the venter, undergoes cell division and develops into a young sporophyte, consisting of three distinct parts: the **foot,** the **seta,** and the **capsule.** Is the chromosomal condition of the sporophyte haploid or diploid?

Examine a prepared slide of a developing sporophyte. Locate the foot embedded in gametophytic tissue. What are some functions of the foot?

Note the large, terminal capsule in which spores are found. By what type of division were the spores formed?

Are the spores haploid or diploid?

___

Into what structure will the spores develop?

___

Among the spores, locate elongated cells with spirally thickened walls. These are **elaters,** which help to disperse the spores when the capsule breaks open. Note that the sporophyte matures within the archegonium. Before the release of the spores, the capsule is pushed free of the gametophyte by the elongation of the stalk.

B. CLASS MUSCI (MOSSES)

Mosses differ from the liverworts in that the gametophyte of the moss begins as a filamentous, branching structure, a **protonema,** that differentiates into an upright, radially symmetric gametophyte consisting of "stems" and "leaves." In addition, the capsule of the sporophyte contains a series of "teeth" around its opening, which regulate the dispersal of the spores contained within.

1. Sporophyte of *Polytrichum*

Examine living or preserved specimens of the common moss *Polytrichum,* which consists of both gametophytic and sporophytic plants. The sporophyte is easily distinguishable because it consists of a terminal capsule (often covered by a hairy cap, the **calyptra),** a slender stalk (the **seta**), and a **foot** that is embedded in the top of the "leafy" gametophyte (Fig. 18-3).

1. Carefully pull the sporophyte out of the gametophyte. Remove the calyptra and save it for later examination.

2. Gently remove the lid of the capsule and examine the exposed surface of the capsule with a hand lens or dissecting microscope. Describe what you see.

___

___

Suggest a function for these structures.

___

___

3. Crush the capsule in a drop of water on a slide. Examine the contents of the capsule microscopically. If the sporophyte is mature, you will see numerous spores. These spores represent the first cells of the gametophyte generation. Are the spores haploid or diploid?

___

What type of division, meiosis or mitosis, is involved in the development of these spores?

___

2. Gametophyte

After its release from the capsule, if environmental conditions are suitable, the spore germinates and develops into a filamentous structure called a protonema. Note the similarity of the protonema to filamentous algae. What evolutionary significance, if any, is indicated here?

___

___

Look for small, budlike structures along the length of the filament. These will later develop into mature, "leafy" gametophytes (Fig. 18-3).

1. Examine the leafy part of *Polytrichum* that remains after removal of the sporophyte. The gametophyte is photosynthetic. The mature gametophyte consists of "leaves," a "stem," and rhizoids. How could you determine if there are true stems and leaves?

___

___

2. *Polytrichum* is a dioecious moss, with the reproductive organs located at the tips of separate male and female plants. With forceps, carefully remove as many leaves as possible from the apex of a female plant. Then cut off the stem tip and mount it in a drop of water on a slide. With a probe, gently tease the stem tip into small fragments. Examine microscopically. Locate the archegonia (Fig. 18-3). Note the canal that leads through the neck and terminates in the venter of the archegonium.

3. In a similar manner, mount a stem tip of a male plant. Locate the antheridia, which are the

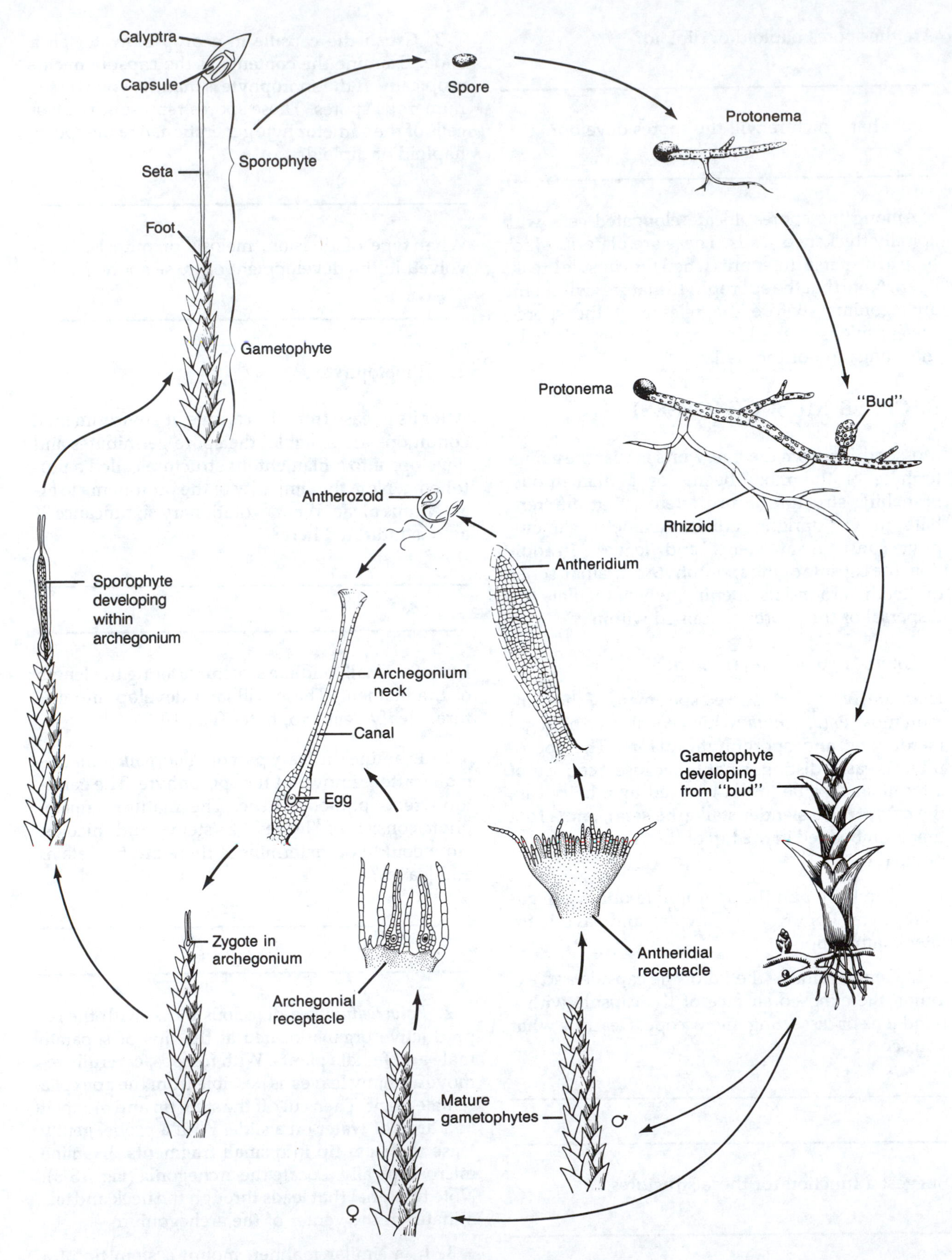

FIGURE 18-3 Life cycle of *Polytrichum*

TABLE 18-1 Increases in complexity shown by the bryophytes over the algae

Characteristic	Algae	Liverworts	Mosses
Morphology			
Habitat			
Asexual reproduction			
Sexual reproduction			

elongated, saclike structures consisting of an outer jacket and an inner mass of cells destined to become **antherozoids** (sperm). If available, examine prepared slides of moss antheridia and archegonia.

3. Fertilization

The nonmotile egg is retained in the venter of the archegonium. When the archegonium is mature, the cells lining the inside of the stalk disintegrate and a canal is opened leading to the female gamete. The male gamete swims through this canal and unites with the egg. What must be present for fertilization to occur?

____

The zygote undergoes a series of divisions and becomes a mass of undifferentiated cells, still held in the venter. The developing embryo differentiates into three parts: foot, seta (stalk), and capsule. The remains of the archegonium are retained as the calyptra (cap) and can be found enclosing the capsule. Is the calyptra sporophytic or gametophytic tissue?

____

Label the various structures in Fig. 18-3 with *N* or 2*N* to indicate whether they are haploid or diploid.

C. INCREASES IN COMPLEXITY SHOWN BY THE BRYOPHYTES

Complete Table 18-1. Indicate the increases in complexity shown by the bryophytes over the algae in morphology, habitat, and reproduction. The following suggested terms can be used.

- Morphology: unicellular, filamentous, spherical colonies, prostrate thallus, erect thallus, radial symmetry, bilateral symmetry
- Habitat: fresh water, salt water, ditches, stagnant pools, damp woods, terrestrial
- Asexual reproduction: fragmentation, cell division, zoospores, nonmotile spores, gemmae
- Sexual reproduction: unicellular or multicellular reproductive structures, motile or nonmotile gametes, alternation of generations, isogametes or anisogametes, necessity of water for fertilization, dependency or nondependency of sporophyte, antheridia, archegonia

REFERENCES

Conrad, H. S., and P. L. Redfern, Jr. 1979. *How to Know the Mosses and Liverworts.* 2d ed. Wm C Brown.

Margulis, L., and K. V. Schwartz. 1987. *Five Kingdoms.* 2d ed. W. H. Freeman.

Raven, P. H., R. F. Evert, and S. E. Eichhorn. 1986. *Biology of Plants.* 4th ed. Worth.

Richardson, D. H. S. 1981. *The Biology of Mosses.* Wiley.

Exercise

19

Kingdom Plantae: The Vascular Plants

Vascular plants, or **tracheophytes,** have a system of specialized tissues that carry water, dissolved minerals, and organic products throughout the plant. These specialized tissues are called **vascular tissues.** Vascular plants can transport great quantities of water and minerals over long distances and to great heights in relatively short periods of time. Vascular plants were able to colonize the land because they were no longer restricted to water environments.

The vascular plants consist of several divisions and many classes. Only the following divisions will be studied in this exercise.

- Division Pterophyta (ferns): The gametophytes are small, free-living, and photosynthetic. Multicellular sex organs called gametangia are present. Most are homosporous (producing one type of spore) but some are heterosporous (producing more than one type of spore, usually two).
- Division Coniferophyta (conifers): These plants have needlelike leaves. The sporophyte develops in specialized structures called seeds, which are contained in a cone.
- Division Anthophyta (flowering plants): The flower is the reproductive structure containing the male (stamen) and female (pistil) parts. Pollination is effected by wind and insects. Double fertilization (of egg and polar nuclei) is a common feature. Ovules are typically enclosed in carpels of the ovary. Mature seeds are contained in fruits, which may be fleshy or hard.

In this exercise, you will become familiar with the structures and reproductive patterns of ferns, conifers, and a flowering plant.

A. DIVISION PTEROPHYTA (FERNS)

1. Sporophyte

Examine the mature sporophyte of the fern *Polypodium.* Locate the horizontal stem, which may lie on or just under the surface of the soil. What is the term given to this type of stem?

____

Note the large, deeply lobed leaves called **sporophylls** (Fig. 19-1). On the undersurface of some of the leaves, locate small yellow-brown spots called **sori** (singular **sorus**), each of which contains many **sporangia.** What is produced in the sporangia?

____

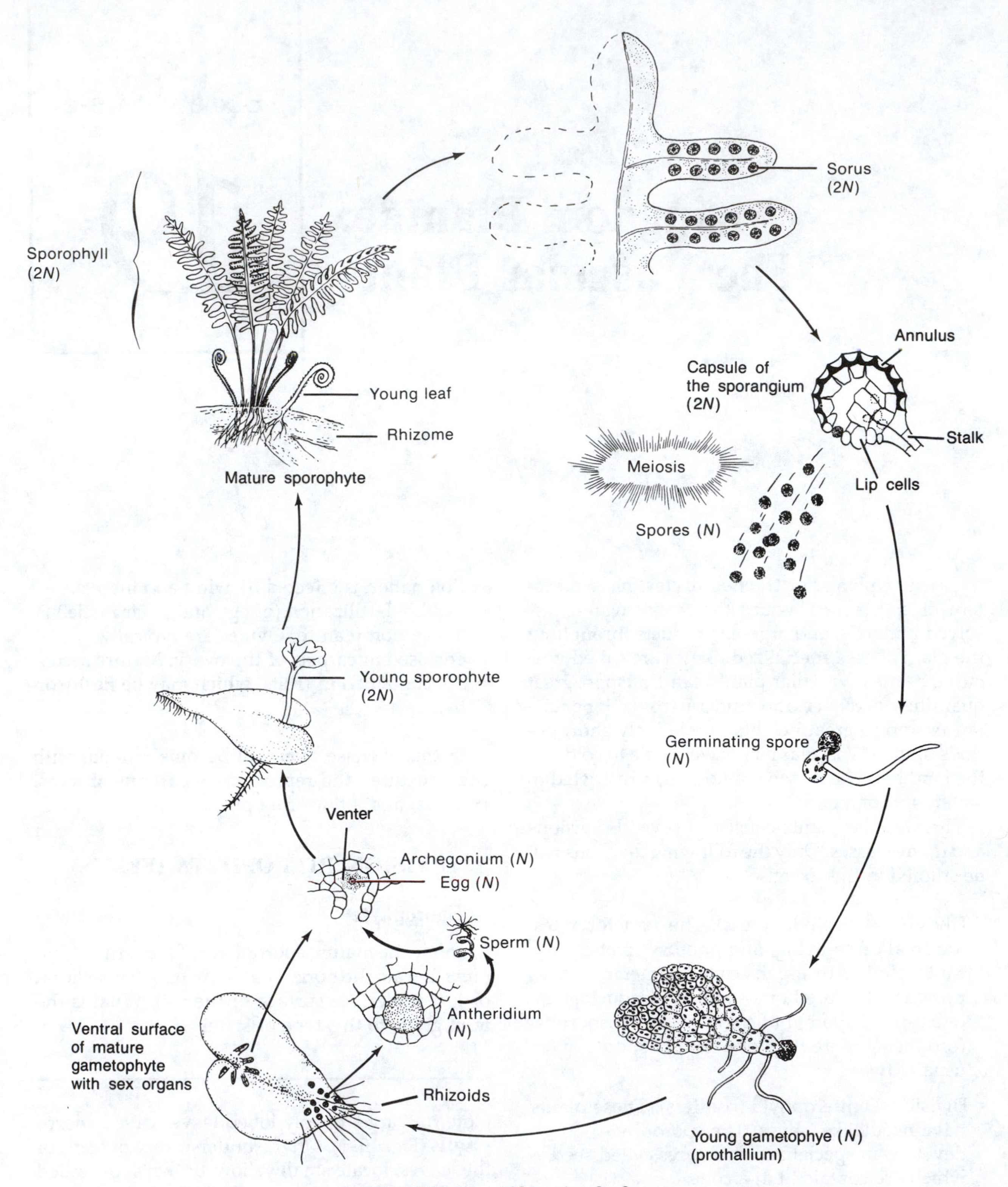

FIGURE 19-1 Life cycle of a fern

Obtain part of a fern sporophyll that bears sori. After examining the sorus with a hand lens or stereoscopic microscope, scrape the contents of a sorus into a drop of water on a slide and add a coverslip. Examine with the low power of the microscope. Locate a sporangium and note that it is composed of a stalk and an enlarged capsule. Examine the capsule under high power. Note a ridge of cells **(annulus)** that extends around the capsule. These cells have thickened inner and radial walls. As the capsule matures, the cells of the annulus lose water and the thin outer walls shrink. This results in a considerable amount of tension on the lateral, thin-walled **lip cells,** which then rupture, releasing the haploid spores.

Remove the coverslip from the slide and blot the excess water with a piece of filter paper. Examine the preparation (without the coverslip), and describe what happens to the sporangia as they dry.

____

____

2. Gametophyte

On germination, the fern spore develops into a short filament of cells that grows into a platelike young **gametophyte.** The mature gametophyte is a heart-shaped structure with a notch, containing the growing point, at one end. With a dissecting microscope, examine living fern gametophytes. Locate **rhizoids** on the lower, or ventral, surface.

Among the rhizoids of mature gametophytes are found numerous sex organs called **antheridia** (♂) or **archegonia** (♀) (Fig. 19-1). Examine the gametophyte with a dissecting microscope. Locate archegonia on the ventral surface just posterior to the apical notch. Only the neck of the archegonium is visible. The **venter** containing the egg is embedded in the gametophytic tissue. The antheridia, located at the opposite end of the plant, appear as small, rounded protuberances. Why do you think antheridia and archegonia are produced on the ventral rather than the dorsal surface of the gametophyte?

____

____

____

Also examine prepared slides of a mature gametophyte and locate the structures discussed.

Following fertilization, the **zygote** undergoes a series of mitotic divisions and develops into an embryo **sporophyte.** The young sporophyte, still held in the venter of the archegonium, differentiates into four lobes: One develops into an anchoring foot, one grows down into the soil and becomes a primary root, the third develops into the primary leaf, and the last becomes the stem. Until the primary root and leaf become functional, the sporophyte is dependent on the gametophyte for nourishment and water, but as soon as the root begins to absorb water and the leaf to photosynthesize, the sporophyte becomes independent. Soon after this, secondary leaves and adventitious roots are formed, and the primary root, leaf, and gametophyte die.

B. DIVISION CONIFEROPHYTA (CONIFERS)

The conifers include pines, spruces, cedars, and firs. Their seeds are exposed (another term for conifer is **gymnosperm,** "naked seeds") on scalelike structures. Seed plants have a well-defined alternation of generations. Unlike the ferns, their gametophytes are microscopic and completely dependent on the large free-living sporophyte, which has stems, leaves, and roots.

1. Pine Sporophyte

Examine branches from various species of pine, and note the two kinds of leaves. Locate the needlelike photosynthetic leaves borne on short spur shoots and inconspicuous scale leaves found at the bases of the spur shoots (Fig. 19-2). Is the number of leaves the same for all species?

____

Describe any variation among the species in the number of leaves in a cluster.

____

____

Would you consider this difference a means of separating one species of pine from another?

____

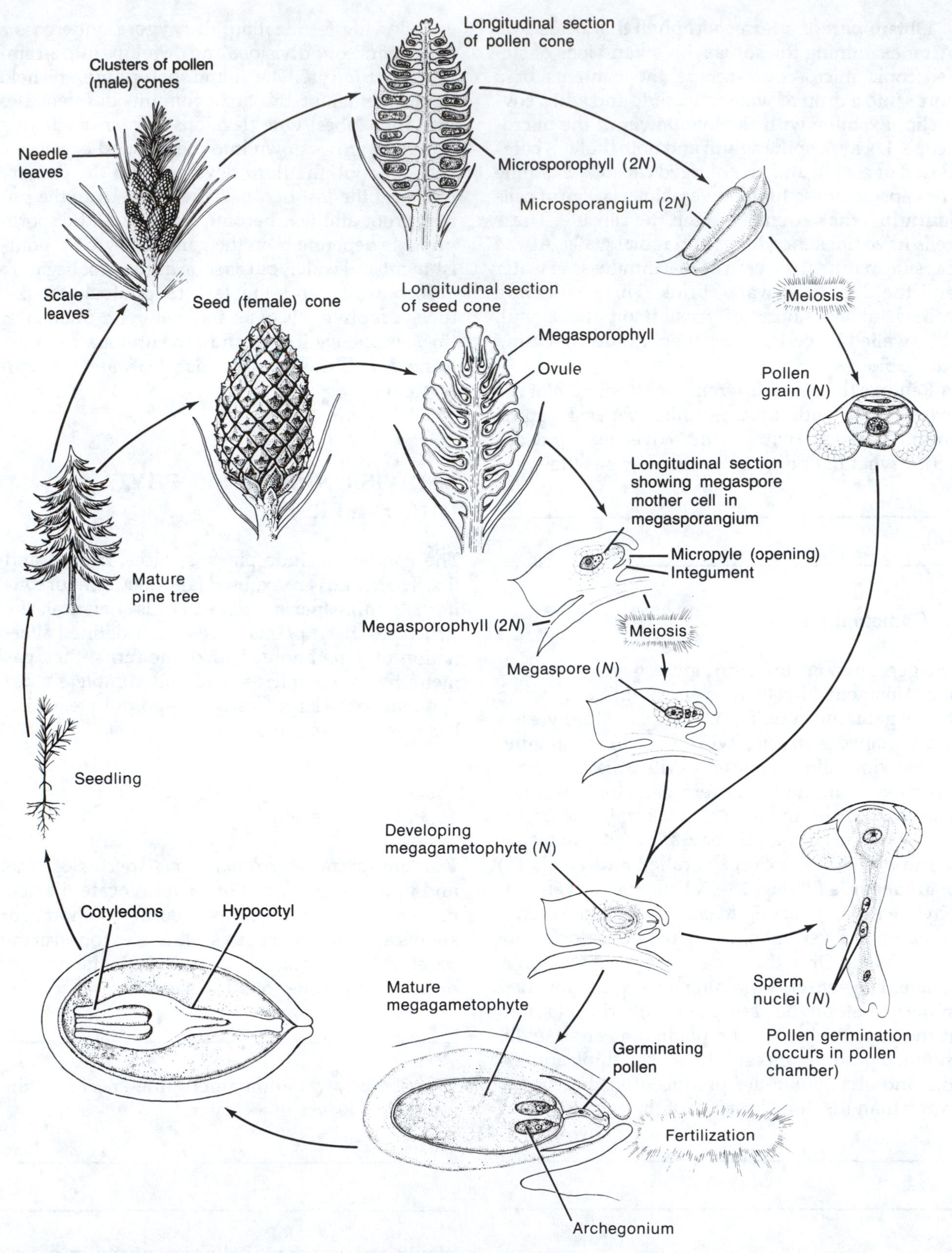

FIGURE 19-2 Life cycle of a pine

Why are pines called evergreens?

___

___

2. Development of the Pine Gametophyte

A pine tree produces two kinds of cones, pollen cones (♂) and seed cones (♀). Examine a branch bearing **pollen cones,** which are usually in clusters at the end of the branch. Dissect one of the pollen cones and locate the central axis, to which are attached spirally arranged scalelike structures called **microsporophylls** (Fig. 19-2). *Note:* The suffixes *micro* and *mega* are attached to the names of the reproductive organs of the higher plants to distinguish the smaller male organs from the larger female organs (Fig. 19-3). Remove a microsporophyll and examine it with a stereoscopic microscope. Note the two saclike **microsporangia.** Break open the microsporangia and examine the contents under the high-power lens of the compound microscope. Depending on the stage of development, you will observe either haploid **microspores** (consisting of a single cell) or **microgametophytes** (consisting of two prominent cells and two degenerate cells). What is the common name for the microgametophyte of pine?

___

Note the lateral, winglike appendages. How might these "wings" be advantageous?

___

The other kind of cone found on the pine tree is the female **seed cone,** which is borne on short lateral branches near the apex of young branches. They are partially hidden by the terminal bud. Ex-

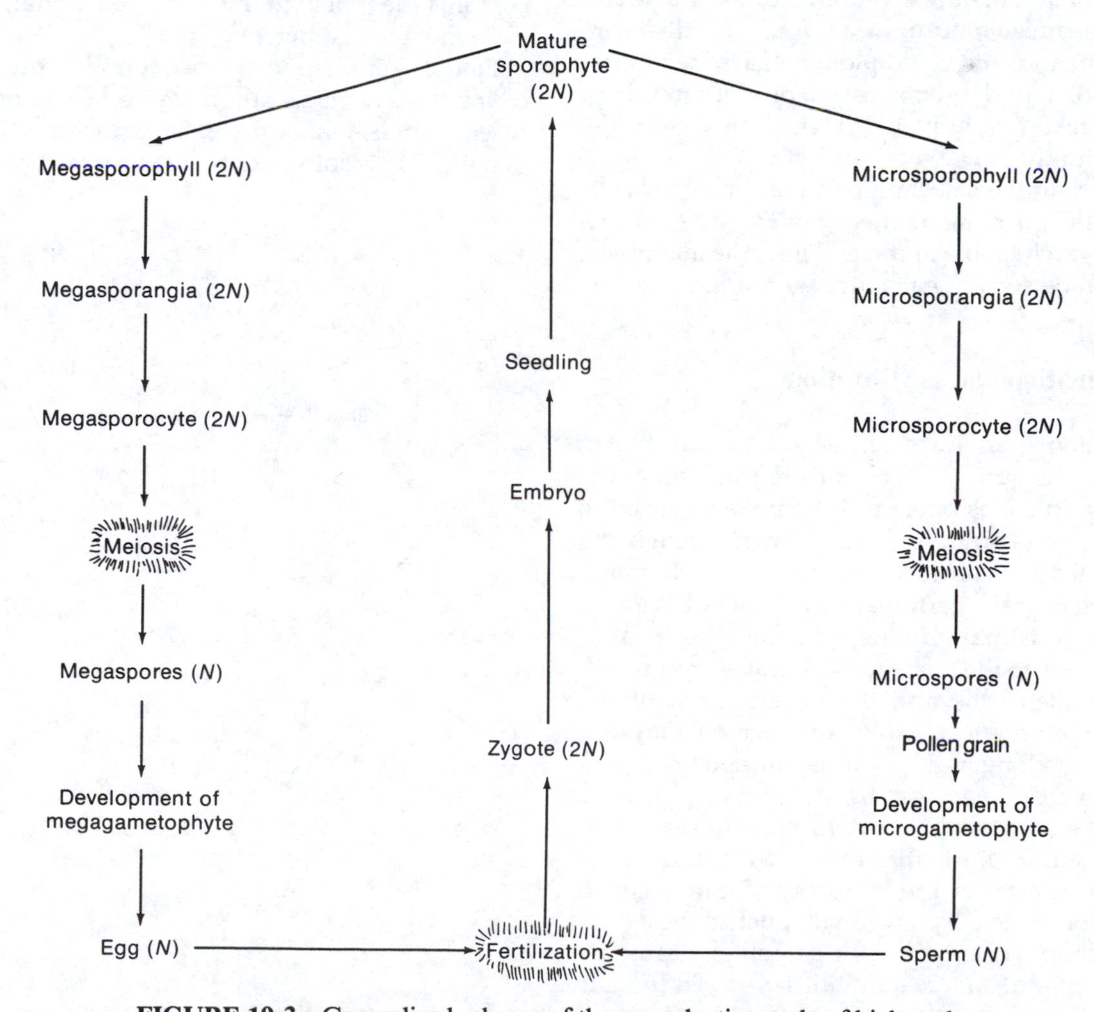

FIGURE 19-3 Generalized scheme of the reproductive cycle of higher plants

amine the seed cones and compare them with the pollen cones on the basis of size, shape, and location. Also examine older seed cones that have opened and discharged their seeds.

Dissect a seed cone and locate the overlapping, bractlike **megasporophylls** (Fig. 19-3) attached to the central axis. On the upper surface of each megasporophyll, locate two small white structures, the **ovules.** Note that the ovules are not attached directly to the megasporophyll but are borne on paper-thin scales, which in turn are attached to each bract. Why are the seeds of conifers considered to be naked?

____

____

Using a stereoscopic microscope, examine prepared slides of a longitudinal section of a young pine ovule (Fig. 19-2). Locate the **integument,** an outer layer of tissue that encloses the inner tissue except for an opening at one end, called the **micropyle.** The **megasporangium** is found inside the integument and contains diploid **megaspore mother cells,** which undergo meiosis to produce four haploid **megaspores,** only one of which survives to develop into the **megagametophyte.**

Next examine a section of a mature pine ovule. Locate the prominent megagametophyte, which contains archegonia at the end near the micropyle. Eggs can be seen in each archegonium.

3. Pollination and Fertilization

The pollen grains are released from the pollen cones in the spring. At about the same time, the young seed cones open and the pollen, carried to the seed cones by wind, sifts down through the cone scales and comes to lie in the pollen chamber. The pollen grain germinates, produces a slender pollen tube containing the male gametes, and digests its way to the archegonia. Because the growth of the pollen tube and the development of the megagametophyte is slow, fertilization may not occur for as long as 13 months after pollination.

Remove the seed coat from a soaked pine seed and make a longitudinal cut through the seed to locate the embryo. Identify the **epicotyl** and **hypocotyl.** At the tip of the hypocotyl, locate a coiled suspensor, which, by its growth, pushed the developing embryo deeply into the gametophytic tissue, which contained large quantities of stored food. If available, examine stages in the germination of a pine seed. Note the peculiar manner in which the embryo comes out of the seed.

C. DIVISION ANTHOPHYTA (FLOWERING PLANTS)

In flowering plants, the gametes are produced in modified branches that we call flowers. Examine a flower of a lily and locate the following structures (Fig. 19-4).

Sepals are modified leaves and are the outermost structures of the flower. They are typically green though they may be other colors, and in some flowering plants they are absent. Collectively, the sepals make up the **calyx.**

Petals are located inside the sepals and may be white or colored. Collectively, the petals make up the **corolla.**

Stamens are structures inside the petals. Each stamen consists of a stalk (the **filament)** and a terminal capsule (the **anther),** which when mature contains the pollen grains. The pollen grains produce the male gametes.

Pistils—one or more—are found at the center of the flower. Pistils are made up of one or more fused **carpels,** which are leaflife structures bearing ovules. The pistil consists of three parts: an en-

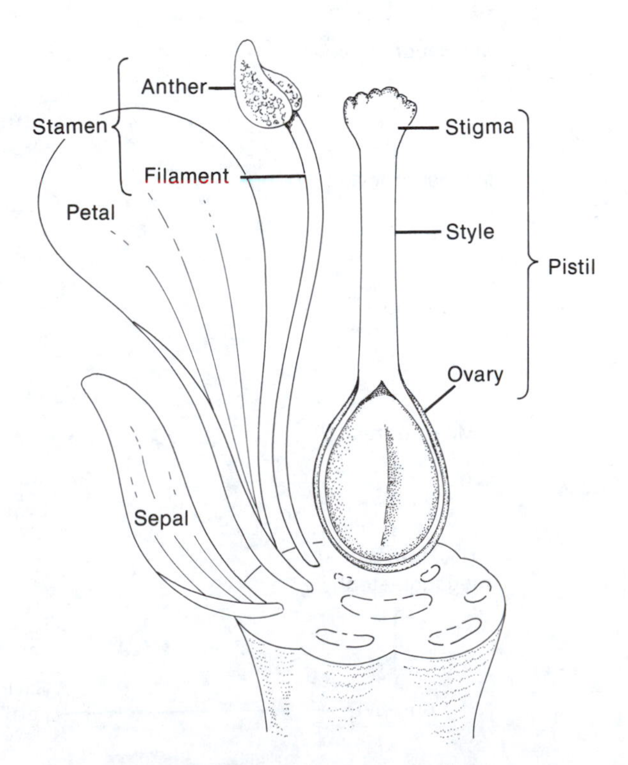

FIGURE 19-4 Diagram of a flower

larged basal region (the **ovary**), a slender stalk (the **style**), and a somewhat flattened tip (the **stigma**). The ovary contains seedlike structures called **ovules**, which, when mature, contain the female gametes. In the process of pollination, the pollen grains are transported to the stigma, where they germinate and send a long tube down to the ovules. How does pollination in the flowering plants differ from that in conifers?

The pollen tube penetrates the ovule and releases two male gametes, one of which unites with the egg to form the **zygote**.

The lily plant is the sporophyte stage in the life cycle of the plant, during which spores are produced in specialized structures called sporangia. The sporangia are formed on or within modified leaves called sporophylls. Which flower parts described earlier can be considered sporophylls?

Spore mother cells, or sporocytes, undergo meiotic division in the sporangium to form the spores, which undergo morphogenetic changes and develop into microscopic gametophyte plants that produce the gametes. In flowering plants, the gametophytic plants are parasitic on the sporophyte.

Recall that the stamen (microsporophyll) consists of an anther and a filament. The anther is made up of four microsporangia in which are found the microsporocytes. Examine slides of a young lily anther (Fig. 19-5). Locate the microsporocytes. What is the chromosomal condition of these cells, *N* or 2*N*?

What indication is there that some of the microsporocytes have divided?

What kind of division has occurred?

The cells formed as a result of the division of the microsporocytes are called **microspores.** They will develop into mature microgametophytes or pollen grains. Remove an anther from a flower, crush it in a drop of water on a slide, and examine it microscopically. For a more detailed examination of the pollen grain, obtain slides showing a section through an older lily anther. Locate the numerous pollen grains. The pollen grains are shed from the anther and are transferred—by wind or insects, for example—to the stigma of the same or a different flower. In the process of pollination, a pollen tube is formed that is chemotropically oriented to grow down through the style to the ovules. During the growth of the pollen tube, gametes are formed as a result of the division of one of the nuclei in the pollen tube. Examine cultures of germinating pollen and observe the pollen tubes. If possible, locate the two male gametes (Fig. 19-5).

In the ovule, a centrally located mass of tissue enlarges to form the **megasporangium,** or **nucellus** (Fig. 19-5). Concomitant with the development of the megasporangium, adjacent cells grow up and around the nucellus to form the integuments, which later develop into the outer coverings of the seed. The integuments do not fuse but leave an opening (the micropyle) at the apex of the nucellus, through which the pollen tube enters the ovule. As the nucellus grows, the nucleus of the centrally located megasporocyte undergoes meiosis to form four megaspore nuclei.

One of the megaspore nuclei migrates to the micropyle end; the other three migrate to the opposite **(chalaza)** end. This 3 + 1 arrangement represents the first four-nucleate stage in the development of the embryo sac.

The three haploid nuclei then fuse to form a 3*N* nucleus. This stage is followed by a mitotic division of the 3*N* and *N* nuclei to form two triploid nuclei and two haploid nuclei. This is the second four-nucleate stage. A final division occurs, resulting in an eight-nucleate embryo sac consisting of four triploid and four haploid nuclei. At this point, three of the triploid nuclei migrate to the chalaza end of the embryo sac, where cell walls are formed around them. These cells, called **antipodals,** are probably nutritive in nature.

Three haploid nuclei migrate to the micropylar end of the embryo sac. The middle nucleus be-

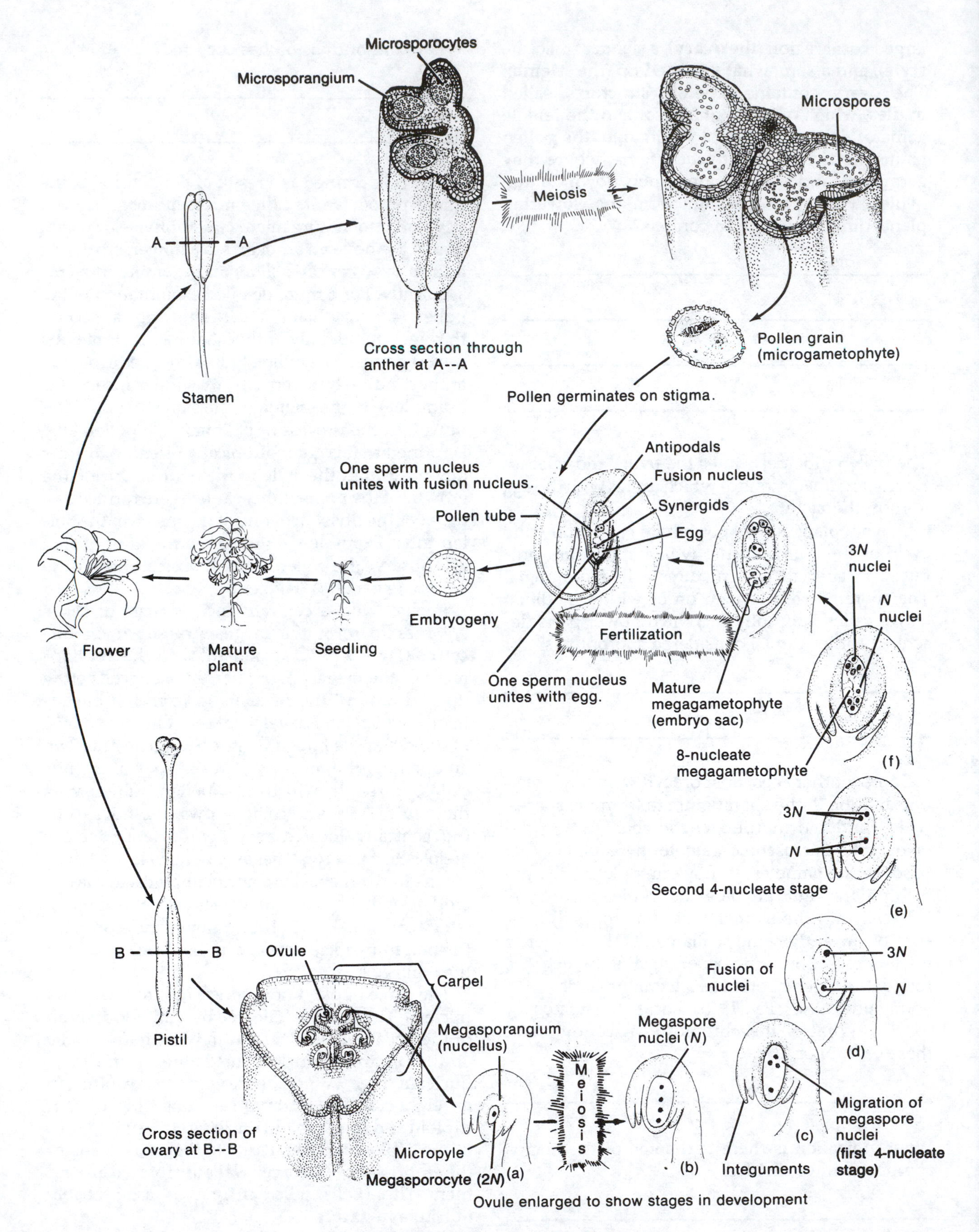

FIGURE 19-5 Life cycle of a lily

TABLE 19-1 Increases in complexity shown by the vascular plants

Characteristic	Liverworts and mosses	Vascular plants
Morphology of sporophyte		
Morphology of gametophyte		
Asexual reproduction		
Sexual reproduction		

comes the egg. The role of the two lateral nuclei (called **synergids**) is not fully understood, though studies have shown that the pollen tube enters the synergid first and then the egg. Thus, the synergids might function in guiding the pollen tube. The two remaining nuclei migrate to the center of the embryo sac and form a **fusion nucleus.** What is the chromosome content of this nucleus?

___

Fertilization is complete when the pollen tube, having reached the micropyle of the ovule, digests its way through the nucellus, reaches the embryo sac, and releases both male gametes. One microgamete unites with the egg to form the zygote, which subsequently develops into an embryo. The second gamete unites with the fusion nucleus to form the endosperm nucleus. What is the chromosomal condition of the endosperm nucleus?

___

Subsequent division of the endosperm results in a mass of nutritive tissue that is used as a food source for the developing embryo. This double fertilization is a phenomenon peculiar to flowering plants and has no counterpart in the lower plants or in the animal kingdom.

D. INCREASES IN COMPLEXITY SHOWN BY VASCULAR PLANTS

Complete Table 19-1. Indicate the increases in complexity shown by the tracheophytes over the bryophytes with respect to morphology and reproduction. The following are suggested terms to use.

- Morphology: true leaves present or absent, true stems and roots present or absent, vascular tissue present or absent, erect or prostrate plant body
- Asexual reproduction: spores, fragmentation, gemmae, death and decay of older parts, and so on
- Sexual reproduction: unicellular or multicellular reproductive organs, motile or nonmotile gametes, alternation of generations, isogametes, anisogametes, water necessary or unnecessary for fertilization, antheridia, archegonia, dependency or interdependence of sporophyte, pollination

REFERENCES

Bold, H. C. 1986. *Morphology of Plants and Fungi.* 5th ed. Harper & Row.

Gifford, E. M., and A. S. Foster. 1989. *Morphology and Evolution of Vascular Plants.* 3d ed. W. H. Freeman.

Raven, P. H., R. F. Evert, and S. Eichhorn. 1986. *Biology of Plants.* 4th ed. Worth.

Exercise

20

Plant Anatomy: Roots, Stems, and Leaves

As you would expect, plants and animals differ significantly in the changes that occur between fertilization and the achievement of their characteristic forms. Development in plants is **indeterminate,** whereas development in animals is **determinate.** That is, development continues throughout the life of a plant, with new organs, tissues, and cells being formed perpetually. For example, in a giant redwood tree growing in California, cells and tissues near the base of the tree may be more than 2000 years old, whereas new cells, tissues, and organs in the tips of its branches continue to form and differentiate into their final structures. Theoretically, there is no limit to the growth of this tree. Conversely, an animal becomes complete (determined) early in its life. In a chicken, for example, differentiation of cells and tissues occurs early in embryonic development. After the animal matures, no further growth and development occur.

Another distinction between plants and animals is that cell division in plants occurs in localized regions called **meristems,** whereas in animals it occurs throughout the immature organism. The three principal growth regions of plants are **apical meristems,** which determine growth in the length of stem and root and are instrumental in the production of leaves, flowers, and branches; **lateral meristems** consisting of the **vascular cambium,** which is involved in the growth in diameter of stem and root; and the **cork cambium,** which is involved in the production of cork, the protective outer covering of the stem and root.

To understand the growth and development of plants and to comprehend their complex structures and the relationships between structure and function, it is necessary to study their anatomy. In this exercise, you will become familiar with the anatomy of the roots and shoots (stems and leaves) of flowering plants.

A. ROOTS

A root is usually the part of the plant that grows beneath the surface of the soil. The principal function of roots is to absorb water and soluble minerals and, through their vascular systems, transport these substances to the above-ground parts: the stems, leaves, flowers, and fruits. Roots also anchor the plant in the soil and store reserves of food for the plant, as in sweet potatoes, carrots, and turnips. Roots may also manufacture food. For example, the aerial roots of orchids contain chlorophyll and supplement the leaves in photosynthesis.

Roots can be broadly classified by form into two groups: **taproots** and **fibrous roots** (Fig. 20-1). A taproot (main root) becomes many times larger than the branch roots (those that arise from the

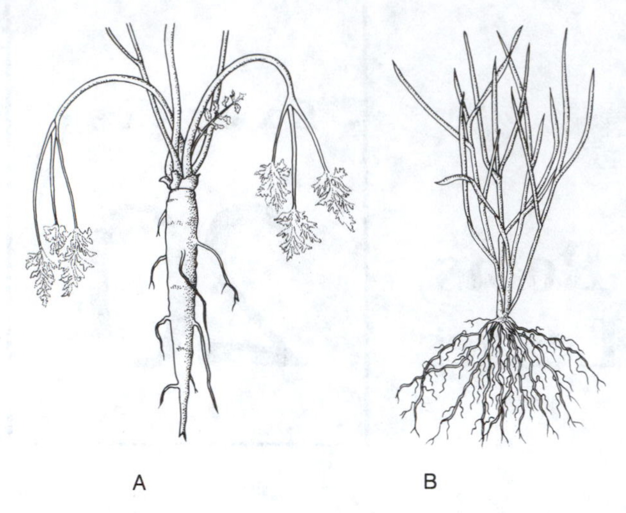

FIGURE 20-1 (A) Taproot of carrot. (B) Fibrous root system of grass.

main root) and penetrates some distance into the soil. In some plants, the taproot may be greater in diameter than the stem. Examine the taproot of a carrot plant for the presence of branch roots.

In fibrous root systems, the primary root and the branch roots are approximately the same length and diameter. Examine the fibrous root system of a grass or bean plant.

1. Root Tip

Obtain a germinating radish or grass seed. Mount it in a drop of water on a slide and examine the young root with a stereoscopic microscope. Locate the mass of cells covering the root tip. This **root cap** covers the apical meristem, a region of active cell division, and protects it from damage as the root grows through the soil. By adjusting the light, you may be able to observe the **root (apical) meristem,** which appears as a dense, opaque region at the tip of the root. Behind the root tip, note the presence of **root hairs**—special absorbing cells on the root. Root hairs generally persist for a short time and then die. New root-hair cells are continually formed near the root tip to replace those that are lost. In what way do root hairs increase the efficiency of absorption of water and minerals by the root?

____

____

____

Examine a prepared slide of a longitudinal section through an onion root tip and locate the structures just described (Fig. 20-2A, B).

2. Primary Tissues of the Root

Using a compound microscope, examine a prepared slide of a cross section of a buttercup (*Ranunculus*) root, cut through a region in which the cells have become differentiated (Fig. 20-2C, D). Starting from the outside, the **primary tissues** (tissues having their origin in cells produced in the root apical meristem) consist of an outer **epidermis,** which is a single layer of cells covering the outside of the root. The **cortex** in most roots consists of thin-walled cells. Note the presence of starch grains in many of the cortical cells and the large intercellular spaces where the cells abut.

Lining the cortex on the inside is the **endodermis,** a single layer of cells separating the cortex from the **vascular cylinder.** The endodermis, present in all roots, is believed to function in directing the flow of water from the root hairs, through the cortex and into the vascular cylinder.

The main pathway of water as it crosses the epidermis and cortex is via the permeable cell walls of these tissues. When water reaches the endodermal layer, it can no longer follow the cell-wall route because of a waxlike strip (called a **Casperian strip)** that surrounds each endodermal cell (Fig. 20-2E, F). Casperian strips act like gaskets, preventing the water from flowing through the cell walls of the endodermal layer. Instead, the water and its dissolved minerals must pass through the cytoplasm of the endodermal cells before reaching the vascular tissue, thereby permitting the root to be more selective about how much and what can enter the vascular system.

Examine the endodermal layer of the buttercup more closely and locate thickenings in the radial walls of the endodermal cell. These thickenings are sections through the Casperian strips.

The vascular cylinder consists of several tissues lying internal to the cortex. The **pericycle,** a unicellular layer of cells internal to the endodermis, has the ability to become meristematic and to initiate the growth of lateral roots. The **xylem,** the primary water-conducting tissue, is represented by three, four, or five radiating arms. Alternating with the xylem arms are groups of **phloem** cells. The phloem conducts various organic molecules, including products of photosynthesis and various hormones. This alternating arrangement of the xylem and phloem is unique to roots and anatomi-

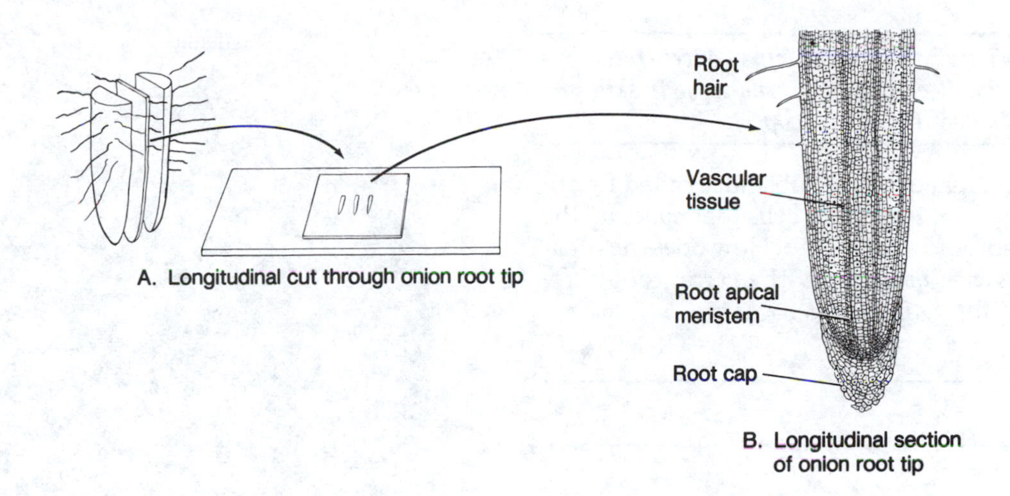

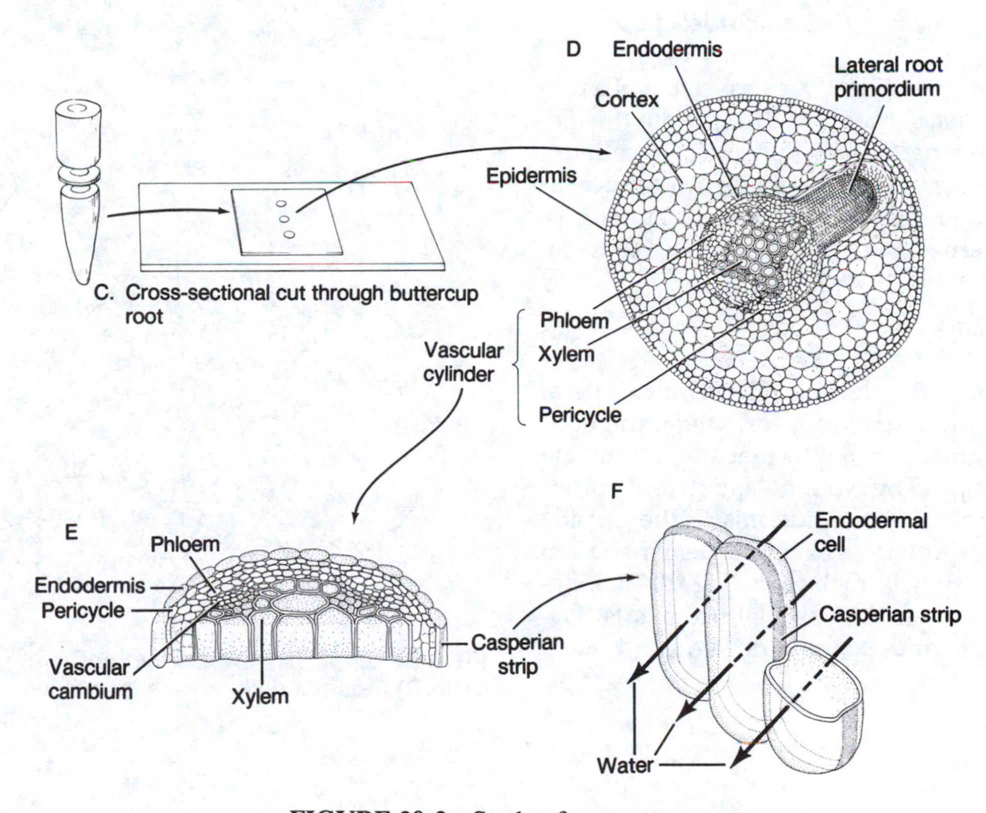

FIGURE 20-2 Study of root anatomy

cally distinguishes roots from stems. Locate the region between the phloem and the xylem where the **cambium** will develop. The cambium is a lateral meristem responsible for growth in the diameter of a root. When the cambium cells divide, they give rise to cells that will differentiate as xylem and phloem; they are called *secondary* xylem and phloem because they originated in meristems other than an apical meristem.

Obtain a carrot. Using a sharp razor, cut a thin cross section from the root and place it on a slide. Add one or two drops of iodine and a coverslip. Examine microscopically. Iodine reacts with starch to form a deep blue-black color. Where is the starch located in this root?

____

Prepare a second cross section of carrot and stain with phloroglucinol-HCl.

Caution: *This solution contains concentrated hydrochloric acid. It can cause burns to your skin and damage the lens of the microscope.*

This chemical reacts with a substance called **lignin** in the cell walls of the xylem cells that make up the vascular tissue and stains it red. How does the organization of vascular tissue of the carrot root differ from that of the *Ranunculus* root?

____

____

3. Primary Tissues of a Monocot Root

Examine a prepared slide of a cross section of a mature corn (*Zea mays*) root (Fig. 20-3). Note the distinctive circular pattern of the vascular cylinder surrounding a central region of mostly undifferentiated tissue called the **pith.** Locate the epidermis, cortex, endodermis, pericycle, xylem, and phloem.

4. Lateral Roots

As noted, the pericycle may give rise to lateral (branch) roots. In lateral root formation, the pericycle cells organize into a new meristem, complete with a root cap. The young lateral root grows through the cortex and epidermis to the outside and continues growing. Examine demonstration slides of developing branch roots (Fig. 20-2D). Remove young roots from water lettuce plants (*Pistia*), if available, and examine developing branch roots.

5. Adventitious Roots

Examine stem cuttings from various plants (e.g., *Coleus*, willow, geranium) that have been rooted in moist sand or vermiculite (Fig. 20-4A). The roots that arise from such cuttings are called **adventitious** roots and generally result from regenerative processes in stem cuttings. What does the term *adventitious* mean?

____

____

Examine the adventitious **prop** roots of a corn plant (Fig. 20-4B). How would you determine whether these prop roots originate in the stem or the root?

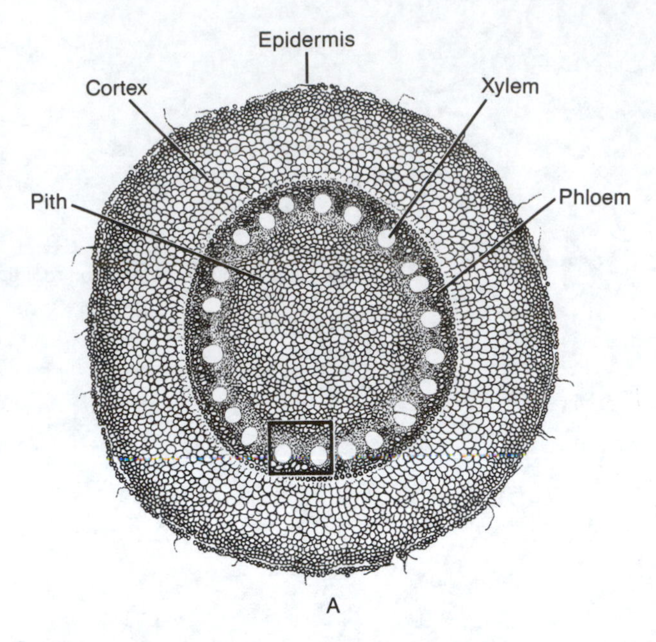

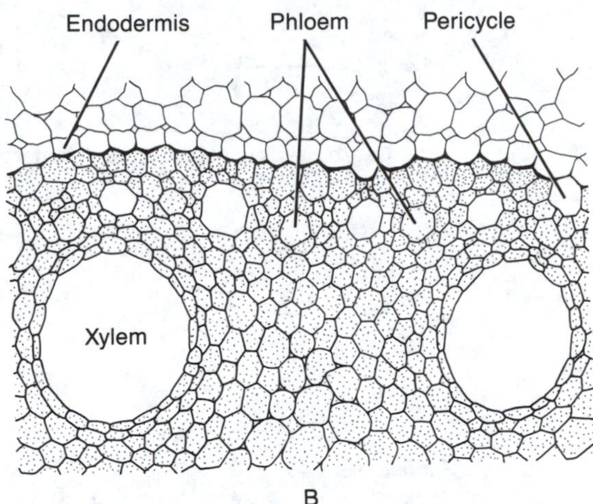

FIGURE 20-3 (A) Cross section of corn (*Zea mays*) root. (B) Detail of mature vascular cylinder.

____

What is (are) the function(s) of prop roots?

____

If available, examine a microscopic section of a prop root. In what way(s) is it similar to the lateral root in *Ranunculus*?

____

____

____

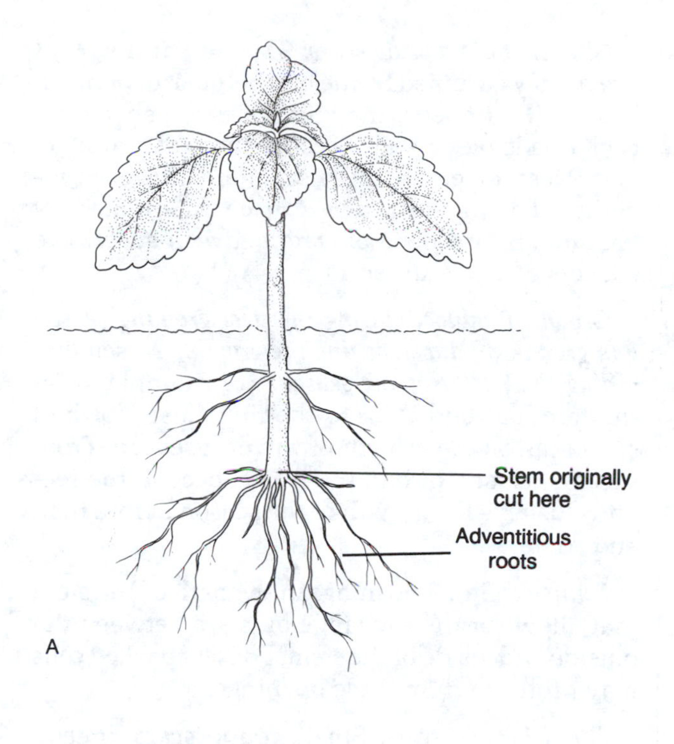

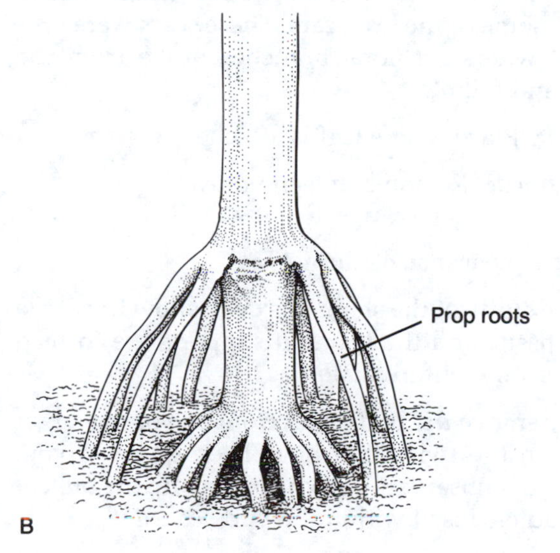

FIGURE 20-4 Types of roots. (A) Adventitious roots on stem cutting. (B) Prop roots of corn.

In what way(s) is it different?

___

___

___

B. THE SHOOT

The **shoot** consists of the stem and its leaves and usually grows upright. Stems are like roots in general structure: they have an epidermis, a cortex, and a vascular cylinder. Stems *differ* from roots in the organization of their vascular structure and in having lateral appendages called **leaves.**

The stem has several functions. It is the framework that supports the leaves, flowers, and fruits. It provides for the transport of the water and solutes absorbed by the roots, and it transports and distributes the sugar manufactured by the leaves during photosynthesis to the places where it is used or stored. Many stems store food—sometimes to the extent of being economically important (e.g., Idaho and Wisconsin potatoes and sugar cane).

1. Buds and Other Structures

In its broadest definition, a **bud** is a compact, undeveloped shoot that may remain small and dormant or may elongate and form a new shoot consisting of one stem and its appendages.

a. Classification of Buds

Buds can be classified in a variety of ways:

By content
- foliage buds
- floral buds
- mixed buds (i.e., flower and foliage primordia)

By position
- terminal buds
- axillary buds
- adventitious buds

By time of development
- active buds
- resting buds
- latent buds

By protective covering
- protected buds
- naked buds

b. Examination of a Woody Perennial

Examine branches from the hickory tree. Locate the following structures and label them on Fig. 20-5.

Terminal bud: Bud located at the tip of the branch and involving the entire stem tip

Axillary bud: Bud located in the axil of a leaf (above the leaf scar if the leaf has fallen from the twig)

Latent bud: Small axillary bud that normally does not develop into a branch during the spring follow-

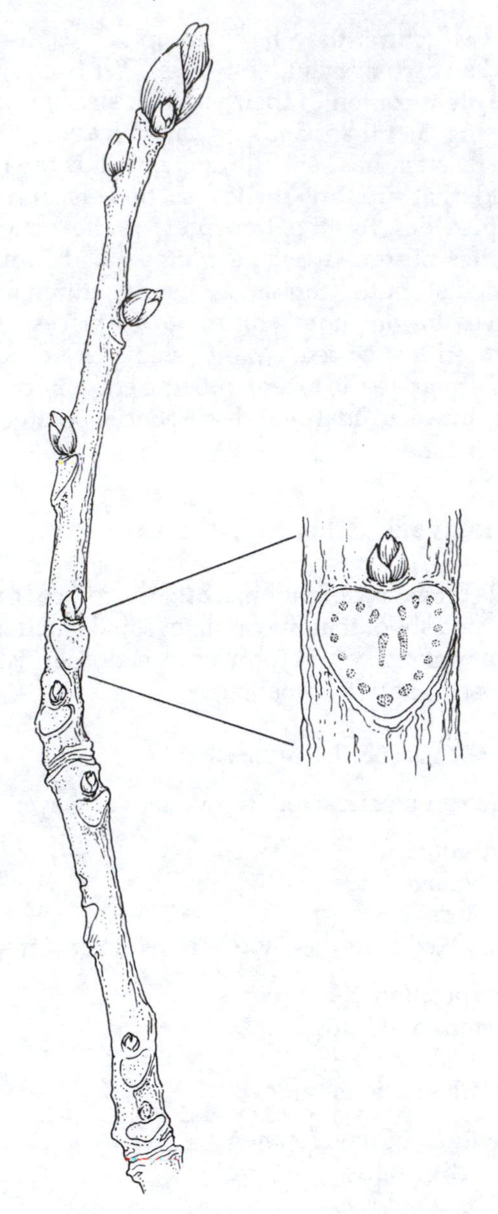

FIGURE 20-5 External features of dormant hickory branch

ing its formation. It may develop into a branch if an accident destroys the terminal or axillary buds.

Leaf scar: Scar on the stem formed when the petiole separated from the stem at the time of leaf fall

Vascular bundle scars: Small "spots" located in the area of a leaf scar. These scars indicate the number of vascular bundles that extended from the stem into the leaf and were severed when the leaf fell.

Bud scales: Protective scalelike leaves covering all the buds

Clusters of bud scale scars: Scars left at the point previously occupied by the bud before it expanded. When a bud opens, the bud scales are shed, and each one leaves a scar. Because the bud scales are very close together, these scars are clustered together. The age of a twig can be determined by counting the sections of stem between successive clusters of bud scale scars.

Growth produced during the past growing season and growth produced during the growing season preceding the past growing season: For example, from the terminal bud back to the first cluster of bud scale scars is the growth of the past season. From this first cluster of bud scale scars back to the second cluster is the growth of the season before that, and so on.

Lenticels: Small openings in the bark of the stem that allow for the exchange of gases between the outside and inside of the stem. Loosely packed cells may protrude from these openings.

Floral branch scars: Small, round scars directly above some of the leaf scars. These scars were produced when the floral branches broke from the stem and fell off.

Node: Place where leaf is attached to stem

Internode: Region between nodes

c. Examination of Shoot Apex

A brief study of the shoot apex will show how new stem tissue and the leaves it supports are formed and become differentiated.

1. Remove the bud scales from one of the larger buds on the twigs that have been given to you. From your observations, what makes up the bulk of the bud enclosed within the bud scales?

Remove the leafy appendages found in the bud and locate the minute tip of the shoot within the bud. How would you describe the shape of the shoot tip?

---

2. Examine a prepared slide of a longitudinal section through a *Coleus* shoot tip (Fig. 20-6). *Coleus* is a common annual plant (dies at the end of summer's growth) used in flower gardens. Locate the shoot apical meristem, which is at the tip of the shoot, enclosed by leaf primordia. Note how the

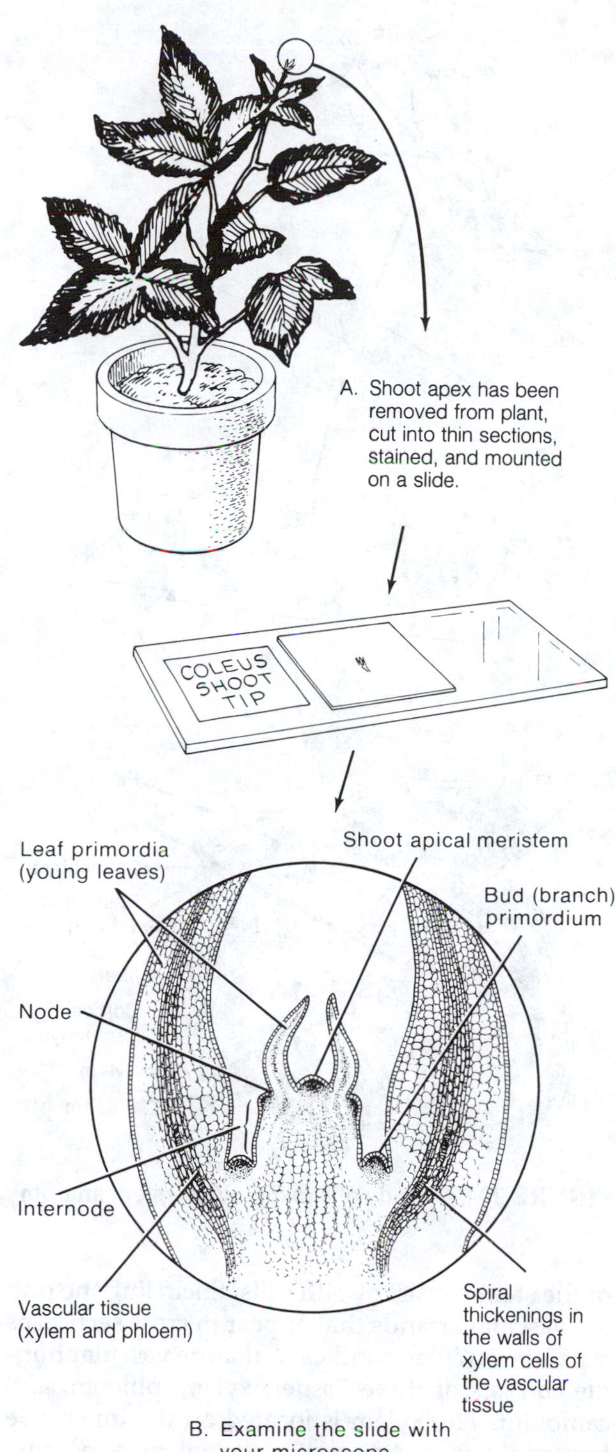

FIGURE 20-6 Study of the *Coleus* shoot tip

leaves progressively increase in size as you move down from the shoot tip. Can you find evidence of cell division in the apical meristem? If not, can you suggest a reason for the absence of divisions?

Are all the cells in the leaf primorida alike? If not, what evidence of differentiation do you find?

What do you find in the angle that each leaf primordium makes with the stem?

---

2. Primary Tissues of Dicot Stem

Flowering plants **(angiosperms)** are divided into two major groups called **dicotyledons** (dicots) and **monocotyledons** (monocots). The differences in the number of cotyledons (leaflike food storage organs) in the seed — one in the monocots and two in the dicots — provide the most familiar distinction between these groups. Other differences are based on

- leaf venation pattern (parallel in monocots, reticulate in dicots)
- vascular bundle arrangement in the stem (peripheral and arranged in a cylinder in dicots; scattered in monocots)
- presence (in dicots) or the absence (in monocots) of a vascular cambium
- numerical arrangement of floral parts (three in monocots; four or five, or multiples thereof, in dicots)

Although you will examine plants that exhibit these classical distinctions in the studies that follow, be aware that there are always exceptions to rules.

Examine various representatives of dicots such as *Coleus*, geranium, bean plants (*Phaseolus*), tomato plants (*Lycopersicon*), and any others provided by your instructor. Identify the following.

Terminal or apical bud: Bud located at the tip of the main stem or at the end of a branch. To what does this structure give rise?

Node: Place on the stem where leaves arise

Internode: Region between two nodes. Stem elongation is primarily the result of elongation of the internodes.

Leaf: Composed of two parts called the **blade** (the expanded flat portion) and the **petiole** (the stalk that attaches the blade to the stem). In some plants, the blade may be subdivided into numerous smaller segments called **leaflets.**

Stipules: Leaflike structures found near the base of the petiole; not always present

Veins: Vascular network located in the leaf blade. In dicots, this pattern is described as **net venation** or **reticulate venation.**

Axil: Angle between the upper side of the petiole and the stem

Lateral or axillary bud: Bud formed in the axil of the leaf. Develops into a leafy branch or flower, sometimes both. You examined these in part B.1 of this exercise.

Obtain a portion of a geranium stem from your instructor. With a razor blade, make a thin cross-sectional cut. Prepare a wet mount and examine it with the low-power objective of the microscope. Locate the outer epidermal layer and the numerous hairs associated with it. Beneath the epidermis is the cortex. Note the presence of numerous chloroplasts in the cells. Suggest a function for this tissue in the living plant.

____

____

The cortex is separated from the centrally located pith by a band of vascular tissue. Note that the vascular cylinder is not of uniform thickness but is composed of more or less separate vascular bundles. The vascular tissue may be observed more easily if stained. Mount the section in phloroglucinol solution.

Caution: *Phloroglucinol is made with a strong acid. It can cause burns to your skin and damage the lens of the microscope.*

Wait about 1 minute and reexamine the section. Phloroglucinol-HCl stains the cell walls of the xylem red. How does the arrangement of the vascular tissue differ from that of the root?

____

To examine the primary tissue in detail, obtain a prepared slide of a cross section of the alfalfa (*Medicago*) stem (Fig. 20-7). The cells in the central part of the stem consist of pith cells. Encircling the pith are vascular strands that appear in cross section as separate vascular bundles. A mature vascular bundle consists of three tissues: xylem, phloem, and cambium. The xylem is located on the inner side toward the pith. Adjacent to the xylem, appearing as small rectangular cells having the long axis at right angles to the radius of the stem, is the cambium. The phloem lies immediately external to the cambium.

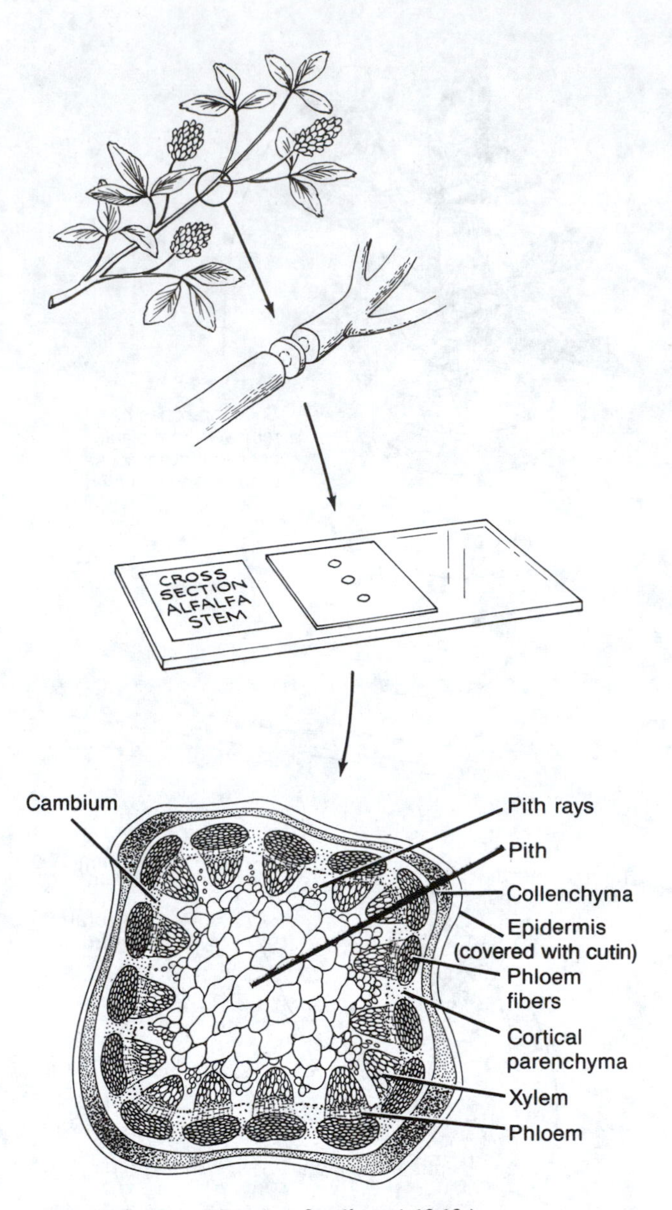

FIGURE 20-7 Study of a dicot (alfalfa) stem anatomy

Capping each vascular bundle is a group of thick-walled phloem fibers. The cells lying between the vascular bundles are **pith rays.**

The cortex lies immediately outside the phloem fibers and consists of two distinguishable tissues,

cortical parenchyma and **collenchyma** (mechanical tissue). The cortical parenchyma consists of loosely packed cells, many of which contain chloroplasts. In many stems, the collenchyma serves a supportive (strengthening) function.

The outermost layer of cells constitutes the epidermis. How many cell layers thick is the epidermis?

____

The outer walls of the epidermal cells usually have become thickened, and during their development a waxy substance **(cutin)** is secreted. Close the diaphragm on your microscope so that you can observe this layer of cutin. It will appear as a faint, pink, noncellular layer covering the epidermis. What effect does this waxy layer have on the loss of water from the stem?

____

____

____

What characteristics of dicots are exhibited by geranium and alfalfa?

____

____

____

3. Primary Tissues of Monocot Stem

Examine a slide of a cross section of a young stem of corn (*Zea mays*) (Fig. 20-8).

	Yes	*No*
Are the vascular bundles located peripherally and in a ringlike pattern as they are in dicots?	___	___
Is any kind of pattern evident with respect to the arrangement of vascular bundles?	___	___

If yes, describe the pattern. ________

____

Are all the vascular bundles the same size?	___	___

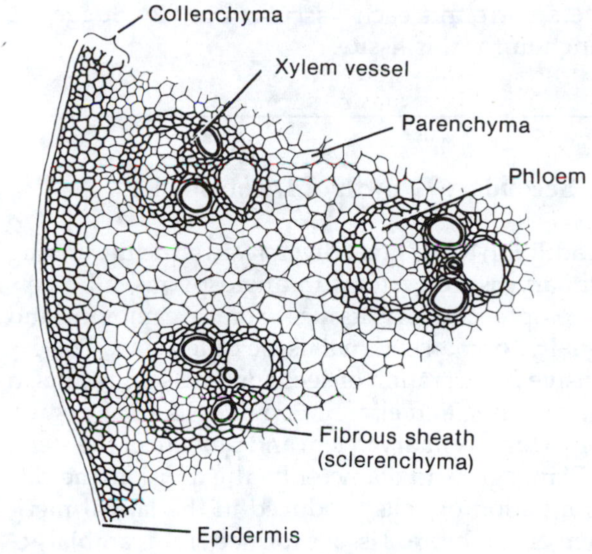

FIGURE 20-8 Study of a monocot (corn) stem anatomy

If not, are they larger in the center or

at the periphery? ____________

Is there any pith?	___	___
Is there any cortical tissue?	___	___

Using the high-power objective lens, study the details of an individual vascular bundle. Locate the following tissues and label them in Fig. 20-8.

Xylem. The xylem tissue consists of xylem **vessels** and xylem **parenchyma.** Vessels are large cells through which water is rapidly transported in the stem. Two or three vessels often look like the eyes and nose of a face. Indeed, the vascular bundle sometimes looks like a skull. Located between the larger vessels are smaller vessels and thinner-walled xylem parenchyma. A large hole (which looks like the mouth in the skull) is created by cells that have been stretched during elongation of the stem and have collapsed.

Phloem. Phloem tissue consists of larger **sieve tubes** and small **companion cells.** In some sections, you may be able to see the perforated **sieve plate** of the sieve tube cell. What is the function of the sieve tubes and the companion cells?

____

Fibrous sheath (sclerenchyma). Thick-walled

fibers surround each vascular bundle. Suggest a function for this tissue.

4. Secondary Growth of Dicot Stem

In addition to the great diversity in structure among primary tissue systems in angiosperms, all dicots develop secondary tissues from a cambium and usually form an external tissue called **cork.** The extensive proliferation of secondary tissues results in the crushing and elimination of epidermal tissues, cortex, and primary xylem and phloem of the stem.

Plants grow in diameter by the division and differentiation of cells produced in the lateral meristem or cambium. Tissues formed from cambial activity are designated as **secondary tissues.** Because monocots typically lack cambial activity and the resulting secondary growth, this exercise will examine the secondary growth of a dicotyledenous plant.

Carefully examine a slide that contains a cross section of one-, two-, and three-year-old basswood stems. Refer to Fig. 20-9 as you locate each of the following tissues.

a. Epidermis

The epidermis usually consists of a single layer of cells that are covered by the **cuticle.** The cuticle is made up of a waxy, waterproof substance called **cutin** that protects the underlying tissue from drying out.

b. Periderm

This region is composed of three layers.

Cork. Cork cells are rectangular and are produced by the cork cambium. Walls of cork cells are impregnated with **suberin,** which makes the cells waterproof. Cork cells die soon after they mature. How do you account for their early death?

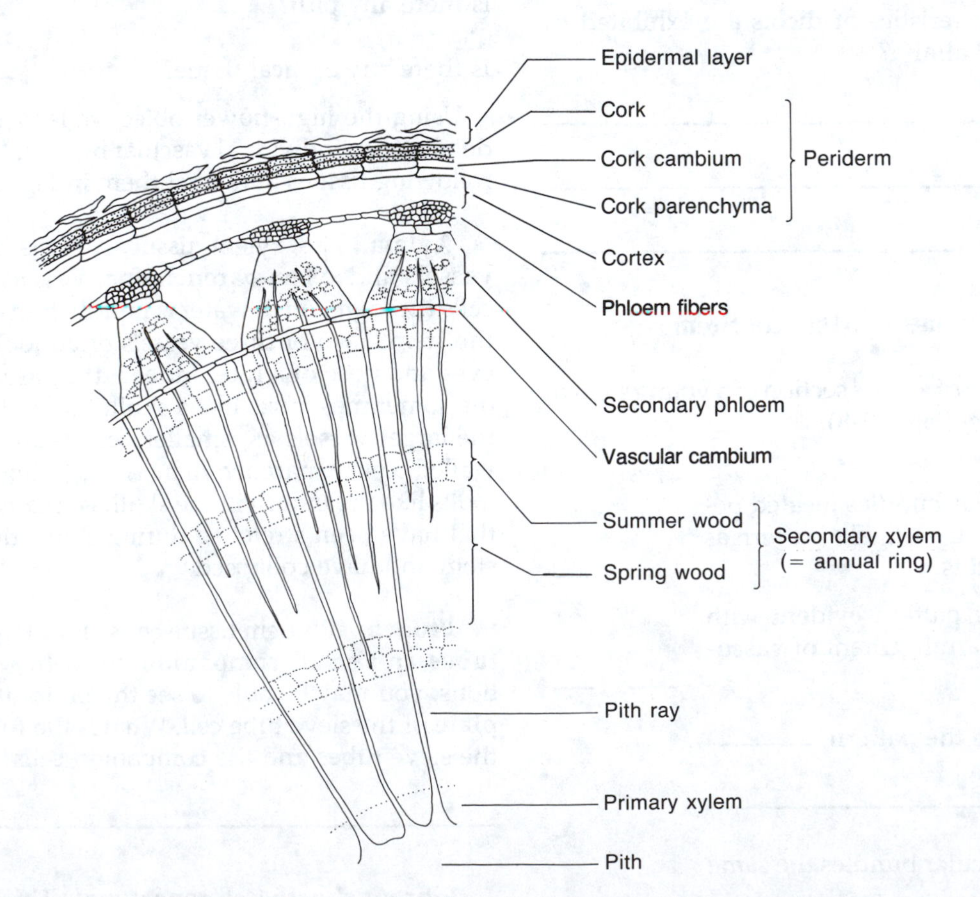

FIGURE 20-9 Cross section of a three-year-old basswood stem

Cork cambium. The cells of the cork cambium are shaped like the cork cells, but nuclei are usually plainly visible in the cork cambium cell.

Cork parenchyma. In addition to producing cork, the cork cambium produces a small layer of parenchyma cells located just inside the cork cambium layer.

c. Cortex

This zone is composed of several layers of thick-walled cells—**mechanical tissue**—and, just interior to the mechanical tissue, an area of thin-walled cork parenchyma cells. Look for crystals in some of these cells.

d. Vascular Cylinder

Phloem. Recall from lectures and previous laboratory work the physiological function of the phloem.

Primary phloem: Thick-walled phloem fibers are located as a cap over the primary phloem. Examination of the primary phloem with the high-power objective of your microscope will show a very thick cell wall surrounding a very small cell cavity.

Secondary phloem: The secondary phloem in the basswood stem looks banded with alternating layers of thick-walled **phloem fibers** and thin-walled cells (the other types of phloem cells such as sieve tubes, companion cells, and phloem parenchyma). The thick-walled phloem fibers here are very similar in appearance to the thick-walled cells of the primary phloem.

Vascular cambium. The cambium consists of a layer of thin-walled cells between the secondary xylem and the secondary phloem that appear rectangular in cross section. The cells of the cambium differ from those of other tissues in that they continue to divide. The planes of division are largely tangential (parallel to the surface). The cambium layer does not increase in thickness because the daughter cells produced mature into xylem and phloem cells. In the vascular tissue, therefore, the cells that flank the cambium have been produced by it. Although primary and secondary phloem differ in origin, they are similar in structure and function and, in the cross section of the stem, cannot be distinguished except by position. The same is true of primary and secondary xylem.

Xylem

Secondary xylem: The secondary xylem of woody plants normally shows a differentiation between the xylem cells formed early in the growing season **(spring wood)** and the xylem elements formed later in the growth period **(summer wood).** The cells formed during the spring are larger. The cells of the secondary xylem produced during the later part of the growing season are smaller and the walls are relatively thick. The zone of spring wood and summer wood produced during one growing season is called an **annual ring.** The annual ring is conspicuous because of the size difference between the small xylem cells produced at the end of one growing season and the large ones formed when growth is resumed in the next growing season.

Primary xylem: The primary xylem is located at the base of the secondary xylem next to the pith. Usually, primary xylem cells are more irregularly arranged and are smaller in diameter than are the secondary xylem cells. The primary xylem makes up only a small portion of the inner part of the first annual ring.

e. Pith

The pith occupies the central area of the stem. In the pith of the basswood stem, there are numerous cells containing tannin and mucilage material, which strongly absorbs the stains used in preparation of slides.

f. Rays

Some of the rays in the stem extend from the pith outward to the cortex. As the stem grows in diameter, the cambium adds new cells to the rays and also initiates new rays.

5. Leaves

Leaves consist of three basic parts: an expanded or flattened portion, the **blade;** a thin, stemlike portion, the **petiole;** and small, paired, lobelike structures at the base of the petiole, the **stipules.** The petiole and stipules are not present in some plants.

Leaves can be classified according to the type of venation or to the arrangement of the leaves on the stem. Thus, leaves are parallel veined when the larger vascular strands traverse the leaf without apparent branching and net veined or reticulate when the main branches of the vascular system form a network. According to their positions on the stem, leaves may be **alternate** or **opposite** (borne singly or in pairs at each node). Examine various plants on demonstration in the laboratory and become familiar with the parts of the leaf and the simple classification given above.

a. Gross Anatomy of Dicot Leaf

Obtain a leaf from a bean or geranium plant. Remove a small piece of the lower epidermis by ripping the leaf, as shown in Fig. 20-10A. Mount the thin, transparent piece of tissue in a drop of water on a slide and then add a coverslip. Observe under low and high power of a compound microscope. Note that many of the epidermal cells have an irregular shape. Scattered throughout the epidermis are openings called **stomata (singular stoma).** Each

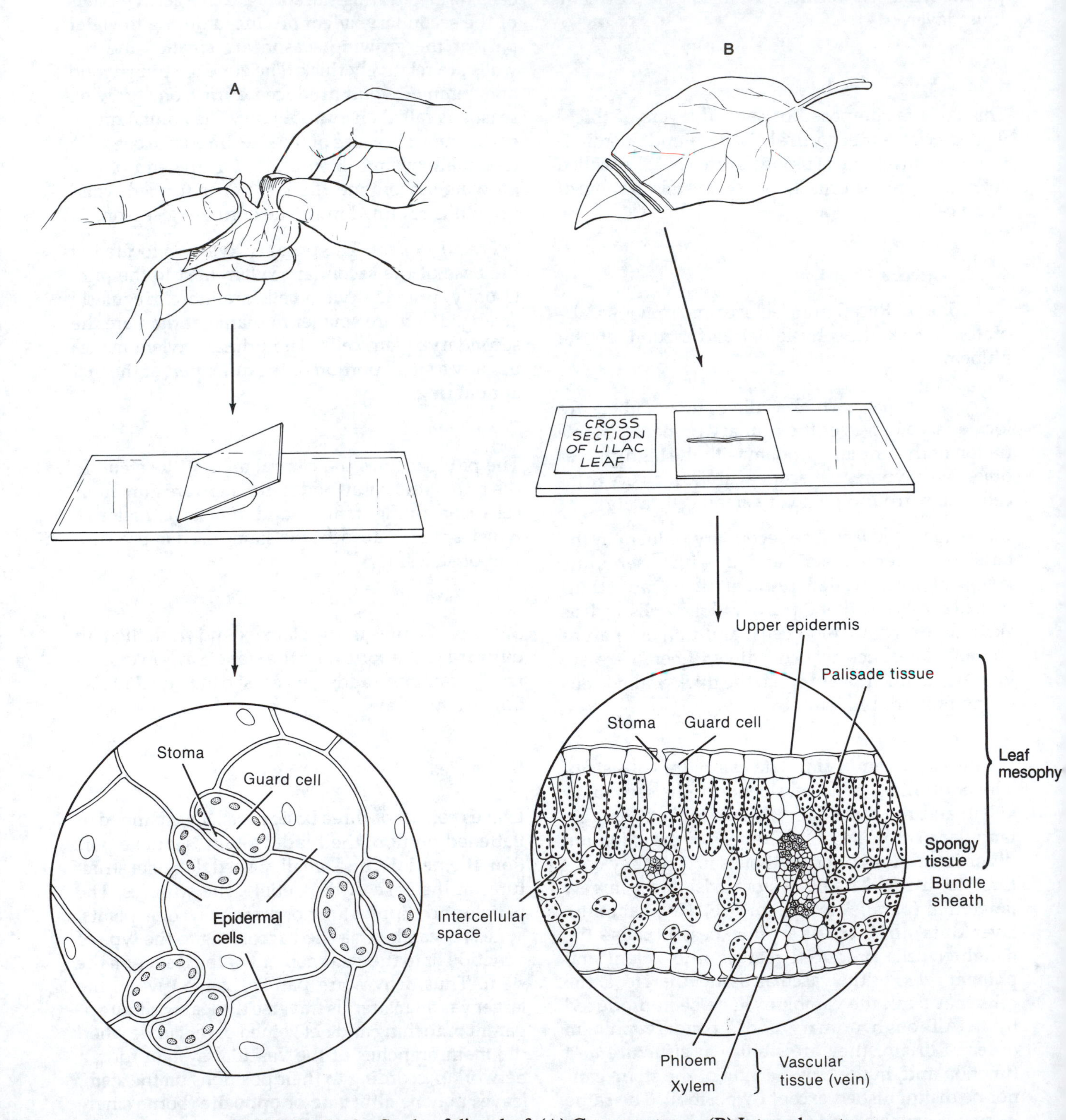

FIGURE 20-10 Study of dicot leaf. (A) Gross anatomy. (B) Internal anatomy.

stoma is surrounded by two bean-shaped **guard cells.** What is the function of the guard cells?

Which of the epidermal cells contain chloroplasts?

What is the function of the chlorophyll in these cells?

b. Internal Anatomy of Dicot Leaf

Obtain a prepared slide of a cross section of a lilac (*Syringa*) leaf for examination of the internal tissues (Fig. 20-10B). Below the upper epidermis is a region of **palisade tissue,** consisting of one or more layers of elongated cells with the long axis of the cell perpendicular to the surface of the leaf. Palisade cells contain numerous chloroplasts. What is the function of this tissue?

Under the palisade tissue, locate a region of rounded cells, which constitute the **spongy tissue.** The palisade and spongy tissue are collectively called **leaf mesophyll.** Note the presence of numerous intercellular spaces in the mesophyll. Locate the stomata. What is the relationship between the stomata and the intercellular spaces of the mesophyll tissue?

The veins of a leaf are vascular bundles (strands) that are continuous with the vascular tissues of the petiole and stem. A vein contains xylem and phloem and has the same cellular elements as the stem. What is the function of the **bundle sheath** surrounding the vascular strand?

c. Internal Anatomy of Monocot Leaf

Examine a prepared slide of a cross section through a monocot (corn) leaf. In the upper epidermis, locate **bulliform cells** on both sides of the midrib (vein) (Fig. 20-11A). When these cells are full of water, the leaves are unrolled and flattened. When these cells lose water under drying conditions, the

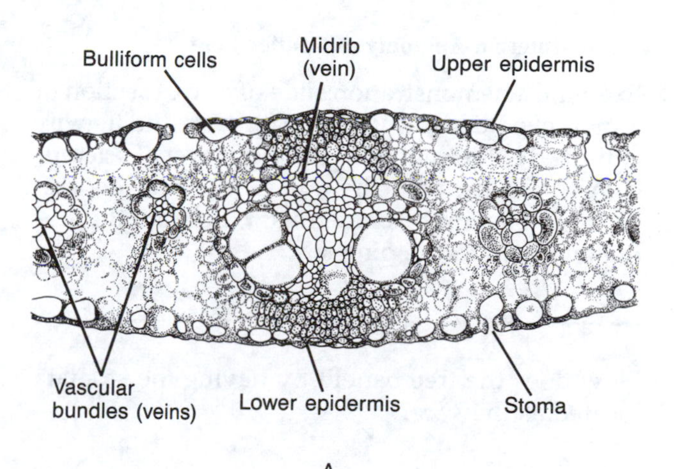

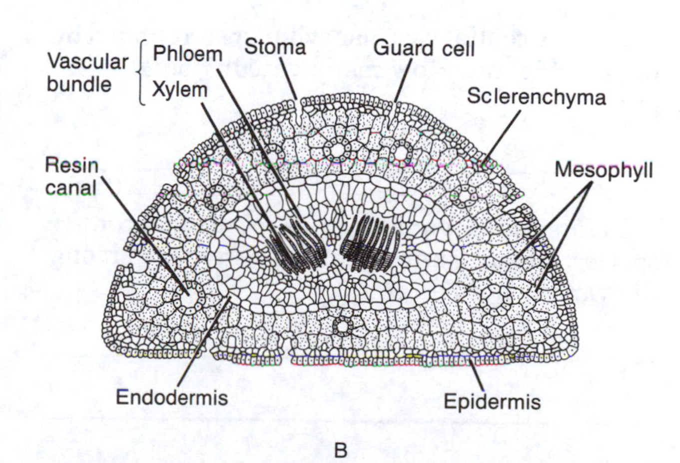

FIGURE 20-11 (A) Study of monocot (corn) leaf. (B) Study of conifer (pine needle) leaf.

leaf rolls up. Of what advantage to the plant is the rolling up of the leaf?

Note that the vascular bundles on either side of the midrib (vein) are rather evenly spaced. This is a reflection of the parallel venation you see when you look at the leaf surface.

Compare the mesophyll in the corn leaf with that of the dicot leaf. Is there a differentiation between spongy and palisade tissue? If so, describe it.

d. Internal Anatomy of Conifer Leaf

Examine a demonstration slide of a cross section of a pine needle leaf (Fig. 20-11B). Note the heavily cutinized epidermal cells. Locate a layer of sclerenchyma just beneath the epidermis. Find depressions, or pits, in the epidermis. Where are the stomata located in the pine leaf?

How does the tree benefit by having the stomata located as they are?

Locate the central vascular cylinder surrounded by the endodermis. How many vascular bundles are present?

Examine cross sections of leaves of other conifer species. Describe any variation in structure among the various needles.

Examine the slides of the various leaves with a stereoscopic microscope. What correlation exists between the number of leaves in a cluster and the shape of an individual needle leaf?

C. MORPHOLOGICAL ADAPTATIONS OF LEAVES TO THE ENVIRONMENT

Water is one of the basic raw materials of photosynthesis. It is the main component of plant tissues, making up 90% of the plant body. Water is the substance that carries most materials into and out of the cells of plants, and it is the solvent for the various biochemical reactions that occur in living cells.

The amount of water used by plants is far greater than that used by animals of comparable weight. The reason for this is that a large amount of the water used by animals is recirculated in the form of blood plasma and tissue fluid. In plants, over 90% of the water taken in by the root system is evaporated into the air. This process, which occurs mostly through the leaves, is called **transpiration.** Plants not only have developed extensive and efficient water-transport systems, but also have evolved numerous morphological adaptations, many of which involve the leaves, to conserve water.

The amount of water in the habitat of a plant is a key to its survival. Many types of habitats exist in nature with respect to water supply. These can be conveniently divided into **xeric, mesic,** and **hydric** habitats. The plants that are adapted for living in these habitats are called *xerophytes, mesophytes,* and *hydrophytes,* respectively.

Xerophytes include species that live in habitats where the supply of water is physically or physiologically deficient (xeric habitat). **Mesophytes** inhabit regions of average or optimum water conditions (mesic habitat) and include the majority of wild and cultivated plants of the temperate regions. **Hydrophytes** are an extensive flora living on the surface of water or at various depths in the water (hydric habitat).

In this part of the exercise, you will study representative examples of hydrophytes and xerophytes, because the structural adaptations are more obvious in these groups. After your examination of these specimens, you will be asked to classify unknown specimens.

1. Hydrophytic Adaptations

The chief structural modifications exhibited by hydrophytes are an increase in leaf surface, the presence of air chambers in the leaf and stem, and a reduction in protective, supportive, and conductive tissues.

a. Dissected Leaves

Examine a hydric habitat. In many aquatic plants, the submerged leaves are finely divided (dissected). Proportionately, there is a much increased surface area in contact with the water. How might this condition be advantageous to the plant?

___

___

b. Air Chambers

The leaves and stems of many plants that are submerged in water have chambers filled with air. Examine prepared slides of cross sections of a leaf of *Potamogeton.* Locate the large air spaces, which are separated from each other by partitions of photosynthetic tissue. List several ways in which these air spaces might be of benefit in a hydric habitat.

___

___

c. Supporting Tissues

Remove one of the hydrophytic plants from the water and note how flaccid (limp) it becomes. This condition is due to the marked reduction of thick-walled supporting tissues or cells. Confirm this by examining prepared slides of *Potamogeton* stems and leaves. Why are large amounts of supporting tissues not needed by hydrophytes?

___

d. Vascular Tissues

Because aquatic plants are submerged in or floating on a nutrient solution, the structures necessary for absorption and transport of mineral nutrients and water are greatly reduced and in some cases absent. The greatest reduction occurs in the xylem. The phloem, though reduced in amount, is fairly well developed. Examine prepared slides of *Potamogeton* stems and leaves. Locate the vascular tissue and note the absence of xylem. By what process does water enter aquatic plants?

___

___

e. Protective Tissues

The epidermis of aerial plants has become modified to prevent or reduce desiccation. Under normal conditions, aquatic plants do not lose water through the epidermis, so the epidermis in hydrophytes is not a protective tissue. In these plants, nutrients and gases can be absorbed directly from the water. The cuticle overlying the epidermis is extremely thin and may be absent. The epidermal cells usually contain chloroplasts and may form a considerable part of the photosynthetic tissue. Would you expect to find guard cells in submerged hydrophytes? If not, why not?

___

___

Where would they be located in floating hydrophytes?

___

Examine slides of *Potamogeton* and locate these modifications.

2. Xerophytic Adaptations

The lack of water that characterizes xeric habitats may be the result of various environmental conditions, such as intense light, heat, and high-velocity winds. Xerophytes have evolved many adaptations to prevent desiccation when exposed to one or any combination of these factors.

a. Stomata

The stomata facilitate the exchange of carbon dioxide and oxygen between the plant and the environment. When the stomates are open, however, water can also leave the plant, which harms the plant as a whole. Consequently, reducing the rate of transpiration is important to xerophytic plants.

The location of the stomates below the level of the surrounding epidermal cells contributes to the reduction of transpiration.

Examine a prepared slide of a cross section of a pine leaf and locate sunken stomates. How does this position of the stomates reduce the rate of transpiration?

____

b. Protective Tissues

In contrast to the hydrophytes, the epidermis of xerophytes is commonly covered by a cuticle made of a waxy material called **cutin.** How does this cut down on the amount of water lost from the plant?

____

____

____

Examine a slide of a pine needle leaf and locate the modifications mentioned here.

c. Supporting Tissues

Xerophytes generally have a large proportion of supportive tissues that prevent water loss and help support the stem or leaf. Examine a slide of a pine needle leaf and locate a layer of thick-walled tissue just below the epidermis. Why is it important that the stems and leaves of aerial plants be supported in some manner?

____

d. Leaf Rolling

During drying conditions, the leaves of many xerophytes—notably, the xerophytic grasses—roll up tightly. The stomates in these plants are more numerous on the upper surface. What effect does this rolling up have on the rate of transpiration? Explain.

____

____

____

____

____

Examine a prepared slide of a corn leaf. In the upper epidermis, locate the large bulliform cells that function in the rolling of the leaves in dry weather.

e. Water Storage

Some xerophytes possess large amounts of water-storage tissue. These plants are called **fleshy xerophytes.** Examine specimens of this group. In some plants, the leaves are fleshy **(leaf succulents).** Plants in which the stem is fleshy are called **stem succulents.** Note the absence or greatly reduced number of leaves on the stem succulents. What is the primary photosynthetic organ in these plants?

____

Cut a thin cross section of a leaf succulent (for example, *Aloe*), and mount it in a drop of water on a slide. Examine it microscopically. Note the large amount of water-storage tissue that makes up the bulk of the leaf. What other xerophytic characteristics can you observe microscopically?

____

____

____

____

____

3. Unknown Specimens

Examine slides of the species of plants listed in Table 20-1. Record whether any of the characteristics listed in this table are "present" or "absent" in each of the plants described, and then decide in what habitat (xeric or hydric) each plant would be found. Base your decision on the information obtained from the examination of the various slides and characteristics cited in this study.

TABLE 20-1 Adaptations of vascular plants to the environment

Species	Characteristics						Habitat (xeric, mesic, hydric) and reason for choice
	Air chambers	Supporting tissues	Protective tissues	Water storage tissue	Vascular tissue	Describe any leaf modifications (stomata, motor cells, other)	
Potamogeton							
Yucca							
Myriophyllum							
Ammophila							
Acorus							
Typha							
Pinus							

REFERENCES

Cutler, D. 1978. *Applied Plant Anatomy.* Longman.

Cutter, E. 1978. *Plant Anatomy: Cells and Tissues.* Part 1. 2d ed. Arnold.

Cutter, E. 1982. *Plant Anatomy: Organs.* Part 2. 2d ed. Addison-Wesley.

Esau, D. 1977. *Anatomy of Seed Plants.* 2d ed. Wiley.

Fahn, S. 1982. *Plant Anatomy.* 3d ed. Pergamon Press.

Foster, A. S., and E. M. Gifford, Jr. 1974. *Comparative Morphology of Vascular Plants.* 2d ed. W. H. Freeman.

Jensen, W. A., and F. B. Salisbury. 1984. *Botany.* 2d ed. Wadsworth.

Raven, P. H., R. F. Evert, and S. Eichhorn. 1986. *Biology of Plants.* 4th ed. Worth.

Saigo, R. H., and B. W. Saigo. 1983. *Botany: Principles and Applications.* Prentice-Hall.

Weier, T. E., C. R. Stocking, and M. G. Barbour. 1982. *Botany: An Introduction to Plant Biology.* 6th ed. Wiley.

Exercise

21

Flowers and Fruits

Flowers are reproductive branches consisting of the shoot axis and lateral appendages called **sepals, petals, stamens,** and **carpels.** The significance of the flower lies in its role in sexual reproduction and its contribution to the formation of the fruit, the ripened ovary of the angiosperm flower, which contains the seeds. A large variety of fruits has evolved during the course of angiosperm evolution. Equally diverse mechanisms have evolved for dispersing the seeds.

A. FLOWERS

Roots, stems, and leaves develop and function to ensure the survival of the plant. Ultimately, however, the plant dies. It is through a constant succession of new individuals that species survive. This is accomplished by a process called **reproduction.**

The two main types of reproduction are **asexual reproduction** (or vegetative) and **sexual reproduction.** Flowering plants can reproduce asexually in many ways. For example, the tips of the branches of strawberry plants arch down, touch the soil, and develop leafy branches that root and form new plants. The most common method of reproduction, however, is sexual. To understand sexual reproduction in plants, it is necessary to study the anatomy of the flower.

Your instructor will provide you with two or three flowers to use for this part of the exercise. With the help of Fig. 21-1, locate the following parts of these flowers.

The **sepals** (modified leaves) are the outermost structures of the flower. They are typically green but can be other colors. In some flowers, the sepals are absent. The petals lie to the inside of the sepals and are often brightly colored. Both the sepals and the petals are attached to the enlarged end of the branch, the **receptacle.**

Carefully remove the sepals and petals. In the center of the flower, locate the stalklike female part of the flower, the **pistil.** Pistils are made up of one or more **carpels,** leaflike structures bearing seedlike structures called **ovules** (Fig. 21-1A, B). It is thought that during evolution the carpels rolled inward, enclosing the ovules. The pistil has a swollen base, the **ovary,** and an elongated **style** that terminates in a **stigma.** A flower may contain more than one pistil.

The ovary contains one or more ovules. You can see the ovules if you cut the ovary crosswise and examine its internal structure with a dissecting microscope. Within the ovule, one of the cells undergoes meiosis. The daughter cells formed have half (haploid) the number of chromosomes of the parent cell. One of these haploid daughter cells develops into a microscopic gametophyte plant that will pro-

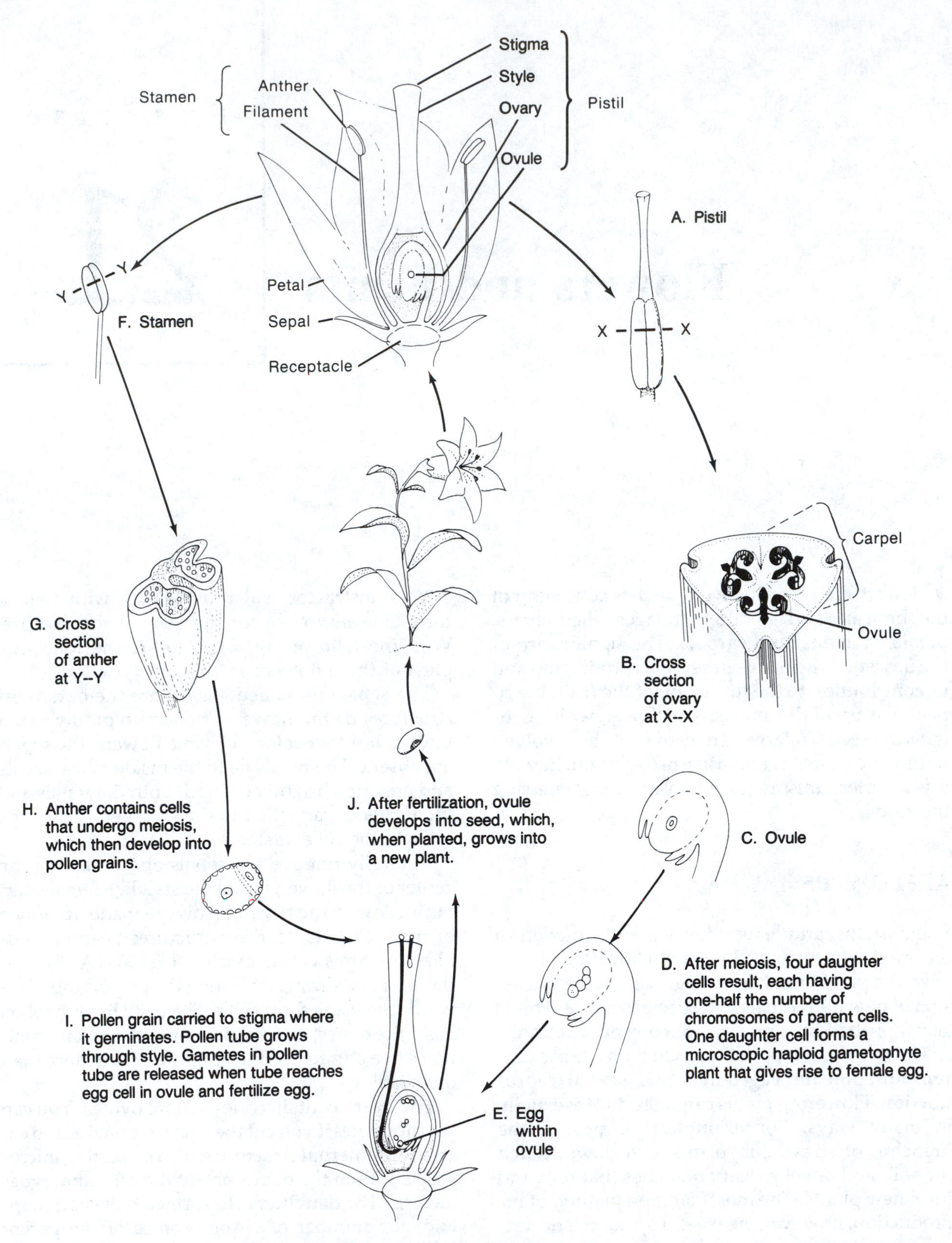

FIGURE 21-1 Study of flower anatomy

duce a female gamete, the egg cell (Fig. 21-1C, D, E). Why is it necessary that meiosis occur during the formation of the gametes?

___

___

___

Locate the **stamens** that surround the pistil (Fig. 21-1F). These male parts of the flower consist of a terminal capsule, the **anther,** attached to a slender **filament.**

The anthers also contain cells that undergo meiosis to produce cells that eventually develop into microscopic gamete-producing plants called **pollen grains** (Fig. 21-1G, H). Crush a small piece of an anther in a drop of water on a slide and add a coverslip. Examine it with a microscope and locate the pollen grains. If available, examine pollen from different plants. Note the diversity in size and in surface markings. During **pollination,** the pollen grains are transferred to the stigma, the sticky surface of the pistil. The pollen grain germinates, and the pollen tube grows through the style to the ovary and enters the ovule (Fig. 21-1I). It is known that many organisms or parts of organisms grow toward or away from various stimuli such as light, chemicals, and gravity. Suggest a mechanism that could direct the growth of the pollen tube toward the ovules.

___

During the growth of the pollen tube, two male gametes (sperm) are produced. These are released when the tube enters the ovule. One sperm fertilizes the egg cell, and the other unites with the nucleus of another cell in the ovule. This second fertilization results in the formation of a special tissue, the **endosperm,** which functions as a nutrient tissue for the developing embryo.

B. FRUITS

Fertilization of the egg initiates extensive changes in the ovary. The ovary enlarges and develops into the fruit. Structurally, the fruit consists of a mature ovary or cluster of ovaries. In some plants, other parts of the flower are modified and incorporated into the ovary to form part or all of the fruit. The seeds, usually located inside the fruit, develop from the ovules.

During the development of the fruit, the wall of the ovary, called the **pericarp,** usually thickens and becomes differentiated into three layers, which may or may not be easy to distinguish visually, depending on the species. These three layers are called the **exocarp** (outer epidermal layer), the **mesocarp** (middle layer), and the **endocarp** (inner layer). As an example, in the peach, the exocarp is the skin, the mesocarp is the fleshy part of the peach, and the endocarp is the stony pit. The seed, containing the embryo, is inside the pit (Fig. 21-1J).

The fruit protects and disperses the seed, which contains the next generation of the plant. Indeed, numerous and varied adaptations have evolved to obtain maximal dispersion of species. For example, some fruits have wings or similar structures so that they are dispersed by the wind. Fleshy fruits with hard, inedible, indigestible seeds are eaten by animals. The seeds are later dispersed in the feces during the animals' wanderings, which not only disperses the seed but also supplies the growing plant with nutrients. Spiny fruits are dispersed by sticking to the coats of animals.

1. Classification of Fruits

A classification of fruits follows.

- **Simple fruits** (derived from a single ovary).
  a. Pericarp fleshy. Examples: drupe, pome, berry, pepo.
  b. Pericarp indehiscent (does not split open when ripe). Examples: akene, nut, caryopsis.
  c. Pericarp dehiscent (splits open when ripe). Examples: legume, silique, capsule, follicle.
- **Aggregate fruits** (derived from numerous ovaries of a single flower that are scattered over a single receptacle and later unite to form a single fruit). Examples: strawberry, blackberry, raspberry.
- **Multiple fruits** (derived from the ovaries of several flowers united into a single mass). Example: pineapple.

2. Some Common Types of Fruit

In this study, you will become familiar with several types of fruit. After examining them, fill in Table 21-1 using the information given in the classification scheme from part B.1 and the characteristics of each fruit type you observed.

TABLE 21-1 Characteristics of some common fruits

Name of plant	Name of fruit	Dry or fleshy	Dehiscent or indehiscent	Structure(s) other than ovary involved in fruit formation	Fruit type

a. Legume (Fig. 21-2A)

Examine a bean or pea pod. Along how many sides does this fruit split open?

Remove a bean (or pea) seed and locate the following structures.

Hilum: Scar on the seed representing the point of attachment of the seed to the wall of the fruit.

Micropyle: Small opening adjacent to the hilum through which the pollen tube entered the ovule. It now serves to admit water to the seed as a preliminary step in the germination process.

Carefully split open the seed. The two fleshy halves, called **cotyledons,** are attached to the embryo, which consists of an embryonic root **(radicle)** and bud **(plumule).**

b. Follicle (Fig. 21-2B)

Examine a milkweed follicle. Open it and remove some seeds. Suggest a function for the large number of "hairs" associated with each seed.

How does the follicle of the milkweed differ from the pod of the legume?

c. Akene (Fig. 21-2C)

Examine the fruit of the sunflower. Crack open the hard fruit coat and examine the seed **(akene).** Remove the seed coats to observe the embryo. In some plants, such as dandelions and lettuce, the akenes are winged (Fig. 21-2D). Of what advantage is this to the species?

d. Samara (Fig. 21-2E)

Examine the winged fruit **(samara)** of the ash or maple tree. Let the fruit drop from a height of sev-

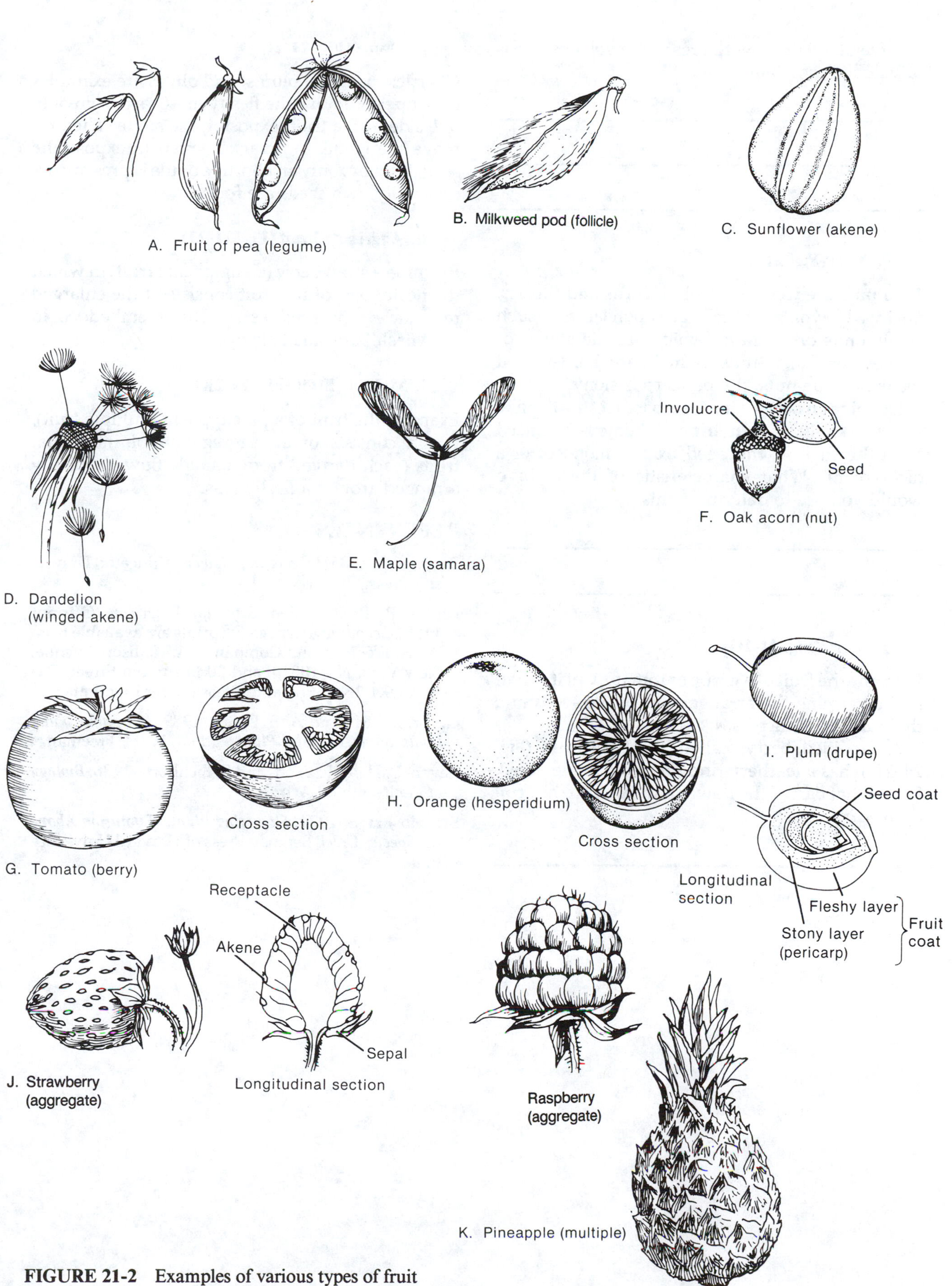

FIGURE 21-2 Examples of various types of fruit

eral feet. Explain how the response you observe is of value to the plant.

____

____

____

e. Nuts (Fig. 21-2F)

True nuts are represented by acorns and filberts, not by what you might call "commercial" nuts such as almonds or walnuts (which are the stones of drupes) or Brazil nuts (which are hard-walled seeds). In true nuts, the pericarp is stony.

Examine an oak acorn. At its base locate a cuplike structure called an **involucre.** Remove the seed from the shell. Is this seed from a monocot or a dicot plant? What characteristic of the embryo would you use to determine this?

____

____

f. Berry (Fig. 21-2G)

Examine the fruit of a grape or tomato. Cut it in half and determine the arrangement of the seeds and the number of carpels.

A modified berry, called a **hesperidum** (Fig. 21-2H), has a leathery rind (oranges and lemons). From what part of the flower is the pulp in this fruit derived?

____

g. Drupe (Fig. 21-2I)

Cherries, peaches, plums, and olives are examples of **drupes.** Remove the fleshy mesocarp from only one side of the fruit, exposing the stone. Then remove the stone and crack it open to expose the seed. Usually, only an abortive ovule is present; occasionally, two seeds are found.

h. Aggregate Fruit (Fig. 21-2J)

Examine a strawberry (an aggregate fruit) in which the fleshy part of the fruit consists of the enlarged receptacle. The seedlike structures embedded in the "flesh" are small akenes.

i. Multiple Fruit (Fig. 21-2K)

Examine the fruit of a pineapple (a multiple fruit), which consists of an aggregation of individual fruits (each derived from a single flower) spirally arranged around a fleshy axis.

REFERENCES

Barth, F. G. 1985. *Insects and Flowers.* Princeton University Press.

Echlin, P. 1968. Pollen. *Scientific American* (Offprint 1105). *Scientific American* Offprints are available from W. H. Freeman and Company, 41 Madison Avenue, New York, NY 10010, and 20 Beaumont Street, Oxford OX1 2NG, England. Please order by number.

Gifford, E. M., and A. S. Foster. 1989. *Morphology and Evolution of Vascular Plants.* 3d ed. W. H. Freeman.

Raven, P. H., R. F. Evert, and S. Eichhorn. 1986. *Biology of Plants.* 4th ed. Worth.

Stebbins, H. S. 1974. *Flowering Plants. Evolution Above the Species Level.* Belknap Press of Harvard University Press.

Exercise

22

Transport and Coordination in Plants

At the cellular level the movement of materials such as water, metabolic wastes, food, hormones, minerals, and gases is accomplished by osmosis, diffusion, and active transport. In higher plants, transport of materials throughout the organism is accomplished by complex vascular systems composed of a continuous network of conducting tissues (phloem and xylem). This network, aided by root pressure, transpiration, and gravity, moves water, minerals, and the products of photosynthesis to all parts of the plant body, where they are used or stored. Some of the substances transported throughout the plant, such as hormones and other growth regulators, are involved in coordinating various biological responses of the plant. The plant responds to external and internal stimuli; generally, the more complex the plant body the more complex are its coordination systems. In addition to various hormones and growth regulators, however, plants have evolved other coordinating mechanisms that respond to such physical factors as radiant energy and electrical gradients.

In the first part of this exercise, you will examine the morphological characteristics of vascular systems of higher plants and analyze some of the factors that influence their function. In the second part of the exercise, you will examine some of the physical and chemical factors that function, in part, to coordinate some of the biological activities of plants.

A. EFFECT OF ENVIRONMENTAL FACTORS ON TRANSPIRATION RATE

Water, one of the basic raw materials of photosynthesis, is the major component of plant tissues. Making up 90% of the plant body, it is the carrier for most materials that enter and leave the cells of plants and it is the solvent for various biochemical reactions. The amount of water used by plants is far greater than that used by animals of comparable weight because a large amount of the water used by an animal is recirculated in the form of blood plasma and tissue fluid. Over 90% of the water taken in by the root systems of plants is evaporated into the atmosphere. This process, which occurs largely through the leaves, is called **transpiration.** Consequently, plants not only have developed extensive and efficient transport systems but also have evolved numerous morphological adaptations to conserve water.

Using the setup shown in Fig. 22–1D, determine the transpiration rate of a plant as follows.

1. Cut a branch from a geranium or *Coleus* plant

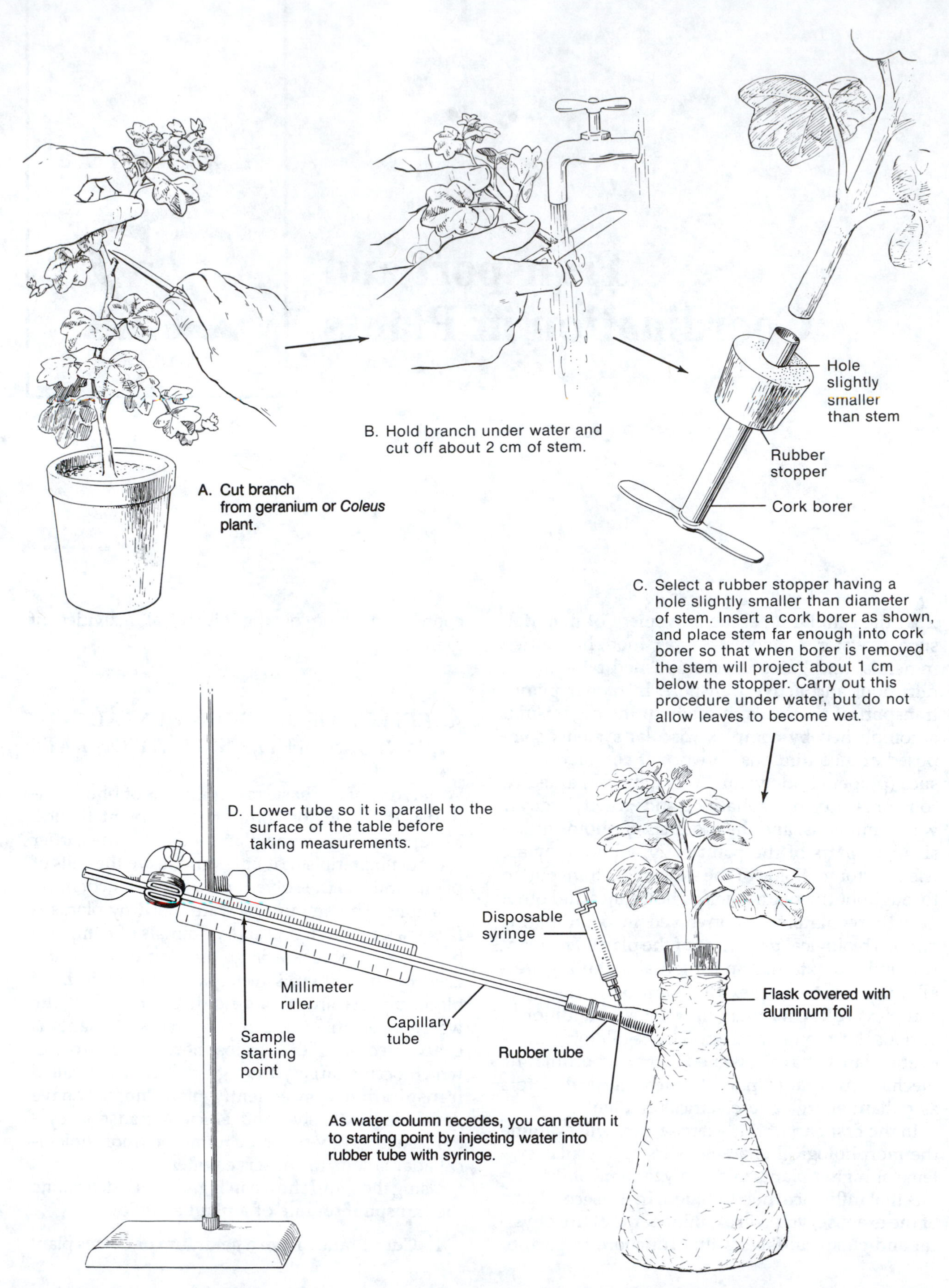

FIGURE 22-1 Procedure for determining the rate of transpiration

and insert it into a rubber stopper as shown in Fig. 22–1.

2. Place the branch and stopper in a flask that was previously filled with distilled water. Insert the stopper slowly to avoid creating bubbles. If you do this properly, water will be forced out of the end of the capillary tube. However, when you release the pressure on the stopper, the fluid in the capillary tube may recede. If this occurs, fill a 5-ml syringe with water and insert the needle into the rubber coupling between the flask and the capillary tube (see Fig. 22–1). Slowly inject water until it comes out of the end of the capillary tube.

3. If the apparatus has been properly set up, the water in the tube will now begin to recede slowly. The rate at which the meniscus moves along the tube is a measure of the rate of water uptake by the branch and can be used as a measure of the rate of transpiration. Determine the transpiration rate by recording the distance the meniscus moves each minute for a period of 10 minutes. If the meniscus goes beyond the graduated scale at the right, return it to the zero mark by injecting water into the rubber coupling as described in step 2.

4. Record your "control" results in Table 22–1. Plot your data in Fig. 22–2.

TABLE 22-1 Effect of environmental factors on transpiration rates in *Coleus* or geranium plants

Environmental factors	Time (min)									
	1	2	3	4	5	6	7	8	9	10
	Distance meniscus moves (mm)									
Control										
Light intensity										
Air movement										
Humidity										
Effect of leaves										

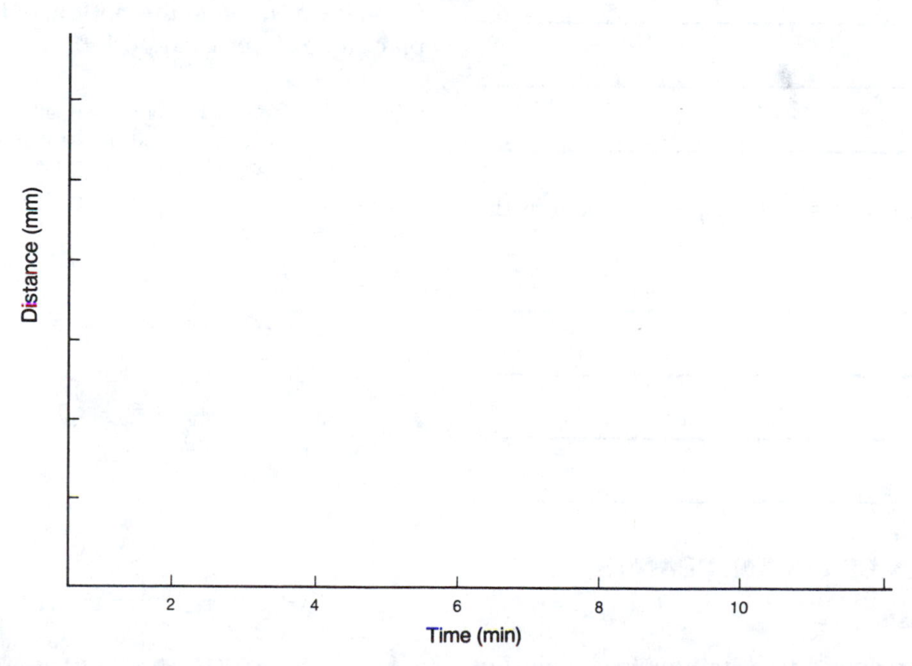

FIGURE 22-2 Comparison of transpiration rates under various environmental conditions

Design and perform experiments to show the effects of light intensity, air movement, humidity, and the leaves on the process of transpiration. Record the experimental conditions and your results in Table 22–1. Plot your data in Fig. 22–2. What is the reason for covering the flask with aluminum foil?

____

____

____

____

All aerial parts of plants can lose water by transpiration. A large portion of the water lost by herbaceous and woody plants is lost through openings on leaf surfaces called stomata. Discuss the effects of the various environmental factors on transpiration in terms of their effects on the stomata.

____

____

____

Why is it important that ornamental evergreen plants be thoroughly watered before winter sets in?

____

____

____

How does the process of transpiration benefit the plant?

____

____

____

____

B. COORDINATION IN PLANTS

1. Phototropism

The ability to respond to a simulus (a chemical or physical change in the environment) is characteristic of living organisms. In this study you will examine the effects of light on the orienting movements of plants. A simple experiment to show this effect is set up as follows.

1. Place 20 radish seeds on the surface of vermiculite in each of four medium- to large-size styrofoam cups. Punch several small holes in the bottom of the cups.

2. Cover the seeds with about ¼ inch (6.5 mm) of vermiculite and water thoroughly. Allow to germinate in the dark (about 2 days).

3. While the seeds are germinating, obtain four boxes that are large enough to accommodate each of the pots and set them up as follows.

Box 1: Seal with masking tape to make it light-tight. Label the box "Dark Control."

Box 2: Cut a rectangular hole in one side of the box (at the level of the top of the pot) and cover the opening with two layers of dark red cellophane. Label the box "Red Light" (Fig. 22–3).

Box 3: Cut a rectangular hole in the box and cover the opening with three layers of dark blue cellophane. Label the box "Blue Light."

Box 4: Cut a rectangular opening in the box and leave the opening uncovered. Label the box "White Light."

4. Place a cup containing seedlings in each box (or you can remove the bottoms of the boxes and place a box over each pot).

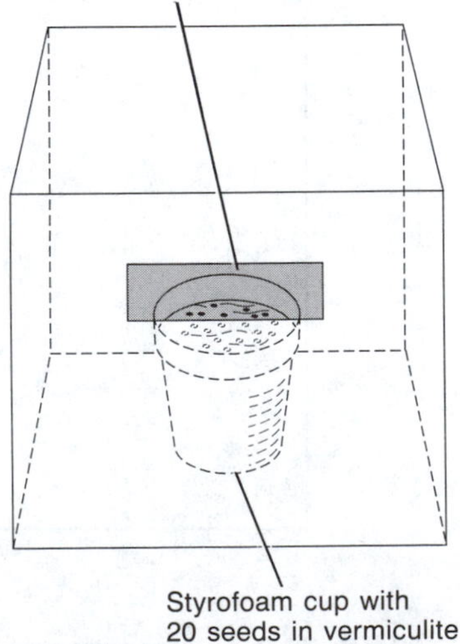

FIGURE 22–3 Phototropism setup

5. Line up the boxes so that they are about 60 cm from a 40-watt fluorescent tube that is lying sideways so that light is directed toward the openings in the boxes.

6. After 4–5 days, remove the boxes and record your observations in Table 22–2. If obvious responses are not observed, replace the boxes and examine the cups again in 2 days.

What is the growth response of the seedlings to light?

Which wavelength of the visible spectrum (red or blue) is effective in producing a phototropic response?

What is the name of the pigment involved in this response and what color would you expect the pigment to be?

Describe the differences in growth between dark-grown and light-grown seedlings.

What is the term for the growth phenomenon that is observed in the dark-grown seedlings?

2. Geotropism

The effect of gravity in determining the direction of growth of various plant parts is called **geotropism.** Most stems grow in the direction that is opposite the force of gravity. Is this negative or positive geotropism?

What do you think is the geotropic response of roots to gravity?

In this study, you will work in teams and will observe the geotropic responses of *Coleus* or *Iresine* shoots. For convenience, each team will be assigned one aspect of the problem. Because the response of the plant usually requires several hours, set the experiment aside for examination during the next laboratory period.

Team 1: What is the effect of gravity on stem orientation? Remove three *Coleus* or *Iresine* branches, keeping their cut surfaces moist during the cutting operation. Set them up as indicated in Fig. 22–4. What is the control in this experiment?

TABLE 22–2 Observations of phototropic responses to red, blue, and white light

Box number and light condition	Growth response
1. Dark control	
2. Red light	
3. Blue light	
4. White light	

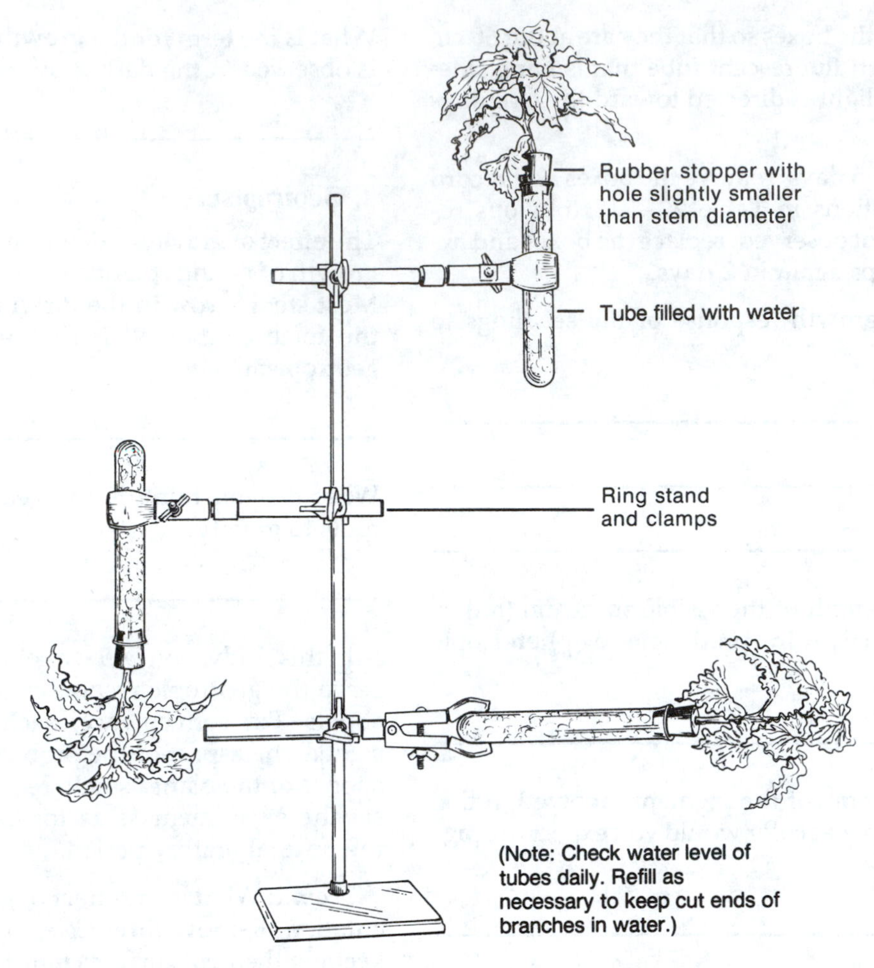

FIGURE 22-4 Experimental setup to show effect of gravity on growth

In Table 22–3 record the results you expect and the observed results.

Team 2: Does defoliation affect the geotropic response of *Coleus* or *Iresine* shoots? Design and perform an experiment to answer this question. Complete Table 22–4.

Team 3: Does the shoot apex exert any effect on the geotropic response of *Coleus* or *Iresine?* Design and perform an experiment to answer this question. Complete Table 22–5.

Team 4: Does "replacing" the shoot apex with auxin have any effect on the geotropic response of

TABLE 22-3 Data on the effect of gravity on stem orientation

Results	Stem position		
	Vertical	Horizontal	Inverted
Expected			
Observed			
Conclusions:			

TABLE 22-4 Data on the effect of defoliation on the geotropic response

Procedure: 1. Experimental treatment	Expected results:	Observed results:	Conclusions:
2. Control treatment			

Coleus or *Iresine?* Design and perform an experiment to answer this question. Complete Table 22-6.

3. Chemotropism

Growth toward or away from a chemical stimulus by nonmotile cells is called **chemotropism,** and we call the effect of a chemical agent on cells *chemotropic activity.* In this study you will observe the chemotropic activity of lily pistils, auxin, and gibberellic acid on lily pollen tubes. Your instructor will provide each team with the chemical agents to be tested for chemotropic activity and the lily pollen tubes upon which to test them.

Testing is done by placing a small amount of the chemical agent on the surface of the nutrient medium in a petri dish and, with a camel's hair brush, positioning the lily pollen tubes in small clumps within 2-3 mm around it. To test the chemotropic

TABLE 22-5 Data on the effect of shoot apex on the geotropic response

Procedure: 1. Experimental treatment	Expected results:	Observed results:	Conclusions:
2. Control treatment			

TABLE 22-6 Data on the effect of replacing the shoot apex with auxin on the geotropic response

Procedure:	Expected results:	Observed results:	Conclusions:
1. Experimental treatment			
2. Control treatment			

activity of liquids, cut a well in the medium with a cork borer and place the liquid in the well; then position the lily pollen tubes in the same way as above.

Team 1: Does the pistil have any chemotropic effect on the pollen tubes? Is chemotropic activity exhibited by all parts of the pistil? Design and perform an experiment to answer these questions. What controls should be used?

____

____

What criteria would you use to determine whether positive or negative chemotropism is indicated?

____

____

____

TABLE 22-7 Data on the chemotropic activity of lily pistils

Procedure:	Results:	Conclusions:
1. Experimental treatment		
2. Control treatment		

TABLE 22–8 Data on the effect of heat on the chemotropic activity of lily pistils

Procedure:	Results:	Conclusions:
1. Experimental treatment		
2. Control treatment		

Complete Table 22–7.

Team 2: Is the chemotropic activity of lily pistils altered by heat? Design and perform an experiment to answer this question. Complete Table 22–8.

Team 3: Is chemotropic activity exhibited by auxin or gibberellic acid? Design and perform an experiment to test the chemotropic activity of each of these plant-growth substances. It will be necessary to know the solvents for each substance so that adequate controls can be established.

If one (or both) of these substances shows chemotropic activity, determine its optimum concentration by testing with a series of concentrations—for example, 1, 10, 50, and 100 parts per million (ppm). Complete Table 22–9.

TABLE 22–9 Data on the chemotropic activity of auxin and gibberellic acid

Procedure:	Results:	Conclusions:
1. Experimental treatment		
2. Control treatment		

REFERENCES

Baker, D. A. 1978. *Transport Phenomena in Plants.* Wiley.

Bowley, J. D., and M. Black. 1985. *Seeds: Physiology of Development and Germination.* Plenum Press.

Raven, P. H., R. F. Evert, and S. Eichhorn. 1986. *Biology of Plants.* 4th ed. Worth.

Salisbury, F., and C. Ross. 1991. *Plant Physiology.* 4th ed. Wadsworth.

Whatley, F. R., and J. M. Whatley. 1980. *Light and Plant Life.* Edward Arnold.

Exercise

23

Plant Growth and Development

The observable changes that take place during the life cycle of a higher plant begin with the embryo in the seed and proceed sequentially through germination of the seed, and the appearance and subsequent enlargement of stems, leaves, and roots to the production of flowers, fruit, and seeds. The locations of these organs and their sizes and shapes are the visible manifestations of the correlated activities of cells that are produced by meristems. This complex, ordered series of events is called **development.**

Development can be divided into phases called growth, cell differentiation, and organogenesis. **Growth** is an irreversible increase in volume, which is usually, though not always, accompanied by an increase in weight. **Cell differentiation** includes all events involved in the creation from generalized cells of cells that are specialized for building the various structures and carrying out the various functions of the mature plant. **Organogenesis** is the development of the various plant organs.

A. COMPARATIVE STUDY OF SEEDS

Seeds develop from ovules and are formed following the sexual union of the sperm and the egg, an event that occurs in the ovary of the flower. If you are not familiar with floral anatomy, refer to Fig. 23-1 and a flower model or living flowers. The ovary and, in some cases, other parts of the flower become the fruit. Within the seed is the embryo, which consists of three parts: the **cotyledons** (sometimes called seed leaves); the **epicotyl,** which is located above the point of attachment of the cotyledons; and the **hypocotyl,** found below the cotyledons. The mature seed may also have an **endosperm,** which functions as nutritive tissue for the developing embryo.

Obtain soaked bean, pea, and corn seeds. (The so-called seed of corn is really a fruit, the seed coat being intimately fused with a hard outer tissue called the **pericarp,** which is part of the fruit.) Carefully remove the outer seed coats from the seeds. Then separate the other parts and compare them with each other and with Fig. 23-2.

1. Cotyledons

Scrape the surfaces of the cotyledons of the bean and pea and add a drop of iodine to the scraped surface. On the basis of your observations, what appears to be the role of the cotyledons in the development of these seeds?

____

____

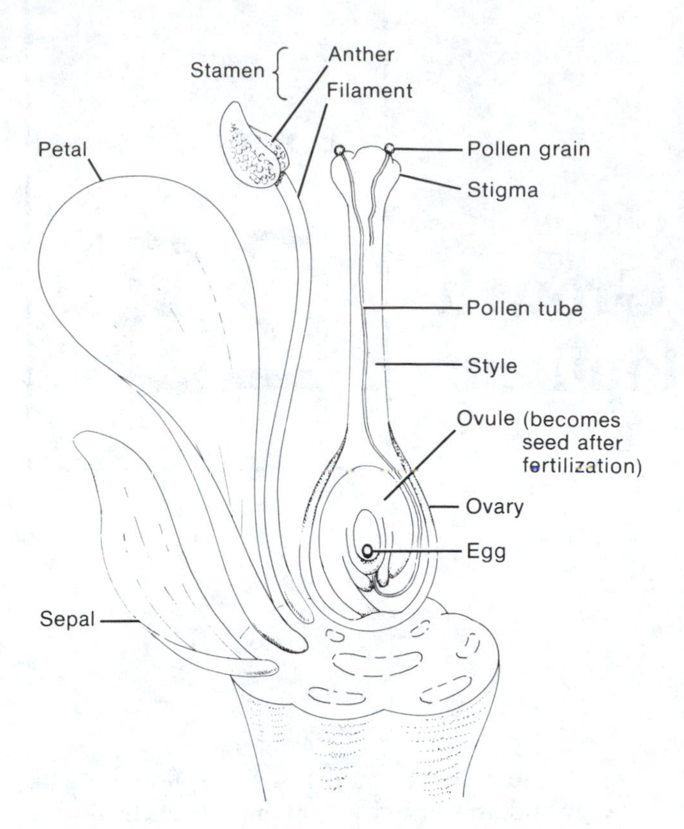

FIGURE 23-1 Diagram of a flower

2. Epicotyl

This part of the embryo gives rise to the shoot system of the mature plant. What indication is there, if any, that this region will produce stem and leaf tissues?

---

3. Hypocotyl

This region gives rise to all root tissues and, in some cases, to the lower part of the stem.

4. Endosperm

The endosperm (although found in many seeds) is present only in the corn "seed" of the group you are examining. Depending on whether the endosperm is composed of starches or sugars, corn tastes starchy or sweet. Remove the endosperm (Fig. 23–2) from several corn seeds and cut it into fine pieces. Then add the pieces to 5 ml of Benedict's solution in a test tube and heat them in a hot-water bath for several minutes. If a reducing sugar is present, the solution will change color from blue (small amount of sugar) to green to orange to red to brown (large amount of sugar), depending on the amount of sugar present.

Cut through a kernel of corn, as shown by the dashed line c--c in Fig. 23–2. Scrape the surface of the endosperm with a razor blade and add a drop of iodine. The scraped surface will turn a dark blue, which indicates the presence of starch. What commercial value does starchy corn have?

B. GERMINATION

The seed is a resting stage in the development of a plant, serving to carry the plant over periods of unfavorable environmental conditions. When provided with optimal growing conditions, the seed, if still living **(viable),** will germinate and produce another plant. The production of viable seeds is required to maintain the existence of any plant species in nature. It is commercially important for seed growers to be able to judge the viability of the seed from any given crop. For biologists who grow research plants from seed, it is important to know that the percentage of germination will be high.

In this part of the exercise, you will determine the degree of viability of a batch of seeds. One way of doing this is to germinate sample batches of seed under standardized conditions. This germination test has one major drawback: it takes 7–10 days for many seeds to germinate. A quicker and easier method is to use the fact that embryos in viable seeds respire and can change colorless forms of dyes, such as tetrazolium, into colored forms.

1. Germination Test

1. Working in teams, obtain 100 bean seeds that have been soaked in water. Boil 50 of these seeds for 10 minutes to kill them.

2. Plant 25 living seeds and 25 killed seeds about 15 mm deep in separate rows in a planting tray. (Save the remaining boiled and unboiled seeds for the tetrazolium test.) Label each row with your section number, team number, date, number of seeds, and treatment (boiled or unboiled).

3. After a week, count the total number of germinated, growing seeds in each treatment and record these data in Table 23–1. Record your team

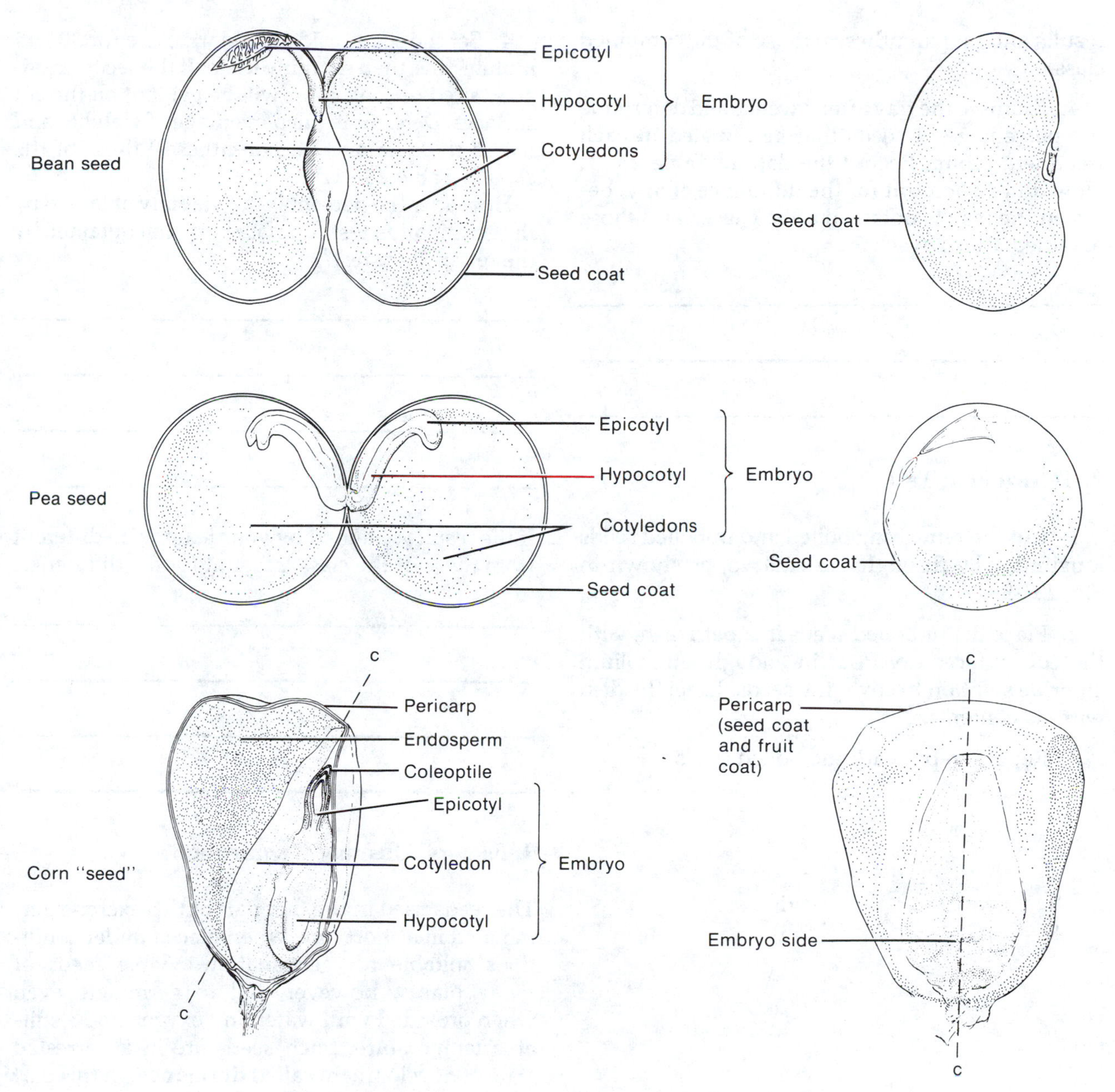

FIGURE 23-2 Comparative anatomy of various seeds

TABLE 23-1 Germination and tetrazolium data

Data	Germination test				Tetrazolium test	
	Unboiled seeds		Boiled seeds		Unboiled seeds	Boiled seeds
	Week 1	Week 2	Week 1	Week 2		
Total number of seeds						
Number of seeds germinated						
Germination (%)						

results and (in parentheses) those of the combined class.

4. Examine the tray after 2 weeks and determine the percentage of seeds that germinated in each treatment group. Record the data in Table 23–1. How do you account for the difference, if any, between the results obtained after 1 week and those obtained after 2 weeks?

____

____

____

2. Tetrazolium Test

1. Cut the remaining boiled and unboiled seeds longitudinally through the embryo, as shown in Fig. 23–3.

2. Place the unboiled seeds in a petri dish, with the cut surfaces up. Pour in enough tetrazolium chloride solution to cover the seeds. Label the dish with its contents.

3. Repeat step 2 with the boiled seeds.

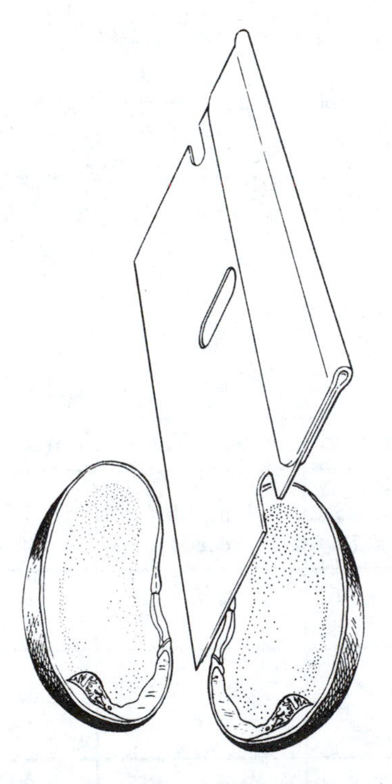

FIGURE 23-3 Procedure for cutting bean seeds for tetrazolium test

4. Set the seeds aside in a dark place for 30–45 minutes and then examine them. If the seeds are viable, a red or pink color will be evident on the cut surfaces. Determine the percentage of viability and record the data and (in parentheses) those of the class in Table 23–1.

How does the percentage of viability obtained by the germination test compare with that obtained by the tetrazolium test?

____

____

____

____

If the results obtained by your team were different from those of the class, account for the difference.

____

____

____

____

3. Factors Affecting Germination

The seeds used in the first part of this exercise germinated in a short time when placed under conditions suitable for germination. Viable seeds of many plants, however, fail to germinate even when provided with water and oxygen and a suitable temperature. Such seeds are in an arrested state of development called **dormancy.** In this part of the exercise, you will work in teams to study one of the factors involved in the dormancy of seeds.

1. Cover the bottoms of three petri dishes with moistened filter paper. Obtain 60 honey locust seeds (other appropriate seeds are sweet clover, okra, and alfalfa), 20 of which have been soaking for 2–3 hours in a 70% sulfuric acid solution in a beaker.

2. With a rubber band, fasten cheesecloth over the beaker and carefully pour off the acid into a sink with the water running.

Caution: *Do not pour the acid into the sink and then turn on the water. A violent splattering of the acid could occur.*

Wash the acid-treated seeds in running water for 5 minutes and place them in one of the petri dishes.

3. Put 20 untreated seeds into a second petri dish.

4. With a razor blade, remove a small chip from each of the untreated 20 seeds, exposing the inner tissue. Place these seeds with their cut surface down in the third dish.

5. Cover all dishes, seal them with tape (to retard evaporation), and label each with its contents. Put the dishes in the dark at room temperature (or wrap them in aluminum foil and take them home for observation).

6. Periodically examine the seeds over the next several days and determine the percentage of germination for each group of seeds. Record your team data in Table 23–2. What is the "factor" that controls dormancy in these seeds?

____

____

Of what advantage to seeds is dormancy?

____

____

C. MEASUREMENT OF PLANT GROWTH

Earlier, we defined growth as an irreversible increase in volume usually accompanied by an increase in weight. Growth, therefore, is quantitative and can be measured. In these experiments, you will determine the locus of growth in a plant organ and plot its growth curve.

1. Localization of Plant Growth

1. Prepare a moist chamber (1000 ml beaker) as shown in Fig. 23–4A. Cover a glass plate (9× 10 cm) with moist toweling, and place it in the chamber. Cover the container to prevent drying.

2. From germinating seeds provided by your instructor, select one with a fairly straight root about 1.5–2 cm long. Blot the root to remove any excess moisture and then lay it against a millimeter ruler. Using the threaded marking device shown in Fig. 23–4D, carefully mark 10 lines, each 1 mm apart, *starting from the tip of the root* (Fig. 23–4D). Make sure you wipe excess ink from the thread or the mark will become smudged, making accurate measurements difficult. Mark five roots. Avoid drying of the roots during marking.

3. Remove the paper-covered glass plate from the chamber and lay the five seedlings on it. Hold the seedlings lightly in place with a rubber band so that the whole length of each root is touching the moist paper (Fig. 23–4F). Place the plate and seedlings in the moist chamber, and then cover and keep in a dark place.

4. After 48 hours, measure the distances between each of the ink marks on each root. Average the lengths for each distance. Record the data in Table 23–3, and plot the data in Fig. 23–5.

Where does most of the growth occur?

____

If the ink lines were initially sharp and clear, how do you account for the smudging of the first, and possibly the second, line?

____

TABLE 23–2 Factors affecting the germination of seeds

Data	Seed treatment		
	Treated with H_2SO_4	Untreated	Seed coat cut
Total number of seeds			
Number of seeds germinated			
Germination (%)			

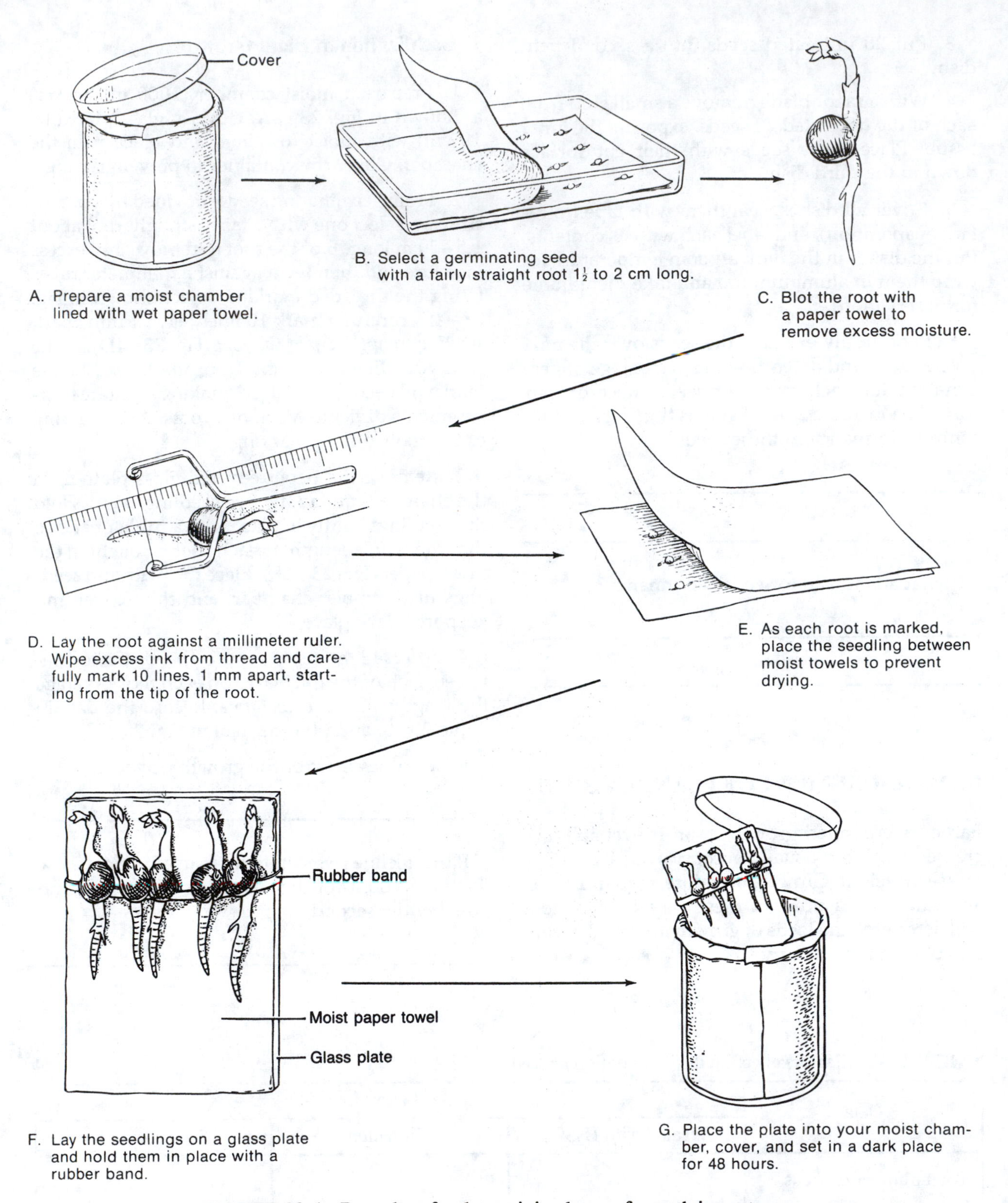

FIGURE 23-4 Procedure for determining locus of growth in roots

TABLE 23-3 Data for locus of root growth

Root tip	Interval									
	1	2	3	4	5	6	7	8	9	10
1										
2										
3										
4										
5										
Total										
Average										
Control										

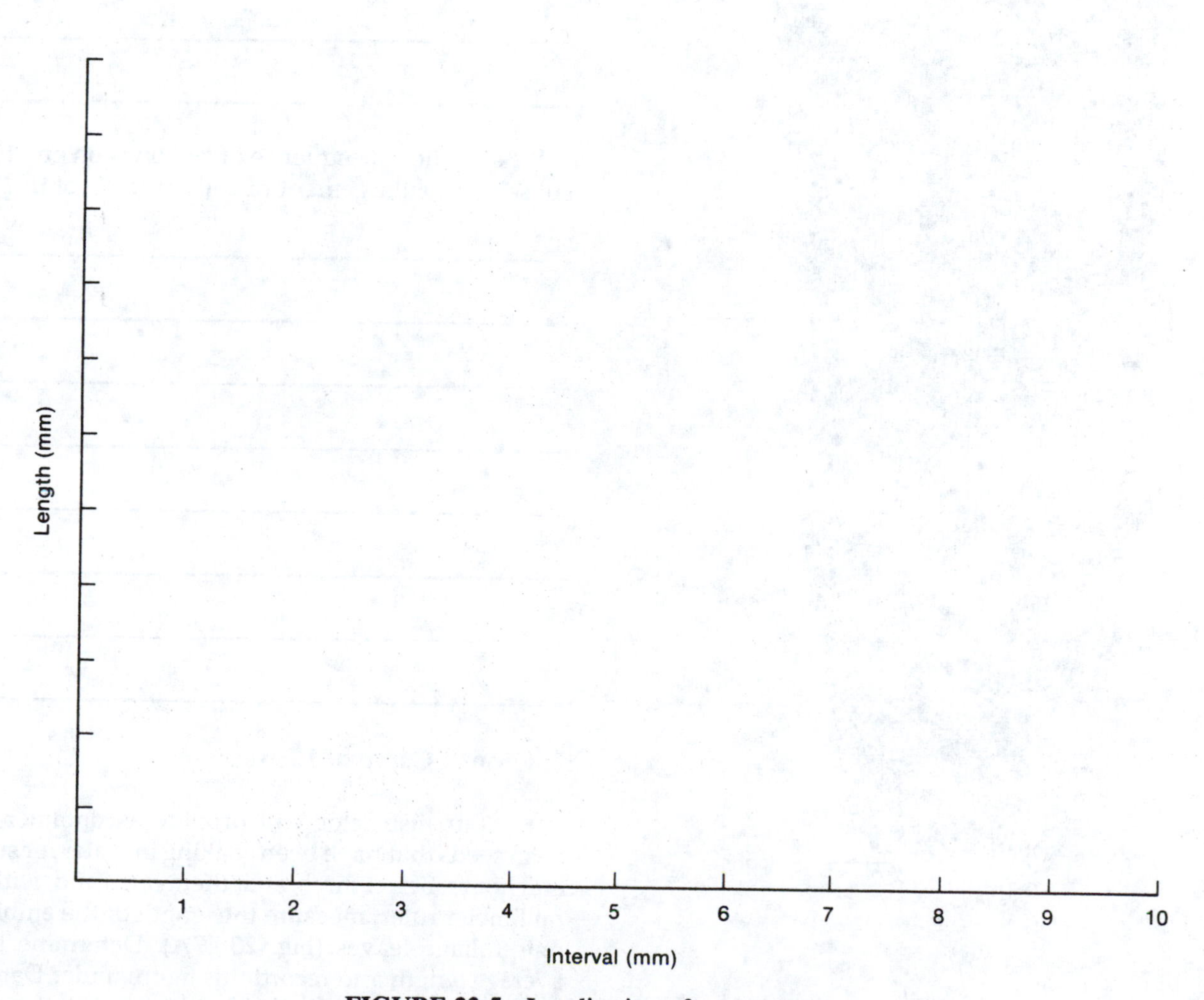

FIGURE 23-5 Localization of root growth

If you were to cut a longitudinal section through a young root and examine it microscopically, what would you expect to see in the first millimeter or so that would account for the results you have observed?

____

____

With a compound microscope, examine a prepared slide of a longitudinal section through an onion root tip (compare with Fig. 23–6). Locate a cone-shaped mass of loosely arranged cells covering the tip of the root. This is the **root cap,** which serves to protect the **meristematic** (embryonic) region of the root in which active cell division is occurring. Closely examine the cells of the meristematic region. What occurred in this region when the tissue was living to account for the results obtained in the growth measurement experiment?

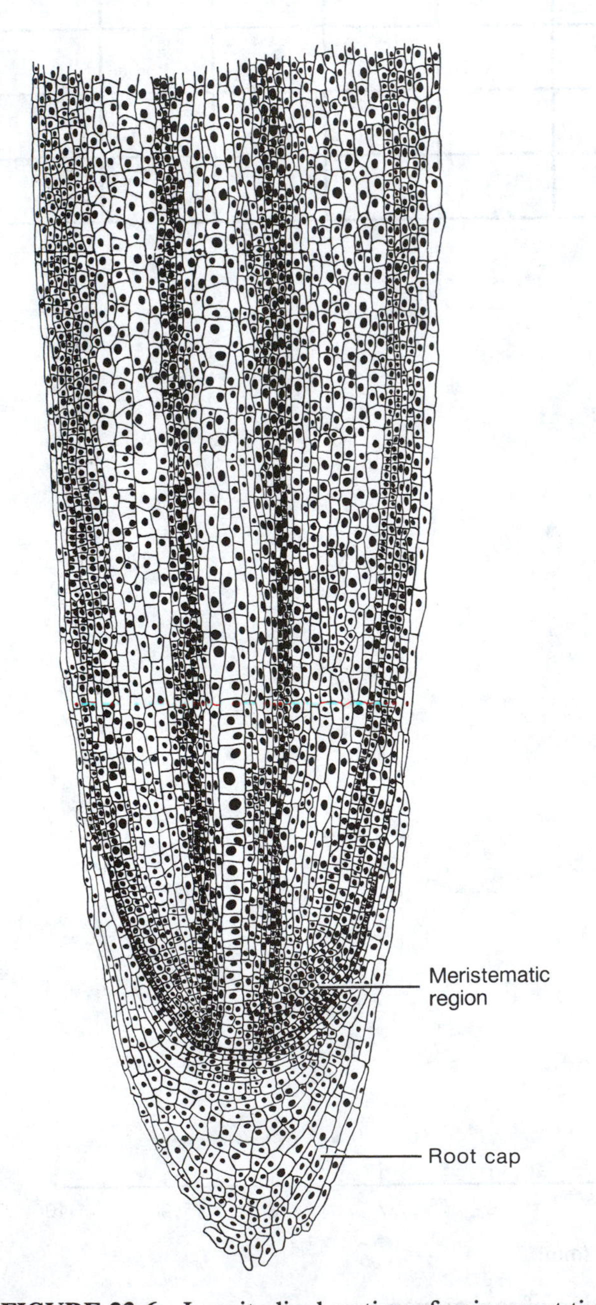

FIGURE 23-6 Longitudinal section of onion root tip

____

____

____

____

Locate the region of the root in which cell differentiation is occurring. Approximately how far back from the tip does this region occur? (The diameter of the low-power field, 10 × objective, is approximately 2 mm.)

____

____

What is the consequence of cell division and the subsequent enlargement of cells at the root tip?

____

____

____

____

____

____

____

____

2. Growth Curve of Leaves

1. Your instructor will provide a container of bean seeds that have been soaking in water for several hours. Select three, split them open and, with a millimeter ruler, measure the length of the embryonic foliage leaves (Fig. 23–7A). Determine the average length and record this figure under Date 0 in Table 23–4. This is the first in a series of mea-

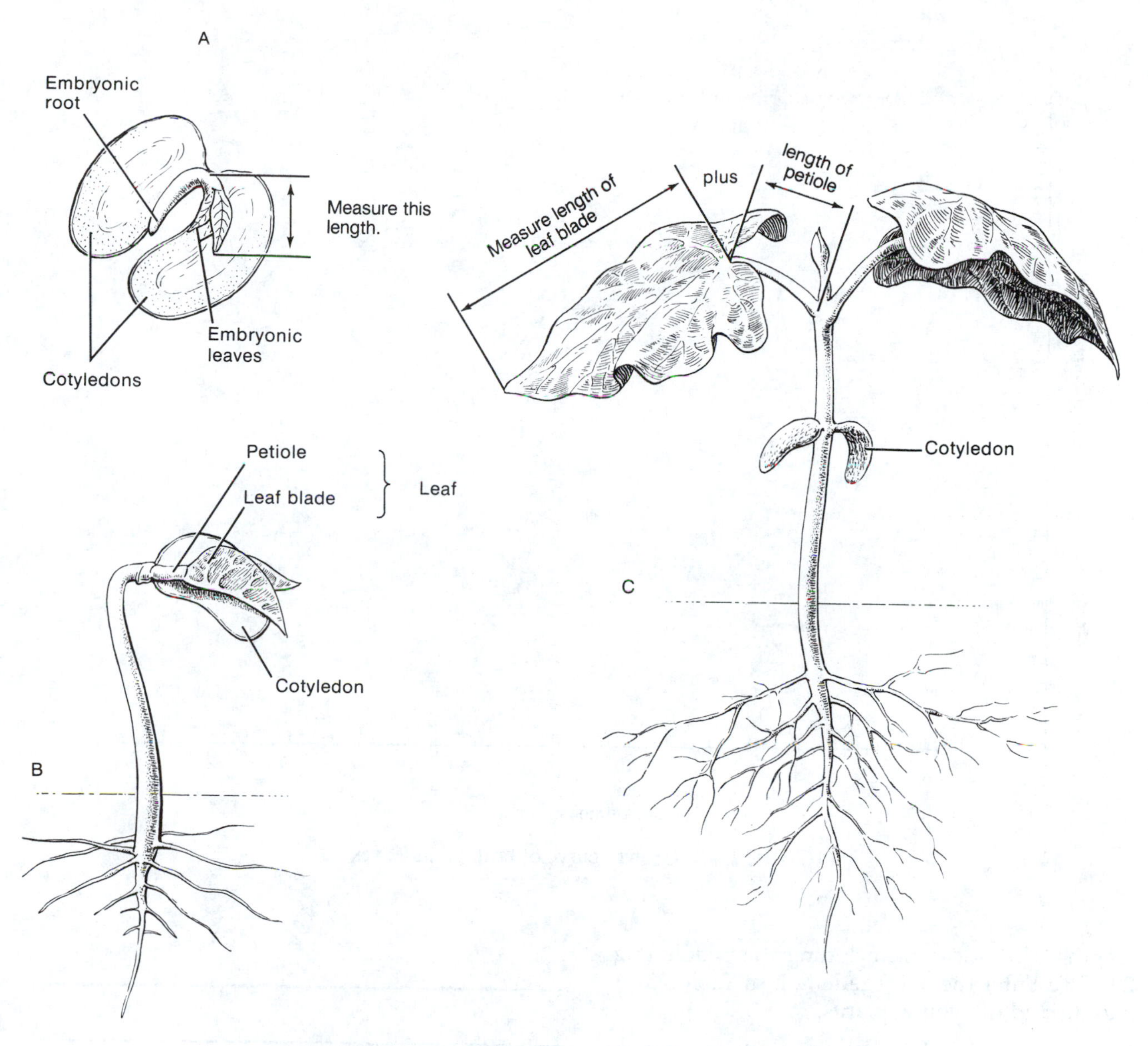

FIGURE 23-7 Measuring the growth of a bean seedling

TABLE 23-4 Average length of leaves

Date of measurement	0						
Average length (mm)							

surements to be made of the growth of the first two foliage leaves.

2. Plant 25 of the soaked bean seeds about 15 mm deep in the container provided. Water thoroughly, and place them in the greenhouse. In 2 or 3 days, dig up three of the seeds and measure the

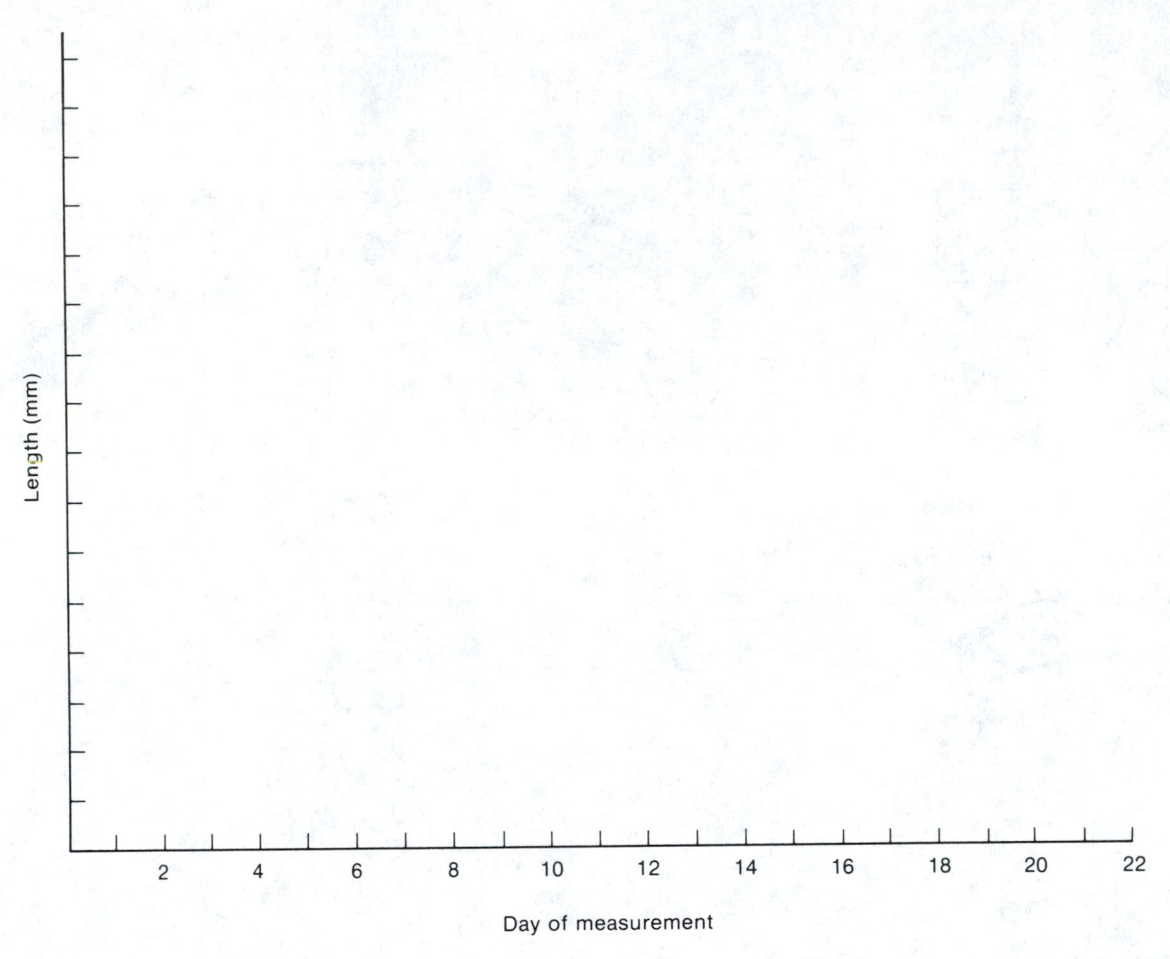

FIGURE 23-8 Growth curve of bean plant leaves

lengths of the leaves, including the petiole (Fig. 23–7B). Enter the average length in Table 23–4, and discard the young plants.

3. In succeeding laboratory periods, select three plants and measure the leaves as in step 2 (Fig. 23–7C). *Do not remove these plants from the container. Use the same three plants for all measurements, beginning with this third measurement.*

4. Plot the data in Fig. 23–8. List several other organisms (or organs) that would show a similar growth curve.

REFERENCES

Raven, P. H., R. F. Evert, and S. E. Eichhorn. 1986. *Biology of Plants.* 4th ed. Worth.

Salisbury, F., and C. W. Ross. 1985. *Plant Physiology.* 3d ed. Wadsworth.

Exercise

24

Plant Development: Hormonal Regulation

The systematic differentiation of millions of cells into various tissues and organs and the development of an organism's specific form require precise coordination.

For example, consider the development of a seed into a mature plant. It consists of growth by cell division and cell elongation; differentiation of organs such as roots, stems, leaves, and flowers; and complex chemical changes. The final form of the plant is a blend of the plant's genetic "blueprint" and the modifying effects of the environment. When the seed begins to germinate, it absorbs large amounts of water, and cells at the meristems of the plant embryo begin to divide. For reasons not yet fully understood, the root almost always begins to develop before the shoot. At both root and shoot ends of the embryo, new cells are formed by the **meristematic regions** (areas of rapid cell division), followed by elongation and differentiation of these cells.

The control of cell growth and differentiation lies in the DNA of the nucleus, which ultimately controls the production of hormones and other regulatory chemicals. Our knowledge of the mechanisms of regulation of growth and differentiation is expanding rapidly as new experimental evidence is reported.

Some of the growth-regulating substances in plants are auxins, gibberellins, cytokinins, and various growth inhibitors.

- **Auxins** which are produced in the meristematic tissues of buds, leaves, embryos, seeds, and fruit, exist in several slightly different chemical forms. The most abundant is **indole-3-acetic acid,** or IAA. Auxins influence cell elongation; inhibit growth of lateral buds; promote the initiation of roots; and, in a few plants, regulate the differentiation of flower buds.
- **Gibberellins** synthesized in meristems and plant seeds, are complex substances that affect cell elongation, cell division, and flowering responses in some plants. More than 70 gibberellins have been identified. Of these, **gibberellic acid** (GA_3) has been most thoroughly studied.
- **Cytokinins** which appear to be related to a component of RNA, promote cell division and, in the presence of auxin, induce the differentiation of roots and shoots.
- **Inhibitors** of various types regulate such responses in plants as flowering, dormancy of buds and seeds, and rates of growth.

In this exercise, you will study the effects of some environmental and chemical factors on the patterns of growth and differentiation in plants.

A. AUXINS

1. Effect of an Auxin on Directional Growth in Plants

That hormones regulate growth in animals has been known for a number of years. By 1930, it was commonly accepted that plant growth is also regulated by special hormonal substances called **auxins.** The principal naturally occurring auxin is indole-3-acetic acid, or IAA (Fig. 24–1).

Before IAA was discovered, *auxin* was used to refer to this growth hormone. The term now describes any one of a large number of compounds having physiological activity similar to that of IAA. Indole-3-acetic acid has several effects on growth; a critical effect, however, is the inducement of cell elongation. In this part of the exercise, you will study some effects of IAA on plant growth and become familiar with some procedures used to assay for its presence in tissues.

1. Select 21 bean plants of uniform size. Divide them into seven groups of three plants each. To groups 1, 2, and 3, apply an IAA-lanolin mixture at the places shown in Fig. 24–2. (*Note:* IAA is mixed with lanolin, which is an inert paste that aids even application of growth substances to the plant.) To groups 4, 5, and 6, apply only lanolin at the same places. Why is only lanolin applied to these plants?

____

____

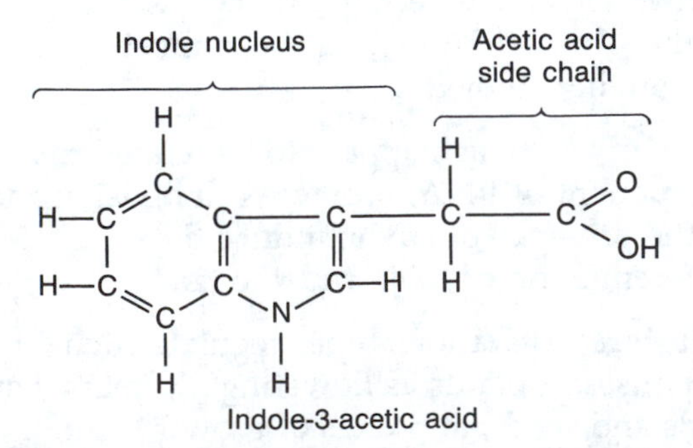

FIGURE 24-1 Structure of indole-3-acetic acid (IAA). (From *An Experimental Approach to Biology*, 2d ed., by Peter Abramoff and Robert G. Thomson. W.H. Freeman and Company ©1976.)

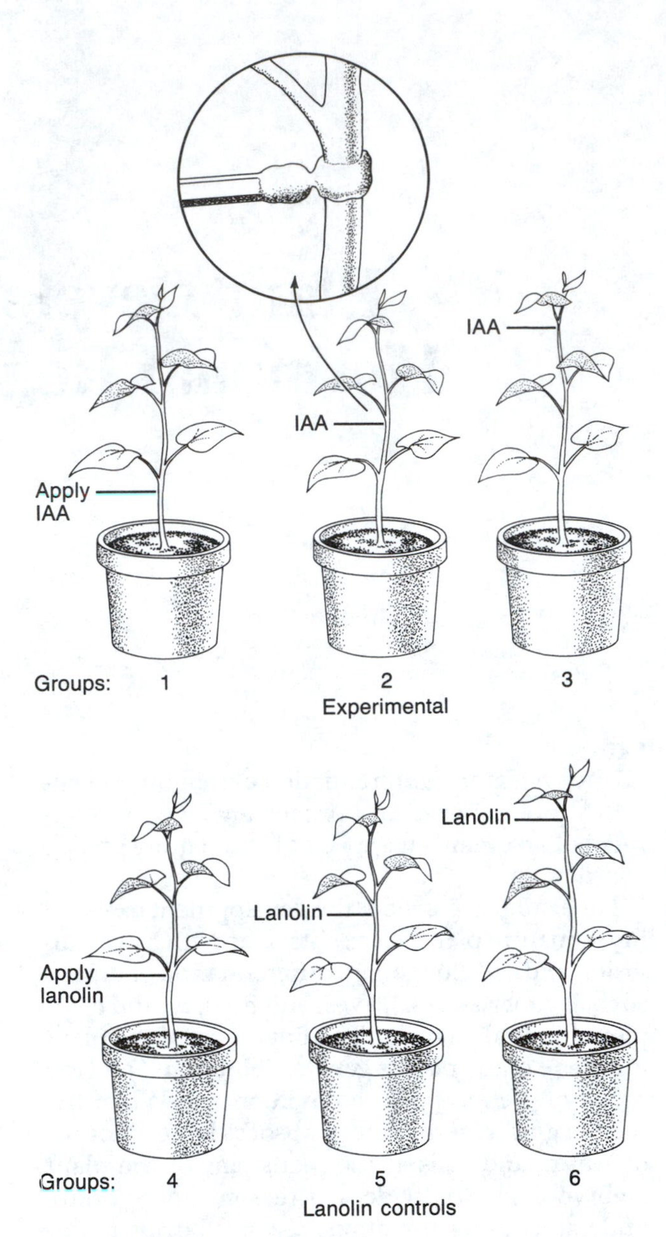

FIGURE 24-2 Method of applying IAA and lanolin to plants

Group 7 plants are untreated controls.

2. Observe the plants daily for the next week. Describe the results (use drawings if necessary). Where would you expect IAA to have its most profound effect, if any, on growth? At the top of the shoot or at the bottom of the plant where it emerges from the soil? Explain.

____

____

____

____

____

____

2. Effect of an Auxin on the Initiation of Root Growth

1. With a small-diameter (5 – 10 mm) cork borer, punch five evenly spaced holes in the covers of four pint-size cottage cheese containers. Fill the containers with the following solutions:

Container 1: Distilled water

Container 2: Hoagland's nutrient solution

Container 3: Hoagland's solution containing IAA at 0.1 mg/liter

Container 4: Hoagland's solution containing IAA at 1.0 mg/liter

Replace the covers and label each container as to its contents (Fig. 24 – 3A).

2. Cut 20 bean plants at ground level (Fig. 24 – 3B). Then remove cotyledons and cut the stems at a point 5 cm below the point of attachment of the cotyledons (Fig. 24 – 3C). Quickly place the plants in the holes of the cottage cheese containers (Fig. 24 – 3D).

3. Place the containers in a greenhouse. Alterna-

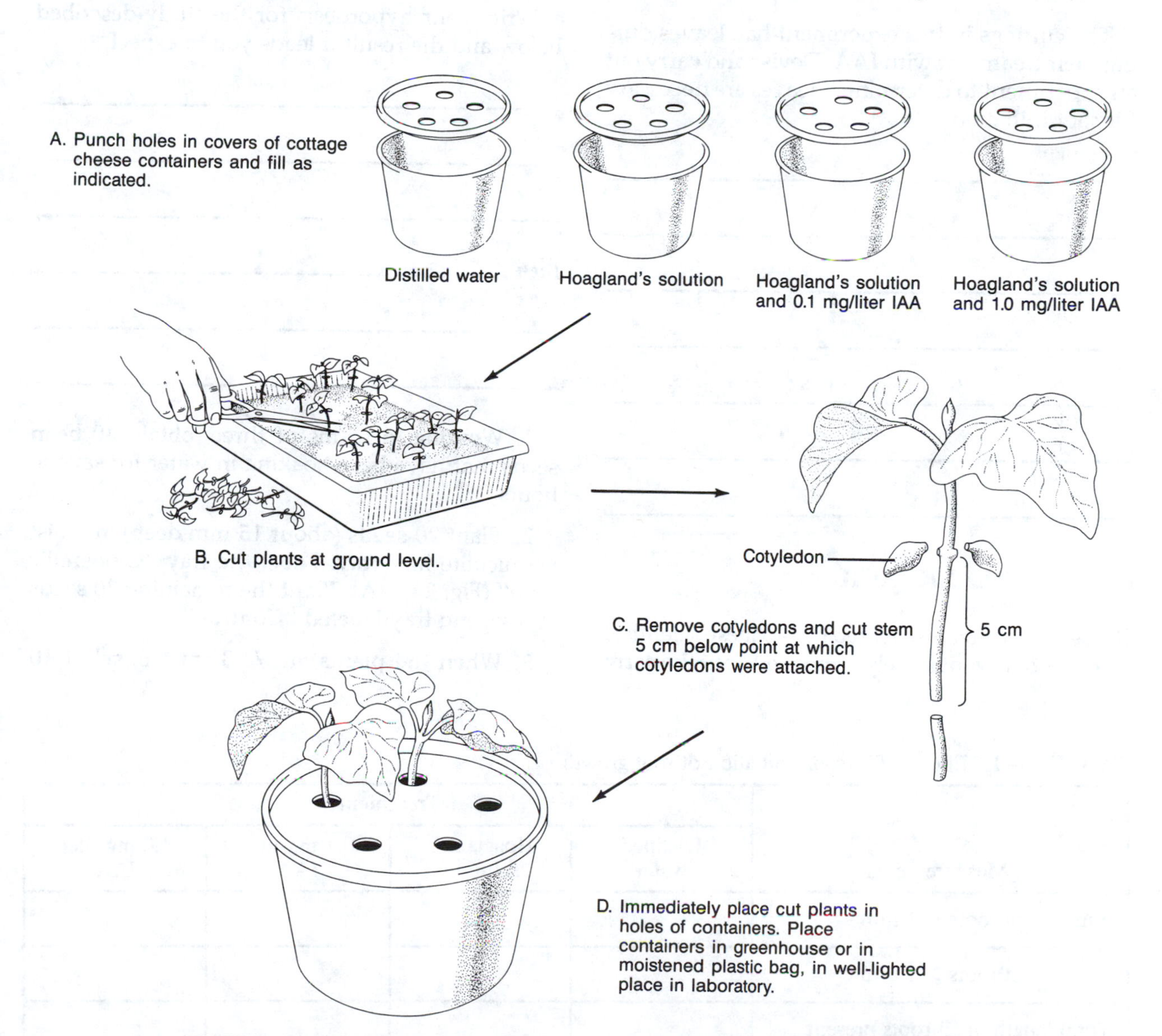

FIGURE 24-3 Procedure for determining the effect of IAA on root initiation

tively, place them in a moistened polyethylene bag, and store them in a well-lighted area of the laboratory.

4. After one week, examine the plants and complete Table 24–1. Describe the effects of IAA on the initiation of root growth.

____

____

____

____

____

The cuttings in this experiment had leaves during their treatment with IAA. Devise and carry out an experiment to determine if leaves are necessary for root initiation.

____

____

____

____

____

____

____

B. GIBBERELLIC ACID

Gibberellins, first discovered by Japanese scientists in the 1920s, went largely unnoticed until the early 1950s, when English and American biologists became interested in them. They were first isolated from the fungus *Gibberella.*

In this part of the exercise, you will determine some of the more obvious effects of gibberellic acid on plant development. Before initiating your study, make a hypothesis about how gibberellic acid will affect plant development and then predict the result that will follow. One way of doing this is to use the *if–then* predictive approach. For example,

If gibberellic acid inhibits growth, *then* plants treated with gibberellic acid should grow less than control plants that have not been treated with gibberellic acid.

Write your hypothesis for the study described below and the result it leads you to expect.

If _______________

____

_______________,

then _______________

____

____

1. Working in teams of three, obtain 40 bean seeds that have been soaking in water for several hours.

2. Plant 20 seeds (about 15 mm deep) in moist vermiculite in a tray. Label the tray "Gibberellic Acid" (Fig. 24–4A). Plant the remaining 20 seeds in a second tray labeled "Control."

3. When the plants are 7–8 cm tall, select 10

TABLE 24–1 Effects of IAA on initiation of root growth.

	Treatment			
Measurements	Distilled water	Hoagland's solution	0.1 mg/liter IAA	1.0 mg/liter IAA
Number of roots < 1 mm				
Number of roots > 1 mm				
Total length of all roots present				

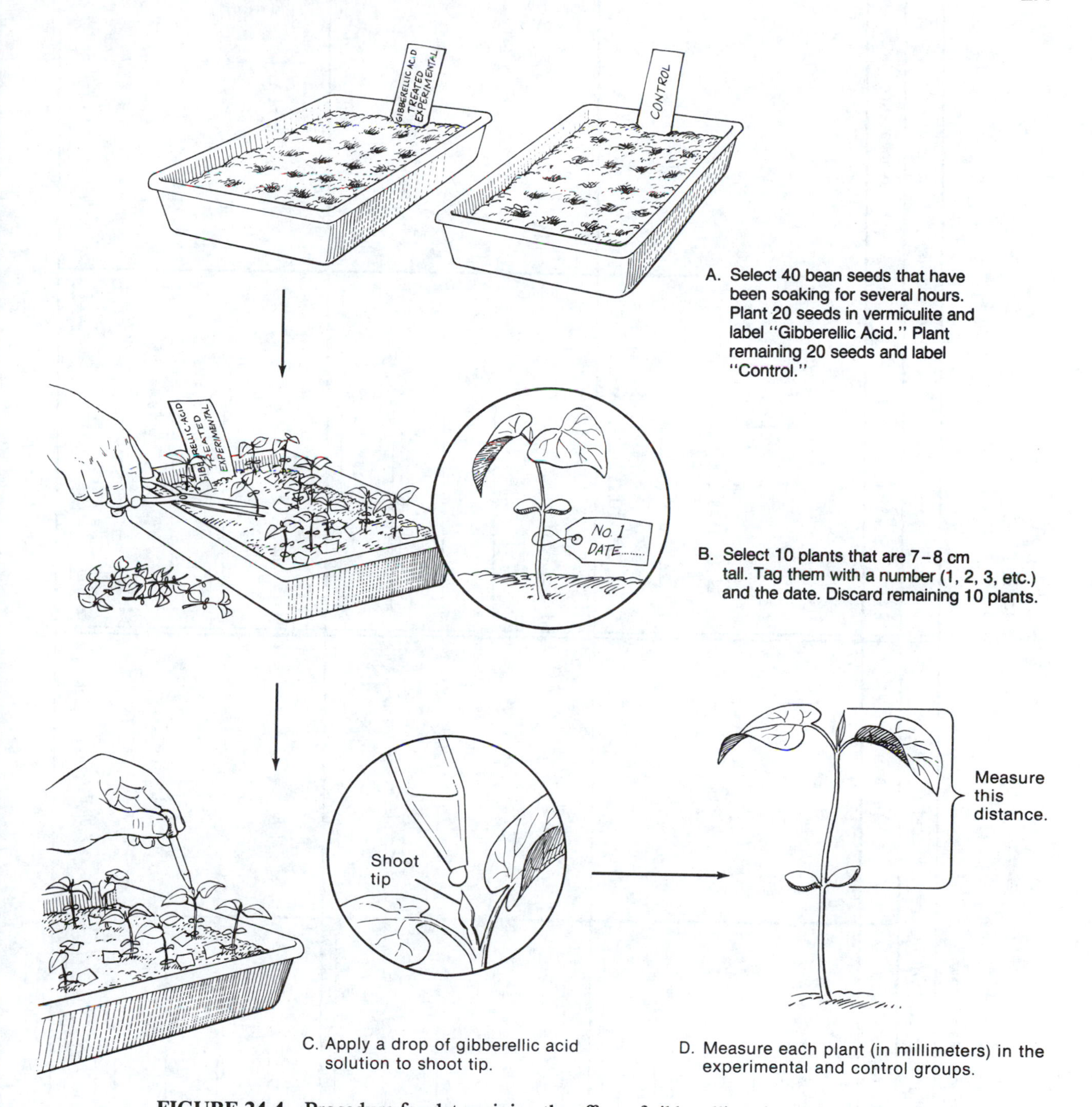

FIGURE 24-4 Procedure for determining the effect of gibberellic acid on plant growth

plants in each tray that are about the same size. Tag each plant with a number and the date. Cut the remaining plants at ground level and discard the parts you have cut off (Fig. 24–4B).

4. Measure the height of each plant (in millimeters) from the cotyledons to the tip of the shoot apex. Can you think of other measurements you might make? If so, use these in place of the one suggested.

5. Record the date, height, and appearance of the plants in Table 24–2 under Day 0.

6. Apply a drop of gibberellic acid to the shoot apex of each plant in the tray labeled Gibberellic Acid. What will you apply to the control plants?

___

7. Apply gibberellic acid to the plants weekly

TABLE 24-2 Effect of gibberellic acid on plant growth

Day	Date	Plants treated with gibberellic acid		Control plants	
		Height (mm)	Appearance	Height (mm)	Appearance
0		1.____ 6.____ 2.____ 7.____ 3.____ 8.____ 4.____ 9.____ 5.____ 10.____ Avg.____		1.____ 6.____ 2.____ 7.____ 3.____ 8.____ 4.____ 9.____ 5.____ 10.____ Avg.____	
		1.____ 6.____ 2.____ 7.____ 3.____ 8.____ 4.____ 9.____ 5.____ 10.____ Avg.____		1.____ 6.____ 2.____ 7.____ 3.____ 8.____ 4.____ 9.____ 5.____ 10.____ Avg.____	
		1.____ 6.____ 2.____ 7.____ 3.____ 8.____ 4.____ 9.____ 5.____ 10.____ Avg.____		1.____ 6.____ 2.____ 7.____ 3.____ 8.____ 4.____ 9.____ 5.____ 10.____ Avg.____	

		1._____ 6._____ 2._____ 7._____ 3._____ 8._____ 4._____ 9._____ 5._____ 10._____ Avg._____		1._____ 6._____ 2._____ 7._____ 3._____ 8._____ 4._____ 9._____ 5._____ 10._____ Avg._____	
		1._____ 6._____ 2._____ 7._____ 3._____ 8._____ 4._____ 9._____ 5._____ 10._____ Avg._____		1._____ 6._____ 2._____ 7._____ 3._____ 8._____ 4._____ 9._____ 5._____ 10._____ Avg._____	
		1._____ 6._____ 2._____ 7._____ 3._____ 8._____ 4._____ 9._____ 5._____ 10._____ Avg._____		1._____ 6._____ 2._____ 7._____ 3._____ 8._____ 4._____ 9._____ 5._____ 10._____ Avg._____	
		1._____ 6._____ 2._____ 7._____ 3._____ 8._____ 4._____ 9._____ 5._____ 10._____ Avg._____		1._____ 6._____ 2._____ 7._____ 3._____ 8._____ 4._____ 9._____ 5._____ 10._____ Avg._____	

and record the height of each plant and the general appearance of all plants in Table 24–2. As the plants grow, it may be advisable to stake them with thin bamboo sticks and twine. Be careful not to crush the stems when tying them to the sticks.

8. At the conclusion of the experiment (as determined by your instructor), plot the data in Fig. 24–5. Analyze the data to determine whether the growth response obtained with gibberellic acid is significantly different from that of the controls. (See Appendix E for a method of statistical analysis of the data.)

Do the results of your statistical analysis of your data support your hypothesis? Explain.

____

____

____

____

____

C. PLANT GROWTH INHIBITORS

The idea that inhibitors are important in regulating plant growth first gained credence when it was discovered that dormant buds of ash trees contain large amounts of inhibitor and dormancy ended as the inhibitor concentration decreased. A substance called **dormin** is believed to cause plants to stop growing and enter their dormant state. Dormin has also been called **abscisin** because it was originally believed to be responsible for causing leaves to fall off (absciss) late in the growing season. Dormin has been shown to be **abscisic acid, ABA.** Thus, a single substance isolated from different plants regulates dormancy and many other growth processes.

In this study, you will determine the effects on growth of an artificial plant growth inhibitor called **phosphon.** Before initiating the study, hypothesize how phosphon affects plant growth and state the result you would expect. Use the if–then format as in part B.

If ______________________________

____

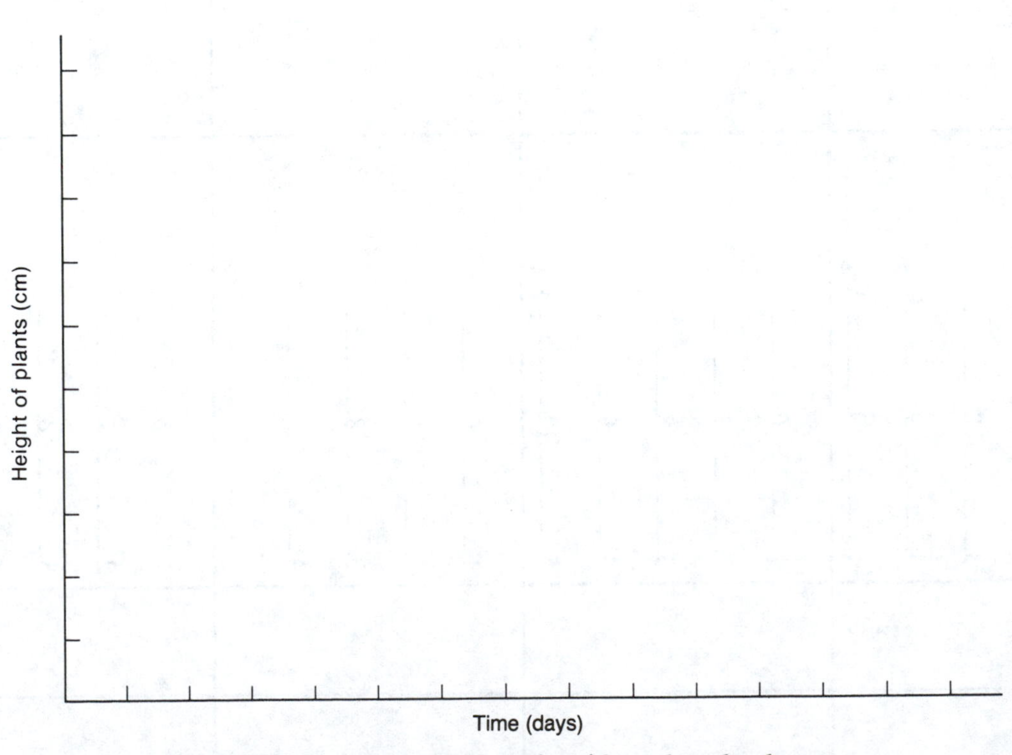

FIGURE 24-5 Effect of gibberellic acid on plant development

____

then ____

____

____

1. Select two bean or sunflower plants that are in about the same stage of development (only the first two leaves should have expanded). Tag one of them "Phosphon" and the other "Control," along with your name and the date (Fig. 24–6B). Record the average height of the plants in Table 24–3. In Fig. 24–7, draw the plants as they appear before treatment.

2. Using a wooden match or similar applicator, apply a ring of phosphon–lanolin paste around the stem of the plant labeled "Phosphon," about 15 mm below the first pair of leaves (Fig. 24–6C). What should be applied to the control plant and why?

____

____

3. Place the two plants in bright sunlight, and examine your plants and those of other students every 2–3 days for the next 3 weeks. Record the average height of the plants and other information in Table 24–3. Draw the plants as they appear at the end of the study in Fig. 24–7 and plot the data in Fig. 24–8.

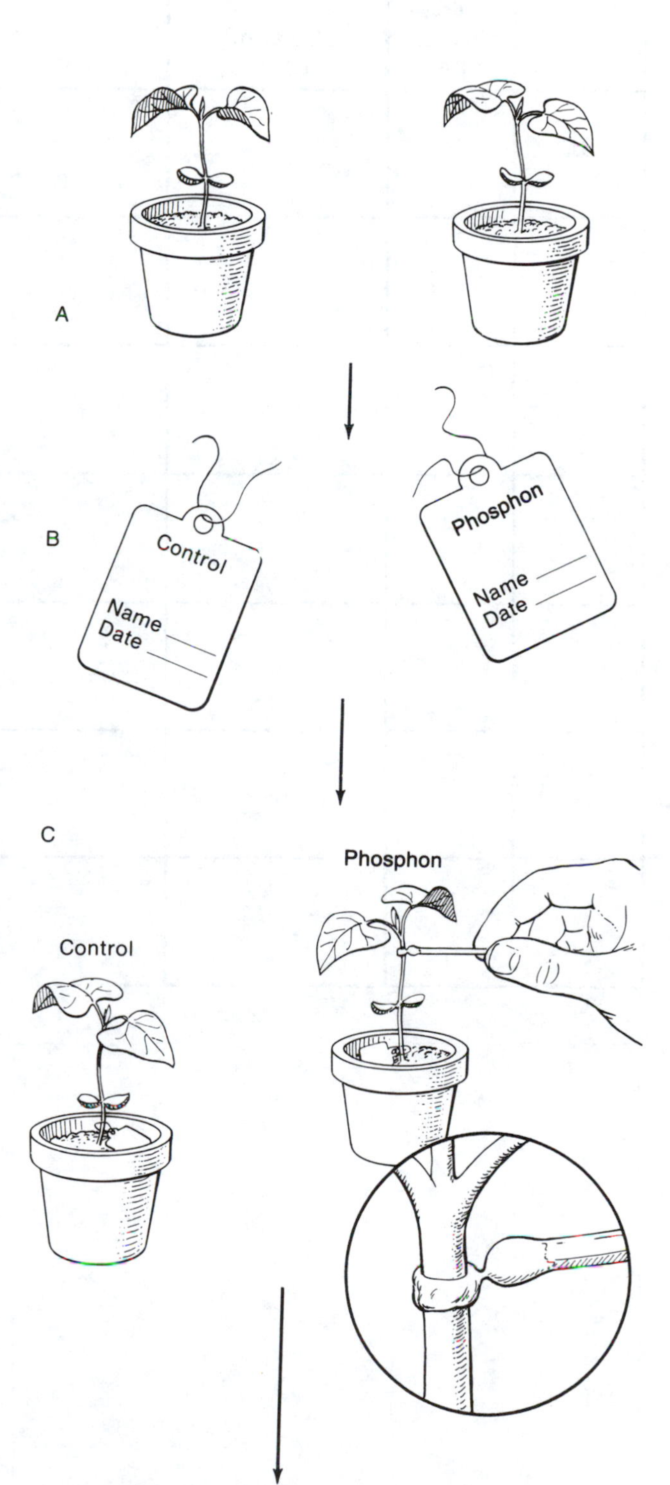

FIGURE 24-6 Procedure for determining the effect of phosphon, a plant growth inhibitor, on plant development

D. ASSAYS FOR IAA

The isolation and identification of IAA have been simplified by a biological testing procedure called a **bioassay.** Numerous biologically active compounds are present in living organisms in amounts so minute that we cannot detect them by the usual chemical procedures. In a bioassay, the whole organism, or some part of it, is used to detect the presence and to measure the amounts of biologically active substances.

1. Bioassay for IAA

A very sensitive and reliable bioassay for IAA is the *Avena* **curvature test,** which makes use of the coleoptile of the *Avena* (oat) seedling (Fig. 24–9). The

TABLE 24-3 Effect of phosphon on plant growth

Date	Plants treated with phosphon			Control plants		
	Average height (mm)	Color of leaves	Other	Average height (mm)	Color of leaves	Other
0						

Before After

Phosphon-treated plant

Before After

Control plant

FIGURE 24-7 Effect of phosphon on growth of bean or sunflower plants

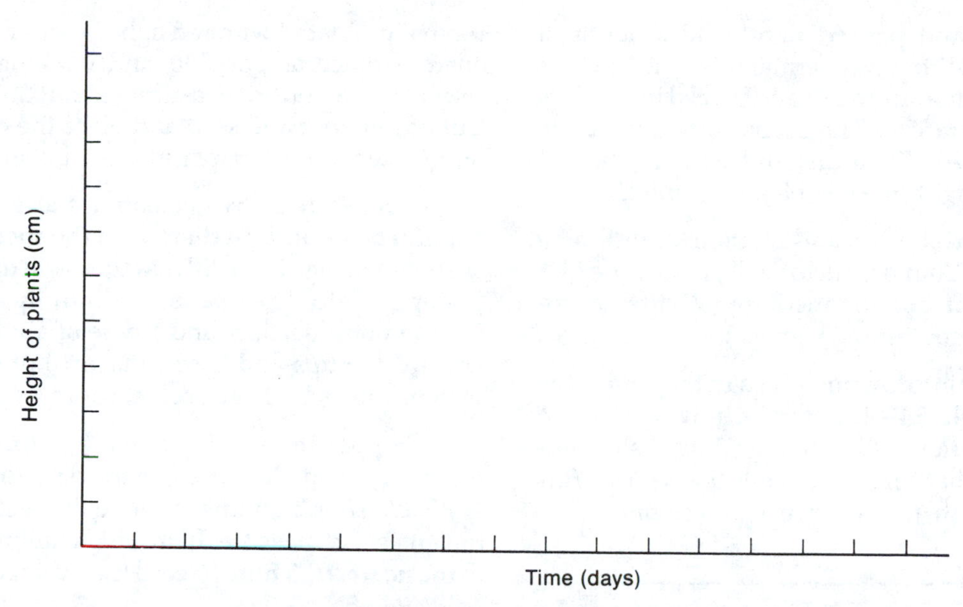

FIGURE 24-8 Effect of phosphon on plant growth

coleoptile is a cellular sheath that surrounds the embryonic leaves of the developing seedling. If oat seeds are germinated in the dark, cells of the coleoptile continue to divide until the coleoptile is approximately 1 cm long. During the next 3–4 days, while the coleoptile is reaching its maximum length

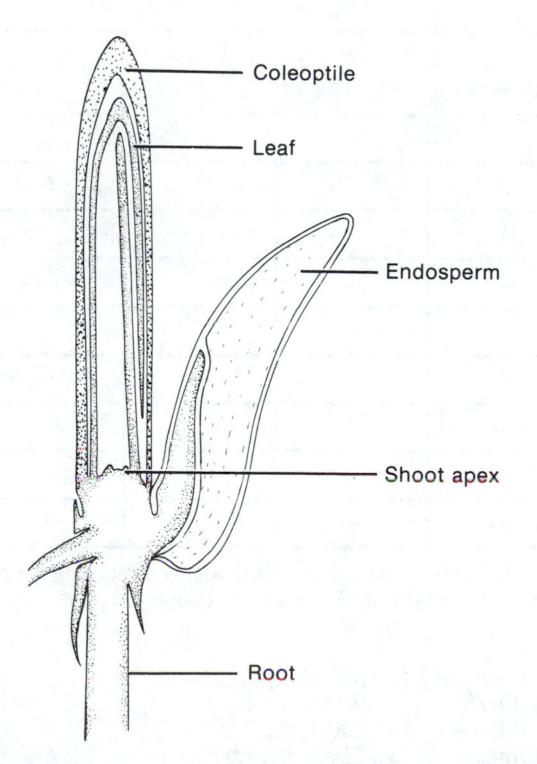

FIGURE 24-9 *Avena* seedling. (From *Plants in Action*, by Leonard Machlis and John G. Torrey. W.H. Freeman and Company © 1956.)

of 5–6 cm, the coleoptile extends solely by cell elongation. This elongation is brought about by IAA that is produced in the tip of the coleoptile. When the seedling is grown in the dark, IAA is directed downward, causing the embryonic cells of the coleoptile to elongate. If the tip is removed, no cellular elongation occurs. If the cut-off tip is replaced by an agar block containing IAA, growth of the coleoptile resumes. If the block of agar is displaced to one side, the cells on that side elongate more rapidly than those on the opposite side, with the result that the coleoptile curves. This is the basis for the curvature test. It has been shown that the degree of curvature is, within limits, proportional to the concentration of IAA in the agar block.

Although the *Avena* curvature test is very sensitive, it requires rigidly controlled conditions of temperature and humidity and accurate measurements of curvature. For these reasons, it is difficult to conduct such a test in most introductory laboratory courses. A simpler bioassay for IAA uses oat coleoptiles and is less sensitive in terms of measurable concentration of auxin. Called the ***Avena* coleoptile straight-growth assay,** it measures straight growth as reflected in an increase in length of coleoptile sections. To save time, step 1 will be carried out before the laboratory meeting.

1. For each laboratory section, 600–700 Brighton hull-less oat seeds were soaked for 2 hours in a liter of distilled water. After soaking, the seeds were rinsed two or three times with distilled water. The oats were then randomly scattered on moist germinating paper (or paper toweling) in a

tray, covered and packed tightly (to a depth of about 15 mm) with moist vermiculite. The tray was covered with aluminum foil and placed in the dark at room temperature. The seeds were allowed to germinate for 70–72 hours, and were exposed to 1 hour of red light in each 24-hour period.

2. Label six test tubes and fill them as indicated in Table 24–4. Your instructor will provide the IAA dilutions and incubating medium (a buffered nutrient solution containing sucrose).

3. Using the information given in the note at the bottom of Table 24–4, describe how you would prepare the various dilutions used in this experiment. (Notice that 1 ml of each dilution series is further diluted with 1 ml of incubating solution.)

____

____

____

____

____

4. Approximately 70 hours after planting, the coleoptiles will be 20–30 mm long. Working in a room illuminated with red light (or in a room darkened as much as possible, and working quickly), select 60 coleoptiles measuring about 30 mm long. Cut off the root and seed and place the coleoptiles on moistened filter paper in a petri dish.

5. Place four or five coleoptiles at a time on a paraffin block and cut them with the special cutter, as shown in Fig. 24–10. This cutter will divide each coleoptile into three parts: a 3-mm tip portion, a 5-mm middle section, and a base of about 20 mm. Discard the tips and bases. Place 10 of the 5-mm sections in each of the six test tubes.

6. Stopper the test tubes with cotton or styrofoam plugs and place them in the dark for 24 hours at 25°C. After 24 hours, remove the sections from the tubes and measure them with a millimeter ruler to the nearest 0.5 mm. Record the average length in Table 24–5.

7. Plot the data in Fig. 24–11 to obtain the standard curve for these concentrations. Does the curve show a proportional increase in growth with increasing IAA concentration? Explain.

____

____

____

TABLE 24–4 Protocol for the *Avena* coleoptile straight-growth assay

	Tube					
Tube contents (ml)	1	2	3	4	5	6
Incubation solution	1	1	1	1	1	1
Distilled H_2O	1	—	—	—	—	—
10^{-7} M IAA solution	—	1	—	—	—	—
10^{-6} M IAA solution	—	—	1	—	—	—
10^{-5} M IAA solution	—	—	—	1	—	—
10^{-4} M IAA solution	—	—	—	—	1	—
Unknown IAA solution	—	—	—	—	—	1

Note: The dilution series uses solutions of varying molar (M) concentration. A molar solution is made by dissolving the molecular weight of a compound, in grams, in a liter of solvent. The molecular weight of IAA is 175.2. Thus a 1.0 M solution of IAA contains 175.2 g/1000 ml of water. In a similar manner:

1/10 M	$= 10^{-1}$M =	17.52000000 g/liter =	17,520.00000 mg/liter
1/100 M	$= 10^{-2}$M =	1.75200000 g/liter =	1752.00000 mg/liter
1/1000 M	$= 10^{-3}$M =	0.17520000 g/liter =	175.20000 mg/liter
1/10,000 M	$= 10^{-4}$M =	0.01752000 g/liter =	17.52000 mg/liter
1/100,000 M	$= 10^{-5}$M =	0.00175200 g/liter =	1.75200 mg/liter
1/1,000,000 M	$= 10^{-6}$M =	0.00017520 g/liter =	0.17520 mg/liter
1/10,000,000 M	$= 10^{-7}$M =	0.000017520 g/liter =	0.01752 mg/liter

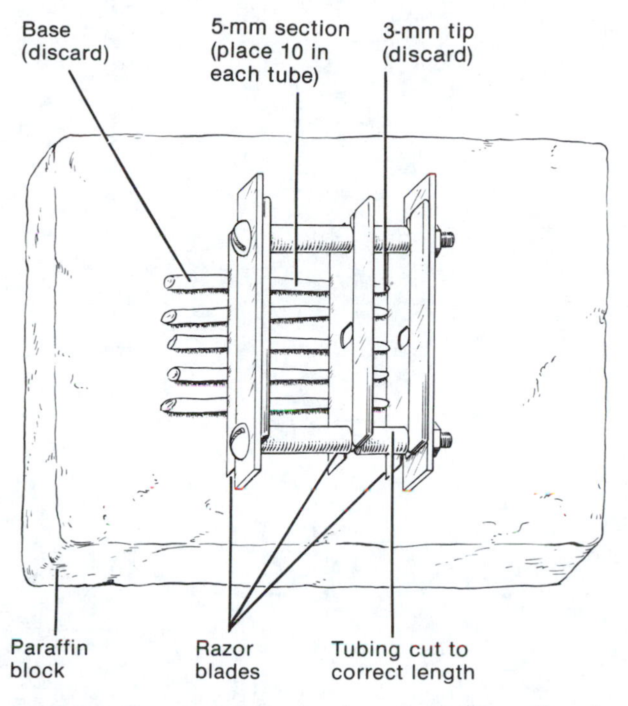

FIGURE 24-10 Method of cutting coleoptiles for *Avena* straight-growth assay

8. Plot the unknown IAA concentration on the graph. On the basis of the position of the unknown, estimate the concentration of the IAA in the unknown solution.

9. What is the reason for having test tube 1?

2. Colorimetric Assay for IAA

There is a simpler chemical assay for detecting a natural auxin (such as IAA) that is much less sensitive than either of the bioassays but is useful for **in vitro** (test tube) studies of IAA in concentrations well above those normally found in plant tissue. For example, an enzyme found in plant tissues, called indole-3-acetic acid oxidase, breaks down IAA. This assay would be useful for studying the action of such an enzyme.

The color assay you will perform makes use of the fact that IAA in the presence of iron chloride in Salkowski reagent forms a red color complex that can be quantitatively measured with a colorimeter (see Appendix B for a discussion of spectrophotometry).

You will be provided with a stock solution of IAA at a concentration of 100 mg/liter. Make 10-ml samples of the following concentrations: 40, 20, 10, 1, and 0 mg/liter. This is done as follows:

1. Collect 10 ml of the stock solution in a test tube. Pipet 4 ml of this solution into another test tube and dilute with distilled water to 10 ml. Label this tube "40 mg/liter."

2. Repeat this procedure using 2 ml, 1 ml, and 0.1 ml of the stock solution, diluting each to 10 ml. Mark these tubes "20 mg/liter," "10 mg/liter," and "1 mg/liter," respectively.

3. For the 0-mg/liter concentration, merely pipet 10 ml of distilled water into a tube.

This dilution series will be used to establish a standard curve. In addition, you will be given a solution containing an unknown concentration of IAA.

To conduct the colorimetric assay, add 2 ml of the solution being tested to a test tube containing 8 ml of Salkowski reagent. (*Note:* Because full color development is a function of time, stagger your tests to allow time for each colorimetric measurement.) Shake the mixture thoroughly (and *carefully* because the reagent contains sulfuric acid!) and set the mixture aside for exactly 30 minutes to allow the color to develop. While waiting for the color reaction, warm up the colorimeter and adjust the

TABLE 24-5 Data for *Avena* coleoptile straight-growth assay

IAA (mole)	0 Incubating medium	10^{-7}	10^{-6}	10^{-5}	10^{-4}	Unknown
Average coleoptile length (mm)						

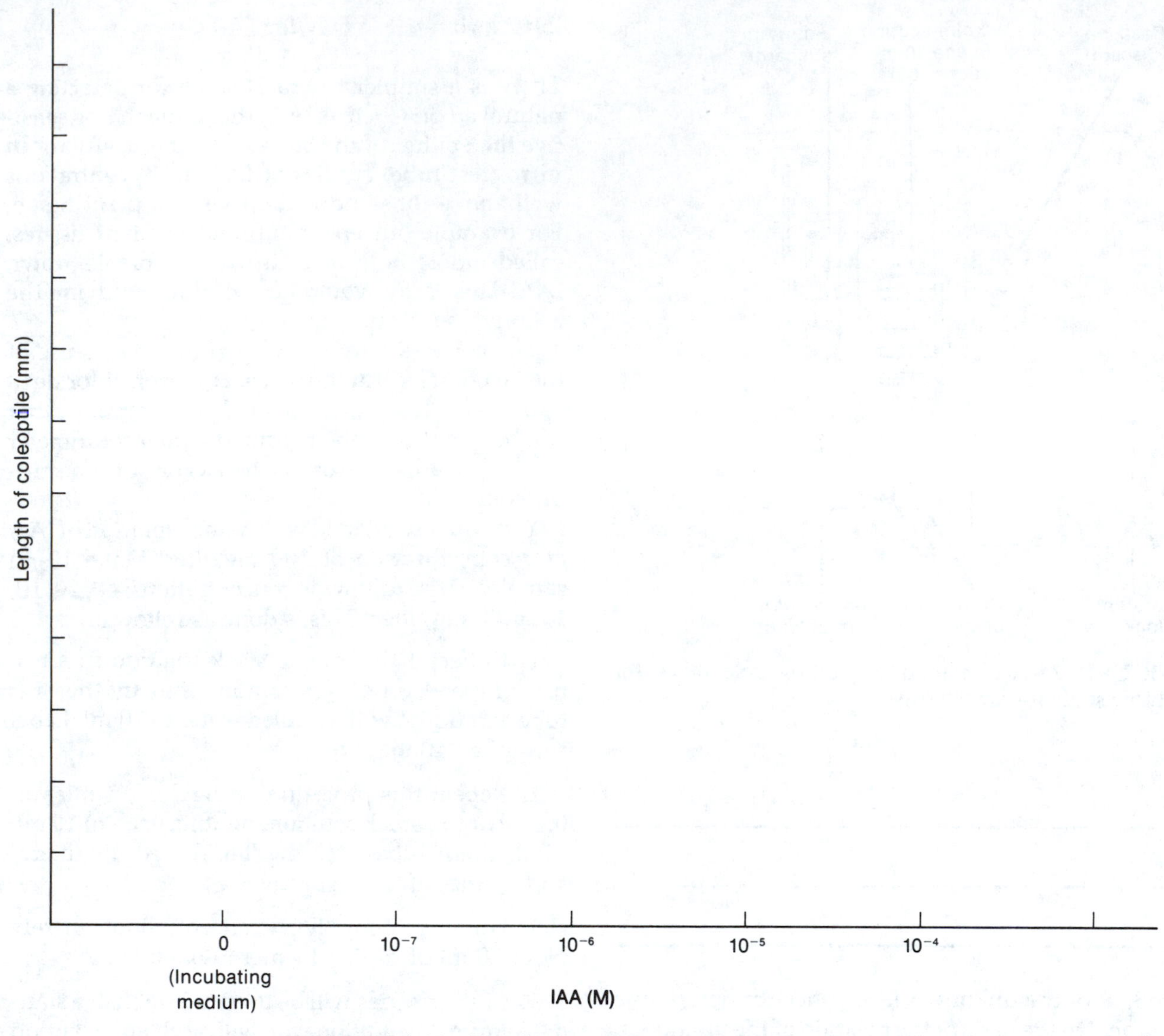

FIGURE 24-11 Standard curve for *Avena* straight-growth assay

wavelength to 510 nm, the wavelength maximally absorbed by the color produced in the Salkowski reaction. (See Appendix A for instructions on the use of the Milton Roy Spectronic 20™ colorimeter.)

Standardize the colorimeter, using a solution consisting of 2 ml of distilled water and 8 ml of Salkowski reagent. Immediately transfer about 3 ml of this mixture to a colorimeter tube and place it in the Spectronic 20. Adjust the instrument to read 100% transmittance. Why is this step required?

____

____

At the end of 30 minutes, record the percent transmittance for each of the IAA concentrations and the unknown; record the data in Table 24-6. Plot your data in Fig. 24-12. What is the concentration of the unknown IAA solution?

TABLE 24-6 Data for colorimetric determination of IAA

Tube contents (mg/liter)	0 Control	1	10	20	40	Unknown
Transmittance (%)						

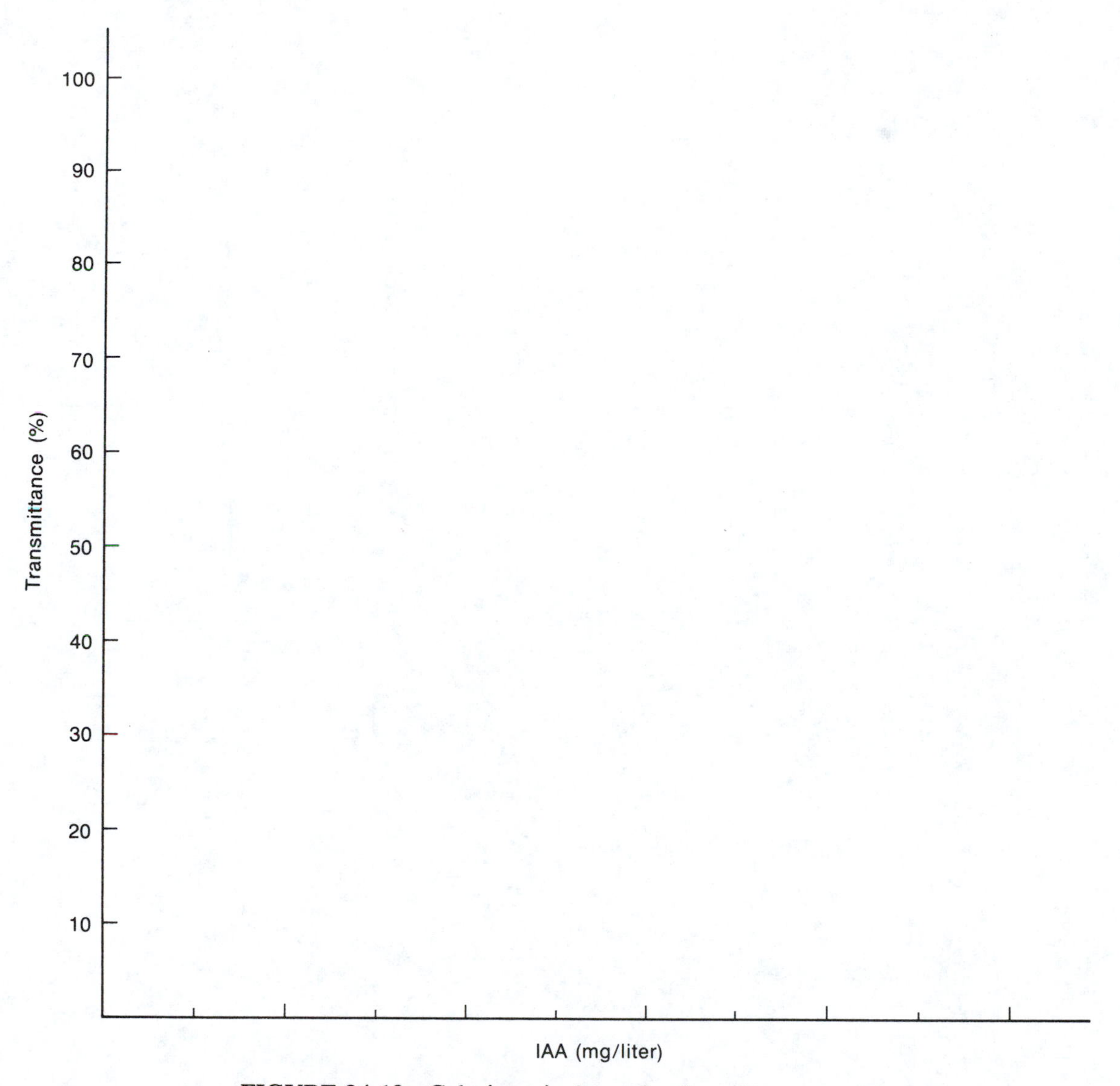

FIGURE 24-12 Colorimetric determination of IAA

Of the three assay methods discussed in this exercise, which would you use if you had to assay accurately for the total amount of IAA in 500 mg of plant tissue? Explain.

REFERENCES

Addicott, F. T., ed. 1983. *Abscisic Acid.* Praeger.

Raven, P. H., R. F. Evert, and S. E. Eichhorn. 1986. *Biology of Plants.* 4th ed. Worth.

Salisbury, F. B., and C. W. Ross. 1985. *Plant Physiology.* 3d ed. Wadsworth.

Sisler, E. C., and S. F. Yang. 1984. Ethylene, the Gaseous Plant Hormone. *BioScience* 33:233–238.

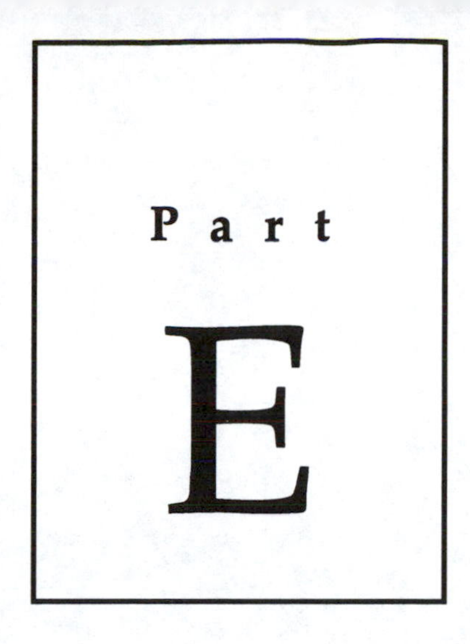

Diversity of Life: Kingdom Animalia

According to Whittaker's classification scheme, the kingdom Animalia includes multicellular organisms that have eukaryotic cells, possess a nuclear membrane and a variety of membranous organelles but lack cell walls, plastids, and photosynthetic pigments. These organisms are heterotrophic, with most taking in nutrients by ingestion; digestion takes place in an internal cavity. Some take in nutrients by absorption and lack an internal digestive cavity.

The more complex forms in the animal kingdom have evolved intricate organization and tissue differentiation, including sensory neuromotor systems and modes of mobility that are based on contractile fibers. Most of the organisms in this kingdom reproduce sexually and, except for some of the more primitive members, haploid cells occur only in the gametes.

Reproduction for most of the organisms in this kingdom is sexual, and, except for some of the lowest phyla, haploid cells occur only in the gametes. In the exercises in this part of the manual, you will become familiar with the great diversity of organisms that comprise the kingdom Animalia.

Exercise

25

Kingdom Animalia: Phyla Porifera, Cnidaria, and Ctenophora

In Whittaker's classification scheme, the kingdom Animalia includes those multicellular organisms possessing eukaryotic cells that lack cell walls, plastids, and photosynthetic pigments. Most members of this kingdom obtain nutrients by ingestion, with digestion taking place in an internal cavity. However, some animals take in nutrients by absorption, and a number of them lack an internal digestive cavity. The higher forms in the animal kingdom have evolved sophisticated levels of organization and tissue differentiation, including sensory neuromotor systems and modes of mobility based on contractile fibers.

In this exercise, you will study representatives of the phyla Porifera (sponges), Cnidaria (coelenterates), and Ctenophora (comb jellies).

A. PHYLUM PORIFERA (SPONGES)

The members of the phylum Porifera (Latin *porus*, "pore"; *ferre*, "to bear") are characterized by having skeletons consisting of minute inorganic spicules or organic fibers. The body surface of sponges is perforated by numerous pores, which are connected to internal canals and chambers lined with flagellated cells.

Based largely on the structural organization of the internal skeleton, this phylum is divided into three major classes:

- Class Calcarea (calcareous sponges): Skeletons are composed of calcium carbonate spicules.
- Class Hexactinellida (glass sponges): Skeletons are composed of siliceous material (chiefly silicic acid) fused into a network.
- Class Demospongiae (natural sponges): Skeletons are composed of organic fibers (spongin) or siliceous spicules or both. Siliceous spicules, when present, are not fused into a network.

1. Class Calcarea

Scypha (formerly called *Sycon* or *Grantia*) is a small, slender sponge that rarely exceeds an inch in height and is found in clusters adhering to various objects in shallow marine waters. Obtain several specimens of *Scyphae* from your instructor and examine them using a stereoscopic microscope. Note their slender, vase-shaped appearance (Fig. 25–1). Observe the basal end by which this organism attaches itself to various objects; at the upper end

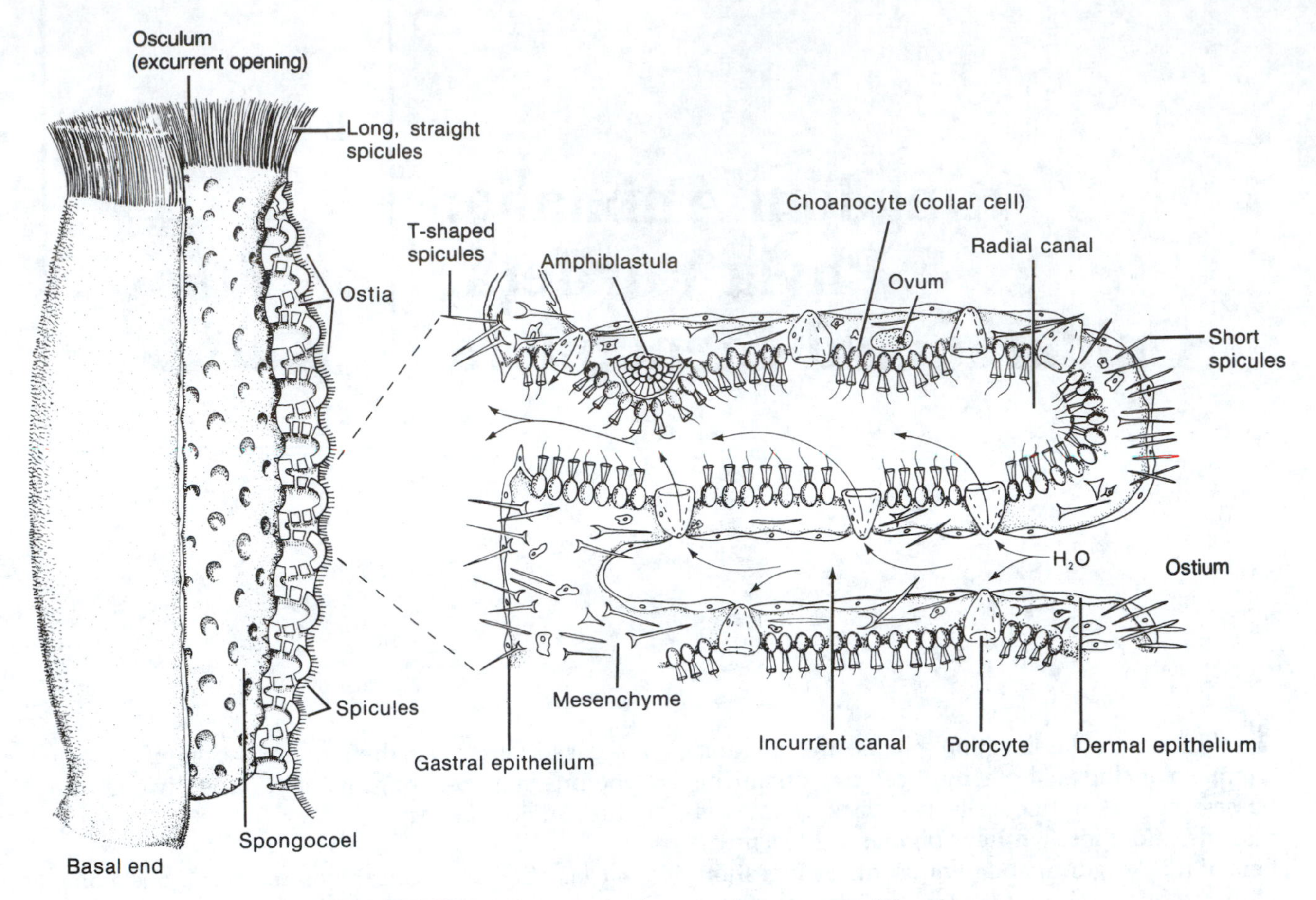

FIGURE 25-1 Gross and microscopic anatomy of *Scypha*

locate the large, single excurrent opening, the **osculum.** Minute pores **(ostia)** along the body wall allow water to enter the single tubular central cavity, or **spongocoel.** Locate bristlelike spicules along the body surfaces which give rigidity to the body and make up the skeleton. The spicules are of four kinds: long, straight spicules surrounding the osculum that protrude beyond the body and produce a bristly appearance; short, straight spicules surrounding the ostia; T-shaped spicules lining the spongocoel; and branched spicules embedded in the body wall. Examine slides showing isolated spicules.

Using a compound microscope, study a slide of a cross section of *Scypha* under low and high magnification (Fig. 25–1). Note the thick body wall and the numerous short radial canals that are lined with small, flagellated collar cells, or **choanocytes.** These cells resemble some of the flagellated protozoans, and indicate a probable evolution of the sponges from protozoans. Water enters the sponge through the **ostia,** which lead into the **incurrent canals.** The incurrent canals are connected to the **excurrent (radial) canals** by a series of small pores through perforated cells called **porocytes.** The excurrent canals open into the spongocoel. The exterior surface of this sponge is covered by a thin **dermal epithelium;** the spongocoel is lined with **gastral epithelium.** Between the dermal and gastral epithelium is a gelatinous **mesenchyme** in which amoebocytes and the various spicules that make up the skeleton are embedded.

Water, taken into the body through the ostia, is passed over the collar cells, which remove tiny bits of food. The food, consisting of plankton (microscopic plants and animals) and bits of organic matter, is digested in the food vacuoles of the choanocytes. Indigestible material is extruded into the canal system and then eliminated through the osculum by way of the spongocoel.

Scypha reproduces both asexually and sexually. Parts of a sponge lost by injury can be replaced asexually by regeneration. Many kinds of sponges also reproduce by simple budding; the buds either separate from the original sponge as growth proceeds or remain attached to form clusters of sponges. *Scypha* is **monoecious,** producing both eggs and sperm (though usually at different times of the year). Sexual reproduction is accomplished by male gametes (sperm) and female gametes (ova) produced by the choanocytes. Sperm leave the sponge by means of water currents and enter other sponges in the same manner. Fertilization is internal. When the embryo reaches an oval-shaped **amphiblastula** stage, it escapes through the osculum. It then swims about for a short time, attaches to a solid object, and grows into a new sponge.

2. Class Hexactinellida

Hexactinellids are beautiful vase- or funnel-shaped deepwater sponges whose skeletons are composed of a network of six-rayed spicules made of siliceous materials. They are often referred to as glass sponges.

Examine specimens of the sponge *Euplectella* (Venus's flower basket) (Fig. 25–2). You are observing merely the skeleton of the sponge because all the protoplasm has dried and decayed away. Note that the skeleton is composed of a delicate network of siliceous spicules fused at their tips.

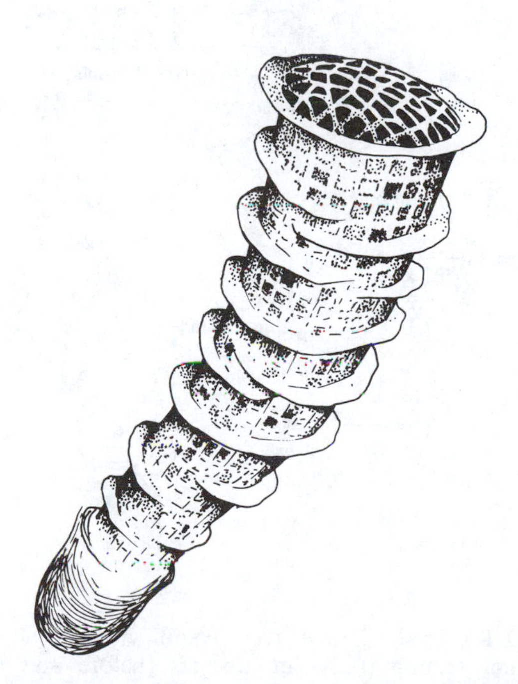

FIGURE 25-2 Skeleton of *Euplectella* (Venus's flower basket)

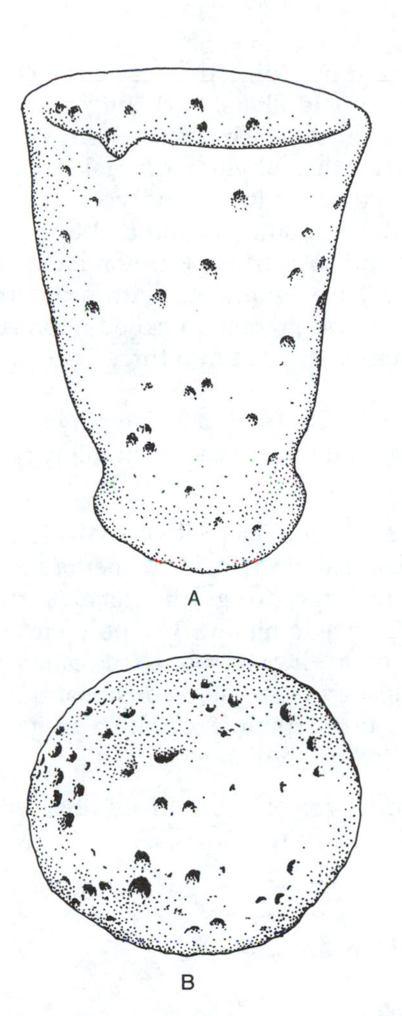

FIGURE 25-3 Some common sponges. (A) *Gelliodes*, the elephant's ear sponge. (B) *Spongia*, the common bath sponge.

3. Class Demospongiae

This class includes the familiar bath sponges and freshwater sponges (Fig. 25–3). Examine the cleaned and dried skeletons of *Spongia*, the common bath sponge. The skeleton holds water because of capillary forces in the fine spaces of the irregular spongin network. In most areas of the world, the natural variety of sponges used by people have been largely replaced by synthetic "sponges."

B. PHYLUM CNIDARIA (COELENTERATES)

The phylum Cnidaria consists of a large and diverse group of radially symmetrical organisms characterized by a digestive **(gastrovascular)** cavity, some muscle fibers, and stinging cells called **cnidocysts,** from which this phylum gets its name. Members of this phylum can take one or two forms, polyp or medusa. A **polyp** is closed at one end and has a mouth surrounded by tentacles at the other end. A **medusa** is umbrella-shaped, jellylike, and free-swimming, with a mouth at the end of a central projection called a **manubrium.** The phylum is divided into three classes:

- Class Hydrozoa (hydras): The polyp form is predominant and gives rise in many species to small medusas.
- Class Scyphozoa (true jellyfish): Includes the larger jellyfish, in which the medusa form, consisting largely of gelatinous **mesoglea** or "jelly," is predominant. The polyp form is minute or lacking. Scyphozoans can be roughly distinguished from the medusae of the hydrozoans by their size, which ranges from 25 mm to 2 m in diameter.
- Class Anthozoa (corals and sea anemones): Only the polyp form exists.

1. Class Hydrozoa

a. Hydra

The common freshwater hydrozoan *Hydra* is an excellent organism for the study of the polyp form. It illustrates a definite, though primitive, level of organization into tissues. Commonly studied species are the brown hydra *(Hydra oligactis)* and green hydra *(Hydra viridissimus)*; the latter are green because of minute algae living symbiotically in their bodies.

Using a hand lens or a stereoscopic microscope, examine living hydras in a Syracuse dish or deep-well slide containing pond water. At the upper end of the animal, observe several elongate, actively moving tentacles that are used to capture food (Fig. 25–4). At the center of this circle of tentacles is the mouth. Tap the edge of your slide or dish, and observe the extreme contractility of this organism. Note that the animal elongates after contracting. Compare the rate of elongation with the rate of contraction.

____

____

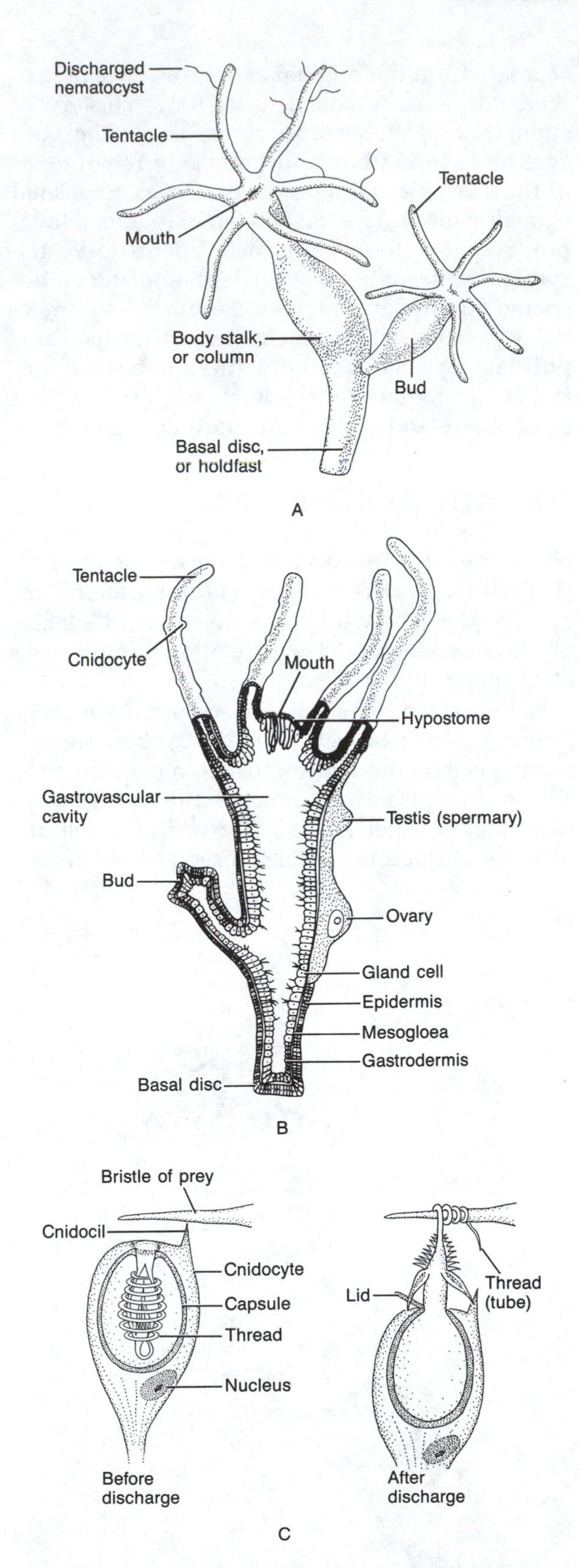

FIGURE 25-4 *Hydra.* (A) Overall view. (B) Longitudinal section. (C) Nematocysts (before and after discharge).

Hydra, like other cnidarians, have **nematocysts,** or "thread capsules" in stinging cells called **cnidocytes** in the epidermis (Fig. 25–4). The nematocysts, when stimulated, release a coiled tube. Some nematocysts capture prey by being sticky and others by coiling around their prey. A third type, tipped with a barb, injects a toxin that paralyzes the prey.

Each cnidocyte has a small projecting trigger called a **cnidocil** on its surface. This structure responds to touch or to chemicals in the water and causes the nematocyst to fire its tube. A nematocyst is used only once. After it has been discharged, a new one is formed by a new cnidocyte.

So that you can observe the hydras feeding and the action of the nematocysts, your instructor will add to your slide or dish several brine shrimp that have been thoroughly rinsed with tap water. Why are the brine shrimp washed before being added to the pond water containing the hydra?

____

____

____

Using a stereoscopic microscope, observe the hydra and describe the feeding process, beginning with the first contact of the hydra's tentacles with the brine shrimp.

____

____

____

It is not unusual for a single hydra to capture and ingest 10 or more brine shrimp in the course of a 2-hour feeding period.

The release of the nematocysts can be observed by placing a hydra in a drop of water on a clean slide and then adding a drop of dilute acetic acid at the edge of the water. Using the low power of your compound microscope observe the tentacles as the acid diffuses toward them. Note the discharge of the nematocysts. The structures of the various types of nematocysts can be observed by adding a drop of dilute methylene blue or safranine stain at the edge of the water.

Hydras reproduce asexually by forming a "bud," an outgrowth from the body wall in which the gastrovascular cavity becomes continuous with that of the parent. After maturing, the bud separates from the parent. Examine specimens of budding hydras.

In sexual reproduction, gametes are produced in organs called **spermaries (testes)** and **ovaries,** which are found in protuberances along the body wall. Both sex organs can be present in a single organism. Microscopically examine prepared slides showing the location and structure of testes and ovaries.

To study the histology of *Hydra,* examine stained cross sections. Note that the body wall surrounding the gastrovascular cavity consists of two distinct layers of cells: an outer layer of **epidermis** and an inner layer of **gastrodermis** (Fig. 25–5). These two layers are held together by a thin, noncellular layer, the **mesoglea.** The epidermis is made up of two principal cell types: the larger **epitheliomuscular cells** containing contractile fibrils at their bases and the smaller **interstitial cells** located between them. Suggest a function for the epitheliomuscular cells.

____

____

The interstitial cells are germinal cells and give rise to the sperm and eggs, as well as to other cell types.

Many of the gastrodermal cells, particularly the flagellated **nutritive muscle cells,** contain food vacuoles in which intracellular digestion takes place. These food vacuoles form as a result of the phagocytosis of food particles by pseudopodia that are extended into the gastrovascular cavity. Many of the digestive cells also contain flagella, whose beating activity directs food toward the pseudopodia. Extracellular digestion in hydra results from the secretion of enzymes directly into the gastrovascular cavity, followed by absorption of the digested food material.

b. Obelia

Unlike *Hydra,* most members of hydrozoans are colonial marine organisms found attached to seaweeds (particularly the kelps), rocks, shells, or pilings in the shallow waters off seacoasts. In *Obelia,* a typical marine hydrozoan, reproduction is by **metagenesis,** in which there is an alternation of sexual and asexual generations. The sexual generation is represented by a medusa form; the asexual generation by the polyp form.

Obtain a small piece of *Obelia,* place it in a Syracuse dish and examine it with a stereoscopic microscope. The plantlike colony consists of numerous

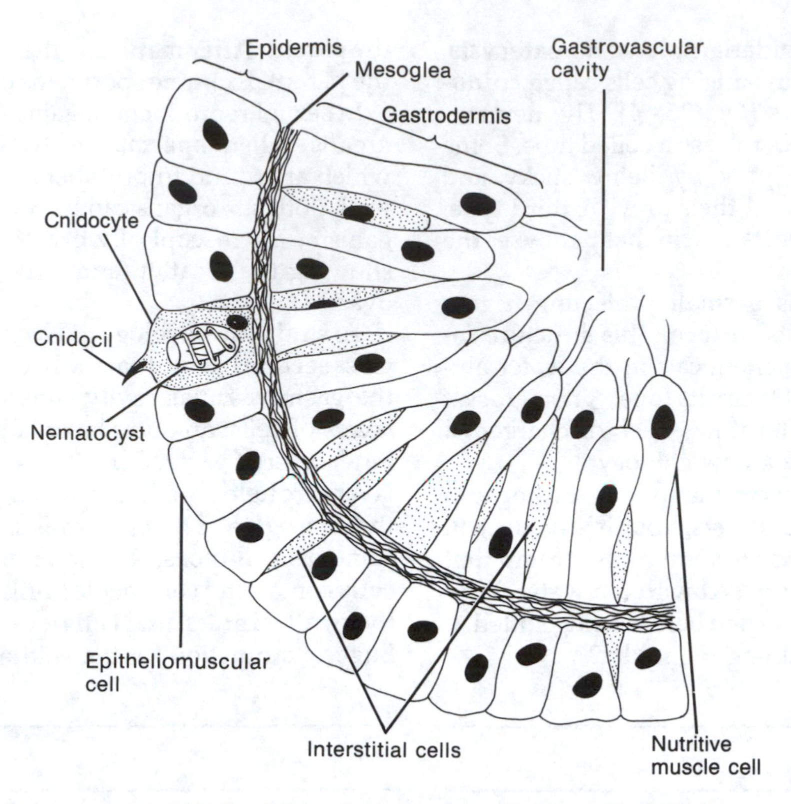

FIGURE 25-5 Cross section of body wall of *Hydra*

branches that terminate in two kinds of polyps. The **feeding,** or **nutritive, polyps (gastrozooids)** possess tentacles and resemble *Hydra.* The **reproductive polyps (gonangia)** are club-shaped and lack tentacles (Fig. 25–6). The branches of the colony are covered by a transparent, noncellular sheath, the **perisarc.** This covering expands around the gastrozooids and gonangia to form vase-shaped protective structures called **hydrothecae and gonothecae,** respectively. The inner cellular core of the colony is called the **coenosarc.**

Because the medusa forms of *Obelia* are too small to study in detail, the large medusa form of *Gonionemus,* a common jellyfish, will be used to illustrate the typical structure of a coelenterate medusa (Fig. 25–7).

c. Gonionemus

Using a hand lens or stereoscopic microscope, examine a specimen of *Gonionemus* in a Syracuse dish (Fig. 25–7). Note the umbrella shape of the medusa. Turn the organism over and note that at the margin of the "umbrella," and extending inward, is a muscular ring of tissue called the **velum.** Suggest a function for the velum.

____

____

The medusa usually swims on its "back" so that the tentacles are facing the surface of the water. Observe the number of tentacles around the velum. Extending into the cavity of the medusa is the **manubrium,** which contains a mouth at its tip surrounded by four **oral lobes.** Extending from the manubrium are four **radial canals** that join with the circular canal at the margin and connect with the cavities of the tentacles. Suggest a function for these canals.

____

____

Observe the numerous rings of nematocysts and the **suctorial pads** on the tentacles.

At the base of each tentacle are round, pigmented structures that are believed to be **photoreceptors.** Between each tentacle is a **statocyst,** a small capsule containing a tiny, solid body, which serves as a balancing organ. The reproductive organs **(gonads)** open into the radial canals. *Gonionemus,* like most coelenterates, is **dioecious;** that is, the male and female organs are in separate

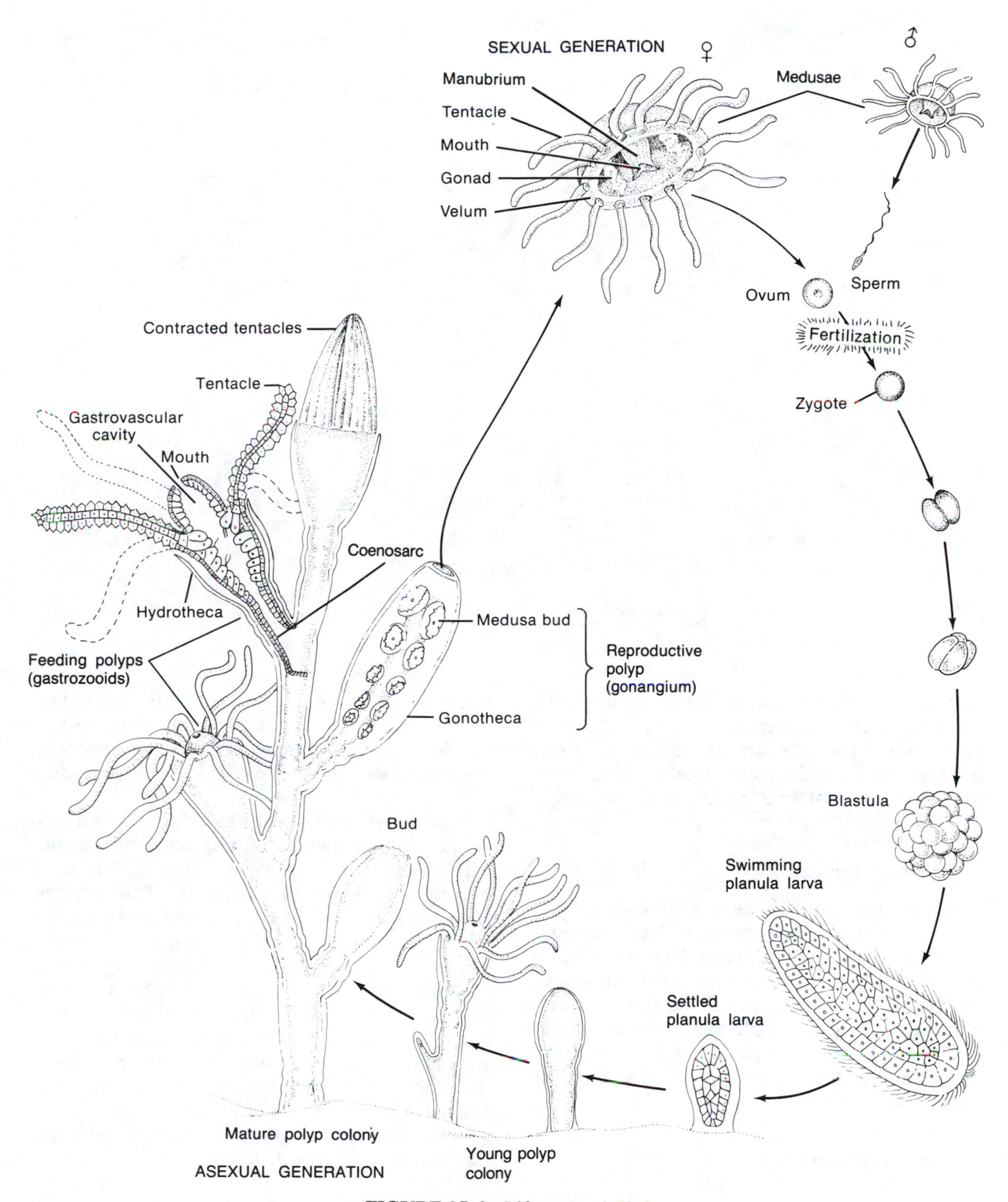

FIGURE 25-6 Life cycle of *Obelia*

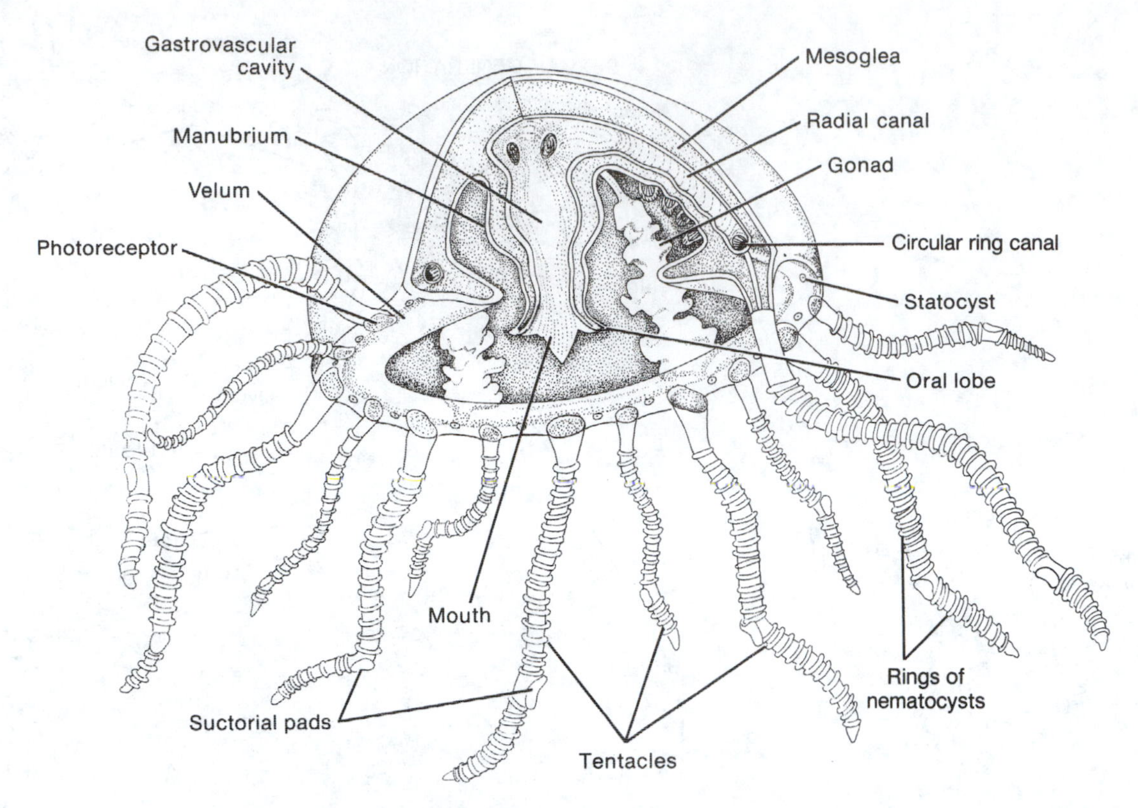

FIGURE 25-7 The internal anatomy of *Gonionemus*

medusae. The eggs and sperm are released into the sea, where fertilization takes place. The zygote develops into a **planula larva,** which swims about and then settles and attaches to a submerged object, where it transforms into a microscopic polyp (Fig. 25–6).

d. Other Hydrozoans

To become familiar with the great diversity of form that hydrozoans exhibit, examine demonstrations of other members of this group. Pay particular attention to *Physalia,* the Portuguese man-of-war, which shows a highly specialized form of **polymorphism** (many forms). In addition to a modified medusa that forms the gas-filled float, there are several kinds of tentacles that paralyze and capture food for the colony.

2. Class Scyphozoa

In scyphozoans (Greek *skyphos,* "cup"; *zoon,* "animal"), the medusa form, commonly known as a jellyfish, is quite large, ranging from 25 mm to 2 m in diameter with tentacles up to 30 m in length. The polyp form is minute or lacking.

Aurelia, one of the commonest scyphozoan jellyfish, is often seen in large groups drifting, or swimming slowly, by rhythmic contractions of the shallow, almost saucer-shaped bell. Great numbers of these jellyfish are sometimes cast on shore during storms.

Examine a specimen of *Aurelia.* Identify as many of the following structural characteristics of this jellyfish as you can (Fig. 25–8). The body is fringed by a row of closely spaced tentacles. The tentacles are interrupted at eight equally spaced intervals by indentations, each of which contains a **sense organ.** The sense organs consist of a pigmented photoreceptor that is sensitive to light; a hollow statocyst, containing minute calcareous particles whose movements set up stimuli that direct the swimming movements; and two sense pits, lined with cells that are thought to be sensitive to food or chemicals in the water. Many circular muscle fibers are present in the margin of the body. What is the function of these muscle fibers?

____

____

The mouth is located in the center of the oral (concave) surface at the end of a short manubrium.

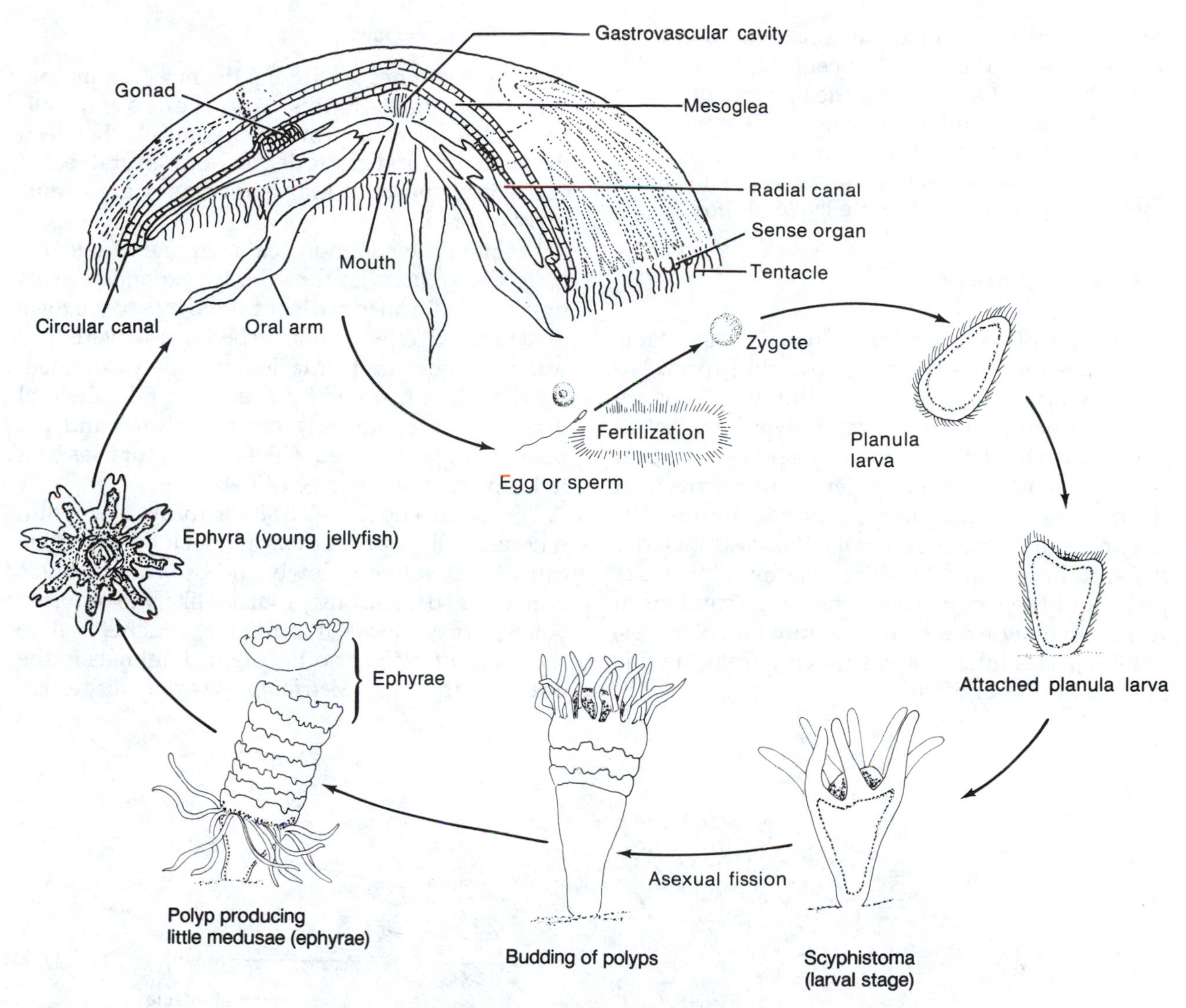

FIGURE 25-8 Life cycle of *Aurelia*

The manubrium lies among four tapering **oral arms,** each of which has a ciliated groove. Nematocysts in the oral arms and in the tentacles paralyze and entangle small animals, which are then swept up the ciliated grooves, through the mouth, and into a **gastrovascular cavity** that extends into the four pouches. The pouches contain tentacle-like projections that are covered with nematocysts that paralyze prey that arrives in the pouches still alive and struggling. Numerous **radial canals** extend through the mesoglea from the pouches to a **ring canal** in the margin of the velum. Flagella lining the entire gastrovascular cavity maintain a steady current of water that brings a constant supply of food and oxygen to, and removes waste from, the animal's internal parts.

The four horseshoe-shaped, colored bodies, by which *Aurelia* is easily identified, are the gonads (testes or ovaries); they are located on the floor of the gastrovascular cavity. In a male medusa, the sperm cells are discharged into the gastrovascular cavity and then to the outside through the mouth. The eggs of the female are discharged into the gastrovascular cavity, where they are fertilized by sperm cells entering the mouth along with incoming food.

The zygotes emerge from the gastrovascular cavity and lodge on the oral arms, where each one develops into a ciliated planula larva. This larva soon escapes, swims about for a while, then settles and attaches to the sea bottom. Losing the cilia, it develops into a larval stage, called a **scyphistoma,** that is about the size of a hydra. This stage lasts for many months, budding off other small polyps like itself.

Usually in the fall and winter, the polyp develops a series of horizontal constrictions that resembles a pile of minute "saucers" **(ephyrae)** with fluted borders. One by one, the ephyrae pinch off from the parent and swim away as little medusae that develop into adult jellyfish.

Examine demonstration specimens and slides showing the stages in the life cycle of *Aurelia.*

3. Class Anthozoa

Anthozoans (Greek *anthos,* "flower"; *zoon,* "animal") are marine polyps of flowerlike form. No medusa forms are present in the life cycle. They are distinguished from hydrozoan polyps by a gastrovascular cavity that is divided by a series of vertical partitions and a surface epidermis that turns in at the mouth to line the **pharynx.** Besides the familiar sea anemones and hard corals, this class includes the soft, horny, and black corals, the colonial sea pens, and the sea pansies. They are abundant in warm, shallow waters but also inhabit polar seas. Other species inhabit areas ranging from the tide lines to depths of 5000 m.

a. Sea Anemones

The sea anemones are among the most highly specialized organisms in this class. They have a well-developed nerve net, mesenchyme cells between the epidermis and gastrodermis, and several sets of specialized muscles. Many anemones are exquisitely colored.

Examine the common sea anemone *Metridium.* Note the stout cylindrical body, expanded at its upper end into an **oral disc** covered by several rows of tentacles. When undisturbed and covered by water, the body and tentacles are widely extended. If irritated, or exposed by a receding tide, the oral disc may be completely turned inward and the body tightly contracted. Cilia on the tentacles beat to keep the oral disc free of debris.

The basal end of *Metridium* forms a smooth, muscular, slimy **pedal disc** on which the anemone can slide about very slowly and by which it holds onto rocks so tenaciously that it is likely to be torn if you try to pry it loose. It commonly attaches itself to the shells of crabs or other shelled animals in the ocean. In this way, *Metridium* is widely dispersed.

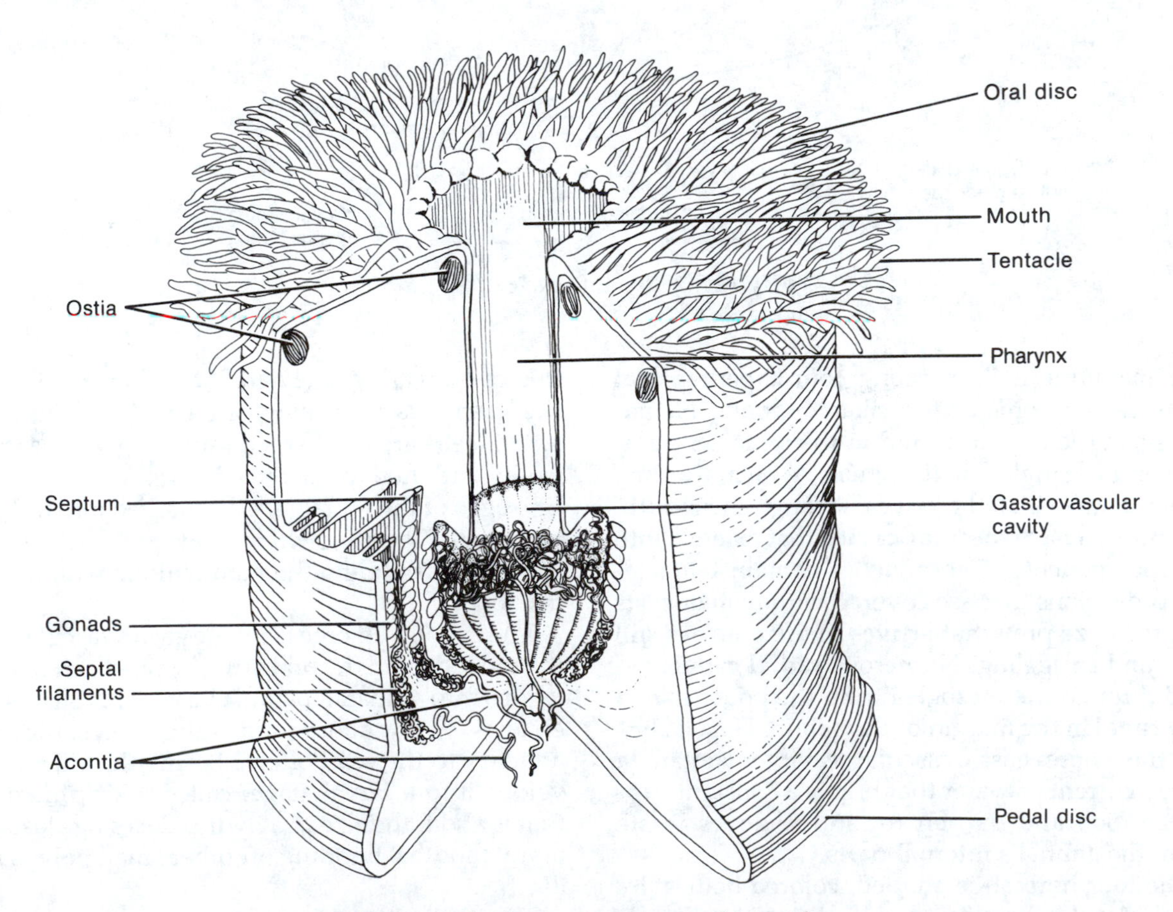

FIGURE 25-9 Internal anatomy of *Metridium*

From the mouth, locate a muscular pharynx that hangs down into the gastrovascular cavity (Fig. 25–9). The pharynx is lined with cilia that beat downward, drawing a current of water into the gastrovascular cavity and steadily supplying the internal parts with oxygen. At the same time, other cilia lining the pharynx beat upward, creating an outgoing current of water that takes carbon dioxide and other wastes with it. When small animals touch the tentacles, the cilia of the pharynx reverse their beat, and the food is swept down the pharynx into the gastrovascular cavity.

By careful dissection you will see that the pharynx is connected with the body wall by a series of vertical partitions **(septa).** What is the function of these septa?

____

____

In the septa, beneath the oral disc, locate openings, or **ostia,** through which water can pass between the internal compartments. The free, inner margin of each septum is a thick, convoluted **septal filament** that becomes a threadlike **acontium** toward the bottom. Both parts bear nematocysts for paralyzing prey and gland cells for secreting enzymes for digestion.

Anemones sometimes reproduce asexually by separating into longitudinal halves; each half then regenerates into a new adult. Sexual reproduction is accomplished by the small, rounded gonads located at the edges of the septa. The sexes are separate. Eggs and sperm leave the gonads through the mouth, and fertilization takes place in the water. The fertilized egg develops into a ciliated planula larva, which finally settles down and grows into an adult anemone.

b. Corals

The polyp form of corals are similar to those of sea anemones, except they usually combine to form huge colonies consisting of millions of individual polyps. Each member of the colony secretes a protective skeleton of limestone containing a pocket into which the polyp partly withdraws when disturbed. New individuals build their limestone skeletons on the skeletons of dead ones, and thus, over many years, huge undersea ledges known as **coral reefs** are built up. Three main types of coral reefs have been described. A **fringing reef** grows in shallow water, bordering the coast or separated from it by a narrow stretch of water that is shallow enough to wade when the tide is out. A **barrier reef** also is parallel to the coast but is separated from it by a lagoon deep enough to accommodate large ships. The best-known of these is the Great Barrier Reef along the northeast coast of Australia. The reef is about 2000 km long and is located from a few to about 150 km offshore. **Atolls** lie above old submerged volcanoes and are more or less circular coral reefs surrounding a central lagoon. Reefs furnish both food and shelter for other marine organisms, including many species of reef fishes and a tremendous variety of invertebrates such as sponges, sea urchins, marine worms, and crustaceans.

Examine samples of coral on demonstration and note the variety of forms. Locate and examine the small pockets that originally contained the living polyps.

C. PHYLUM CTENOPHORA (COMB JELLIES)

The ctenophores are called comb jellies because they have eight longitudinal rows of comblike plates of fused cilia (Fig. 25–10). The beating of

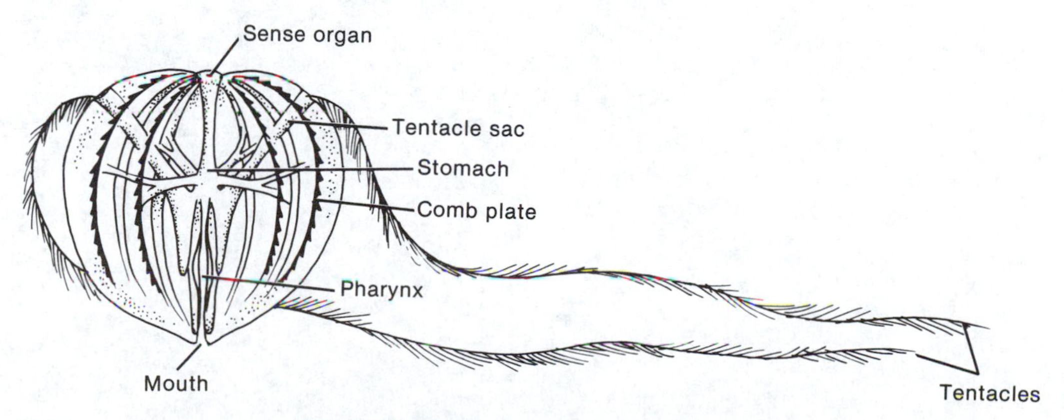

FIGURE 25-10 *Pleurobrachia,* a ctenophore

these cilia propel the animals, mouth end forward, through the sea. Examine preserved specimens of ctenophores and note the jellylike consistency of their body walls, which is due to the gelatinous mesoglea. Observe the two long tentacles, which contain specialized cells that secrete a sticky substance used to catch prey.

Reproduction in the ctenophores is only sexual, each organism having male and female reproductive organs. The fertilized eggs develop into free-swimming larvae that develop into adults.

REFERENCES

Barnes, R. D. 1987. *Invertebrate Zoology.* 5th ed. Saunders.

Buchsbaum, R., M. Buchsbaum, J. Pearse, and V. Pearse. 1987. *Animals Without Backbones.* 3d ed. rev. University of Chicago Press.

Lane, C. E. The Portuguese Man-of-War. 1960. *Scientific American* 202(3):156–168.

Storer, T. I., et al. 1979. *General Zoology.* 6th ed. McGraw-Hill.

Villee, C. A., W. F. Walker, and R. D. Barnes. 1984. *General Zoology.* 6th ed. Saunders.

Kingdom Animalia: Acoelomates (Phylum Platyhelminthes) and Pseudocoelomates (Phyla Nematoda and Rotifera)

Exercise

26

The acoelomates are represented in this exercise by the phylum Platyhelminthes. Platyhelminthes are tripoblastic (their body tissues have developed from three embryonic germ layers), bilaterally symmetrical (their right and left halves are mirror images of each other), lack a body cavity (**coelom**), and lack organs for secreting waste or transporting oxygen to their internal organs. In contrast, the pseudocoelomates have a fluid-filled body cavity containing their internal organs. This cavity, derived from the blastocoel, is a primitive cavity that is found in the embryo and persists as the **pseudocoelom** in the adult.

A. ACOELOMATES

Platyhelminthes, characterized by a dorsoventrally flattened body, are commonly called flatworms. In addition to being bilaterally symmetrical, they also have a dorsal (top) and a ventral (bottom) surface. Most bilateral animals also have distinct anterior (head) and posterior (tail) ends.

The Platyhelminthes are the most primitive of all bilateral animals. However, they are considerably more complex than the coelenterates in possessing well-developed nervous, excretory, muscular, digestive, and reproductive systems. The phylum includes the following three major classes:

- **Class Turbellaria** Free-living flatworms. Most are marine, although some inhabit fresh water or moist places on land. They possess a ciliated epidermis.
- **Class Trematoda** Flukes, which are internal or external parasites of vertebrates. Their epidermis, which is secreted by underlying cells, is ciliated.
- **Class Cestoda** Tapeworms, the adults of which are intestinal parasites of vertebrates.

1. Class Turbellaria (Free-Living Flatworms)

Flatworms are best exemplified by a group of small freshwater organisms called planarians. The common American planarian, *Dugesia*, is found in slow-moving streams or ponds of cool, fresh water, either adhering to or crawling over sticks, stones, leaves, and other debris.

Examine a living planarian in a Syracuse dish with a hand lens or stereoscopic microscope. Observe its general shape and size (Fig. 26-1A). From its movements, determine its anterior, posterior, dorsal, and ventral sides. What is its coloration?

____

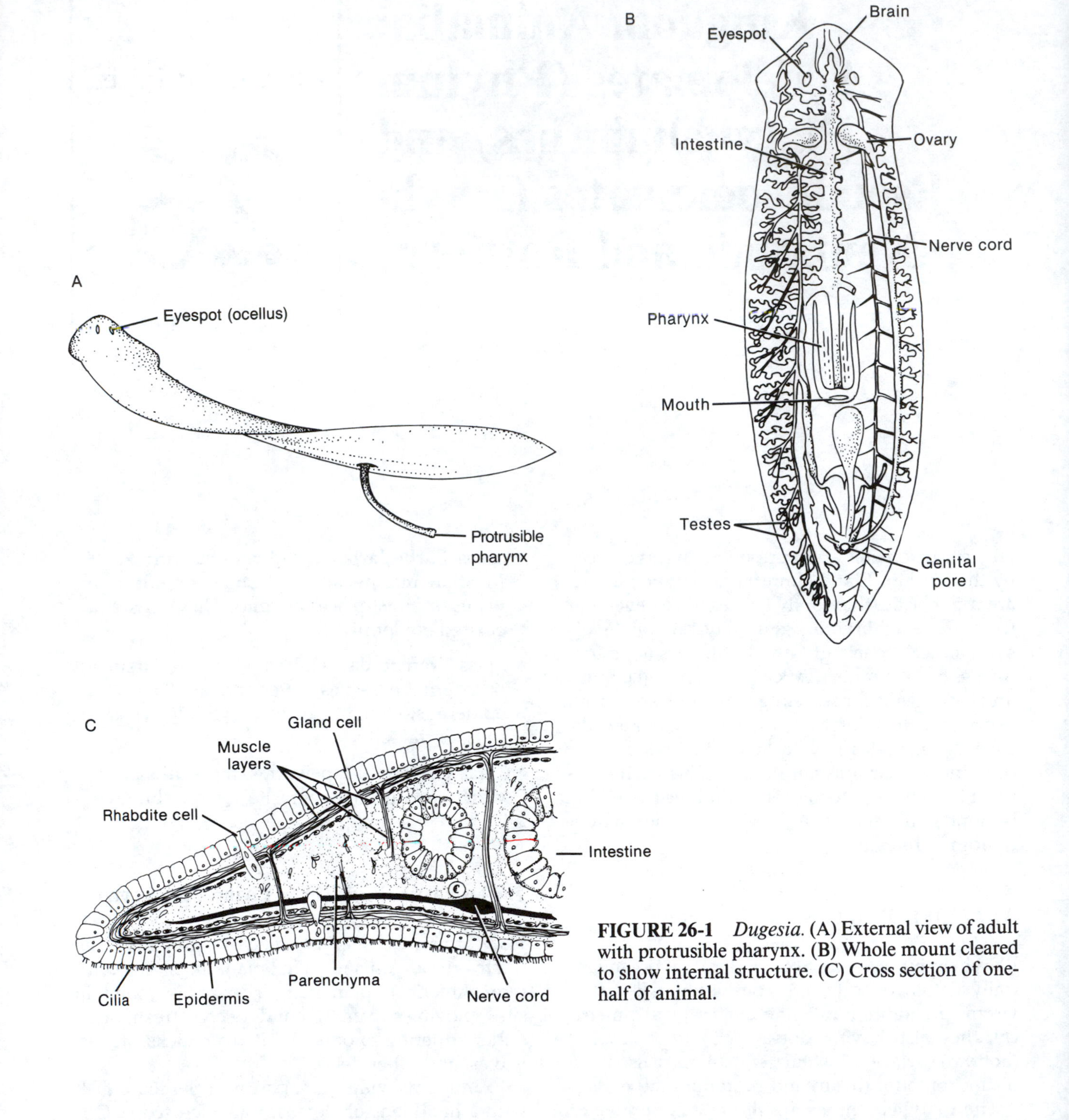

FIGURE 26-1 *Dugesia.* (A) External view of adult with protrusible pharynx. (B) Whole mount cleared to show internal structure. (C) Cross section of one-half of animal.

What are its reactions to such stimuli as jarring the dish, touching it with a toothpick, or turning it over?

____

____

Observe the protrusible **pharynx** on the ventral surface. What is the function of such a structure?

____

____

____

Note the large eyespots **(ocelli)** on the "head." They have no lenses and cannot form an image. However, they can distinguish light from dark. Shine light from a pen light onto the planarian. Describe its response.

___

___

Using a stereoscopic microscope, examine a prepared slide of a whole mount of a planarian cleared to show its internal anatomy (Fig. 26-1B). Locate the intestine leading from the protrusible pharynx. Suggest a function for the numerous lateral projections from each of the three main branches of the intestine.

___

___

Excretory tubes and nerve fibers are difficult to observe because of the thickness of the animal.

Using a compound microscope, examine cross sections of a planarian cut through the pharynx and through regions anterior and posterior to the pharynx. Locate the intestine in each of these regions (Fig. 26-1C). Note that there is no body cavity; the space between the surface epidermis and the linings of the intestine is filled with a mass of cells called **parenchyma.** Observe cilia on some of the epidermal surface cells. On which surface are they most prominent?

___

The epidermis contains **rhabdite cells.** It is thought that the rod-shaped contents of these cells are secreted into the water, where they swell and form a protective gelatinous sheath around the animal.

The muscles are arranged in two layers, one circular and one longitudinal. How does such an arrangement of muscles give the planarian flexibility?

___

___

___

Locate the two nerve cords, one on either side of the animal, just inside the ventral epidermis.

Planarians have been used extensively in studies of **regeneration.** Their regenerative powers are easily demonstrated.

- Add several drops of magnesium sulfate to a Syracuse dish that contains a planarian and allow 15 minutes for it to become anesthetized.
- With a clean, oil-free razor blade, carefully cut the planarian into sections. (Your instructor will indicate the cuts to be made.) Diagram your cuts in Fig. 26-2.

FIGURE 26-2 Regeneration in *Dugesia*

- Add fresh pond water and place your dish, covered to prevent evaporation, in a cool, dark place in the laboratory.
- Over the next few weeks observe your planarian for any regeneration that occurs. *Note:* Periodically add fresh pond water and remove any degenerating tissues that will appear grayish in color.
- Make sketches of the regeneration in Fig. 26-2. Based upon your results, and the observations made by other students, do all parts of the animal regenerate equally well?

____

- If not, which parts of the animal (e.g., head region, tail region, middle region) appear to have the best potential for regeneration?

____

____

____

2. Class Trematoda (Flukes)

The flukes are parasitic flatworms that lack cilia and an epidermis but are covered by a thick integument secreted by the underlying cells. This integument is resistant to the hosts' enzymes and is thus an important adaptation for a parasitic way of life. One or more suckers that attach the flukes to their hosts may also be present on the ventral surface. A large part of the body is occupied by reproductive organs and a two-branched intestine, which unlike the intestine in the turbellarian, does not ramify throughout the body.

a. *Opisthorchis (Clonorchis) sinensis* (Chinese Liver Fluke)

Opisthorchis sinensis is often used to illustrate the typical morphology and life cycle of flukes. It is a common parasite in China, Japan, and Korea, where it has infested millions of human beings. The Chinese liver fluke has a life cycle that requires two intermediate hosts.

The adult fluke, about 15 mm long, lives in the bile ducts of the human liver where it remains firmly attached by means of a pair of suckers. Being hermaphroditic, it is capable of self-fertilization, though cross-fertilization is much more common. What advantage is gained from possessing the capability to self-fertilize?

____

____

____

The fertilized "eggs," called zygotes, are released into the bile ducts and are eventually evacuated in the feces of the host. If the feces get into water, the zygotes may be eaten by snails, the first intermediate host (Fig. 26-3). In the digestive system of the snail, a larval form, called a **miracidium,** emerges from the zygote and makes its way into the tissues of the snail, where it gives rise to other larval forms called **sporocysts** and **redia** (Fig. 26-3). It has been estimated that a single miracidium can give rise to 250,000 infective larvae called **cercariae.** Why is it essential that one miracidium give rise to such a large number of cercariae in the life cycle of the fluke?

____

____

____

The cercariae escape from the snail and swim about until they come in contact with the second intermediate host, one of several species of carp. They burrow through the skin of the fish, lose their tails, and encyst to form **metacercariae.** When raw or partly cooked fish is eaten by human beings, the cyst walls are digested in the stomach, and the metacercariae are released. They then migrate to the bile ducts and mature into adult flukes to complete the life cycle. In light of what you know about the life cycle of the fluke, what would be the most effective methods of controlling the infestation of human beings by this organism?

____

____

____

____

____

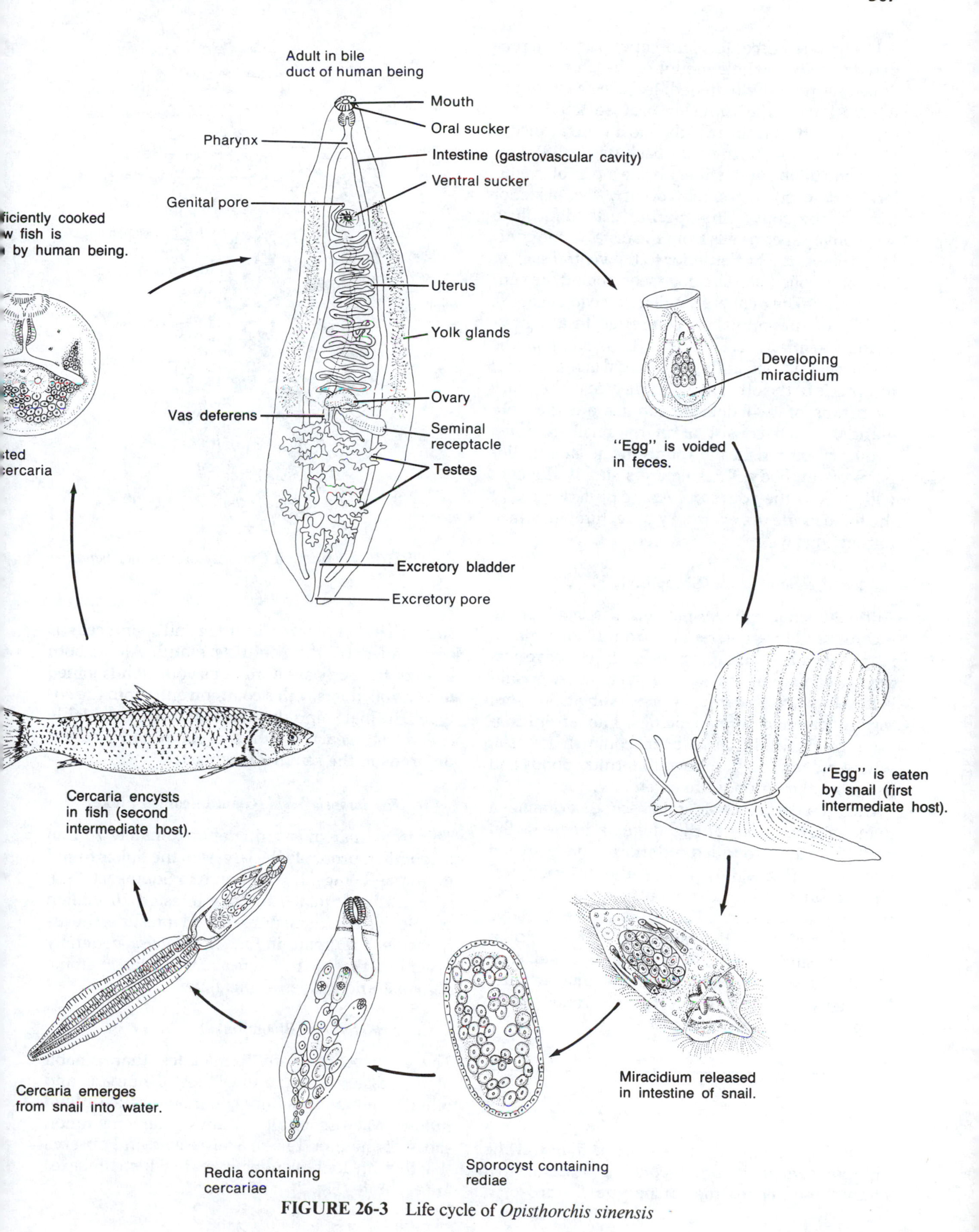

FIGURE 26-3 Life cycle of *Opisthorchis sinensis*

Using a stereoscopic microscope, examine a prepared slide of a whole mount of the Chinese liver fluke and note the flattened, leaflike shape of the worm. Locate the **anterior oral sucker,** in the center of which is the mouth, and a **ventral sucker** about one-third of the way back (Fig. 26-3). Observe the bilobed intestine. Locate a pair of irregularly branched testes that occupy the posterior third of the body. Tiny **sperm ducts** (difficult to see) convey sperm cells from the testes to the **genital pore** located just anterior to the ventral sucker. The long, coiled uterus can be seen behind the ventral sucker. The **ovary,** a single body lying near the middle of the animal, is connected to a lighter-staining **seminal receptacle.** After copulation, the sperm cells of each animal are stored in the seminal receptacle of the other. Also connected to the ovary by means of two delicate tubules are the **yolk glands,** which consist of many small, rounded bodies characteristically found in the lateral midparts of the body. These glands supply the eggs with yolk as they develop. At the posterior end of the fluke locate the **excretory pore,** through which nitrogenous wastes are excreted.

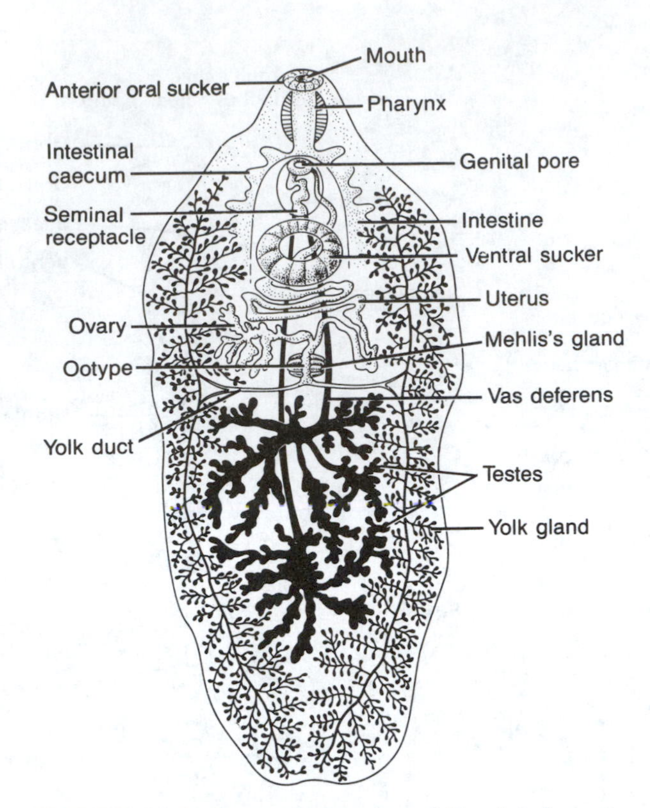

FIGURE 26-4 Internal anatomy of *Fasciola hepatica*

b. *Fasciola hepatica* (Sheep Liver Fluke)

Although similar to *Opisthorchis, Fasciola hepatica* is considerably larger, ranging from 15 to 50 mm in length, and has a more complex reproductive system. It commonly infests the livers of sheep and cattle, but can also infest horses, rabbits, camels, pigs, goats, and human beings. Human infestations of this fluke have been common in Asia, where human feces are used to fertilize ponds and fish from these ponds are eaten raw.

Using a stereoscopic microscope, examine a demonstration slide of this fluke and note its flat body, which is rounded anteriorly and flattened posteriorly (Fig. 26-4). Locate the **anterior oral sucker** surrounding the mouth and the ventral sucker a short distance back from it. The digestive system consists of the mouth, a short muscular pharynx, and a branched intestine that extends along both sides of the fluke. Each of the branches has many smaller lateral branches, called **caeca.** Suggest a function of the caeca?

____

____

The female reproductive system, found in the anterior half of the body, consists of a branched ovary connected to a median **ootype.** The ootype, a small chamber that receives egg and sperm cells, is surrounded by the **Mehlis's gland.** Along both sides of the body are numerous yolk glands joined to two yolk ducts with a common entry into the ootype. The male reproductive system has two highly coiled testes, each of which is connected by a **vas deferens** to the seminal receptacle.

c. *Fasciolopsis buski* (Giant Intestinal Fluke)

This trematode, measuring approximately 75 mm in length, is probably the largest of the flukes found in human beings. It is common in Southeast Asia, where it is estimated to have infested 10 million people. Except for an unbranched intestine, the arrangement of organs in *Fasciolopsis buski* generally resembles that of the Chinese liver fluke. Examine demonstration slides of this fluke.

d. *Schistosoma* (Blood Fluke)

The schistosomes are trematodes that inhabit blood vessels near the intestines of humans and other vertebrates and are commonly called **blood flukes.** Microscopically examine demonstration slides of the blood fluke *Schistosoma* and observe that they are long, slender flukes in which the sexes are separate (Fig. 26-5).

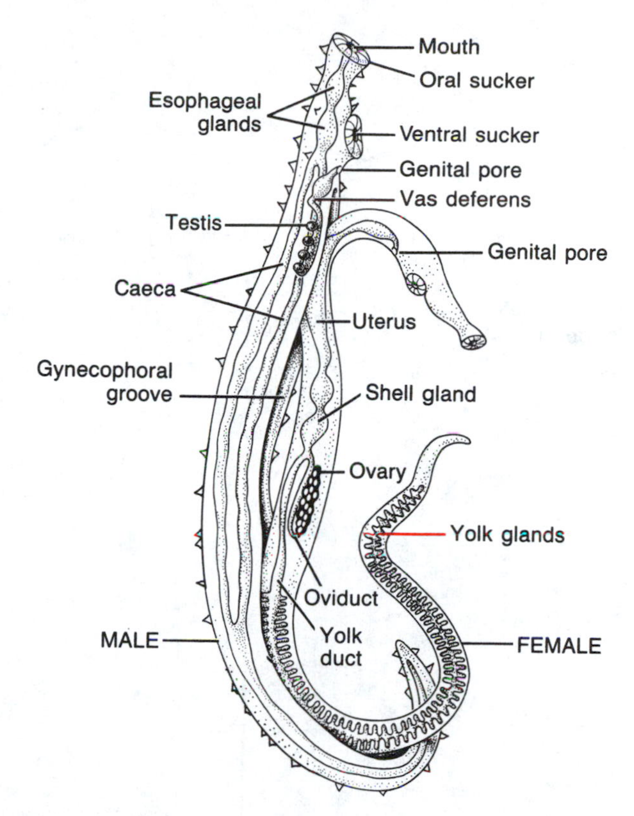

FIGURE 26-5 Longitudinal section of male and female copulating *Schistosoma*

The life cycle of *Schistosoma* is very similar to that of *Opisthorchis* and *Fasciola*, in that the snail is the intermediate host for the various larval stages. However, rather than being ingested with food, the cercariae enter their human hosts by burrowing through the skin, particularly the feet, or by being taken in with drinking water. Cercariae of certain schistosomes that do not normally parasitize human beings can burrow into the skin and produce a "swimmer's itch."

From your study of the anatomy and the life cycles of flukes, describe how they are adapted to their parasitic life.

____

____

____

____

____

3. Class Cestoda (Tapeworms)

Tapeworms are typically long, dorsoventrally flattened ribbonlike organisms. The extent to which the tapeworm has adapted to its parasitic existence is evident in its degenerate digestive and nervous systems and its highly developed reproductive system. The flatness of these worms enhances diffusion so that there is no need for respiratory and circulatory systems.

a. *Dipylidium caninum* (Dog Tapeworm)

The common cat or dog tapeworm, *Dipylidium caninum*, will be studied as a typical tapeworm. Fleas are the hosts for the larval stage of this species, and children are sometimes infected by it if they accidentally swallow fleas while playing with dogs or cats.

Using a stereoscopic microscope, examine a prepared whole-mount slide of the head and several segments of an adult tapeworm (Fig. 26-6). The head, or **scolex** (pl. scolices), with its four suckers and several rows of hooks, is used for attachment to the intestine of its host. Behind the scolex is a short neck region, followed by a long, ribbonlike body consisting of a series of segments called **proglottids.** Mature proglottids contain both male and female reproductive organs. The smaller ducts leading from the lateral **genital pores** are the oviducts, which lead into the diffuse ovaries. Locate the **testes,** which are distributed throughout the proglottid, and the vas deferens, by which the sperm leave the proglottid via the genital pore. Lateral and transverse **excretory canals** are present in the mature proglottids, as are thin, longitudinal nerve cords.

The gravid proglottids near the posterior end of the tapeworm become filled with numerous **ovarian capsules**, each of which contains a large number of fertilized eggs, or zygotes. The gravid proglottids break loose from the tapeworm and are released in the feces of the host. The zygotes (actually embryos because development has already begun) are released from these segments, eaten by the larvae of the dog or cat flea, and develop into infective **cysticercoid larvae.** When swallowed by a dog or cat, the infected flea is digested by the host and the cysticercoid larva is released. It attaches to the intestine wall of the host and develops into an adult tapeworm.

b. *Taenia saginata* (Beef Tapeworm)

The most common intestinal tapeworm of human beings is the beef tapeworm, which ordinarily

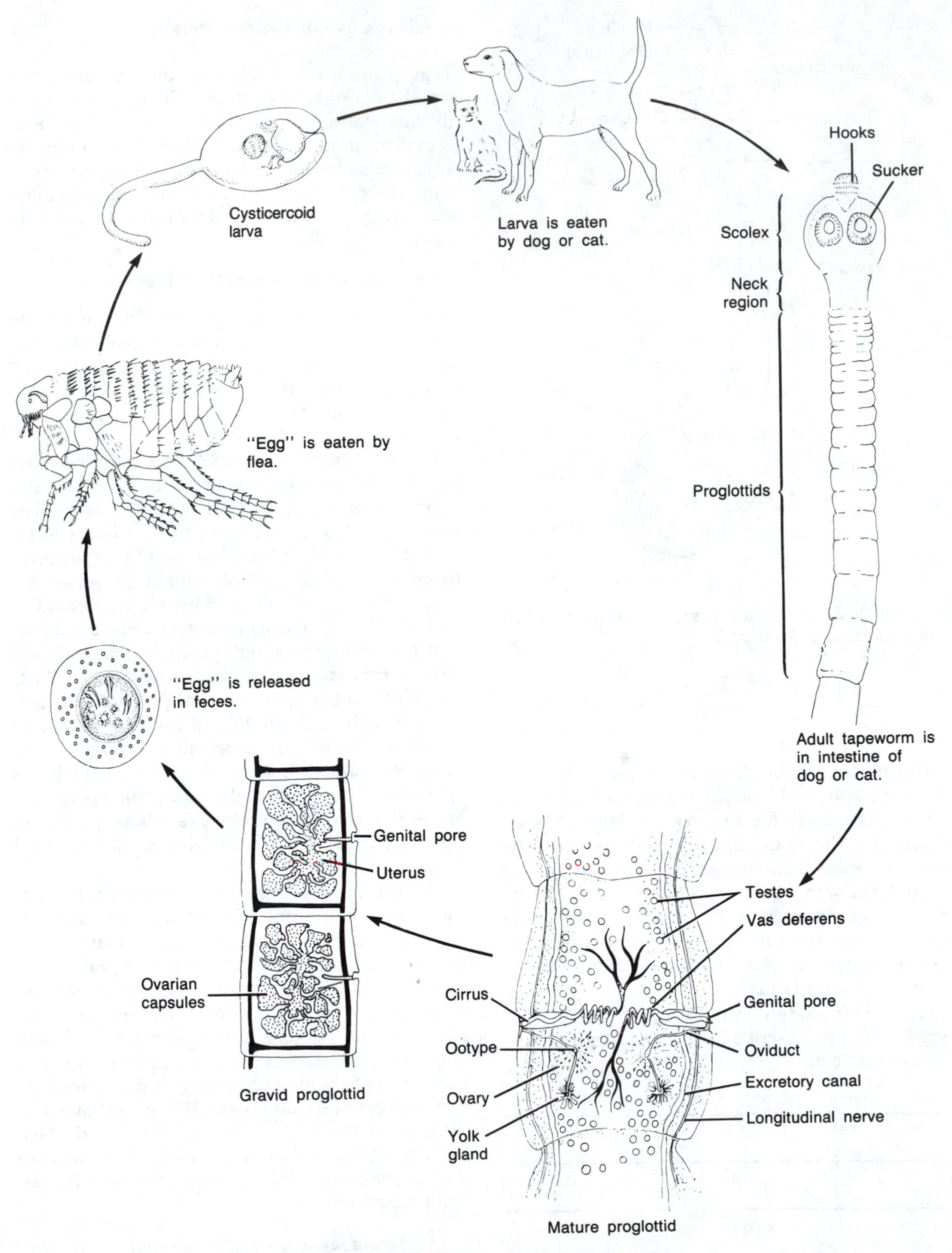

FIGURE 26-6 Life cycle of *Dipylidium caninum*

grows to a length of 5–7 m. Cattle, buffalo, giraffes, and llamas have also been recorded as natural alternative hosts for the cysticercus stage of this tapeworm. Meat inspection has reduced the occurrence of this parasite to about 1% of the cattle in the United States. In countries in which beef is prepared by broiling large pieces over open fires, the incidence of infection is quite high. Why?

____

____

____

Examine preserved specimens of this tapeworm, as well as slides showing scolices and mature and gravid proglottids.

c. *Taenia solium* (Pork Tapeworm)

The pork tapeworm that infests human beings closely resembles the beef tapeworm and has a similar life cycle, except that the larval cysticerci stages, called **bladderworms,** develop in pigs. A mature pork tapeworm can be 2–8 m long. Human beings are infected by eating raw or insufficiently cooked pork. Meat inspection has made it uncommon in the United States.

d. *Dibothriocephalus latus* (Broad Fish Tapeworm)

The broad fish tapeworm, which can grow to a length of 18 m, is the largest tapeworm to parasitize human beings. This tapeworm requires two intermediate hosts to complete its life cycle. The eggs, when released into water, develop into larvae that are eaten by small crustaceans called copepods. These in turn are eaten by fish, in which the larval stage (**plerocercoid**) develops. People ingest the parasite when they eat raw or partly cooked fish. This tapeworm has become well established in the western Great Lakes region of North America, where it commonly infests pike and pickerel.

Using a stereoscopic microscope, examine whole-mount slides of the broad fish tapeworm. Note that the scolex has sucking grooves rather than the suckers and hooks found in the other tapeworms you have studied. Also note that the proglottids are broader than they are long and that the reproductive organs are located in the center of the proglottid.

Do you consider tapeworms to be more adapted or less adapted to parasitism than flukes? Give the reasons for your answer.

____

____

____

____

____

____

B. PSEUDOCOELOMATES

The pseudocoelomates are a heterogeneous group of organisms that includes rotifers, gastrotrichs, kinorhyncha, nematomorphs (horsehair worms), acanthocephala, entoprocts, loricifera, and one of the most successful groups in the animal kingdom, the nematodes, or roundworms. In addition to a pseudocoelom, all of these organisms have a complete digestive tract with a separate musculature and a nonliving body covering called the **cuticle.** The reproductive organs are located in the fluid-filled pseudocoelom.

Of the many phyla of pseudocoelomates, we will study the following two:

- **Phylum Nematoda** Roundworms, including both free-living and parasitic worms having elongate cylindrical bodies tapered at both ends.
- **Phylum Rotifera** Microscopic, free-living animals found primarily in freshwater habitats and characterized by an anterior locomotor and feeding organ and a specialized internal grinding organ (**mastax**).

1. Phylum Nematoda (Roundworms)

Nematodes have long cylindrical bodies that are usually tapered at both ends and covered with a tough cuticle. They exhibit bilateral symmetry, a complete digestive tract, three definite tissue layers derived from the embryonic ectoderm, mesoderm, and endoderm, and definite organ systems. They lack a circulatory system. The sexes are usually separate, with the male nematode being smaller than the female.

Among multicellular animals, roundworms are probably second only to insects in number. Most of the thousands of species of nematodes are free-living. Some are found in fresh water; some in salt water; and some in mud, field, or garden soils. Other nematodes are parasites of roots, stems, leaves, and seeds of many horticultural and agricultural plants, causing inestimable damage to crops. Many thousands of species of roundworms are parasites of invertebrate and vertebrate animals; about a dozen are parasitic in human beings. Most of the species that are either free-living or parasitic in plants or invertebrates are barely visible to the eye. The species that are parasitic in vertebrates, on the other hand, can be several meters in length and have more complicated life cycles.

a. *Ascaris lumbricoides* (Pig Roundworm)

Ascaris is a common parasite in the intestine of hogs. The sexes are separate, with the male worm being shorter and more slender than the female and easily distinguished by a sharply curved posterior end. In addition, the male has a pair of hairlike structures **(spicules)** that extend from the anal opening and are used during copulation. Both animals are several centimeters in length. Examine preserved specimens of both sexes. Using a hand lens or stereoscopic microscope, locate the terminal **mouth** surrounded by three lobelike **lips.** Locate the **excretory pore** just behind the mouth.

Using a scalpel, carefully slit a female worm along the whole length of the body. Expose the internal organs by pinning the body wall back in a dissecting pan (Fig. 26-7). To prevent drying during your dissection, cover the worm with water. Locate the long, flat digestive tract extending the length of the body cavity. The Y-shaped female reproductive system can be seen to occupy most of the body cavity. Identify the single **vagina,** which divides into a pair of large, straight, and uncoiled **uteri.** The uteri narrow down into the **oviducts,** which in turn continue as the long, thin, highly coiled **ovaries.**

The male reproductive tract consists of a single long, highly coiled tube that is regionally divided into a long slender **testis,** a **sperm duct,** an enlarged **seminal vesicle,** and a small **ejaculatory duct** that opens into the terminal end of the digestive tract.

Using a compound microscope, examine a prepared slide of a cross section of the female (Fig. 26-8). The body wall consists of the epidermis covered by the noncellular cuticle. Suggest a function of the cuticle.

____

____

____

The epidermis has four extensions into the body cavity. Two of these contain the **dorsal** and **ventral nerve cords,** and the other two contain the **excretory tubes.** The greater part of the body wall is composed of muscle tissue. The digestive tract is centrally located in the pseudocoelom. Observe the thin-walled intestine and note that the intestinal wall consists of a single layer of large **gastrodermal** cells. Locate the two large cross sections of the bilobed uterus and note that the ova have shells surrounding them. Sections through the oviducts and

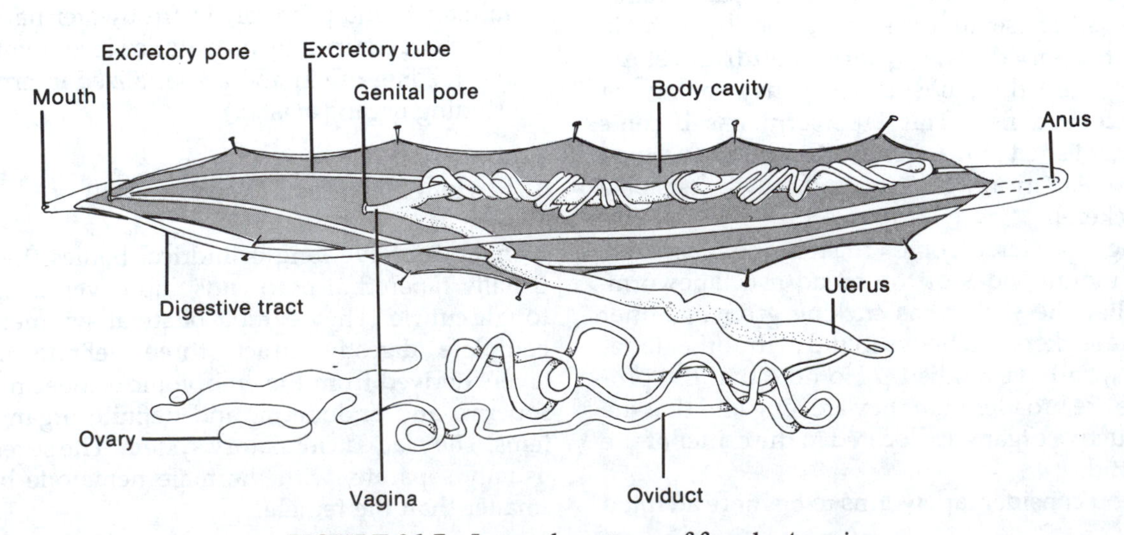

FIGURE 26-7 Internal anatomy of female *Ascaris*

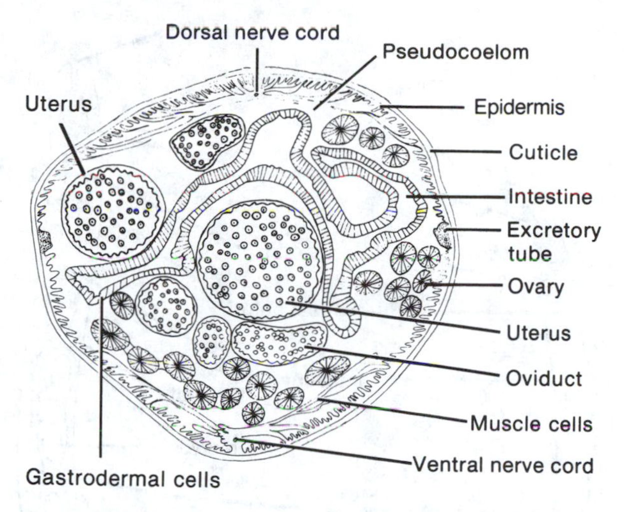

FIGURE 26-8 Cross section of female *Ascaris*

ovaries can be distinguished from sections through the uteri because the former are smaller and the ova lack shells.

Look for meiotic stages in the oviduct (e.g, primary and secondary oocytes, and cells in metaphase I and II).

b. *Trichinella spiralis*

Infection by this roundworm, which causes **trichinosis** in human beings, results from eating undercooked pork that contains the encysted larvae of this parasite. In the human intestine, the larvae are released from their cysts and develop into adult worms. The mature female worm releases larval worms that migrate through the human body and encyst in the muscles of the diaphragm, ribs, and tongue. The greatest damage to tissues is done during this migration phase. Why?

___

___

___

The incidence of human infection in the United States is astonishingly high: approximately 17% of the population.

Examine prepared slides of adult worms and encysted larvae.

c. *Necator americanus* (Hookworm)

No group of nematodes causes more injury to human beings or greater economic loss through attacks on domestic animals than the hookworms. These tiny parasites have highly developed mouths that contain plates by which the worm holds onto the intestine while it sucks blood and tissue. Larvae develop from eggs deposited on moist soil in warm climates. Infestation of human beings occurs when the larvae penetrate the skin, and is very often a result of going barefoot in areas where the parasite is prevalent.

Examine prepared slides of hookworms. Locate the **buccal capsule** at the anterior end of this worm and note the cutting plates inside the capsule. At the posterior end, locate the **bursa,** an umbrellalike expansion of the cuticle that is used in copulation.

d. *Enterobius vermicularis* (Pinworm)

This roundworm is one of the most common parasites of children, and its incidence in Canada and the United States is estimated to be 30–60% of the population. After initial infection, the 2–12 mm worms attach to the wall of the appendix and large intestine. At night, females migrate down the intestine, crawl outside and deposit their eggs on the skin around the anus. Since the most common symptom is **anal pruritus** (itchiness in the anal region), the eggs may be carried back to the mouth of the child on his or her fingertips. Alternatively, the eggs may get into the bedding and ultimately be spread to other members of the family. Control of infection involves treatment of the entire family and thorough observance of personal hygiene.

Examine slides of adult pinworms. These small, spindle-shaped worms are characterized by a lateral expansion of the cuticle at their anterior ends. The posterior end of the male is strongly curved.

e. Free-Living Nematodes

1. In Soil.

Free-living nematodes can be found in almost any loose soil containing organic matter. Your instructor has placed some soil in a petri dish. Using a stereoscopic microscope or the low power of a compound microscope, examine the soil for the presence of nematodes. Scrape several of the worms you find onto a slide, add a drop of water, and examine them under the low- and high-power magnifications of the microscope. Because these nematodes are relatively transparent, the various organs, with the exception of the nervous system, can be observed. Describe and account for the type of movement exhibited by these worms. *Note:* If the movements of the worms are too rapid, add a drop of dilute hydrochloric acid to your slide.

____

____

____

2. In Vinegar.

A common free-living nematode is the vinegar eel (*Anguillula aceti*, formerly *Turbatrix aceti*), which is frequently found on the bottom of a vinegar barrel where it feeds on the bacteria and yeasts that have settled there. Mount a drop of vinegar eel culture on a slide and examine it microscopically. Because this nematode is transparent, its internal organs can readily be seen. Further, being viviparous (i.e., giving birth to living young), all stages of development can be studied. For better observation, add a drop of 0.2% neutral red stain to the drop of vinegar and add a coverslip. If the animals are moving too vigorously for observation, slow them down by withdrawing some of the solution with a small piece of filter paper or paper toweling placed at the edge of the coverslip.

Use Fig. 26-9 to identify the sexes and study the internal anatomy of this free-living nematode.

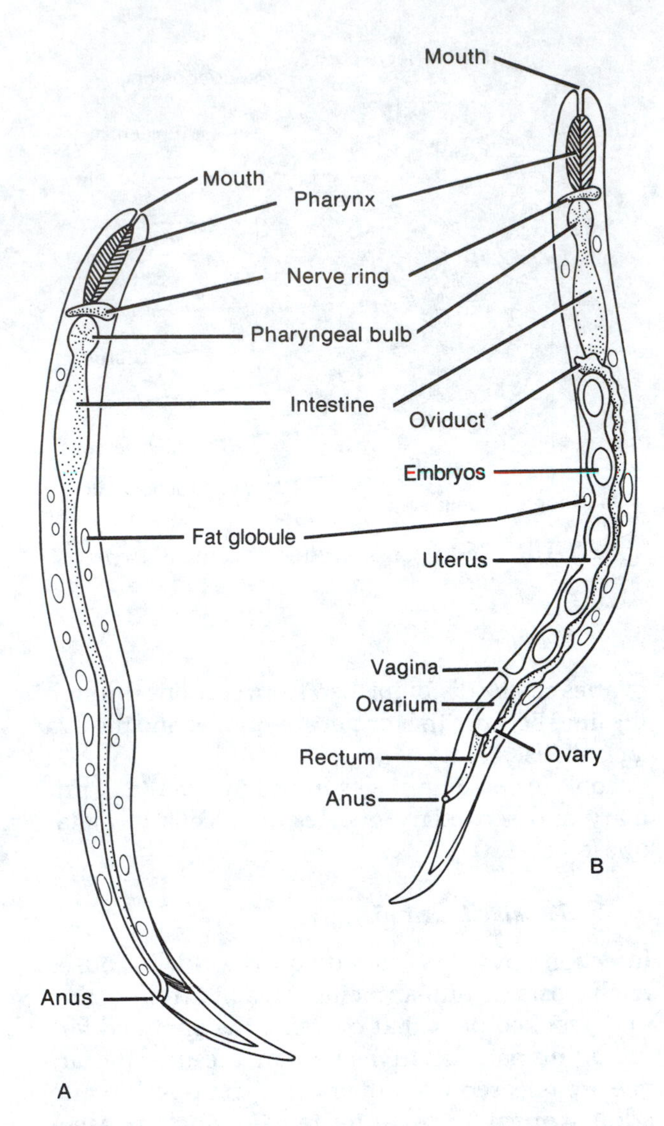

FIGURE 26-9 Adult *Anguillula aceti*. (A) Male, (B) Female.

2. Phylum Rotifera

Rotifers are microscopic organisms that are abundant in freshwater lakes, ponds, quiet streams, and wayside ditches. They are even found in street gutters and eave troughs of buildings. They constitute an important component of the plankton that is a source of food for many fish species.

Prepare a wet mount of a culture of rotifers that contains one or more of the genera shown in Fig. 26-10. Observe the elongate cylindrical body, which is divided into a somewhat cylindrical trunk and foot. Although the cuticle is divided into several segments,the internal organs are not; therefore, this is not true segmentation. You may observe, in a rotifer that is moving about, a telescoping of these segments as the animal expands and contracts.

The anatomy of a typical rotifer is shown in Fig. 26-11. Observe the long, tapering **foot** and its two **spurs** at the posterior end of the body. The anterior end of the body is expanded into a crown of cilia, called a **crown** or **corona,** which is used primarily for locomotion and for gathering food. The cilia create a current of water that draws food particles toward the mouth. These animals bear the name rotifer, or wheel animals, because the beating motion of the cilia gives the impression of a turning wheel. The mouth leads into a muscular pharyngeal apparatus called a **mastax.** This structure can be observed in living rotifers because of its constant grinding movement. Posterior to the mastax is the narrow **esophagus,** surrounded by the **salivary glands** and saclike **stomach.** The digestive tract is lined with cilia that move food into the **intestine** and waste products to the **cloaca,** where they are then eliminated through the **anus,** located near the base of the foot.

The female reproductive system of rotifers consists of a single ovary combined with a yolk gland. After fertilization, the eggs pass from the oviduct

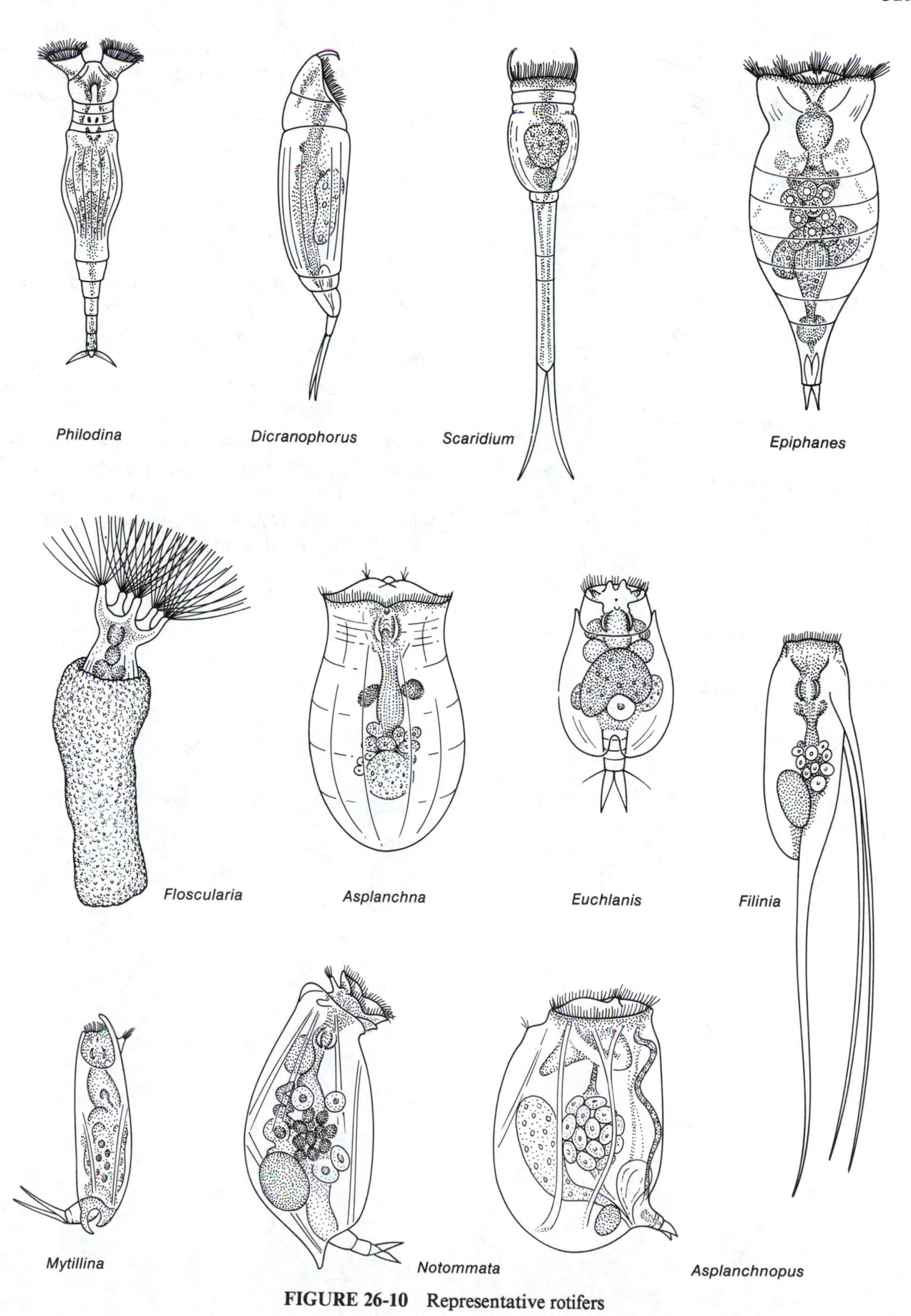

FIGURE 26-10 Representative rotifers

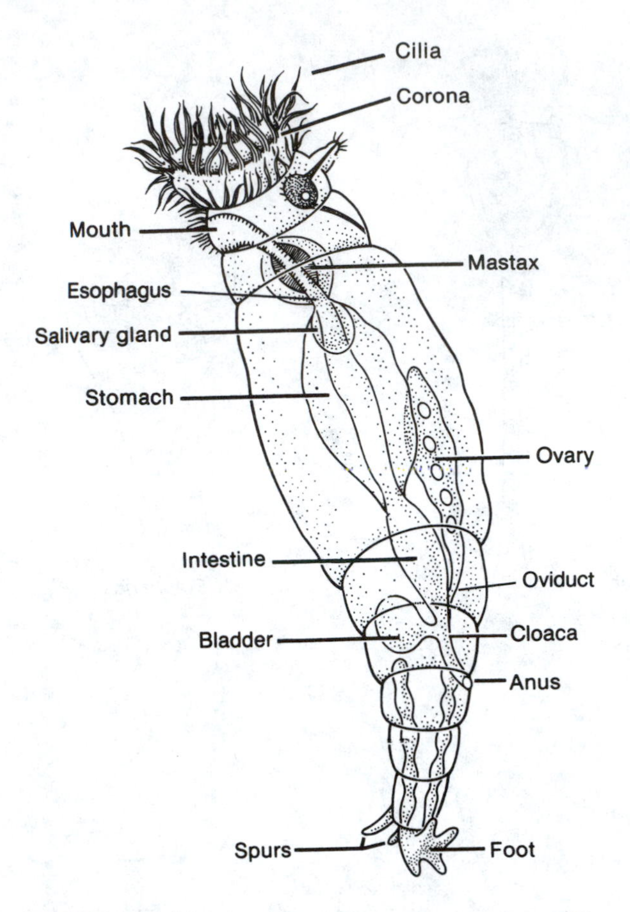

FIGURE 26-11 Anatomy of a representative rotifer

into the cloaca and then to the outside. Males are smaller than the females and have a single testis and a penis, which is inserted into the female oviduct during copulation.

REFERENCES

Barnes, R. D. 1987. *Invertebrate Zoology.* 5th ed. Saunders.

Buchsbaum, R., et al. 1987. *Animals Without Backbones.* 3d ed. rev. University of Chicago Press.

Croll, N. A., and B. E. Matthews. 1977. *Biology of Nematodes.* Wiley.

Noble, E. R., and G. A. Noble. 1989. *Parasitology.* 6th ed. Lea and Febiger.

Pearse, V., et al. 1987. *Living Invertebrates.* Blackwell.

Storer, T. I., et al. 1979. *General Zoology.* 6th ed. McGraw-Hill.

Villee, C. A., W. F. Walker, and R. D. Barnes. 1984. *General Zoology.* 6th ed. Saunders.

Exercise

27

Kingdom Animalia: Phylum Mollusca

Mollusks (Latin *molluscus*, "soft") are soft-bodied animals that have an internal or external shell. Included in the phylum are snails, oysters, slugs, clams, octopuses, and squids. Most mollusks are **bilaterally symmetrical** and have well-developed respiratory, excretory, digestive, and circulatory systems with a heart.

Mollusks are similar to annelids (segmented worms) in their development; both have **trochophore** larvae. Mollusks differ from annelids, however, in the absence of segmentation. Further, the coelom, so prominent in the annelids, is greatly reduced in the mollusks and is generally restricted to the cavity surrounding the heart.

Most mollusks are slow moving (snails and slugs), but the bodies of several species are modified for rapid locomotion (squid and octopus). Although primarily marine organisms, some mollusks are found in fresh water (clams) and on land (slugs).

The mollusks are characterized by three main body regions: a **foot,** the locomotive part of the body (in squid, the **head-foot** also contains sensory organs); a **visceral mass** containing the excretory, digestive, and circulatory organs; and the **mantle,** a sheet of specialized tissue that covers the internal organs and that secretes the **shell.** The **gills,** which function in respiration, are located interior to the mantle.

In this exercise, you will study representatives of the five major classes of mollusks.

- **Class Bivalvia** Visceral mass is contained in a shell having right and left valves that are hinged. A hatchet-shaped foot extends out between the valves during locomotion.
- **Class Polyplacophora** The elliptical body is covered by a shell consisting of eight transverse plates, although some species lack these plates.
- **Class Scaphopoda** Visceral mass is enclosed in an elongated, tapered, and toothlike shell. Foot is cone-shaped. Gills are absent.
- **Class Gastropoda** Visceral mass is contained in a spirally coiled shell (though the shell in some is greatly reduced or absent). Distinct head with one or two pairs of tentacles. Large, flat foot.
- **Class Cephalopoda** Have either an internal or an external shell and a large and prominent head with conspicuous, complex eyes. A mouth is surrounded by 8 – 10 or more tentacles.

A. CLASS BIVALVIA (MUSSELS, CLAMS, SCALLOPS, OYSTERS)

Members of this class, characterized by a shell consisting of two **valves,** include common freshwater clams and mussels and marine varieties of clams and oysters. Freshwater clams, which you will study in this section, are abundantly distributed on the bottoms of lakes, rivers, ponds, and streams where they feed on **plankton** (microscopic plant and animal life).

1. General Structure

Examine the shell of the freshwater clam *Anodonta* and note that it consists of two valves hinged along the dorsal side (Fig. 27-1A). On the anterior part of each valve is a protruding region, the **umbo.** The

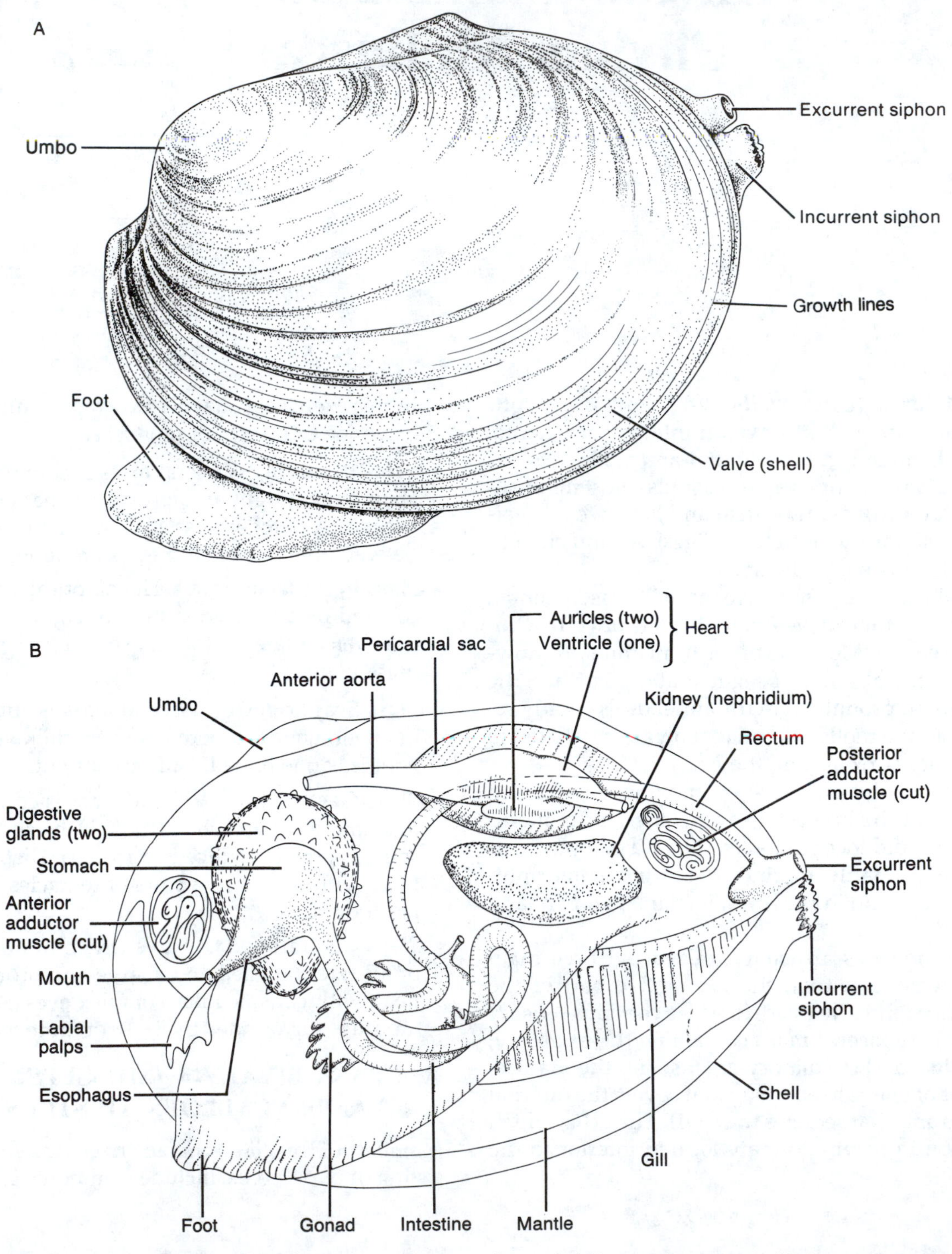

FIGURE 27-1 Morphology of the freshwater clam *Anodonta.* (A) External anatomy. (B) Internal anatomy.

concentric lines that extend outward from the umbo are lines of growth. The valves are held together by two large **adductor muscles** located at opposite ends of the shell (Fig. 27-1B). To get to the inside of the clam, cut these muscles by inserting a knife or scalpel between the two valves of the shell and drawing it toward the hinge where the valves are joined. Open the valves and observe that they are lined by a glistening **mantle.** The outer epithelial layer of the mantle secretes the shell, which is made up of three layers (Fig. 27-2). The thin outer layer, the **periostracum,** functions to protect the underlying parts of the shell from acids in the water. It also gives the shell its color. The middle **prismatic layer** is composed of calcium carbonate. The inner **nacreous layer,** called mother-of-pearl, is made up of many layers of calcium carbonate and has an iridescent sheen.

In *Anodonta*, the edges of the mantle are partially fused to form incurrent and excurrent siphons for the circulation of water through the clam. In many bivalves, the openings are extended as tubular structures called **siphons** so that the animal can remain in contact with the water when it burrows into the mud. Water enters the clam through the ventral **incurrent siphon,** circulates through the mantle cavity and over the gills, and then leaves through the dorsal **excurrent siphon.**

Remove one of the valves and detach the mantle from the inner surface of the shell. Observe the large muscular foot extending down from the visceral mass. Locate the gills, which hang down into the mantle cavity. How many gills are there and what is their function?

____

____

Locate two pairs of flaplike **labial palps** on the anterior edge of the visceral mass, near the anterior muscle. These palps surround the mouth. Dorsal to the gills is the **pericardial sac,** which encloses the heart. Carefully cut open this sac and locate the heart. As shown in Fig. 27-3, the heart consists of three chambers, two lateral **auricles** and one **ventricle.** Carefully cut away the gills and locate the kidney (**nephridium**), a dark-colored organ lying

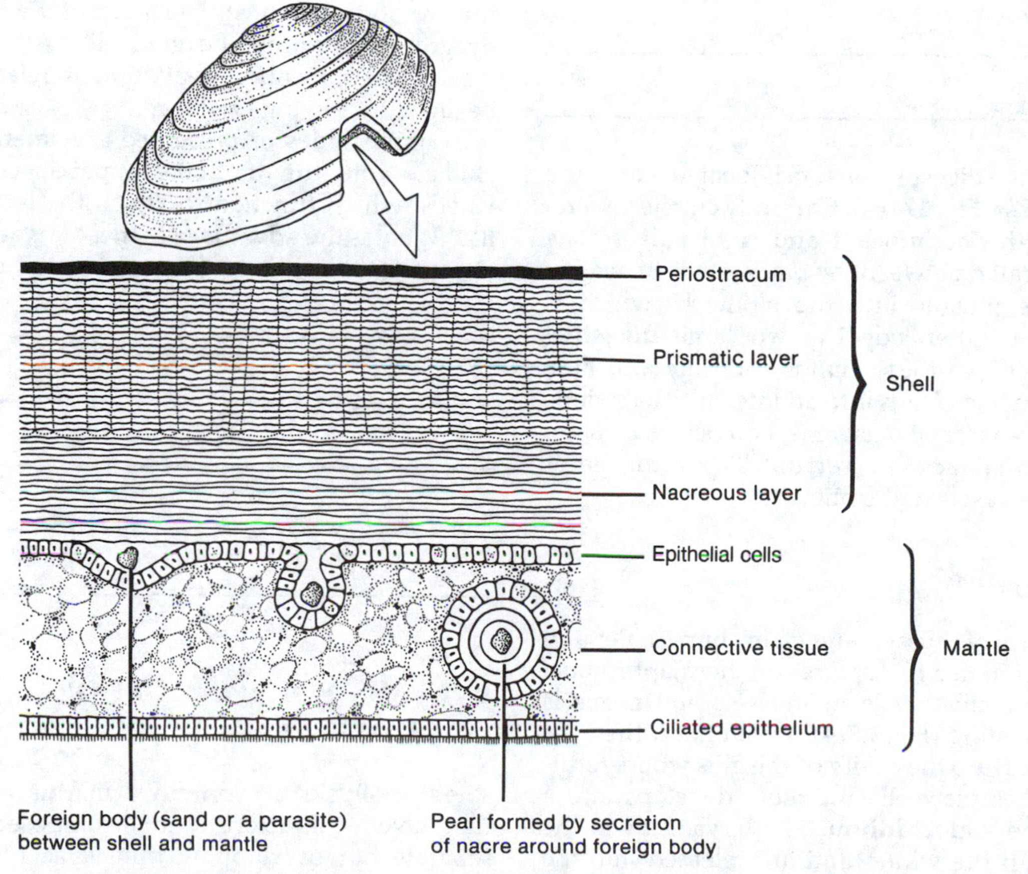

FIGURE 27-2 Diagram of structure of shell and mantle of *Anodonta*

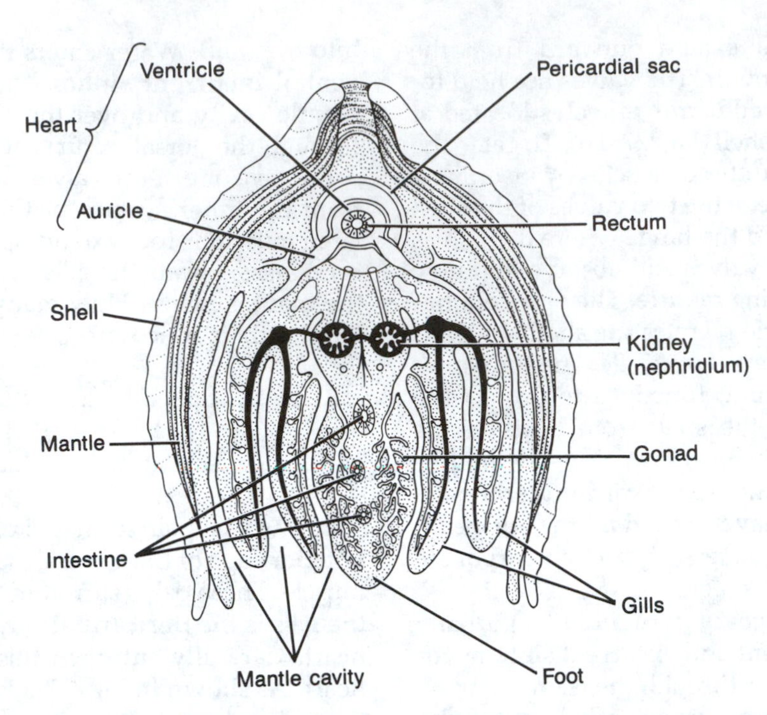

FIGURE 27-3 Cross section through shell, mantle, and visceral mass of a clam

near the gills and just below the pericardial sac. What is the function of this organ?

____

____

Most of the digestive system is located within the visceral mass (Fig. 27-1B). Carefully cut the visceral mass lengthwise into left and right halves. The mouth, located between the palps, leads by way of a short esophagus into the stomach, which is flanked on either side by two large **digestive glands.** Remove these glands carefully and note that the stomach leads into an intestine that winds through the visceral mass and then passes through the pericardial sac as the **rectum.** The rectum empties into the excurrent siphon.

2. Reproduction

Most species of mussels and clams have males and females, though a few species are hermaphroditic. The reproductive cycle in mussels and clams is quite interesting (Fig. 27-4). In summer, the eggs are released into the cavity of the gills where fertilization takes place. Each zygote develops into a larva called a **glochidium.** The larvae stay in the gill through the winter and are released into the water the following spring. If they come in contact with a fish, a contact stimulus causes them to close their valves and become attached to the gills and the fins of the fish host. The tissue of the fish reacts by growing around the glochidia. After several weeks, the parasitic larval form is released and begins a free-living existence.

Examine slides or preserved specimens of glochidia. Note the toothlike appendages on the valves, which function to attach the larva to the fish. What is the advantage to these larvae of being attached to the gills or fins instead of the body wall?

____

____

____

____

____

B. CLASS POLYPLACOPHORA (CHITONS)

These mollusks are primitive marine organisms that have a characteristic shell composed of eight separate but overlapping transverse plates (Fig. 27-5A). A large, broad, flat muscular foot occupies

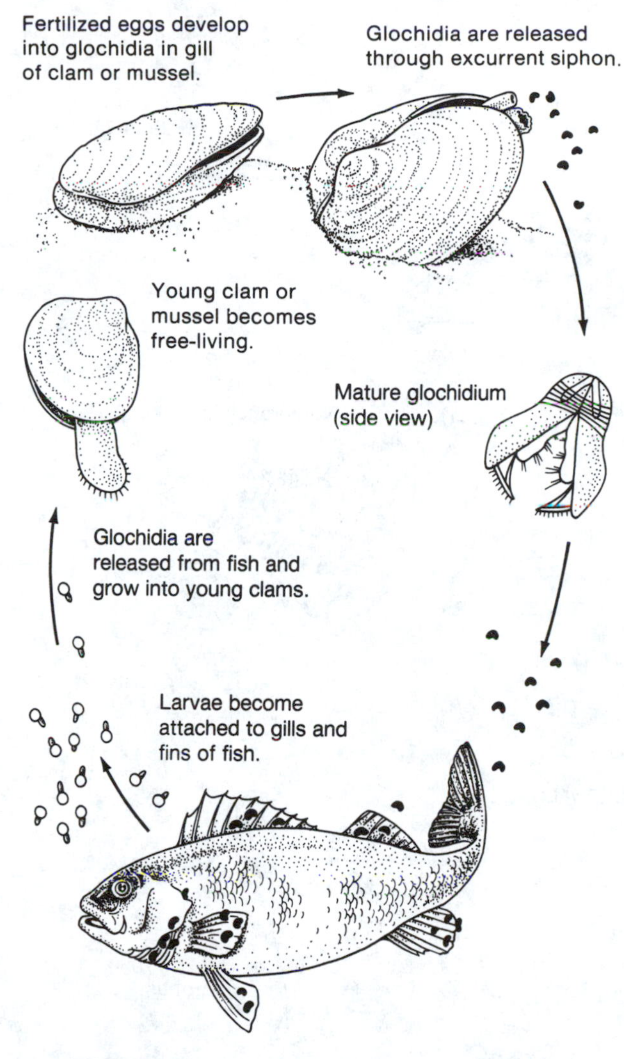

FIGURE 27-4 Life cycle of a clam or mussel

the greater part of the ventral surface. Chitons are usually found on rocky seashores or in water less than 25 fathoms deep. They attach tightly to rocks by suction produced by the foot. When pulled off, they tend to roll up, much like an armadillo or a pill bug, with their soft parts covered by the hard shell. Examine the chitons on demonstration.

C. CLASS SCAPHOPODA (TOOTH, OR TUSK, SHELLS)

The visceral mass of this marine mollusk is enclosed in a long, tubular, toothlike shell that is open at both ends (Fig. 27-5B). The foot is typically cone-shaped and is used for burrowing into mud or sand head first, leaving the narrow end of the shell exposed above the surface. Examine the tooth shell on demonstration.

D. CLASS GASTROPODA (SNAILS, SLUGS, AND NUDIBRANCHS)

This is by far the largest and most diverse class of mollusks (Fig. 27-5C). Most gastropods are marine, but some live in fresh water and others have become adapted to land. In most species of gastropods, adults retain the coiled shell; in slugs, the shell has been lost.

Snails have an unusual mode of locomotion. A **slime gland,** located in the forward part of the foot, secretes a thin film of mucus over which the snail moves by means of wavelike contractions of the muscular foot.

Slugs, because they lack a protective shell, must remain in moist areas to avoid desiccation. Consequently, they are not very active in the daytime, feeding at night instead.

Sea slugs (nudibranchs) live among the seaweeds. Some species so resemble the plants that they are difficult to see. Others have warning coloration and are protected by stinging cells.

Examine demonstrations of various gastropods.

E. CLASS CEPHALOPODA (SQUIDS AND OCTOPUSES)

The cephalopods are considered the most advanced and highly developed class of mollusks (Fig. 27-5D). In contrast to other mollusks, the cephalopods are active, free-swimming animals.

All cephalopods are marine organisms that are characterized by the modification of a foot to form tentacles and a head with prominent, highly developed eyes. The eye is remarkably similar to the vertebrate eye in that it has an eyelid, iris, pupil, lens, cornea, and retina. In some species the shell is external, and in others it is internal. Some species reach sizes of several meters in length. The giant squid of the North Atlantic ocean is the largest living invertebrate.

Examine a squid. What morphological characteristics can you see on this organism that are adaptations to a predatory existence?

____

____

____

____

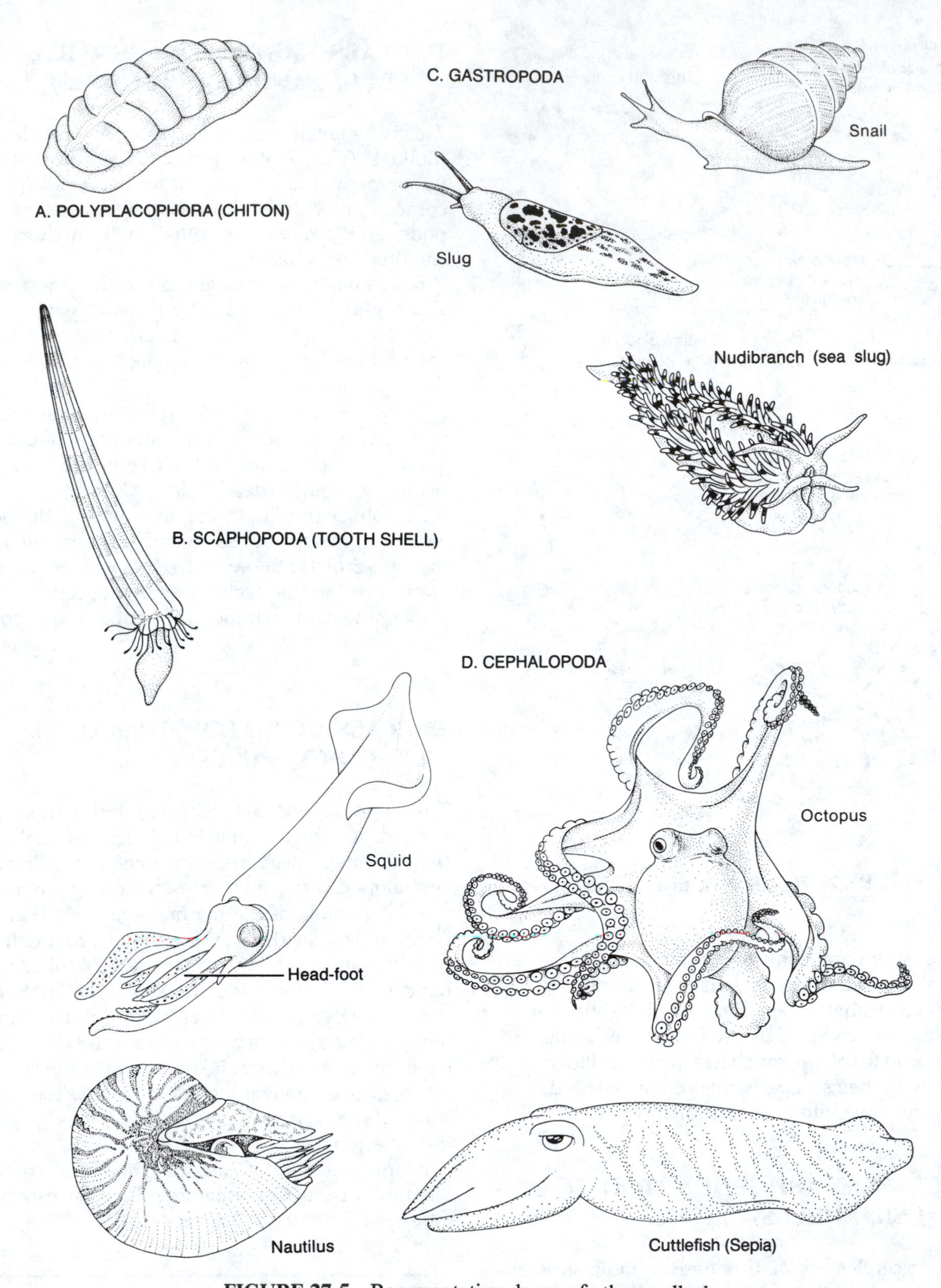

FIGURE 27-5 Representative classes of other mollusks

Because the squid relies on its ability to swim rapidly for protection, it has no need of a cumbersome external shell. Consequently, the shell is a vestigial structure, consisting of a horny plate buried in the visceral mass. Interestingly, the shell, called a **cuttlebone,** is sold in pet stores as a source of calcium for pet birds. It is attached to a bird's cage to be used by the bird to "sharpen" its beak.

Nautilus is the only present-day cephalopod that has a well-developed shell (Fig. 27-5D). Its flat, coiled shell consists of many chambers separated by transverse septa. The animal occupies only the outermost chamber. By secreting air into its inner chambers, the *Nautilus* is able to float.

In the cuttlefish (*Sepia*), the shell is greatly reduced to an internal stiffening support (cuttlebone) overgrown by the mantle.

Examine an octopus. How are the squid and octopus similar?

____

____

____

____

How are they different?

____

____

____

____

REFERENCES

Barnes, R. D. 1987. *Invertebrate Zoology.* 5th ed. Saunders.

Buchsbaum, R., et al. 1987. *Animals Without Backbones.* 3d ed. University of Chicago Press.

Pearse, V., et al. 1987. *Living Invertebrates.* Blackwell.

Purchon, R. D. 1977. *The Biology of the Mollusca.* 2d ed. Pergamon.

Villee, C. A., W. F. Walker, and R. D. Barnes. 1984. *General Zoology.* 6th ed. Saunders.

Exercise

28

Kingdom Animalia: Phylum Annelida

The organisms in the phylum Annelida (Latin *anellus*, "ring") are referred to as **segmented worms** to distinguish them from the nonsegmented flatworms and roundworms. This phylum includes the common earthworm, leeches, and a large number of freshwater and marine species that are unfamiliar to most people.

The distinguishing characteristic of this phylum is **metamerism,** the division of the body into similar segments that are arranged linearly along the anterior–posterior axis. Annelids also exhibit **cephalization;** that is, the nervous system is concentrated in a dorsal ganglionic mass, or brain, at the anterior end in structures called **cerebral ganglia.** A ventral nerve cord arises from this brain and passes posteriorly. The excretory and circulatory systems are well developed. The digestive tract is straight and tubular and, in contrast with that of the roundworms, is supplied with its own musculature, which enables it to function independently of the muscular activity of the body wall. Annelids, like roundworms, have a fluid-filled body cavity separating the digestive tract from the body wall. However, in annelids this cavity is a **true coelom** that is formed by splitting of the middle embryonic germ layer, the mesoderm, during development.

In this exercise, you will study the following three classes.

- **Class Polychaeta** Conspicuous segmentation; pairs of lateral projections called **parapodia;** a head that has tentacles; separate sexes in most species; **trocophore** larva. Most are marine organisms (clamworms).
- **Class Oligochaeta** Conspicuous segmentation; lack a well-developed head; few setae per segment; hermaphroditic. Usually found in fresh water or moist soils (earthworms).
- **Class Hirudinea** Dorsoventrally flattened bodies with a large posterior sucker and inconspicuous segmentation; lack appendages and setae; hermaphroditic. Found in fresh and marine waters (leeches).

A. CLASS POLYCHAETA

The polychaete worms are common marine animals whose secretive habits cause them to be overlooked by the casual observer. Many are found in the intertidal zone, and a few have been found at depths of more than 4500 m. Most polychaetes are strikingly beautiful, colored green, red, or pink. Some exhibit a combination of colors, and some are iridescent.

Because of their abundance (thousands have

been found in a square meter), these animals are important in the marine food chain. They are eaten by flatworms, starfish, fish, and even other marine annelids.

Even though numerous and varied modifications have enabled polychaetes to adapt to a wide variety of habitats, all have retained their fundamental organization. This basic body structure can be observed in *Neanthes virens*, the marine clamworm, which lives in burrows in the sand at tide level (Fig. 28-1A). It remains in its burrow throughout the day, coming out at night to feed.

Examine a specimen of *Neanthes.* Note that the long, slender body is somewhat compressed and

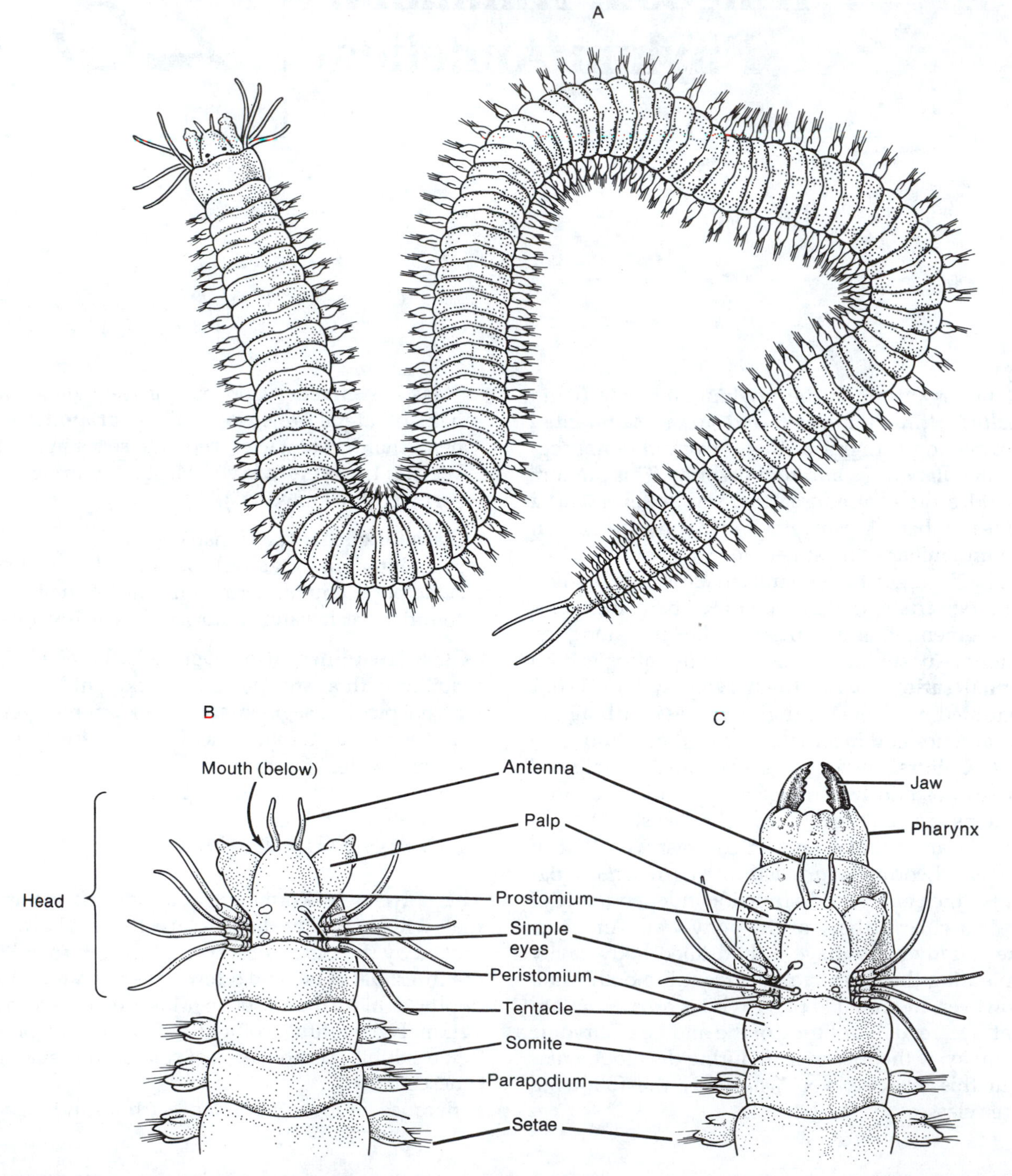

FIGURE 28-1 *Neanthes virens.* (A) External anatomy. (B) Dorsal view of head with pharynx retracted. (C) Dorsal view of head with pharynx everted.

perfectly metameric. All of the body segments are identical and bear a pair of lateral **parapodia.** Each parapodium contains several terminal, bristlelike structures called **setae.** Suggest a function for the setae.

___

___

___

The head is generally well developed and contains the mouth, which is retracted except when the animal is feeding, at which time the pharynx is everted and the jaws, used in capturing small animals, are exposed (Fig. 28-1B, C). Over the mouth region is the highly modified segment called the **prostomium,** which has a pair of **antennae,** two pairs of **simple eyes,** and laterally placed **palps.** The **prostomium** bears four tentacles. What function might these tentacles serve?

___

___

___

B. CLASS OLIGOCHAETA

You will study the common earthworm, *Lumbricus terrestris,* in detail to learn the main characteristics of the oligochaetes. The bodies of most earthworms are divided externally into segments that are separated internally by membranous partitions. Except for the head and tail regions, all segments are essentially alike. The earthworm hunts food at night and thus has been called a "night crawler." It extends its body from the surface opening of a small tunnel, which it makes by "eating" its way through the soil. The rear end of the worm's body remains near the opening while the head end forages for nearby decaying leaves and animal debris.

It has been estimated that an acre of good soil contains more than 50,000 earthworms. By their continual foraging and tunneling, earthworms turn over from 18 to 20 tons of soil per acre and bring more than 25 mm of rich soil to the surface every 4–5 years.

1. External Anatomy

Obtain a specimen of *Lumbricus* and, using Fig. 28-2 as a guide, study its external anatomy. A hand lens or stereoscopic microscope is helpful in identifying the smaller features.

At the anterior end, locate the **prostomium,** a small fleshy projection over the **mouth.** It is not considered a segment of the worm. At the posterior end is the **anus,** the opening from the digestive tract through which solid wastes are expelled. About one-third of the way back from the mouth region is a thick cylindrical collar, the **clitellum,** which functions in reproduction.

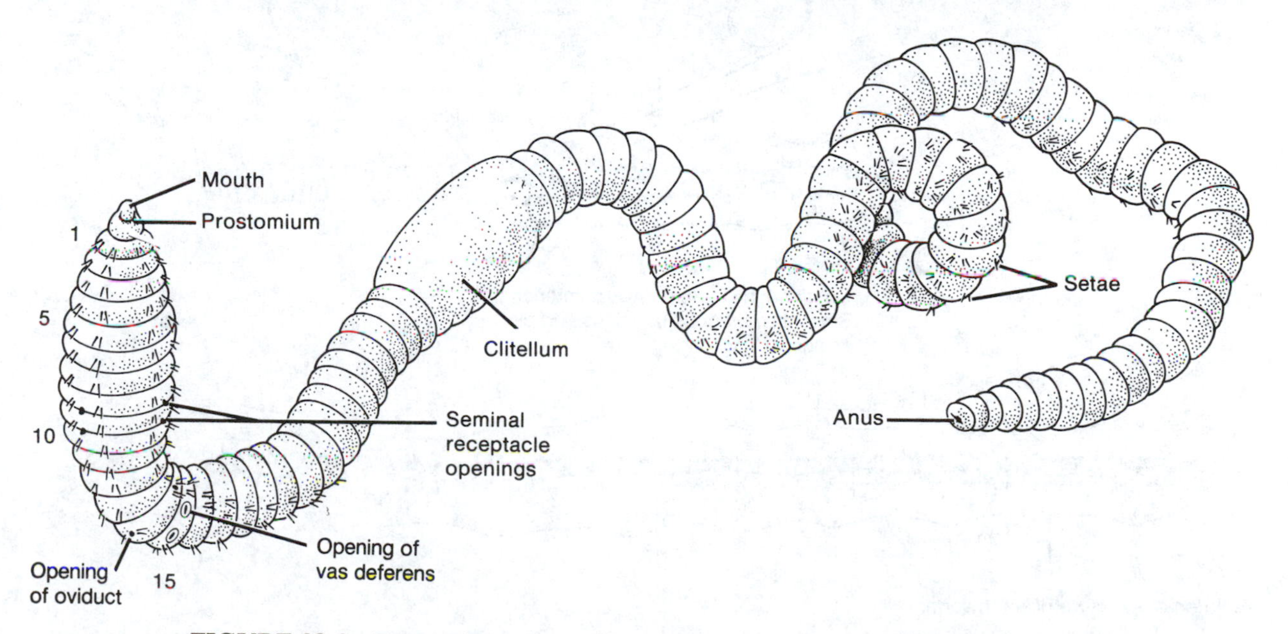

FIGURE 28-2 External anatomy of the common earthworm, *Lumbricus terrestris*

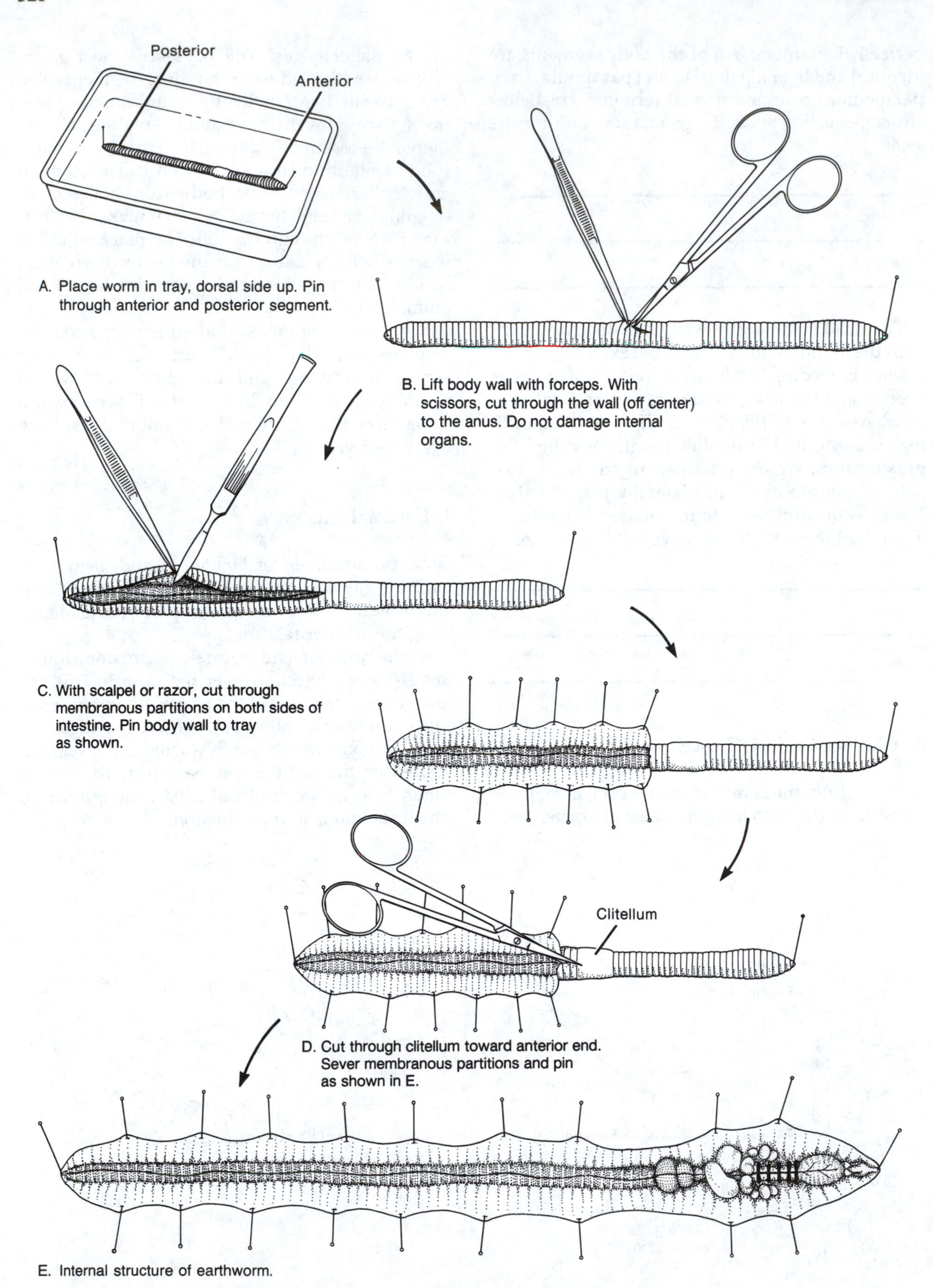

FIGURE 28-3 Procedure for dissecting earthworm

Position the worm so that the ventral, lighter colored side is uppermost. With your finger, lightly stroke the ventral surface in an anterior direction. The bristles you feel are the setae and are used by the worm in movement. Using a stereoscopic microscope, determine how many pairs of setae there are in each segment of the worm.

___

___

Which segments do not have setae?

___

Every segment (except the first three and the last one) contains **nephridiopores** (Fig. 28-4). These small openings connect with the **nephridia,** which are the primitive kidneys of the earthworm. Liquid wastes, which collect in the body cavity, are excreted through the nephridiopores. A pair of rather large openings of the **vas deferens** (male) are located on each side of segment 15. Sperm are released from the worm through these openings. Eggs produced in the ovaries are released through openings of the **oviducts** located in segment 14.

2. Internal Anatomy

Place the earthworm on the dissecting tray, dorsal side up, and pin in position. To expose the internal organs, dissect the worm as outlined in Fig. 28-3.

a. Digestive System

Locate the mouth at the anterior end just below the overlapping prostomium (Fig. 28-4). The mouth leads to a slightly expanded and muscular **pharynx.** Food taken in by the animal is passed on by muscular contractions in the pharynx through the **esophagus,** which is covered dorsally by three pairs of whitish **seminal vesicles,** to the **crop** where it is temporarily stored.

The crop opens into a thick-walled, highly muscular **gizzard** where, with the aid of small soil particles taken in during feeding, food is ground. The food then passes into the **intestine,** where it is digested and the nutrients absorbed. The dorsal wall of the intestine has an infolded **typhlosole,** which increases its surface area for food absorption. Solid waste products of digestion are passed to the exterior through the anus.

b. Circulatory System

The circulatory system is a "closed" system in which the blood circulates in a series of blood ves-

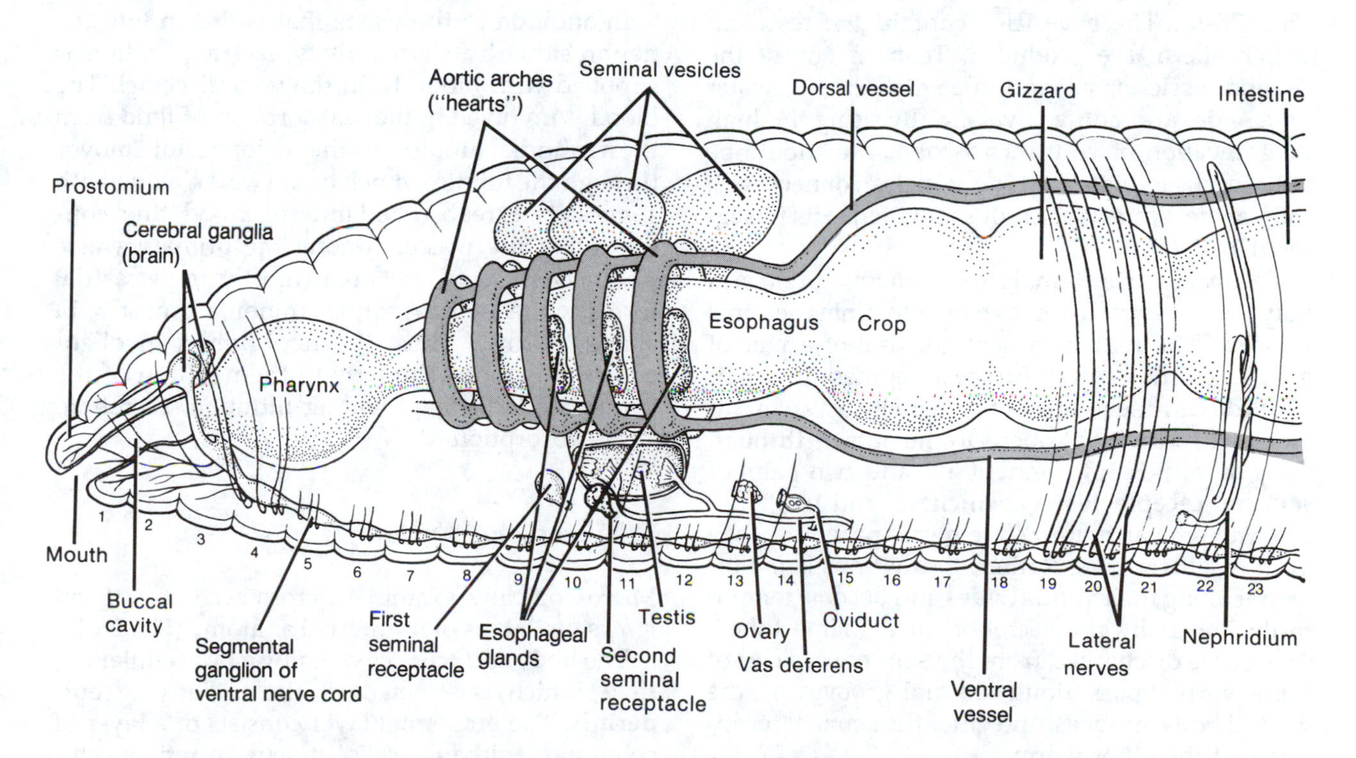

FIGURE 28-4 Internal structure of *Lumbricus*: lateral view

sels. The blood is red because it contains hemoglobin, the same pigment that gives the red color to human blood. The hemoglobin, however, is not contained in cells but is dissolved in the plasma.

Locate the following major vessels of the circulatory system. The **dorsal vessel** lies on top of the digestive tract and a **ventral vessel** lies below it. These two vessels are connected by five pairs of vessels passing around the esophagus. These are larger than the other blood vessels and constitute the **aortic arches** or the "hearts." Although the hearts contain valves and contractile muscle tissue, and serve as pumps to circulate blood through the animal, they are thought to play a minor role in circulation. The dorsal and ventral vessels are the major circulatory pumps. Associated with the hearts and surrounding the esophagus are two pairs of **esophageal** (calciferous) **glands,** which are excretory organs that function in controlling the amount of calcium in the blood.

c. Reproductive System

To obtain a clear view of the reproductive organs, cut through the intestine near the clitellum. Carefully lift the intestine and tease it free as far forward as the pharynx (Fig. 28-5). Then cut it out.

Earthworms are **hermaphroditic,** having complete sets of male and female reproductive organs. The male system consists of a pair of trilobed **seminal vesicles** located between segments 9 and 13 (Fig. 28-5). These vesicles contain the **testes** in which sperm are produced. Tear off one of the seminal vesicles, smear it in a small drop of water on a slide, and add a coverslip. By using the high magnification of your microscope, you should be able to observe sperm. The vesicles connect with the **vas deferens** (sperm duct), which exits in segment 15.

To observe the female reproductive organs, it may be necessary to remove the remaining seminal vesicles. The female system consists of a pair of **ovaries** on the ventral surface in segment 13, a pair of **oviducts** in segment 14 that lead by way of a ciliated **egg funnel** that opens to the outside through the genital pore of segment 14, and two pairs of **seminal receptacles** in segments 9 and 10.

Although earthworms are hermaphroditic, they do not self-fertilize. Rather, two worms come together along their ventral sides and become temporarily joined by the secretion of a "slime tube." Sperm are discharged from the seminal vesicles of both worms, pass along seminal grooves on the ventral body surfaces, and enter the seminal receptacles of the other worm.

When eggs leave the ovaries, glands of the clitellum secrete a tube of mucus that slides over the anterior segments and picks up eggs released from the oviducts in segment 14 and sperm from segments 9 and 10. The tube of mucus finally slips over the anterior end of the worm to form the egg cocoon, from which the young eventually hatch.

d. Nervous System

The major component of the earthworm nervous system is the **ventral nerve cord,** which runs the length of the worm on the inner ventral surface. At its anterior end, the cord divides and passes around the front part of the pharynx where it enlarges to form two swellings, the **cerebral ganglion** or brain. Along the length of the cord, lateral nerves extend from segmental ganglia to the muscles of the body wall.

e. Excretory System

In *Lumbricus*, every body segment except the first three or four and the last contains two excretory organs called **nephridia** (Fig. 28-6). Each nephridium opens into the coelomic cavity, just anterior to the segment on which it is located, through a ciliated funnel (**nephrostome**). The nephrostome continues as a finely coiled **tubule**, surrounded by capillaries, that leads to a **bladder,** which discharges to the outside through a **nephridiopore** located near the ventral surface of the body wall.

In addition to the waste that is drawn into the nephrostome by ciliary activity, each nephridium is supplied with blood from the ventral vessel. The blood is involved in the reabsorption of fluid from the nephridial tubules. As the coelomic fluid moves through the tubules, much of the water, along with some salts, is reabsorbed into the blood, thus concentrating the waste. Among the primary waste products secreted are urea (mostly in terrestrial forms of Annelida) and ammonia (mostly in aquatic forms). Thus, the nephridium in oligochaetes functions similarly to the nephron of the vertebrate kidney, providing filtration, secretion, and reabsorption.

3. Microscopic Anatomy

Microscopically examine a cross section of *Lumbricus* for details of its internal anatomy (Fig. 28-7).

The body surface is covered by a noncellular **cuticle,** which is secreted by the underlying **epidermis.** The epidermal layer consists of a layer of columnar epithelial cells, mucus-secreting cells,

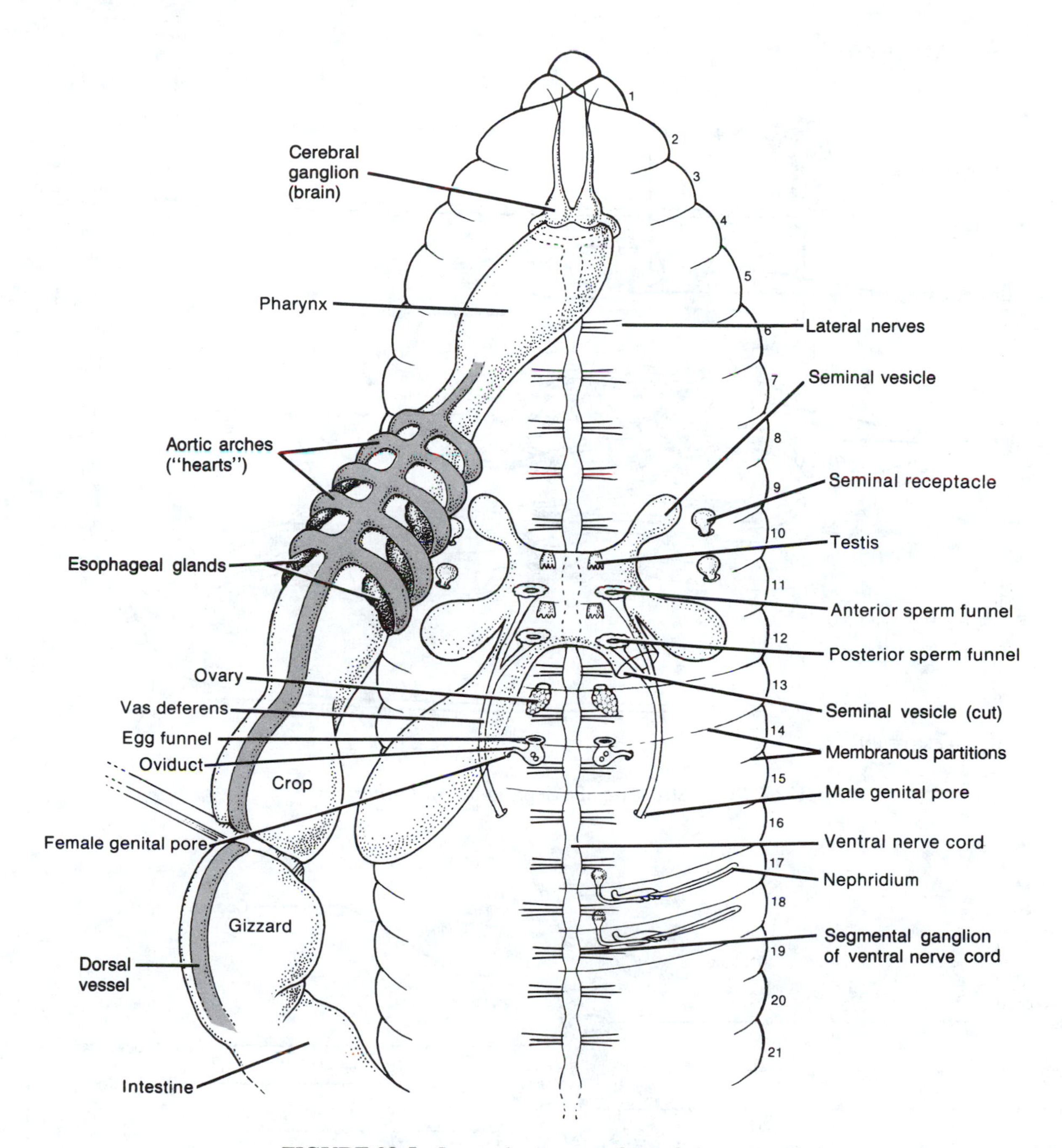

FIGURE 28-5 Internal structure of *Lumbricus*: dorsal view

and sensory cells, and photoreceptor cells that help orient the animal in its environment.

Under the epidermis is a layer of **circular muscles.** What effect would contraction of this muscle layer have on elongation or contraction (and thus locomotion) of the worm?

___

___

___

Under the circular muscle layer is a thick band of **longitudinal muscles** that are arranged like barbs in a feather. What effect would contraction of this muscle layer have on locomotion?

___

___

The innermost layer of the body wall is covered with thin, flattened cells that form the **peritoneum.** The space between the body wall and the digestive

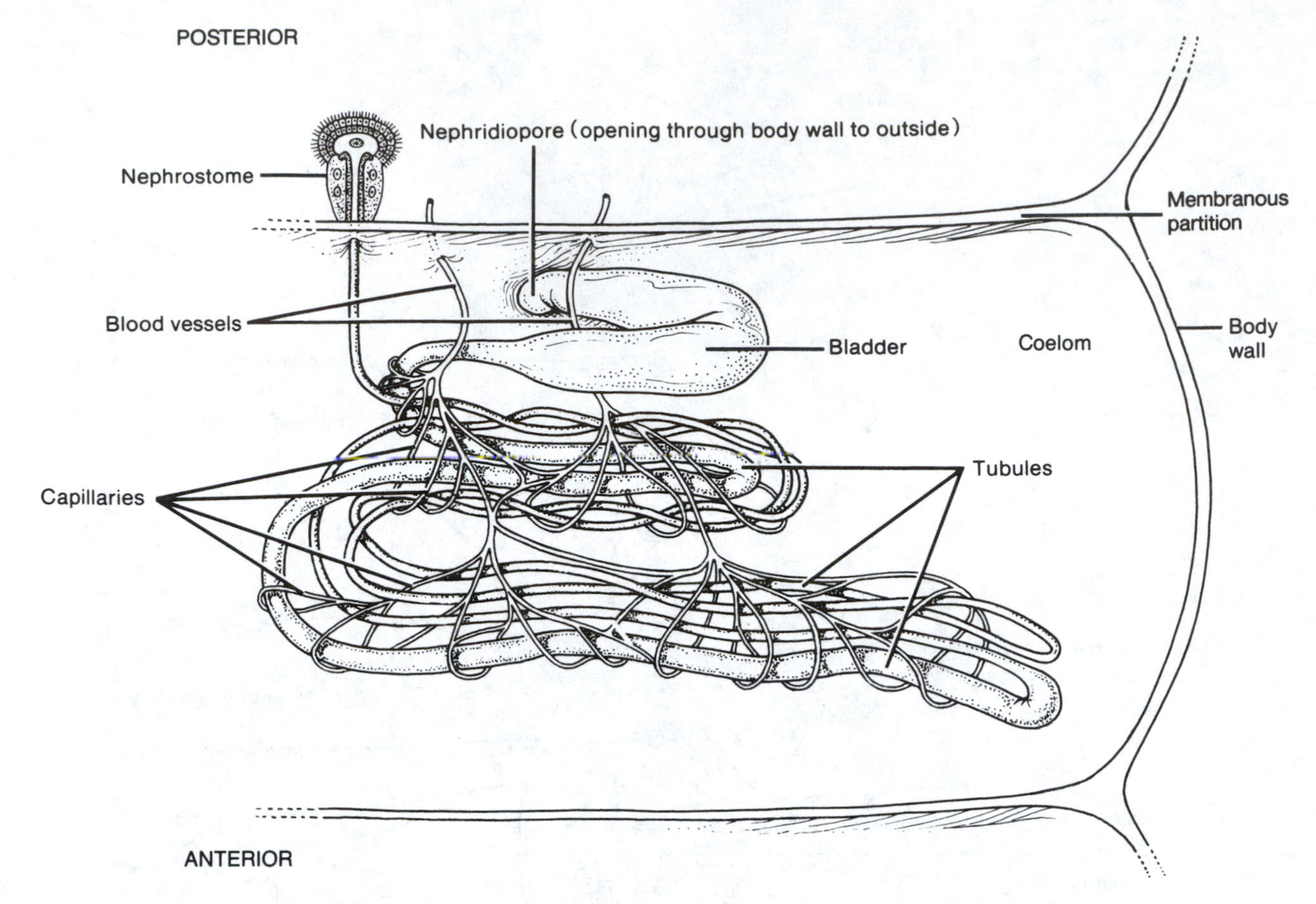

FIGURE 28-6 Structure of a nephridium in *Lumbricus*

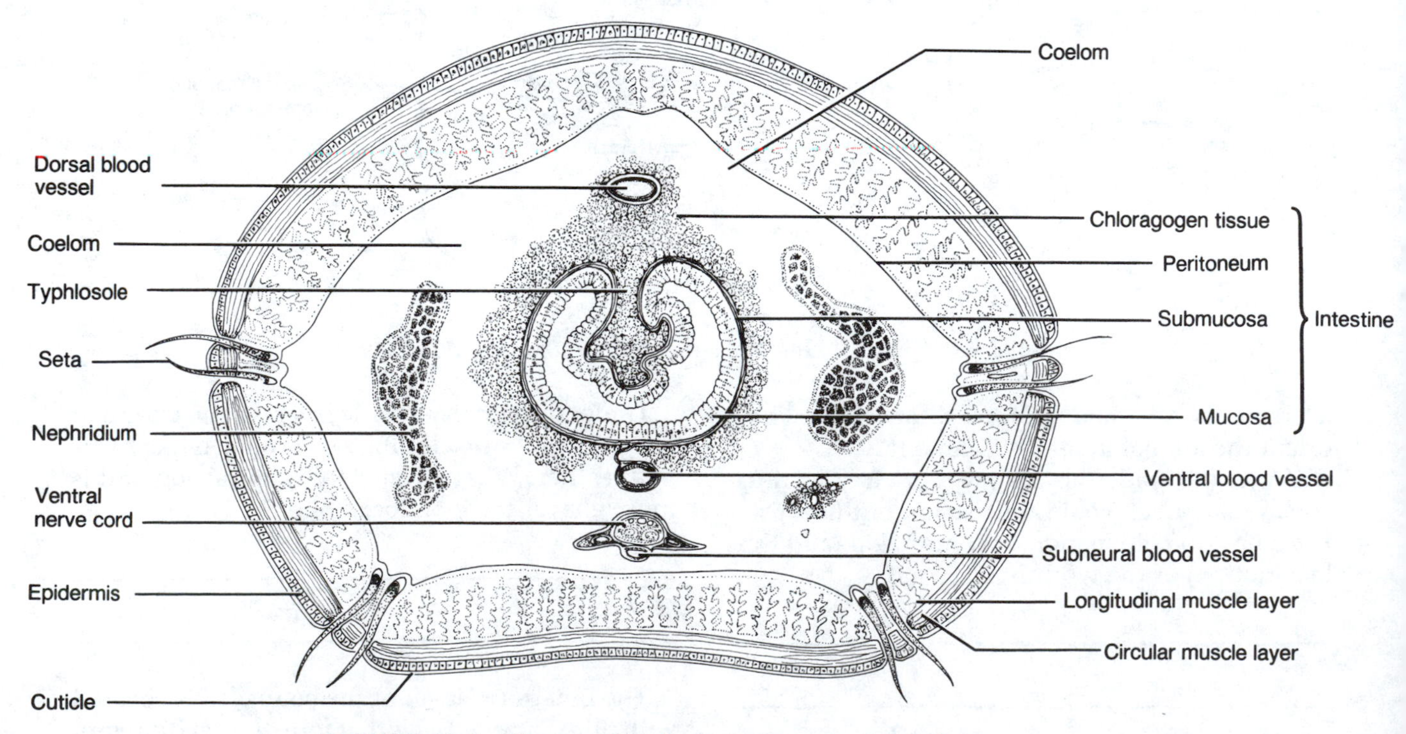

FIGURE 28-7 Cross section through segment of *Lumbricus*

cavity is the **coelom.** The coelomic fluid contains protein, metabolic wastes, and various types of phagocytic cells. Locate the dorsal and ventral blood vessels and the ventral nerve cord in the coelomic cavity. You may be able to observe nephridia, located in the coelomic cavity on either side of the intestine, in your cross section.

The wall of the intestine is composed of three layers. The innermost layer is the **mucosa,** a layer of narrow, ciliated, columnar cells. The intermediate layer is the **submucosa,** which contains circular and longitudinal muscle fibers and numerous small blood vessels. What is the function of these muscle layers and blood vessels in the intestine?

____

____

____

The outermost layer, surrounding the intestine and the dorsal blood vessel, is a layer of **chloragogen tissue.** This tissue plays an important role in intermediary metabolism similar to the function of the liver in vertebrates. Chloragogen tissue is the primary site for fat and glycogen synthesis and storage. It is also involved in the deamination of proteins, the formation of ammonia, and the synthesis of urea.

C. CLASS HIRUDINEA

Leeches are highly specialized annelids; most live in freshwater habitats, though a few are marine organisms, and some have become adapted to a terrestrial existence in tropical climates.

Contrary to common belief, leeches are not parasites. Rather, they can be considered predators that feed on the blood of various invertebrates and vertebrates.

Examine various leeches on display. Note that they are dorsoventrally flattened and tapered at both ends (Fig. 28-8). Most species are 20–60 mm long. The largest is said to be as long as a half meter when it is fully extended while crawling. Note the **anterior** and **posterior suckers,** which are used for attachment. The anterior sucker contains the mouth, which has jaws covered with chitinous teeth for biting. Blood that is sucked up is stored in an enormous crop, enabling the animal to ingest

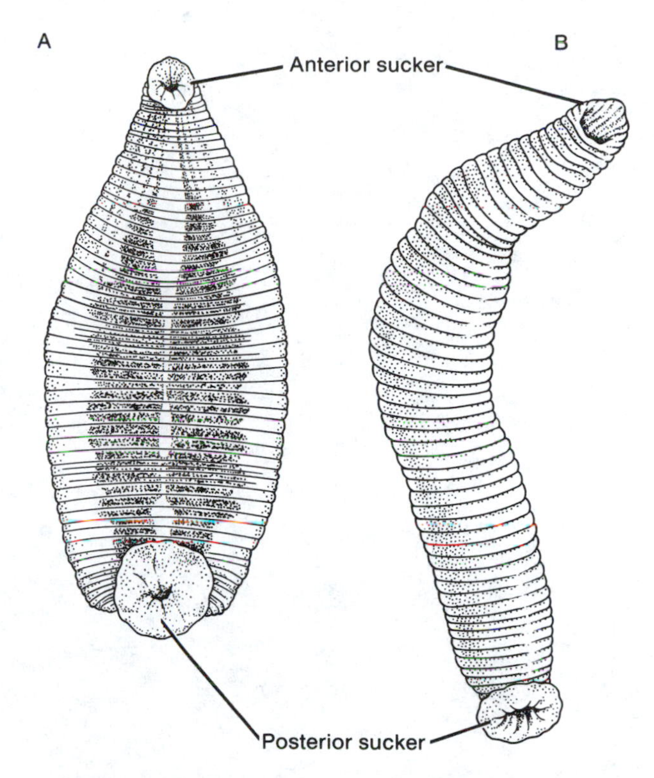

FIGURE 28-8 Leeches. (A) *Placobdella,* commonly found on turtles. (B) *Hirudo medicinalis,* a leech used for bloodletting.

three times its weight in blood. This permits the leech to go as long as 9 months between feedings.

In most leeches, respiration takes place directly through the moist body surface, though some leeches have gills for this purpose. Waste products and removed from the coelomic fluid and blood by nephridia.

Leeches are hermaphroditic, but reproduction is by cross-fertilization. Copulation and the formation of a **cocoon** are adaptations to terrestrial life that are similar to those of earthworms.

REFERENCES

Barnes, R. D. 1987. *Invertebrate Zoology.* 5th ed. Saunders.

Buchsbaum, R., et al. 1987. *Animals Without Backbones.* 3d ed. rev. University of Chicago Press.

Edwards, C. A., and J. R. Lofty. 1977. *Biology of Earthworms.* 2d ed. Chapman and Hall.

Pearse, V., et al. 1987. *Living Invertebrates.* Blackwell.

Villee, C. A., W. F. Walker, and R. D. Barnes. 1984. *General Zoology.* 6th ed. Saunders.

Exercise

29

Kingdom Animalia: Phyla Onychophora and Arthropoda

The phylum Onychophora (Greek *onychus*, "claw"; *phorus*, "bearing") includes a group of terrestrial organisms that have a distinct head, a somewhat cylindrical, soft, externally unsegmented body, a pair of nephridia in each internal segment, and a number of short, unjointed legs. They are found primarily in moist land habitats, under stones, leaves, and logs in forests. In this exercise, you will study *Peripatus*, a representative genus of this phylum.

The phylum Arthropoda (Greek *arthros*, "joint"; *podos*, "foot") is the largest of all animal phyla. Of the million or so known species of animals, more than three-fourths are arthropods. The arthropods are considered to have attained the greatest "biological success": they comprise the largest number, not only of species but also of organisms; they occupy the greatest variety of habitats, consume the largest amounts and kinds of food, and are capable of defending themselves against their enemies.

Members of the phylum Arthropoda are characterized by their rigid, chitinous body covering, called an **exoskeleton.** The body is segmented externally and internally to varying degrees, depending on the species, and the paired appendages are jointed and modified for a large variety of functions. Arthropods have an open circulatory system with a dorsal "heart."

The classification of the Arthropoda is complex because of its large number of species and the tremendous diversity in form among its members. It is divided into four subphyla, which are further subdivided into classes.

- **Subphylum Trilobita** Primitive marine arthropods, now extinct, characterized by a distinct head, a segmented body, and appendages covered by a hard, segmented shell.
- **Subphylum Chelicerata** Chiefly terrestrial arthropods with the first two body parts (head and thorax) combined in a single cephalothorax; lack antennae and mandibles but have six pairs of jointed appendages.

  Class Merostomata (horseshoe crabs): Most species are extinct. Living merostomes are aquatic arthropods having lateral compound eyes, a cephalothorax, and an abdomen with six pairs of appendages.

  Class Pycnogonida (sea spiders): Small marine organisms having a short, slender body, a mouth on a long proboscis, and four pairs of long legs.

  Class Arachnida (spiders, scorpions, mites, ticks): Terrestrial arthropods having simple eyes, no gills, and four pairs of walking legs.

- **Subphylum Crustacea** The most common aquatic arthropods. They have two pairs of antennae, one pair of jaws, and two pairs of maxillae. Appendages are **biramous** (double branched).
- **Subphylum Uniramia** The largest group of arthropods. They have three distinct body parts and three or more pairs of legs, one or two pairs of antennae, and one pair of mandibles (jaws). Appendages are **uniramous** (unbranched).

  Class Chilopoda (centipedes): Terrestrial arthropods having long, flattened bodies with many segments, each with one pair of appendages.

  Class Diplopoda (millipedes): Terrestrial arthropods having a long, usually cylindrical body with many segments, each with two pairs of appendages.

  Class Insecta: Mainly terrestrial arthropods. They have a distinct head, thorax, and abdomen and one pair of antennae, three pairs of legs, and two pairs of wings.

A. PHYLUM ONYCHOPHORA

The members of this phylum possess features that are common to both the annelids and the arthropods and thus come closer than any others to being the "missing link" between the two phyla.

Examine a preserved specimen or plastic mount of *Peripatus*, which resembles the annelids in the structure of the nephridia, the ciliated reproductive ducts, and the simple gut. It resembles the arthropods in having a **hemocoel** (a body cavity that functions as part of an open circulatory system) and a dorsal "heart" and in the general structure of the reproductive organs. Note the absence of external segmentation, although there is a pair of legs for each internal segment of the body (Fig. 29-1). The legs terminate in claws that resemble those of arthropods but differ in not being jointed. The head bears three pairs of appendages: two short **antennae,** a mouth flanked by a pair of clawlike **mandibles,** and two blunt **oral papillae** that are used in defense. When disturbed, onychophorans use a special pair of glands that secrete an adhesive. These glands open into the oral papillae, and the secretion is ejected as two streams for a distance

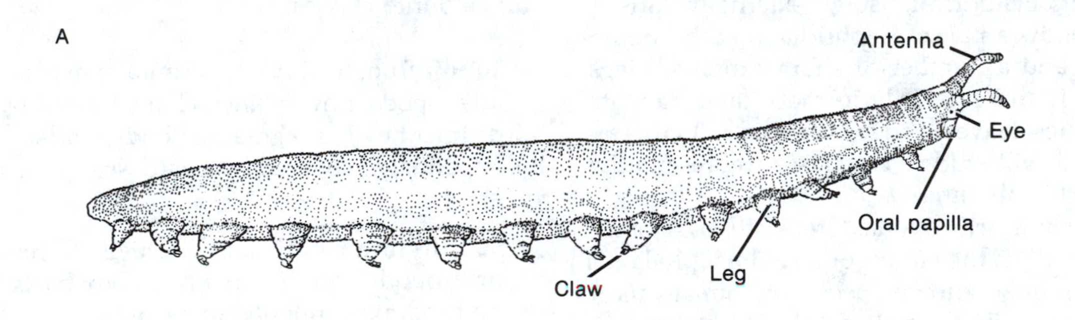

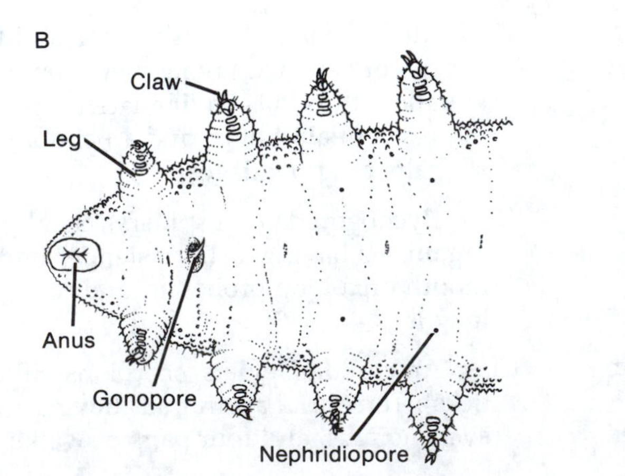

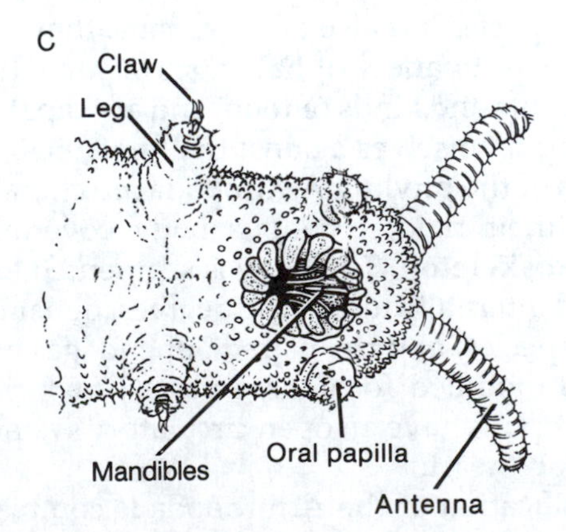

FIGURE 29-1 *Peripatus.* (A) Lateral view of whole organism. (B) Posterior ventral surface. (C) Anterior ventral surface.

up to 50 cm. It hardens almost immediately on contact and entangles the intruder in a mesh of sticky threads. Like the annelids, *Peripatus* possesses an internal system of excretory **nephridia** with external openings (**nephridiopores**) at the base of each appendage. The **anus** opens at the blunt posterior end and is preceded by a single genital opening, the **gonopore.**

B. PHYLUM ARTHROPODA

1. Subphylum Chelicerata

The body of most chelicerates is divided into two regions, a fused **cephalothorax** and an abdomen. The first pair of appendages on the cephalothorax are pincerlike and are called **chelicerae.** There are usually five other appendages on the cephalothorax, which in some groups are all walking legs. In others, the first pair are modified as feeding appendages, called **pedipalps.**

a. Class Merostomata

This class includes many fossil forms and only four living species of horseshoe crabs. Examine a preserved specimen of *Limulus polyphemus,* a horseshoe crab that is common in the shallow marine waters off the Atlantic coast from Nova Scotia to Yucatan. The dark brown outer covering, or **carapace,** is shaped like a horseshoe, thus giving the animal its common name. Posterior to the carapace is an **abdominal shield** with short, lateral, movable spines, to which is attached a tail spine (**telson**). The telson is used to right the body when it is overturned and to push it forward when the animal is burrowing into the ocean floor.

Turn the specimen over and observe the six pairs of jointed appendages of the cephalothorax (Fig. 29-2). The first pair consists of the chelicerae, which function to some extent in locomotion but are used mainly in capturing and macerating food. The next four pairs are walking legs that end in small pincers, or **chelae.** The last pair are leglike

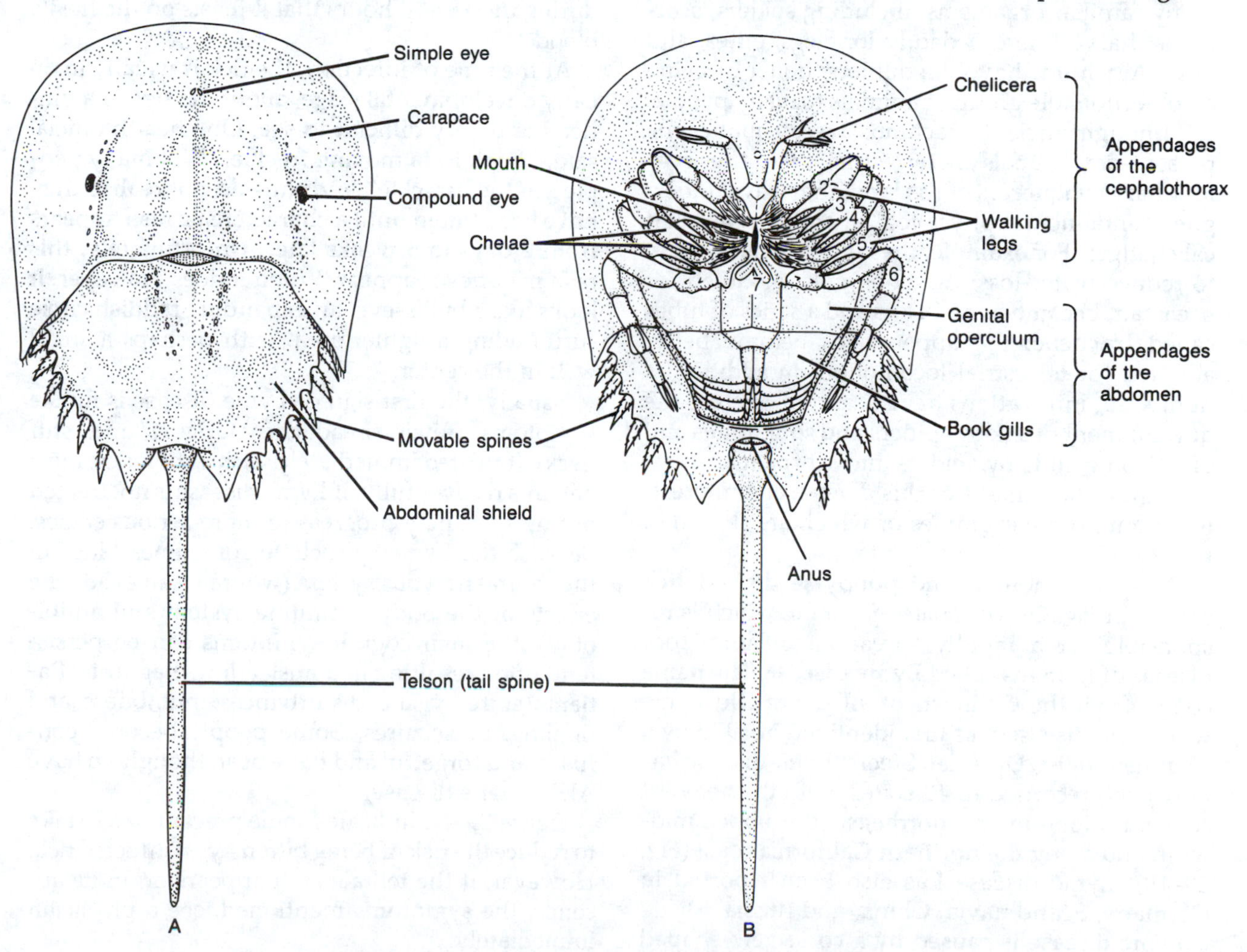

FIGURE 29-2 *Limulus.* (A) Dorsal view. (B) Ventral view.

and end in leaflike tips used to sweep away mud that clings to the body when burrowing.

The abdomen also bears six pairs of appendages. The first pair are fused at the midline to form the genital **operculum.** Two genital pores are located beneath this flap. Posterior to the genital operculum are five pairs of appendages that are modified to function as **gills.** Like the genital operculum, these are flaplike and fused along the midline. On the undersurface of each flap are numerous leaflike folds called **lamellae.** Each gill contains about 150 lamellae. The arrangement of the lamellae has resulted in these appendages being called **book gills.** Gaseous exchange between the blood and the water that is constantly moving past these organs takes place across the lamellae as the book gills wave back and forth.

b. Class Arachnida

The arachnids comprise the largest and, from our point of view, the most important of the chelicerate group of arthropods. Included in this class are many familiar organisms, including spiders, scorpions, harvestmen or daddy longlegs, mites, and ticks. Arachnids have the dubious honor of being an objectionable group of animals to many people.

Although ancient arachnids were aquatic, the present-day animals are terrestrial. Like other evolutionary conquerors of the land, they have undergone fundamental physiological and morphological changes. For example, the cuticle became waxy to reduce water loss; the book gills, modified for use in air, became **book lungs** and a series of tubes called **tracheae;** the appendages became better adapted for terrestrial locomotion. In addition, a number of innovations have occurred, such as the development of silk by spiders and some mites and of poison glands by spiders and scorpions.

Examine specimens of this diverse and interesting group, some examples of which are shown in Fig. 29-3.

Of special note is the poppy-seed-sized tick shown in Fig. 29-4A, *Ixodes dammini,* which is responsible for a rapidly spreading inflammatory disease of humans called **Lyme disease.** The name comes from the Connecticut village of Old Lyme where the disease was first identified in 1975 by a rheumatologist, Dr. Allen Steere. Lyme disease has now been recorded in 43 states, with the heaviest concentrations in the northeast, the upper midwest, and along the northern California coast (Fig. 29-4B). Lyme disease has also been reported in Germany, Scandinavia, China, and Russia.

Lyme disease is caused by a corkscrew-shaped bacterium, called a **spirochete,** and is spread by infected ticks living in or near wooded areas, tall grass, and brush. These ticks lead a parasitic life, feeding on a variety of mammals including mice, deer, and people. If a mouse, for example, is harboring the spirochete, it is picked up by the tick as it feeds on the mouse. The bacterium in turn is transmitted to humans when the tick comes in contact with the skin and feeds.

Lyme disease is not fatal, but it can be debilitating physically and emotionally. Its symptoms include arthritis, severe headaches, loss of sensation, facial palsy, and heart arrhythmias. Unless caught early, the joint and neurological damage is sometimes irreversible.

These variable symptoms make the disease a challenge for physicians to diagnose because it does not follow any predictable pattern. All the patients have in common is a bite from an infected tick, which may crawl around for several hours on the victim's body looking for a suitable place to feed. The tick deposits the spirochete in a capillary during the 18–24 hours that it feasts on the host's blood.

At the time of infection, the tick is usually in an early developmental stage, about the size of a pinhead and very difficult to see. One health official said, "Look for a moving freckle." Not many people see the "freckle" and thus do not realize they have been bitten until a characteristic rash appears from 2 days to 5 weeks later. Unfortunately, this rash may never appear. When it does, however, it looks like a bull's-eye: an expanding reddish circle surrounding a lighter area, with perhaps a small welt in the center.

Usually, the first sign of Lyme disease is flulike symptoms (fever, headache, fatigue, and stiff neck). If treated immediately with antibiotics, most patients recover fully. If Lyme disease is not treated promptly, it may progress to more serious stages. Because the Lyme spirochete sometimes hides in the central nervous system (where it can evade the effects of the body's immune system and antibiotics), the neurological symptoms can be persistent, often resulting in a misleading diagnosis. Patients suffer visual disturbances, numbness and tingling, or seizures. Some people become confused and forgetful and have been thought to have Alzheimer's disease.

Figure 29-4C indicates some precautions to take to reduce the risk of being bitten by an infected tick. However, if the telltale rash appears or, in its absence, the symptoms mentioned, see a physician immediately.

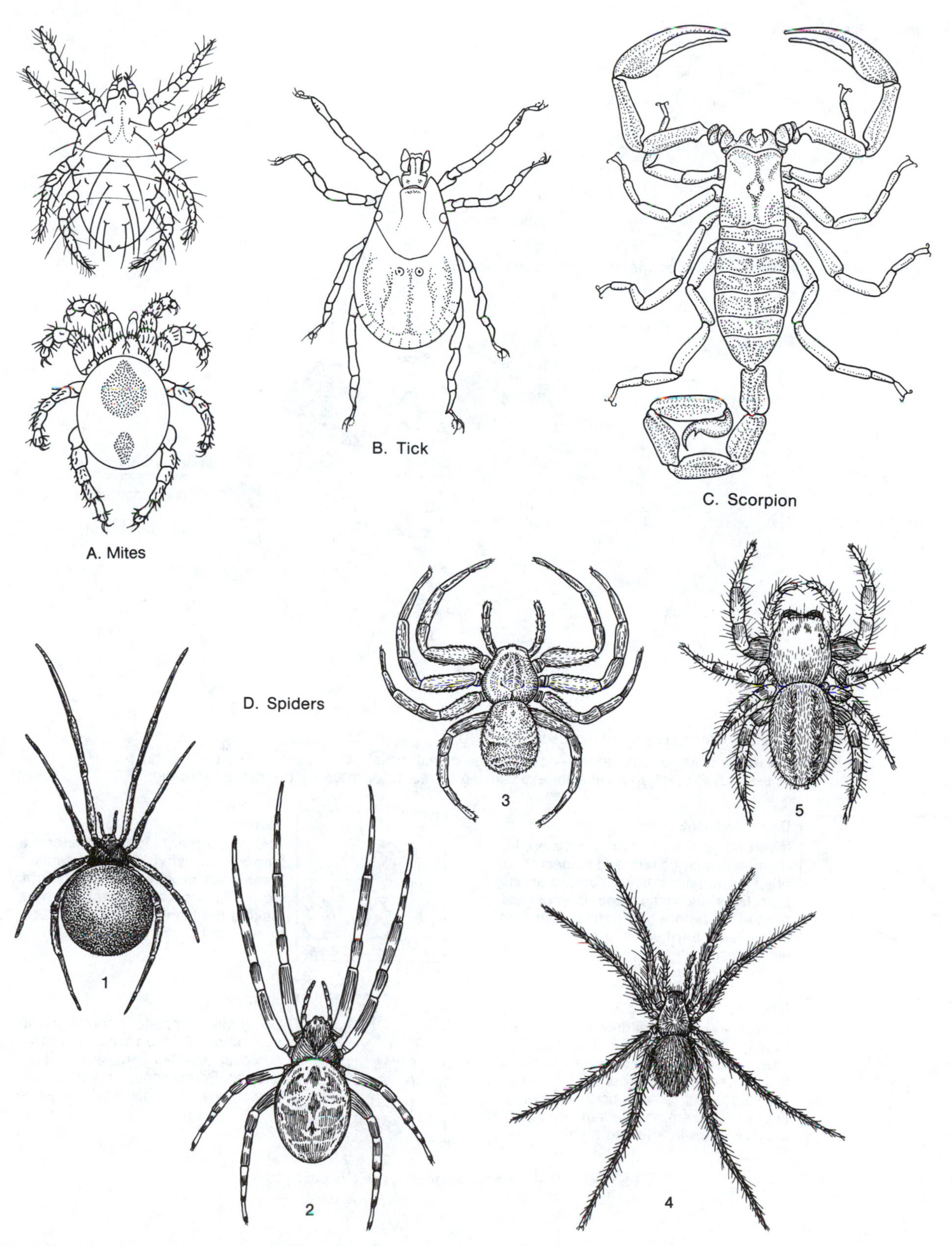

FIGURE 29-3 Representative arachnids. The spiders shown are (1) black widow, (2) common garden, (3) crab, (4) jumping, and (5) wolf.

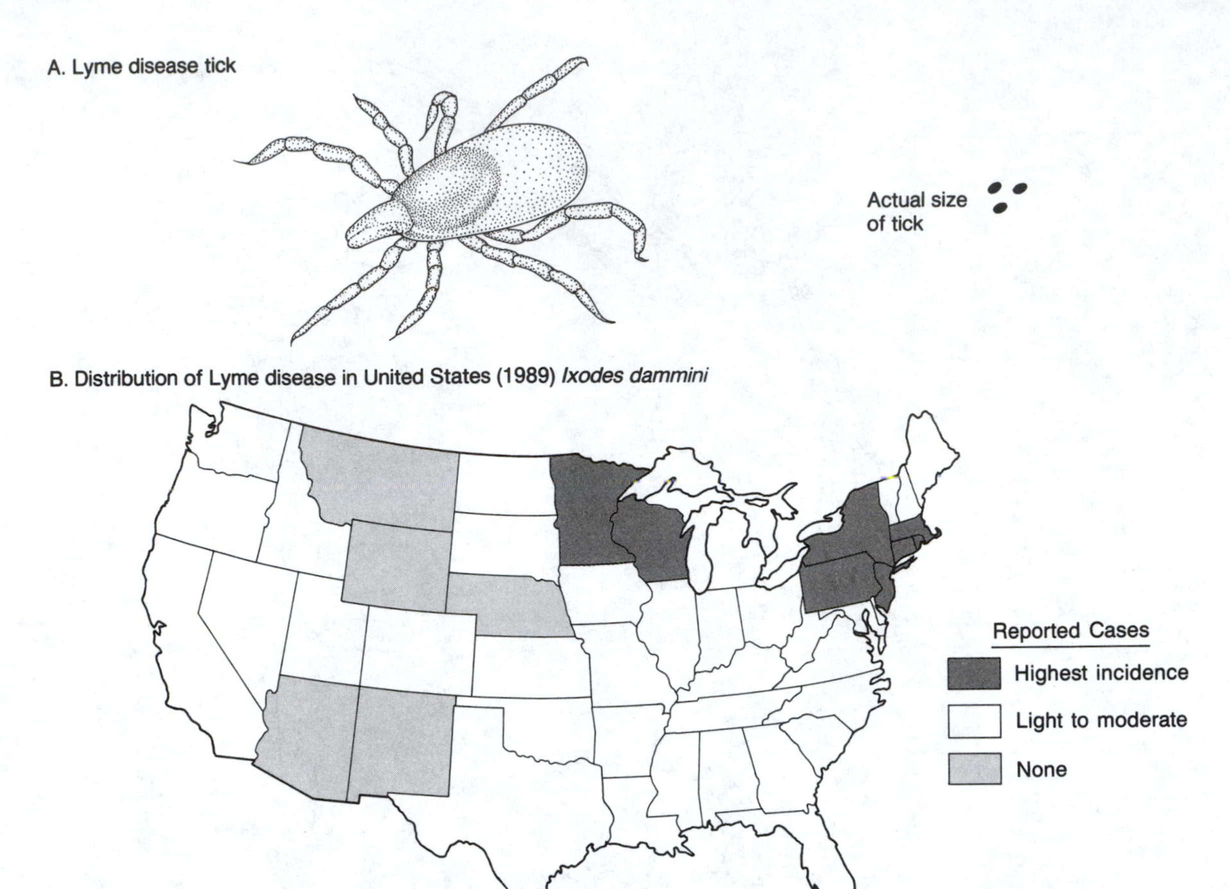

C. How to minimize the risk of infection by Lyme disease

To remove ticks, use tweezers—not matches or nail polish, which may leave part of the tick behind. Protect yourself from ever having to reach for them by doing the following:

Dogs and Cats
If you let your dogs or cats outside, fit them with tick collars and inspect them often for attached ticks. *You* are at risk from ticks that hitch rides on your pets but fall off before biting them. Don't let animals on furniture and don't sleep with them.

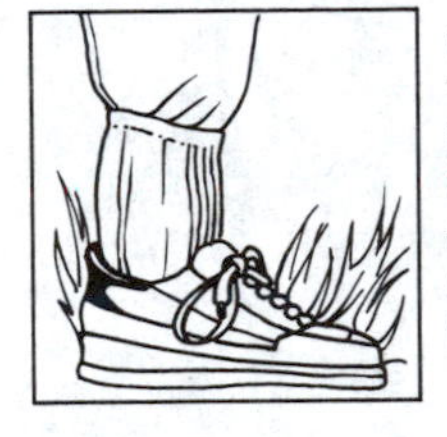

Clothing
Tuck trousers into long socks and a long-sleeved shirt into your pants. Wear white or light colors and tightly woven fabrics. Check often for ticks making the climb from your lower legs to open skin.

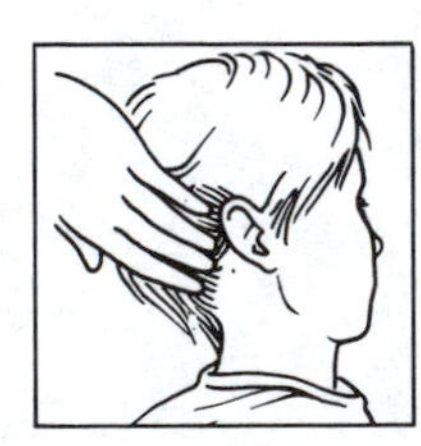

Children
Unless you keep children indoors, you may not be able to keep them tick-free. Teach them to avoid tall grass and low brush, check each other for "moving freckles," and remove ticks only with tweezers. Examine children closely after they've been in infested areas.

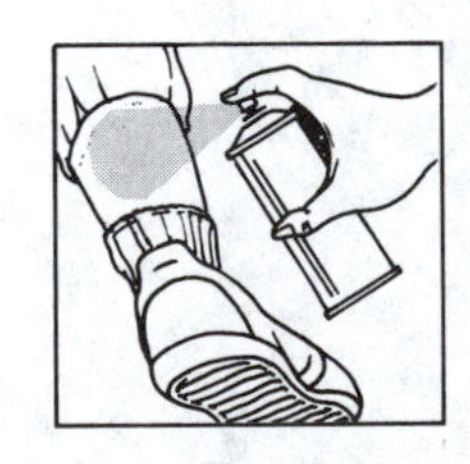

Repellents
Spray insect repellent containing the ingredient DEET on your skin. If you do cover up, also spray permethrin (sold as Permanone) on your clothes, particularly on pants legs and socks.

FIGURE 29-4 Facts about Lyme disease

2. Subphylum Crustacea

The subphylum Crustacea, which has more than 31,000 species, includes some of the most familiar arthropods: crabs, lobsters, shrimp, crayfish, and wood lice. In addition, there are a great number of tiny crustaceans that live in ponds, lakes, and oceans and fulfill a basic position in aquatic food chains. Although a few crustaceans (e.g., wood lice and pill bugs) are terrestrial, most are aquatic. Many are marine, but there are numerous freshwater species.

Crustaceans are extremely diverse in structure, but they are unique among the arthropods in having two pairs of antennae on the head as well as one pair of mandibles and two pairs of maxillae. Although there is much variation in the trunk, a carapace that covers part or all of the body is common.

The appendages of crustaceans are typically **biramous** (they have two jointed branches at their ends). Depending on the group, the appendages have become adapted for many functions. Gills are typically associated with the appendages, though the number, location, and form of the gills vary greatly among the species.

a. Class Crustacea

You will study the common crayfish (*Cambarus* or *Procambarus*) as a representative of this class. It is a cannibalistic scavenger that lives on the muddy bottoms of freshwater lakes, streams, and ponds. It emerges at night to feed on decaying matter, insect larvae, worms, and other crayfish. It can be as long as 15 cm, and its appendages are differentiated for various functions.

If available, observe a living specimen of a freshwater crayfish in a shallow pan of water or in an aquarium. Study its manner of walking and swimming. In which direction does the crayfish normally walk?

____

Hold the crayfish firmly against the bottom of the container and then introduce a drop of india ink at the posterior end of the abdomen. From your observation, describe the direction of water flow and the mechanism used to set up such water currents.

____

____

____

Feed a live insect (cockroach or cricket) to the crayfish and note the coordinated activity of the mouth.

External anatomy. Examine a preserved specimen of the crayfish. The body is divided into an anterior cephalothorax (fused head and thorax) and a posterior abdomen (Fig. 29-5A). The chitinous exoskeleton effectively protects the crayfish from predators. To grow, however, the crayfish periodically **molts** (sheds it exoskeleton). During the period in which a new skeleton is being formed, the crayfish is defenseless and hides to escape its enemies.

The **carapace** is a saddlelike covering of the cephalothorax. A transverse cervical groove separates the fused head from the thoracic region. Laterally, the carapace covers the gills. The **rostrum** is an anterior, pointed extension of the head. The eyes are located on either side of the rostrum.

The abdomen consists of several segments and is terminated by the **telson,** an extension of the last abdominal segment. To escape its enemies, the crayfish spreads the telson and the wide **uropods** like a fan, rapidly drawing the telson forward under its body. This motion causes the crayfish to dart backward into the muddy bottom, which clouds the water because of the flipping movements. Each abdominal segment contains a dorsal plate, a ventral plate, and two lateral plates. Examine the appendages on the lower side of the abdomen (Fig. 29-5B). These appendages are the **swimmerets.** What is the function of these appendages?

____

____

____

____

Although male and female crayfish have an equal number of swimmerets, in the male those adjoining the thorax have been modified: they are elongate and can be brought together to form a troughlike channel, which is used to transfer sperm to the seminal receptacles of the female. In the female, the eggs (which look like clusters of grapes) are attached to the abdominal swimmerets and are aerated by gentle, waving movements of the swimmerets.

Examine the four pairs of walking legs and count the number of segments in each leg. Locate the male genital pores, which open into the base of the

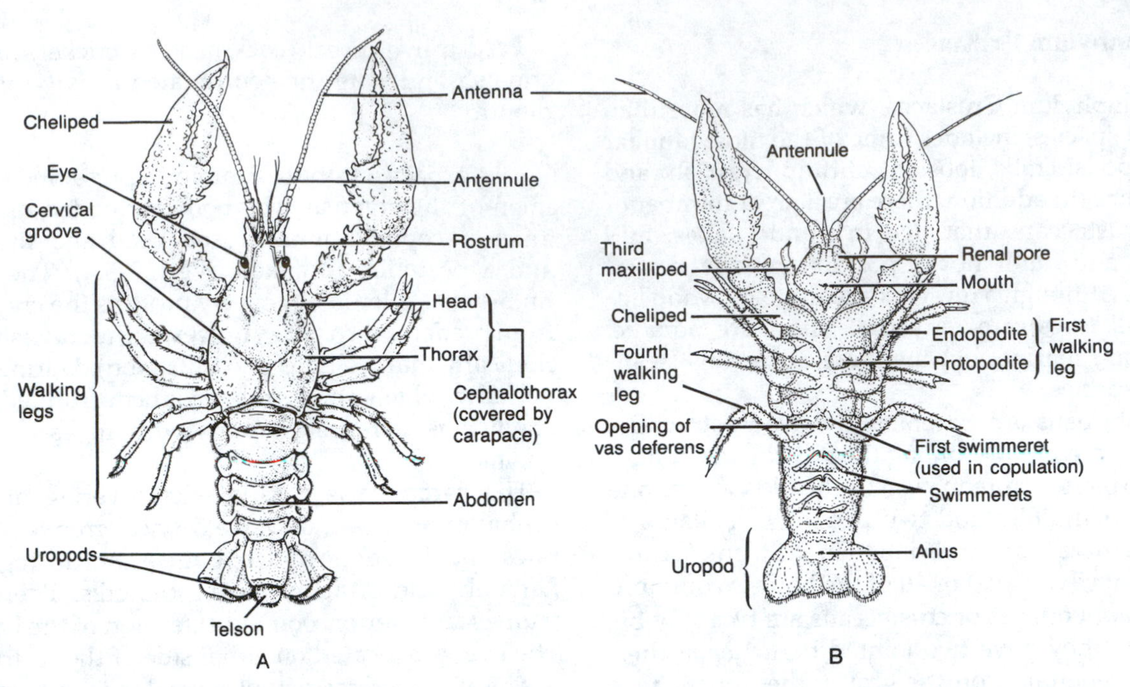

FIGURE 29-5 Crayfish (*Cambarus* or *Procambarus*). (A) Dorsal view. (B) Ventral view.

last pair of legs. The openings of the oviducts in the female are located at the base of each third walking leg. Carefully cut the membrane at the base of each leg and remove the right walking legs, starting with the most posterior one and working forward. Arrange these appendages in order on a piece of paper. How many of these legs have feathery gills attached to them?

____

What is the advantage of the featherlike structures in these gills?

____

____

____

____

The first pair of legs, the **chelipeds,** are much larger than the walking legs.

The appendages of the crayfish and other crustaceans are **homologous** structures; that is, all are fundamentally similar in structure and arise from a similar embryonic rudiment. The structural adaptations of the appendages in different regions are correlated with their functions (Fig. 29-6). When corresponding structures in different segments of the same animal are homologous, it is called **serial homology.** On the other hand, **analogous** organs are similar in function but not necessarily in structure. For example, the eyes of annelids and vertebrates are similar in function but not in structure. Give examples of some other analogous structures or organs.

____

____

____

____

Internal anatomy. Remove the walking legs and swimmerets from the left side of the animal. Then, using a small sharp scissors, make a midline anterior-to-posterior cut through the carapace on the dorsal and ventral surfaces. Be careful not to cut into the internal organs. Remove the left side of the shell to expose the internal organs.

Using Fig. 29-7 as a guide, locate the diamond-shaped heart perforated by three pairs of openings called **ostia.** Identify the following blood vessels: the **ophthalmic artery** extending anteriorly from the heart toward the eyes; the **hepatic arteries** going to the digestive region of the stomach; the

LOCATION	APPENDAGES OF CEPHALOTHORAX	FUNCTION
In front of mouth	Antennule	Senses other organisms and helps to balance crayfish
In front of mouth	Antenna	Senses other organisms
In mouth	Mandible, or jaw	Crushes food
Behind mandibles	First maxilla	Moves food to mouth
Behind mandibles	Second maxilla	Bails water in gill chamber
At anterior and ventral part of thorax region	First maxilliped	Holds food, touches, and tastes
At anterior and ventral part of thorax region	Second maxilliped	Holds food, touches, and tastes
At anterior and ventral part of thorax region	Third maxilliped	Holds food, touches, and tastes
Posterior to maxillipeds at ventral part of thorax	Cheliped	Grasps food
Posterior to maxillipeds at ventral part of thorax	Walking leg (Gill)	For locomotion

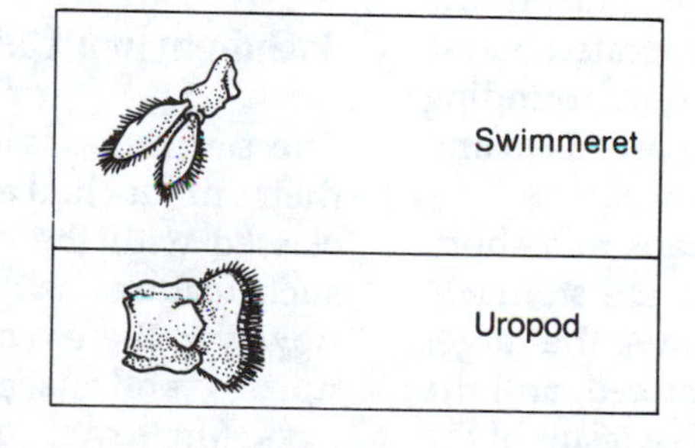

LOCATION	APPENDAGES OF THE ABDOMEN	FUNCTION
On ventral side of abdomen	Swimmeret	First swimmeret in male transfers sperm to female, which uses 2d, 3d, 4th, and 5th swimmerets to hold eggs and young
At posterior end	Uropod	For swimming

FIGURE 29-6 Serial homology of crayfish appendages

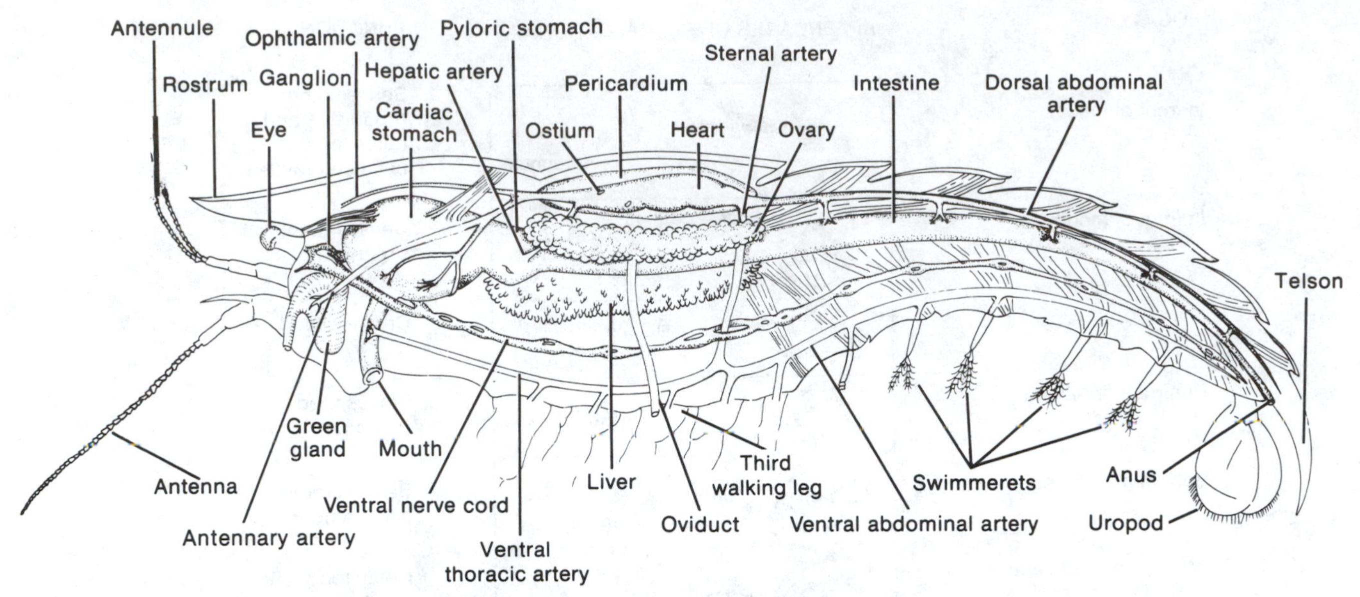

FIGURE 29-7 Internal anatomy of the crayfish

antennary arteries going to the antennae and green glands; the **dorsal abdominal artery** leading posteriorly from the heart to the dorsal part of the abdomen; the **sternal artery** leading from the heart to the ventral part of the crayfish where it joins the **ventral thoracic artery** anteriorly and the **ventral abdominal artery** posteriorly.

There are no veins in the crayfish; the blood flows from the arteries into open spaces or **sinuses** in the tissues instead of into capillaries as it does in the earthworm. The blood, after being oxygenated in the sinuses of the gills, returns through channels to the heart. This type of circulatory system is known as an **open system** to distinguish it from the **closed system** found in the annelids and many higher forms, including human beings.

Remove the heart and locate the gonads. In the female crayfish, the two ovaries are located below the **pericardium,** the saclike structure surrounding the heart. In the male, the two testes are fused and are also located below the pericardial sac.

Locate the mouth. The mouth leads to a short, tubular **esophagus,** which leads to the stomach. The stomach is made up of two parts, the larger **cardiac stomach,** in which food is stored, and the small, posterior **pyloric stomach.** As a result of the grinding action and the enzymes that are secreted into this region by the digestive glands, most of the digestion takes place in the pyloric stomach. The stomach empties into the intestine, a straight tube leading to the anus. Anterior to the stomach and just behind each antenna are the excretory structures, commonly called the **green glands.** Find their external openings at the bases of the antennae.

The nervous system is similar to that of annelids, except for the fusion of several (originally separate) nervous elements. Remove the main organs from the thoracic and abdominal regions of the crayfish and locate the large **ganglion,** or "brain," in front of the esophagus. Expose the brain by careful dissection and find the nerves passing to the eyes, antennae, and antennules. The brain is connected to a **ventral nerve cord** by a pair of nerves that pass around the esophagus. Follow the ventral nerve cord as it passes through the thorax and abdomen and count the number of ganglia. Each ganglion gives off pairs of nerves to the appendages and internal organs of the segment in which it lies.

3. Subphylum Uniramia

These arthropods are called uniramians because of their unbranched appendages. They were formerly classed with the crustaceans because they share such features as mandibles and compound eyes; however, there is no evidence that centipedes, millipedes, and insects ever had branched appendages. Moreover, uniramians have unjointed mandibles (unlike crustaceans), and they use them to handle food, not to grind the food before it passes into the mouth.

a. Class Chilopoda (Centipedes)

Centipedes are distributed throughout the world in temperate and tropical regions where they are found in soil and humus and under logs, stones, and bark on decaying trees. The common house centipede *Scutigera coleoptrata*, which is found in Europe and North America, is frequently seen when it becomes trapped in bathtubs and sinks (Fig. 29-8A). The largest centipede is the tropical *Scolopendra gigantea*, which grows to 26 cm long (Fig. 29-8B). Other tropical forms range from 18 to 24 cm long.

Examine specimens of centipedes. Note that they are elongated, somewhat flattened, and wormlike in shape. Each body segment bears a pair of legs, with the last pair directed backward. The typical head (Fig. 29-8C) bears paired antennae and mandibles. Beneath the mandibles is a pair of maxillae that function as a lower lip. A pair of maxillae are located above the first pair. A large pair of maxillipeds (poison claws) covers most of the mouth. Each claw bears a terminal pointed fang, which contains the outlet of the poison gland located in the appendage.

Centipedes are predators. Smaller arthropods make up the bulk of their diet, though some feed on earthworms, roundworms (nematodes), and snails. The prey is located with the antennae or legs and is killed or stunned with the poison claws. The venom of large tropical centipedes has been known to be painful but not sufficiently toxic to cause death in humans. Indeed, the bite of these dreaded forms has been likened to that of a severe hornet or yellow jacket sting. The prey, once captured, is held by the poison claws and second maxillae while the first maxillae and mandibles carry out the various manipulative actions that prepare the food

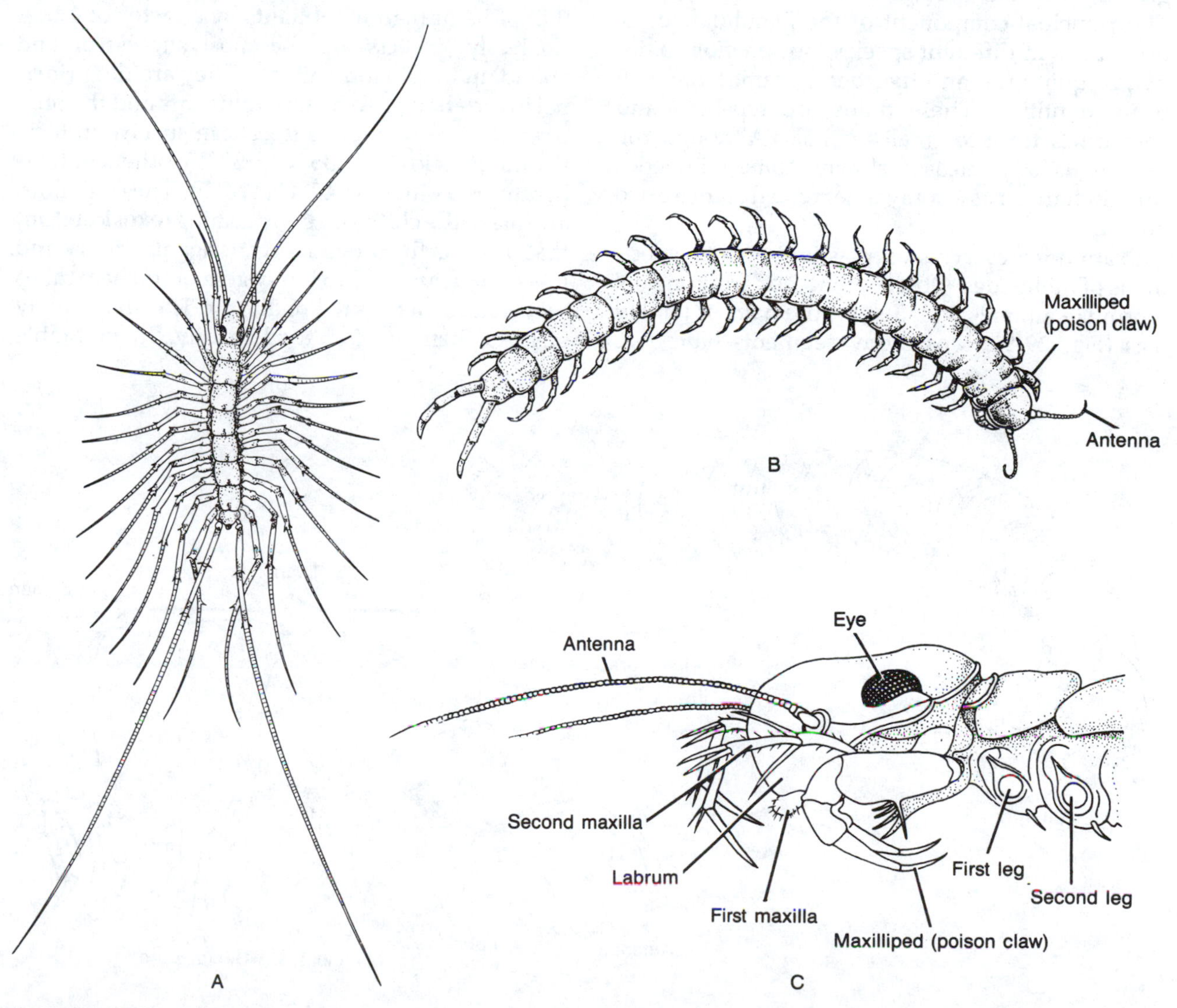

FIGURE 29-8 (A) *Scutigera coleoptrata*, the common house centipede. (B) *Scolopendra gigantea.* (C) Lateral view of head of *Scutigera coleoptrata.*

for ingestion. Salivary secretions are provided by salivary glands associated with the feeding appendages.

b. Class Diplopoda (Millipedes)

Millipedes, or "thousand leggers," are secretive and shun light. Like the centipedes, these wormlike animals live under leaves, stones, boards, and logs and in the soil. Some millipedes take over the old burrows of other animals, such as earthworms. Quite a large number live in caves.

To compensate for their lack of speed, a number of protective mechanisms have evolved in millipedes. The exoskeleton offers some protection from predators, and many millipedes can roll up into a ball to protect their softer ventral surfaces. Perhaps the most unusual device involves "stink" glands, which are found in every body segment. The principal component of the glandular secretion varies in different species, but phenols, aldehydes, quinones, and hydrogen cyanide have all been identified. These fluids are repellent and sometimes toxic to small animals. Although the fluid is usually released slowly, some millipedes can discharge it as a spray or jet for a distance up to 30 cm.

Examine preserved or plastic-mounted specimens of millipedes. Observe that they are elongate, wormlike animals with 30 or more pairs of jointed legs (Fig. 29-9). Most body segments have two pairs of legs. The head bears two clumps of many simple eyes (**ocelli**) and a pair each of mandibles and antennae. The thorax is short, consisting of four single segments, all but the first having a pair of legs.

c. Class Insecta

It has been estimated that there are as many as 5 million insect species, of which fewer than a million have been identified and classified (there are, for example, more than 300,000 species of beetles alone). Insects range in size from a species of winged hairy beetle that can crawl through the eye of a needle to the Atlas moth of India, which has a 30-cm wingspan. It has been estimated that the insect population of the world is at least 1×10^{18} and, taking the average weight of insects as 2.5 mg, the weight of the earth's insect population exceeds that of its human inhabitants by a factor of 12.

Clearly, insects are the most successful and abundant of all land animals. They are the principal invertebrates in dry environments and the only ones able to fly. Some insects can survive in temperatures as low as −35°C (−30°F), others in temperatures as high as 49°C (120°F). These abilities are due to the chitinous body coating **(exoskeleton)** that protects the internal organs against injury and loss of moisture and to the system of tracheal tubes that enables insects to breathe air. The ability to fly enables them to find food readily. In favorable

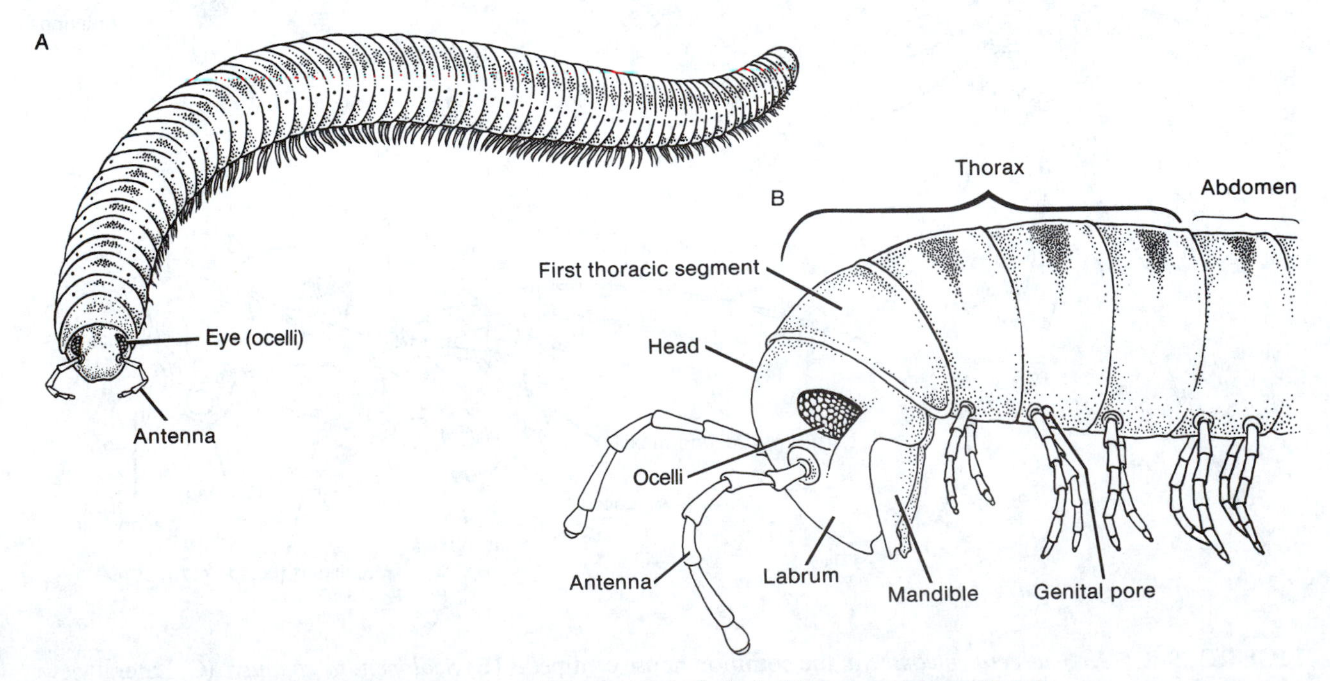

FIGURE 29-9 (A) Millipede (*Spirobolus*). (B) Lateral view of head.

conditions, insects can multiply rapidly. The insect's small size frees it from competition with larger animals for a place in the environment.

Insects have great ecological impact in terrestrial environments. For example, three-fourths of all flowering plants depend on insects for pollination. The principal pollinators include flies, wasps, bees, butterflies, and moths.

Insects are also of enormous importance to humans. Fleas, lice, bedbugs, mosquitos, and a variety of flies contribute directly to our misery. More important, these and other insects serve as vectors (carriers) of human disease and of diseases of our domesticated animals; examples are the tsetse fly (sleeping sickness), lice (typhus and relapsing fever), the housefly (dysentery and typhoid fever), and mosquitos (yellow fever, malaria, and encephalitis).

Although domesticated and other plants depend on insects for pollination, it is also true that many plants are destroyed by insects. Large amounts of money are being spent to control insect pests and thus to prevent the loss of foods needed by people. However, we also find that overuse of pesticides is exceedingly hazardous to the environment.

Insects are distinguished from the other arthropods by three pairs of legs and one or two pairs of wings attached to the middle (thoracic) region of their body. In addition, the head characteristically has a single pair of antennae and one pair of compound eyes. Simple eyes (ocelli) may also be present. A tracheal system provides the mechanism for exchange of respiratory gases. The openings of the reproductive system are typically at the posterior end of the abdomen.

Although insects are exceedingly diverse organisms, their basic internal and external organization is well illustrated by a study of the large black lubber grasshopper *Romalea microptera*. The following description, though it pertains to the short-winged lubber grasshopper, will serve for any common species of insect.

External anatomy. The body of the grasshopper is divided into a head that consists of six fused segments, a thorax of three segments to which the legs and wings are attached, and a long, segmented abdomen that terminates with the reproductive organs (Fig. 29-10). The exoskeleton consists largely of chitin, which is secreted by the epidermis. The young, growing grasshopper periodically sheds this exoskeleton (molts); adults do not molt. Pigment in and under the cuticle, which covers the exoskeleton, gives the grasshopper coloration resembling that of its surroundings, thus affording protection from predators.

The **head** (Fig. 29-11) has one pair of jointed antennae, two **compound eyes,** and three simple eyes or **ocelli.** The mouth parts are of the chewing type and include a broad upper lip or **labrum;** a tongue-

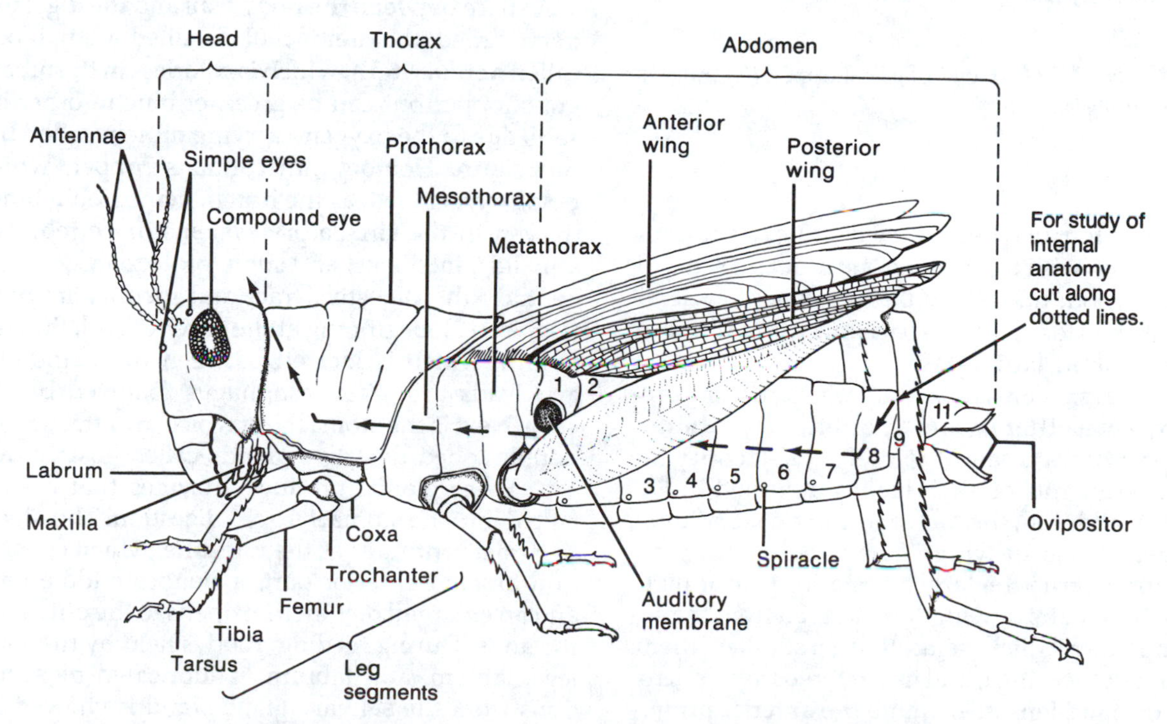

FIGURE 29-10 External anatomy of a female lubber grasshopper

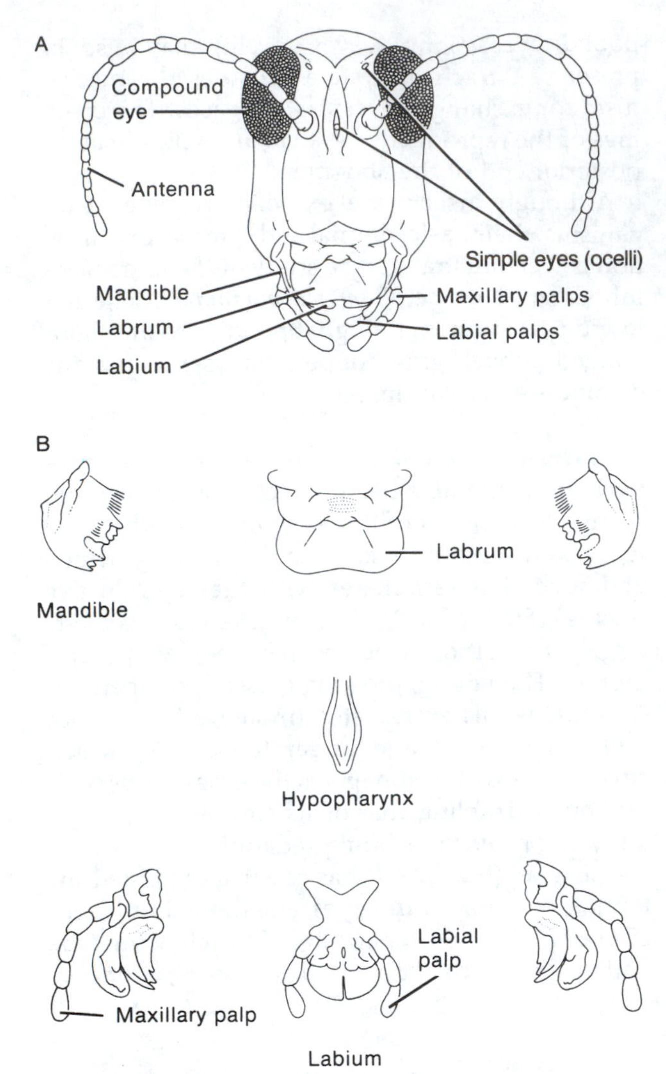

FIGURE 29-11 (A) Head of grasshopper. (B) Individual mouth parts.

like **hypopharynx;** two heavy lateral jaws or **mandibles,** each with teeth along the inner margins for chewing food; **maxillary palps** (sensory appendages) at the side; and a broad lower lip or **labium** with two short **labial palps.**

The **thorax** consists of a large anterior **prothorax,** a **mesothorax,** and a posterior **metathorax.** Each part bears a pair of jointed legs. Identify the various segments of each leg, as indicated in Fig. 29-10. In addition, the mesothorax and metathorax each bear a pair of wings. The anterior wings are thick and overlie the larger posterior pair of flight wings. Both pairs of wings are derived from the cuticle and have thick veins that strengthen them. Stretch out the wings, and use a stereoscopic microscope or hand lens to examine the anterior protective wings and the flight wings.

The abdomen consists of 11 segments, the posterior ones being modified for reproduction. The male has a blunt terminal segment, whereas the female has four sharp conical prongs, the **ovipositors,** which are used in egg laying (Fig. 29-10). Along the lower sides of the thorax and abdomen are 10 pairs of **spiracles,** the small openings of the **tracheae,** which branch to all parts of the body and constitute the respiratory system of insects. This system of air tubes, which open and close to regulate the flow of air, brings atmospheric oxygen directly to the tissues of the body. The three most anterior pairs of spiracles are inhalatory; the other spiracles are exhalatory.

Internal anatomy. It is difficult to preserve the internal organs of the grasshopper because the preservative often fails to penetrate the exoskeleton. Careful dissection is therefore necessary to study the internal anatomy.

After removing the wings, use a small scissors or a scalpel to make two lateral cuts toward the head, one on either side of the body, as indicated in Fig. 29-10. Remove the dorsal wall. Locate the muscles on the inside of the body wall and note their arrangement. What is their function?

____

____

A space between the body wall and the digestive tract, called the **hemocoel,** is filled with blood called **hemolymph,** which is colorless in the lubber grasshopper but can be green or blue in other insects due to the oxygen-carrying protein called **hemocyanin.** Hemocyanin contains copper, which binds oxygen, just as the iron in hemoglobin binds oxygen in the circulatory systems of vertebrates, annelids, mollusks, and some protozoans.

Study the digestive tract and identify its parts (Fig. 29-12). Beginning at the anterior end, find the mouth, which is located between the mandibles and leads to a short esophagus followed by the **crop.** Next is the stomach, to which are attached six double-lobed digestive glands called **gastric caecae.** These glands produce enzymes that are secreted into the stomach to aid digestion. The digestive tract continues as the intestine, which consists of a tapered anterior part, a slender middle part, and an enlarged rectum that opens to the outside at the anus. During feeding, food is held by the forelegs, labium, and labrum, is lubricated by secretions from the salivary glands, and is chewed by the mandibles and **maxillae.** Chewed food is

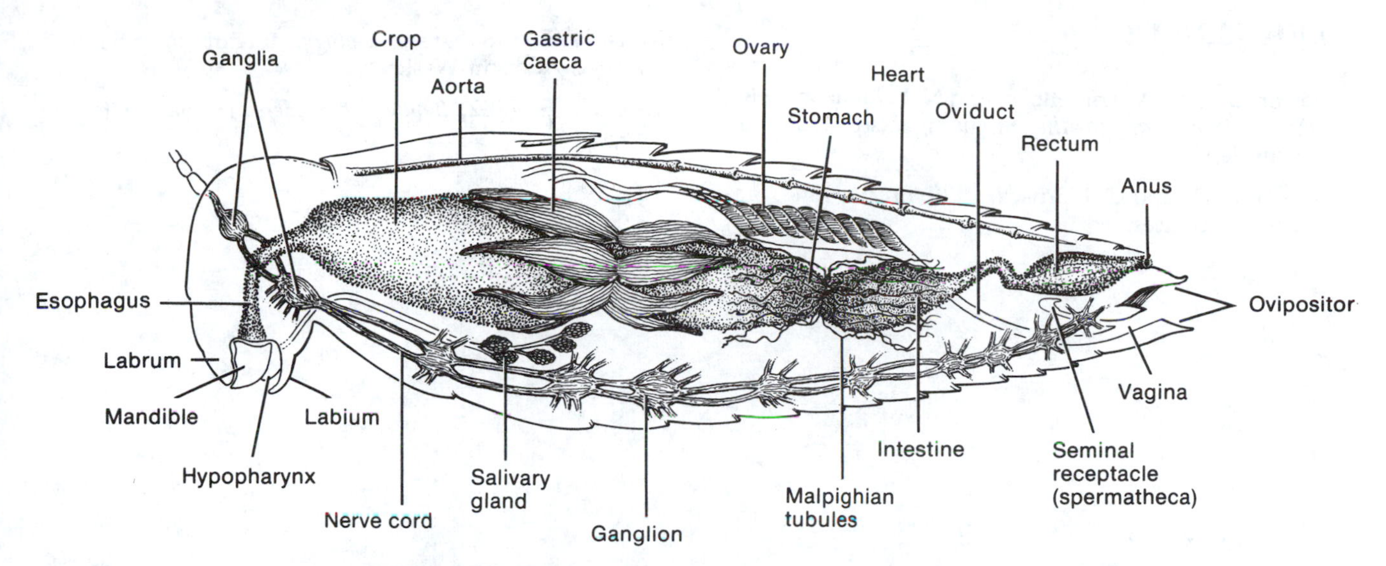

FIGURE 29-12 Internal anatomy of a female lubber grasshopper

stored in the crop. Because most of the digestive tract, except for the stomach and crop, is lined with chitin (which prevents the absorption of digested food), digestion and absorption take place in the stomach. Excess water from undigested food is absorbed in the rectum.

The excretory system is made up of numerous **Malpighian tubules,** which empty their products into the anterior end of the intestine. These tubules remove uric acid and salts from the hemolymph.

The reproductive organs of the separate sexes are in the terminal abdominal segments. In the male, each of the two **testes** is composed of a series of slender tubules and is located above the intestine. Each testis is joined to a **vas deferens** (Fig. 29-13). The vasa deferentia are joined to seminal vesicles that empty into a single ejaculatory duct, to which **accessory glands** are attached.

In the female, each ovary is composed of several **ovarioles,** which produce the ova. Each ovary is joined to an **oviduct** leading to the **vagina,** to which a pair of accessory glands and a single **spermatheca** (seminal vesicle) are attached. The latter organ is used to store sperm received at copulation.

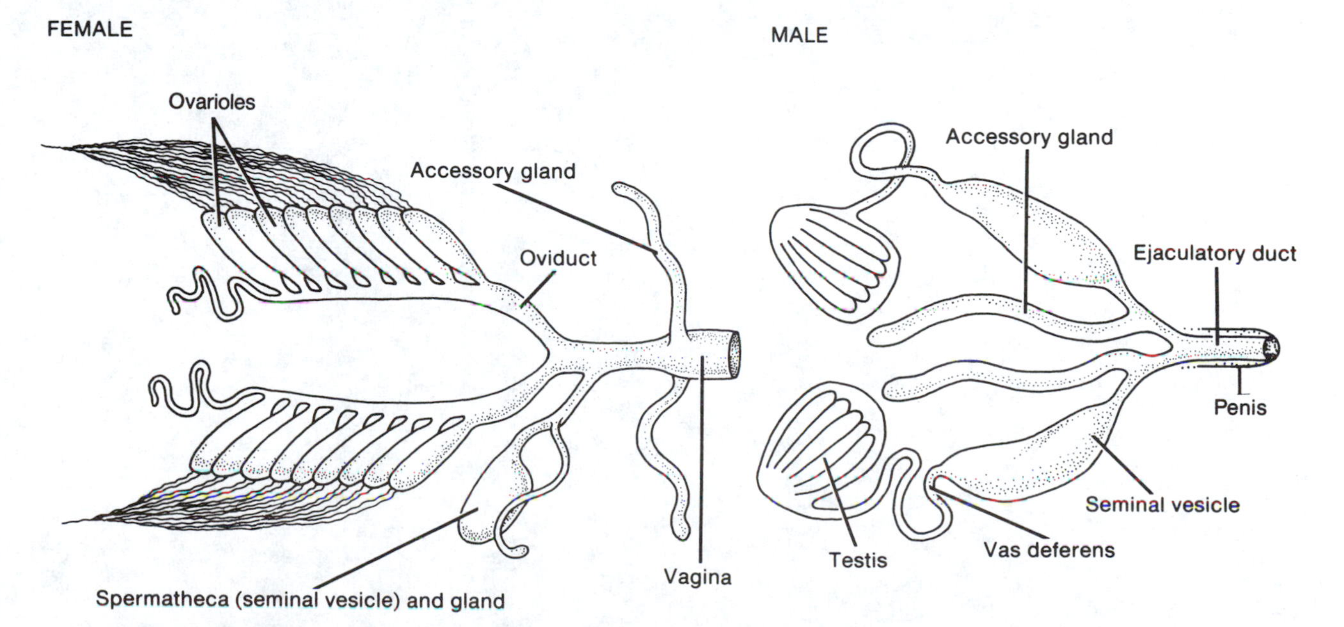

FIGURE 29-13 Structures of female and male reproductive systems of the lubber grasshopper

REFERENCES

Borror, D. J., C. A. Triplehorn, and N. F. Johnson. 1989. *An Introduction to the Study of Insects.* 6th ed. Saunders.

Brusca, R. C., and G. J. Brusca. 1990. *Invertebrates.* Sinauer Associates.

Evans, H. E. 1984. *Insect Biology: A Textbook of Entomology.* Addison-Wesley.

Foelix, R. F. 1982. *Biology of Spiders.* Harvard Univ. Press.

Exercise 30

Kingdom Animalia: Phylum Echinodermata

Members of the phylum Echinodermata (Greek *echinos*, "hedgehog"; *derma*, "skin") are exclusively marine, bottom-dwelling animals commonly known as starfish, sea urchins, sand dollars, sea cucumbers, and sea lilies. The phylum is so named because of the presence of spiny plates (**calcareous ossicles**), which form a dermal skeleton. Most adult echinoderms are radially symmetrical (though the larva is bilaterally symmetrical), and they have true coeloms arising as outpocketings from embryonic mesoderm of the gut. Because of this last characteristic and because the bipinnaria larva more closely resembles the chordate larva, the echinoderms are thought to be closely related to the chordates.

The most distinctive feature of the echinoderms is a unique coelomic canal system, and surface appendages made up of a water vascular system. Although a circulatory system is present, it is greatly reduced. Thus, the coelomic fluid acts as the principal medium for the transport of food and respiratory gases.

The phylum includes many classes of extinct echinoderms but only the following six include living species.

- **Class Stelleroidea (Asteroidea) (starfish or sea stars)** Have a star-shaped body, with 5–25 arms, that is covered by a flexible, spiny skeleton.
- **Class Crinoidea (feather stars and sea lilies)** Have a flowerlike body with many slender, branched arms.
- **Class Ophiuroidea (brittle stars)** Have a body with a central disc and five distinct slender, jointed arms.
- **Class Echinoidea (sea urchins and sand dollars)** Have a cylindrical or disc-shaped body in a shell of fused plates that bear movable spines.
- **Class Holothuroidea (sea cucumbers)** Have a soft, wormlike body with no arms or spines.
- **Class Concentricycloidea (sea daisies)** Only recently discovered. Have round bodies with a ring of marginal arms but no spines.

A. CLASS STELLEROIDEA (ASTEROIDEA) (STARFISH OR SEA STARS)

The simplest and perhaps the most familiar of all echinoderms is the starfish. The common starfish

Asterias, found along the Atlantic coast of North America, is a typical example. Starfish crawl on the shallow bottom or in tide pools among the rocks and sand of the seashore and coral reefs. They have been serious predators of oysters. At one time, oyster fishermen caught starfish, cut them up, and threw them back into the ocean. Then it was discovered that each piece could regenerate and grow into another starfish. Today, "sea mops" made of cloth are dragged over the oyster beds to entrap the starfish. They are then exposed to the sun to dry.

1. External Anatomy

Examine your preserved specimen and note that the body is composed of a central **disc** from which radiate five **arms** (Fig. 30-1A). Some of the specimens may have fewer arms, but this is probably because the arms have broken off in handling. Some starfish have more than five arms; rare specimens with as many as 25 have been found. The ventral or oral surface of each arm contains grooves extending outward from the centrally located mouth. The **aboral,** or dorsal, surface is spiny. The spines are extensions of small calcareous plates (**ossicles**) that lie buried beneath the surface. These plates form the **endoskeleton.** Surrounding each spine are numerous minute pincerlike **pedicellariae** and tiny **skin gills,** which function in respiration (Fig. 30-1B). Each pedicellaria has two jaws, moved by muscles, that open and shut when touched. They keep the body surface clean of debris and may also help to capture food.

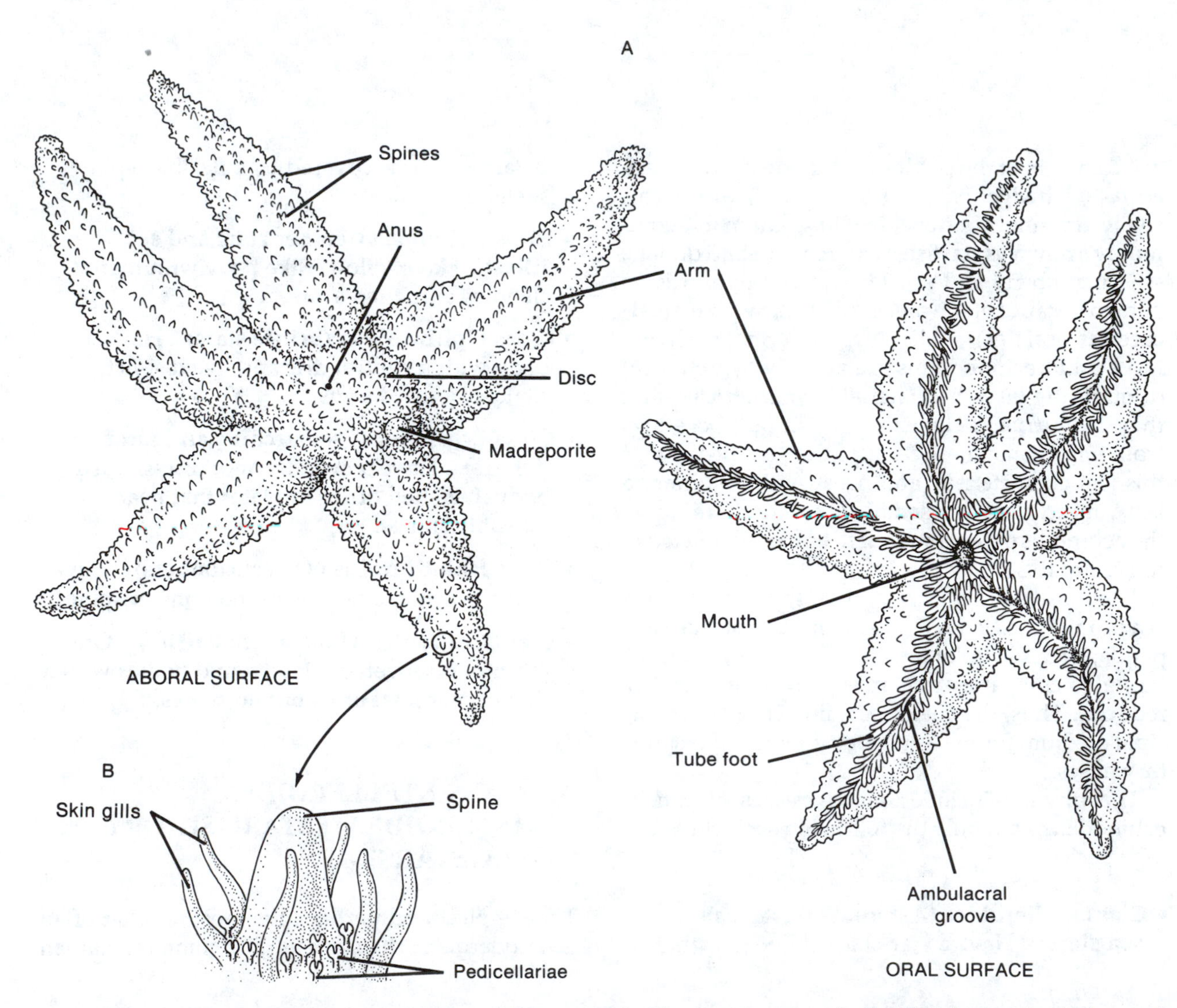

FIGURE 30-1 Starfish (*Asterias*). (A) External features of the aboral and oral surfaces. (B) Spine, pedicellariae, and skin gills.

The groove in the oral surface of each arm is called the **ambulacral groove.** Along the sides of the groove are a series of flexible **spines** that lie across the groove and protect rows of small, finger-like **tube feet,** which are organs of locomotion. If you separate the tube feet, you may be able to see a thick, white **radial nerve cord** that runs down the center of each arm. At the tip of each arm are small, light-sensitive **eye spots.** These light-sensitive tips are thrust upward during locomotion.

2. Internal Anatomy

Cut off about 15 mm from the tip of one of the arms of your starfish, and then make longitudinal cuts on both sides of this arm to the central disc. Carefully remove the aboral surface to expose the internal organs (Fig. 30-2A). Note that most of the coelom in the arm is taken up by two highly branched digestive glands, the **hepatic caecae.** Examine the glands with a hand lens or stereoscopic microscope and note the numerous lobes that secrete digestive enzymes. The ducts of the hepatic caecae join at the base of the arm to form the **pyloric duct,** which enters the centrally located, saclike **stomach.** The ventral **mouth** and a short **esophagus** lead directly into the stomach, which consists of a multilobed lower **cardiac stomach** and an upper **pyloric stomach.** The food is partly digested in the cardiac stomach and passed into the pyloric stomach where digestion is completed. The pyloric stomach then empties into the anus, located in the center of the aboral disc. Two small **rectal caecae,** usually found near the anus, function as temporary storage areas for waste products.

Cut the pyloric duct where it enters the stomach and remove the hepatic caecae to expose the **gonads** (reproductive organs). If the starfish was caught during the breeding season, the gonads will fill the arms. At other times the gonads are very small.

The male and female gonads look alike. To determine the sex of the starfish, the contents of the gonads must be examined microscopically. To do this, remove a small piece of a gonad and mince it in a drop of water on a slide. Add a coverslip and examine under the low and the high power of the microscope. The testes of the male have flagellated sperm. The ovaries of the female produce spherical eggs that are considerably larger than the sperm. Eggs and sperm are discharged into the water through openings called **gonopores** on the oral surface of each arm. Fertilization takes place in the water. The fertilized eggs develop into bilaterally symmetrical, ciliated larvae. A similar larval stage is formed during the development of the hemichordates. This larva may pass through several distinct stages before it develops into an adult. The similarity between the larval stage of Echinodermata and Hemichordata (primitive chordates), as well as similarities in their early development, suggest that both groups arose from a common ancestor at some remote time.

The **water vascular system,** which is unique to the echinoderms, consists of a series of interconnected canals and appendages associated with the body wall (Fig. 30-2B). This system is well-developed in the Stelleroidea and functions as a mechanism of locomotion. To study the anatomy of this system, carefully remove the reproductive organs from one arm and the digestive system, including the stomach and anus.

The internal canals of the water vascular system are connected to the outside through a button-shaped **madreporite** located on the aboral (dorsal) surface. This sievelike structure opens into the **stone canal,** which descends to the oral side of the animal. The stone canal is so named because of calcareous deposits in its walls. On the oral side, the stone canal joins the circular **ring canal.** The inner sides of the ring canal give rise to several pairs of pouches called **Tiedemann's pouches,** which are believed to produce amoeboid cells that circulate in the system. From the ring canal, a long **radial canal** extends into each arm. **Lateral canals** arise from both sides of each radial canal along its entire length. Each lateral canal contains a valve and terminates in a bulb called an **ampulla** and a **tube foot.** Typically, the tip of the tube foot is flattened, forming a sucker.

The entire water vascular system is filled with a fluid that is similar to sea water except that it contains some protein, a high potassium-ion content, and amoeboid cells. This system functions as a hydraulic system during locomotion. When the ampulla contracts, the valve in the lateral canal closes and water is forced into the tube foot, which then elongates. When the foot touches the surface on which the animal is moving, the center of the terminal sucker is withdrawn. This produces a vacuum, thereby causing the foot to adhere to the surface. A thick secretion produced at the tip of the foot also aids adhesion.

After adhesion of the foot, longitudinal contractile fibers of the foot contract, shortening the foot and forcing fluid back into the ampulla. Thus, during locomotion, each foot performs a sort of stepping motion. The foot elongates, swings forward, adheres, contracts, and moves backward.

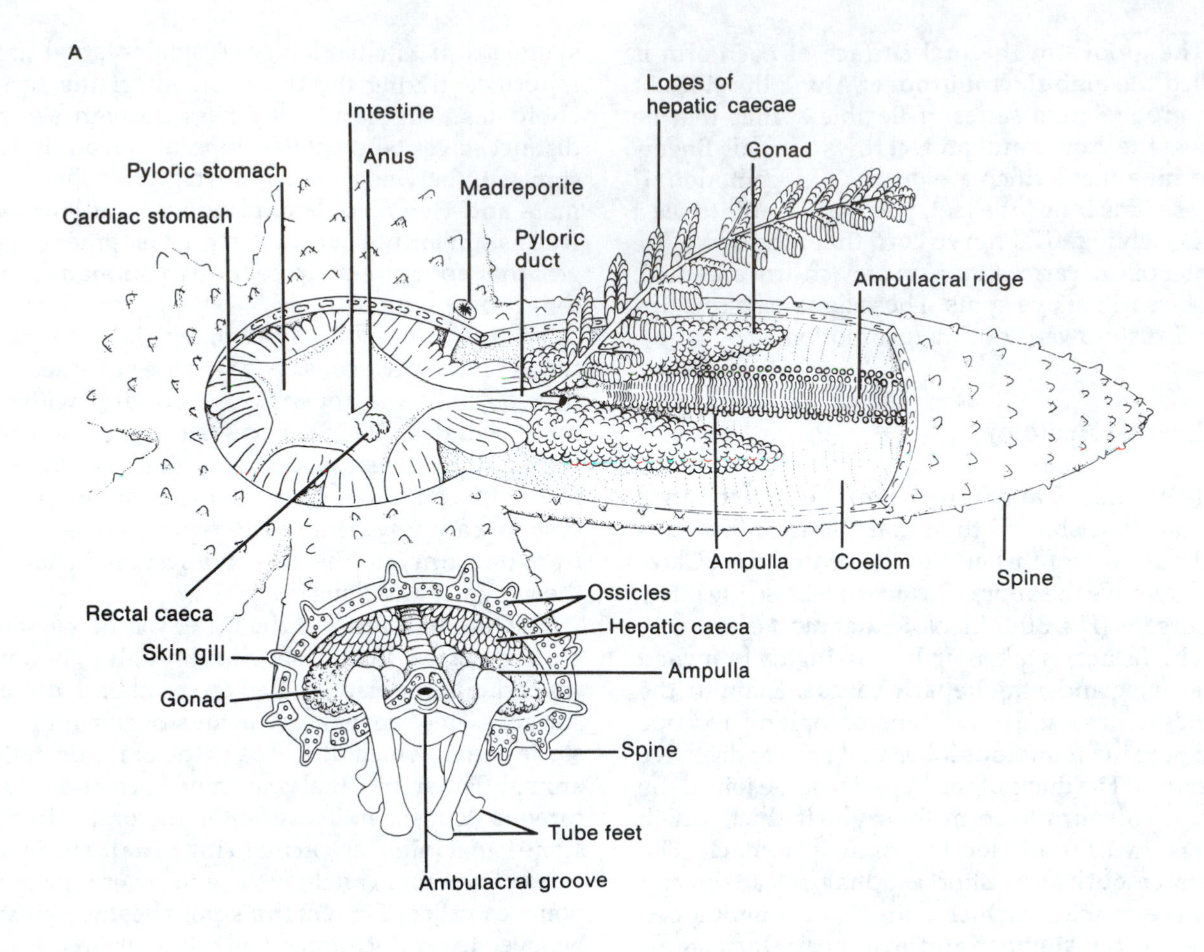

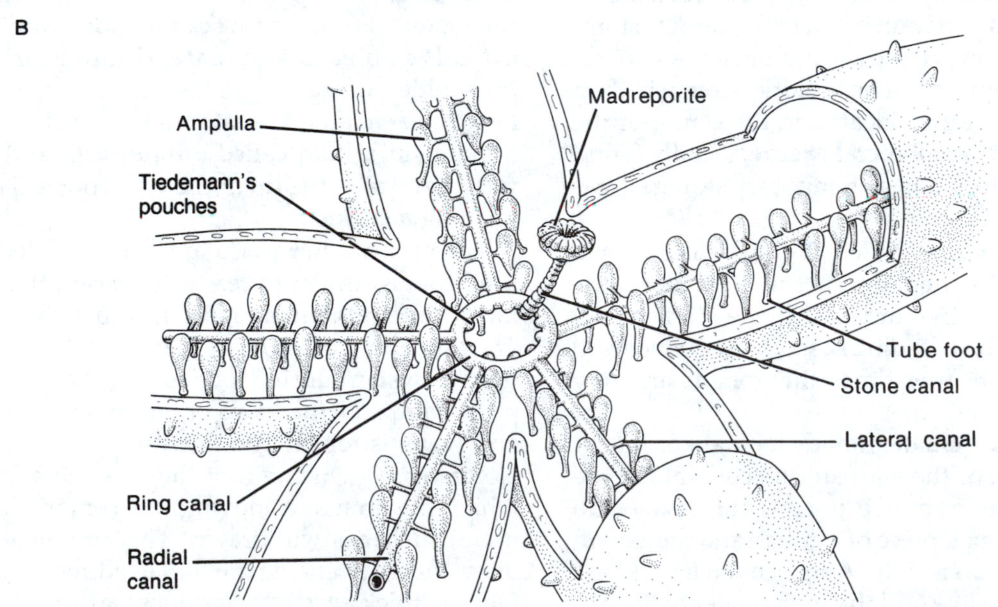

FIGURE 30-2 (A) Internal organs of the starfish. (B) Water vascular system of the starfish.

The net result of many feet "walking" is that the animal moves forward.

B. CLASS CRINOIDEA (FEATHER STARS AND SEA LILIES)

Examine preserved or plastic-mounted specimens of feather stars and sea lilies, the oldest and most primitive of the present-day echinoderms. The habitat of these flowerlike animals ranges from just below the tideline to depths of more than 3600 m. The body of a feather star (*Antedon*) consists of a small, cup-shaped **calyx** of calcareous plates, to which are attached several flexible arms that bear many slender lateral **pinnules,** arranged like barbs on a feather, thus giving the animal its common name (Fig. 30-3A).

The sea lily (*Metacrinus*) has a long, jointed **stalk** that attaches to the sea bottom by rootlike outgrowths called **cirri** (Fig. 30-3B). Both mouth and anus are on the oral surface of the calyx. In the crinoids, the oral surface is oriented upward, which is different from all other modern echinoderms. Each arm has an ambulacral groove that is lined with cilia and contains tentaclelike tube feet.

FIGURE 30-3 Crinoidea. (A) Feather star (*Antedon*). (B) Sea lily (*Metacrinus*).

C. CLASS OPHIUROIDEA (BRITTLE STARS)

Examine preserved or plastic-mounted specimens of brittle stars. These echinoderms have five arms like the sea stars, but the arms are longer, more slender, and more flexible (Fig. 30-4). The skeleton consists of an outer, superficial endoskeleton and a deeper, internal, articulated series of vertebral ossicles. This arrangement permits the solidly armored arm to move quite freely, which enables this animal to crawl rapidly or swim. The arms, which break easily, are quickly regenerated.

D. CLASS ECHINOIDEA (SEA URCHINS AND SAND DOLLARS)

Members of this class have globose, oval, or disc-shaped bodies that lack arms or rays but are covered with slender movable spines and tube feet.

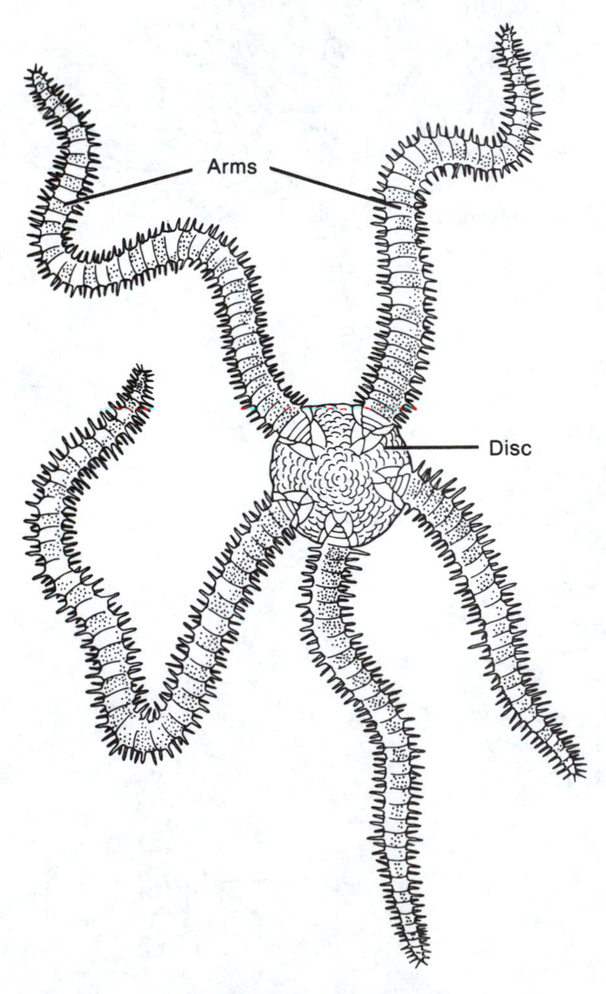

FIGURE 30-4 Brittle star

1. Sea Urchin

Examine a preserved specimen of the sea urchin *Arbacia* or *Strongylocentrotus*. Study the surface of this animal and observe the sharp, movable spines that are attached to the solid shell, or **test** (Fig. 30-5A). On the shell are the rounded tubercles to which the spines are attached. Among the spines are pedicellariae on long, flexible stalks. Some echinoids have several kinds of pedicellariae; a few bear poison-producing glands. Locate the long, slender tube feet and note that they are restricted to regions of the shell known as the ambulacra.

Hold the sea urchin so that the oral side faces you (Fig. 30-5B). In the center of the oral surface is the mouth, which contains a highly developed scraping apparatus called **Aristotle's lantern.** This apparatus is made up of five plates, or "teeth," each looking like a barbed arrowhead with its point directed toward the mouth. The lantern can be projected and retracted through the mouth by special muscles. Sea urchins, most of which are grazers, scrape the substrate over which they move by opening and closing the lantern plates. They thereby obtain a diverse diet of plant and animal material.

2. Sand Dollar

Examine the concave aboral surface of the sand dollar *Echinarachnius* (Fig. 30-6). Observe the arrangement of the ambulacra on the surfaces. In the center of the aboral surface, you will find the madreporite, at the periphery of which are five genital pores (gonopores).Turn the specimen over and locate the ambulacral grooves, the mouth in the center of the disc, and the anus at the edge.

E. CLASS HOLOTHUROIDEA (SEA CUCUMBERS)

Examine preserved specimens of the sea cucumbers *Cucumaria* or *Thyone* (Fig. 30-7). Note that the body surface has no spines. The endoskeleton is reduced to microscopic spicules, thus giving the body wall a tough, leathery texture. The mouth is located at the center of a conspicuous crown of tentacles that are modified tube feet. In *Cucumaria*, you also observe lengthwise zones of tube feet that are tactile and respiratory in function. In *Thyone*, the tube feet are distributed over the whole body.

The body wall of the sea cucumber is composed of a cuticle over a nonciliated epidermis, a dermis, a

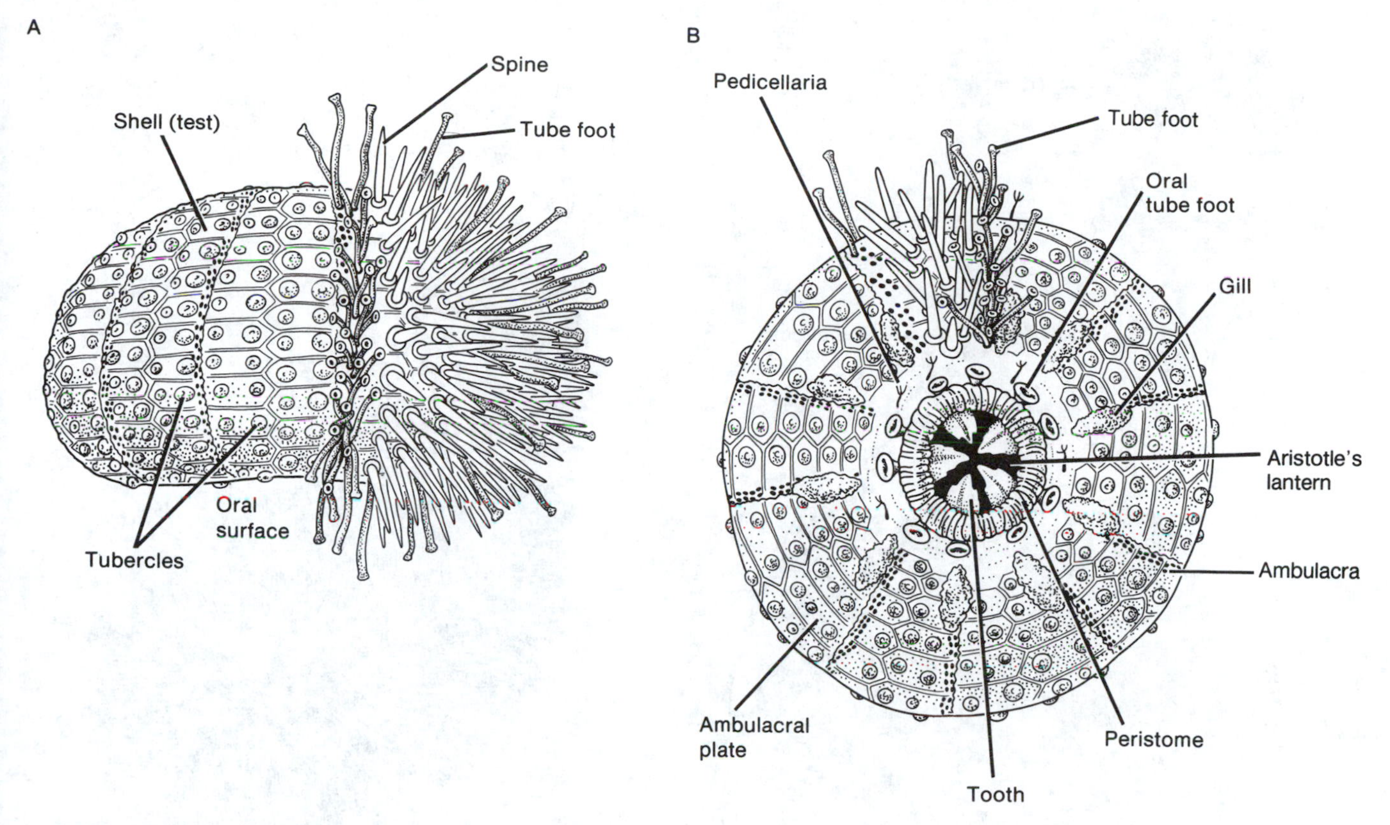

FIGURE 30-5 Sea urchin. (A) Lateral view. (B) Oral surface; spines and tube feet have been removed from the left side to show structure of the test.

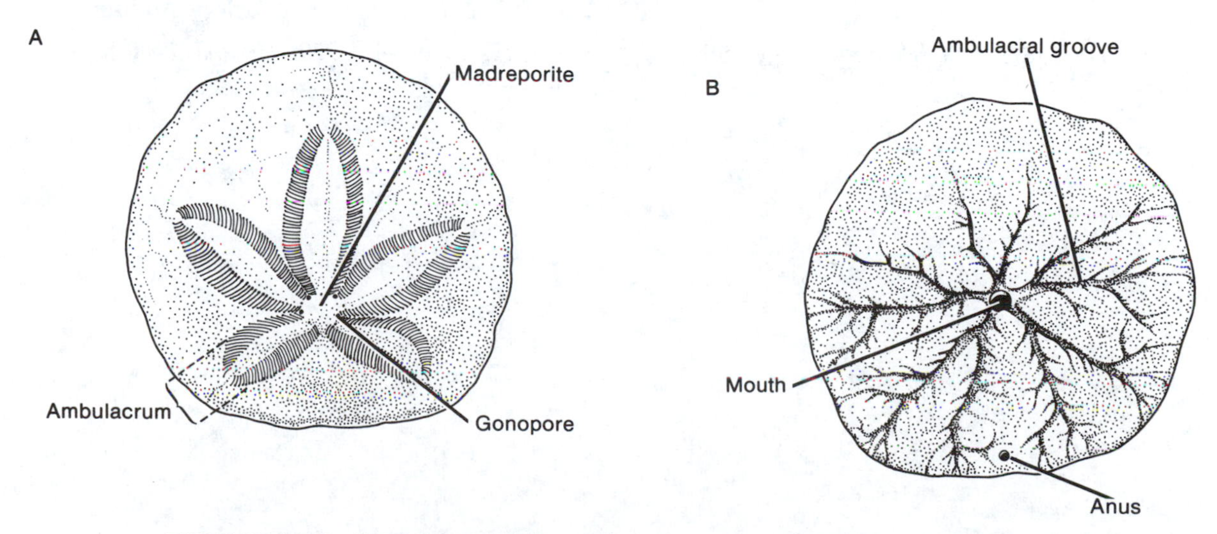

FIGURE 30-6 Sand dollar (*Echinarachnius*). (A) Aboral surface. (B) Oral surface.

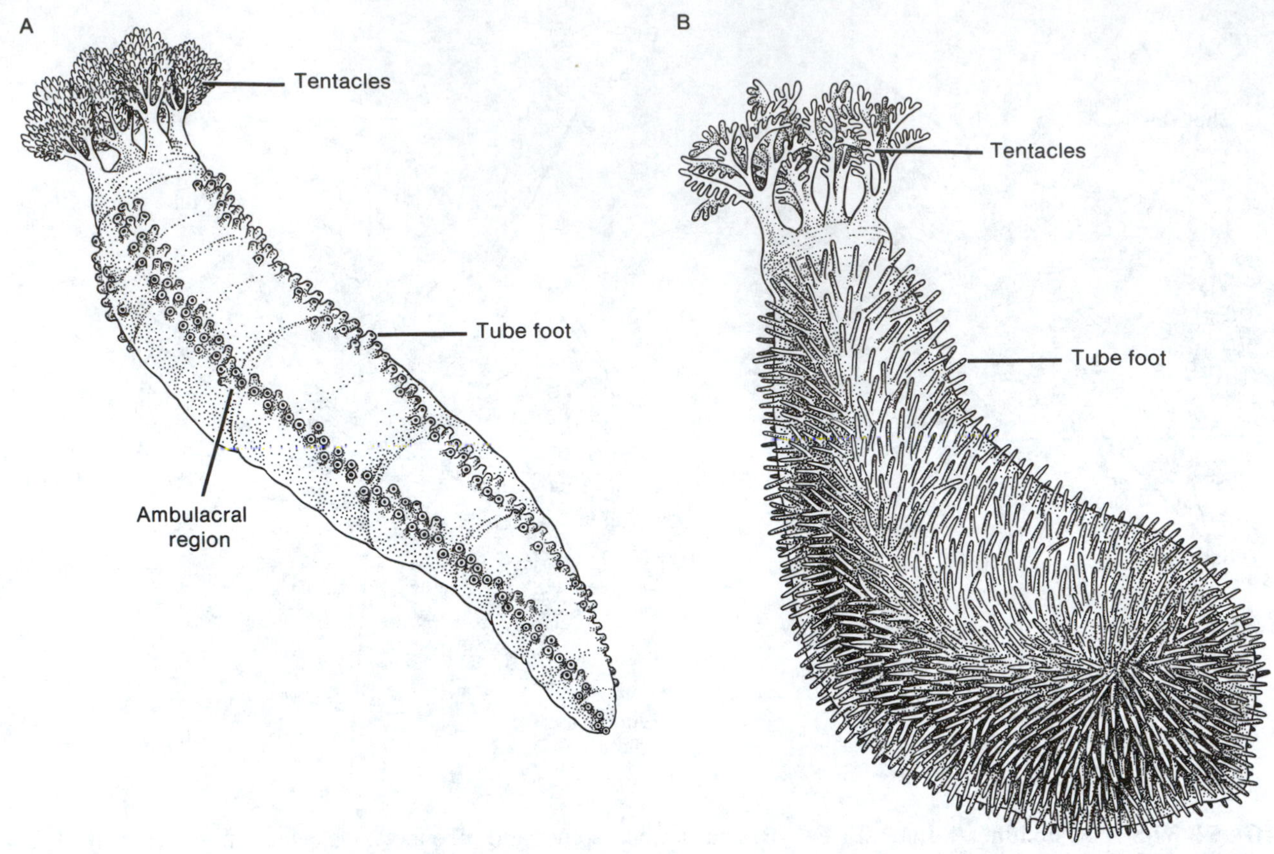

FIGURE 30-7 Sea cucumbers. (A) *Cucumaria*. (B) *Thyone*.

layer of circular muscles, and five double bands of longitudinal muscles. The action of these muscles enables the sea cucumber to extend or contract its body and to move by wormlike movements.

REFERENCES

Barnes, R. D. 1987. *Invertebrate Zoology.* 5th ed. Saunders.

Brusca, R. C. and G. J. Brusca. 1990. *Invertebrates.* Sinauer Associates.

Buchsbaum, R., et al. 1987. *Animals Without Backbones.* 3d ed. rev. University of Chicago Press.

Hyman, L. H. 1955. *The Invertebrates. Vol 4: Echinoderms.* McGraw-Hill.

Lutz, P. E. 1986. *Invertebrate Zoology.* Addison-Wesley.

Storer, T. I., et al. 1979. *General Zoology.* 6th ed. McGraw-Hill.

Exercise

31

Kingdom Animalia: Phyla Hemichordata and Chordata

The phylum Hemichordata (Greek *hemi,* "half"; *chorda,* "string") is a group of small, soft-bodied animals that are entirely marine, often found in U-shaped burrows on sandy or muddy sea bottoms. Because they possess both echinoderm and chordatelike characteristics, hemichordates presumably represent an evolutionary link between echinoderms and chordates. In this exercise, you will study the acorn worm *Saccoglossus* as a representative of this phylum.

The phylum Chordata is ecologically the most significant phylum, in large part because of the negative impact of humans on the environment. It is divided into three subphyla, the Urochordata (tunicates), Cephalochordata (lancelets), and Vertebrata, all of which, at some stage in their development, share three important characteristics: a notochord, pharyngeal gill slits (or pouches), and a dorsal tubular nerve cord.

You will study representatives of the following phyla, subphyla, and classes in this exercise.

- Phylum Hemichordata (acorn worms): Wormlike animals characterized by bilateral symmetry, a well-developed enterocoelom, pharyngeal slits, and a primitive dorsal nervous system.
- Phylum Chordata (chordates): Possess at some stage in their life cycles well-developed pharyngeal gills or gill slits, a dorsal tubular nerve cord, and a notochord.

Subphylum Urochordata or Tunicata (tunicates): Have a larval stage in which chordate characteristics are present. The neural tube and notochord are lost in the sedentary adult, though it does possess a primitive circulatory system.

Subphylum Cephalochordata (lancelets): Have a well-developed coelom, a circulatory system without a discrete heart, and a fusiform body that has prominent muscle segments (myotomes).

Subphylum Vertebrata (vertebrates): Have a cranium (skull), visceral arches, and a spinal column of vertebrae that are cartilaginous in lower forms and bony in higher forms; have a notochord that extends from cranium to base of tail, an enlarged brain, and a head region with specialized sense organs.

Class Agnatha (cyclostomes): Have a long, slender, cylindrical body with median fins, a mouth located ventrally, 5–16 pairs of gill arches, and a persistent notochord; lack true jaws and scales (hagfish and lampreys).

Class Chondrichthyes: Have a cartilaginous skeleton with notochord, tough skin covered with scales, median and paired lateral fins, a ventrally located mouth with upper and lower jaws, and pectoral and pelvic girdles (sharks, skates, and rays).

Class Osteichthyes: Have a bony skeleton, a terminal mouth, gills covered by an operculum, median and paired fins, and (usually) skin that is covered with scales (perch), though some are scaleless (trout).

Class Amphibia: Have a moist, glandular skin that lacks scales, two pairs of limbs but no fins, a bony skeleton, a terminal mouth with upper and lower jaws, and a tongue that is often protrusible. Aquatic in the larval stage but usually terrestrial as adults (frogs, toads, salamanders, and newts).

Class Reptilia: Have a body that is dry and covered with scales; two pairs of limbs (absent in snakes) with digits adapted to running, crawling, climbing, or swimming; and a bony skeleton (lizards, snakes, turtles, crocodiles, and alligators).

Class Aves: Warm-blooded animals that have a body covered with feathers, forelimbs modified as wings, a bony but light skeleton, and a beak (includes all birds).

Class Mammalia: Have mammary glands that secrete milk for nourishing the young, hair in varying quantities, and young that are born alive (human beings, dogs, cattle, and mice).

A. PHYLUM HEMICHORDATA

Hemichordates are common marine animals of broad distribution. They are small, soft-bodied organisms that live singly or in colonies on sandy or muddy sea bottoms or in open ocean waters. The most common representatives are the acorn worms *Balanoglossus* and *Saccoglossus.*

Examine a preserved specimen of *Saccoglossus.* Note the softness of the body and its wormlike form (Fig. 31-1). The epidermis is ciliated and well supplied with mucus-secreting cells, which are important in burrowing and feeding. The body is divided into three regions: an anterior **proboscis,** a short **collar,** and a long **trunk.** The **mouth** opening is ventral and located at the base of the proboscis. The trunk is divisible into three regions: an anterior **branchial region** containing pharyngeal slits; a **genital region,** which, in other species, may be enlarged into large genital ridges; and, posteriorly, an **abdominal region** containing the intestine and lateral pouches of the hepatic caecae. The **anus** is terminal.

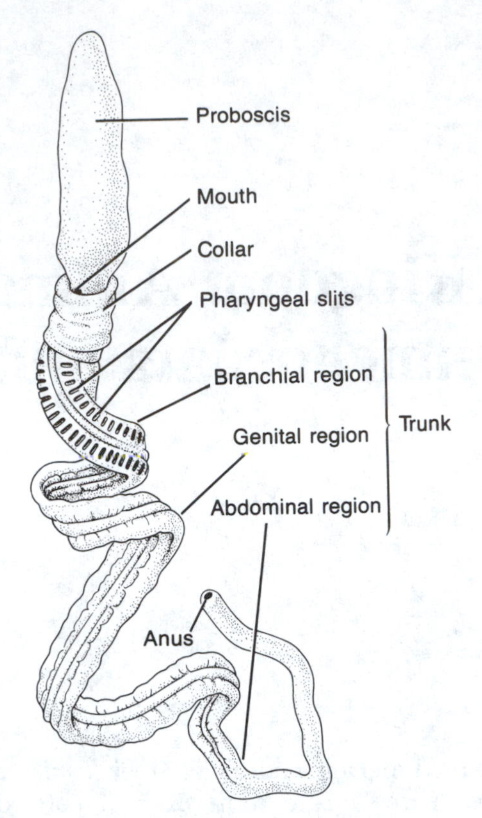

FIGURE 31-1 External anatomy of the acorn worm *Saccoglossus.*

During development, a ciliated larva stage **(tornaria)** is formed that looks very much like the bipinnaria larvae of some of the echinoderms. The similarity of the larvae in these two phyla and the similarities in their early development suggest that the two groups probably arose from a common ancestor.

B. PHYLUM CHORDATA

Chordates are characterized by three major features: the dorsal, tubular **nerve cord,** which in mammals becomes the brain and spinal cord; the **notochord,** a cartilaginous rod that develops dorsal to the primitive gut in the early embryo (in the lower chordates, the notochord persists throughout life, whereas in the vertebrates it persists as the soft center of the intervertebral discs); the presence, during some stage in the life cycle, of **slits** in the pharynx or throat.

1. Subphylum Urochordata or Tunicata

These exclusively marine and sessile organisms are covered by a firm protective tunic, from which they get their common name, tunicates.

Examine a preserved or plastic-mounted specimen of the tunicate *Ciona* or *Molgula* (Fig. 31-2A). The adult animal is saclike and is usually attached by its base to rocks, seaweeds, shells, wharf pilings, or ship hulls. At its free end are two openings: an incurrent siphon, and an excurrent siphon. The body wall consists of a firm tunic, under which lies the mantle, which secretes the tunic and contains the muscles that can alter the body shape. Carefully cut away the body wall and locate the atrial cavity containing a large pharynx that has many gill slits, and a visceral cavity containing the major organs (gonads, stomach, and heart).

Examine a prepared slide of the larval stage of *Molgula.* This transparent, free-swimming larva

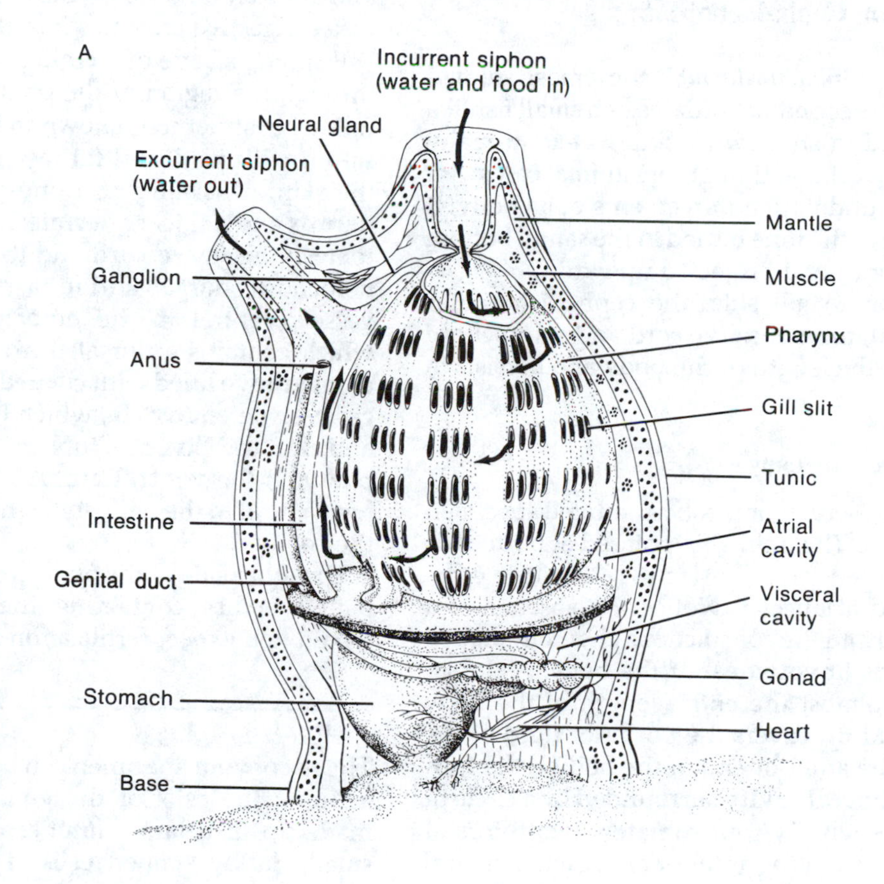

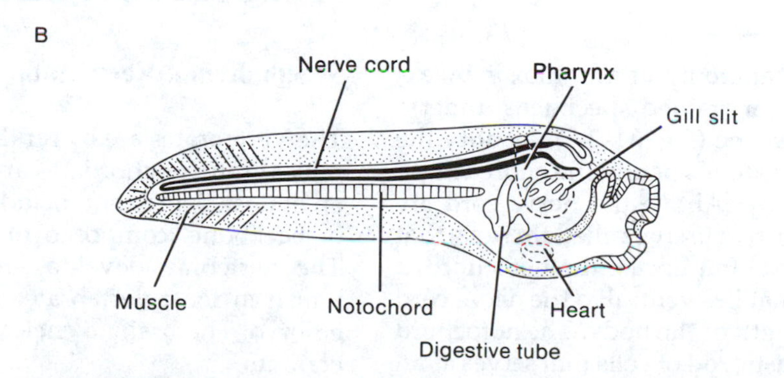

FIGURE 31-2 *Molgula.* (A) External anatomy of an adult. (B) Larva.

resembles an amphibian tadpole. The larva has all three chordate characteristics. Its tail contains a supporting notochord, a dorsal tubular nerve cord, serial pairs of lateral, segmental muscles, and gill slits in the pharynx (Fig. 31-2B). It swims about like a tadpole, settles down on some object, and then transforms into a sedentary adult. During this transformation, two of the chordate features (the notochord and nerve cord) are lost.

2. Subphylum Cephalochordata

Amphioxus *(Branchiostoma),* the most-studied member of the cephalochordates, is a small fishlike animal found in shallow marine waters in many parts of the world. Although the animal can swim with lateral, undulating movements of its body, it spends most of the time buried in the sandy bottom with its anterior end projected upward.

In addition to gill slits, the cephalochordates have a dorsal, tubular nerve cord and a notochord that extends the length of the body and persists in the adult.

a. External Anatomy

Examine a preserved or plastic-mounted specimen of amphioxus. The animal is pointed at both ends and compressed laterally (Fig. 31-3A). It is commonly called a lancelet. Notice the absence of a distinct head and the conspicuous chevron-shaped muscle bands **(myotomes)** of the body. A **dorsal fin** extends almost the entire length of the body and a **ventral fin** covers the posterior third of the animal. At the anterior end of the body is the funnel-shaped **buccal cavity** surrounded by a circle of oral tentacles **(cirri).** Approximately two-thirds of the way back from the anterior end, on the ventral side, locate the **atriopore,** through which water leaves after having been pumped through the pharynx.

b. Internal Anatomy

Study the internal anatomy of amphioxus by examining preserved or stained specimens under a stereoscopic microscope (Fig. 31-3A). Identify the **nerve cord,** which extends nearly the entire length of the animal's body. Above the nerve cord are short rods of connective tissue, called **fin rays,** that strengthen the dorsal fin. Locate the notochord, a cartilaginous rod that lies ventral to the nerve cord and extends the length of the body. The notochord is a longitudinal elastic rod of cells that serves as an internal skeleton. Ventral to the notochord is the digestive tract. The mouth, located in the buccal cavity, leads into the relatively large pharynx. Observe the gill slits in the wall of the pharynx. The pharynx is lined with cilia that beat inward to produce a steady current of water, which enters the mouth, passes over the gill slits (leaving behind suspended food particles), and is then eliminated through the atriopore. Posteriorly, the pharynx joins the intestine. Close to the point at which they join is a ventral outgrowth, the "liver," which secretes digestive enzymes into the intestine.

Examine a slide of a cross section of amphioxus through the region of the pharynx and locate the following structures, shown in Fig. 31-3B: the dorsal fin with its dorsal fin ray; **metapleural folds;** the skin, consisting of a one-celled layer of **epidermis** and a thicker **dermis;** muscle bands (myotomes); the **nerve cord** and its **central canal;** the notochord; **dorsal aortae,** a pair of small blood vessels ventral to the notochord; the pharynx, which contains a dorsal furrow, the **hyperbranchial groove** lined with ciliated cells, and a ventral groove, the **endostyle,** which has both gland cells and cilia; the "liver," a tube on the right side of the pharynx; the **ventral aorta,** a small blood vessel just ventral to the endostyle; the **atrial cavity;** and the **coelom.**

On either side of the pharynx, locate the **gonads,** paired bodies containing the sex cells. Where would you expect fertilization to take place?

____

The sex of your specimen can be determined by examining the cells of the gonads. The testes are made up of a large number of densely packed, small, darkly stained cells. The ovaries contain fewer, larger cells with vesicular nuclei. What is the sex of your specimen?

____

3. Subphylum Vertebrata

The Vertebrates are by far the largest and most familiar group of chordates and are characterized by an endoskeleton that includes a vertebral column, or backbone, composed of a series of vertebrae. The vertebrae develop around the notochord, which in most vertebrates is present only in the embryo. The brain is enclosed and protected by a cranium.

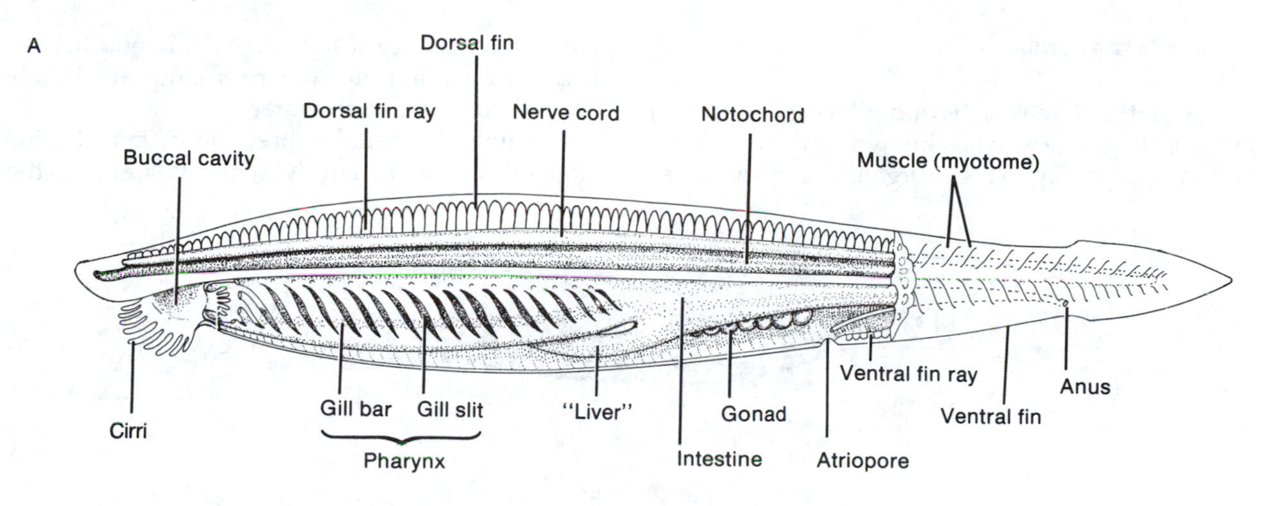

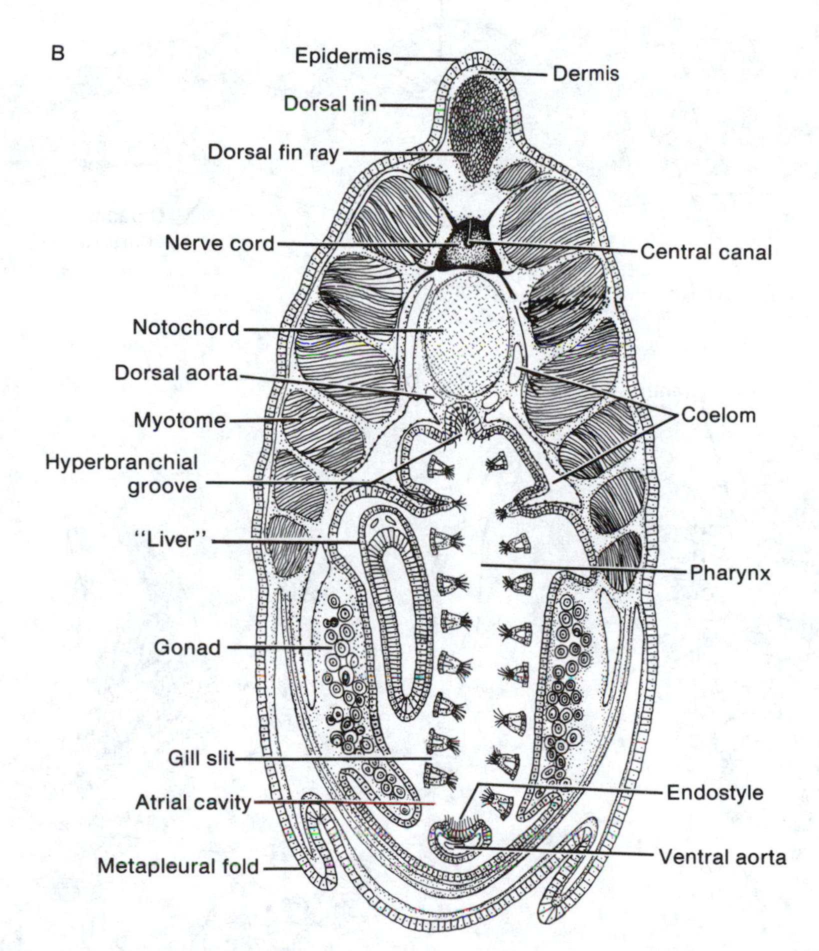

FIGURE 31-3 Amphioxus *(Branchiostoma)*. (A) Lateral view of internal anatomy. (B) Cross section.

a. Class Agnatha

The Agnatha (Greek *a*, "without"; *gnathos*, "jaw") include the jawless fishes known as the hagfishes and lampreys (Fig. 31-4). Organisms in this class are also called *cyclostomes* (round mouths). All hagfishes are marine, whereas lampreys inhabit both marine and fresh water.

Examine the sea lamprey on demonstration. Note the shape of its body, which is similar to that

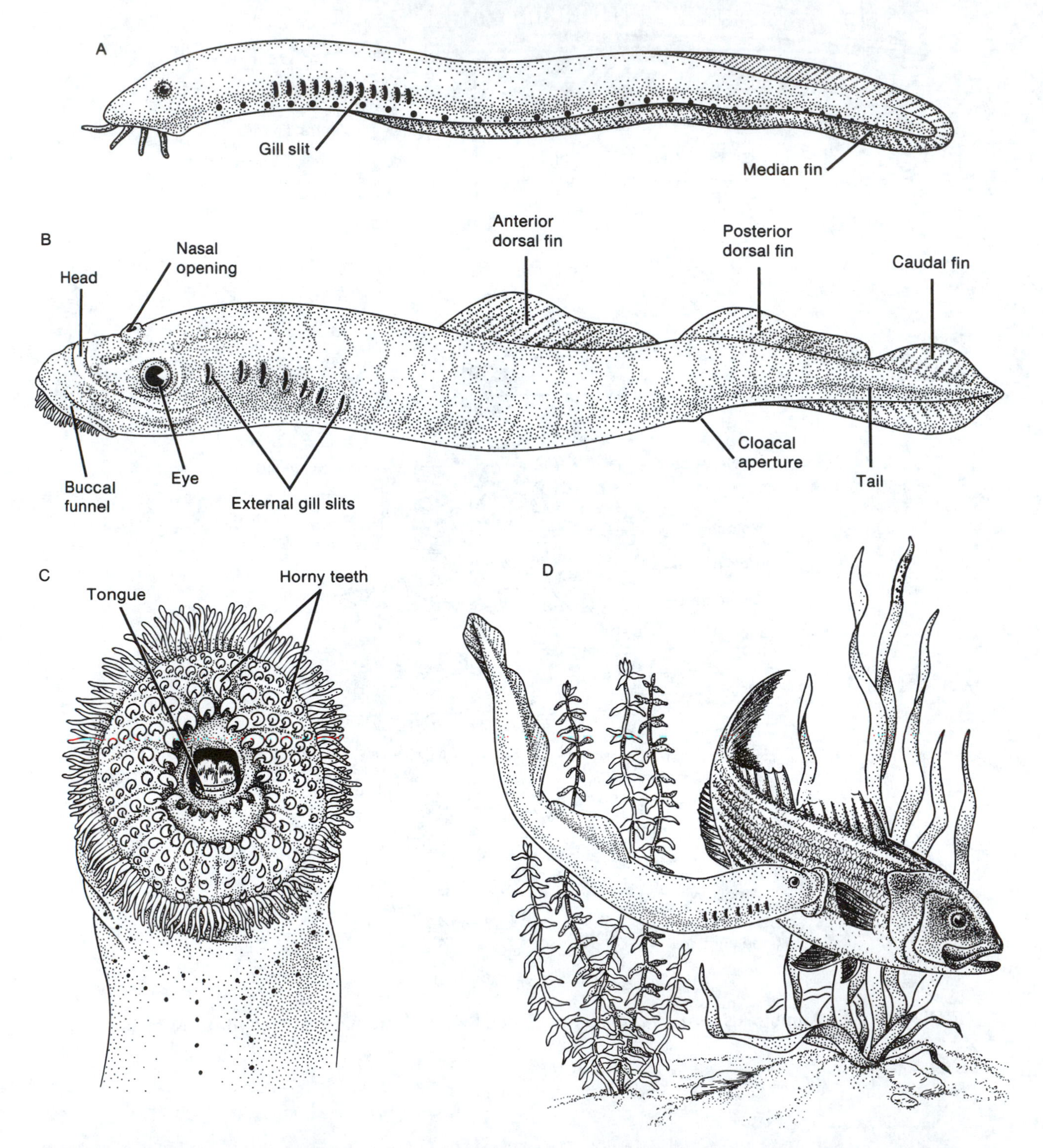

FIGURE 31-4 Agnatha. (A) Hagfish. (B) Sea lamprey. (C) Buccal funnel of lamprey. (D) Lamprey attached to fish. (Parts B–D after *Atlas and Dissection Guide for Comparative Anatomy,* 5th ed., by Saul Wischnitzer. W.H. Freeman and Company © 1993.)

of an eel, and the dorsal fins. How many dorsal fins are there?

___

How many ventral fins?

___

At the anterior end of the body, locate the seven external gill slits, which open into the gills. Just anterior to the gill slits are the eyes. Between the eyes, on the dorsal surface, is a single nasal opening that functions as an olfactory (smell) organ. The ventrally located mouth lies in a suctorial disc called the **buccal funnel** (Fig. 31-4C). Lampreys attach themselves to their prey by the suction generated by the buccal funnel. Then, using the pointed horny teeth inside the buccal funnel and the rasp-like tongue, the lamprey penetrates the flesh of the animal and feeds on the blood (Fig. 31-4D). Some years ago, the sea lamprey nearly eliminated the commercial fish (whitefish, lake trout, lake perch) of the Great Lakes. Successful eradication measures, including chemical control and electric barriers to block migration, have virtually eliminated the lamprey, and the population of commercial fish has steadily increased in these waters.

Examine the **ammocoete larva** of the lamprey. To which organism already studied does it bear a striking resemblance?

___

In terms of evolutionary relationships, what is the significance of this similarity?

___

___

___

b. Class Chondrichthyes

Chondrichthyes (Greek *chondros,* "cartilage"; *ichthys,* "fish") include sharks, skates, and rays. All are predators, and most live in the ocean, though a few are found in tropical rivers and lakes and in other fresh and brackish waters.

This class is characterized by skeletons that are made of **cartilage,** a softer, more flexible material than bone. The skin is covered by small, pointed scales made of plates of dentine covered by enamel, which structurally resemble vertebrate teeth.

Sharks, which are active swimmers, are usually found in the open oceans. Their typical diet includes fish, squid, and small crustaceans. Some of the larger sharks prey on sea lions and seals. The principal diet of rays and skates is small invertebrates.

In many countries, including the United States, sharks and rays are used for food, though their meat is usually not identified as such. In Asia, the shark's fins are boiled to yield a gelatinous substance that is used as a flavoring for soups.

External anatomy. Examine the dogfish shark *(Squalus acanthias),* a small species of shark that frequents the eastern shore of the United States (Fig. 31-5A). Locate the following external features.

The blunt head contains a broad **mouth** that has several rows of sharp teeth. The teeth, unlike those of the bony fishes and higher vertebrates, are not attached to the jaw but are embedded in the flesh. Teeth are continuously being formed and move forward to replace those that are lost.

The two **nostrils** contain olfactory organs, enabling the shark to smell chemicals dissolved in the water. The **eyes** lack lids and are adapted for use in dim light.

Anterior to each **pectoral fin** are six external **gill slits.** Five of them look like slits. The sixth, located just behind the eye, is the **spiracle,** a specialized structure through which, along with the mouth, oxygen containing water enters the gill chamber.

The **anus** is located between the pelvic fins. In the male, the pelvic fins possess claspers, which are brought close together during mating and inserted into the cloaca of the female. Seminal fluid containing sperm flows down the channel formed by the claspers and enters the female.

Separate anterior and posterior **dorsal fins** each have a characteristic spine just anterior to the fin. The **caudal** (tail) **fin** is bilobed and is used to propel the animal through the water.

The **lateral line,** a fine groove along each side of the body, contains a canal that has numerous openings to the surface. In the canal are sensory hair cells that connect to the tenth cranial nerve. These sensory cells respond to low-frequency pressure stimuli in the water, serving as a sense of touch at a great distance.

The body itself is covered evenly by diagonal rows of **placoid scales.** Each scale is covered by enamel and has an inner dentine layer.

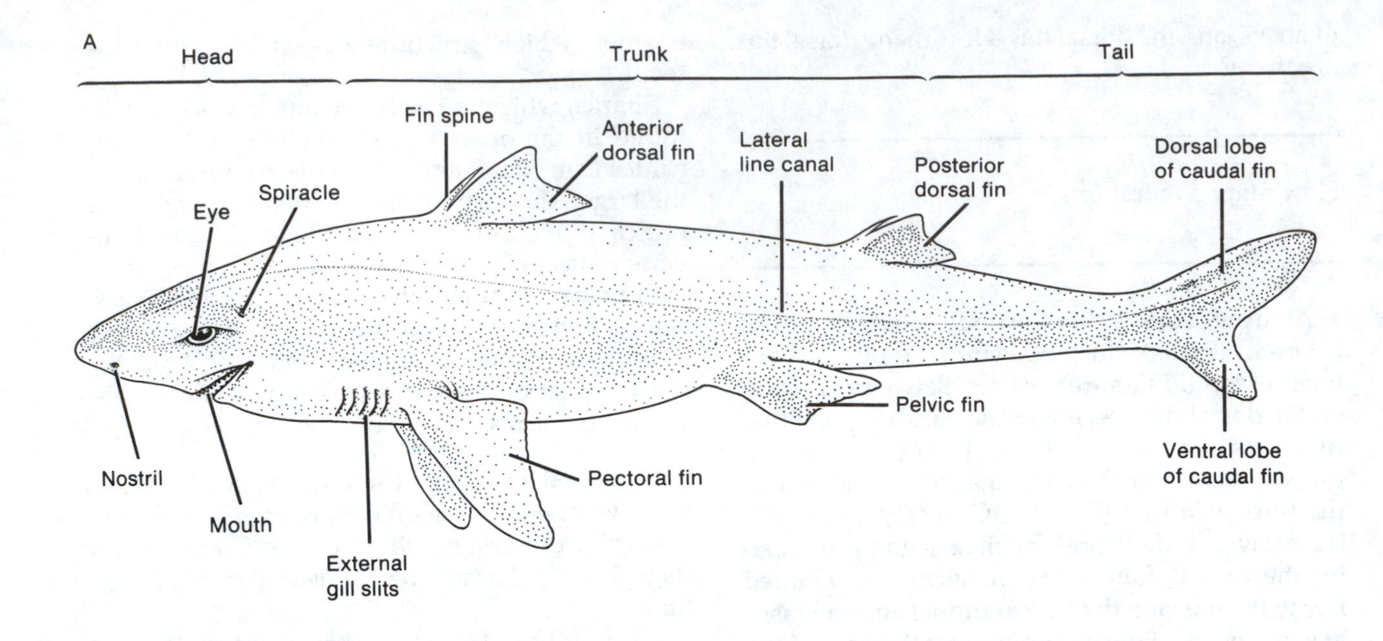

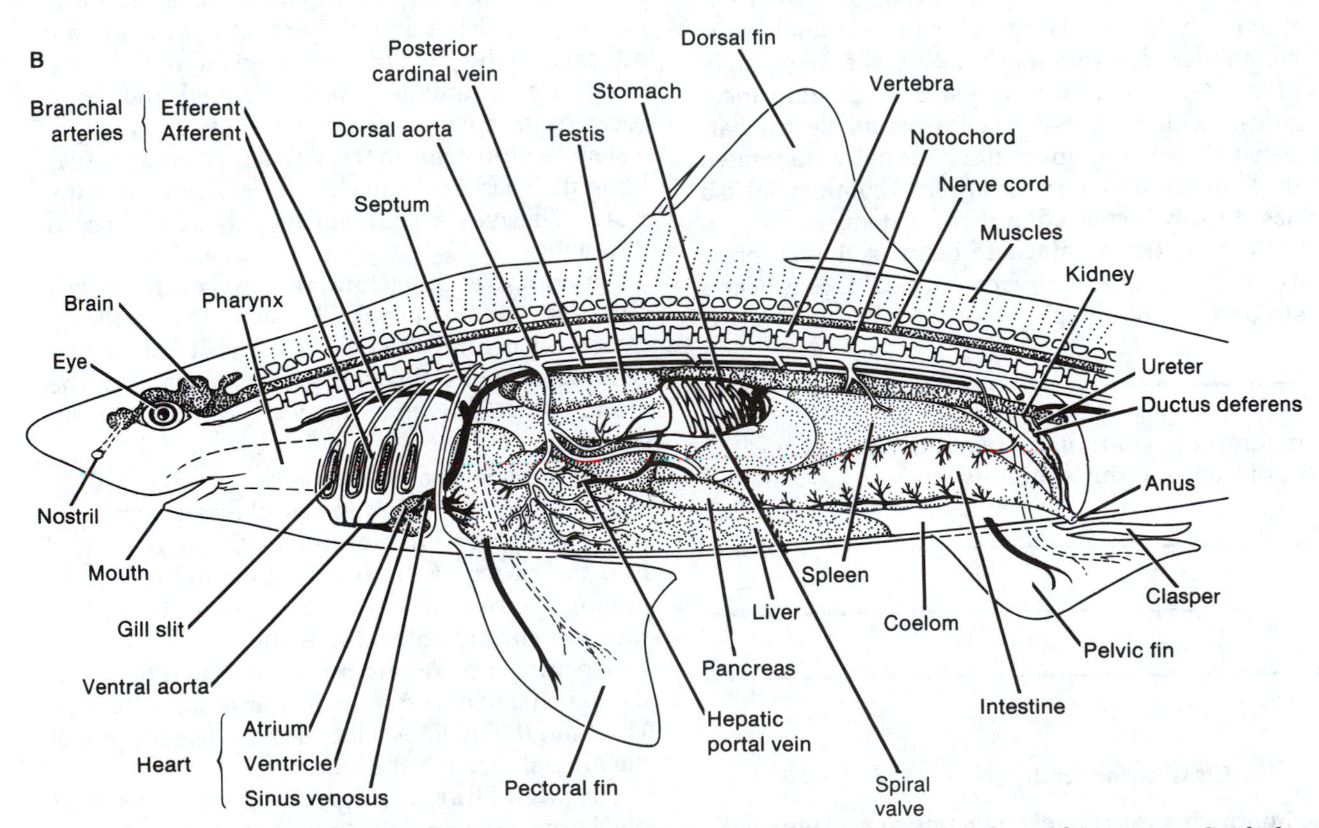

FIGURE 31-5 Dogfish shark. (A) External anatomy. (B) Internal anatomy. (After *Atlas and Dissection Guide for Comparative Anatomy,* 5th ed., by Saul Wischnitzer. W.H. Freeman and Company © 1993.)

Internal anatomy. The internal anatomy of the shark is shown in Fig. 31-5B. A peculiar feature of the digestive system is the **spiral valve,** a spirally arranged partition covered with a mucous membrane that increases the area for absorption.

c. Class Osteichthyes

The Osteichthyes (Greek *osteon,* "bone"; *ichthys,* "fish"), or bony fish, are characterized by skeletons of bone, a skin covered with dermal scales, and gills covered by an operculum. This class includes most of the animals we think of as fish (Fig. 31-6), from the common yellow perch to the more unusual lungfish. Fish are a successful and adaptable group that inhabit virtually all types of water habitats, including fresh, brackish, and salt. Although most fish are streamlined, usually having spindle-shaped bodies, they have developed a variety of forms for various habitats. Fish are vitally important as a source of food for people throughout the world. Many species are caught for sport.

The common yellow perch *(Perca flavescens)* is typical of the bony fish and thus is a good example to study (Fig. 31-7).

External anatomy. The main body regions of the perch are the head, trunk, and tail (caudal fin). The head extends from the tip of the snout to the hind edge of the operculum, the trunk from this point to the anus, and the tail is the remainder (Fig. 31-7A). Examine the head and observe the large mouth, which has distinct upper and lower jaws, the upper jaw being the **maxilla** and the lower jaw the **mandible.** Dorsally on the snout are two double external **nares** (nostrils); they are the openings to the olfactory sacs, which are highly sensitive to dissolved chemicals in the water. There is no external ear, but the internal ears contain semicircular canals, which function as balancing organs that enable the fish to maintain the proper orientation in the water. The lateral eyes are without lids. Behind each eye is a bony operculum, which is a protective covering for the four comblike respiratory gills.

Cut away the operculum from the left side and examine the gills. Cut away one gill, place it in water, and examine it with a stereoscopic microscope. The spaces between the gills are called gill slits or gill clefts. Attached to the lower edge of the operculum is the **branchiostegal membrane,** supported by rays of cartilage. This membrane serves as a one-way valve that allows water to pass out of the opercular opening but prevents its return.

The lateral line, which extends along each side of the whole body of the perch, is essentially a system of water-filled canals that communicate by means

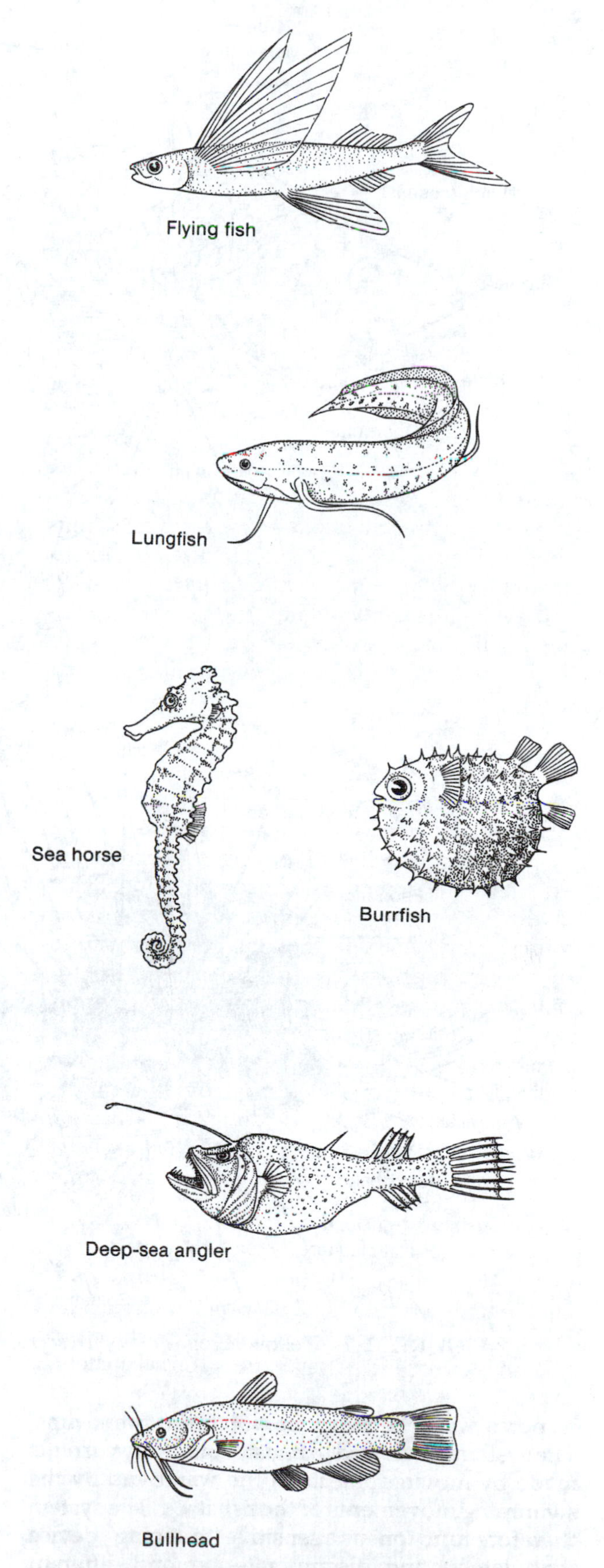

FIGURE 31-6 Representative bony fish

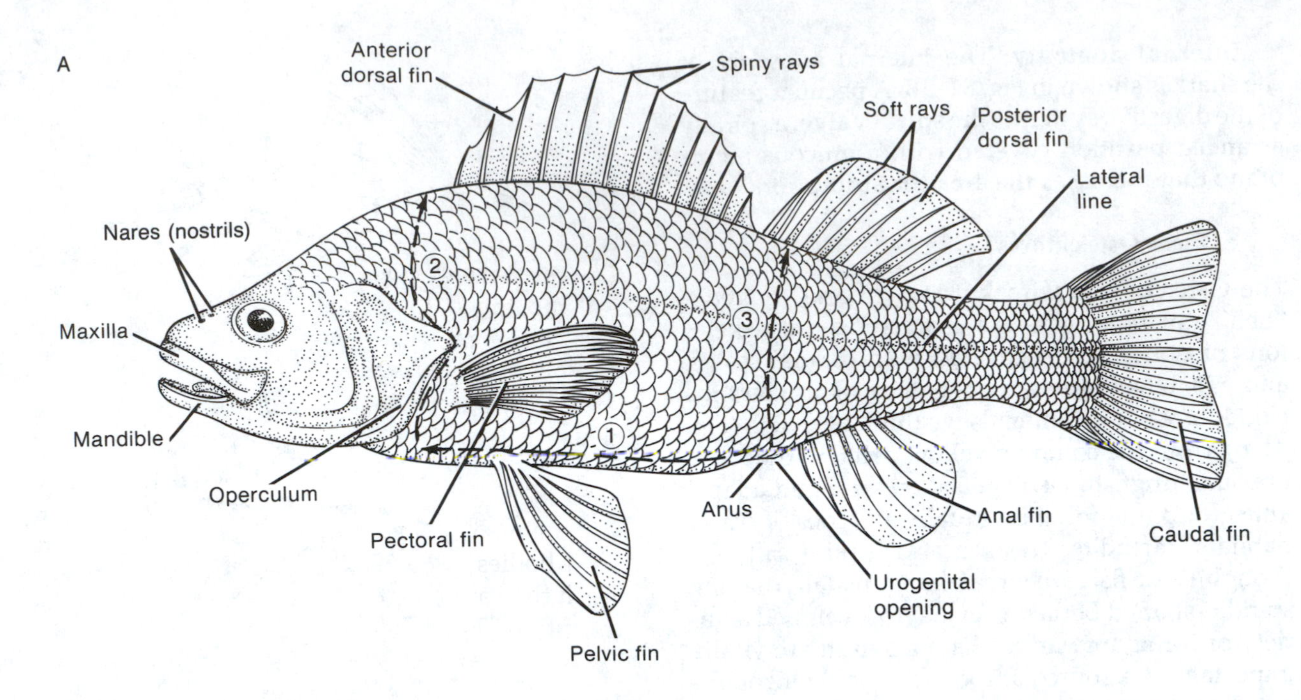

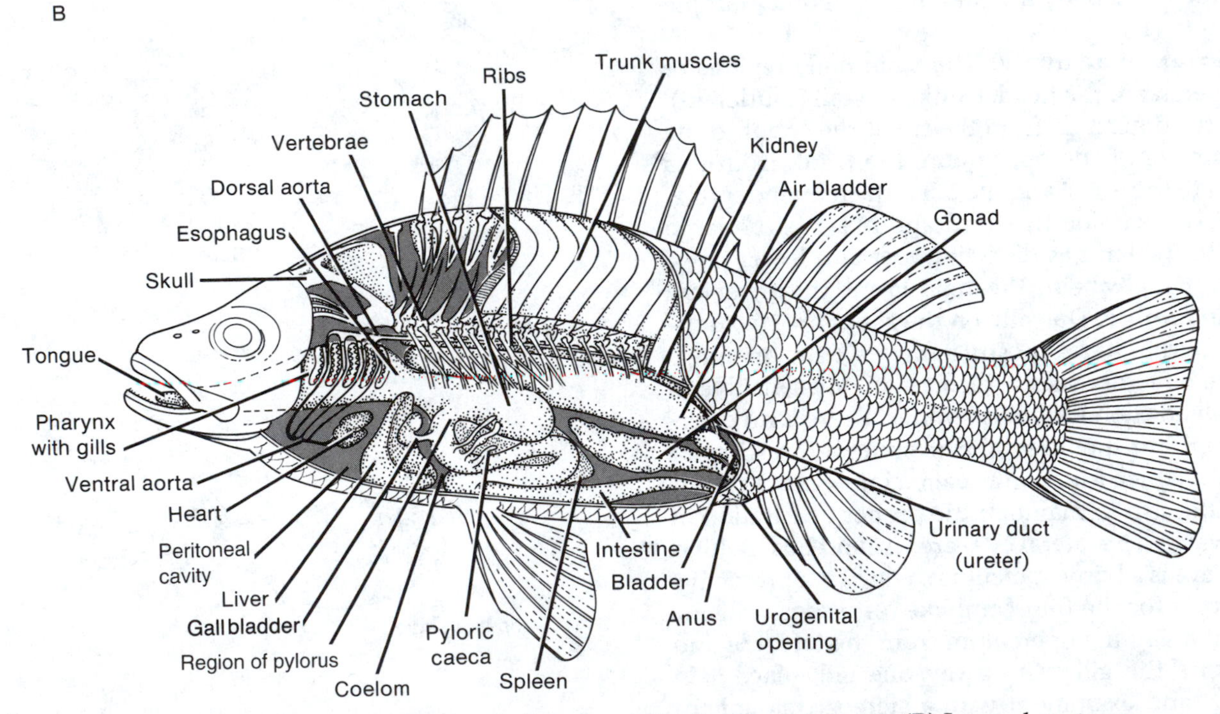

FIGURE 31-7 Yellow perch *(Perca flavescens).* (A) External anatomy. (B) Internal anatomy.

of pores with the water in which the fish swims. The system is believed to register vibratory currents made by moving objects in the water and by the swimming movements of the fish itself. The system therefore functions as a sensitive "listening" device that detects and discriminates among different kinds of turbulence.

Two separate dorsal fins are on the back of the perch. A caudal fin is on the end of the tail, and an anal fin is on the ventral side of the tail region. Just anterior to the anal fin are the anus and the **urogenital opening.** The lateral paired pectoral fins are behind the opercula, and the ventral pelvic fins are just below them. The fins, membranous extensions of the integument, are supported by **fin rays.** Each anterior dorsal fin has 13–15 solid calcified spines, and there are one or two similar rigid spines at the anterior edge of the other fins. The fins are used in swimming, steering, and maintaining equilibrium.

The trunk and tail are covered by thin, rounded scales. Remove a scale, mount it in a drop of water on a slide, and examine it under the low power of a microscope. Note the annual growth rings. As a fish grows, the size of each scale increases, rather than the number of scales. As a scale grows, concentric rings are formed. Rings formed in the fall and winter are closer together than those formed in the spring and summer. By counting the number of regions of closely spaced concentric rings, you can determine the number of winters your specimen lived. What is the approximate age of your fish?

____

The whole body of the fish is covered by a soft mucus-producing epidermis that facilitates its movement in water and protects against entry of disease organisms.

Internal anatomy. Hold the fish in the dissecting pan, ventral side up with the head pointing away from you. Insert the point of your scissors through the body wall in front of the anus and cut up the midline of the body to the space between the opercula (line 1 in Fig. 31-7A). Now lay the fish on its right side (with the head on your left). Continue to cut around the back edge of the gill chamber to the dorsal surface (line 2 in Fig. 31-7A). Make another incision from the starting point of the ventral incision close to the anus, and cut upward to the dorsal surface (line 3 in Fig. 31-7A). Be careful not be disturb the internal organs. Remove the whole lateral body wall by cutting along the top of the body cavity. This procedure will expose the internal organs in their normal positions.

Digestive system. Locate the reddish brown liver in the anterior end of the body cavity (Fig. 31-7B). Raise the lobes of the liver and find the **gallbladder** attached to the lower side. Cut the liver free and remove it. This will expose the short esophagus and the stomach. Locate the **pylorus** where the stomach and intestine join. Three tubular **pyloric caecae,** secretory or absorptive in function, attach to the intestine. Examine the loops of the intestine and trace the intestine to the anal opening. Long masses of fat lie along the intestinal loops. Observe the **spleen,** attached to an internal mesentery near the stomach. It is a dark red gland that has no duct and no functional connection with the digestive system. Cut the esophagus at its anterior end, and carefully remove the entire digestive tract, from the mouth to the anus.

Reproductive system. Having removed the digestive tract, you should be able to see the gonads and the urinary bladder.

In the female, the ovary lies between the intestine and the air bladder. It is a membranous sac filled with eggs. The posterior end of the ovary is tapered, and the eggs pass to the outside through the urogenital opening just behind the anus. The ovaries are paired in early stages, as they are in other vertebrates, but during development they fuse into a single organ.

In the male, the testes are a pair of white, elongated bodies lying just below the air bladder, to which they are joined by a thin sheet of tissue, the mesentery. They fuse toward the posterior end, and the sperm are passed to the outside through the urogenital opening.

Air bladder (swim bladder). Locate the **air bladder** along the top of the body cavity. The bladder is filled with gases (oxygen, nitrogen, and carbon dioxide) and acts as a hydrostatic organ to adjust the specific gravity of the fish at different depths of water. By secretion or absorption of gases through blood vessels in the bladder wall, a fish makes the adjustment slowly as it moves from one depth to another.

Excretory system. Remove the air bladder from the body. Lying in the roof of the bladder, near the middle, is a large blood vessel, the dorsal aorta. Parallel to it is a pair of long, narrow, dark-colored kidneys. The urinary ducts **(ureters)** run along the edges of each kidney, join at their posterior ends, and empty into the urinary bladder. Fluid nitrogenous wastes removed from the blood are emptied by means of this excretory system through the urogenital opening.

Circulatory system. Using the points of a pair of scissors, make a horizontal cut through the anterior wall of the body cavity in front of the liver. This will expose the **peritoneal cavity,** in which lies the two-chambered heart. The heart consists of a single, light-colored **ventricle** and a larger thin-walled **auricle.** Behind the auricle is the **sinus venosus.** The large vessel carrying blood from the ventricle is the **ventral aorta.** It is greatly enlarged just anterior to the ventricle, forming the **conus arteriosus.** Trace the ventral aorta forward toward the gill region and identify some of the branchial arteries. Rhythmic contractions of the ventricle force the blood through the conus arteriosus and short ventral aorta into four pairs of **afferent**

branchial arteries, which distribute blood to capillaries in the gill filaments for oxygenation. In the gill filaments, the blood is collected into correspondingly paired **efferent brachial arteries** leading to the dorsal aorta, which has branches to all parts of the head and body.

Nervous system. Hold the fish dorsal side up with the head pointing away from you. Using a scalpel, cut the skin from the skull and scrape the skull carefully to wear away the bone. When the bone is very thin, remove the pieces with forceps to expose the brain. Locate the **olfactory lobes** in front, the larger lobes of the **cerebrum** behind them, and the very thin **optic lobes** posterior to the cerebrum. The **cerebellum** is posterior to the optic lobes, and the **medulla** is the enlargement where the **spinal cord** joins the brain.

d. Class Amphibia

The amphibians include toads, frogs, mud puppies, and salamanders (Fig. 31-8). As the name of this class implies (Greek *amphi,* "dual"; *bios* "life"), these animals live both on the land and in the water. Indeed, their position in the evolutionary scale is between fish and reptiles in that they were the first chordates able to live on land. They exhibit several modifications that have enabled them to adapt to a terrestrial existence. These modifications include legs, lungs, nostrils, and sense organs that function in both water and air. Amphibians do not rely only on their lungs for respiration. Their moist glandular skin, which is extensively supplied with blood vessels, also serves as a respiratory surface.

Most amphibians lay their eggs in the water, and have an aquatic larval stage called a **tadpole.** Some amphibians, particularly the salamanders and newts, resemble lizards. Reptiles, however, have scales, whereas salamanders and newts have smooth, slimy skin.

Amphibians are very beneficial to human beings. For example, frogs are used extensively for laboratory dissections, pharmacological experiments, and fish bait. Extracts of frog skin are used to make the glue used in book bindings. In Asia, toad skin is used medicinally; certain glands contain digitalislike secretions that increase blood pressure. In France and in tropical countries, frogs and toads have been introduced as insect controls.

The common leopard frog, *Rana pipiens,* will be studied in detail in Exercises 32 and 33. Examine other representatives of this class, noting particularly the external adaptations that have enabled

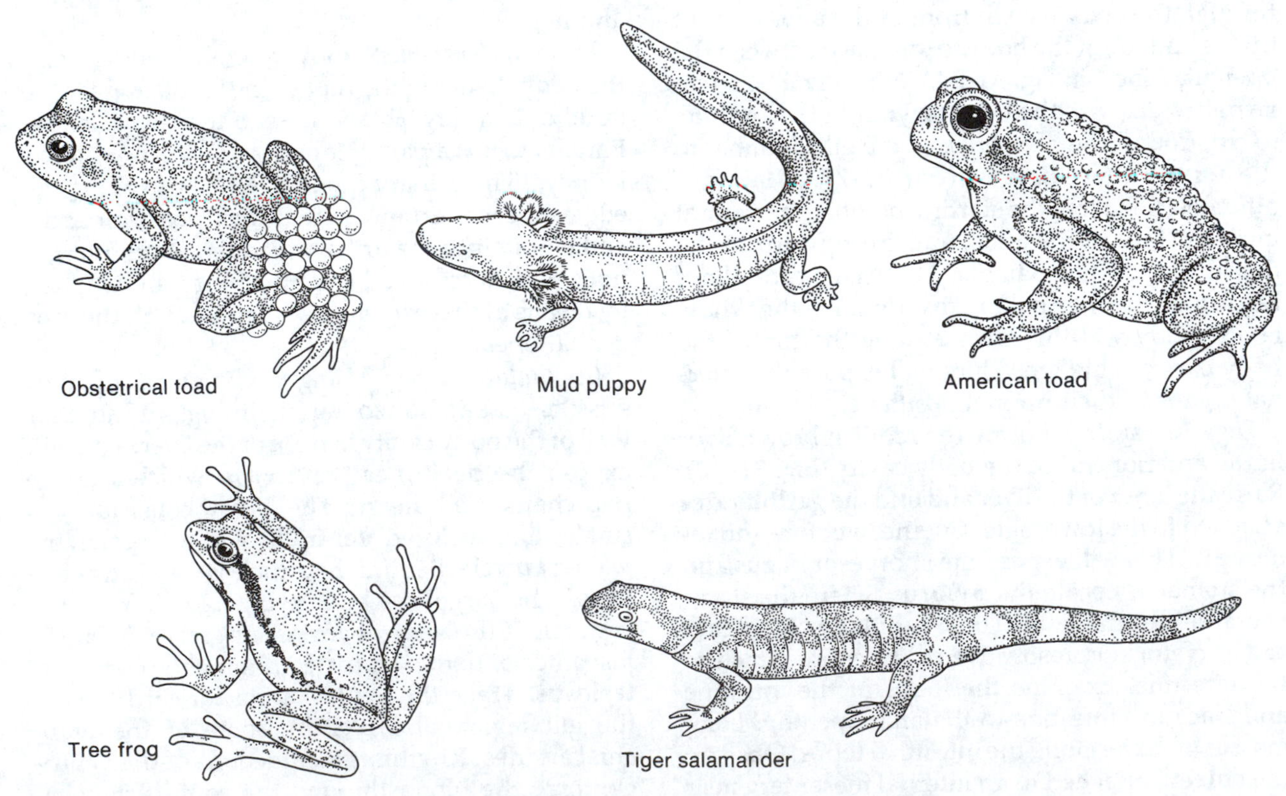

FIGURE 31-8 Representative amphibians

them to make the transition from the water to the land.

e. Class Reptilia

Reptiles include some of the most interesting and diverse chordates, such as the turtles, snakes, lizards, crocodiles, and alligators (Fig. 31-9). They are **poikilothermic** (have variable body temperature), and most are covered with hard, dry, horny scales or bony plates. They respire by means of lungs. They were the first vertebrates to adapt to dry places, the dry skin and scales retarding moisture loss from the body and enabling the animal to occupy rough land surfaces. The reptiles are true terrestial animals and do not need to return to water to reproduce, as do most of the amphibians. The name of the class indicates its mode of locomotion (Latin *reptur,* from *repere,* "to creep").

Examine available representatives of the various modern reptiles and note their great diversification in form. Note particularly the ways in which reptiles are morphologically more complex than amphibians—ways that have adapted them for life on land, such as a dry, scaly skin to prevent desiccation, and appendages (when present) that are suited for rapid locomotion on land or in the water.

Modern reptiles are only a fraction of the known orders that flourished during the Mesozoic era, the age of reptiles. During that time, reptiles were the dominant vertebrates and occupied most of the habitats available, from dry uplands and deserts to marshes, swamps, and oceans.

f. Class Aves

Birds are probably the best known and most easily recognized of all nonmammalian vertebrates because they have feathers and are capable of flight (Fig. 31-10). No other animals have feathers, though bats (which are mammals) can fly. These animals have very efficient respiratory systems that provide the oxygen for the high metabolic rate necessary for the tremendous muscular activity of flying.

Birds are considered to have a reptilian ancestry. Indeed, their early embryonic development parallels that of reptiles, and scales, a reptilian characteristic, persist on the legs of birds. Fossils of some of the earliest birds have reptilelike teeth.

Examine the demonstration of various birds. These animals have a number of striking adaptive features that enable them to occupy a variety of habitats. Coloration, for example, is varied and striking. Although some birds show only one color, the feathers of most are marked with spots, stripes,

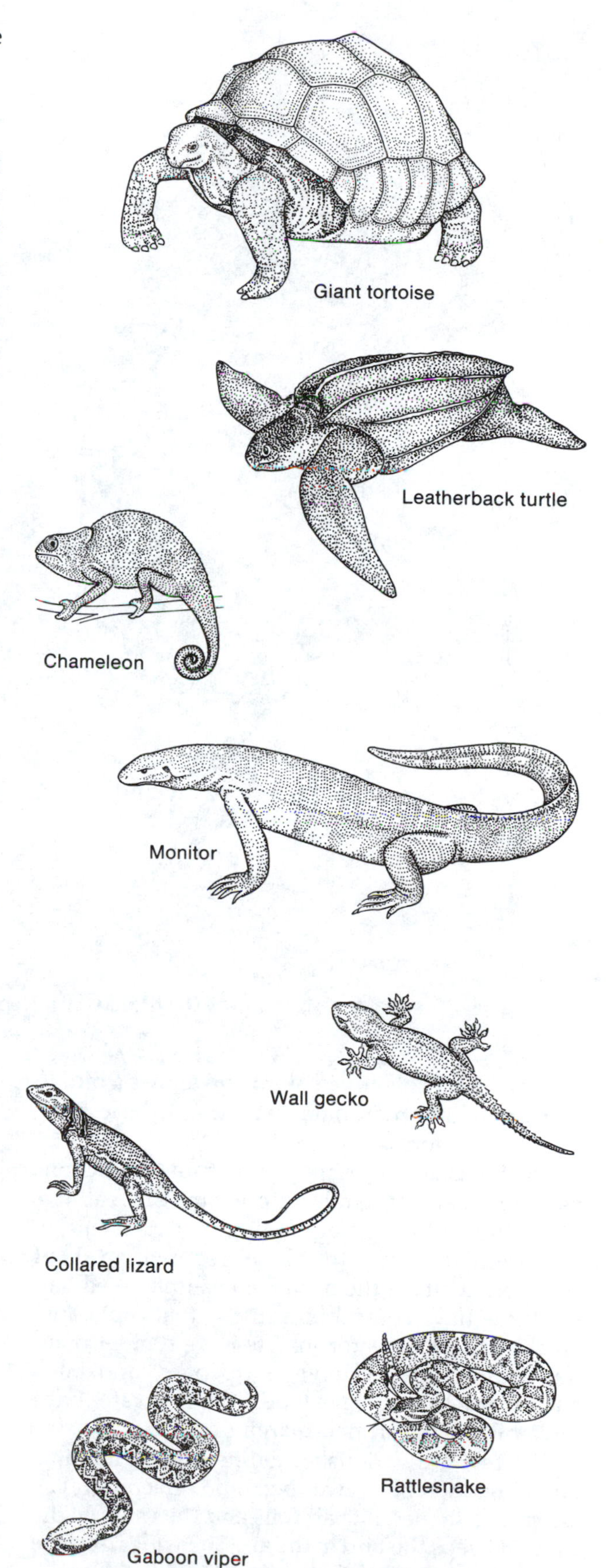

FIGURE 31-9 Representative reptiles

FIGURE 31-10 Representative birds

or bars. This **protective coloration** allows a bird to blend with the environment and thus to be less visible to predators.

The bill is a multipurpose structure. It functions as a mouth and as hands for preening the feathers, obtaining and arranging nesting materials, and in defense. The form of the bill tells something about the food habits of the bird. For example, seed-eating birds have conical bills, those that probe into cracks and crevices for insects have slender ones, flesh-eating species have sharp ones, and birds such as ducks that sieve food from the water have wide bills with serrated margins.

The feet are variously modified for swimming, climbing, perching, wading, and grasping. The wings are shaped like air foils and thus supply the lift that keeps the bird in the air. The wings of some birds (penguins) are further modified for "flying" under water. In some species, such as the ostrich and kiwi, the wings have so degenerated that the bird is no longer capable of flight.

g. Class Mammalia

Mammals include rodents, monkeys, bats, horses, whales, cattle, deer, and human beings (Fig. 31-11). All are covered to varying degrees with hair and are warm-blooded. They are called *mammals* because the young are nourished by milk from the mammary glands of the females. The evolution of mammalian care of their young is complex and interesting. "Primitive" mammals, such as the duckbill platypus, resemble reptiles in that they lay eggs. In more complex mammals, such as the marsupials (opossum and kangaroo), the young are born in a very immature condition and transfer to a pouch where they suckle until they are more mature. The young of the most complex mammals are

FIGURE 31-11 Representative mammals

retained in the uterus until they reach an advanced stage of development. What is the advantage of retaining the embryo in the uterus during its development?

____

____

____

What advantage is there in nourishing the young with milk from the mammary gland rather than with the food in the yolk of a reptilian egg?

____

____

____

____

Mammals live in all kinds of habitats, from the tropics to the poles (musk ox and polar bear) and from the oceans to the driest deserts (kangaroo rat, camel). They have been able to move into these niches and survive because of the wide diversity in their morphological, physiological, and behavioral features.

REFERENCES

Applegate, V. C., and J. W. Moffett. 1965. The Sea Lamprey. *Scientific American* 192(4):36–41.

Gilbert, P. W. 1962. The Behavior of Sharks. *Scientific American* 207(1):60–68 (Offprint 127). *Scientific American* Offprints are available from W. H. Freeman and Company, 41 Madison Avenue, New York, NY 10010, and 20 Beaumont Street, Oxford OX1 2NQ, England. Please order by number.

Romer, A. S., and T. S. Parsons. 1986. The Vertebrate Body. 6th ed. Saunders.

Storer, T. I., et al. 1979. *General Zoology.* 6th ed. McGraw-Hill.

Villee, C. A., W. F. Walker, and R. D. Baines. 1984. *General Zoology.* 6th ed. Saunders.

Exercise

32

Vertebrate Anatomy: External Anatomy, Skeleton, and Muscles

Frogs are probably the animals most studied in introductory biology courses. Their structure clearly shows the major features of the organ systems characteristic of the phylum Vertebrata to which they belong. In addition, the frog is one of the most widely used animals in vertebrate physiology and other experimental studies. Vertebrates are of particular interest because humans are members of the phylum.

Contemporary amphibians are grouped into three orders: frogs and toads, salamanders, and legless, worm-shaped caecilians of the tropics. All reproduce in water or in very moist places on land. Most go through an aquatic larval stage, called the **tadpole,** which is followed by metamorphosis to a terrestrial adult. Adult amphibians have only a rudimentary ability to conserve body water, so they must live in damp habitats on land and must often return to water. As a result of their phylogenetic position between fishes and reptiles and their double mode of life in water and on land, amphibians have a mixture of aquatic and terrestrial attributes.

The most widespread of North American frogs is the leopard frog, also known as the grass or meadow frog. Its scientific name is *Rana pipiens.* Another frog that is often studied, because of its large size, is the bullfrog, *Rana catesbeiana.*

A. ANATOMICAL TERMINOLOGY

Beginners in the study of any science are often confused by its jargon, and students of anatomy are no exception. However, the specialized terminology of anatomy is necessary to achieve precision. For example, when referring to a point on an animal, what do we mean when we say *above, behind, over, on top of,* or *below?* These words may be interpreted differently by different persons. To eliminate ambiguity, anatomists have developed a set of well-defined terms that are used universally to locate and identify body structures and features (see Appendix H, Glossary of Common Anatomical Terms).

1. Orientation and Direction

The following terms describe direction and the position of body parts with respect to each other. In studying these terms, refer to Fig. 32-1. Note that some terms have a different connotation when referring to a four-legged animal than they do when referring to a human.

- **Dorsal/ventral** (back side/belly side): These terms are generally applied to four-legged

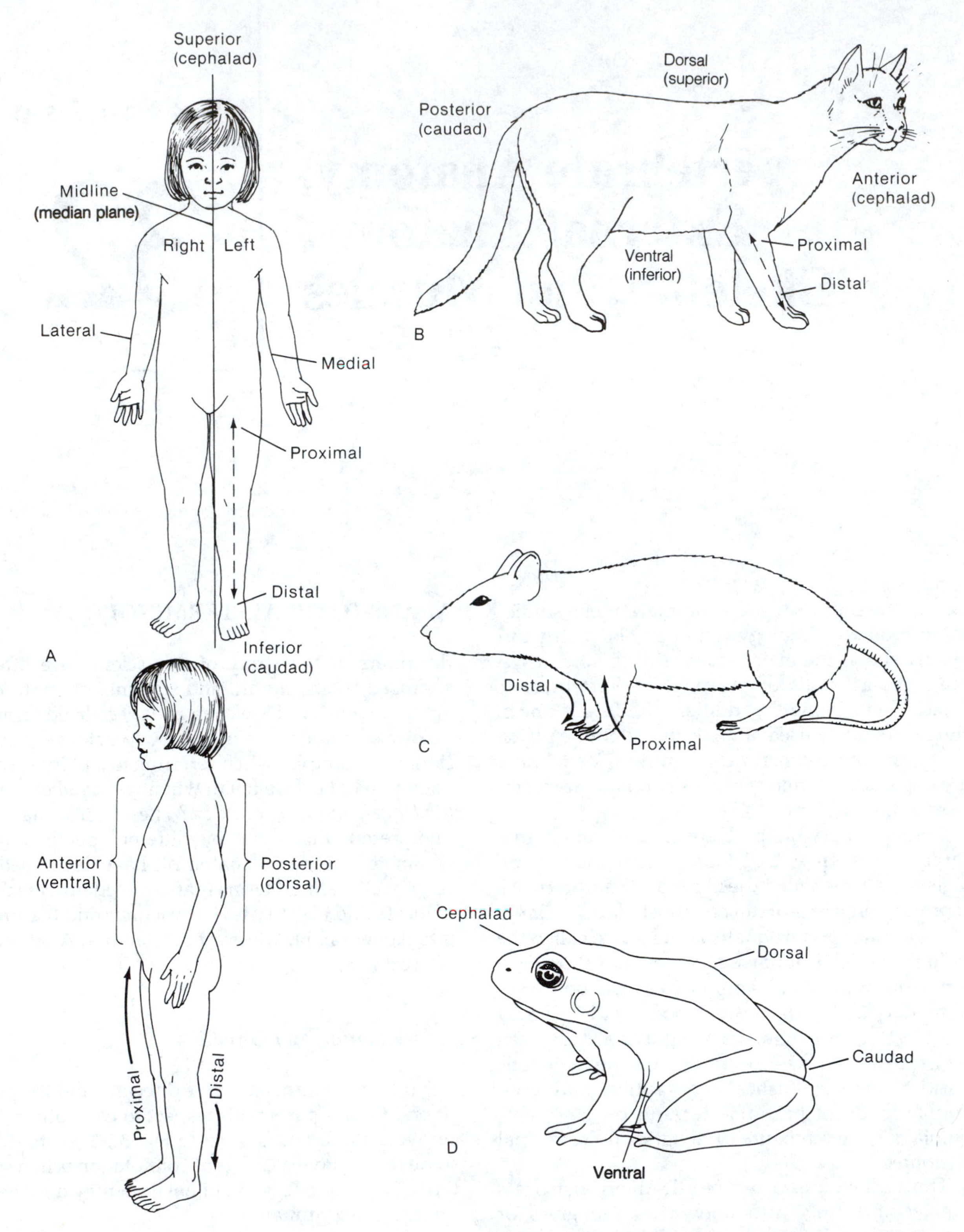

FIGURE 32-1 Terms used for anatomical orientation and direction in (A) human, (B) cat, (C) mouse, and (D) frog

animals and are used interchangeably with the terms *superior* and *inferior*, respectively. *Dorsum* is Latin for "back," so dorsal refers to the back side of the body or other structure. The back side of the arm, for example, is its dorsal surface. The Latin term *venter* means "belly," so ventral refers to the belly side of an animal. In humans, dorsal and ventral are synonymous with the terms *posterior* and *anterior*, respectively.

- **Anterior/posterior** (front/back): When these terms are used in reference to humans, they describe surfaces that are facing forward or backward. The abdomen, the chest, and the face are on the anterior surface. Posterior surfaces (the back and buttocks) are on the back side of the body. These terms can also be used to describe the position of one structure in relation to another. For example, one structure can be anterior or posterior to another (e.g., the molar teeth are posterior to the bicuspids).
- **Superior/inferior** (above/below): The Latin *super* means "above." Thus, a structure located above another is said to be superior; the eyes are superior to the nose, for example. *Inferus* is Latin for "low" or "below." A structure is inferior to another structure if it is underneath or below that structure; the abdomen is inferior to the chest.
- **Cephalad/caudad** (Toward the head/toward the tail): For four-legged animals, these terms are interchangeable with anterior and posterior. For humans, they can be used alternatively with superior and inferior.
- **Proximal/distal** (nearer, or toward the body or attached end/farther, or away from the body or attached end): These terms locate various parts of the limbs. For example, the fingers are distal to the wrist; the knee is proximal to the ankle; the elbow is distal to the shoulder.
- **Medial/lateral** (toward the midline, or median plane, of the body/away from the midline of the body): The midline of the body is an imaginary line on the plane that divides the body into left and right halves (Fig. 32-1A). Medial (Latin *medius*, meaning "middle") refers to surfaces or structures closest to the midline. The inner surface of an arm or leg is its medial surface because it is closest to the body's midline. The term *lateral* is the opposite of medial (Latin *lateralis*, meaning "side"). The outside surface of an arm is its lateral surface.

2. Planes and Sections

To see internal structures and their positions relative to each other, it is necessary to make a cut, or **section**, through the body. When a section of the body is made, it is through an imaginary line called a **plane.** Three sections have been classified by anatomists (Fig. 32-2).

- **Longitudinal (sagittal) section:** A cut that is made parallel to the long axis of the body and divides the body into right and left sections is a longitudinal, or sagittal, section. If the cut divides the body into equal left and right halves, it is referred to as the midsagittal, or **median, section.** Cuts made parallel to the median (midsagittal) plane are called **parasagittal sections.**

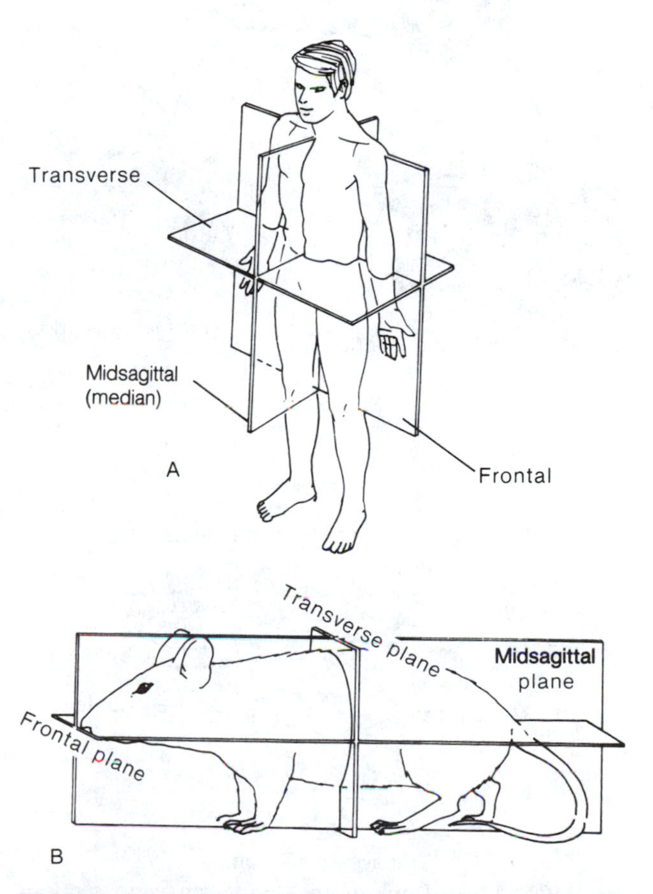

FIGURE 32-2 Planes and sections used in anatomical terminology shown in (A) a human and (B) a mouse

- **Frontal section:** This section is cut along a longitudinal plane that divides the body into anterior (ventral) and posterior (dorsal) regions.
- **Transverse (cross) section:** Any cut of the body that is made perpendicular to the longitudinal or frontal plane is a transverse section. It divides the body into superior (cephalad) and inferior (caudad) parts.

  Longitudinal and transverse sections are frequently made of various organs and tissues for gross and microscopic observation. As shown in Fig. 32-3, there is quite a difference in appearance between an organ or tissue cut in a transverse section and one cut in a longitudinal section.

The vertebrate body has two main cavities (Fig. 32-4).

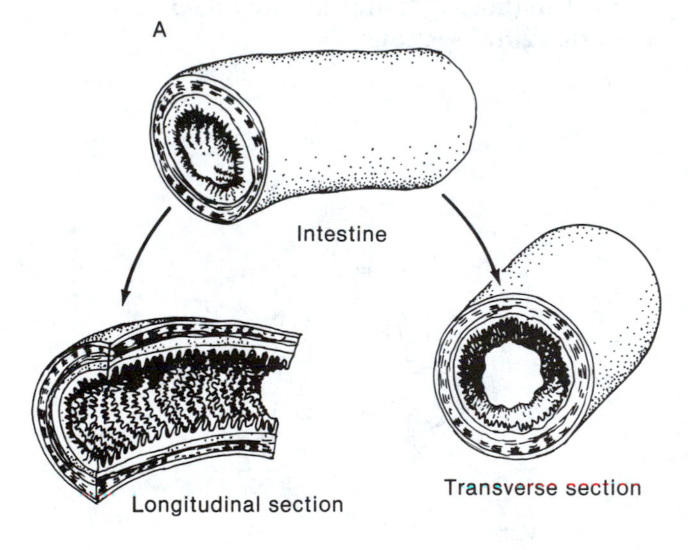

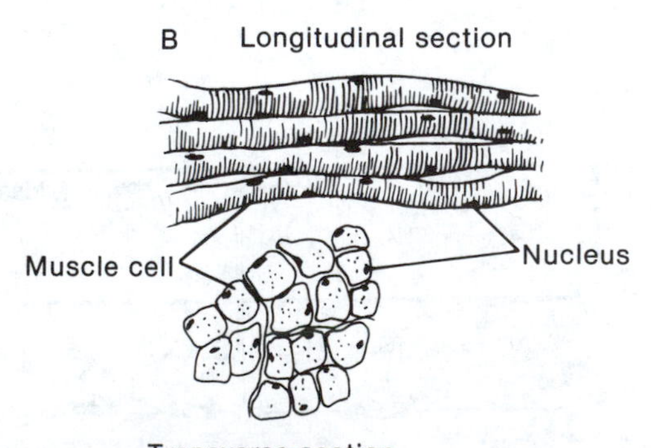

FIGURE 32-3 Longitudinal and transverse sections made of (A) an organ, the human intestine, and (B) a tissue, human muscle

- **Dorsal body cavity:** This cavity consists of the **cranial cavity** and the **vertebral** or **spinal canal.** The cranial cavity is surrounded by the bones of the skull and contains the brain. The vertebral canal, surrounded by the vertebrae, contains the spinal cord, parts of the spinal nerves, and the spinal fluid.
- **Ventral body cavity (coelom):** This cavity is located in the anterior (or ventral) part of the body and contains the organs **(viscera).** The coelom is subdivided into two regions by the muscular diaphragm. The upper region is called the **thoracic** (or chest) **cavity.** This is further partitioned into two **pleural cavities,** which contain the lungs, and the **pericardial cavity,** in which the heart is located.

  The lower region, the **abdominopelvic cavity,** consists of two parts. The **abdominal cavity,** which is immediately inferior (caudad) to the diaphragm, contains the stomach, liver, spleen, pancreas, gallbladder, kidneys, ureters, small intestine, and most of the large intestine. Caudad to the abdominal cavity is the **pelvic cavity,** which contains the urinary bladder, sigmoid colon, rectum, and the reproductive organs.

B. EXTERNAL ANATOMY OF THE FROG

Examine a living or preserved frog and locate the anatomical features described here.

The body of a frog is divided into the **head,** which extends posteriorly to the shoulder region, and the **trunk** (Fig. 32-5). The adult frog moves by means of its powerful hind legs and webbed feet. The much smaller front legs are used primarily to raise the front of the body above the ground, but the male also uses them to clasp the female during mating. Notice the lack of a distinct neck; independent motion of the head and trunk would be disadvantageous for swimming.

The **cloacal aperture,** the opening for both the digestive and the urogenital tracts, is at the posterior end of the trunk just above the attachment of the hind legs. Undigested food wastes (feces), liquid waste from the kidneys (urine), and the gametes (sperm or eggs) from the reproductive organs are all discharged from this opening. The anus of human beings and other mammals is the opening of only the digestive tract and so is only partly comparable to the cloacal aperture.

A large **mouth,** a pair of **nostrils,** or **external nares** (singular **naris**), and the **eyes** will be recognized on the head. The upper **eyelid** is a simple

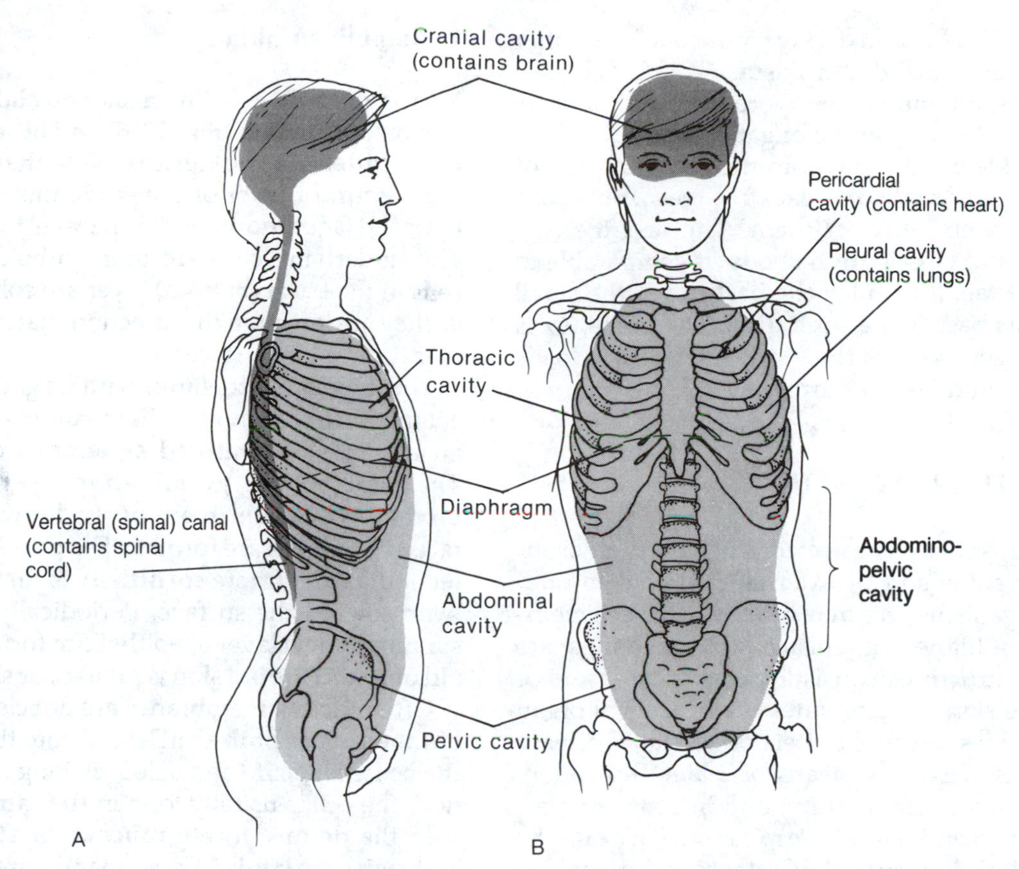

FIGURE 32-4 (A) Lateral view and (B) frontal view of the human showing the location of the dorsal and ventral body cavities

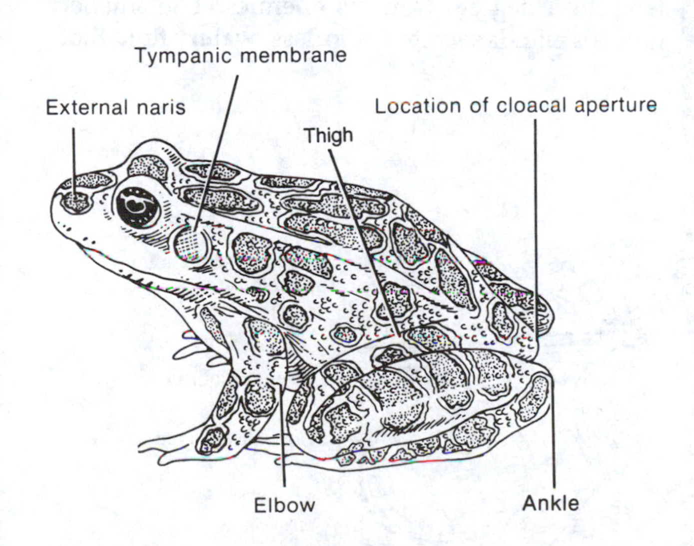

FIGURE 32-5 Lateral view of a leopard frog, *Rana pipiens.* (From *Dissection of the Frog,* 2d ed., by Warren Walker, Jr. W.H. Freeman and Company © 1981.)

fold of skin. The lower lid is a transparent **nictitating membrane,** which can be drawn across the eyeball. The flat disc-shaped area posterior to each eye is the eardrum, or **tympanic membrane.** It is the same size in male and female leopard frogs but is considerably larger in male than in female bullfrogs. This is one example of **sexual dimorphism** (structural differences between the sexes).

Look carefully at the top of the head between the eyes; you may see a small, light **brow spot,** the diameter of a pinhead. This is the remnant of a light-sensitive eye that characterized primitive groups of fishes and amphibians.

The appendages of a frog have the same parts as your own arms and legs. On the front appendages, locate the upper arm, elbow, forearm, wrist joint, hand, and four fingers. Although the finger closest to the body is comparable to our second digit, it is often called the *thumb* because it is stouter than the others. During the breeding season, the male's thumb is swollen and darkly pigmented—another example of sexual dimorphism.

In the hind appendages, locate the thigh, knee, shank, ankle joint, and elongated foot. Two elongated ankle bones lie within the proximal part of the foot: the distal part bears five toes with a conspicuous web between them. The first toe, the smallest and closest to the body, is comparable to our great toe. It is called the **hallux** and the small spur at its base is the **prehallux.** The prehallux is much larger in toads than in frogs, because toads use their hind feet in burrowing.

C. VERTEBRATE SKIN

All organisms are protected from their environment by some sort of external covering. In one-celled organisms, this may be only a thin cell membrane; in higher organisms, the coverings are complex structures consisting of several layers of cells. The skin of vertebrates is a functional organ that provides physical protection, excludes disease organisms, acts as a means of water absorption, helps regulate temperature, and in some animals serves an essential role in respiration. For example, in both the adult and tadpole stages of the frog, oxygen is absorbed and carbon dioxide is eliminated through the skin. This **cutaneous respiration** occurs both in water and in air. During the winter, when the frog lies buried in the mud, the skin becomes the only respiratory organ. Adult frogs also absorb oxygen and give off carbon dioxide through the lining of the mouth **(buccopharyngeal respiration)** and through the lungs **(pulmonary respiration).**

1. Amphibian Skin

Microscopically examine a stained slide of a cross section of frog skin (Fig. 32-6) and observe the distinct **epidermis** and **dermis.** Note that the cells of the outermost layer of the epidermis are thin and flattened **(squamous)** and lie parallel to the surface. The underlying cells are more cuboidal, and the cells in the basal (bottom) layer are columnar; that is, they are longer in the direction that is perpendicular to the skin surface.

This type of epithelium, with its gradation from columnar to squamous cells in successive layers, is an example of **stratified squamous epithelium.** The basal (bottom) **germinative layer** continually produces cells, which are pushed toward the surface as new cells are formed. The cells become flatter and harder (more **cornified)** as they move outward toward the surface. Periodically during the summer, a new layer of epithelium forms under the old one, and the old skin is molted, or sloughed off.

Often the cell membranes are not clearly visible, but the shapes of the nuclei indicate the shapes of the cells. Thus, if the nucleus is long in one direction, the cell is usually long in the same direction.

In the dermis, locate **mucous** or **slime glands** and **poison glands.** These glands are lined by a layer of secretory cells and produce fluids that are secreted onto the surface of the epidermis through ducts. The poison gland can be identified by its larger size and the granular-appearing material in the lumen (cavity). If a frog is roughly handled, the poison glands discharge a thick, whitish secretion that, by causing a burning sensation, protects the frog (to a degree) from its enemies. The smaller mucous glands secrete a colorless, watery fluid that

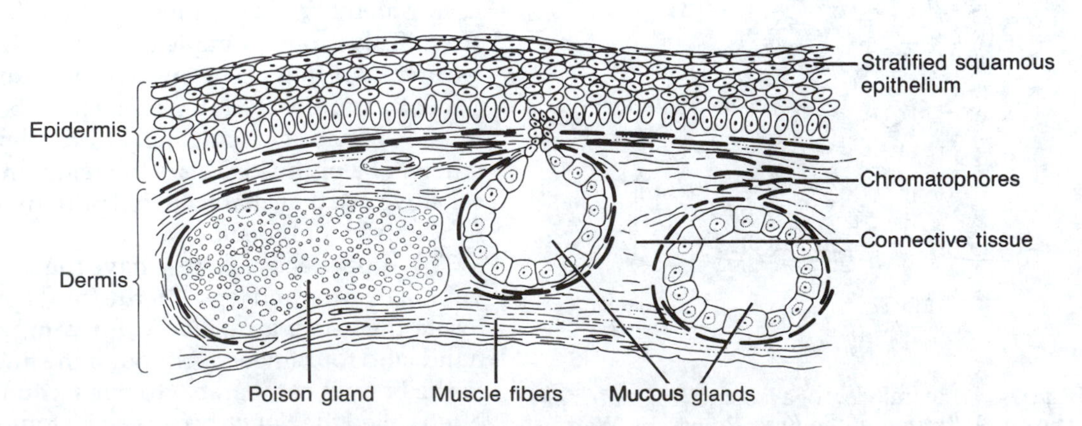

FIGURE 32-6 Vertical section through frog skin

keeps the skin moist, glistening, and sticky. The surface openings of the mucous glands can be widened or narrowed by the contraction of a **stoma** cell, which regulates the amount of mucus discharged. Examine a slide containing a "peel" of frog skin. Find a stoma cell and note its star-shaped opening.

The dermis also contains a large number of small blood vessels that transport food and carbon dioxide to the skin and take back oxygen absorbed through surface layers from the air. Nerve and muscle fibers are also present in the dermis but will not be easily seen in your slides.

Located between the epidermis and the dermis is a prominent layer of pigment cells, or **chromatophores.** Grass frogs, usually protectively colored to resemble their surroundings, are green on the dorsal and lateral surfaces and pale green to whitish on the ventral surface. The color patterns of most amphibians are stable, but the colors of some frogs undergo marked changes. The skin darkens when the pigment granules in the chromatophores spread out and cover other elements in the skin, and it lightens when they shrink. Changes in color are caused by both external conditions and internal states; low temperature produces darkening, whereas high temperature, drying of the skin, or increased light intensity causes the color to lighten.

2. Mammalian Skin

Examine a slide of mammalian skin taken from the scalp. Compare and contrast it with the skin of the frog. The **epidermis** consists of stratified squamous epithelium, usually composed of several more layers of cells than that of the frog and having more flattened keratinized layers at the surface (Fig. 32-7A). **Keratin** is a tough proteinaceous material that serves several functions. It is a relatively waterproof substance and therefore prevents water from entering the body through the skin; more important, it prevents water loss. Because it is a tough material, it protects the underlying epithelial layers from damage through ordinary wear. It is particularly protective on the palms of the hands and soles of the feet. The predominant pigment in human skin is melanin, which is produced by **melanocytes** located in the basal layers of the epithelium. The relative concentration of melanin in the epidermis accounts for the variability of human skin color.

The **dermis** is composed mainly of densely interwoven connective tissue. Note the abundance of small blood vessels. Suggest a mechanism by which they regulate body temperature.

____

____

Hair follicles and **sweat** and **sebaceous (oil) glands** are located at various levels in the dermis. Locate a hair follicle with its enclosed hair. The root of the hair develops into the hair shaft, the free end of which protrudes from the surface of the skin. One or more sebaceous glands (Fig. 32-7B) are located in the dermis and open into the hair follicle. These glands secrete an oily substance, called **sebum,** which lubricates the hair and the surface of the epidermis. It may also prevent evaporation of moisture during cold weather and thus aid in conserving body heat.

Distributed over the surface of the skin are the openings of the sweat glands (Fig. 32-7C). The long, tubular, highly convoluted ducts of the glands penetrate deeply into the dermis and connect with the secretory parts of the gland. Because of the highly coiled structure of a sweat gland, it is difficult to see an entire duct or gland in one section. Rather, you will see cross or oblique sections of the various parts (Fig. 32-7C). The secretions produced by sweat glands flow onto the surface of the skin, where they cool the surface by evaporation. These glands help regulate body temperature.

The **subcutaneous layer** beneath the dermis is composed of fat cells **(adipose tissue);** the number of such cells present varies according to the body part and the nutrition of the organism. The fat cells on your slide will look empty because the method used to prepare the slide dissolved the fat droplet in each cell, leaving only a thin film of cytoplasm with its compressed nucleus. The fat cells are held together by fine, fibrous, connective tissue. The subcutaneous layer is especially important as an insulating layer.

D. VERTEBRATE SKELETON

One of the more significant evolutionary advances of the vertebrates over the invertebrates is an internal skeleton or **endoskeleton.** The external skeleton or **exoskeleton** of invertebrates limits the size of an organism and in several animals is heavy enough to restrict movement. The endoskeleton of vertebrates permits relatively unrestricted size, as demonstrated in such vertebrates as the whale and the elephant.

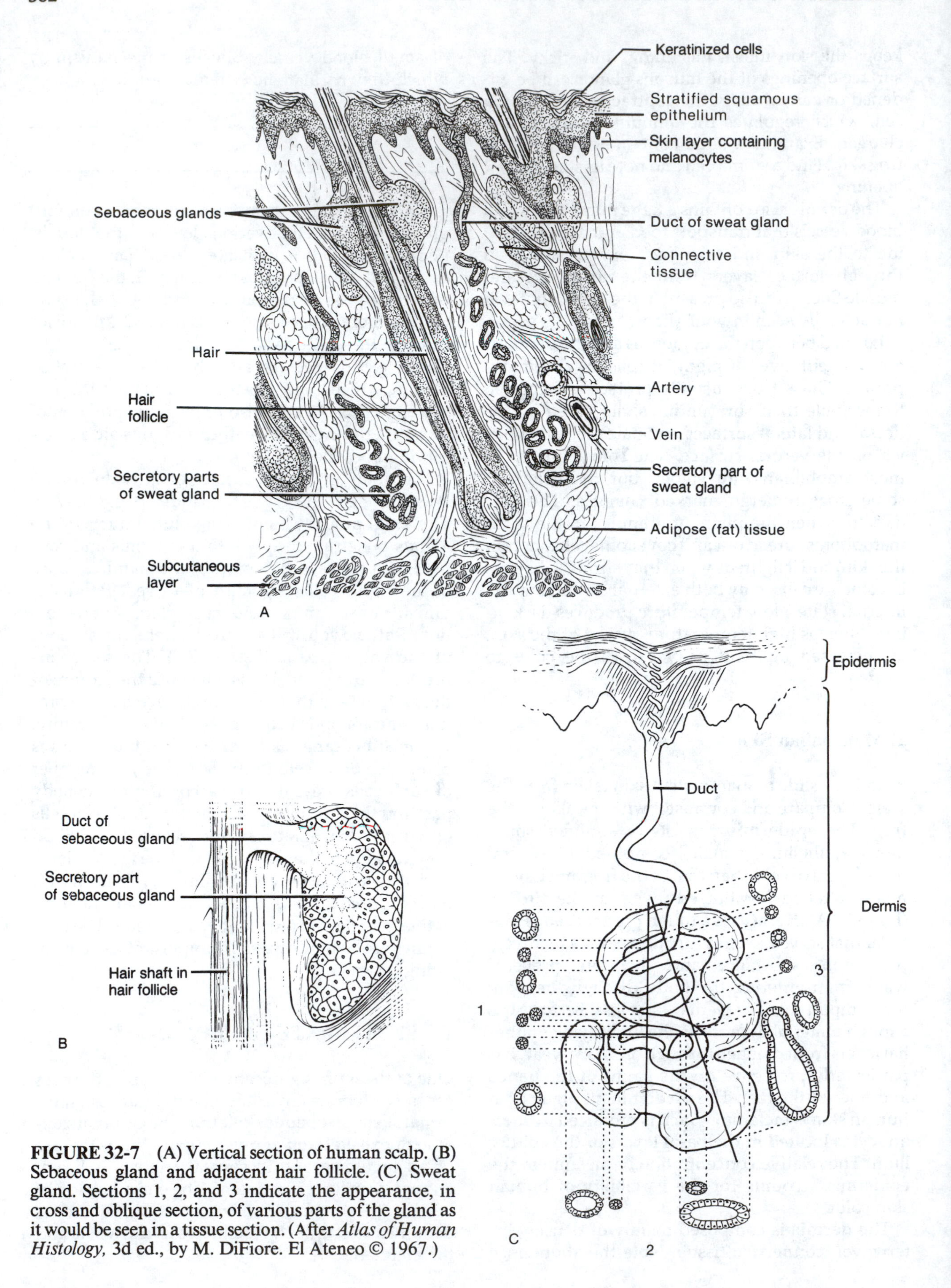

FIGURE 32-7 (A) Vertical section of human scalp. (B) Sebaceous gland and adjacent hair follicle. (C) Sweat gland. Sections 1, 2, and 3 indicate the appearance, in cross and oblique section, of various parts of the gland as it would be seen in a tissue section. (After *Atlas of Human Histology,* 3d ed., by M. DiFiore. El Ateneo © 1967.)

The vertebrate skeleton, in addition to protecting the internal organs, is an efficient supporting structure for the attachment of muscles. Furthermore, the flexibility allowed by the large number of separate bones, coupled with the strength and lightness of vertebrate skeletons, enables even the largest animals to have great mobility.

All endoskeletons have a basic pattern with two major parts: the **somatic skeleton,** consisting of the major bones of the body and appendages, and a less conspicuous **visceral skeleton,** the bones of which are located chiefly in the wall of the pharynx. The somatic skeleton is subdivided into the **axial skeleton,** which consists of the skull, the vertebral column, and the sternum (breastbone), and the **appendicular skeleton,** which consists of the pelvic (hip) and pectoral (shoulder) girdles and the limbs.

As you study the parts of the vertebrate skeleton, examine a mounted specimen of a frog skeleton and, if available, a human skeleton to see their similarities and differences (Fig. 32-8).

1. Axial Skeleton

a. Skull

The skull is the anterior part of the axial skeleton and houses the brain and the olfactory, optic, and auditory capsules for the organs of smell, sight, and hearing (Figs. 32-8A, 32-9, and 32-10). It can be divided into the **cranial region,** containing the brain and the inner ear, and the **facial region,** which forms the jaws and encloses the eyes, nose, and part of the ear.

Examine the dorsal surface of a frog skull (Fig. 32-9). The skull is attached to the anterior end of the vertebral column. Locate the small brain case, or **cranium,** which is roofed by the **frontoparietal** bones, the **nasal** bones (which cover the nasal capsules), the **prootics** (which house the inner ears), and the **exoccipital** bones (each of which has rounded **occipital condyles**). The two condyles fit depressions in the first vertebra, permitting slight movements of the head on the spinal column. Between the condyles is a large opening, the **foramen magnum,** through which the spinal cord passes. Between the prootics and the **maxilla** are the **squamosal** and **pterygoid** bones, which form the lateral borders of the skull. The upper jaw consists of the small **premaxillary** bones in front and the larger **maxillary** bones that extend posteriorly to join with the pterygoid.

On the ventral surface of the skull, locate the **maxillary teeth** on the margin of the upper jaw and the **vomerine teeth** on the roof of the mouth (Fig. 32-10A). The **sphenethmoid** bone, which connects with the frontoparietal bone dorsally and the **parasphenoid** bone ventrally, make up the anterior part of the cranium.

The lower jaw (mandible) consists of a rodlike **Meckel's cartilage** encased by the **dentary** and **angulare** bones, which are united with the **quadratojugal bone** by means of the **quadrate cartilage** (Fig. 32-10B). Notice that the bones of the lower jaw do not have teeth.

b. Vertebral Column

The vertebrae vary greatly among animals of different species and in different regions of the vertebral column in an individual animal. The vertebrae in fish are differentiated into trunk vertebrae and caudal vertebrae. In many other vertebrates, they are differentiated into neck **(cervical),** chest **(thoracic),** back **(lumbar),** pelvic **(sacral),** and tail **(caudal)** vertebrae. In birds and human beings, the caudal vertebrae are reduced in number and size and the sacral vertebrae are fused. The number of vertebrae also varies in different species of animals. The python has the largest number: 435. Human beings have 7 cervical, 12 thoracic, and 5 lumbar vertebrae, plus 5 that have fused to form the **sacrum** and the **coccyx.**

The first cervical vertebra in the frog, the **atlas** (Fig. 32-8A), is modified for articulation with the skull; the vertebral column terminates in a long bone, the **urostyle.** How many vertebrae does the frog have?

Note that the frog's vertebrae are much alike. The lateral projections are called **transverse processes.** In humans and other higher vertebrates, the ribs are attached to these processes.

There are many variations in the ribs of vertebrates. The basic plan seems to have been a pair of ribs for each vertebra from head to tail, but the evolutionary trend has been to reduce the number of pairs from the simpler to the more complex vertebrates. Ribs, however, are not universal among vertebrates; many, including the frog, do not have them at all. Human beings have 12 pairs of ribs.

2. Appendicular Skeleton

Most vertebrate animals have some form of paired appendages supported by **pectoral** (shoulder) and **pelvic** (hip) **girdles.** Among vertebrates, there are

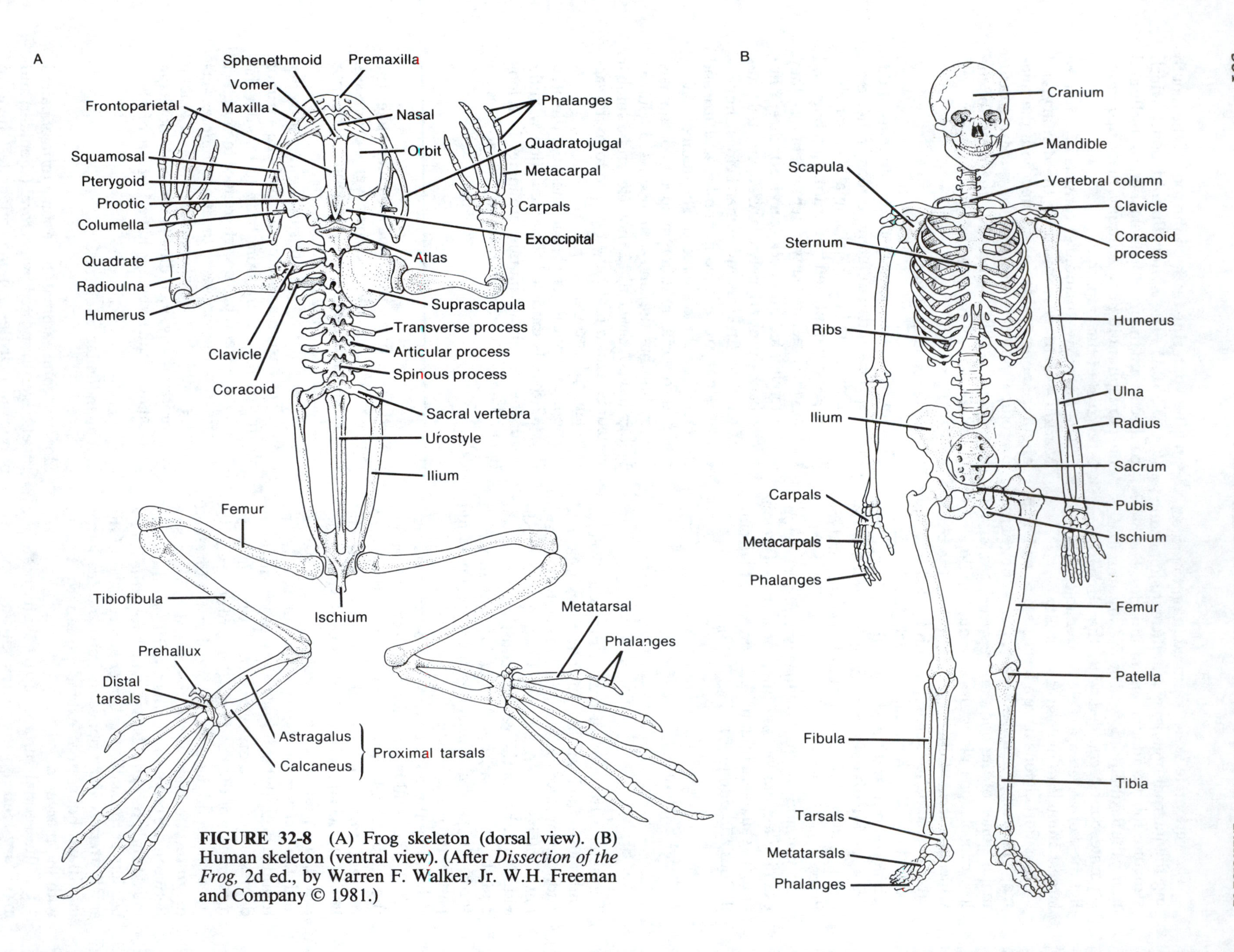

FIGURE 32-8 (A) Frog skeleton (dorsal view). (B) Human skeleton (ventral view). (After *Dissection of the Frog*, 2d ed., by Warren F. Walker, Jr. W.H. Freeman and Company © 1981.)

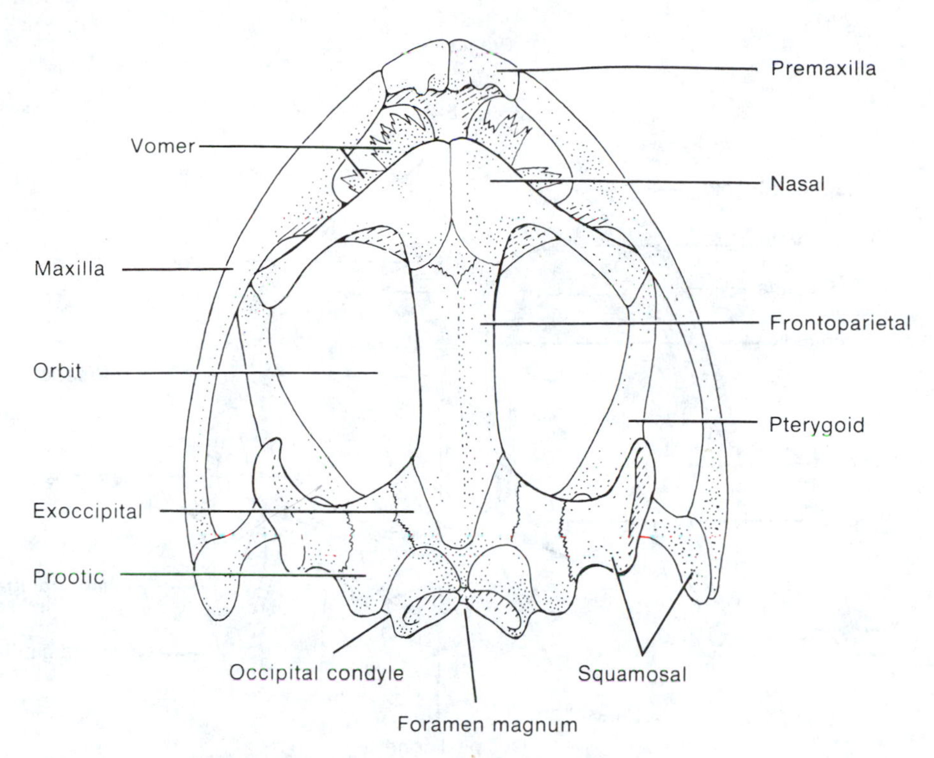

FIGURE 32-9 Frog skull (dorsal view)

many modifications in the girdles, limbs, and digits that enable the animals to meet the requirements of their special modes of life. Whatever the modification, in forms higher than the fish the girdles and appendages have the same basic structure.

The single strong bones closest to the body in the legs of a frog are the **humerus** in the anterior limb (Fig. 32-11) and the **femur** in the posterior limb (Fig. 32-12). Distal to these are the **radioulna** (a fusion of the **radius** and the **ulna**) in the forelimb and the **tibiofibula** (a fusion of the **tibia** and **fibula**) in the hind limb. The humerus is attached to the pectoral girdle at the **glenoid fossa** by means of ligaments. The bone passing dorsally, the **scapula,** has a broad extension called the **suprascapula.** Ventrally and posterior to the **clavicle** is the **coracoid.** At the point of junction of the two clavicles is the **sternum,** which continues anteriorly as the **omosternum** and **episternum** and posteriorly as the **mesosternum** and **xiphisternum.**

The femur attaches to the pelvic girdle (Fig. 32-12) in a socket, the **acetabulum,** that is formed by the fusion of three bones of the pelvic girdle, the **ilium,** the **ischium,** and the **pubis.** The feet and hands are built according to a common pattern, with a number of **carpal** (wrist) or **tarsal** (ankle) bones followed by a group of elongated **metacarpal** (hand) and **metatarsal** (foot) bones, and then the **phalanges** (bones of the fingers or toes). The **astragalus** (inside the foot) joins the tibiofibula to the tarsal bone.

E. VERTEBRATE MUSCULATURE

1. Joints

Before studying the muscles of the frog, it is advisable to learn something about the joints between bones, because they allow the muscles to bend, twist, and turn the various parts of the body. Become familiar with the following classification of joints before proceeding with the muscle dissection.

- **Ball-and-socket** joints allow movement in any direction, including rotation, as in the shoulder and hip.
- **Condyloid** joints allow movement in any direction except rotation, as between metacarpals and phalanges.
- **Hinge** joints allow bones to bend in only one direction, as in the knee, and between flat surfaces of bones, as between the vertebrae.

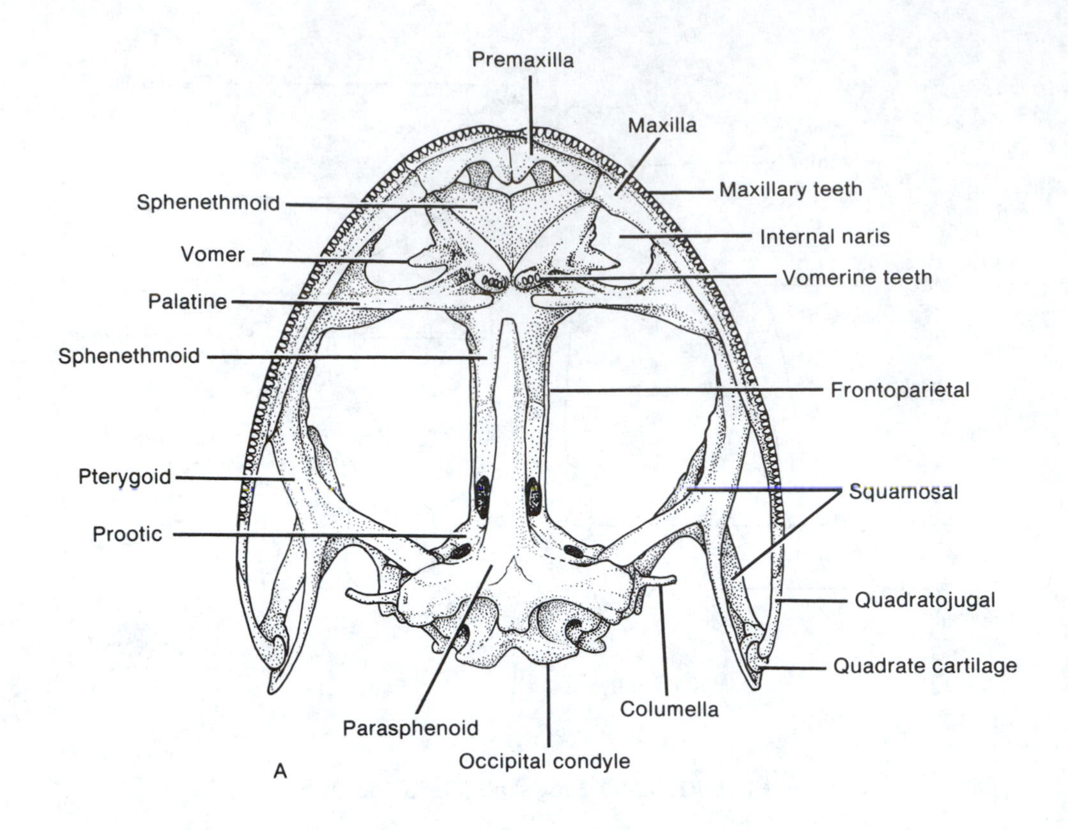

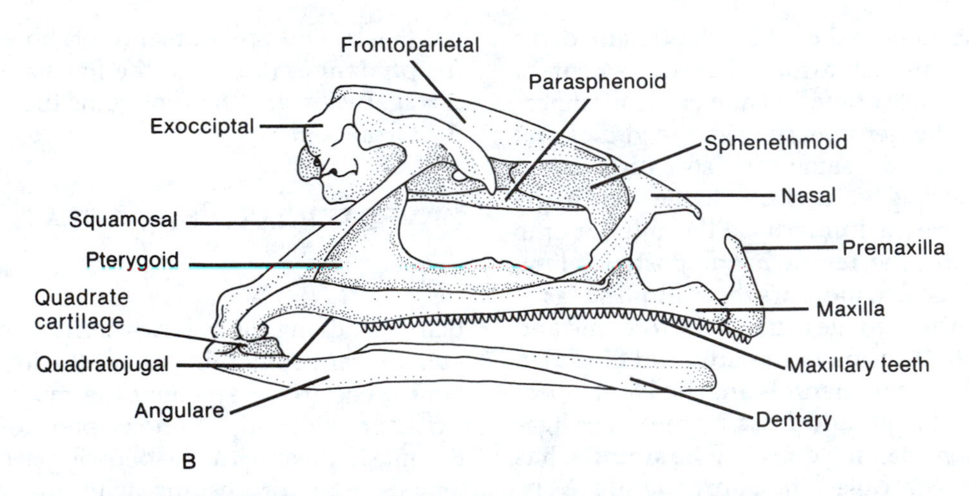

FIGURE 32-10 Frog skull. (A) Ventral view. (B) Lateral view.

- **Plane** joints allow sliding movements between flat surfaces of bones, as between the vertebrae.
- **Radial** joints permit rotation of one bone on another, as between the proximal ends of the radius and ulna.

2. Muscle Terminology

Muscles have several functions. They support the body by holding the bones in proper relationship to each other. They are responsible for the movement of the body as a whole, the passage of food through

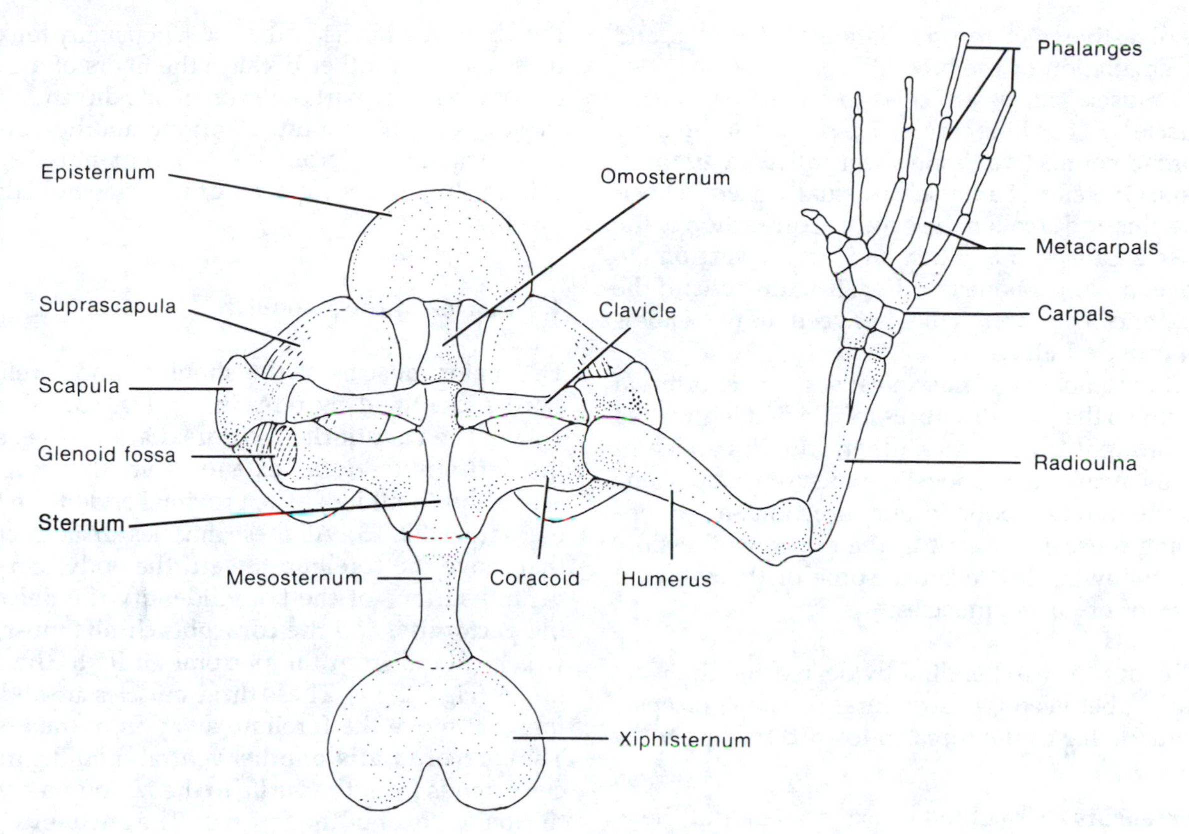

FIGURE 32-11 Frog pectoral girdle (ventral view)

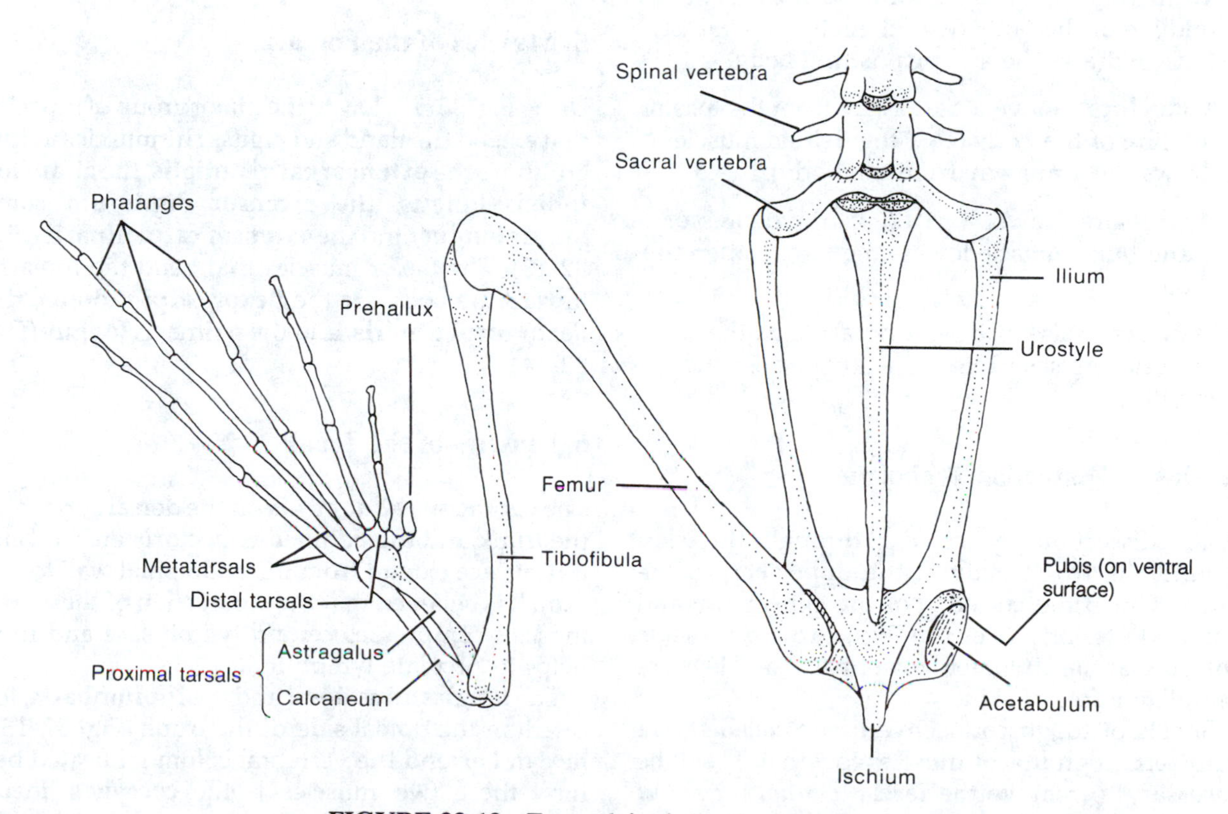

FIGURE 32-12 Frog pelvic girdle (dorsal view)

the digestive tract, the ventilation of the lungs, and the circulation of the blood.

A muscle can be attached to a bone or another muscle by a cordlike connective tissue **tendon** or by a broad connective tissue sheet called an **aponeurosis.** One end of a muscle is usually fixed in position; this end is called the **origin.** Contraction of the muscle causes the other end, the **insertion,** to move, pulling a bone or other structure toward the fixed end. The part that lies between the two ends is the **muscle belly.**

Contraction of a muscle causes it to shorten in length so that the structures to which it is attached are brought toward each other. Muscles are commonly arranged in opposing pairs (or groups): one muscle moves a bone in one direction and an opposing muscle moves it in the opposite direction. The following list includes some of the common types of opposing muscles:

- **Flexors** cause bending by decreasing the angle between two structures (e.g., the biceps muscle flexes the forearm toward the upper arm).
- **Extensors** straighten or extend a part of the body (e.g., the triceps muscle extends the forearm away from the upper arm).
- **Adductors** move a part toward the axis or midline of the body (e.g., the latissimus dorsi muscle draws the arm against the body).
- **Abductors** move a part away from the axis or midline of the body (e.g., the deltoid muscle draws the arm away from the body).
- **Depressors** lower a part (e.g., the depressor mandibulae muscle lowers the jaw to open the mouth).
- **Levators** raise or elevate a part (e.g., the masseter muscles raises the jaw to close the mouth).

3. Muscle Dissection Technique

Begin dissection by cutting through the skin around the frog's body at the upper end of the trunk. Grasp the cut edge of the skin firmly and pull it posteriorly over the legs, turning it inside out, so that the abdomen, lower back, and legs are completely free of skin.

Sheets of tough connective tissue called **fascia** bind certain groups of muscles together. It will be necessary to remove the fascia. *Do not try to cut muscles apart.* First, observe the direction in which the fibers of a muscle run. You can usually tell one muscle from another because the fibers of a muscle, or the major part of it, run in one direction and those of an adjacent muscle run in another. Separate each muscle from the adjacent muscles by using a blunt probe to part the fascia that hold them together.

4. Muscles of the Shoulder

The major muscles of the shoulder and forelimb can be identified by referring to Figs. 32-13 and 32-14. The **cucullaris,** the **dorsalis scapulae,** and the **latissimus dorsi** extend from the scapula across the shoulder to the proximal end of the humerus (Fig. 32-13). All these muscles are adductors that move the forelimb toward the body. On the ventral surface of the body, identify the **deltoid,** the **pectoralis,** and the **coracobrachialis** muscles, which extend from the pectoral girdle to the humerus (Fig. 32-14). These three muscles are abductors that move the forelimb away from the body. The **coracoradialis,** another ventral shoulder muscle, extends from the girdle to the radioulna and is the major flexor of the forearm. The **anconeus,** situated on the dorsal surface of the humerus, is the major extensor of the forearm.

5. Muscles of the Forearm

Using Fig. 32-13, locate the major group of muscles that extend the hand and digits. The muscles in this group are the **extensor carpi radialis,** the **abductor indicis longus,** the **extensor digitorum communis longus,** and the **extensor carpi ulnaris** (Fig. 32-13). The flexor muscles that bend the forearm toward the body are the **flexor carpi radialis,** the **flexor carpi ulnaris,** and the **palmaris longus** (Fig. 32-14).

6. Muscles of the Trunk

The **cutaneous abdominis** on the dorsal surface of the trunk and the **cutaneous pectoris** on the ventral surface extend from the abdominal wall to the skin. It is believed that the contraction of these two muscles compresses certain lymph sacs and thus helps to circulate lymph fluid.

The **longissimus dorsi** and the **iliolumbaris,** located on the dorsal side of the trunk (Fig. 32-13), flex and extend the vertebral column. Located behind these two muscles is the **coccygeoiliacus** muscle, which is attached at one end to the urostyle

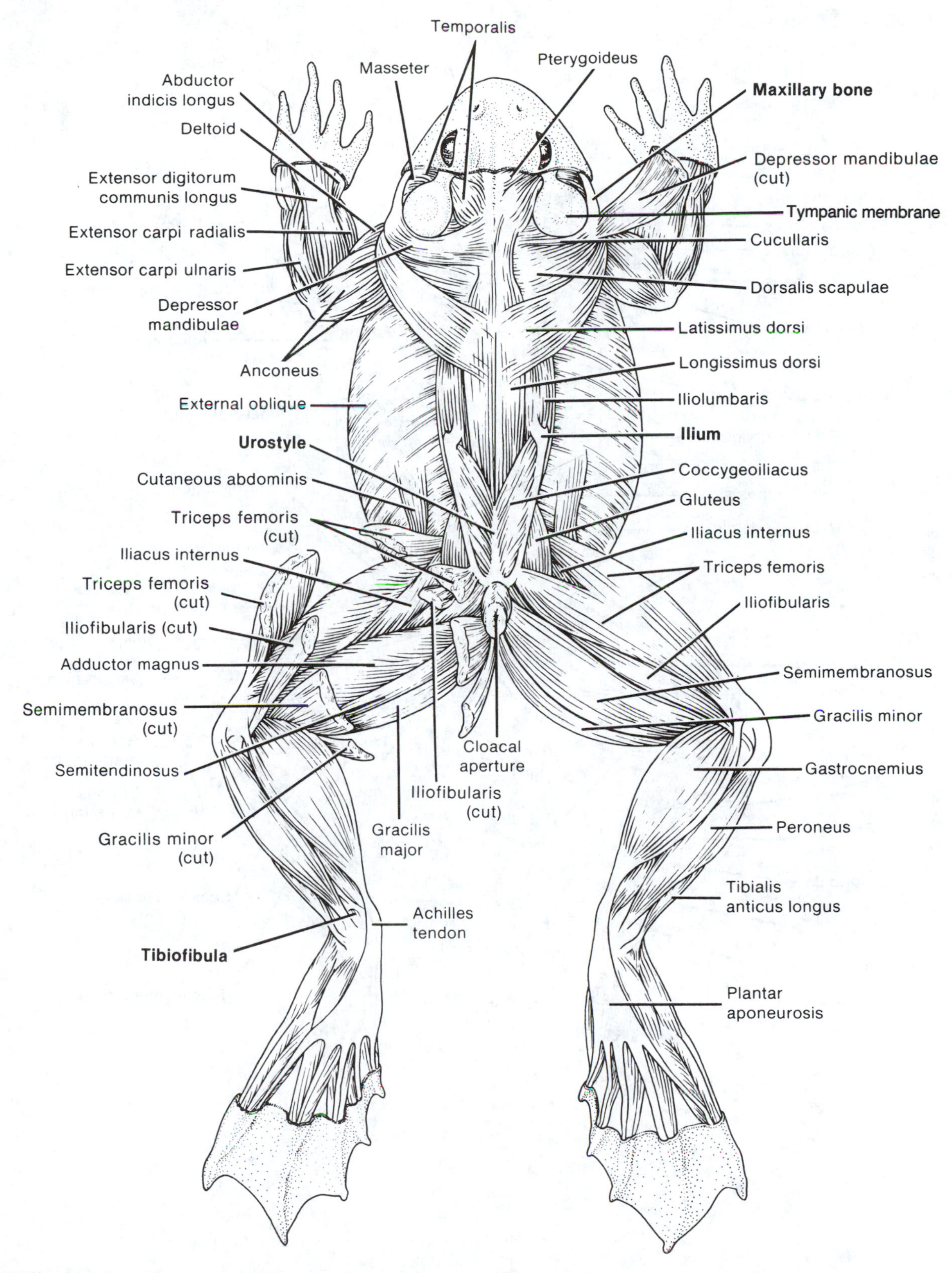

FIGURE 32-13 Muscles of the frog as seen in a dorsal view. Superficial muscles are shown on the right side of the drawing, deeper muscles on the left side. Skeletal parts are in boldface. (After *Dissection of the Frog,* 2d ed., by Warren F. Walker, Jr. W.H. Freeman and Company © 1981.)

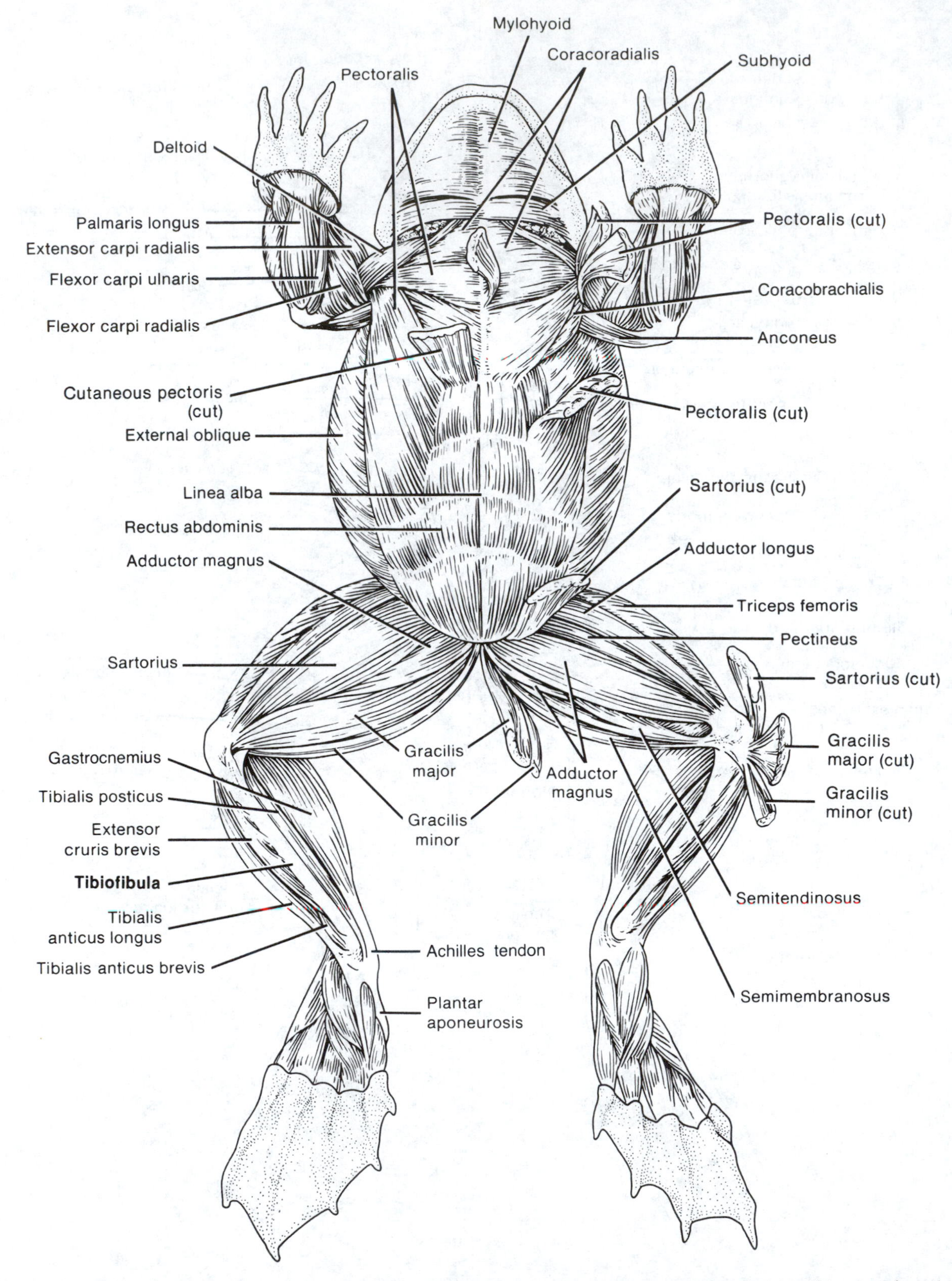

FIGURE 32-14 Ventral view of frog muscles. Superficial muscles are shown on the left side of the drawing, deeper muscles on the right side. (After *Dissection of the Frog,* 2d ed., by Warren F. Walker, Jr. W.H. Freeman and Company © 1981.)

and at the other to the ilium bones. These three muscles are thought to extend the urostyle and pelvic girdle when the frog is making strong leaps.

The ventral body wall of the frog is largely supported by the **rectus abdominis** (Fig. 32-14). The rest of the abdominal wall is covered on the outside by an **external oblique muscle** and an inner **transverse muscle,** whose fibers run nearly perpendicular to each other. These muscles support and tense the abdomen.

7. Muscles of the Head

The **temporalis, pterygoideus,** and **masseter** muscles, located on the dorsal surface of the head (Fig. 32-13), elevate the mandible. Extending transversely on the ventral surface of the head are the **mylohyoid** and **subhyoid** muscles, forming the ventral surface of the lower jaw. These muscles raise the floor of the mouth and are therefore important in the swallowing and breathing movements of the mouth.

8. Muscles of the Thigh

Located on the anterior part of the thigh is the **triceps femoris** (Fig. 32-13), which extends the shank during jumping and swimming. The rest of the dorsal surface of the thigh is composed of the large **semimembranosus,** the **gracilis minor,** and a slender **iliofibularis** muscle (Fig. 32-13). The origins of all three of these muscles are on the bones of the pelvic girdle; their insertions are on the tibiofibula.

On the ventral surface of the thigh is the **sartorius** (Fig. 32-14); its origin is on the pubis and its insertion on the tibiofibula. This muscle flexes the thigh at the hip. The **adductor longus** and **adductor magnus** muscles lie on either side and partly beneath the sartorius. Their main function is to move the thigh toward the body. Lying posterior to the adductor magnus is the **gracilis major,** which flexes the knee and extends the thigh.

To observe the short muscles that extend from the pelvic girdle to the femur, it is first necessary to cut through some of the superficial thigh muscles, as shown in Fig. 32-13. When this is done, you can identify the **gluteus** and the **iliacus internus** (Fig. 32-13), which extend from the ilium bone of the pelvic girdle to the proximal end of the femur. These two muscles flex and rotate the thigh.

To observe the deep muscles on the ventral side of the thigh, cut the gracilis major, gracilis minor, and sartorius muscles as shown in Fig. 32-14. Locate the **pectineus** muscle between the adductor longus and adductor magnus. This muscle helps to move the thigh toward the body. Posterior to the adductor magnus is the **semitendinosus,** which acts both to reflect the shank and to extend the thigh.

9. Muscles of the Shank of the Hind Limb

The largest muscle of the shank is the **gastrocnemius.** This muscle arises by means of two large tendons from the distal end of the femur and ends in the Achilles tendon, which passes around the ankle and spreads into a sheet of connective tissue called the **plantar aponeurosis.** The function of this muscle is to flex the knee and toes. Anterior to the gastrocnemius, as viewed from the dorsal surface, are the **peroneus** and **tibialis anticus longus** muscles, both of which assist in foot extension.

From the ventral view of the shank, identify the **extensor cruris brevis** (Fig. 32-14), which inserts on the tibiofibula and is an extensor of the flank. The **tibialis posticus,** which arises from the shaft of the tibia and inserts on the astragalus bone of the foot, assists in foot extension. The **tibialis anticus brevis,** which arises from the distal part of the tibiofibula and inserts on the tarsal bones, is a flexor of the foot.

REFERENCES

Hammersen, F. 1976. *Histology, A Color Atlas of Cytology, Histology and Microscopic Anatomy.* Lea and Febiger.

Romer, A. S., and T. S. Parsons. 1986. *The Vertebrate Body.* 6th ed. Holt/Saunders.

Underhill, R. A. 1988. *Laboratory Anatomy of the Frog.* 5th ed. Wm C Brown.

Walker, W. F. 1981. *Dissection of the Frog.* 2d ed. W.H. Freeman.

Vertebrate Anatomy: Digestive, Respiratory, Circulatory, and Urogenital Systems

Exercise

33

This exercise deals with the digestive, respiratory, circulatory, and urogenital systems of the frog.

A. DIGESTIVE SYSTEM

1. Oral (Buccal) Cavity

Holding a frog as shown in Fig. 33-1, make a small cut at the angle of the jaw on each side so that the jaws can be opened widely. The **oral (buccal) cavity** narrows into the **pharynx,** which connects to the **esophagus** (or gullet) (Fig. 33-2). Just anterior to the opening into the esophagus is the **glottis,** a short longitudinal slit in the floor of the pharynx. The glottis opens for breathing but closes when food is being swallowed. Posterior to the bulge of the eyeball into the oral cavity, and near the corners of the mouth, are small openings leading into the **eustachian tubes,** each of which connects to the chamber of the middle ear beneath the tympanic membrane. In many species of frog, the males have openings into two **vocal sacs** located at the posterior corners of the oral cavity. When the vocal sacs are inflated, which happens during the breeding season, they amplify the croaking of the male frog.

Locate the **maxillary teeth** on the margin of the upper jaw and the **vomerine teeth** on the roof of the mouth. Because the frog swallows its food whole, these teeth are used not for chewing but for holding food. Near the vomerine teeth are two small openings, the **internal nares,** which lead to the **external nares** (nostrils) through which air passes to and from the oral cavity during breathing.

Lying on the floor of the oral cavity and attached to the mouth anteriorly is the large tongue. The forked end of the tongue is covered with a sticky substance secreted by glands in the roof of the mouth and can be quickly thrust out of the mouth to capture small insects or other live prey. This sticky substance holds the prey as the tongue is retracted into the mouth.

2. Body Cavity

Make a longitudinal incision through the abdominal skin and muscle layers just left of the midsagittal line extending from the pectoral to the pelvic girdle (X–Y in Fig. 33-3). Now make transverse incisions (A–B and A′–B′) through the skin and muscle just anterior to the hind legs and posterior to the forelegs. Continue the cuts to the dorsal surface, turning the flaps of the body wall back and pinning them to the dissecting pan. Finally, cut through the pectoral girdle bone to expose the cavity that contains the heart.

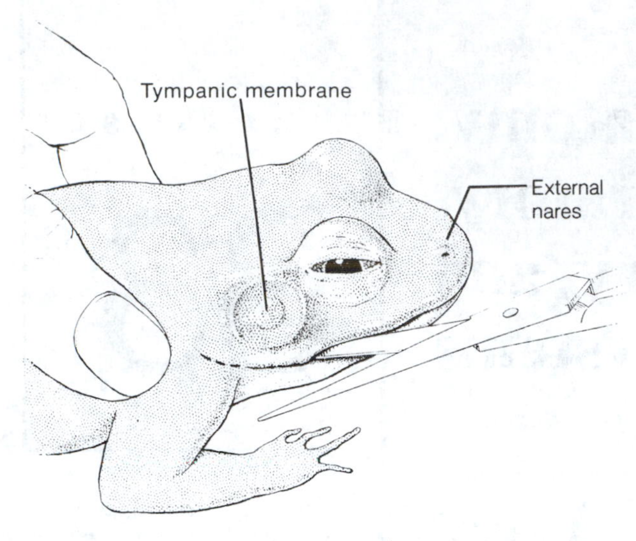

FIGURE 33-1 Procedure for dissection of frog mouth

The body cavity, or **coelom,** in which the internal organs **(viscera)** are located, is completely lined by a shiny layer of epithelium, called the **parietal peritoneum.** The epithelium covering the viscera is called the **visceral peritoneum.** Membranous sheets of this epithelium, called the **mesenteries,** also extend from the dorsal body wall to the organs and between many of the organs. The mesenteries, which serve as pathways for blood vessels and nerves, also limit the movement of various internal organs.

The coelom of the frog is divided into two cavities, the **pleuroperitoneal cavity** containing the lungs and most of the other viscera and the **pericardial cavity** containing the heart. These cavities are separated from each other by the **pericardium,** a tough, connective-tissue membrane that closely invests the heart. Thus, the pericardial cavity is a very narrow, fluid-filled space.

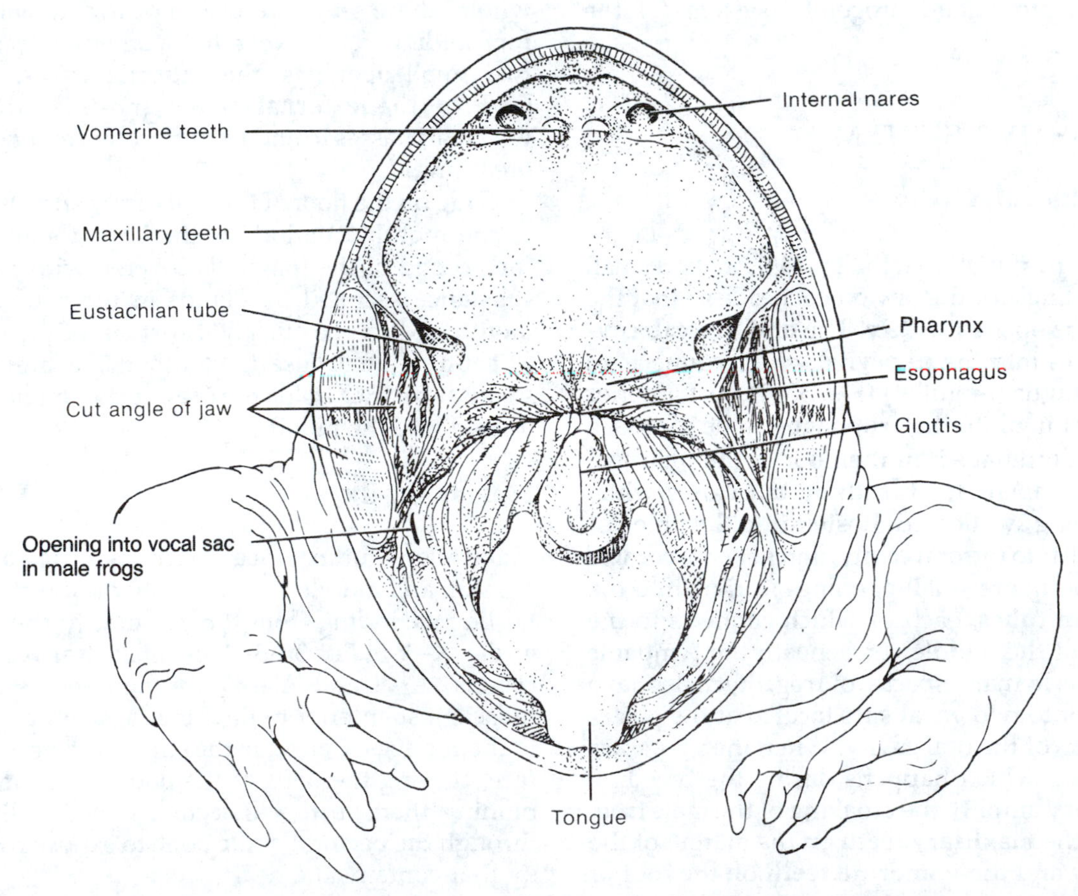

FIGURE 33-2 Interior view of the oral (buccal) cavity of a male frog. The angles of the jaw have been cut so that the mouth can be opened wide. (After *Dissection of the Frog,* 2d ed., by Warren F. Walker, Jr. W.H. Freeman and Company © 1981.)

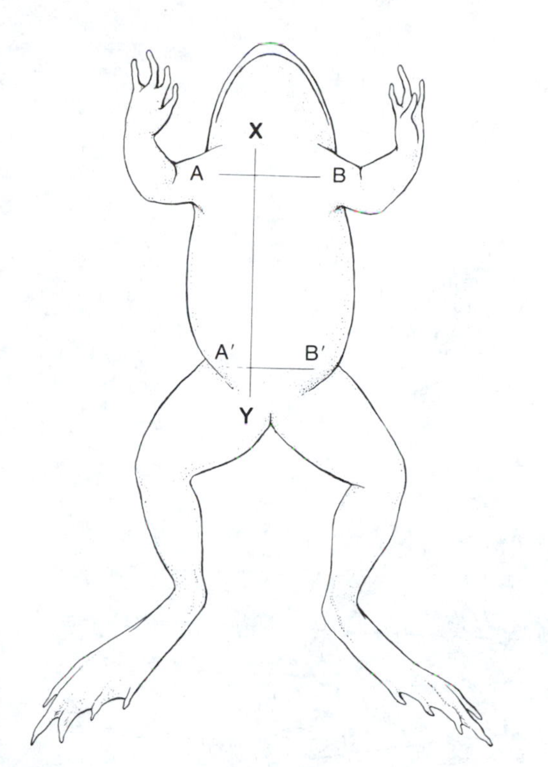

FIGURE 33-3 Procedure for dissection of the abdominal body wall

3. The Viscera

First locate the **liver,** a large, trilobed organ occupying much of the anterior part of the **pleuroperitoneal cavity** (Fig. 33-4). Locate the **stomach,** which lies dorsal to the liver and to the right as you view the body cavity. The stomach, which is connected to the pharynx by the short **esophagus,** is a large, saclike organ in which food is stored while digestion begins. Use a scalpel to open the stomach lengthwise. Observe the longitudinal folds **(rugae)** on the inner walls, which help to grind the food. At the posterior end of the stomach is the **pyloric sphincter,** a circular band of muscle tissue that regulates the passage of partially digested food from the stomach into the small intestine, where digestion is completed and the products of digestion are absorbed.

The intestine is divided into a short **duodenum,** which curves upward toward the liver, and a much-coiled **jejunoileum,** which leads into the **colon,** where water and ions are absorbed and undigested waste products are stored as feces before elimination through the **cloacal opening.**

In the small intestine, digestion is aided by secretions from two large digestive glands, the liver and the **pancreas.** One function of the liver is to produce **bile,** an alkaline secretion that aids the digestion of fatty materials. Bile is stored in the **gallbladder** until food enters the intestine, at which time the bile is released through the **cystic duct** into the **common bile duct** and then into the duodenum. Locate the gallbladder. It is a small, dark sac attached to the dorsal surface of the liver.

Locate the slender, irregular, and whitish **pancreas,** which lies in the mesentery between the duodenum and the stomach. This gland secretes an alkaline digestive fluid that also empties into the common bile duct. The pancreas is also an endocrine gland, containing the **islets of Langerhans,** which produce the hormones insulin and glucagon. These hormones regulate glucose levels in the blood.

Lift up the stomach and locate a small, dense, round structure—the **spleen**—on the left side of the mesentery supporting the intestine. The spleen is intimately related to the circulatory system, and in adult frogs, it is an important site for the production of red and white blood cells and the storage and destruction of senescent (aging) red blood cells. It is also a major site for the production of antibodies, protein molecules produced as a defense mechanism in response to foreign organisms or substances.

In the male frog, locate the **testes,** small, oval bodies, one on each side of the mesentery. The fingerlike lobes attached to the testes are **fat bodies.** The **ovaries,** which are located in the same position in the female as the testes are in the male, vary greatly in size depending on the female's reproductive state. Just before ovulation, they are filled with ripe eggs and take up all the available space in the pleuroperitoneal cavity. Dorsal to each **gonad** (testis or ovary) is the elongated **kidney.** A long, coiled white tube—the **oviduct**—is lateral to each kidney in the female. Males may have a small vestigial oviduct. The bilobed sac lying ventral to the large intestine is the **urinary bladder.**

B. RESPIRATORY SYSTEM

The primary functions of the respiratory system are to take up oxygen from the air and eliminate carbon dioxide from the blood. In the frog, as noted in Exercise 32, such gas exchanges take place not only in the lungs **(pulmonary respiration)** but also through the skin **(cutaneous respiration)** and the epithelial lining of the mouth and pharynx **(buccopharyngeal respiration).**

In pulmonary respiration, air passes through the nostrils (external nares) into the nasal cavity and

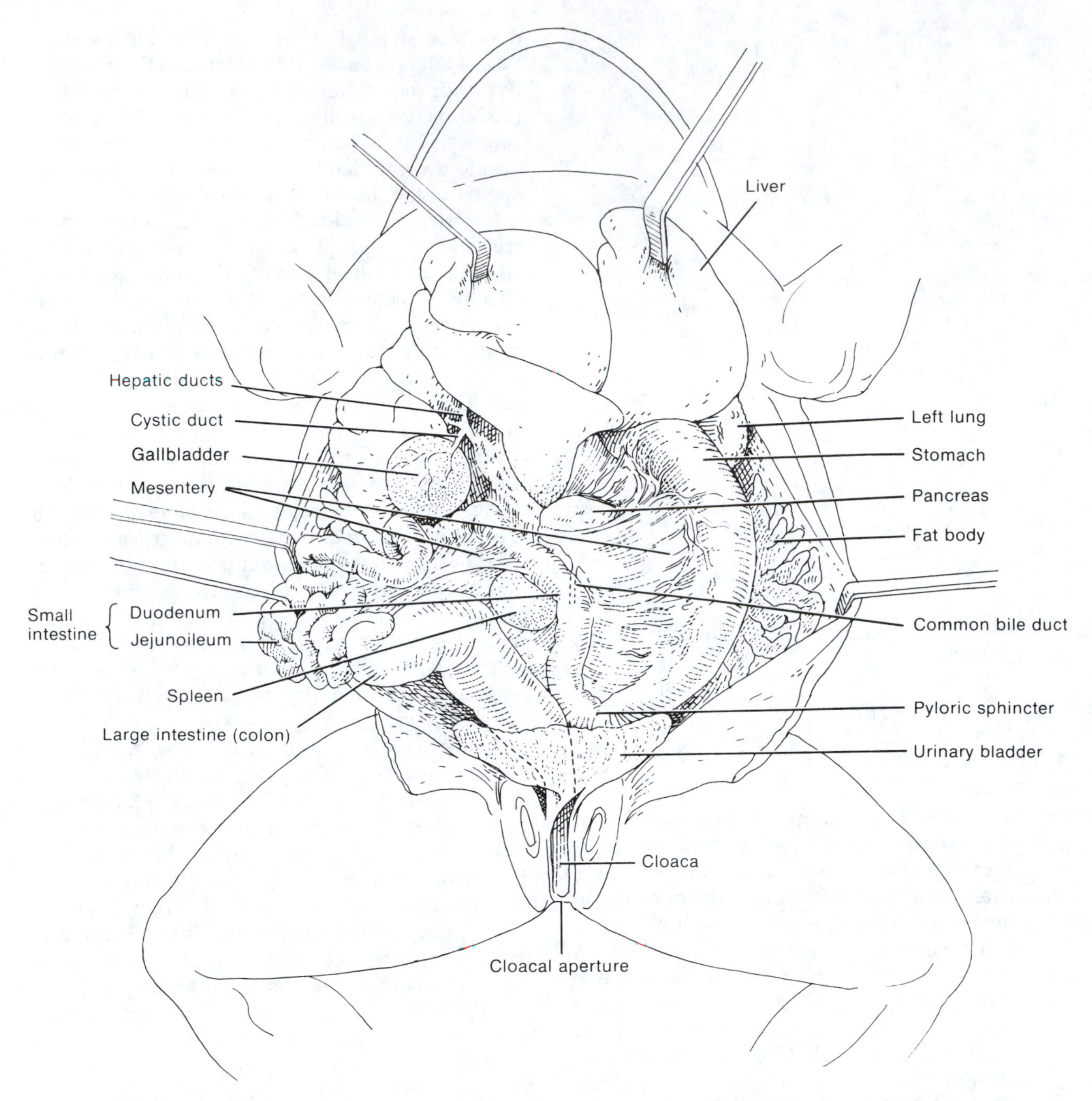

FIGURE 33-4 Ventral view of the digestive tract and associated organs. The liver has been pulled forward and the pelvic region cut open. (After *Dissection of the Frog*, 2d ed., By Warren F. Walker, Jr. W.H. Freeman and Company © 1981.)

then through the internal nares into the oral cavity. The external nares are then closed, the floor of the mouth is raised, and air is forced through the glottis into the **larynx** (see Fig. 33-2). The larynx is reinforced by cartilage and contains two elastic bands, the **vocal cords,** that vibrate and produce croaking sounds when air is forced vigorously from the lungs. The larynx is connected to each lung by a very short tube, the **bronchus.** In most preserved specimens, the lungs are contracted because of the shrinking effect of the preservative. Remove part of the lung and slit it open to observe the network of partitions that divides the lung into many minute chambers called **alveoli.** The alveoli, with their extensive blood capillary network, provide the large surface area necessary for gas exchange.

C. CIRCULATORY SYSTEM

The circulatory system of vertebrates can be divided into two parts, **lymphatic** and **cardiovascular** (or blood vascular). The lymphatic system performs several functions. Its lymph nodes serve as filters to trap and prevent the spread of microorganisms through the body and are involved in the production of lymphocytes. The smallest vessels in this system, the lymphatic capillaries, penetrate the villi in the intestinal wall, where they function in the uptake of fatty acids.

The cardiovascular system is a **closed system,** in which the circulating blood is confined to the heart and a network of blood vessels (the veins, arteries, and capillaries). It is also a **double system:** the heart is divided into right and left parts. The right side of the heart pumps unoxygenated blood to the lungs where exchange of respiratory gases occurs, and oxygenated blood is returned to the left side of the heart; this is the **pulmonary circuit.** The blood is then pumped from the left side of the heart, via the **systemic circuit,** to the tissues and cells of the body. Figure 33-5 is a schematic drawing depicting the closed and double vertebrate circulatory system.

1. The Frog Heart

To study the structure of the frog heart, you will first have to cut away the pericardial membrane that encloses it. Using Fig. 33-6, identify the thin-walled left and right **auricles** and the thick-walled **ventricle.**

The frog heart is different from that of most other vertebrates because it has one ventricle instead of two. Arising from the base of the ventricle on its ventral side is a stout, cylindrical vessel, the **conus arteriosus.** The **sinus venosus,** on the dorsal side, receives the **posterior vena cava** and the two **anterior vena cavae.** Locate the **pulmonary veins,** which come from the lungs and enter the left atrium.

The action of the heart can be studied by examining a pithed frog that has been dissected to expose the beating heart. (Refer to Appendix I for the procedure for pithing a live frog.) As you observe the heart (and refer to Figs. 33-6 and 33-7), note that

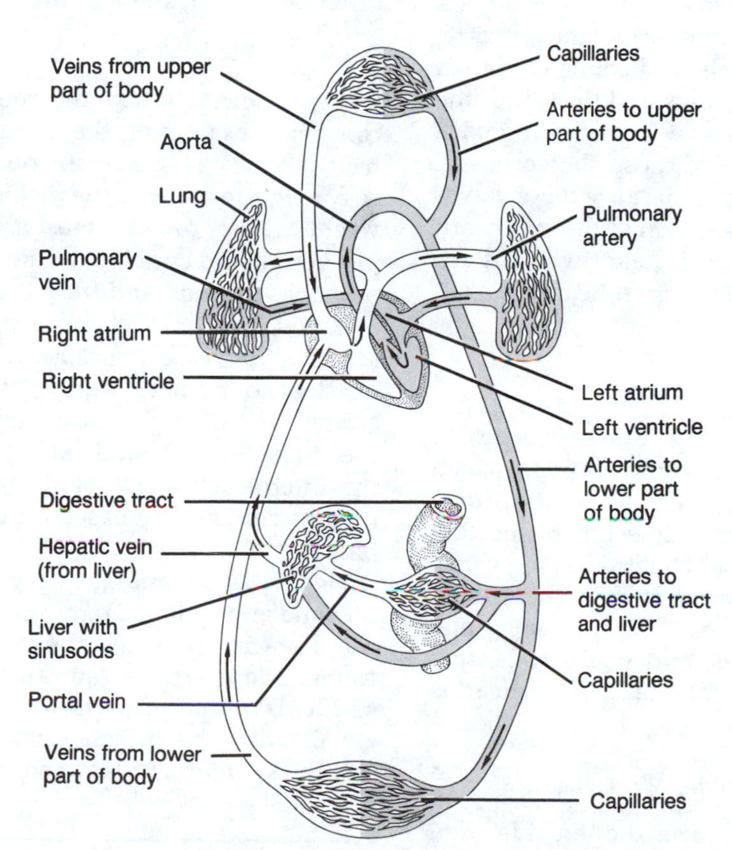

FIGURE 33-5 Double and closed system of circulation. Oxygenated blood circulation shaded; unoxygenated blood circulation in white.

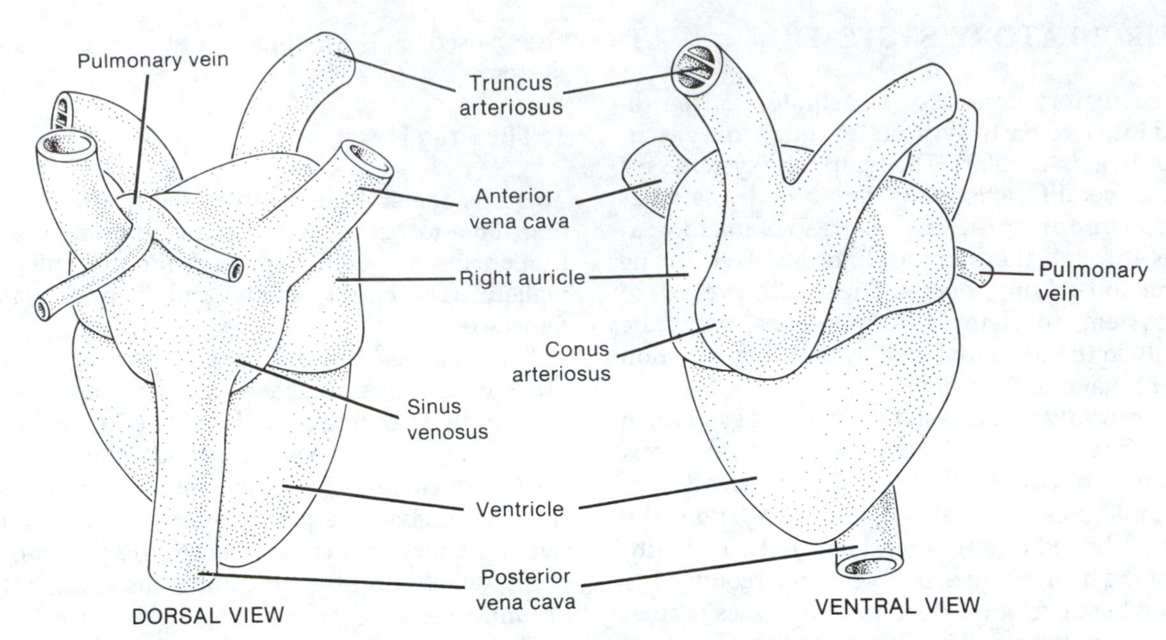

FIGURE 33-6 Dorsal and ventral views of frog heart

the sinus venosus contracts first and sends unoxygenated venous blood into the right auricle. Oxygenated blood from the lungs passes into the left auricle via the pulmonary veins. Then both auricles contract and pump their contents into the ventricle. Contraction of the ventricle and conus arteriosus next pumps blood to the lungs and the rest of the body. The blood is prevented from flowing back into the heart by valves like those shown in Fig. 33-8, which is a sheep heart. Because these valves are difficult to find in a dissection of the frog heart, you will next study the sheep heart, which is approximately the size of the human heart.

2. The Sheep Heart

The heart provided to you may be an isolated organ or one that is still attached to the lungs, in which case the entire structure is referred to as the **pluck.** Begin your examination and dissection by locating the auricles. In the preserved heart, they appear as small, collapsed caplike structures on either side of the superior end of the heart. They usually look darker than the ventricles, and you may be surprised by their small size. The bulk of the heart is made up of the ventricles.

a. The Right and Left Sides of the Heart

Before proceeding, distinguish the left side from the right side of the heart. The pointed end (apex) is composed entirely of the left ventricle and the left ventricle feels firm and muscular when you squeeze or compress it whereas the right ventricle feels soft and flabby. The difference is due to the difference in thickness of the muscular walls.

b. Surface Features

Once you can tell the right from the left side, you can begin examining the surface features of the heart, shown in Fig. 33-8. If you have the pluck, it is easy to locate many of the major vessels associated with the heart because most are intact. For example, you should be able to locate and follow the pulmonary vessels to and from the lungs.

If you have an isolated heart, it is difficult to identify the vessels because they are usually cut off close to the heart during its removal from the animal. If this is the case, save identification of the major vessels until later when you examine the internal structure of the heart. The following descriptions are based on examination of the pluck.

Fig. 33-8A, an anterior view, shows the heart as it would appear if you removed the anterior wall of the thoracic cavity and the pericardium that surrounds the heart. As you can see, the bulk of the surface is composed of the wall of the **right ventricle.** Confirm this by squeezing the wall. Should it feel thick and muscular or soft and flabby?

____

At the superior end, you should be able to locate

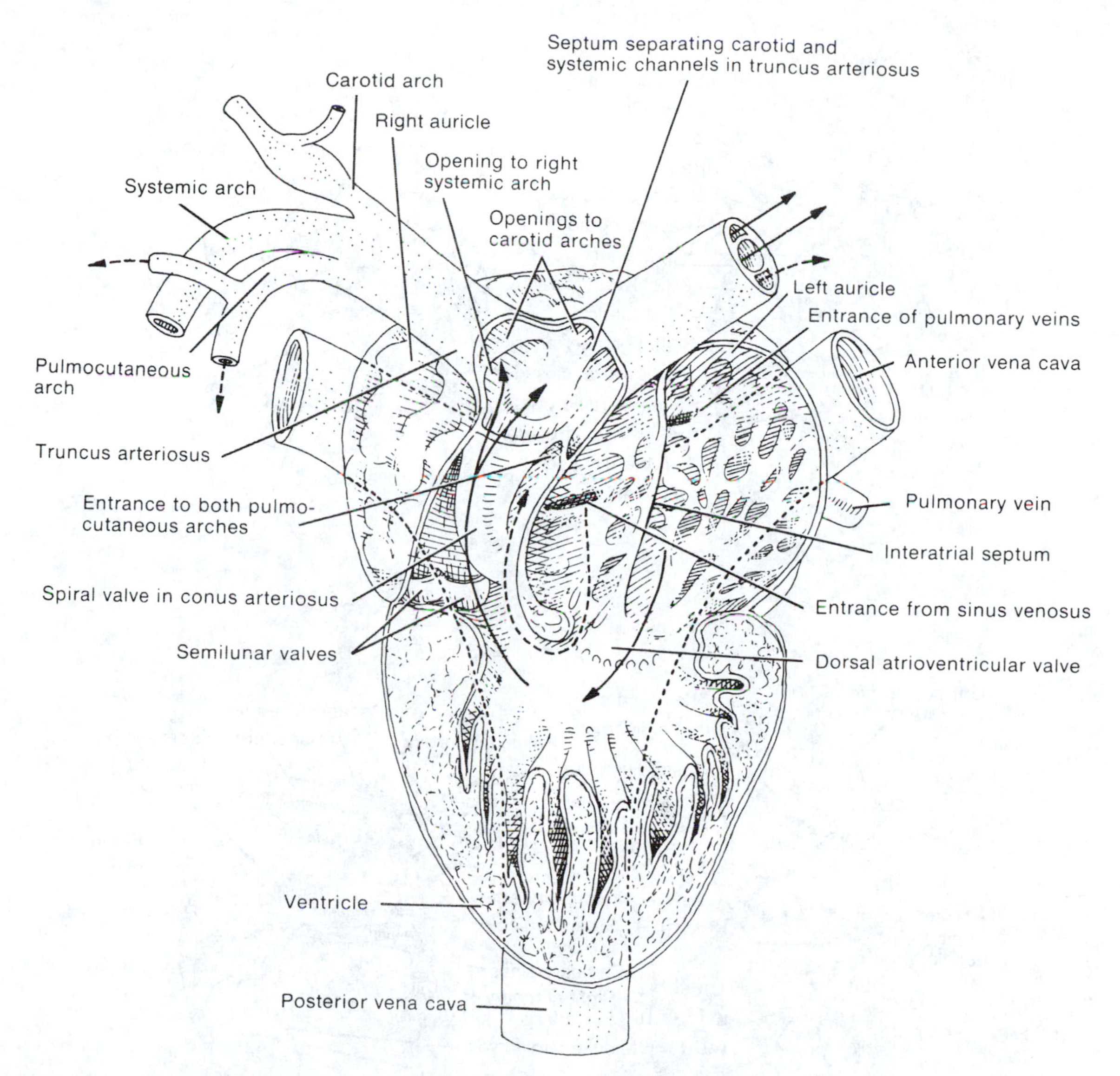

FIGURE 33-7 Longitudinal section of a frog heart. The position of the sinus venosus on the dorsal surface of the heart is shown by broken lines. (After *Dissection of the Frog,* 2d ed., by Warren F. Walker, Jr. W.H. Freeman and Company © 1981.)

the **pulmonary trunk,** which divides into the **pulmonary arteries.**

The surface of the right ventricle is separated from the left ventricle by the **interventricular groove,** in which lies the **great cardiac vein** and the **left anterior descending branch** of the left coronary artery. This groove also marks the position of the interventricular septum, which internally separates the right and left ventricles. (*Note:* A fairly large amount of fat (adipose tissue) usually covers the coronary blood vessels. Take the time to pick away the fat to locate the two blood vessels mentioned.)

Try to locate the cordlike **arterial ligament** that connects the pulmonary trunk to the arch of the aorta. This ligament is the remnant of the fetal **ductus arteriosus.** What role did the ductus arteriosus have in the fetus?

____

____

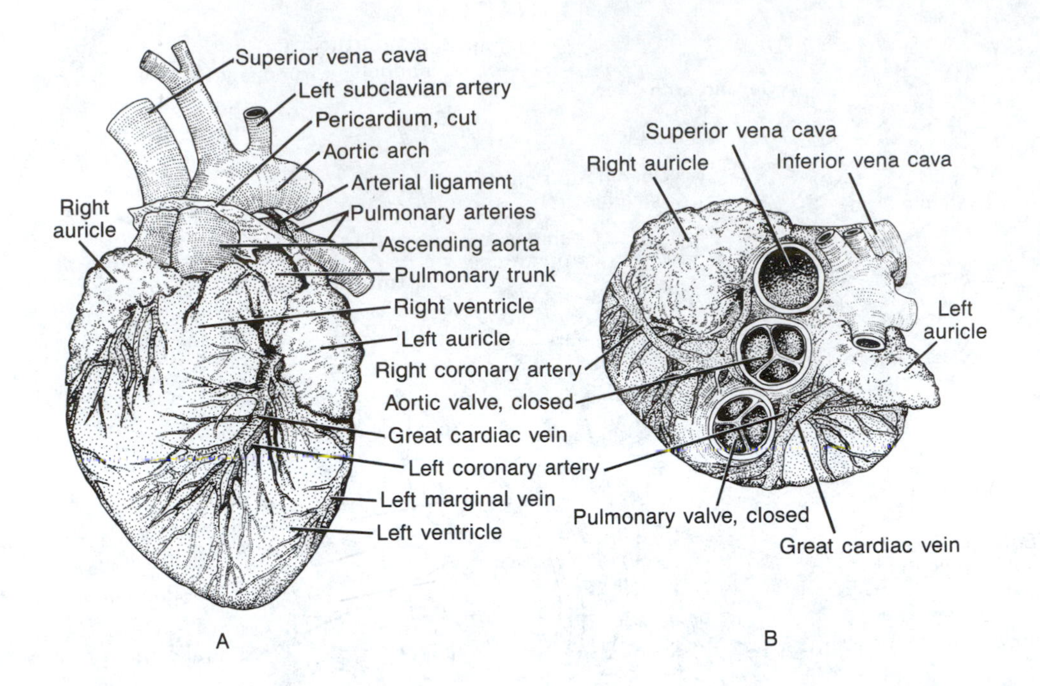

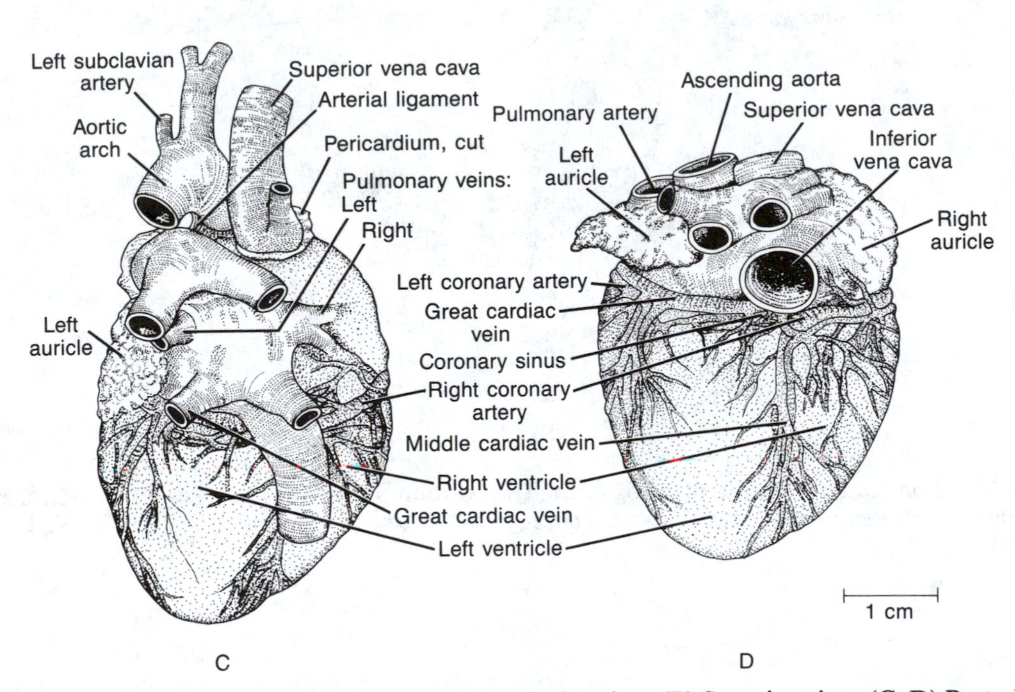

FIGURE 33-8 Surface features of the sheep heart. (A) Anterior view. (B) Superior view. (C, D) Posterior views.

What problems would occur if this duct did not close following birth?

____

____

Entering the right auricle is the **superior vena cava.** From what part of the body is this vessel returning blood to the heart?

____

____

What is the oxygen and carbon dioxide content of the blood in this vessel (high or low)?

____

____

The **aortic arch** can be seen where it leaves the left ventricle.

Fig. 33-8B shows the superior (cranial) view of the heart with the major vessels cut close to the heart. You can make these cuts later when you examine the interior of the heart. For now, try to locate the **superior and inferior vena cavae** where they enter the right atrium.

Figs. 33-8C, D are posterior views of the heart. Locate the superior and inferior vena cavae, the pulmonary trunk and the right and left pulmonary arteries, and the right and left auricles and ventricles.

Before you cut through the major blood vessels in preparing to study the internal anatomy, try to locate the coronary groove, which is the outward evidence of the position of the atrioventricular septum.

c. Internal Structure

The major internal features of the heart are shown in Fig. 33-9. In preparation for this examination,

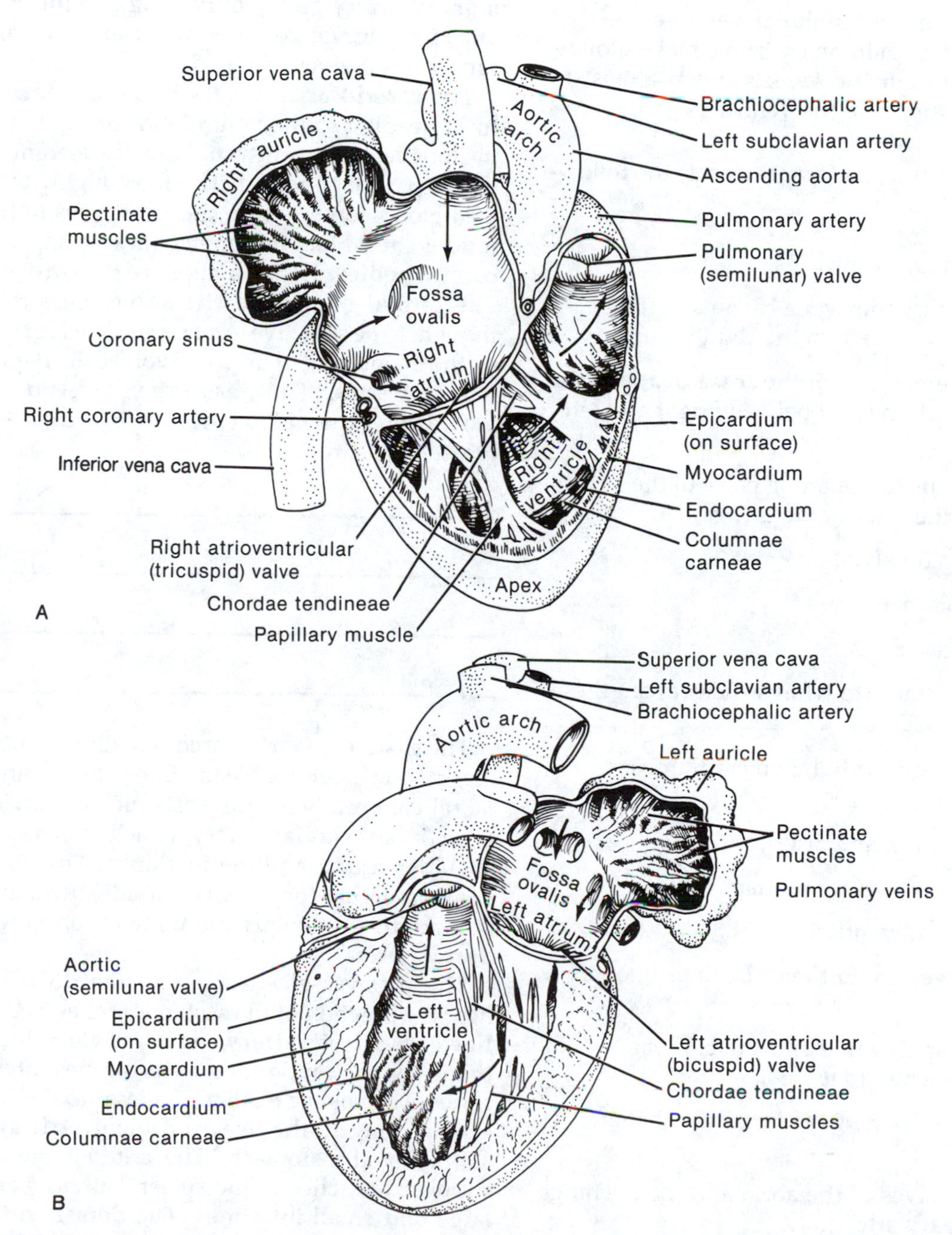

FIGURE 33-9 Schematic longitudinal section of the mammalian heart. (A) Anterior view. (B) Posterior view.

remove the heart from the lungs by cutting the vessels about 1 inch (2.5 cm) from where they enter or leave the heart. To examine the internal anatomy, open the heart as follows.

1. Make a midlateral incision of the right side of the heart, beginning at the top of the auricle and proceeding downward through the wall of the right ventricle.

2. Make a similar incision through the left auricle and ventricle.

3. To examine the semilunar valves associated with the aorta and pulmonary trunk, make a longitudinal cut through the vessels down almost to where the vessels leave the ventricles.

Using Fig. 33-9 as a reference, locate the following structures.

- Right atrium and ventricle
- Superior and inferior vena cavae and their respective openings into the right atrium
- Approximate position of the **fossa ovalis** in the interatrial septum (wall between right and left atria)
- Coronary sinus (an opening close to the opening of the inferior vena cava)
- Right tricuspid valve
- Trabeculae carnae
- Papillary muscles
- Chordae tendinae (to what structures are these attached?) ______________________
- Ventricular endocardium, epicardium, and myocardium
- Pulmonary trunk and arteries
- Semilunar valves of pulmonary trunk
- Left atrium and ventricle
- Pulmonary veins and their openings into left atrium
- Mitral (bicuspid) valve and the chordae tendinae attached to it
- Aorta and its opening leading from the left ventricle

Semilunar valves of the aorta and the openings into the coronary arteries

3. The Frog Arterial System

The arterial system consists of vessels that carry blood away from the heart to the capillaries and tissues. These vessels have thick, muscular walls that do not collapse when empty.

The conus arteriosus of the heart divides into two branches. Each branch is a **truncus arteriosus** and gives rise to three arteries (Fig. 33-10) the pulmocutaneous arch, the carotid arch, and the systemic or aortic arch.

The **pulmocutaneous arch** divides into the **pulmonary artery** going to the lung and the **cutaneous artery** to the skin: this artery and its branches carry nonoxygenated blood.

The **carotid arch** divides into the **external carotid,** leading to the ventral part of the head and the tongue, and the **internal carotid,** leading to the dorsal part of the head. The most highly oxygenated blood coming from the heart goes to the carotid arches. Note the small oval swellings, called **carotid bodies,** at the junctions of the two carotids. The carotid bodies are **chemoreceptors** that are thought to be sensitive to changing levels of oxygen in the blood and that are involved in regulating blood pressure. Thus, as oxygen levels decrease, would you expect blood pressure to increase or decrease? Explain.

____

____

____

____

The **systemic (aortic) arch** first divides into several blood vessels that lead off to the skull and vertebral column. Next, the systemic arch divides to form the **subclavian artery,** which branches to the shoulder region and the forelimbs. Then the two systemic arches (one from each side) bend dorsally around the esophagus and unite to form the large **dorsal aorta.**

The first vessel arising from the dorsal aorta, near the junction of the systemic arches, is the **coeliacomesenteric artery,** which divides into two branches, the **coeliac** and **anterior mesenteric arteries.** The coeliac further divides into the **hepatic artery** going to the liver and the **gastric arteries** supplying the stomach. The anterior mesenteric supplies branches to the spleen **(splenic)** and the large and small intestines. The dorsal aorta next

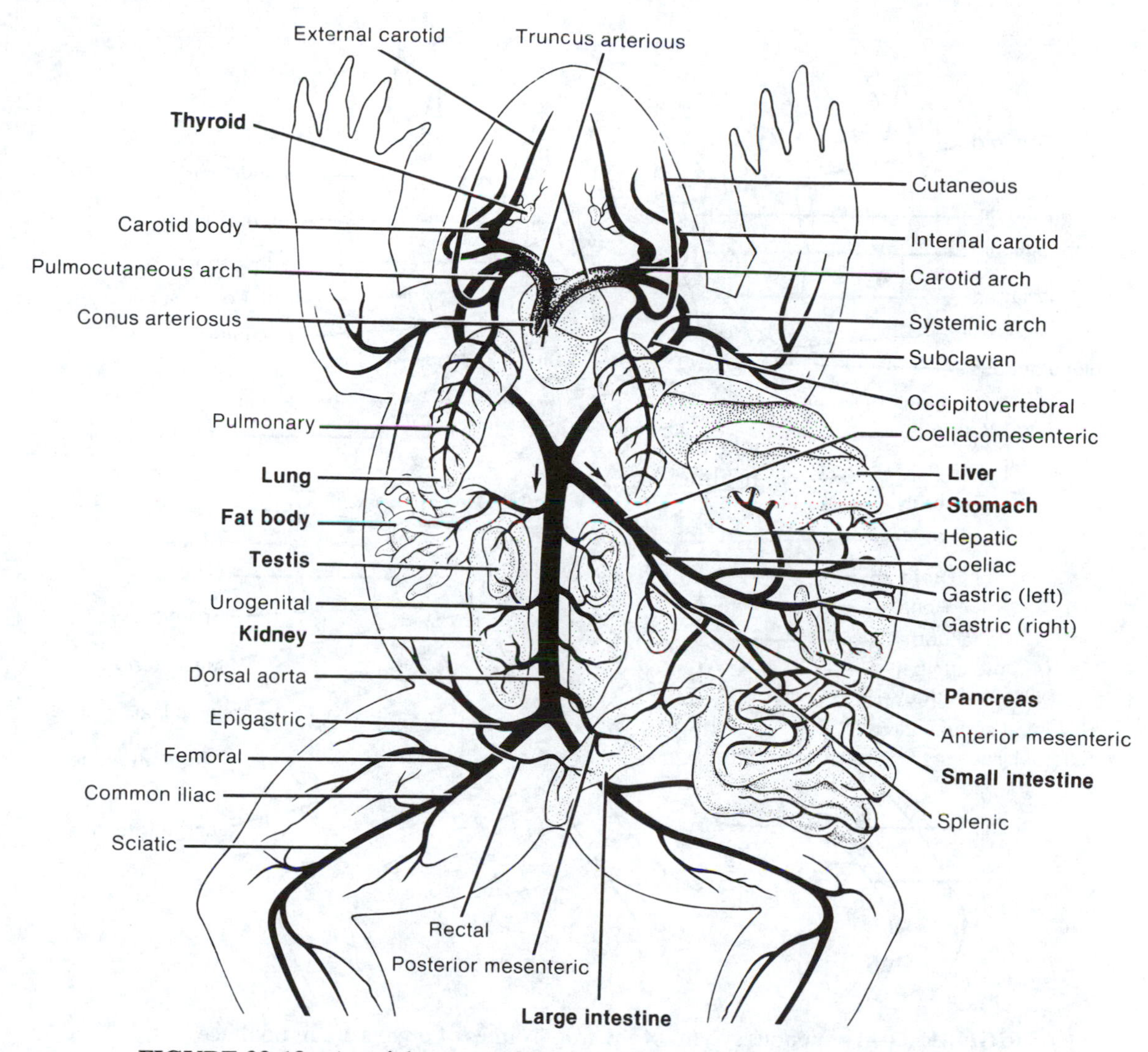

FIGURE 33-10 Arterial system of the frog. Names of organs are in boldface.

gives off four to six **urogenital** arteries to the kidneys, gonads, and fat bodies. The **posterior mesenteric artery** arises from the dorsal aorta to supply the colon. Next, the aorta divides into two **common iliac arteries,** which supply the hind limbs. As you trace these vessels down the legs, you will see the **epigastric arteries,** which extend to the urinary bladder, large intestine, and body wall; the **femorals,** which supply the hip and outer part of the thigh; and the **sciatics,** which serve the thigh, shank, and foot.

4. Frog Venous System

The vertebrate venous system, which consists of the vessels that carry blood back toward the heart, is somewhat more complex than the arterial system. Anteriorly, the head, forelimbs, and skin are drained by a pair of anterior venae cavae, which enter the heart by way of the sinus venosus (Fig. 33-11). Each anterior vena cava receives blood from three veins. The first of these is the **external jugular,** which returns blood from the tongue and floor of the mouth **(lingual vein)** and from the lower jaw **(mandibular vein).** The **innominate veins** collect blood from the head by means of the **internal jugulars** and from the shoulder and back of the forelimb by means of the **subscapular veins.** The **subclavian veins** collect blood from the forelimbs by means of the **brachial veins** and from the muscles and skin on the lateral and dorsal part of the head and trunk by means of the **musculo/cutaneous veins.**

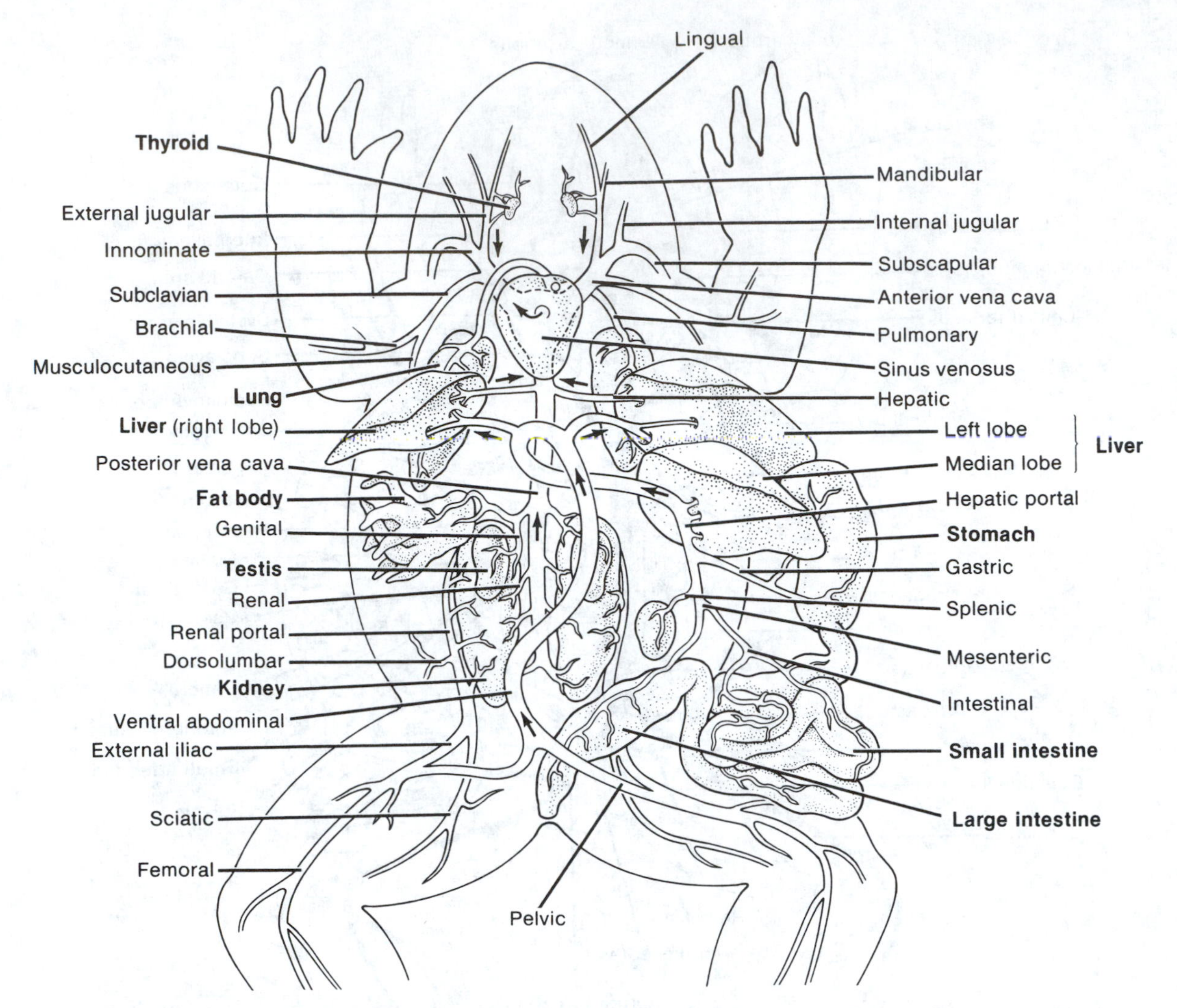

FIGURE 33-11 Venous system of the frog. Names of organs are in boldface.

The single posterior vena cava enters the sinus venosus posteriorly and collects blood from the testes or ovaries through the **genital vein,** from the kidneys through four to six pairs of **renal veins,** and from the liver by means of a pair of **hepatic veins.**

Veins that do not pass from organs or tissues directly to the heart, but instead enter another organ and subdivide within it to join its capillary system, are termed **portal veins** and constitute a **portal system.** The frog has two such portal systems, the renal and the hepatic.

The **renal portal system** is found only in lower vertebrates. It consists of a **renal portal vein,** which receives blood from the hind limbs by means of the **sciatic, external iliac,** and **femoral veins** and from the body wall by means of the **dorsolumbar vein.** It carries blood to the dorsal border of the kidney.

The **hepatic portal system** consists of a large number of veins that carry blood into the liver from the stomach, intestine, spleen, and pancreas. Such blood is heavily laden with recently digested food products absorbed by these organs. As this blood passes through the liver before entering the main circulation, food products are removed and stored and other substances are added to the blood. In addition, bacteria that may have entered the blood through the digestive tract are phagocytized by the Kupffer cells in the liver. The main vein draining into the liver is the **hepatic portal vein,** which collects blood from the spleen via the **splenic vein,** from the stomach via the **gastric vein,** from the small intestine via the **intestinal**

vein, and from the large intestine via the **mesenteric vein.**

D. UROGENITAL SYSTEM

Because the urinary and reproductive systems of the frog are closely connected, they are termed the **urogenital system.**

1. Urinary System

The main organs of the urinary system are two elongate, reddish brown kidneys that lie close together on the dorsal wall at the posterior end of the body cavity. The urine from each kidney is drained by a **ureter,** which can be seen along the kidney's lateral edge. The ureters empty into the dorsal surface of the cloaca, and so the urine must flow across the cloaca to enter the bilobed urinary bladder, which is attached to its ventral surface. The yellow or orange stripe running along the ventral side of each kidney is an adrenal gland, which secretes adrenaline into the blood. This hormone increases the heart rate, raises blood sugar levels, and helps the body react to stress.

2. Female Reproductive System

The ovaries in most frogs are conspicuous, occupying much of the body cavity. However, they vary greatly in size according to the season of the year. They are largest during the fall and winter, when they are filled with thousands of ripe eggs, and smallest after ovulation in the spring.

To identify the rest of the female reproductive system, remove one of the ovaries. Using Fig. 33-12, locate the oviduct, a highly coiled duct

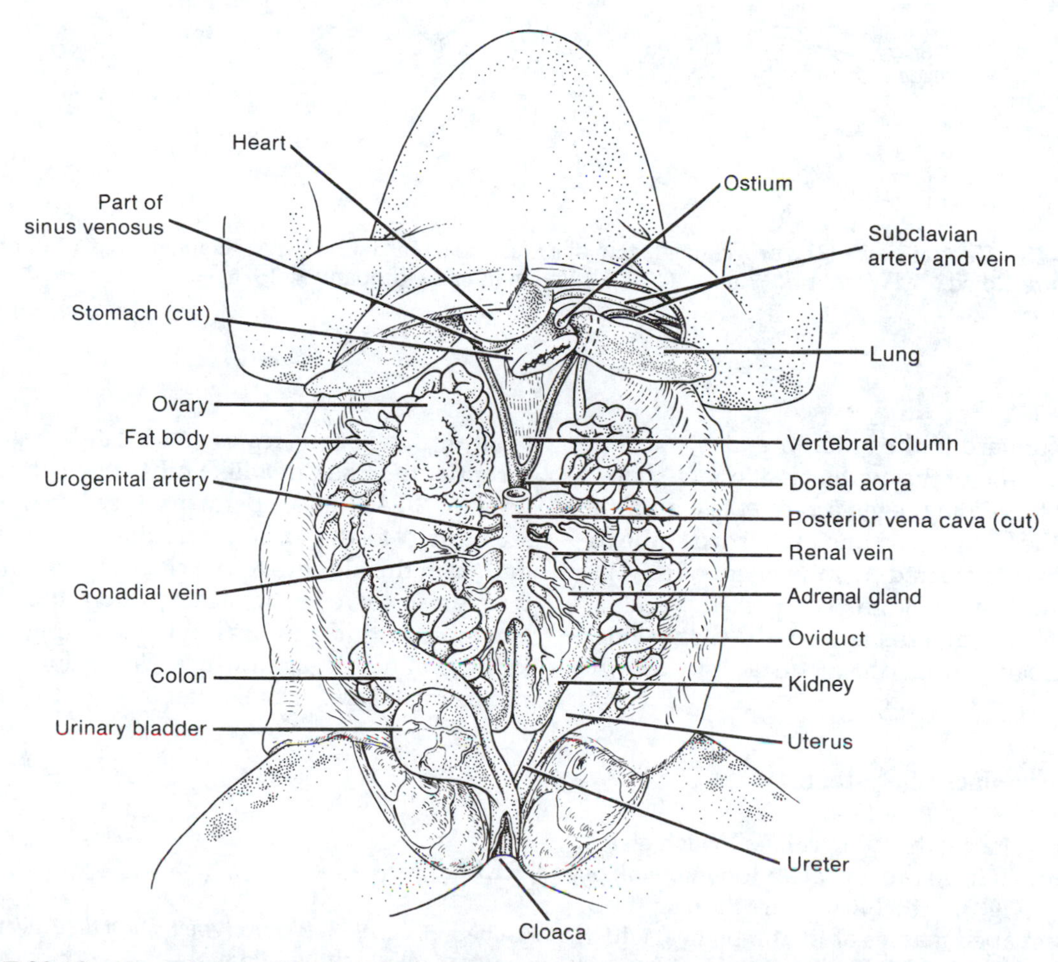

FIGURE 33-12 Ventral view of the female frog urogenital system. One ovary and one fat body have been removed. Because this drawing is based on a specimen taken in midsummer, the ovaries and fat bodies have not reached their full size. (After *Dissection of the Frog,* 2d ed., by Warren F. Walker, Jr. W.H. Freeman and Company © 1981.)

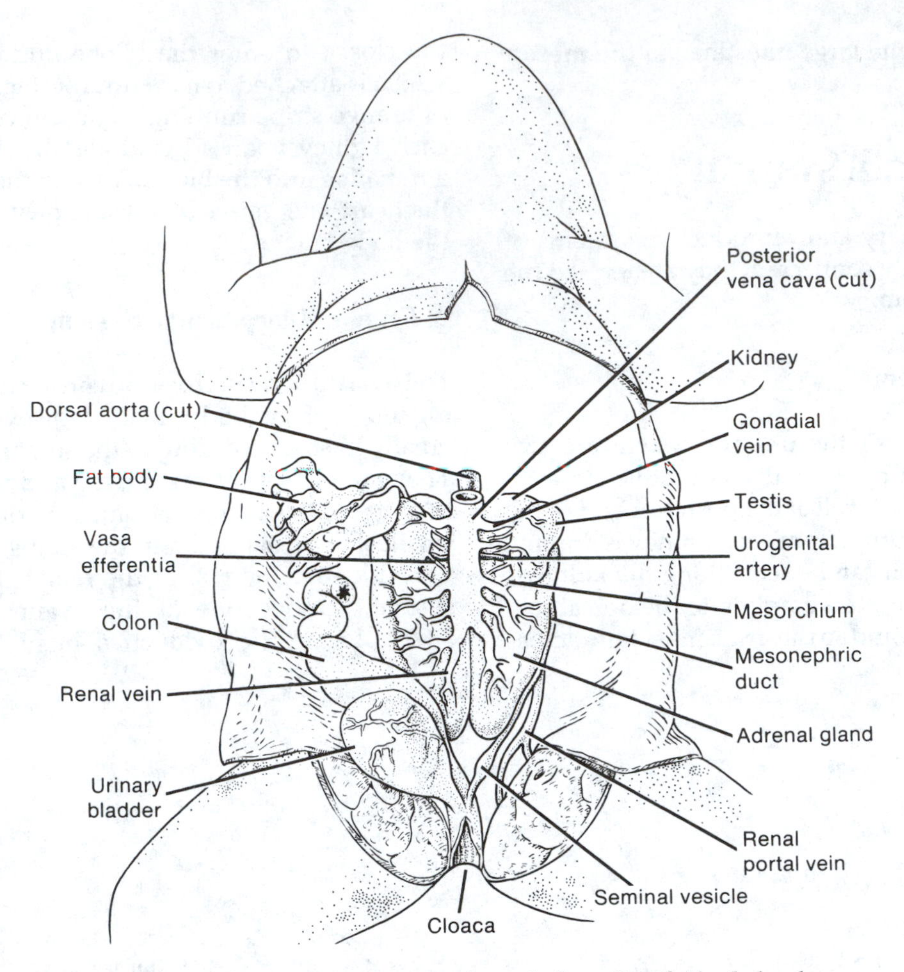

FIGURE 33-13 Ventral view of the urogenital system of a male frog. One fat body has been removed. (After *Dissection of the Frog,* 2d ed., by Warren F. Walker, Jr. W.H. Freeman and Company © 1981.)

extending forward to the anterior end of the body cavity. At the end of the oviduct is a funnel-shaped **ostium.** In the spring, eggs are released from the ovary into the body cavity and are carried to the ostium by currents created by movement of the cilia of the coelomic epithelium. The posterior ends of the oviducts are enlarged to form the **uteri,** which enter the cloaca on its dorsal surface next to the ureters.

3. Male Reproductive System

In a male frog, locate the two oval, yellowish **testes,** each suspended from the dorsal abdominal wall by a mesentery (Fig. 33-13). Nearby are the bright yellow, finger-shaped masses of the fat bodies, which are largest during the fall. Some of the food stored in them is used during hibernation, and the rest is absorbed during the spring breeding season.

Lift up one of the testes and locate the **vasa efferentia,** the fine threadlike tubes that connect the testes to the kidneys. Sperm produced by the testes are carried to the kidneys via the vasa efferentia and from the kidneys to the cloaca via the ureters, which also serve as the urinary ducts. It is not uncommon to find vestigial oviducts alongside the kidneys in male leopard frogs. These ducts should not be mistaken for the ureters.

REFERENCES

Gilbert, S. G. 1965. *Pictorial Anatomy of the Frog.* University of Washington Press.

Romer, A. S., and T. S. Parsons. 1986. *The Vertebrate Body.* 6th ed. Holt/Saunders.

Underhill, R. A. 1988. *Laboratory Anatomy of the Frog.* 5th ed. Brown.

Walker, W. F., Jr. 1981. *Dissection of the Frog.* 2d ed. W.H. Freeman and Company.

Wiggers, C. J. 1957. The Heart. *Scientific American* 196(5):74–87 (Offprint 62). *Scientific American* Offprints are available from W.H. Freeman and Company, 41 Madison Avenue, New York, NY 10010, and 20 Beaumont Street, Oxford OX1 2NQ, England. Please order by number.

Wood, J. E. 1968. The Venous System. *Scientific American* 218(1):86–99.

Zweifach, B. W. 1959. The Microcirculation of Blood. *Scientific American* 200(1):54–74.

Exercise

34

Biological Coordination in Animals

The nervous system is distributed throughout the body. It functions, basically, to collect stimuli from the external and internal environments, transform the stimuli into nervous impulses, and pass the stimuli to highly organized reception and correlation areas where they are interpreted and transmitted to various effector structures where appropriate responses to the stimuli result. All of these functions are carried out by highly specialized cells called **neurons.** These cells, along with a large number of **neuroglia** (or supporting cells) and the associated extracellular materials, make up a highly integrated coordinating network.

Anatomically, the nervous system consists of two major subdivisions, the **central nervous system (CNS)** and the **peripheral nervous system (PNS).** The central nervous system lies deep within the body and is surrounded and protected by bone. It is composed of the **brain,** enclosed by the skull, and the **spinal cord,** which extends from the brain downward through the vertebral canal where it terminates between the first and second lumbar vertebrae. Functionally, the CNS receives thousands of pieces of "information" from different sensory structures located inside and outside the body and then integrates the information to determine the appropriate response. More than 90% of the sensory information received by the CNS is considered to be unimportant and is not acted on. For instance, you are not generally aware of the pressure on your seat when you are sitting down, and you typically are not aware of your clothes touching your body. Even the noise in your surroundings is normally relegated to the background. On the other hand, "important" sensory information is channeled to the spinal cord, and frequently to the brain, to bring about desired responses. For example, when you touch a hot stove, the CNS determines whether you pull your hand away from the stove or move your entire body away and whether you shout out from the pain.

The peripheral nervous system serves to connect various sensory structures and effector organs (muscles and glands) with the CNS. The connecting structures are the cordlike **nerves** that emerge bilaterally from the brain (the **cranial nerves**) and spinal cord (the **spinal nerves**). The PNS is composed of an **afferent system** of sensory nerves (these carry impulses *from* various receptors *to* the CNS) and an **efferent system** of motor nerves that transmit information (or impulses) *from* the CNS *to* various muscles and glands. The efferent component is subdivided into **somatic** and **autonomic** parts. The somatic subdivision carries nerve impulses from the CNS primarily to skeletal muscles and is under conscious or voluntary control. The autonomic subdivision contains efferent (motor) nerves that transmit impulses to numerous glands and the smooth (involuntary) muscles of the cardiovascular, digestive, respiratory and excretory systems. These impulses are transmitted through **sympathetic** or **parasympathetic** nerve fibers. The sympathetic and parasympathetic divisions are anatomically and functionally distinct and generally antagonistic in their action to each other.

In this exercise you will study the anatomy of the brain and the interrelationships of the central and peripheral nervous systems.

A. THE SHEEP BRAIN

Because of its large size and ready availability, you will study the sheep brain for a basic understanding of brain anatomy.

1. Meninges

The living brain is covered by three layers of membranous tissue called the **meninges** (meninges may be absent in the preserved brain). The outer membrane, the **dura mater** (Latin *dura,* "hard"; *mater,* "mother"), is a tough, opaque membrane that is bound to the surface of the skull. Under the dura is a delicate, semitransparent middle membrane called the **arachnoid** (Greek *arachnoeides,* "cobweblike") **membrane** because of its spidery lacelike appearance. The **pia mater** (Latin *pia;* "tender"; *mater,* "mother"), a very thin, vascular membrane, forms an intimate covering of the brain and dips into all of its clefts and fissures. Between the arachnoid membrane and pia mater is the **subarachnoid space** containing delicate strands that connect the dura mater and pia mater. This space is filled with **cerebrospinal fluid.**

In human beings, inflammation of the meninges, caused by bacteria, fungi, or viruses, is called **meningitis** and can affect the brain or the spinal cord or both. One of the more common agents of this disease is the bacterium *Neisseria meningitidis,* frequently called *meningococcus.*

As you examine the sheep brain, note that the dura mater has already been removed and the arachnoid membrane has been destroyed during its removal from the sheep cranium. Only the pia mater remains. During your study of the sheep brain, examine specimens and/or models of the human brain, if available, and note the many similarities between the two.

2. Brain

Locate the following major parts, some of which can be seen from the dorsal and ventral surfaces and some only from the ventral surface.

a. Cerebrum

The cerebrum forms the greater part of the brain (Fig. 34-1A). It is divided into two **cerebral hemispheres** by a deep longitudinal **cerebral fissure.** Small ridges **(gyri)** and grooves (**fissures,** or **sulci**) cover the dorsal surfaces of the cerebral hemispheres. The anterior region of these hemispheres are called the **frontal lobes;** the posterior regions are the **occipital lobes.** The lateral and widest parts of these hemispheres are the **temporal lobes.**

Turn the brain over so that its ventral surface is uppermost and locate a pair of elongated **olfactory bulbs** at the anterior end of the cerebral hemispheres (Fig. 34-2). Extending from the posterior end of each of these bulbs are the medial, lateral and middle **olfactory tracts.** The **olfactory nerve** from the nasal mucosa (inside lining of nose) enters the brain through the olfactory bulbs.

Examine a half brain cut lengthwise through the cerebral tissue (Fig. 34-3A). The prominent curved band seen on the medial surface in the middle of the cerebral hemispheres is the **corpus callosum.** This broad band of tissue connects the two cerebral hemispheres, and forms the roof of the lateral ventricles, one of several fluid-filled cavities in the brain that communicate with each other and with the central canal of the spinal cord.

b. Diencephalon

The diencephalon is made up of the structures that surround the **third ventricle** (a fluid-filled cavity in the brain shown in Fig. 34-3A). To see the dorsal regions of the diencephalon, spread apart the occipital lobe areas of the cerebral hemispheres and, using a small scalpel, cut through the corpus callosum slowly and carefully until the structures shown in Fig. 34-1B are visible. Dorsally, the diencephalon includes a knoblike prominence, the **pineal gland,** an endocrine gland that is involved in regulating biological rhythms.

From the ventral surface of the brain, you can observe three major parts of the diencephalon (Fig. 34-2). Most prominent is the **optic chiasma,** an X-shaped structure from which the optic nerves extend to the eyes. Just behind the optic chiasma is the **pituitary gland** (Fig. 34-3A), which is attached to the brain by the **infundibulum.** The pituitary gland is usually missing because it is easily torn away when the brain is removed from the cranium. Just posterior to the infundibulum is the **mammillary body,** a pair of rounded elevations that form the posterior end of the **hypothalamus.**

Examine a sagittal section of the brain and locate the **third ventricle,** the cavity of the diencephalon (Fig. 34-3A). The floor of the third ventricle forms

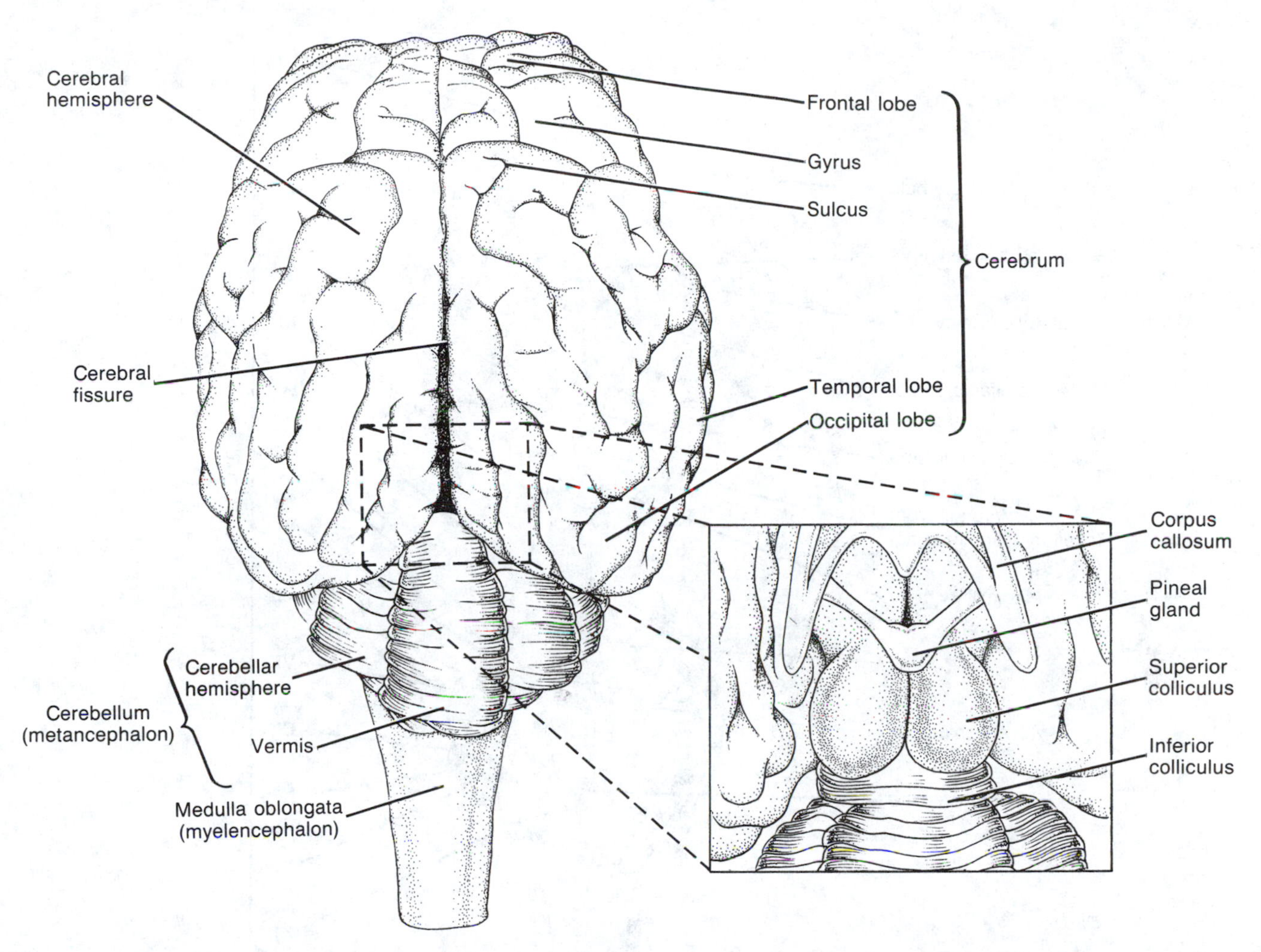

FIGURE 34-1 Sheep brain. (A) Dorsal view. (B) Diencephalon and mesencephalon.

the **hypothalamus,** that part of the brain that controls various physiological functions (temperature regulation, coordination of the autonomic nervous system, control of the pituitary gland). The **thalamus,** which appears as a larger ovoid structure on the lateral walls of the third ventricle, is an important relay center in which sensory pathways from the brain and spinal cord form synapses on their way to the cerebral cortex.

c. Mesencephalon

Four elevations form the roof of the mesencephalon, or midbrain (Fig. 34-1B). The larger anterior pair of elevations **(superior colliculi)** are rounded and form a cradle for the pineal gland. The posterior pair of transverse elevations, the **inferior colliculi,** are involved with auditory reflex activity.

The ventral surface of the mesencephalon is called the **cerebral peduncles.** They emerge from under the optic chiasma and disappear under the pons.

d. Metencephalon

The fourth part of the brain, the metencephalon, can readily be seen lying behind the cerebral hemispheres. When viewed from the dorsal surface (Fig. 34-1A), it consists of the **cerebellum,** which is divided into a median **vermis** and a pair of **cerebellar hemispheres.** The cerebellum has a series of transverse ridges. Viewed from the ventral surface of the brain, you can see the **pons,** the other major portion of the metencephalon (Fig. 34-2).

e. Myelencephalon

The fifth and last part of the sheep brain is the myelencephalon, also known as the **medulla oblongata,** or simply the medulla. It surrounds the

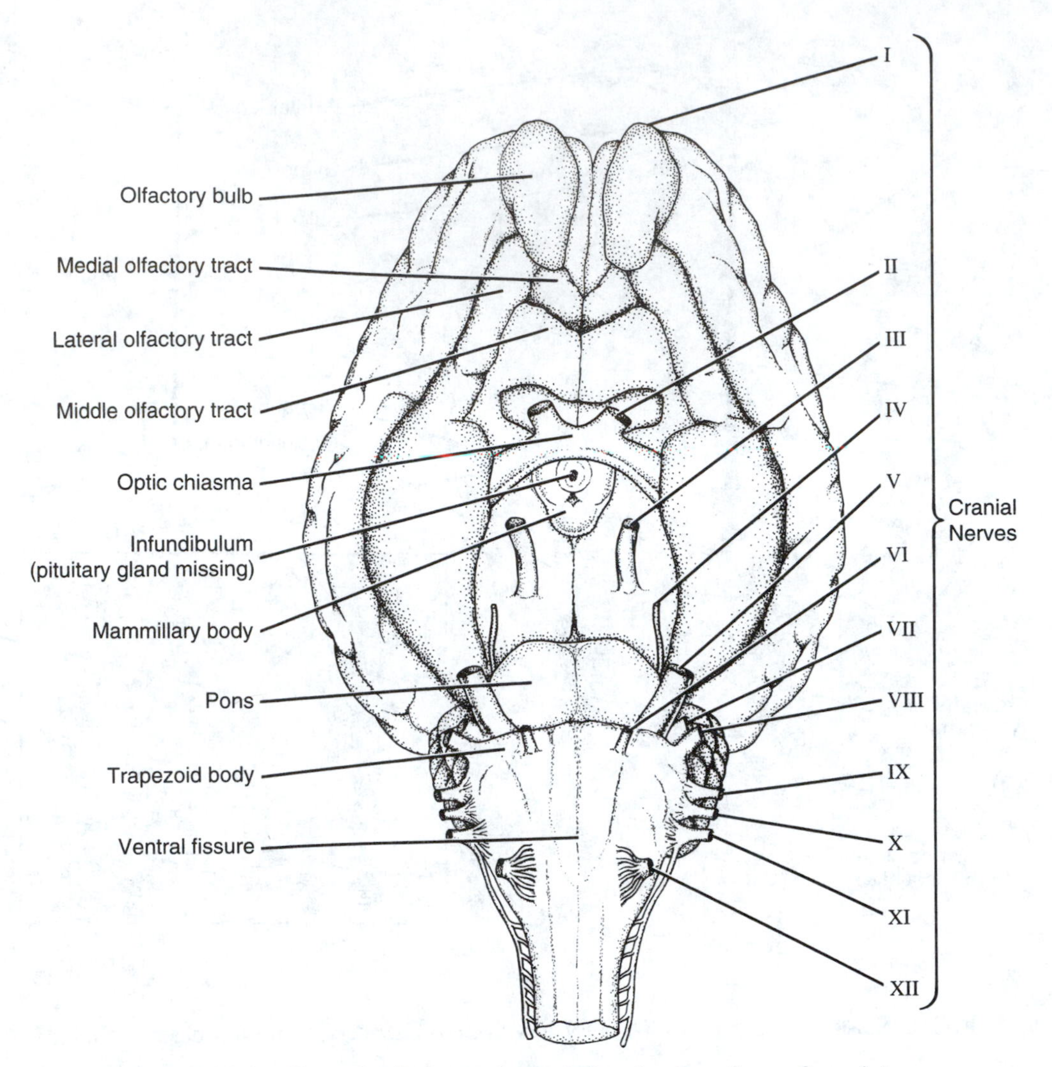

FIGURE 34-2 Sheep brain (ventral view) showing locations of cranial nerves

large, diamond-shaped **fourth ventricle** (Fig. 34-3A).

When viewed from the ventral surface, you can observe a median **ventral fissure** that extends the length of the medulla (Fig. 34-2). The triangular-shaped area of the medulla just behind the pons is called the **trapezoid body.**

2. Cranial Nerves

The sheep brain has 12 pairs of cranial nerves that emerge from the ventral surface and leave the skull through holes called **foramina.**

Some cranial nerves carry impulses from various sensory structures to the brain and thus are sensory in function. Others carry impulses from the brain to effector structures, such as muscles or glands, and are called motor nerves. Still other nerves carry both sensory and motor impulses.

Using Figs. 34-2 and 34-3B, locate as many of the 12 cranial nerves as you can. In some cases, you will only find stubs of the nerves depending on how carefully the brain was removed. In Exercise 35, you will test for the activity of the cranial nerves using yourself or a partner as subjects for various tests of cranial nerve function.

B. HISTOLOGY

The functional unit of the nervous system, the **reflex arc,** will be used to understand the structure and function of the somatic subdivision of the pe-

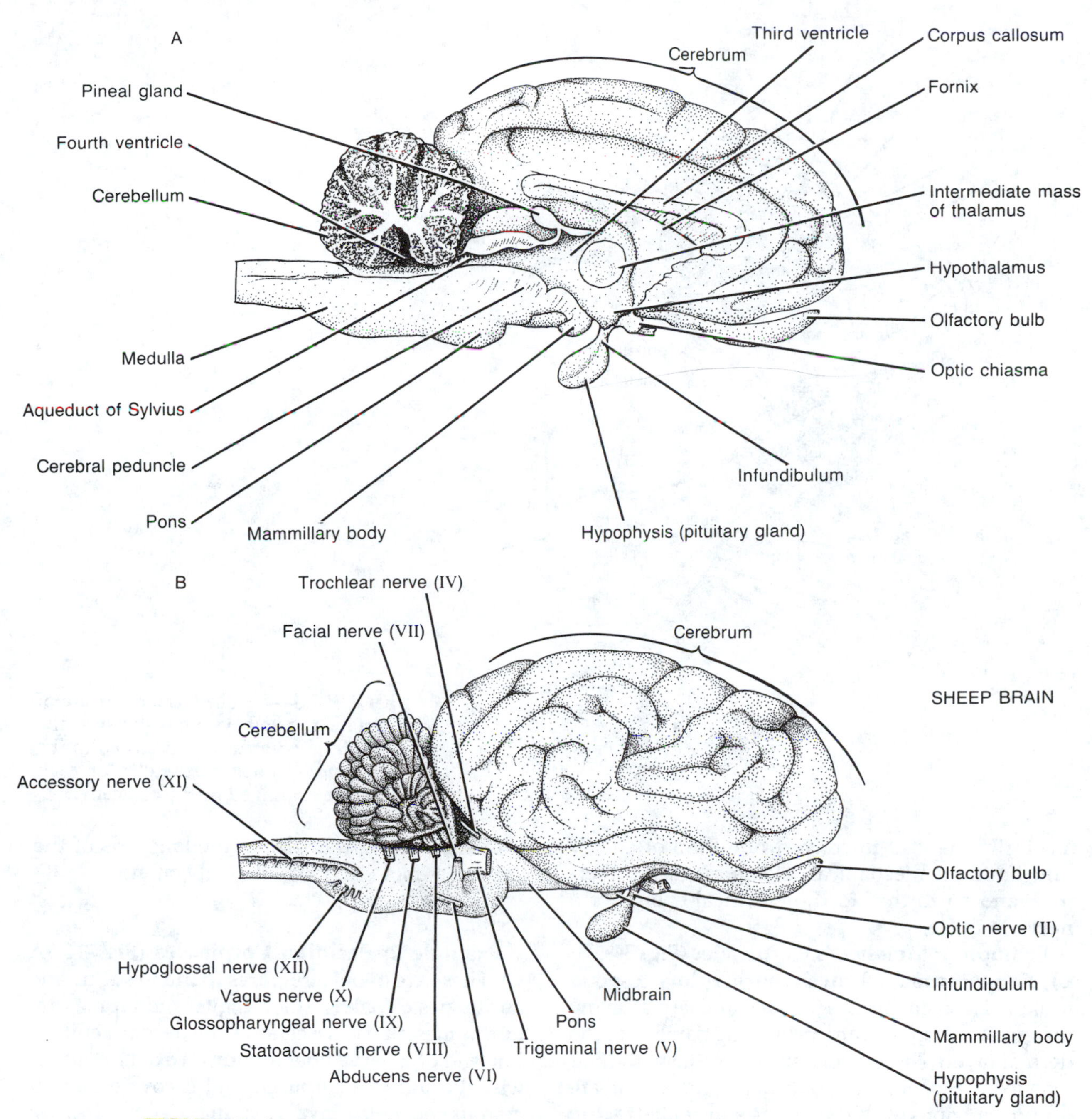

FIGURE 34-3 Sheep brain. (A) Saggital section. (B) Lateral view of cranial nerves.

ripheral nervous system. A reflex arc is one type of **conduction pathway** by which nerve impulses, such as those from a pain or pressure receptor are transmitted to the spinal cord and then to an effector structure, such as a skeletal muscle. The functional cell in such a conduction pathway is the **neuron** and all conduction pathways consist of circuits of neurons. The basic components of a reflex arc follow and are shown in Fig. 34-4. Each component discussed will be examined microscopically to determine its microanatomy.

1. Receptors (Fig. 34-4A)

Receptors are specialized structures that detect changes in the external or internal environment.

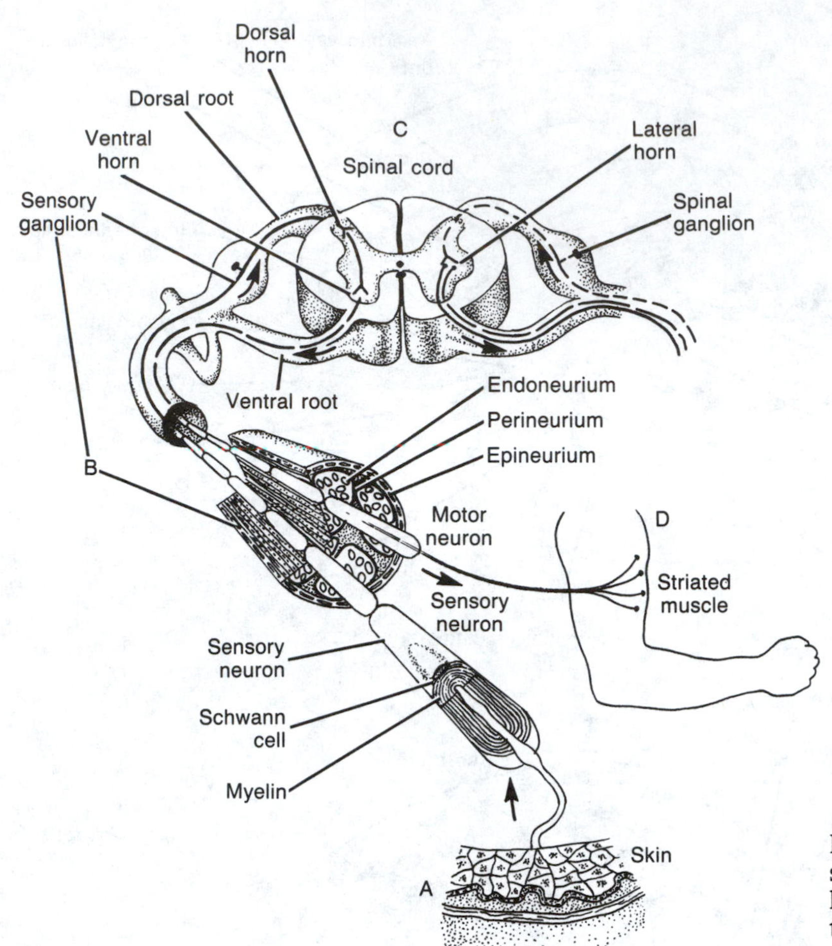

FIGURE 34-4 The basic reflex arc of somatic nervous system is shown at the left. Lettered areas are discussed in the text and histologic examples of each area are examined microscopically.

Typically they act to convert one form of energy into another: mechanical, electromagnetic, thermal, and so forth into the electrical energy of a nerve impulse.

Example 1: Meissner's Corpuscles (Fig. 34-5A, B) Nonencapsulated **mechanoreceptors** respond to skin displacement due to touch and lie just below the epidermis in upward-projecting papillae of the dermal layer. These ovoid structures lie with their long axis perpendicular to the skin surface and are found in regions of skin showing substantial **tactile sensitivity** (e.g., lips, eyelids, external genitalia, and nipples).

Examine a tissue section of skin containing these receptors and locate the following structures.

- The outer or **epidermal layer** of the skin. *Note:* If this is a section of thick skin, there may be a layer of dead cornified cells above the living epithelial layer.
- Fingerlike **papillae** projecting up into the epidermis. Look at these individually under high power (40X or oil immersion) and look for cells and other supporting structures that are oriented at right angles to the long axis of the papilla. Not every papilla will contain a Meissner's corpuscle.

Example 2: Pacinian Corpuscles (Fig. 34-5A, C) These are ovoid structures found in subcutaneous tissues of palms, soles, digits, and nipples and in the mesentery, viscera, and external genitalia. The end of a nerve fiber (dendrite) extends lengthwise through the corpuscle and is covered by numerous concentric layers of flattened cells. In a section, this receptor has the appearance of a cut onion. The turgidity of the corpuscle, due to fluid between the layers, allows a response to pressure and vibration, which then generates a nerve impulse in the dendrite ending.

Examine a tissue section from the skin and locate a Pacinian corpuscle. How could you determine whether this sensory structure has been cut in a transverse or longitudinal section?

____

____

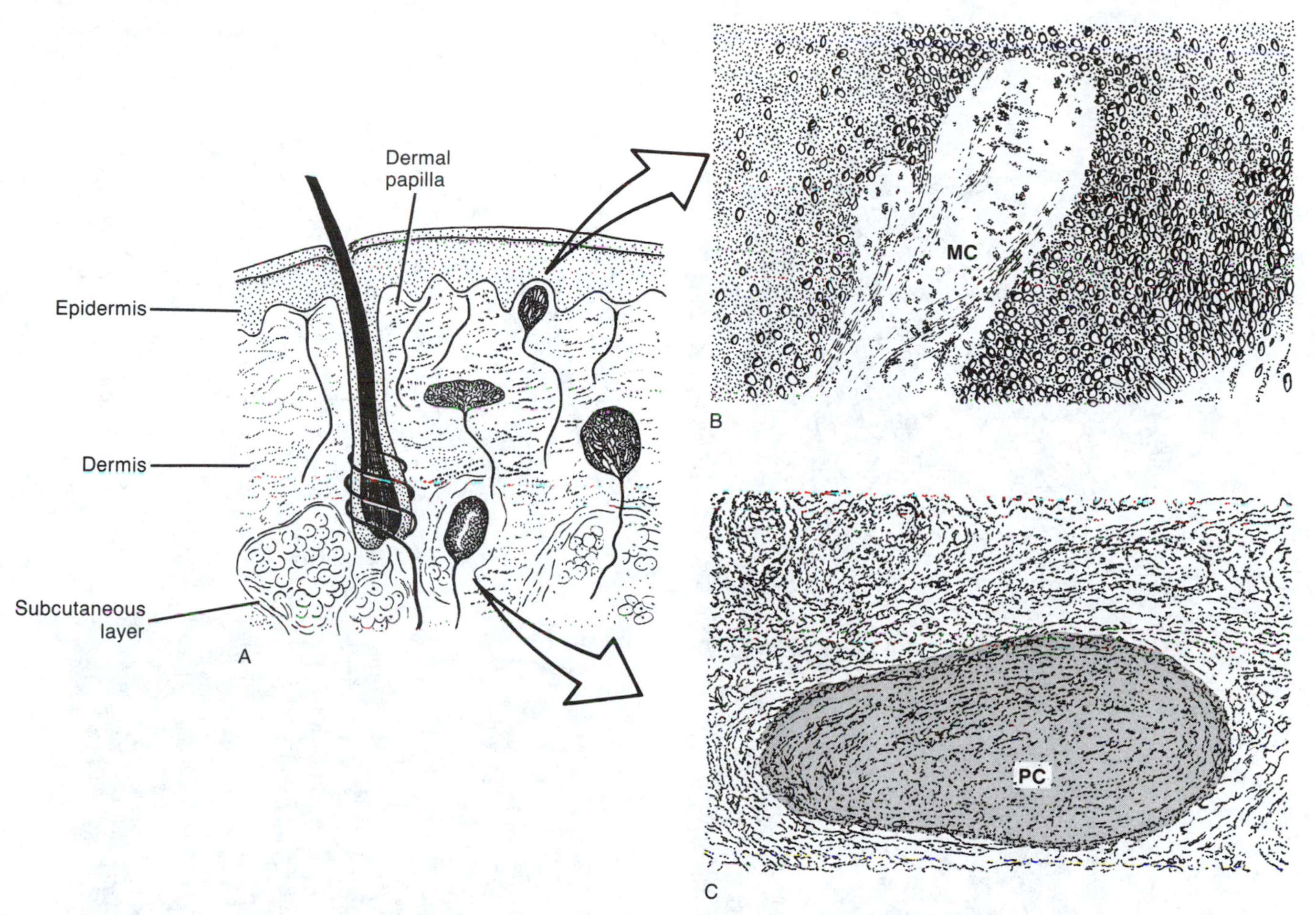

FIGURE 34-5 Sensory receptors. (A) Diagram of section through skin. (B) Meissner's corpuscle (MC) located in upward-projecting papilla on boundary between epidermis and dermis. (C) Pacinian corpuscles (PC) located in subcutaneous tissue just below dermis.

2. Sensory Neurons (Fig. 34-4B)

Typically, large numbers of individual neurons are clustered together to form a **nerve,** which may be a **spinal** or **cranial** nerve depending on whether it emerges from or enters the spinal cord or the brain. In the reflex arc, the bilaterally paired spinal nerves involved are connected to the spinal cord by two points of attachment, called **roots** (see Fig. 34-7A). The **dorsal** root contains only sensory (afferent) nerve fibers transmitting impulses toward the CNS. The other point of attachment is called the **ventral** root and contains only motor neurons carrying impulses away from the CNS.

Thus, once the sensory dendrite associated with a receptor is stimulated, the nerve impulse passes through the sensory neuron to its axonal end in the spinal cord. The cell bodies of these neurons, however, are located in swellings, called the dorsal root ganglion of the dorsal root.

Example 1: Transverse section of a myelinated nerve (Fig. 34-6A, C) Recall that a nerve is a collection or aggregation of individual nerve cell fibers. There are, in addition, other tissue components in nerves such as blood vessels, other types of nervous tissue cells, and connective tissue, which covers the nerve and may partition the nerve fibers into smaller bundles within the larger nerve.

Examine a transverse section of a nerve and locate the following structures.

- The **epineurium,** a relatively strong connective tissue sheath surrounding the entire nerve. It is composed chiefly of collagenous fibers, mostly longitudinal in orientation, and a small number of elastic fibers. Blood vessels that supply the nerves are typically found in this layer, but very difficult to observe.
- **Bundles** or **fascicles,** groupings of nerve fibers within the epineurium. Typically the larger the nerve the greater the number of fascicles. Fascicles are covered by a connective tissue

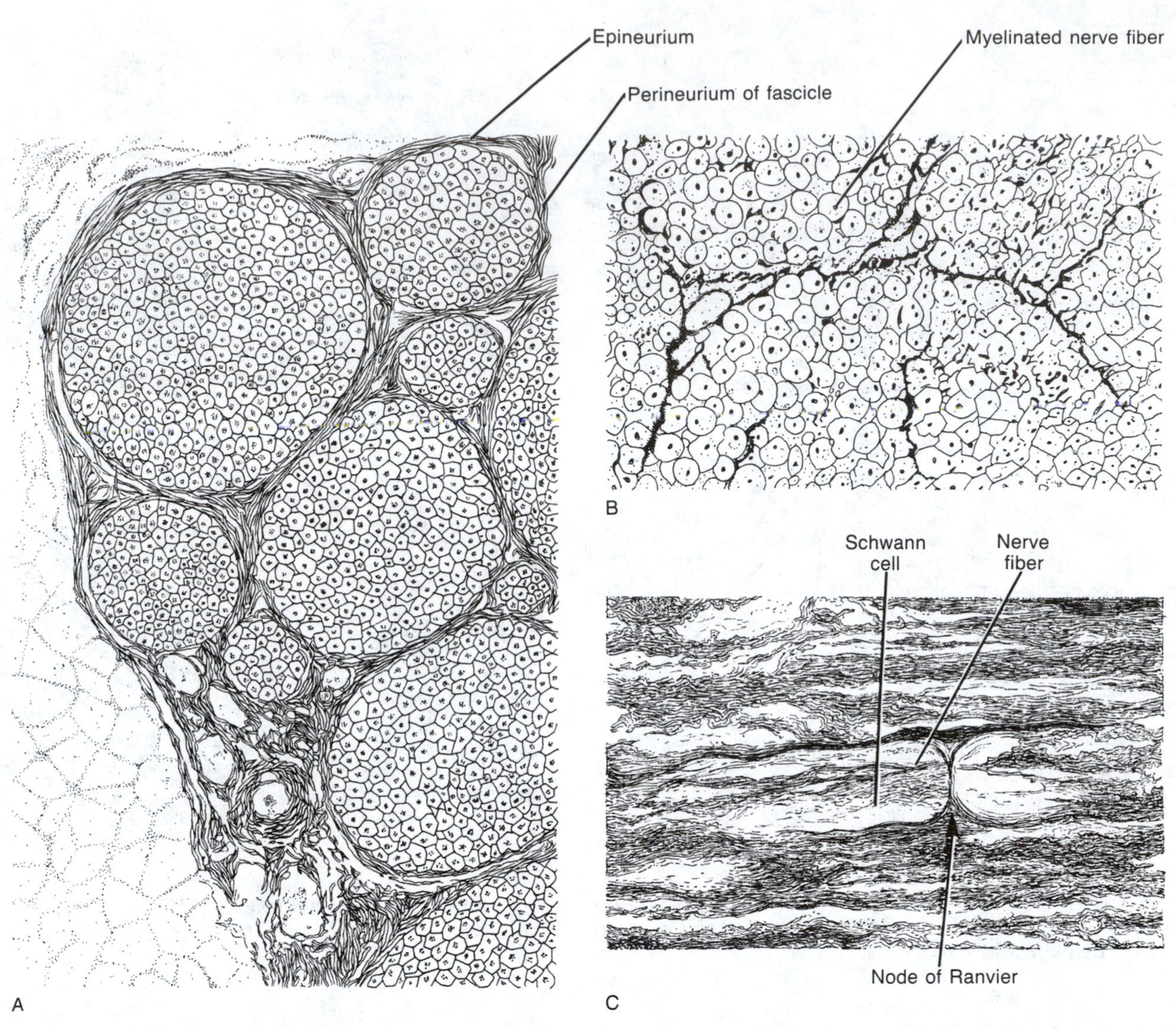

FIGURE 34-6 Transverse section of a spinal nerve. (A) Cross-section of nerve (low power). (B) Cross section of nerve showing several fasciles, or bundles (high power). (C) Light microscopic view of macerated, myelinated nerve fibers showing nodes of Ranvier, nerve fiber, and Schwann cell.

layer that is much thinner than the epineurium and is called the **perineurium.**

- Myelinated **nerve fibers,** which are the bulk of the fascicle. These appear as white circles containing small, roughly circular, dark-staining structures that you might mistake for nuclei but are actually axons or dendrites (Fig. 34-6B). The white circles are the myelin sheaths, which, in the living condition, contain myelin. The myelin, which is lipid-soluble, has been removed during the preparation of the tissue for staining.

The myelin sheath is composed of the spiral layering of the plasma membranes of **Schwann cells** or **oligodendrites,** depending on whether the fiber is in the peripheral, or central nervous system. Since the section above is a nerve, it is a component of the PNS, and Schwann cells provide the myelin sheath.

Example 2: Macerated myelinated nerve fibers (Fig. 34-6C) Examine a segment of myelinated nerve that has been stained with osmic acid and teased apart so that individual nerve fibers can be seen. The osmium preserves and blackens the myelin.

With careful focusing at high power you should be able to locate the following structures.

- A **Node of Ranvier,** which is an interruption of the myelin sheath.
- The **nerve fiber** itself within the myelin sheath.

- The **Schwann cell** that envelopes the nerve fiber and produces the myelin sheathing.

Example 3: Dorsal root ganglion The dorsal root ganglion is a swelling of the dorsal root of a spinal nerve close to where the nerve is attached to the spinal cord (Fig. 34-7A). This ganglion is largely a collection of nerve cell bodies, which accounts for its swollen configuration.

Examine a dorsal root ganglion and, with the aid of Fig. 34-7, locate the following structures.

- The large numbers of **unipolar sensory nerve cell** bodies that are the bulk of the ganglion (Fig. 34-7B). These cells contain large nuclei and prominent nucleoli.
- **Satellite cells** with rounded nuclei forming a single, layered capsule around each nerve cell body (Fig. 34-7C). They may appear to be separated from the cell body by a clear space or cleft. This is an artifact of shrinkage during tissue preparation.
- Bundles of myelinated **nerve fibers** running between the nerve cell bodies. The larger bundles can be seen running in a longitudinal direction.

3. Spinal Cord (Figs. 34-4C, 34-8)

The spinal cord is a cylindrical structure that is flattened somewhat on its dorsal (or posterior) and ventral (or anterior) surfaces. It is protected by the meninges, connective tissue coatings that cover both the spinal cord and brain, and by its location inside the vertebral column.

a. Gray and White Matter

Two main regions can be recognized in fresh slices through the spinal cord. These are distinct areas of **gray matter** and **white matter** that extend through the entire cord and into the brain. The gray matter roughly resembles an H or a butterfly, when viewed in transverse section. The arms of the H (or wings of the butterfly) comprise two **anterior horns** and two **posterior horns.** The region in between may be referred to as the **lateral horns.** The crossbar of the H, connecting the right and left horns, is called the **gray commissure.**

Gray matter is grayish in color because it consists primarily of nerve cell bodies and unmyelinated axons and dendrites of motor and connecting neurons. The white matter, which surrounds the gray matter, is whitish because the majority of the nerve fibers, which extend up and down the cord, are myelinated. White matter does not contain any cell bodies or neurons. White matter fibers originate from cell bodies lying in either the gray matter of the brain or spinal cord, or in spinal ganglia. The fibers are organized into **tracts,** each of which contains fibers from neurons having similar roles. Thus, there are motor tracts and sensory tracts. A tract in the central nervous system is comparable to a nerve in the peripheral nervous system.

Example 1: Transverse section of spinal cord Microscopically examine a cross-section of the spinal cord and, with the aid of Fig. 34-8, locate the following structures.

- The **central canal** found in the center of the gray commissure. This canal runs the length of the spinal cord and is continuous with the fourth ventricle of the brain. It is filled with cerebrospinal fluid, as are all cavities within and around the CNS.
- The H- or butterfly-shaped **gray matter.** (You might want to reduce the amount of light to provide more contrast.)
- The **white matter.** Trace the white matter completely around the gray matter to determine the H configuration of the gray matter.
- The rather large, deeply staining **cell bodies** of **motor (efferent) neurons.** These are readily visible at 4X magnification and are typically found in large numbers in the ventral (or anterior) horn. Reposition the slide so that the ventral horn is at the bottom facing you. It is difficult in a stained section to determine which processes are axons or dendrites. Select a large cell body that is somewhat multiangular in shape and examine it under high power. The cytoplasm should appear to be rather deeply stained and granular. The granules are **Nissl bodies** composed of endoplasmic reticulunar and associated ribosomes. Look for a corner that appears rather light and without any Nissl bodies. This region is the **axon hillock** and is at the origin of the axon where it leaves the cell body.

As well as the large motor neurons, notice intermediate-size cell bodies (about half the size of motor neurons) and numerous small nuclei throughout the gray matter. The intermediate-size cells are called **association (adjuster) neurons** because they take impulses from the incoming sensory neurons and relay them to other "adjusters" or up to the brain via long nerve fibers or directly to motor

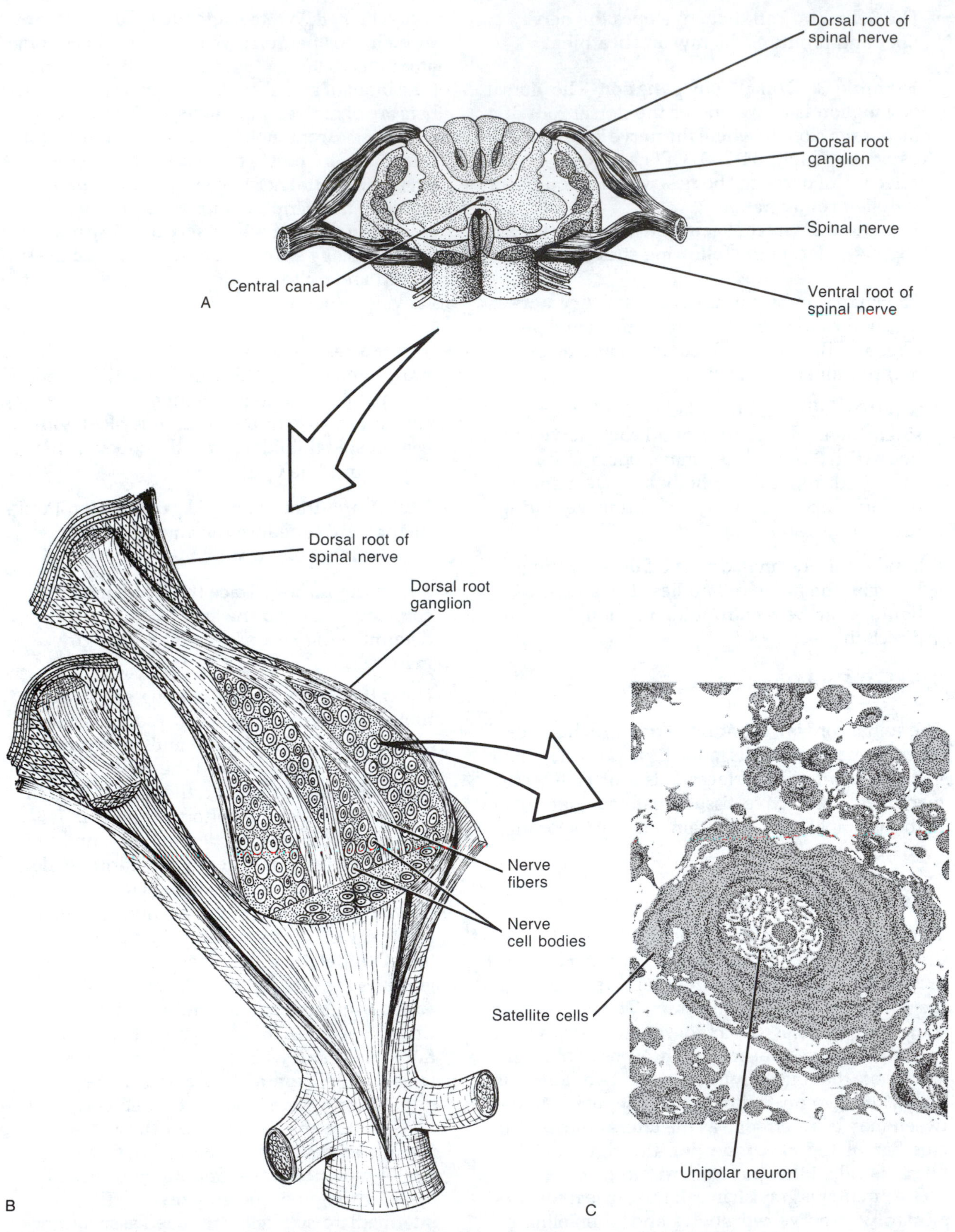

FIGURE 34-7 (A) Diagram of spinal cord showing dorsal and ventral roots of spinal nerve and dorsal root ganglion. (B) Diagram of section through dorsal root ganglion showing sensory neuron cell. (C) Cell body of sensory neuron surrounded by satellite cells.

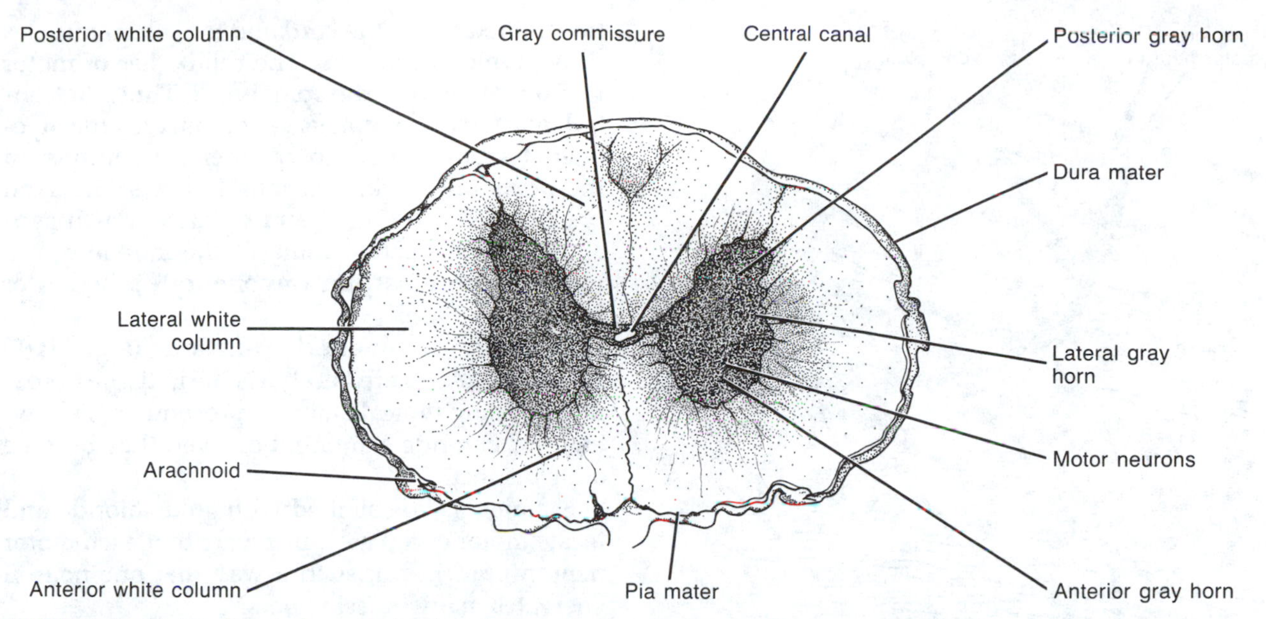

FIGURE 34-8 Sectional view of spinal cord. Organization of gray and white matter in spinal cord as seen in microscopic section.

neurons. Axons of adjuster neurons frequently branch. Thus, one incoming impulse can be relayed to a number of other cells and these, in turn, to numerous others.

- **Myelinated nerve fibers** in the white matter. When viewed under low power, the white matter appears to consist entirely of a lacy network of small, somewhat circular, white spaces containing dark-staining "dots." Reexamine under higher magnifications. The "dots" are transverse sections through the nerve fibers of ascending (sensory) and descending (motor) nerve tracts. *Note:* There are also numerous nuclei present (containing one or more nucleoli). Most of these are associated with neuroglia cells.

- The three **meninges** surrounding the spinal cord (Fig. 34-8). The **pia mater** is the innermost membrane and is tightly applied to the spinal cord. It is relatively thin and consists of collagen and elastic fibers covered with an outer layer of squamous epithelial cells. Numerous small blood vessels are present.

  The **dura mater** is the thick, outermost membrane. It consists mostly of dense connective tissue with large amounts of collagen fibers and smaller amounts of elastic fibers.

  The **arachnoid** is the middle layer (it is frequently torn during preparation of the tissues for sectioning, resulting in large, clear spaces between the dura and pia). The intact arachnoid has the appearance of a cobweb of delicate trabeculae extending between the dura and pia membranes. The spaces between the trabeculae are filled with cerebrospinal fluid and are called the subarachnoid spaces.

- **Sensory** and **motor tracts,** if your tissue section also has part of the spinal nerve and the dorsal root ganglion. You should be able to follow large bundles of sensory nerve fibers from the dorsal root ganglions into and through the dorsal white matter and then into the dorsal horn of gray matter.

  You should also see thin bundles of motor nerve fibers coming from the ventral horn and transversing the white matter. At the margin of the cord, you may be able to see these fibers aggregate and form the ventral root of the spinal nerve, which then passes close to the dorsal root ganglion.

4. Effectors (Fig. 34-4D)

Effectors are the component of the reflex arc that ultimately responds to an environmental stimulus detected by a receptor organ. Effectors are skeletal muscles, which, by their contraction or relaxation, provide for an almost unconscious and rapid response to potentially hazardous changes in the environment.

Sensory impulses relayed to the spinal cord are

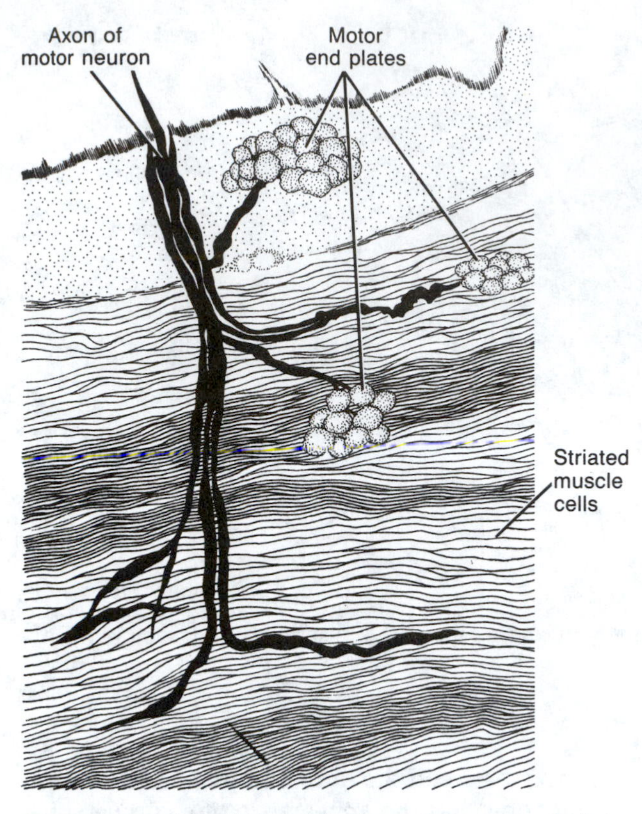

FIGURE 34-9 Microscopic view of motor end plate attached to striated muscle cells.

transmitted out of the cord and to skeletal muscles through motor neurons. The cell bodies of motor neurons are within the spinal cord. Thus, the ventral root and the spinal nerve contain only the myelinated axons of the motor (efferent) neurons. In the proximity of skeletal muscle fibers, the axon branches. The terminal end of each branch synapses with the sarcolemma of the muscle cell to form a **neuromuscular (myoneural) junction** or **motor end plate.**

Example: Myoneural junction (Fig. 34-9) Motor end plates are barely visible in tissue preparations using typical staining procedures. Following gold chloride staining, however, they become quite distinct.

Examine a slide stained with gold chloride and locate motor end plates. Be aware that each motor neuron branches in such a way that one neuron enervates many muscle fibers.

REFERENCES

Kuffler, S. W., J. E. Nicholls, and A. R. Martin. 1984. *From Neuron to Brain.* 2d ed. Sinauer Associates.

Llinas, R. R. 1988. *The Biology of the Brain: From Neurons to Networks.* W. H. Freeman.

Llinas, R. R. 1990. *The Workings of the Brain: Development, Memory, and Perception.* W. H. Freeman.

Shepherd, G. 1988. *Neurobiology.* 2d ed. Oxford University Press.

Exercise

35

Nervous System Physiology

To function effectively, organisms must be "aware" of their external environment. This is accomplished by various cues that stimulate peripheral sensors which produce impulses that are transmitted to the central nervous system (CNS). Many peripheral receptors are clustered together to form **sense organs** (for example, the eye), but not all receptors are grouped (for example, touch receptors). Receptors also possess differential sensitivities; that is, they respond more readily to one form of energy than another. Although receptors may respond to different stimuli (light, heat, touch, and so on), they all produce the same response: a series of action potentials along the afferent neurons—which, however, are interpreted differently in the CNS. Quality, intensity, and localization of the stimulus can be extracted from the information: *quality* from the sensitivity of the different receptors to the different stimuli, *localization* from the sensory map projected in the cortex of the brain, and *intensity* either by the number of sensory units activated or by the frequency of firing of the sensory unit.

Stimuli originating within or outside the body often elicit stereotyped responses of the CNS, called reflexes, that have proved to be especially appropriate to the physiological well-being of the organism. These reactions occur *before* the subject becomes consciously aware of them and, in fact, with some reflexes, such as those associated with the cardiovascular system, are never consciously experienced.

In this study you will examine some common human reflexes to various stimuli and test for cranial nerve function.

A. SWALLOWING REFLEX

Swallow the saliva in your mouth and then *immediately* swallow several times again and again. Describe what happens.

___

___

___

Next, rapidly drink a glass of water. Compare the results of trying to swallow your saliva and swallowing the water.

___

___

___

What do you think is involved here—must moisture of some sort (e.g., water or saliva) be present in order to swallow? Or must some "mass" be perceived by sensory structures in the esophagus to bring about the swallowing reflex? Indeed, is swallowing a reflex activity at all?

____

____

Can you consciously initiate the act of swallowing?

____

Once initiated, can you stop yourself from swallowing?

____

B. PUPILLARY RESPONSES

The size of the pupil of the eye, like the aperture setting of a camera, determines the amount of light that enters the eye (Fig. 35-1). And indeed, many of the changes in pupil diameter are mediated by *light* **(pupillary light responses).** The diameter of the human pupil also depends on the *distance* from the eyes of the object being viewed. This response, called **near point reaction,** is also referred to as the convergence response because the axes of the two eyes converge as the object being viewed is brought closer to the eyes.

Constriction of the pupil is effected by the contraction of a ring-shaped sphincter muscle in the iris. Contraction of a dilator muscle results in pupillary expansion. These muscles are under the control of the autonomic nervous system, which also accounts for variation of pupil diameter due to physiological factors, fatigue, or the consumption of various drugs such as nicotine, caffeine, and alcohol.

1. Direct Light Response (work in pairs)

Note: Pupil responses to light are best observed away from direct or bright light.

Close your eyes for 2 minutes. Then open your eyes and have your partner shine light from a pen light into your eye. Describe what happens to the pupil.

____

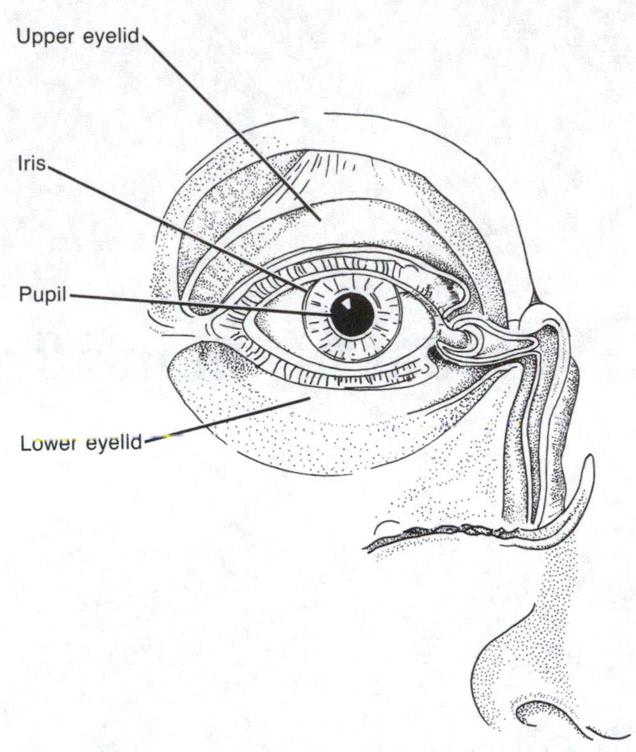

FIGURE 35-1 Frontal view of eye showing pupil and iris

____

Do the pupils of both eyes respond in the same way?

____

Next, have your partner observe your pupils when you shade your eyes (or move to a more dimly lit area). What is the value of such a pupillary response?

____

____

____

2. Consensual Light Response (work in pairs)

(This study should be carried on in a dimly lit or shaded area.)

Using a pen light, have your partner shine the light into your *right* eye, but observe what happens to pupil diameter of *both* eyes. The response of the pupil in the illuminated eye is referred to as a **direct**

light response. The response observed to the contralateral (left) eye is called a **consensual light response.**

3. Convergence Response (work in pairs)

(This study should be carried out in a moderately lit area.)

Focus on the eraser of a pencil held about 36 inches away from your eyes by your partner. Have your partner note the position of your eyeballs and the diameter of your pupils. Then have your partner slowly bring the pencil closer until it nearly touches your nose (keep your eyes on the eraser!). What change is observed in the position of your eyeballs?

____

____

What change occurs in your pupils?

____

____

Several pairs of eye muscles control various movements associated with the eyes. Their action makes it possible for you to fix your gaze on moving or stationary objects and to coordinate your eyes so they both look at the same thing at the same time.

4. Other Visual Activities

a. Near Point Accommodation (work in pairs)

As one becomes older, there is a progressive decrease in the elasticity of the lens. This phenomenon, called **near point accommodation,** results in the inability of older persons to clearly see close objects and is reflected, for example, in difficulty in reading without wearing corrective lenses (Fig. 35-2).

To demonstrate this response, hold a 12-inch ruler pointing forward near the lateral edge of one eye. With your other hand, position a pencil at the end of the ruler farthest from the eye. Then slowly move the pencil toward your eye, stopping when the image of the pencil becomes unclear. Have your partner note the distance from your eye to the pencil.

____

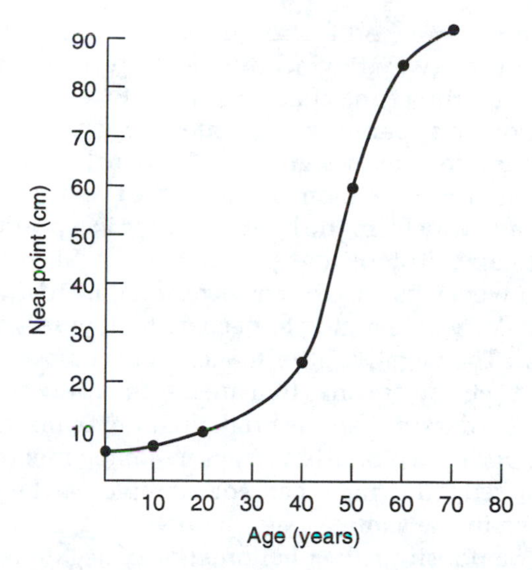

FIGURE 35-2 The near point recedes with advancing age because the lens of the eye gradually loses its ability to accommodate for near vision. The most dramatic changes are usually noted in humans beyond 40 years of age.

This is your near point distance.

Repeat this test on the other eye. Is the near point value the same for both eyes?

____

If you wear glasses or contact lenses, repeat the test without the lenses. Is the near point value the same for the "corrected" and "uncorrected" eyes?

____

b. Afterimages

You probably have had the experience of looking at a bright light for a few seconds and then, when you look away, "seeing" an image of the bulb in your visual field for some time afterward. Because the original object and the image are both about the same color, this is called a *positive afterimage.* However, if you were to prolong looking at the bulb, you would experience a *negative afterimage,* where the light parts of the original image now appear darker against a light background and vice versa.

Further, if you were to view the light through a piece of red cellophane, the negative afterimage would appear in the complementary color of green. If you were to view the light through green cellophane, what color would you expect the afterimage to be?

____

The occurrence of afterimages appears to depend on the way the incoming light rays are processed by the retina. Brief exposure to an intense light source apparently stimulates the cone cells of the retina to convey signals to the visual cortex of the brain for a few moments after the eye has been directed away from the light, resulting in a positive afterimage. Prolonged exposure to the light stimulus, however, causes the cones responding to those particular wavelengths to become temporarily fatigued. The negative afterimage, produced by continued activity of cones that are less intensely stimulated, is darker than the bright light that caused it. The appearance of afterimages in complementary colors (red following green, for example) can be explained in the same way.

To demonstrate the phenomenon of negative afterimages, carry out the following studies.

1. Focus with one eye on the white dot in the center of the pattern on the right for about 30 seconds. Then fix the eye on the dot in the middle of the circle on the left. Describe the appearance of the afterimage.

____

____

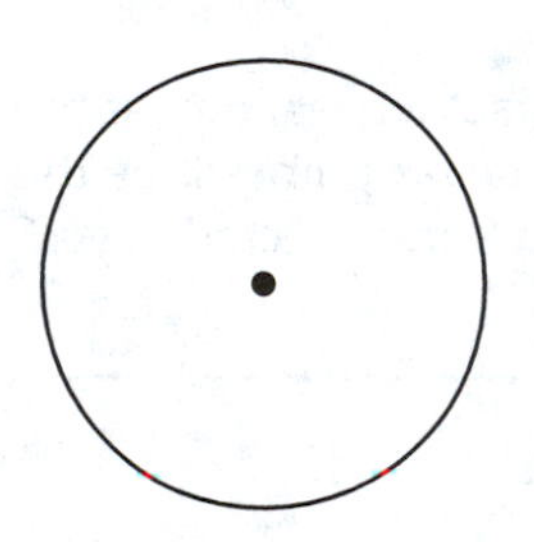

2. Using red and green pens or pencils, color the cross as indicated.

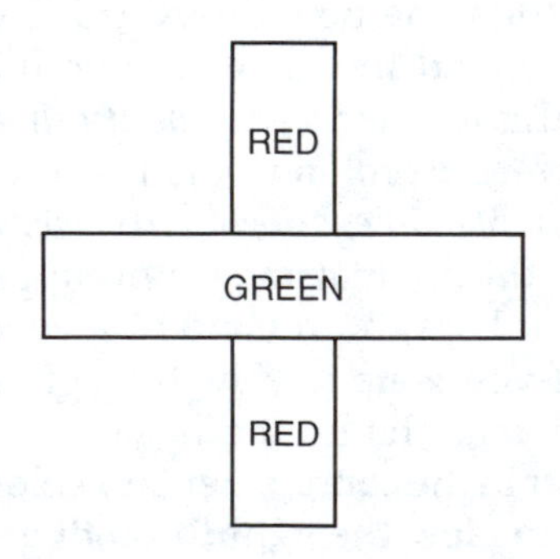

Stare at the center of the cross for thirty seconds and then look at a blank white surface. Describe what you "see" in the afterimage.

____

____

c. The Blind Spot

Except where the optic nerve leaves the rear of the eye, the inner surface of the eye is covered by the **retina.** The retina contains pigment cells that absorb light and trigger specialized neurons, called **rods** and **cones.** The sensory signals from all parts of the retina ultimately leave the eye through the optic nerve and pass to the primary visual area of the cerebral cortex.

Due to the lack of retinal tissue in this area, called the **papilla,** nothing is "seen" there. This area is thus called the "blind spot."

1. To demonstrate the presence of your blind spot (work in pairs), cover your left eye and, holding the figure below about half a meter from your face, fixate your right eye on the X.

Slowly bring the figure toward your eye. You will find that the black disk on the right disappears when its image falls on the papilla of your eye. When the image disappears, tell your partner to measure the distance of the disk from your eye. Is the distance the same for your left eye?

____

(*Note:* When you test your left eye, what should you fixate on—the X or the spot?)

2. To determine the size of your blind spot (you will need a partner), prepare a 15-mm by 75-mm strip of paper and draw a 4-mm spot at one end: This will be the "movable" spot. Next, close your right eye and position your head so that your left eye is about 30 cm directly above the black "stationary" spot below. (Use a ruler or 30-centimeter stick to maintain this distance.) Your partner should position the strip of paper so that the "movable" spot is directly next to and to the left of the "stationary" spot. Then, while fixating with your left eye on the "stationary" spot, move the strip of paper slowly to the left until the "movable" spot disappears. Mark this point on the page (A in the following example).

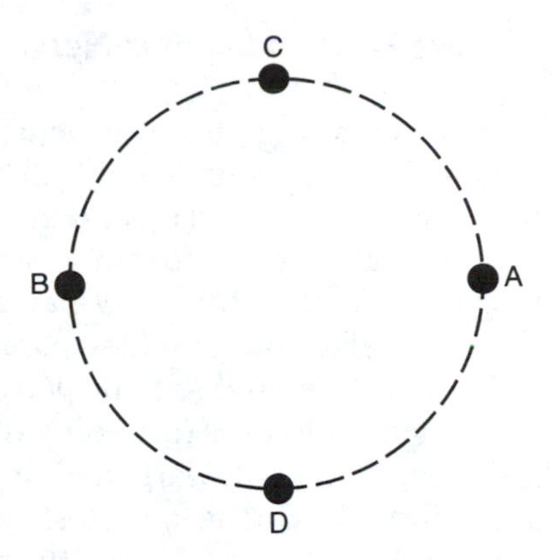

Continue to move the strip to the left until the spot reappears (B in the example.) Mark this point, too. Now position the movable spot midway between A and B. Move it up and then down and mark the points at which it reappears each time (C and D).

A line connecting the four points should form a rough circle or oval. Using these results, calculate the diameter of your blind spot using the following relationships:

$$\frac{a}{b} = \frac{c}{d}$$

where

a = diameter of circle or oval (distance between A and B in diagram)

b = diameter of blind spot (unknown)

c = distance from eye to spot (30 cm)

d = distance from retina to lens of eye (assume 3 cm)

Once you have calculated the diameter of your blind spot, determine its area. Compare the size of your blind spot with that of other students. Would you expect them to be the same size? Suggest reasons for any differences.

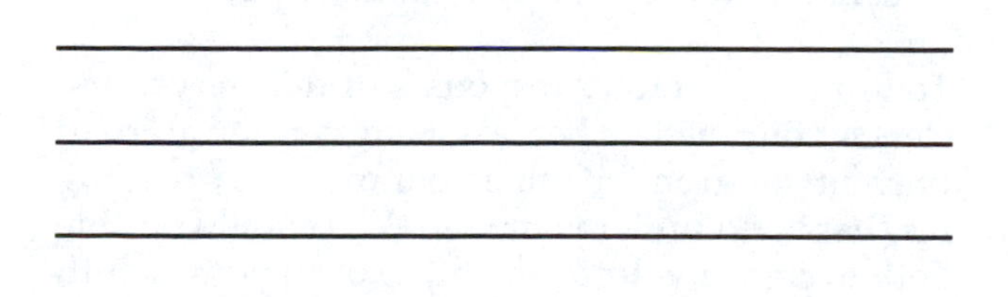

C. SNEEZING REFLEX

Using a piece of thread or a small camel's hair brush, gently touch the lining of your nose just inside the nostrils. Describe what happens.

____

____

What advantage is there, if any, for this type of response?

____

____

D. QUADRICEPS REFLEX (also called patellar reflex, knee jerk, knee reflex)

This reflex is routinely tested during a physical exam as follows: the patient (you) sits on the edge of a table with the legs hanging loosely. The patellar ligament (Fig. 35-3) is tapped briskly with a

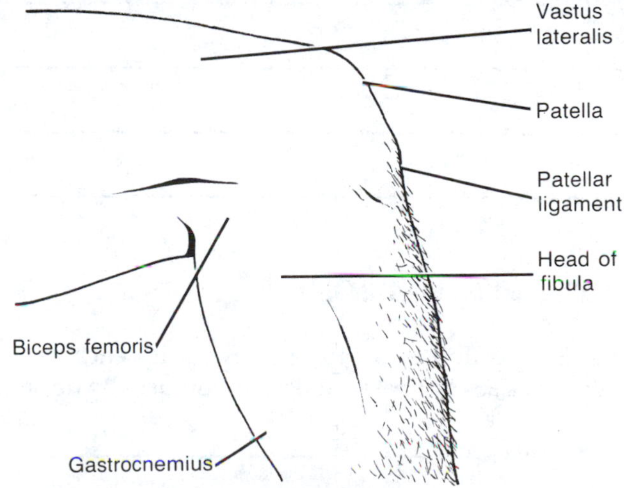

FIGURE 35-3 Location of patellar ligament. This ligament is a strong, thick band that is a continuation of the tendon of the quadriceps femoris muscle. It inserts onto the superior end of the tibia.

small rubber mallet (you can use the heel of your hand) until the quadriceps femoris muscle contracts. This results in extension of the leg at the knee joint.

Tapping the patellar ligament activates sensory structures called **muscle spindles** in the quadriceps femoris muscle. Sensory impulses from these spindles travel via the femoral nerve to lumbar segments of the spinal cord. From here, efferent impulses travel via motor neurons in the femoral nerve to the quadriceps femoris muscle group, resulting in the contraction (jerk) of the muscle and extension of the lower leg.

The knee reflex is blocked by damage to the femoral nerve or by damage to the reflex centers in spinal cord.

E. TESTS FOR CRANIAL NERVE FUNCTION

A rather superficial testing of cranial nerve function can be done using the following simple procedures. Work in pairs and record your results for each test.

1. Cranial Nerve I (Olfactory Nerve)

Identify the odor of various "known" materials provided by your instructor. (*Note:* This test may be invalid if you have a cold. You should be able to discriminate between various common odors.)

Results: ______________________________

____

____

____

2. Cranial Nerve II (Optic Nerve)

Read a portion of a printed page with each eye. Wear glasses or contact lenses if you usually do so.

Results: ______________________________

____

____

3. Cranial Nerve III (Oculomotor Nerve)

Follow your partner's fingertip or a pencil eraser with both eyes, keeping your head still as your partner slowly moves the object up and then down.

As well as regulating eye movements through the rectus ocular muscles, this nerve also innervates the upper eyelid and provides parasympathetic stimulation to the pupils. Your partner is to observe you for signs of *ptosis* (drooping of one or both eyelids) and, with lights darkened, for pupillary reaction to light. To do the latter test, bring the penlight in from the side and observe the reaction of the pupils (they should constrict).

Results: ______________________________

____

____

4. Cranial Nerves IV (Trochlear) and VI (Abducens Nerves)

Have your partner hold a pencil or finger several centimeters in front of your nose. Then follow your partner's fingertip or the pencil eraser, keeping your head still while s/he slowly moves the object laterally in each direction. Both of your eyes should follow the object as it is moved from side to side.

Results: ______________________________

____

____

____

5. Cranial Nerve V (Trigeminal Nerve)

To test for the motor responses of this nerve, first clench your teeth. Then ask your partner to try to prevent you from opening your mouth by holding his/her hand under your chin. You should be able both to clench your teeth and to open your mouth against resistance.

Results: ______________________________

____

____

To test the sensory responses for this nerve:

1. Close your eyes. Then have your partner *lightly* "whisk" a dry piece of cotton over the mandibular (jaw), maxillary (cheek), and ophthalmic (eye) areas on *both* sides of your face.

2. Repeat, having wet the cotton with cold water. You should be able to discriminate temperature as well as touch in the three areas.

Locate the various branches of the trigeminal nerve (Fig. 35-4) involved in the responses elicited in this study.

Results: ______________________________

____

____

6. Cranial Nerve VII (Facial Nerve)

As you are testing for this nerve, have your partner watch you and note your reactions.

1. Look at the ceiling, wrinkle your forehead, frown, puff your cheeks, raise your eyebrows, smile showing your teeth. All the muscles controlling these movements are controlled by the facial nerve. Inability to perform these movements or any asymmetry in the movements indicates potential damage to the nerve.

2. Close your eyes tightly and try to keep them closed while your partner attempts (gently and carefully!) to open them. The strength of the sphincter muscles of the eye are controlled by the facial nerve. When this nerve is damaged, the subject cannot close the eyelid tightly on the affected side.

3. To test the sensory responses of this nerve, have your partner touch a cotton-tipped applicator moistened with a 10% sugar solution to the tip of your tongue. Repeat using a 10% salt solution on the sides of the tongue just back from the tip. You should be able to discriminate between these solutions.

Results: ______________________________

____

____

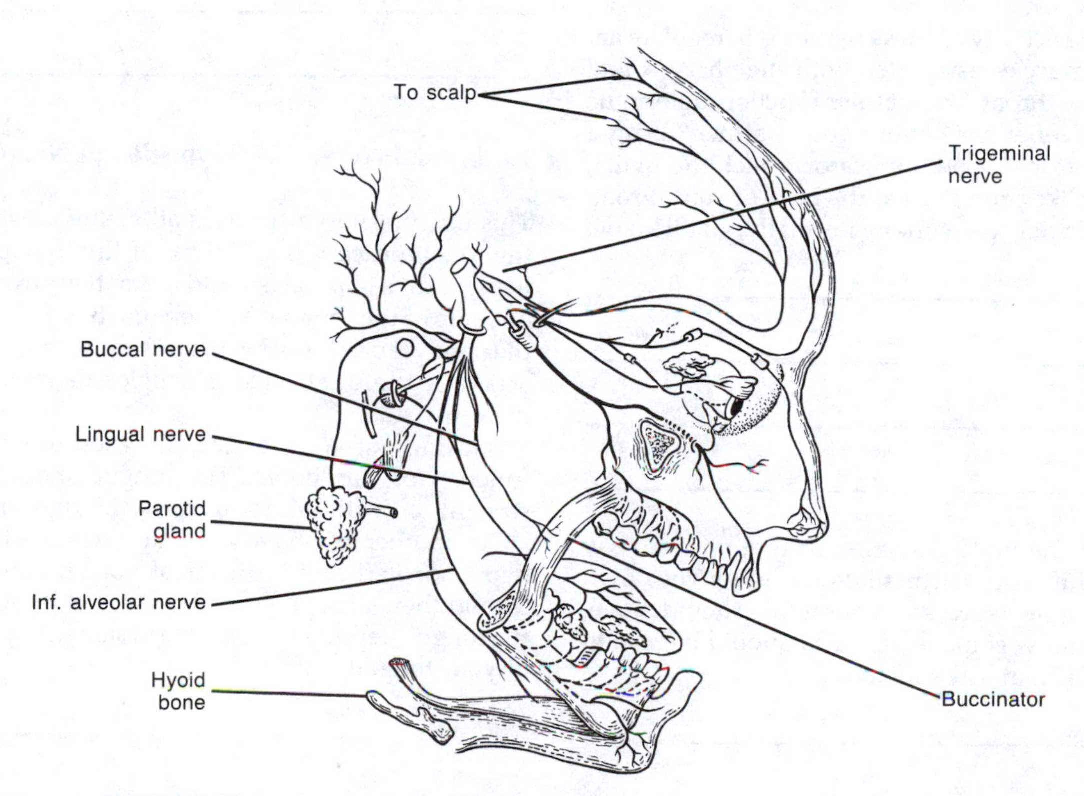

FIGURE 35-4 Cranial nerve V (trigeminal nerve) and its branches.

7. Cranial Nerve VIII (Vestibulocochlear Nerve)

To test the cochlear (hearing) portion of this nerve, have your partner determine your ability to hear a ticking watch and repeat a sentence whispered behind your back.

To test the vestibular (balance) portion of this nerve, sit on a swivel stool and have your partner turn you 10 times in about 20 seconds and then stop the chair. When stopped, your partner should observe your eyeballs for rapid movement or quivering. This is a *normal* response (called **nystagmus**) when one is dizzy.

In general, if you are able to keep your balance while walking under normal conditions, the vestibular branch of this nerve is normal.

Results: ______________________________

____

____

____

8. Cranial Nerves IX (Glossopharyngeal) and X (Vagus) Nerves

Part of the activity of these nerves is to regulate activity of muscles associated with the pharynx and back of the throat. To test their function, determine your gag reflex by having your partner (using a sterile, cotton-tipped applicator) touch the uvula, the fingerlike projection at the back of your throat. If your gag reflex is working normally, you'll know!

Results: ______________________________

____

____

____

To test the motor function of these nerves, say "aah" while your partner holds your tongue down with a tongue depressor. Your uvula should move. Then swallow some water. You should be able to do this without any difficulty.

Results: ______________________________

____

____

____

9. Cranial Nerve XI (Accessory Nerve)

This procedure tests the strength of the trapezius and sternocleidomastoid muscles of the shoulder and neck. To test the strength of the trapezius muscles, have your partner place his hands on your shoulders. Can you raise your shoulders against the resistance s/he applies to your shoulders trying to keep them down?

____

To test the strength of the sternocleidomastoid muscle, have your partner place his/her hands on each side of your head and try to prevent you from turning your head from side to side.

Results: ______________________________

____

____

____

10. Cranial Nerve XII (Hypoglossal Nerve)

This efferent nerve supplies all of the muscles of the tongue except one. Injury of the hypoglossal nerve results in paralysis and ultimately atrophy of one side of the tongue. The tongue thus deviates to the paralyzed side during protrusion because of the action of the unaffected geneoglossus muscle on the other side.

Sticking out your tongue is a good test for the function of this nerve. The tongue should stick straight out with no deviation to the right or left.

For another test, have your partner hold a tongue depressor vertically in front of your mouth. You should then stick your tongue out and try to push the tongue depressor from side to side using the tip of your tongue.

Results: ______________________________

____

REFERENCES

Camhi, J. M. 1984. *Neuroethology: Nerve Cells and the Natural Behavior of Animals.* Sinauer Associates.

Hubel, D. H. 1988. *Eye, Brain, and Vision.* Scientific American Library Series, no. 22. W. H. Freeman.

Jacobs, G. 1983. Color Vision in Animals. *Endeavour* 7(3):137–140.

Parker, D. 1980. The Vestibular Apparatus. *Scientific American* 243(5):118–130.

Stryer, L. 1987. The Molecules of Visual Excitation. *Scientific American,* 257(1):42–50.

Exercise

36

Fertilization and Early Development of the Sea Urchin

The development of a fertilized egg into the complex, coordinated, and interdependent systems of tissues and organs that make up an adult animal is one of the more fascinating studies in developmental biology. In this exercise and the two that follow, you will study the patterns of early development in an echinoderm (the sea urchin), an amphibian (the frog), and a bird (the chick).

The type of development an embryo undergoes is strongly affected by the amount, the position, and the distribution of the yolk in the egg, which markedly affects the patterns of cleavage and subsequent developmental events. In the **isolecithal** egg, found in the sea urchin, the amount of yolk is small and is evenly distributed throughout the cytoplasm. In the **telolecithal** egg, found in the frog and chick, the amount of yolk is large and is concentrated at one end of the egg, called the **vegetal pole;** the cytoplasm being concentrated at the **animal pole.**

In this exercise, you will observe fertilization and the early stages of embryonic development of a marine echinoderm, the sea urchin. This animal, related to the starfish, can be induced to shed its gametes either by injecting potassium chloride into the body cavity or by stimulating the animal with a weak electric current. Each female lays approximately one billion eggs, and each male ejects several billion sperm.

A. FERTILIZATION

Before this laboratory session, your instructor obtained living eggs and sperm from female and male **sea** urchins, using the procedure shown in Fig. 36-1, and made egg and sperm suspensions.

Obtain a depression slide and coverslip (see Fig. 36-1). With a clean pipet, transfer a drop of the egg suspension to the slide and examine it microscopically.

Add one drop of dilute sperm suspension to the eggs on the slide and examine it immediately with the high power of your compound microscope. Note the time so that you will know when to expect the cleavage stages (Table 36-1). Because the rate of development varies with temperature, keep the eggs at or close to 22 °C. How do the sperm respond to the eggs?

____

____

What might be the cause of this response?

____

____

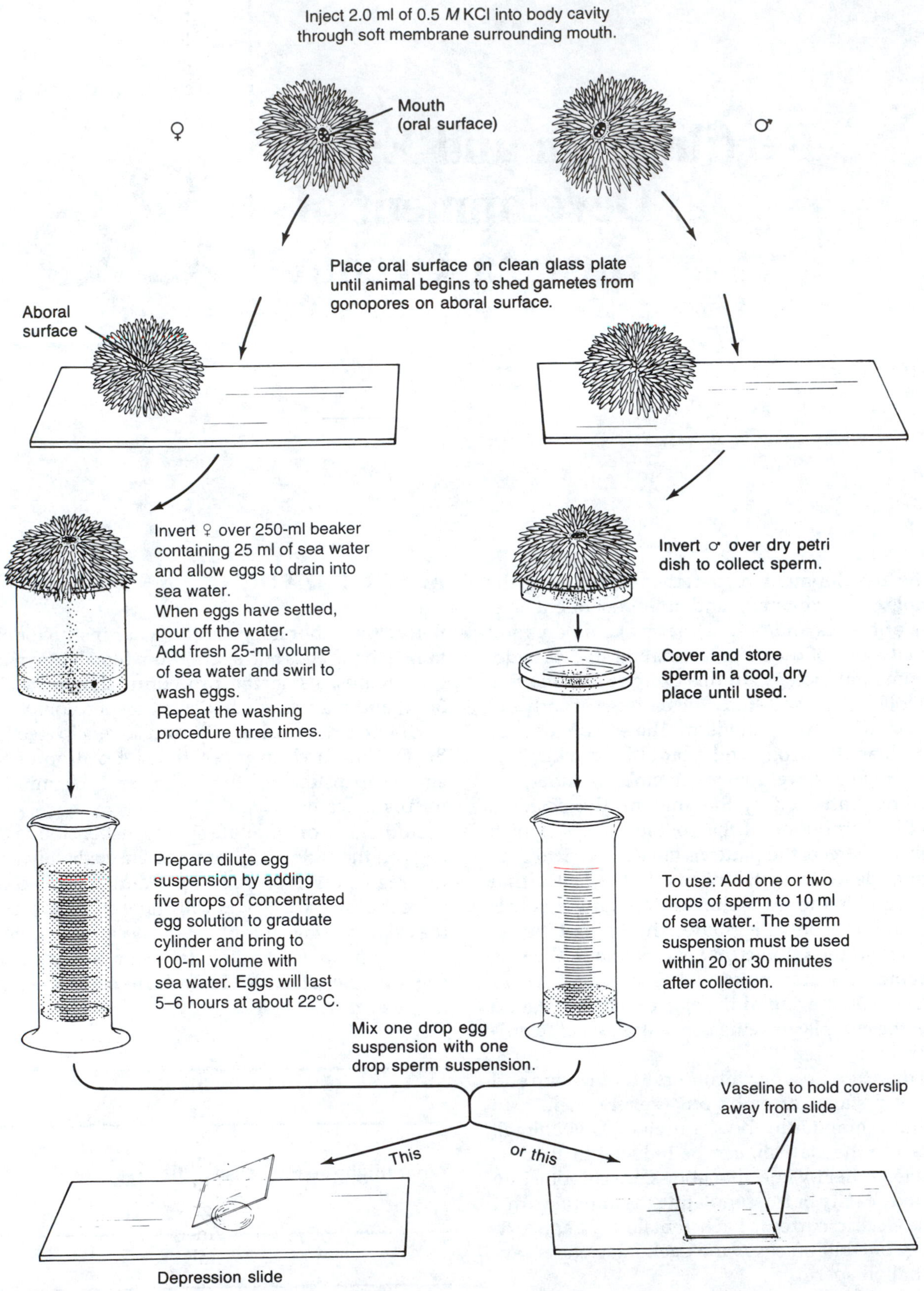

Inject 2.0 ml of 0.5 M KCl into body cavity through soft membrane surrounding mouth.

♀

Mouth (oral surface)

♂

Place oral surface on clean glass plate until animal begins to shed gametes from gonopores on aboral surface.

Aboral surface

Invert ♀ over 250-ml beaker containing 25 ml of sea water and allow eggs to drain into sea water.
When eggs have settled, pour off the water.
Add fresh 25-ml volume of sea water and swirl to wash eggs.
Repeat the washing procedure three times.

Invert ♂ over dry petri dish to collect sperm.

Cover and store sperm in a cool, dry place until used.

Prepare dilute egg suspension by adding five drops of concentrated egg solution to graduate cylinder and bring to 100-ml volume with sea water. Eggs will last 5–6 hours at about 22°C.

To use: Add one or two drops of sperm to 10 ml of sea water. The sperm suspension must be used within 20 or 30 minutes after collection.

Mix one drop egg suspension with one drop sperm suspension.

This

or this

Vaseline to hold coverslip away from slide

Depression slide

FIGURE 36-1 Preparation of sea urchin egg and sperm suspensions for observing fertilization and early development

TABLE 36-1 Approximate time sequence for the development of fertilized sea urchin eggs at 22°C

Formation of fertilization membrane	2–5 minutes
First cleavage	50–70 minutes
Second cleavage	78–107 minutes
Third cleavage	103–145 minutes
Blastula	6 hours
Hatching of blastula	7–10 hours
Gastrula	12–20 hours
Pluteus larvae	24–48 hours

Soon after a sperm penetrates an egg, a **fertilization membrane** begins to lift from the egg's surface and surrounds the egg. The entire process—from the time the sperm touches the egg to the formation of the fertilization membrane—takes 2–5 minutes.

The first cleavage does not occur until 50–70 minutes after the formation of the fertilization membrane. After the sperm enters the egg, the sperm nucleus migrates toward the egg nucleus and unites with it, reestablishing the diploid number of chromosomes. Subsequently, the chromatin is organized into chromosomes, and the first nuclear division, followed by cell division, occurs.

B. CLEAVAGE

The early divisions of the fertilized egg is called **cleavage.** During cleavage, other events occur in the nucleus and cytoplasm, but you will be concerned only with the more easily observable external changes.

While waiting for the first division, examine prepared slides of sea urchin or starfish eggs. How can you tell if an egg on a slide is fertilized or unfertilized?

____

____

Next, prepare a slide of the living sperm. By using the diaphragm of the compound microscope to increase contrast, you may be able to observe the flagellum by which the sperm moves. *If, at this point, the fertilized eggs have not reached the first cleavage stage, continue with part C.*

If the eggs are beginning to divide, note that the first cleavage begins at one end of the egg, called the animal pole. If you carefully observe the egg, you should see it begin to elongate slightly before it divides. The second cleavage occurs approximately 15–20 minutes after the first cleavage. Describe the plane of this second division with respect to the first cleavage.

____

____

Describe the plane of the third cleavage with respect to the first two cleavages.

____

____

If your preparation does not attain this stage of development, examine fertilized eggs that were prepared earlier by your instructor or examine your fertilized eggs later in the day.

C. LATER STAGES OF DEVELOPMENT

Before the eight-cell stage, the cleavages are uniform. After the eight-cell stage, the four cells of the vegetal pole divide unequally, resulting in four large cells **(macromeres)** and four small cells **(micromeres).** Then the four cells of the animal pole divide into eight cells of equal size **(mesomeres)** (Fig. 36-2).

1. Blastula

As the cells of the embryo undergo additional cleavage, a spherical mass containing a fluid-filled central cavity, the **blastocoel,** if formed. With further cleavages, the blastocoel increases in size until the embryo, now called a **blastula,** consists of several hundred cells that arrange themselves into a single-layered hollow ball. At this stage, the embryo begins to rotate within the fertilization membrane. Approximately 8 hours after fertilization, the blastula breaks through the membrane by secreting a "hatching enzyme," which enables it to digest the fertilization membrane. If you are unable to observe a living blastula (Fig. 36-3A), examine slides of this stage in the sea urchin or in related forms such as the starfish.

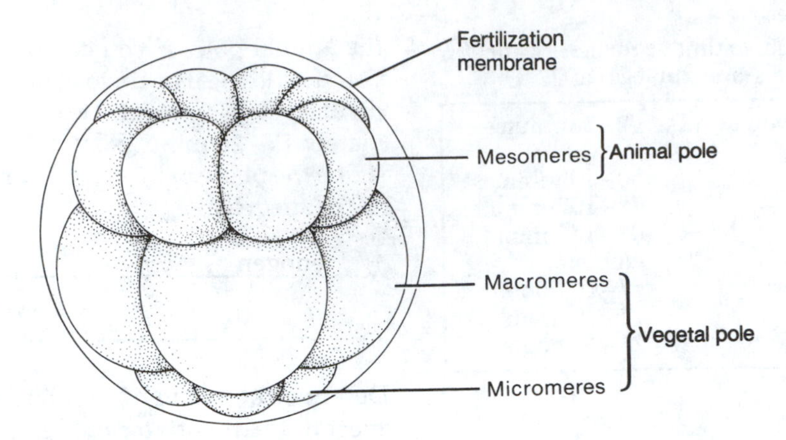

FIGURE 36-2 Unequal cleavage in the sea urchin (fourth division)

2. Gastrula

For several hours after "hatching," the blastula swims actively. About 15 hours after fertilization, the single-layered blastula changes into a double-layered **gastrula** (Fig. 36-3C). In the sea urchin, this process begins when cells at the vegetal pole begin to pulsate intensely, particularly on the inner surfaces facing the blastocoel. This forces some of the cells into the blastocoel, forming a slight indentation called the **blastopore.** These cells then send out long cytoplasmic strands that become attached to the opposite wall near the animal pole. Contraction of these strands deepens the blastopore, ultimately forming an invagination that will become the **archenteron,** or primitive gut, of the embryo. The blastocoel remains but is soon invaded by other cells. The archenteron continues to invaginate toward the animal pole, where it eventually meets the opposite wall. At this junction the mouth will form. The opening of the blastopore will become the anus in the adult.

At the conclusion of gastrulation, the embryo consists of an outer layer that will become the ectoderm. An inner layer will become the endoderm. A third layer, the mesoderm, develops between the ectoderm and endoderm from the proliferation of cells arising in the endodermal layer (Fig. 36-3D). These embryonic germ layers develop into the various tissue and organ systems of the adult animal. The skin, sense organs, and nervous system arise from the ectoderm. The muscles, skeletal elements, and blood originate in the mesoderm. The endoderm gives rise to the digestive system and its various derivatives; the pharynx, esophagus, stomach, liver, intestine, pancreas, and endocrine glands.

Examine living gastrulae or slides of this stage of development.

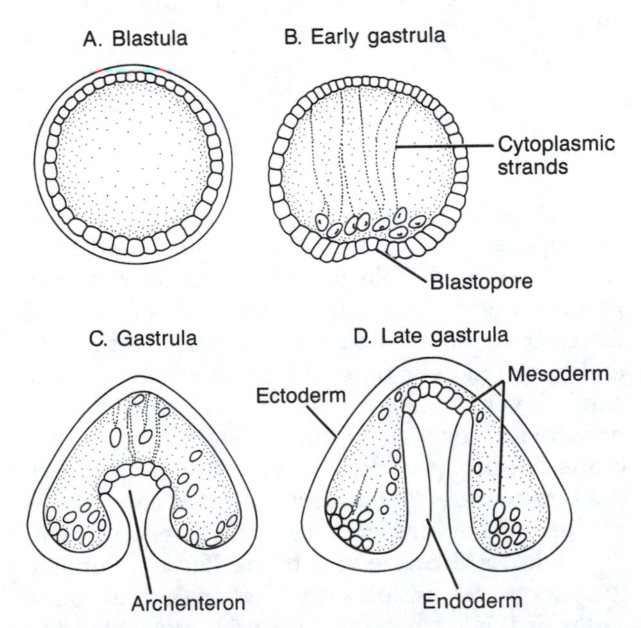

FIGURE 36-3 Blastula (A) and gastrula (B–D) stages of sea urchin development

3. Pluteus Larva

In the later stages in the development of the gastrula, skeletal elements begin to form as pairs of rodlike structures on either side of what will become the anus. Twenty-four hours after fertilization, the larva is well developed. Examine slides or living specimens of the **pluteus larva** (Fig. 36-4A). When 2 or 2½ months old, the larva metamorphoses into the adult sea urchin (Fig. 36-4B). Examine specimens of the adult animal. Echinoderms

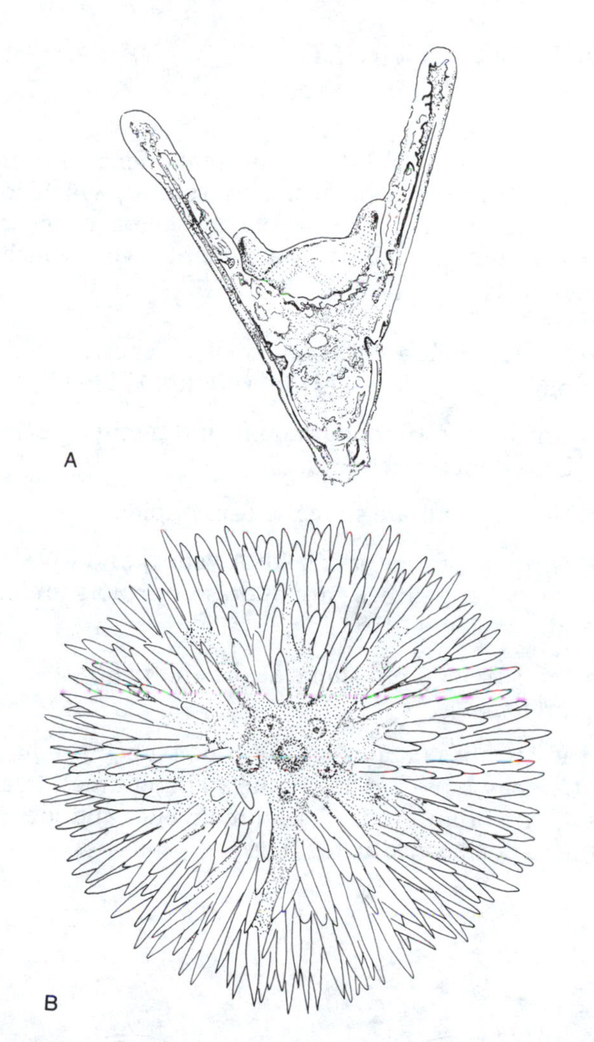

FIGURE 36-4 Sea urchin. (A) Pluteus larva. (B) Adult.

are characterized by numerous "tube feet." Where on the sea urchin are these feet located?

____

What seems to be their function?

____

____

____

____

D. ARTIFICIAL PARTHENOGENESIS

Eggs of sea urchins and of many other organisms have the ability to develop **parthenogenetically** (Greek *parthenos,* "virgin," and *genesis,* "birth"), that is, without being fertilized by sperm. Sea urchins, in particular, are susceptible to the action of external substances. For example, various concentrations of electrolytes such as magnesium chloride ($MgCl_2$) and potassium chloride (KCl), as well as nonelectrolytes such as urea and sucrose, induce parthenogenetic cleavage in sea urchin eggs. In the following study, you will attempt to determine if hypertonic seawater induces cleavage in nonfertilized eggs.

1. Prepare four petri dishes, each containing one of the following concentrations of synthetic seawater.

Undiluted (normal)
1.5 × normal
2.0 × normal
4.0 × normal

2. Transfer samples of washed, unfertilized eggs (as prepared in part A) to each of the dishes.

3. After 5 minutes, wash the eggs in four changes of normal seawater and then leave them immersed in the fourth change.

4. Examine the eggs for fertilization membranes, which appear more slowly in parthenogenetically induced eggs than in eggs that have been fertilized sexually. Check further development, noting particularly if the development of any of the eggs is arrested at a cleavage stage.

E. DEVELOPMENT OF EMBRYOS FROM ISOLATED BLASTOMERES

In 1892, Hans Dreisch separated the cells (blastomeres) of sea urchin blastulae and observed their development in isolation. He was able to do this because, immediately after fertilization, the fertilization membrane is quite fragile and can be ruptured by vigorously shaking the test tube containing the eggs. Without the membrane, the two cells can then be separated. You will attempt to duplicate this procedure.

1. Place a sample of unfertilized eggs into a test tube half full of seawater. Add a drop of sperm suspension and seal the test tube with a clean rubber stopper. Mix the contents by gently inverting the test tube once or twice.

2. After 1 minute, vigorously shake the test tube for 2 minutes. This should rupture a large number of fertilization membranes.

3. Transfer the contents of the test tube to a petri dish. After an hour, examine the eggs for the appearance of the first cleavage. If cleavage has begun, watch carefully until the first cleavage is just completed. Then, using a clean pipet, vigorously squirt seawater into the egg mass in the petri dish several times. Allow the eggs to settle to the bottom and then remove a sample of eggs. Examine the sample microscopically for isolated blastomeres.

4. Using a clean pipet transfer the isolated blastomeres to finger bowls of seawater and observe their development for the next several days.

Compare "normal" development with that of isolated "half embryos." Consider such things as the rate of development and the size of pluteus larvae that develop.

____

____

____

____

F. VITAL STAINING OF SEA URCHIN EGGS

Many chemical and structural components of cells can be "visualized" by using **vital stains,** which do not kill the cells. Listed below are several of these stains and the cellular constituents with which they react.

- Toluidine blue stains mucopolysaccharides pink and nucleic acids (mostly RNA) blue.
- Janus green B, rhodamine B, and methylene blue stain mitochondria.
- Nile blue sulfate stains phospholipids.
- Acridine orange stains DNA yellow and RNA deep orange (these are fluorescent colors under ultraviolet light).
- Brilliant vital red stains acidic and neutral proteins.

Using these stains in the following procedure (you may have to modify the staining time), treat samples of unfertilized sea urchin eggs and locate the various constituents listed.

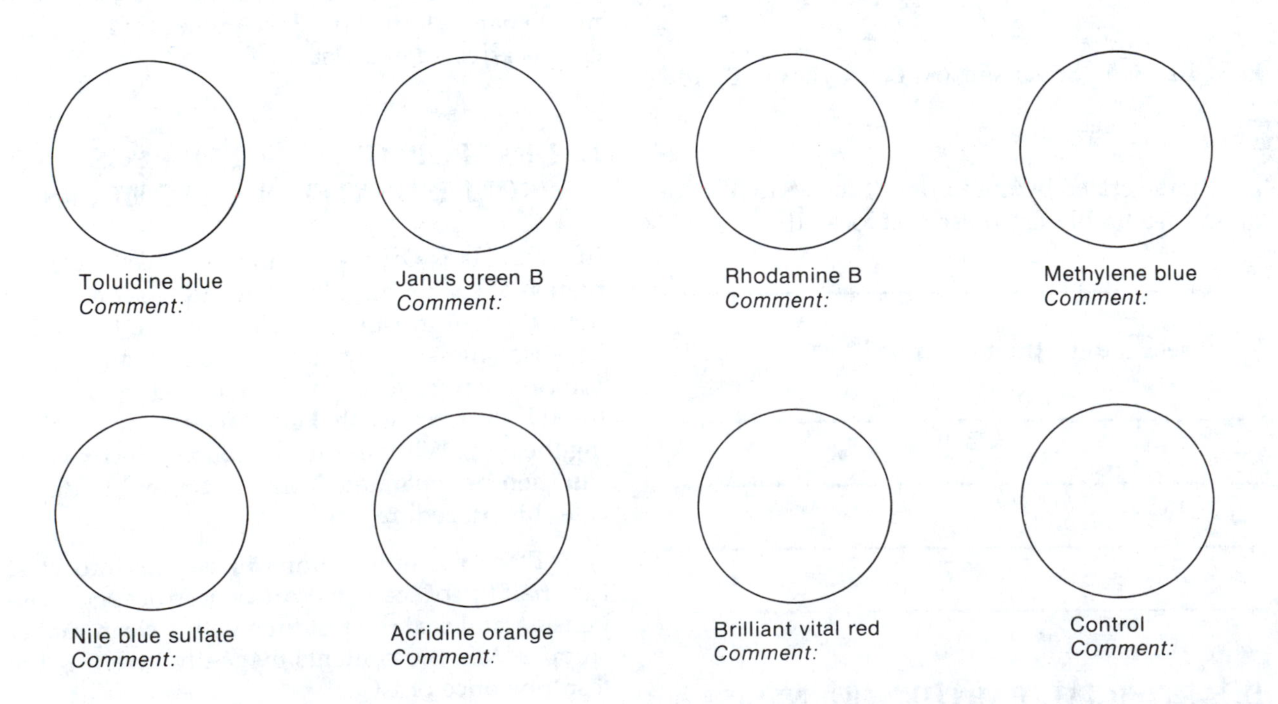

FIGURE 36-5 Vital staining of sea urchin eggs

1. Using a marking crayon, label the depressions in a spot plate to correspond to the various stains. Use one depression as a seawater control.

2. Add 1 ml of seawater and four or five unfertilized sea urchin eggs to each depression.

3. Add one drop of stain to each depression. Add no stain to the seawater control.

4. Stain for 1–2 hours. Then carefully pipet off the staining solution and add fresh seawater. Examine the eggs microscopically, and "describe" your results with drawings in Fig. 36-5.

REFERENCES

Browder, L. W. 1991. *Developmental Biology.* 3d ed. Saunders.

Costello, D. P., et al. 1957. *Methods for Obtaining and Handling Marine Eggs.* Marine Biological Laboratory, Woods Hole, Massachusetts.

Edelman, G. M. 1984. Cell Adhesion Molecules: A Molecular Basis for Animal Form. *Scientific American* 250(4):118.

Harvey, E. B. 1954. Electrical Method of Determining the Sex of Sea Urchins. *Nature* (London) 176:86.

Trinkaus, J. P. 1984. *Cells into Organs.* 2d ed. Prentice-Hall.

Walbot, V., and N. Holder. 1987. *Developmental Biology.* Random House.

Exercise

37

Fertilization and Early Development of the Frog

The eggs of most frogs begin to mature during the summer months when the frogs are feeding heavily. By fall, the eggs are fully developed but are usually retained in the ovaries until the following spring, at which time they are released en masse. After their release from the ovaries, in a process called **ovulation,** the eggs pass into the oviducts and then out of the body.

In nature, ovulation results from stimulation of the ovaries by gonadotropic hormones produced in the pituitary gland. A mature female frog can be induced to ovulate earlier than usual, however, by injecting pituitary glands or a pituitary extract into the body cavity. The number of glands needed varies with the time of year and the sex of the frog supplying the glands. Table 37-1 outlines the number of female pituitaries needed at various times of the year to induce ovulation. Because male pituitaries are about one-half as potent as female pituitaries, twice the number of male glands must be used.

In this exercise, you will induce ovulation in mature female frogs and prepare a frog sperm suspension. You will use the sperm suspension to fertilize the eggs so that you can examine the patterns of cleavage and early development of the frog embryo.

Your instructor will demonstrate how to remove and inject the pituitary glands. To have frogs that are ovulating at the time of the laboratory meeting, you should inject the female frogs 48 hours ahead of time. (*Note:* A powdered pituitary extract is available that can be substituted for pituitaries extracted from living frogs.)

A. ARTIFICIAL INDUCTION OF OVULATION

1. Anesthetize the number of frogs necessary to obtain the quantity of pituitaries shown in Table 37-1 by placing the frogs in a sealed container in which you have suspended a wad of cotton saturated (but not dripping) with ether. Perform this anesthetization in a fume hood or a well-ventilated room. When the frogs stop moving as you gently shake the container, remove them and dissect out the glands as illustrated in Fig. 37-1 and demonstrated by your instructor. Place the glands in a petri dish containing Holtfreter's solution (a solution used for maintaining amphibian eggs and embryos).

2. When you have collected the required number of pituitaries, remove the needle from a 2-ml syringe and draw the glands into the barrel along with 1 or 2 ml of Holtfreter's solution. Attach an 18-gauge needle to the syringe. Hold the syringe,

TABLE 37-1 Number of female pituitaries needed to induce ovulation in frogs

September–December	January–February	March–April
5–6	3–5	1–3

needle down, so that the glands settle to the bottom of the barrel. If any pituitaries stick to the sides, tap the syringe until all collect on the bottom.

3. Select the female to be injected and, holding her as shown in Fig. 37-2C, insert the needle into the body cavity. Keep the needle parallel with the body wall to avoid injuring any internal organs. When the needle is deeply inserted, check to see that the pituitaries are at the needle end of the syringe, and then quickly inject the pituitary glands. Leave the needle in position for a minute and then slowly remove it, pinching the skin to prevent any loss of fluid or glands.

4. Draw more Holtfreter's solution into the syringe to determine if any pituitary material is lodged in the needle. If so, eject all but about 0.5 ml of solution and then inject the preparation containing the remaining pituitaries.

5. Place the frog in a ventilated covered container with about 50 ml of water and put the container in a cool room, at about 68°F (20°C). Check for ovulation after 24 hours by gently squeezing the abdomen, as shown in Fig. 37-2E and demonstrated by your instructor.

If, when you gently squeeze the frog, eggs come out of the cloaca, ovulation has taken place. **Do not squeeze overly hard.** The frog can then be placed in a container of fresh water and kept in a refrigerator until needed. Eggs will remain viable in the oviduct for 3–4 days if the animal is kept at a temperature of 10–15°C. If ovulation has not occurred, put the frog in fresh water and try again 24 hours later.

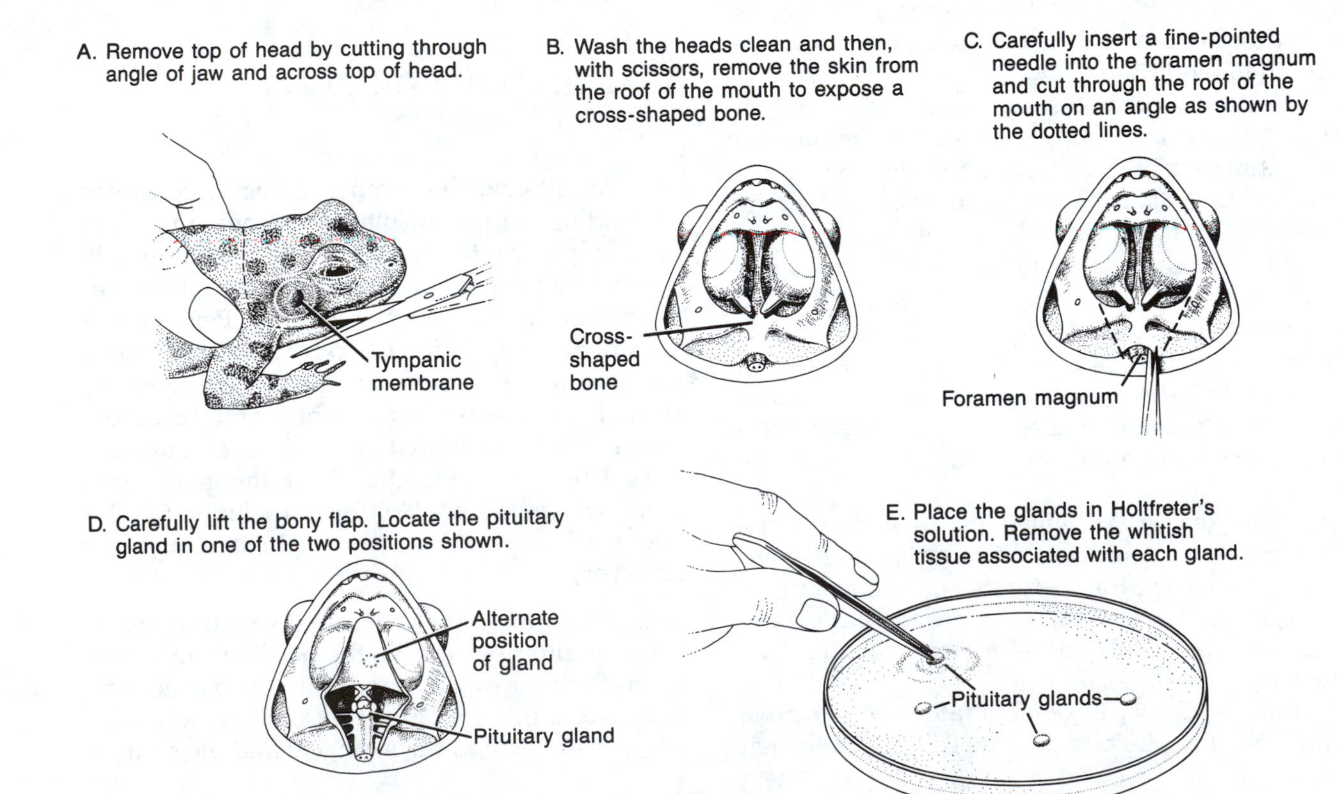

FIGURE 37-1 Procedure for removing female pituitary glands

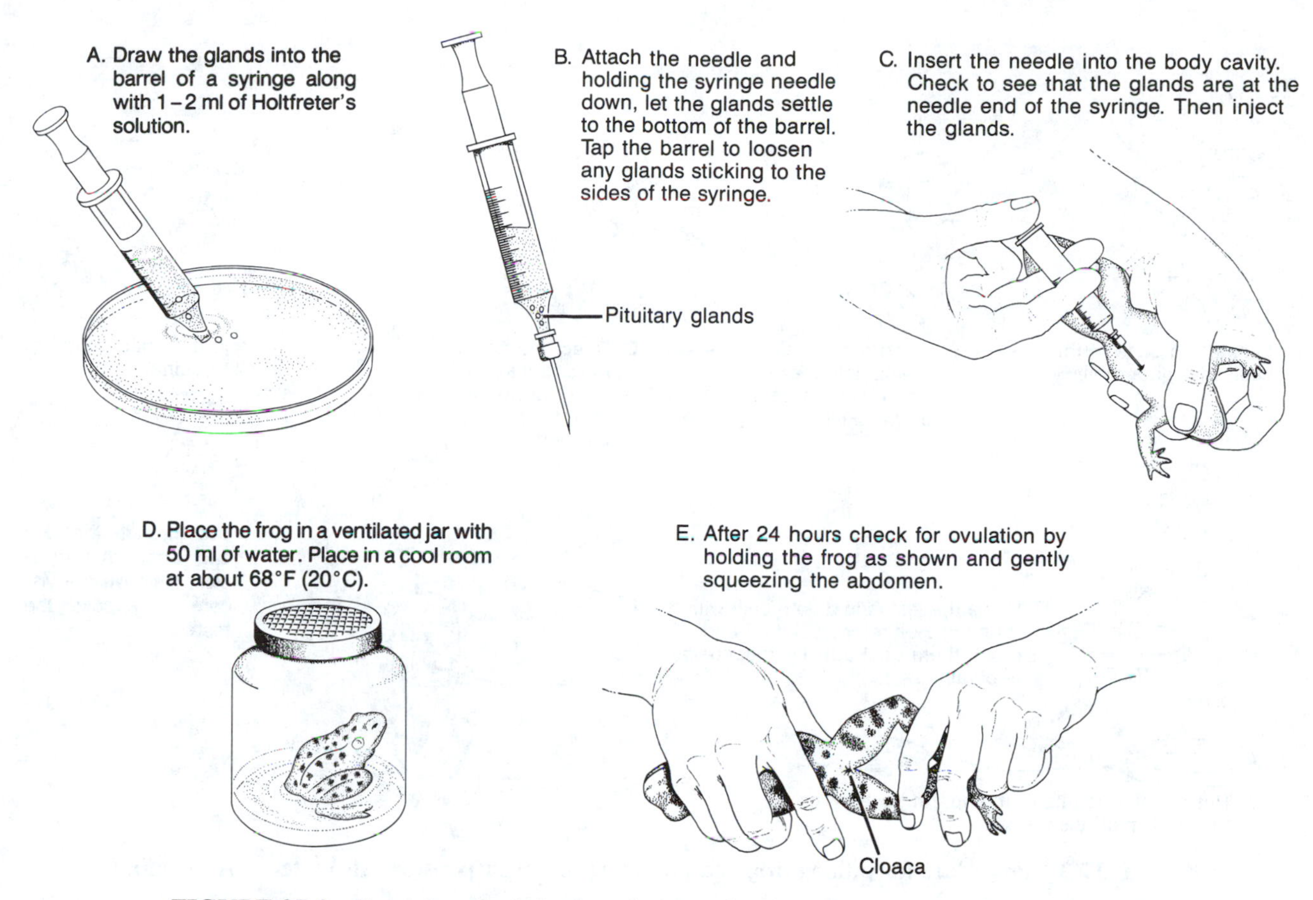

FIGURE 37-2 Procedure for injecting pituitary glands and checking for ovulation

B. PREPARATION OF SPERM SUSPENSION

The sperm suspension should be prepared about 30 minutes before the laboratory session so that active sperm are available at the time the eggs are stripped from the female.

1. Prepare a petri dish containing 20 ml of 10% Holtfreter's solution.

2. Pith a mature male frog and remove the paired testes as shown in Fig. 37-3. As described in Appendix I, clean away any adhering blood and tissue and, using the blunt end of a clean glass probe, thoroughly macerate the testes in the petri dish until a milky suspension is obtained.

3. Set the sperm suspension aside for 15–20 minutes to allow the sperm to become motile.

4. Pipet a drop of the sperm suspension onto a glass slide and examine the sperm under the high power of the compound microscope. Describe the shape of the sperm.

___

C. FERTILIZATION

Divide the motile sperm suspension among two or three *clean* petri dishes so that the bottom of each dish is just covered. Holding the female frog as shown in Fig. 37-4A, strip the eggs directly into the sperm suspension in each dish. Line the egg masses in rows or in a spiral so that all the eggs are in contact with the sperm. *Do not place them in a heap in the sperm.* Shake the dishes gently. Why?

___

Note the random orientation of the poles of the eggs. The pigmented animal pole of some will be uppermost; in others the light-colored vegetal pole will face up.

When the eggs are shed, the first polar body has been formed; the formation of the second polar body can be observed shortly after fertilization. To see this process, remove two or three eggs after they have been in the sperm suspension for several minutes and place them in a Syracuse dish. Cover them completely with Holtfreter's solution. Remove their jelly coats (Fig. 37-5) and then examine

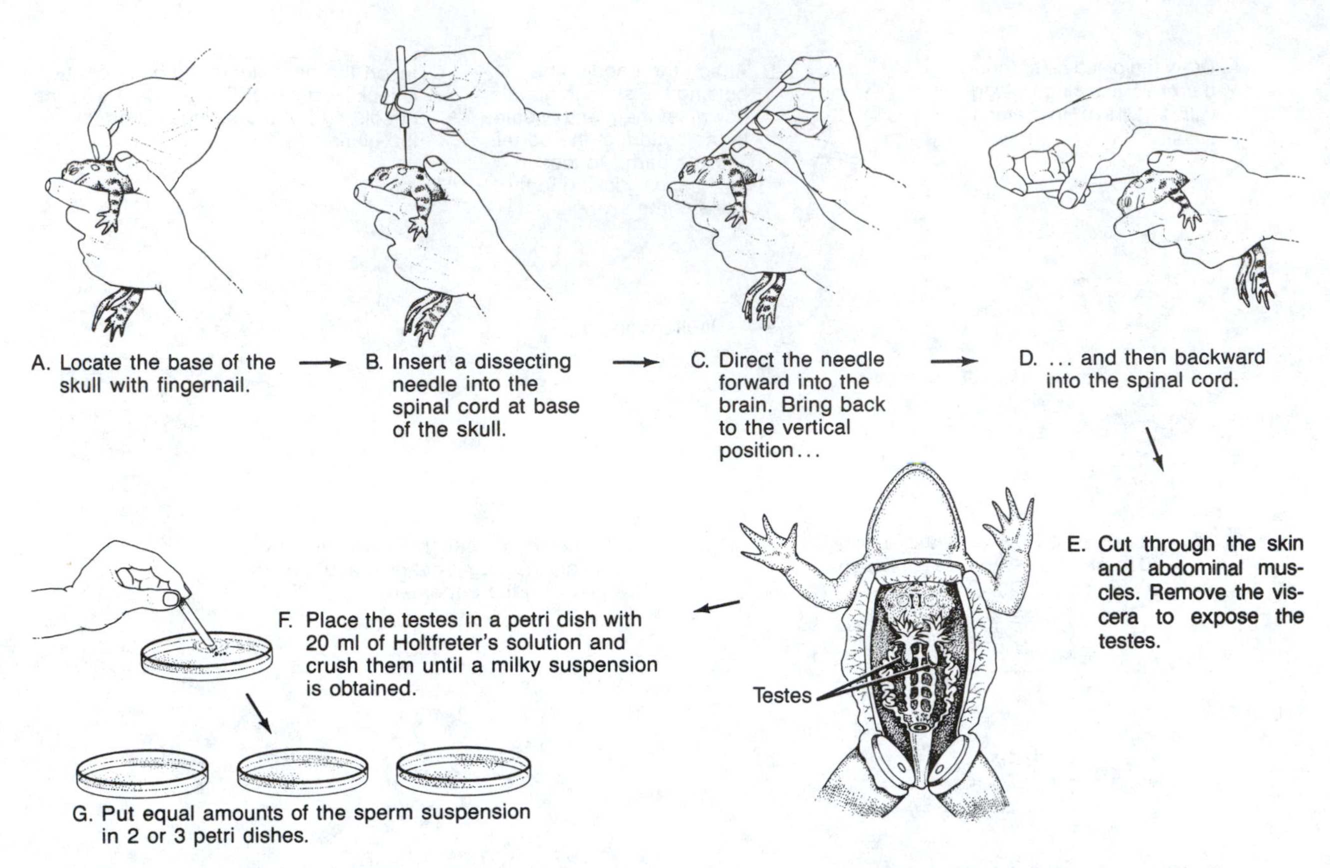

FIGURE 37-3 Procedure for pithing frog and preparing sperm suspension (also refer to Appendix I).

the eggs with a stereoscopic microscope. Adjust the light so that it strikes the egg surface at an oblique angle. With careful focusing, you should see a small, light, circular area in the animal pole. By adjusting the light, you may observe a pit in this light area (Fig. 37-6). This is where the second polar body will be expelled. With patience and careful observation, the formation of this polar body can be seen. It can sometimes be seen more easily if you rotate the egg so that the pit area is at right angles to your field of vision.

After the remaining eggs have been in the sperm suspension for 10 minutes, flood them with 10% Holtfreter's solution to remove the remaining sperm suspension, and then completely cover the eggs with fresh Holtfreter's solution. The first cleavage will occur in about 2 hours. While waiting, go on to part D or E as directed by your instructor.

D. UNFERTILIZED FROG EGGS

While waiting for cleavage to occur, gently squeeze about a dozen eggs from the injected female frog into a petri dish containing Holtfreter's solution. The solution should completely cover the eggs. Note that the eggs are clustered when they first leave the cloaca but tend to spread apart in the Holtfreter's solution. The eggs separate because the jelly that surrounds each egg swells considerably as the protein in the jelly absorbs liquid. Why is the swelling and spreading apart of the eggs important in their development?

____

____

Place one or two eggs on a glass slide, cover with Holtfreter's solution, and examine them with the low power of the compound microscope. How many jelly layers are present?

____

Examine the remaining eggs with a stereoscopic microscope. Adjust the spot lamp so that the light is

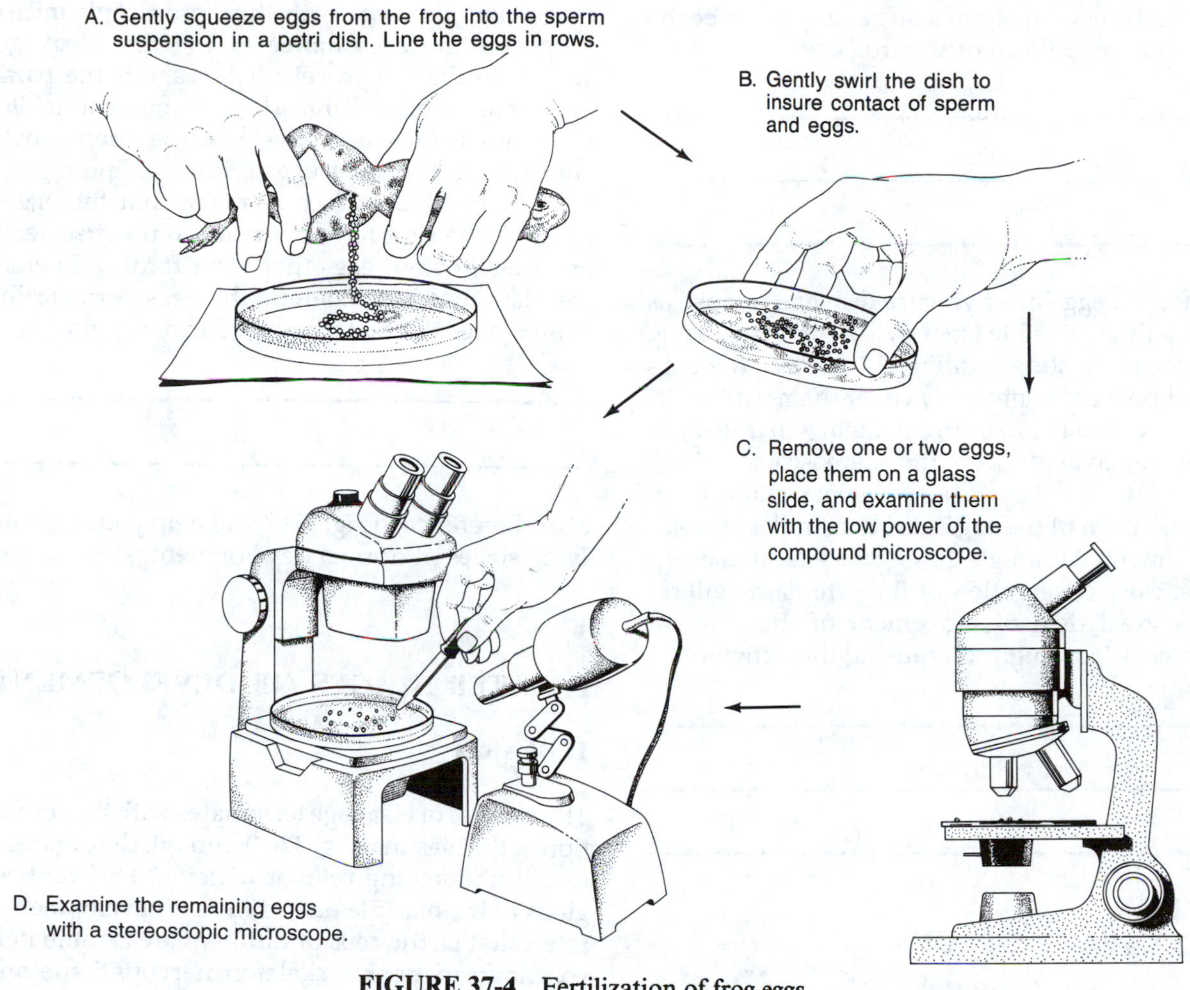

FIGURE 37-4 Fertilization of frog eggs

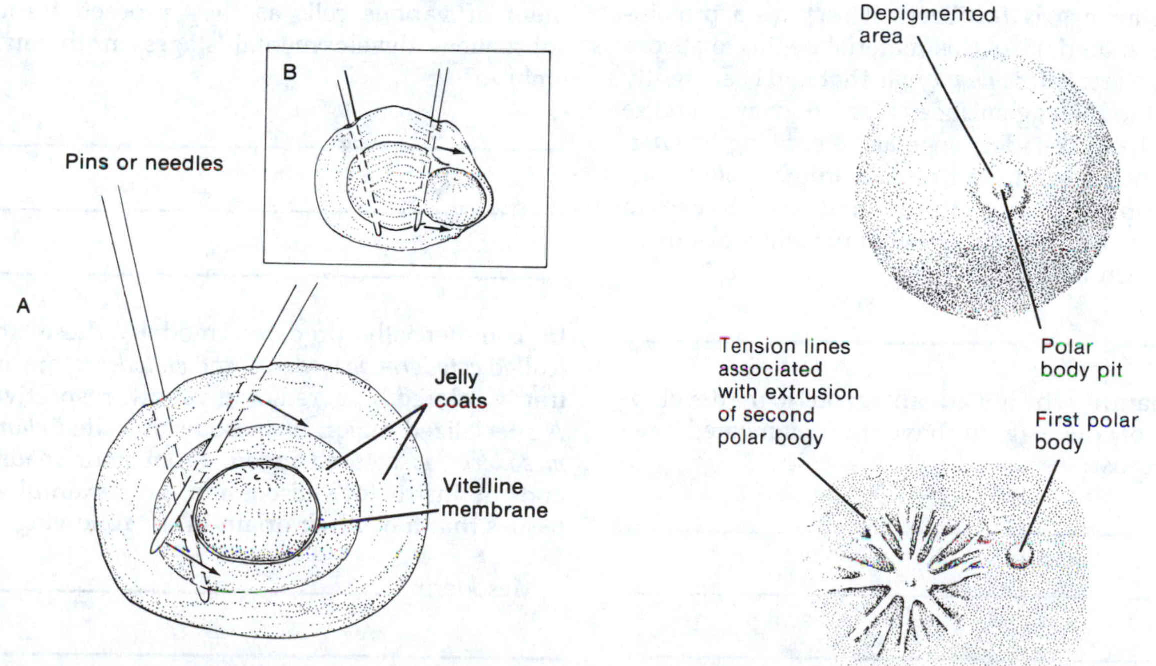

FIGURE 37-5 Procedure for stripping jelly coats from frog eggs

FIGURE 37-6 Location of second polar body

striking the eggs at about a 45° angle. Describe the pigmentation pattern of the frog egg.

______

______

______

Place the eggs in a Syracuse dish and remove as much jelly as possible from two or three eggs using the procedure shown in Fig. 37-5. Place the eggs animal pole up, completely cover them with Holtfreter's solution, and examine them with the highest power available on the stereoscopic microscope. Adjust the spot lamp for maximum light. The dark color of the egg is due to granules of melanin pigment. Although the egg may seem to be inactive, close observation of the cytoplasm will reveal a great deal of movement of the pigment granules. What might be causing this activity?

______

______

______

E. CLEAVAGE STAGES

Examine the fertilized eggs in the Syracuse dish. When an egg is fertilized, it secretes a proteinaceous material. As this material begins to absorb water, the *vitelline membrane* that had been tightly bound to the egg surface is forced away from the egg. The yolk-laden vegetal pole, being heavier, shifts downward, so that the animal pole is oriented upward. Check the orientation of the eggs in the dish. Approximately what percentage of them have been fertilized?

______

In nature, why is it advantageous to the development of the egg to have the pigmented area uppermost?

______

______

______

Examine the eggs with the stereoscopic microscope for the appearance of the first cleavage, which results in two cells. Cleavage is the partitioning of the egg cell into a large number of smaller cells with no increase in the mass. (Be careful not to confuse degenerating eggs with dividing eggs; a broken, mottled surface is a sign that the egg is dead.) If the eggs have not reached the first cleavage stage, obtain eggs that were fertilized before the class meeting. Examine the eggs periodically. Where does the first cleavage furrow begin?

______

______

Note: Reference to Fig. 37-7 will help you in identifying stages of normal development.

F. LATER STAGES OF DEVELOPMENT

1. Blastula

The process of cleavage terminates with the formation of the blastula (Fig. 37-7), though the organism is still undergoing cellular division. Different regions of the blastula have different developmental fates: that is, the cells of different areas ultimately contribute to the skin, skeleton, nervous tissue, and so on (Fig. 37-8). How could you follow the movement of various cells as they proceed through subsequent developmental stages in the living embryo?

______

______

______

In commercially prepared models, these areas (called *ectoderm, mesoderm,* and *endoderm*) are arbitrarily colored blue, red, and yellow, respectively. A specialized region of mesoderm, called *chordamesoderm,* is colored green. From your readings and the instructor's discussion, give examples of tissues that have their origin in the following:

Mesoderm: ______

______

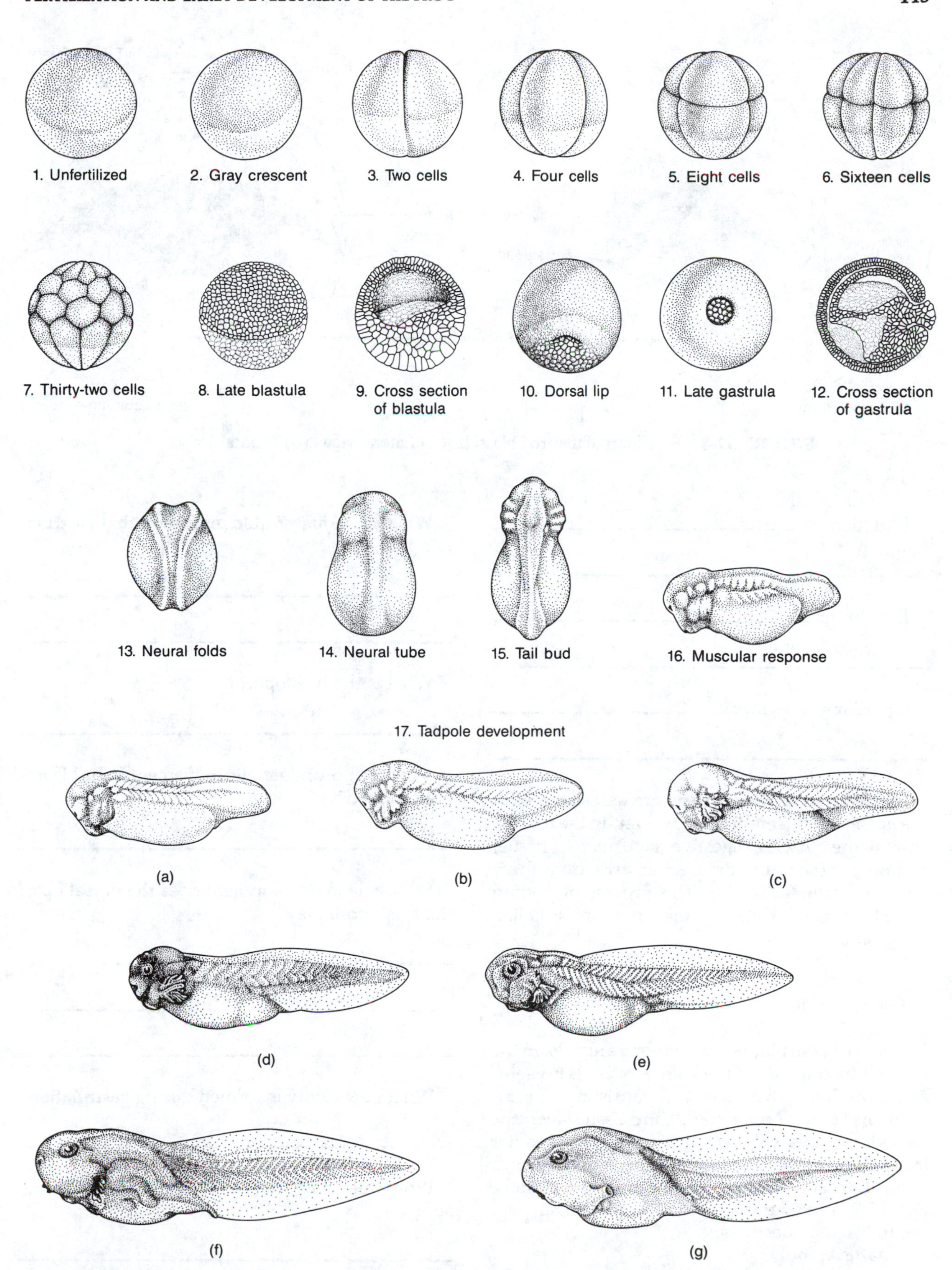

FIGURE 37-7 Normal embryonic developmental stages of the frog from unfertilized egg through tadpole

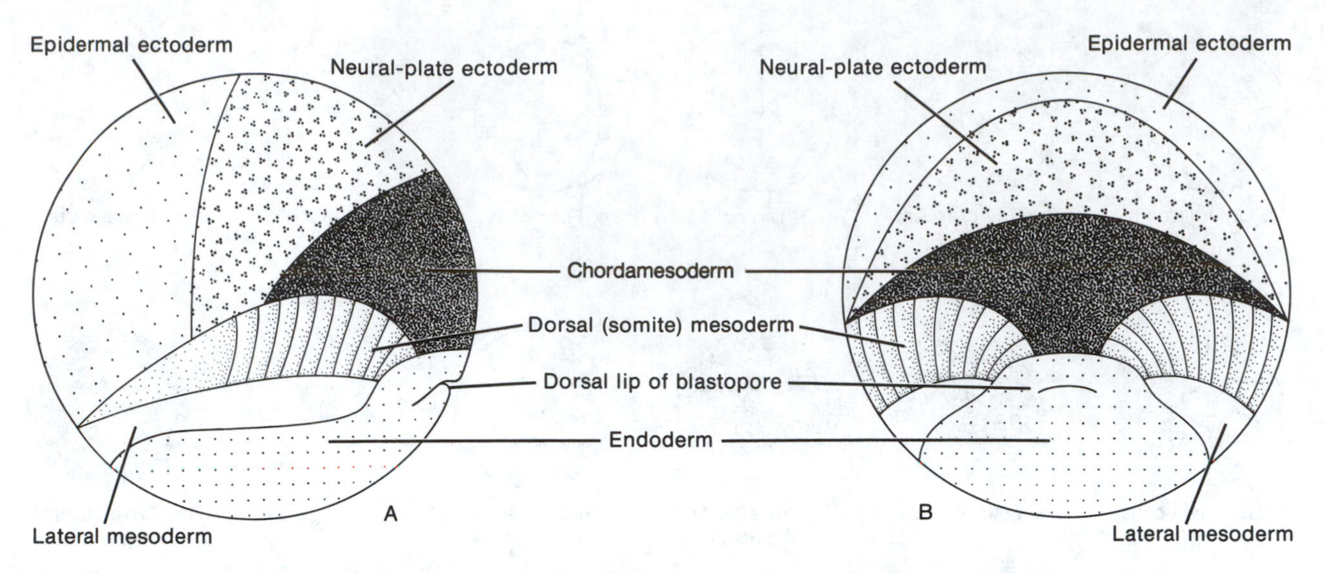

FIGURE 37-8 Fate map of the frog blastula. (A) Lateral view. (B) Frontal view.

Endoderm: ______________________

____

Ectoderm: ______________________

____

Chordamesoderm: ______________________

____

In the blastula, these areas are associated with the *outside* of the embryo. However, in the mature animal, the muscles, digestive tract, nervous tissue, and other tissues and organs that arise from these areas are situated *inside.* The process of getting these areas of cells to the interior is called *gastrulation.*

2. Gastrulation

Patterns of gastrulation vary among animals. In the sea urchin, one side of the hollow blastula invaginates, forming a two-layer cup consisting of ectoderm and endoderm. Later a third tissue layer, the mesoderm, arises between the inner and outer layers. In the frog, gastrulation is accomplished in a quite different manner. From the use of models, Fig. 37-9, and your instructor's discussion, answer the following questions.

What is epiboly?

____

What is the first visible indication that gastrulation has started?

____

____

What is the blastopore?

____

By what morphogenetic movement is the blastopore formed?

____

What role in development does the dorsal lip of the blastopore play?

____

____

____

What new cavity is formed during gastrulation?

____

What does this cavity ultimately become in the adult?

____

____

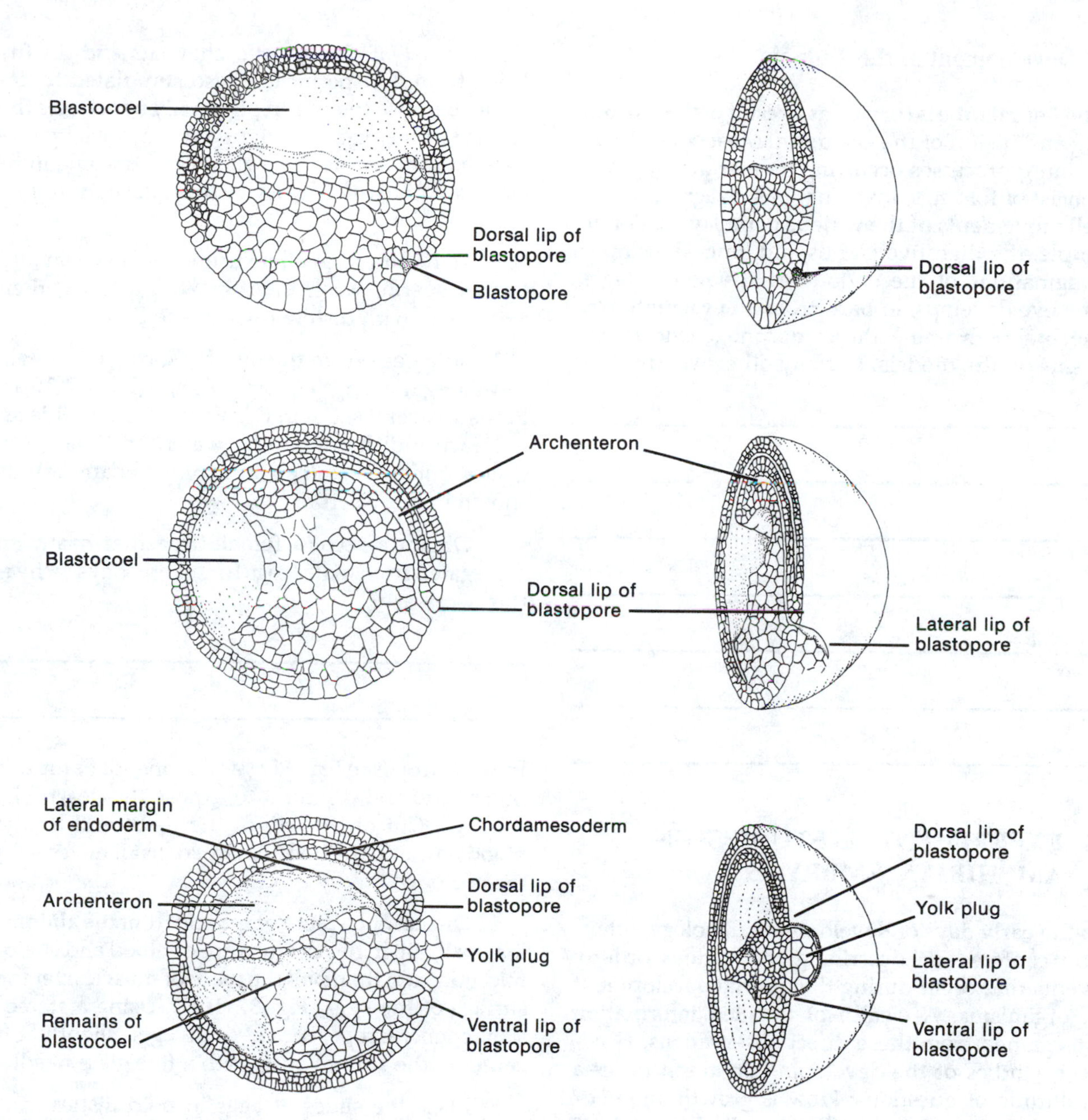

FIGURE 37-9 Gastrulation in the frog. (After *An Introduction to Embryology*, 5th ed., by B. I. Balinksy. Holt, 1981.)

The chordamesoderm, which will form the notochord, plays what other important role in the development of the frog?

___

___

3. Formation of the Neural Tube

When gastrulation is finished, the embryo is completely covered, except for the yolk plug, by ectoderm. The endoderm and mesoderm have moved inside. Soon, two ectodermal ridges form on the dorsal surface of the gastrula. These grow upward and toward each other, eventually fusing in the midline. When the edges fuse, a tube of ectoderm is formed, running from the anterior to the posterior end of the embryo. This tube, covered by an outer ectoderm layer, develops into the brain and spinal cord. Follow these events using Fig. 37-7 (stages 13 and 14) and models. If available, examine living or preserved embryos at these stages of development.

4. Development of the Tadpole

The neural tube is formed as a result of the outfolding and fusion of the ectoderm. Most other tissue-forming processes occurring in later development consist of foldings, invaginations, evaginations, or cell movements of the various germ layers. For example, the digestive glands and lungs develop as evaginations of the endoderm at various stages. The eye develops, in part, from an evagination of the brain. Examine these morphogenetic movements on the models. List any others you find.

____

____

____

____

____

____

____

G. EXPERIMENTAL STUDIES ON AMPHIBIAN EMBRYOS

In the early days of developmental biology, attention centered on describing the various orderly events that occur during the normal development of organisms. A wealth of detailed information was gained from these direct observations. However, studies of the developing organism raises a multitude of questions: How is growth initiated? Why does growth stop? Why do cells differentiate? What regulates form? To answer these questions developmental biologists turned from observation and description to experimental studies designed to shed light on these and other developmental phenomena.

In this part of the exercise you will use the amphibian embryo to study two developmental phenomena: initiation of development of the frog egg by artificial means and the reaggregation of the dissociated cells of a blastula.

1. Parthenogenesis

In the normal sequence of development, a mature egg is penetrated by a sperm. As a result, the egg receives a set of the genetic characteristics in the DNA from the sperm. It is also stimulated to develop. Because the sperm physically penetrates the surface of the egg, we might ask if artificial penetration of the surface by a glass needle would initiate development. To answer this question, follow this procedure.

1. To be sure that all instruments are clean and free of sperm, wash them in 70% alcohol, then rinse them with distilled water and air dry.

2. Strip eggs from the ovulating frog in a single row along the length of a glass slide (Fig. 37-10A). Prepare several slides in this way. Set each slide on a Syracuse dish containing water, and place the dishes under a jar (i.e., in a moist chamber), as shown in Fig. 37-10B.

3. Obtain a second female frog that has been segregated from male frogs for several days. Why is this necessary?

____

____

Pith the frog (see Fig. 37-3 and Appendix I for this procedure) and dissect it to expose the heart (Fig. 37-10C). Cut off the tip of the heart and let the blood flow into the body cavity so that it mixes with the coelomic fluid.

4. Dissect out a piece of muscle from the abdominal wall, dip it into the mixture of blood and coelomic fluid, and then pull the piece of muscle over the surface of the eggs (Fig. 37-10D). Using a stereoscopic microscope, prick each egg, slightly off center in the animal pole, with a fine glass needle.

5. Place the slides of eggs in petri dishes containing Holtfreter's solution so that the eggs are covered by the fluid. Cover the dishes, label them Experimental, and keep them in a cool place at about 68°F (20°C).

6. Describe the control group that should be tested.

____

____

____

Use the remaining eggs for your controls.

7. Examine eggs after 2 hours and periodically

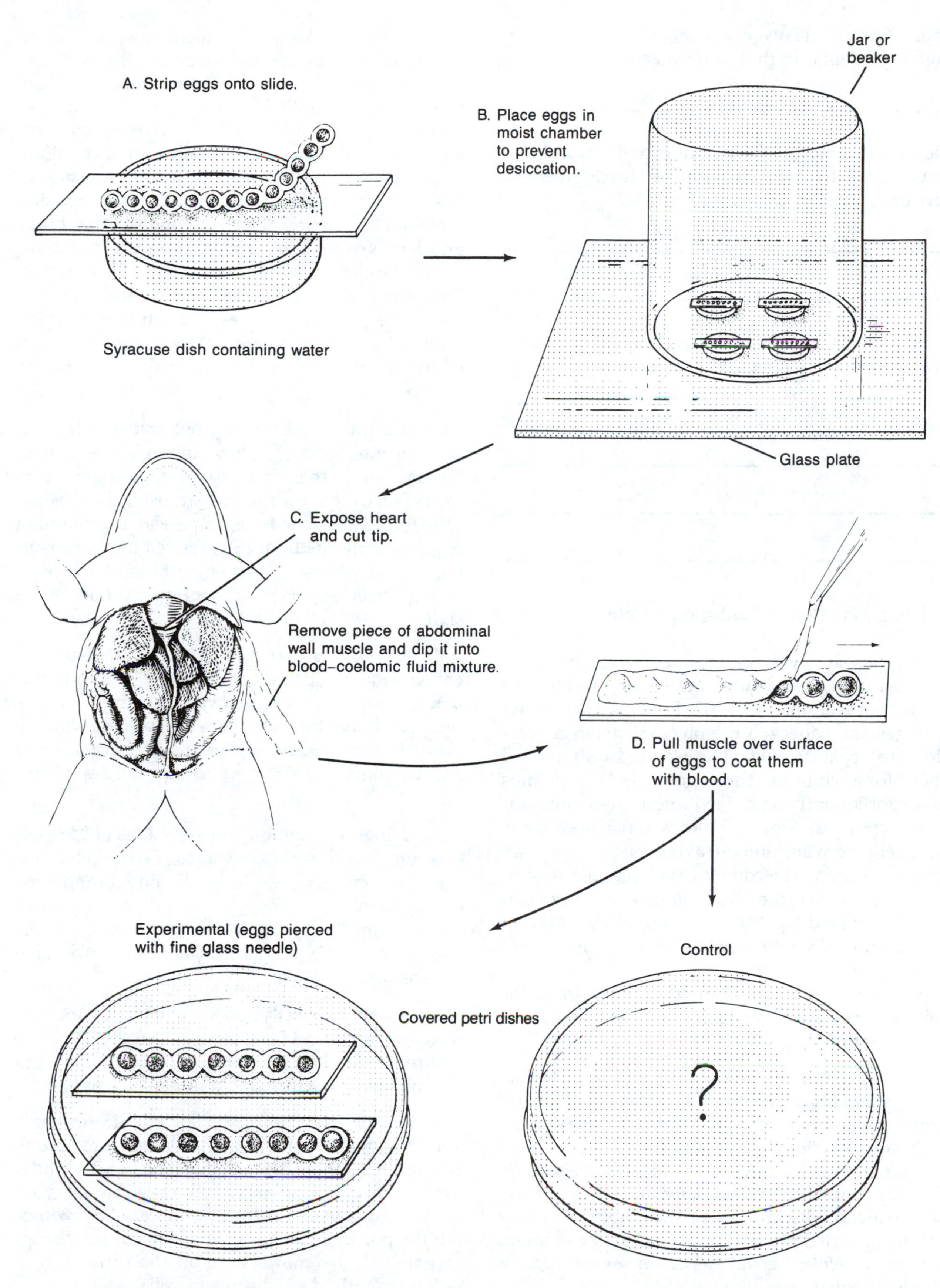

FIGURE 37-10 Procedure for inducing parthenogenetic development of frog eggs

thereafter. If cleavage occurs, is the pattern of cleavage similar to that in fertilized eggs?

____

Describe the stages of parthenogenetically induced development compared with the development of fertilized eggs, as shown in Fig. 37-7.

____

____

If no cleavage occurs, does this prove that sperm are necessary for development? Explain.

____

____

____

____

2. Reaggregation of Embryonic Cells

Numerous studies have centered on the behavior of cells isolated from their normal relationships in a tissue. For example, in 1907, H. V. Wilson cut up sponges and squeezed the pieces through cheesecloth to separate cells. The isolated cells moved about for a while and then aggregated into clumps that subsequently organized themselves into miniature sponges. Similar studies have been done with cells from amphibian, avian, and mammalian species. Aaron Moscona carried out a study in which chick cartilage and liver cells were isolated and then mixed together. These cells sorted themselves into discrete aggregations of liver and cartilage.

In this study, you will dissociate (separate) the cells of a frog blastula and determine the pattern of reaggregation.

1. Obtain several frog embryos in late blastula or early gastrula stage (Fig. 37-7, stages 8, 9, or 10). Using a watchmaker's forceps, transfer one of the embryos to 70% ethyl alcohol *for not more than 5 seconds.* This will reduce the number of bacteria on the jelly coat without damaging the embryo. Rinse off the alcohol by immersing the embryo in sterile Steinberg's medium and then place it in a separate container. (*Note:* Repeat step 1 on several eggs to obtain adequate results.)

2. Using a stereoscopic microscope and clean watchmaker's forceps, remove the jelly coat and vitelline membrane from each embryo. Then, using a pipet supplied by your instructor, transfer each embryo to a separate small petri dish containing 2–3 ml of dissociating medium. This medium lacks magnesium and calcium ions (important in intercellular binding) and contains ethylene diamine tetraacetate (EDTA), a compound that binds and removes calcium or magnesium from the intercellular "cement" that holds the cells together. Thus, in a short time the cells of each embryo should begin to dissociate. You can speed up the process by gently aspirating the embryo in and out of the pipet or by gentle manipulation with the forceps.

3. Using a clean Pasteur pipet, transfer the cells of each embryo to fresh Steinberg's medium to "wash" any adhering dissociating solution from the cells. After a few minutes, transfer the washed cells to flat-bottomed well-type slides containing reaggregating medium (Steinberg's medium supplemented with serum or egg albumin). Transfer enough cells to sparsely scatter over the floor of the slide.

4. Add a thin coat of vaseline to the rim of a coverslip and dip the entire coverslip into Steinberg's medium. Place the coverslip over the well of the slide so that the vaseline makes contact with the glass. This procedure will retard evaporation and reduce fogging of the coverslip during observations.

5. Using the scanning objective lens (4× –5×) of a compound microscope, focus on the cells. Disturb the cultures as little as possible during observations.

(**Caution:** *To avoid overheating the cells, turn the microscope lamp off when you are not making observations.*)

6. Observe the cells periodically throughout the next few hours and for several days thereafter. The cultures should be kept at room temperature and the microscope covered with a plastic bag.

7. Record your observations. Look for such events as cyclosis (internal cytoplasmic streaming), protrusions from the cell surface, and cellular movements. You might consider recording your observations in the form of annotated drawings. For the purpose of this experiment, consider reaggregation to be complete upon the formation of rather smooth, dark clusters of cells.

REFERENCES

Browder, L. W. 1984. *Developmental Biology.* 2d ed. Saunders/Holt, Rinehart and Winston.

Bryant, P. J., S. V. Bryant, and V. French. 1977. Biological Regeneration and Pattern Formation. *Scientific American* 237:66–81.

Epel, D. 1977. The Program of Fertilization. *Scientific American* 237:128–138 (Offprint 1372). *Scientific American* Offprints are available from W. H. Freeman and Company, 41 Madison Avenue, New York, NY 10010, and 20 Beaumont Street, Oxford OX1 2NQ, England. Please order by number.

Hamburger, V. 1960. *A Manual of Experimental Embryology.* Rev. ed. University of Chicago Press.

Jones, K. W., and T. R. Elsdale. 1963. The Culture of Small Aggregates of Amphibian Embryonic Cells in Vitro. *Journal of Embryological and Experimental Morphology* 11:135–154.

Moscona, A. 1961. How Cells Associate. *Scientific American* 205(3):142–162 (Offprint 95).

Rugh, R. 1951. *The Frog: Its Reproduction and Development.* McGraw-Hill.

Rugh, R. 1962. *Experimental Embryology: Techniques and Procedures.* 3d ed. Burgess.

Shumway, W. 1940. Stages in the Normal Development of *Rana pipiens.* I. External Form. *Anatomical Record* 78:139–147.

Wolper, L. 1978. Pattern Formation in Biological Development. *Scientific American* 239:154–164 (Offprint 1409).

Exercise

38

Early Development of the Chick

The chick embryo is used in developmental studies because its early embryonic development is markedly similar to that of the mammalian embryo. Many of the very early events (cleavage, blastulation, and gastrulation) are not as easily observed as similar stages in the frog embryo. One reason for this is that the chick "egg" (popularly called the *yolk*), is surrounded by various accessory coverings secreted by the female reproductive tract.

A. FERTILIZATION AND EGG LAYING

Direct observations of fertilization in birds and mammals are very difficult because the process takes place inside the body of the female. After the egg is liberated from the ovary, it undergoes changes that can be characterized as aging. These changes progress rapidly to a point at which the egg, though technically still alive, can no longer be fertilized. If, however, the egg is fertilized soon after liberation, these aging processes are checked. Fertilization in the chicken takes place just as the egg enters the oviduct (Fig. 38-1A).

Accessory coverings are secreted around the egg in the course of its passage toward the cloaca. In the part of the oviduct adjacent to the ovary, a mass of stringy **albumen** is produced, which adheres to the **vitelline membrane** and extends along the oviduct in both directions from the egg. Because of spiral folds in the oviduct wall, the egg rotates as it moves toward the cloaca, and this rotation twists the albumen into spiral strands known as the **chalazae,** one on either side of the egg. More albumen is added in concentric rings as the egg moves down the first half of the oviduct. These layers of albumen can be seen in boiled eggs because the heat coagulates the albumen.

The **shell membranes,** which consist of matted organic fibers, are added farther along the oviduct. The **shell** is secreted as the egg passes through the shell gland (the uterus). The entire passage of the egg, from its discharge from the ovary until it reaches the end of the oviduct and is ready for laying, takes about 22 hours.

To examine the internal organization of the chicken egg, remove the shell from a hard-boiled egg and then carefully remove the egg white surrounding the yolk.

On the surface of the yolk, locate a small, whitish spot, the **blastodisc,** which normally faces up. This was the living part of the egg. The remainder of the yolk is composed of inert food material and water.

As the fertilized egg passes through the upper region of the oviduct, mitotic cleavage of the blastodisc quickly begins and proceeds as the egg passes through the oviduct, including the part that

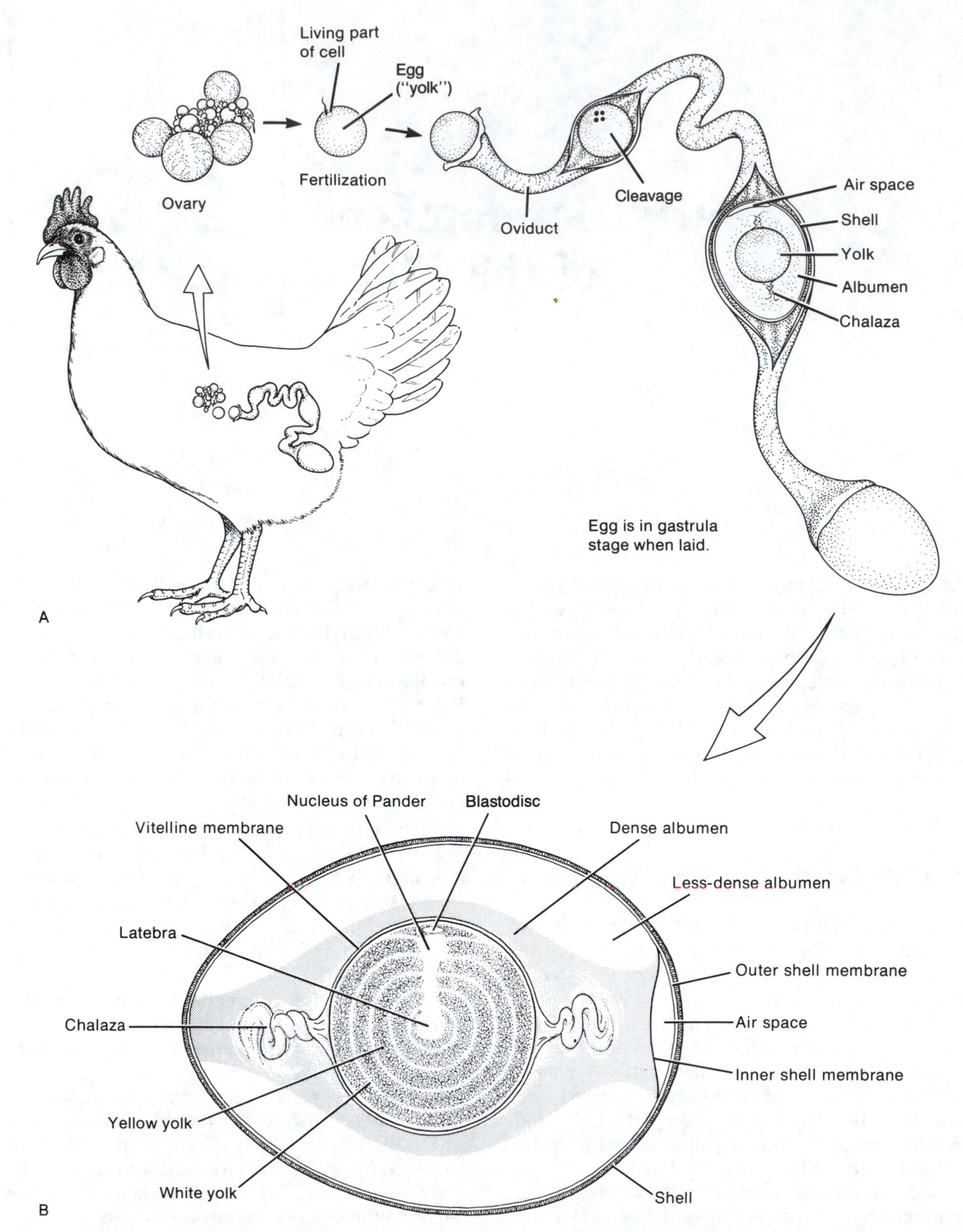

FIGURE 38-1 (A) Stages of fertilization and early development in the hen oviduct. (B) Structure of the egg when it is laid.

contains the shell gland. Cleavage in the chick is termed *partial* or *discoidal* because the irregular cleavage divisions, restricted to the blastodisc, are incomplete (Fig. 38-2), and the large mass of yolk does not divide. As cleavage continues, the blastodisc becomes several cell layers thick, and a fluid-filled **subgerminal space** develops beneath it (Fig. 38-3A).

Locate, on the surface of the yolk, a small, light, circular area. With a sharp razor blade, cut the yolk into two equal parts through this area. Close examination of the cut surface of the yolk will reveal **white yolk** and **yellow yolk,** the colors of which are due to differences in the size and shape of the yolk granules (Fig. 38-1). The granules and globules in white yolk are generally smaller and less uniform than those in yellow yolk. These concentric layers of yolk are thought to indicate daily accumulations of cytoplasm in the final stages of the formation of the egg. The principal accumulation of white yolk lies in a central flask-shaped area, the **latebra,** which extends toward the blastodisc and flares out under it in a mass known as the **nucleus of Pander.**

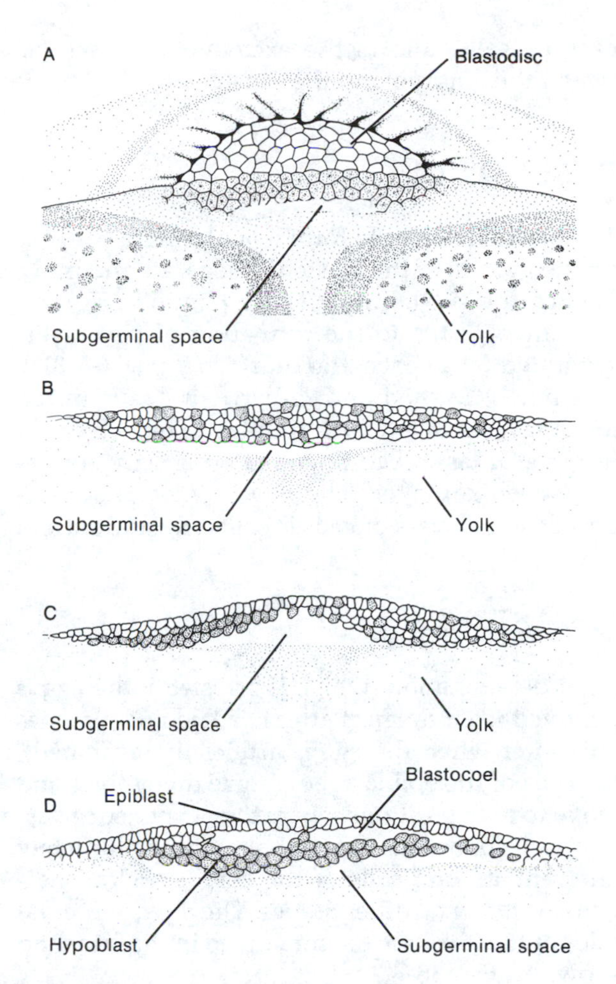

FIGURE 38-3 (A) Cross section of the blastodisc. (B) Delamination of the blastodisc to form the epiblast and hypoblast.

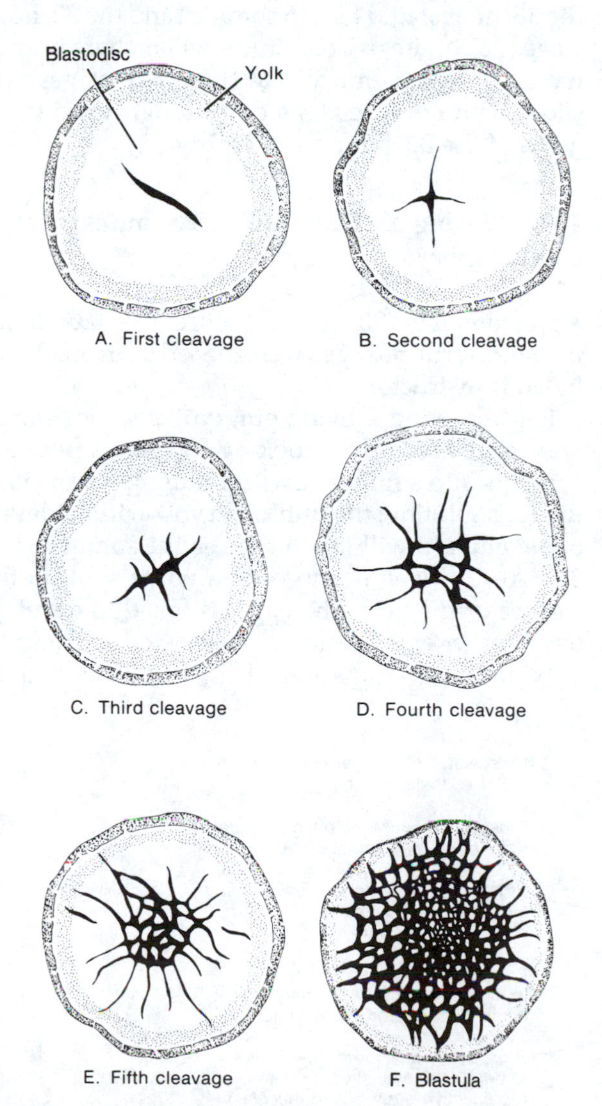

FIGURE 38-2 Successive cleavages as observed looking down at the surface of the blastodisc

The albumen, except for the chalazae, appears nearly homogeneous. Two layers of shell membranes lie in contact with the albumen everywhere except at the large end of the egg, where the inner and outer membranes separate to form an air space. This space is thought to form because the egg contracts as it cools after being laid and because of the evaporation of water. The air space increases in size over time because of continued evaporation of water. The familiar method of testing the freshness of eggs by floating them in water is based on this fact.

The porous egg shell, composed largely of calcareous salts, allows the exchange of gases between the embryo and the outside air.

B. BLASTULA

The blastula begins to form when some of the cells on the bottom of the blastodisc separate and move toward the subgerminal space (Fig. 38-3B). This movement leads to the formation of two layers separated by a space, the **blastocoel** (Fig. 38-3D). The newly formed and rather loosely organized lower layer is called the **hypoblast.** The cells of the upper **epiblast** are arranged in an orderly epithelial (sheetlike) layer. The developmental separation of such layers of cells is called **delamination.**

C. GASTRULA

After delamination (which is arrested if the egg is removed from the nest after it is laid and resumes only later when the egg is artificially incubated), the cells of the epiblast reorganize themselves and move to new positions where they experience various developmental fates; that is, they ultimately differentiate into cells of the skin, skeleton, nervous system, and other tissues. The developmental roles of these cells are summarized in the **fate map** shown in Fig. 38-4.

At this time, the **primitive streak,** a thickening of the blastodisc, develops into the **primitive groove,** a narrow groove that traverses the surface of the pear-shaped blastodisc from near its center to near its more pointed end (Fig. 38-5). Cells of the epiblast then begin moving toward the primitive groove, into it, along it, and then away from it (Fig. 38-5B, C) to form a **mesodermal** layer and an **endodermal** layer. The remaining surface layer becomes the **ectoderm.**

When the mesoderm is fully established, gastrulation is complete. Eventually, the surface ectoderm is folded under the embryo. When this folding is completed the ectoderm forms a complete outer covering over the dorsal and ventral surfaces of the embryo (Fig. 38-5D).

D. LATER STAGES OF DEVELOPMENT

The morphological stages of development of the chick embryo are usually identified by the number of hours of incubation (at 38°C) that are required to reach each stage. Stages are also identified by the number of **somites,** blocklike cell masses of mesoderm that begin to appear after about 21 hours of incubation.

You will study two stages of chick development: the 33-hour stage (12–15 somites) and the 72-hour stage (36 somites). Your study will include examination of living embryos at these two stages and slides with cross sections cut through various regions of the 33-hour chick embryo.

1. The Living 33-Hour (12–15 Somites) Chick Embryo

Approximately 33 hours before the laboratory meeting, fertilized eggs were placed in an incubator by your instructor.

For observing a living embryo, crack an egg as you usually would for cooking and gently pour the contents into a finger bowl containing warm chick Ringer's solution; the unbroken yolk with its developing embryo will float in the saline solution (Fig. 38-6A–C). (If you find only a white spot on the surface of the yolk, the egg has failed to develop; use another egg.) Remove the blastodisc from the yolk mass as shown in Fig. 38-6D–F and as

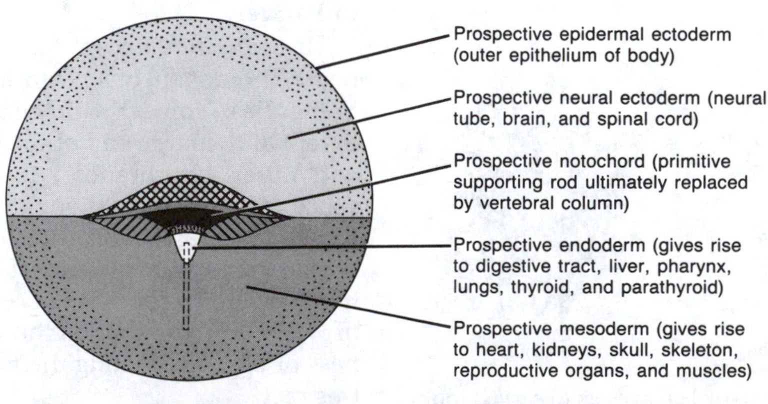

FIGURE 38-4 Fate map of chick embryo

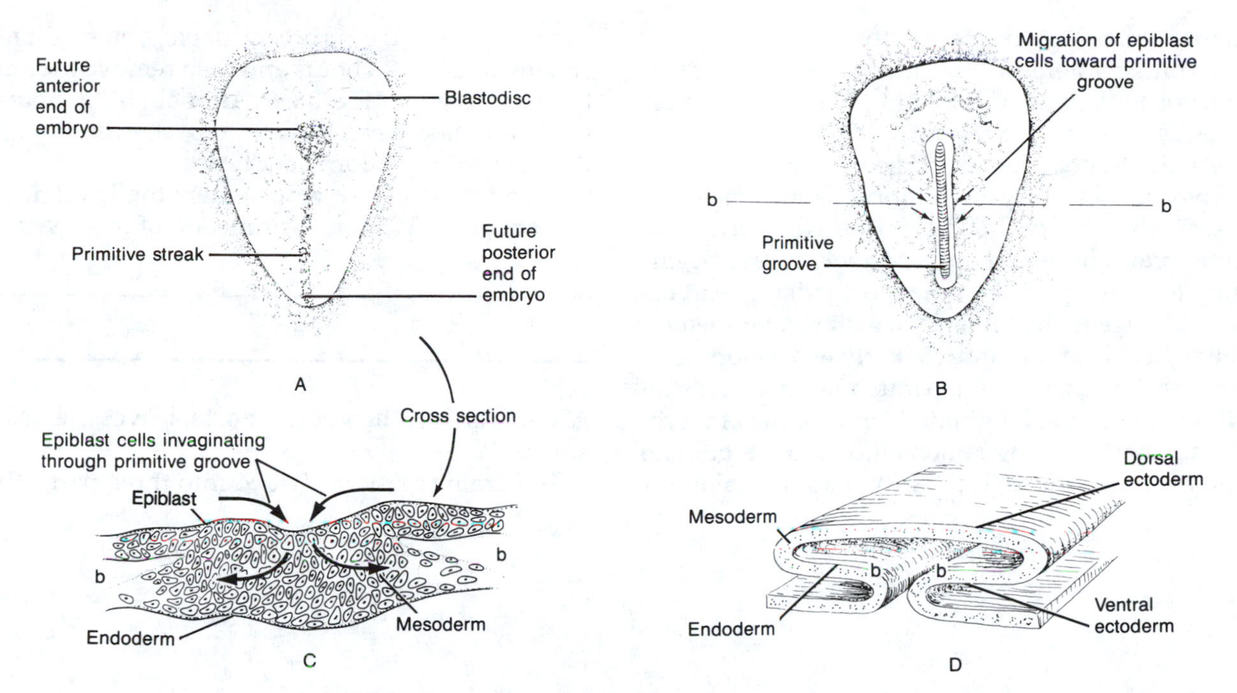

FIGURE 38-5 Gastrulation in the chick

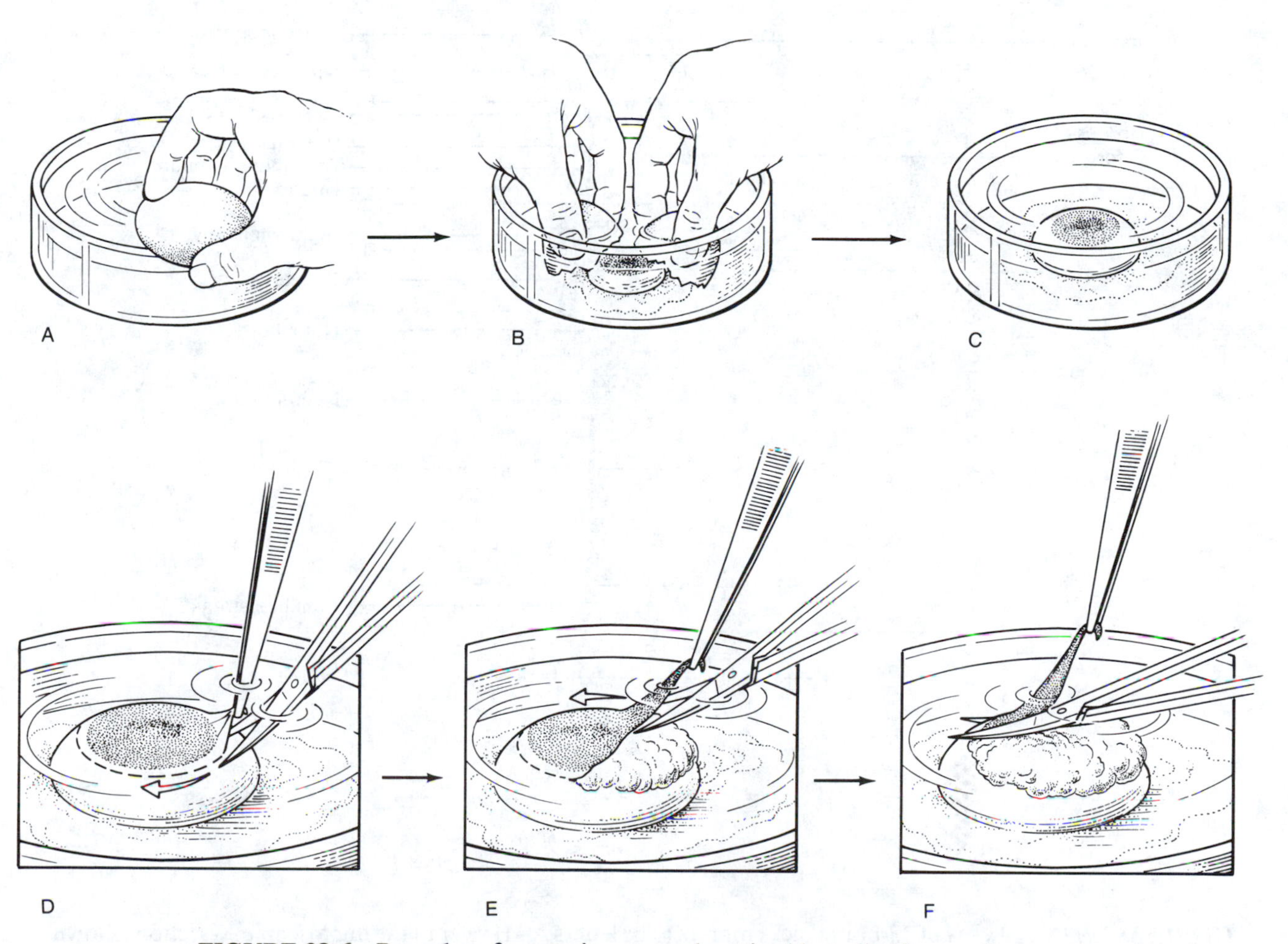

FIGURE 38-6 Procedure for opening egg and cutting blastodisc from yolk

demonstrated by your instructor. Examine the embryo under a stereoscopic microscope. Supplement your observations of the living embryo with prepared slides and models (Fig. 38-7).

At 33 hours, the blastodisc is about 20 mm across. Surrounding the embryo is a transparent region, the **area pellucida.** Outside this area is the **area opaca,** in which the blood vessels are beginning to develop. One of the most striking features of this stage is the "heart," which can be seen on the right side of the embryo. Early in development, the heart originates as separate masses of cells on either side of the longitudinal axis of the embryo. These heart-forming regions move to the midline and eventually fuse (Fig. 38-8). If additional unincubated fertilized eggs are available, you might incubate them for 26 hours and then remove the embryo as shown in Fig. 38-6. After about 26 hours, the heart masses can be seen to beat, even though they have not yet formed a heart.

A pair of **vitelline veins** enters the heart at its posterior end. What is the function of these veins?

____

____

A single artery, the **ventral aorta,** leaves the heart anteriorly.

The brain region is divided into three parts: the

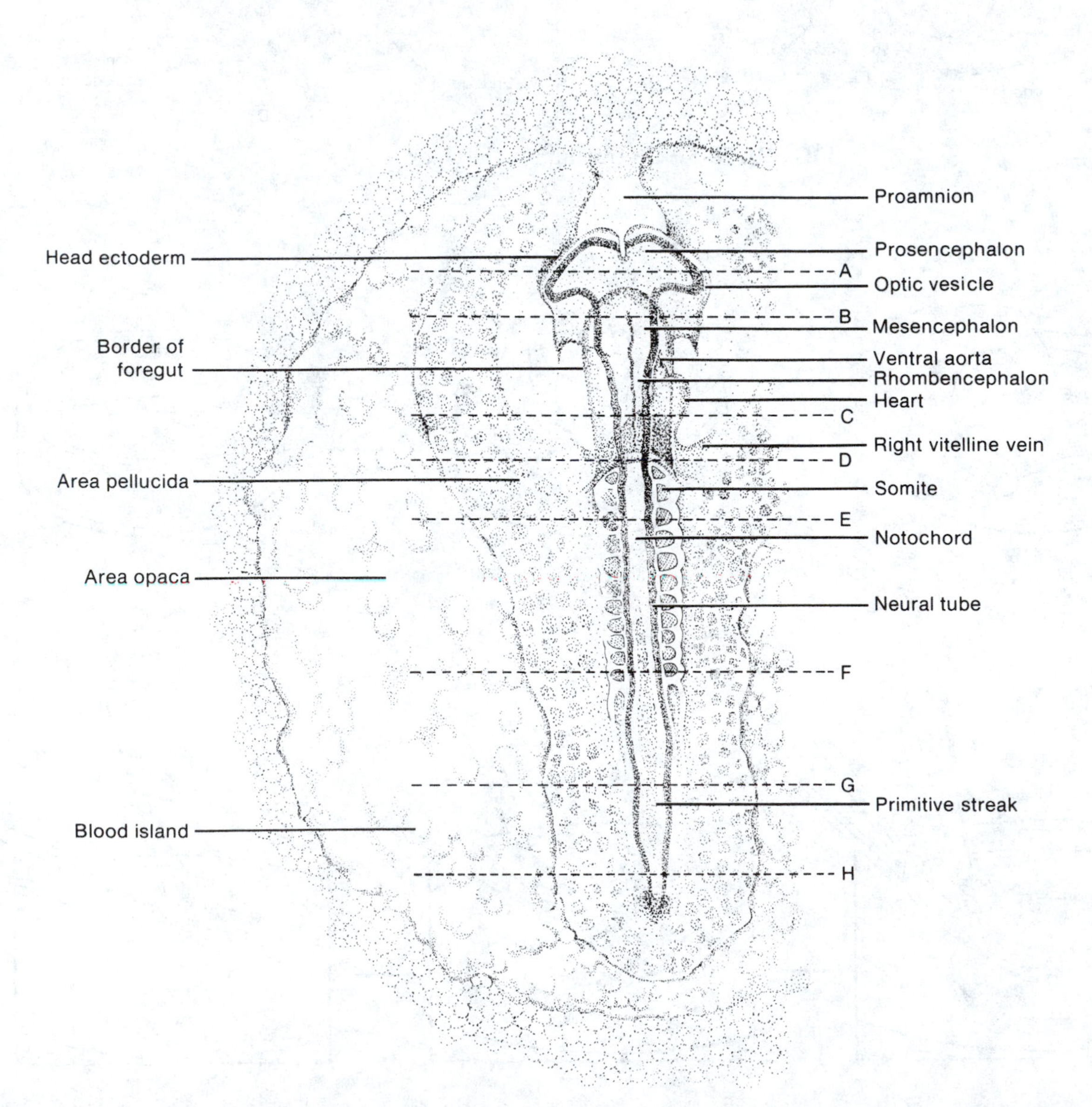

FIGURE 38-7 Dorsal view of 33-hour chick embryo. (The letters A – H refer to the microscopic X-sections shown in Fig. 38-9.)

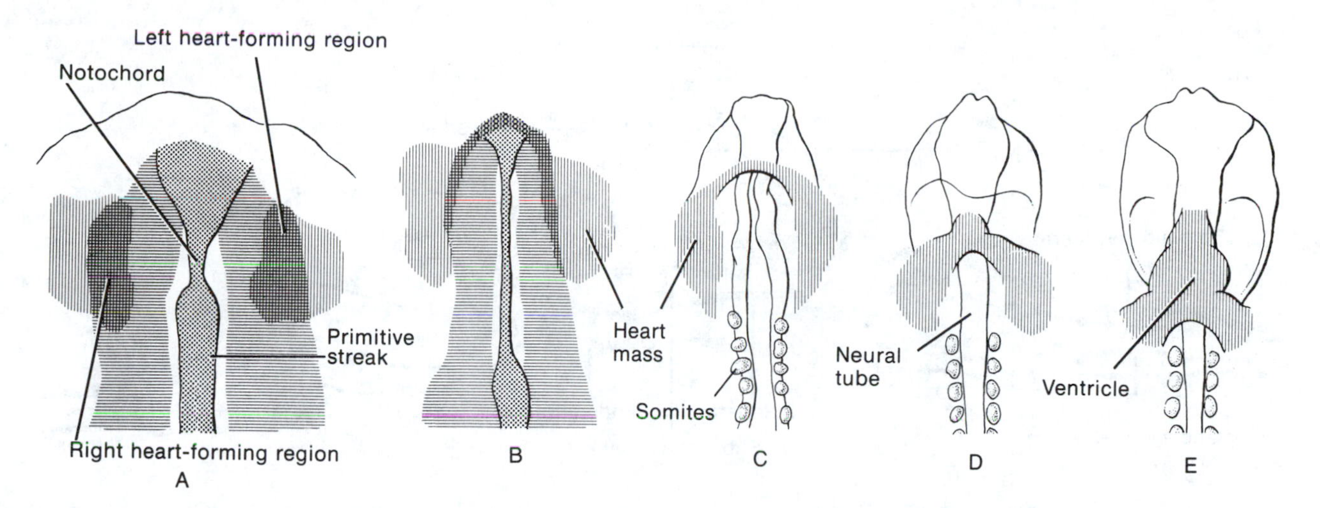

FIGURE 38-8 Formation of the heart in the developing chick embryo

anterior **prosencephalon (forebrain),** the **mesencephalon (midbrain),** and the **rhombencephalon (hindbrain),** which continues for the length of the embryo as the **neural tube.** The lateral evaginations of the forebrain are the **optic vesicles.**

The **notochord,** which lies ventral to the neural tube, can be seen in the midline of the embryo, extending anteriorly toward the forebrain as a solid rod. Lying on either side of the neural tube are 12–15 somites that ultimately will form the vertebral column and the musculature on the dorsal side (they may also contribute to the ribs and musculature of the body wall).

2. Serial Sections of the 33-Hour Embryo

Refering to Fig. 38-9, study demonstration slides of cross sections cut through the various regions of the 33-hour embryo shown in Fig. 38-7. The relationships of various body layers to one another and their differentiation into organs and tissues of the embryo can be reconstructed by study of these sections.

- Section A is through the prosencephalon and optic vesicles, which will produce major parts of the eyes. The head region has raised up and is separated from the rest of the blastodisc by a space, called the **subcephalic pocket.** Note that the blastodisc immediately under the free head region consists only of ectoderm and endoderm.
- Section B is through the head posterior to the optic vesicles. You can see the notochord, the dorsal and ventral aortas, and the pharynx (foregut).
- Section C is through the region of the heart. Note the **epimyocardium,** the thick outer wall of the heart, and the **endocardium,** the thin inner lining. In this region, the embryo is attached to the blastodisc.
- Section D is through a region a short distance posterior to the heart. Observe the **midgut,** which is that part of the digestive tract just below the pharynx. Note the paired dorsal aortae, which will fuse into a single dorsal aorta.
- Section E is through a region that includes the third mesodermal somite. Note the absence of the digestive tract.
- Sections F and G are through regions near the posterior end of the embryo and show the origin of the neural tube, which forms by a folding of the neural plate (section G).
- Section H is through the region of the primitive streak. Note the three germ layers that meet and fuse along the longitudinal axis of the embryo. A depression of the ectoderm forms the primitive groove bounded by two primitive ridges.

3. The 72-Hour Embryo

Remove the embryo from the yolk as shown in Fig. 38-6 and locate the various regions described in this study. At this stage, the anterior part of the embryo has turned to lie on its left side, whereas the posterior half remains dorsal side up (Fig. 38-10).

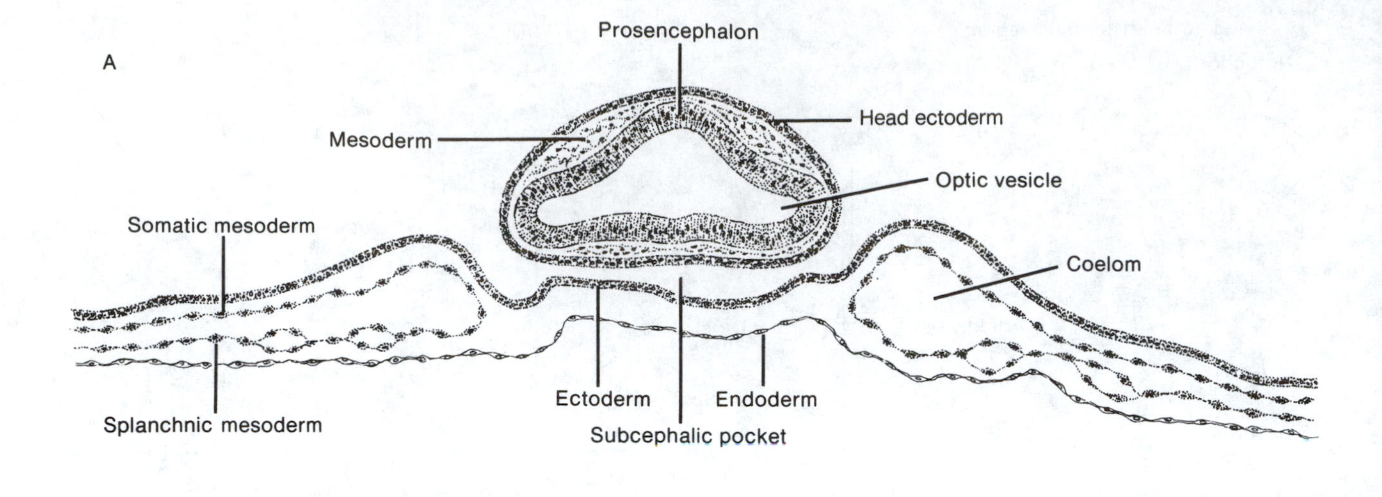

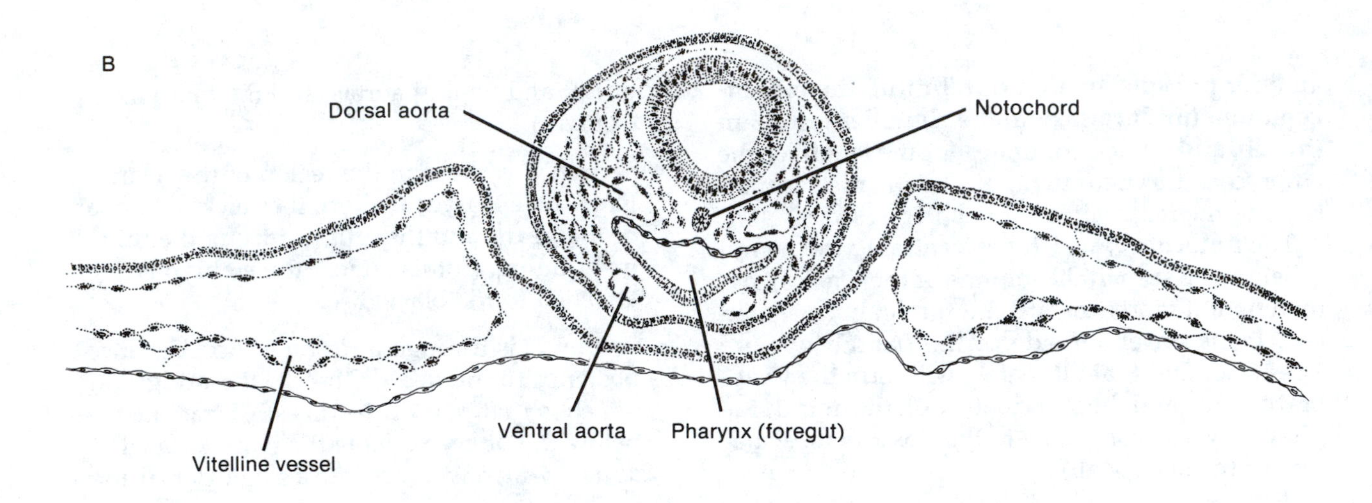

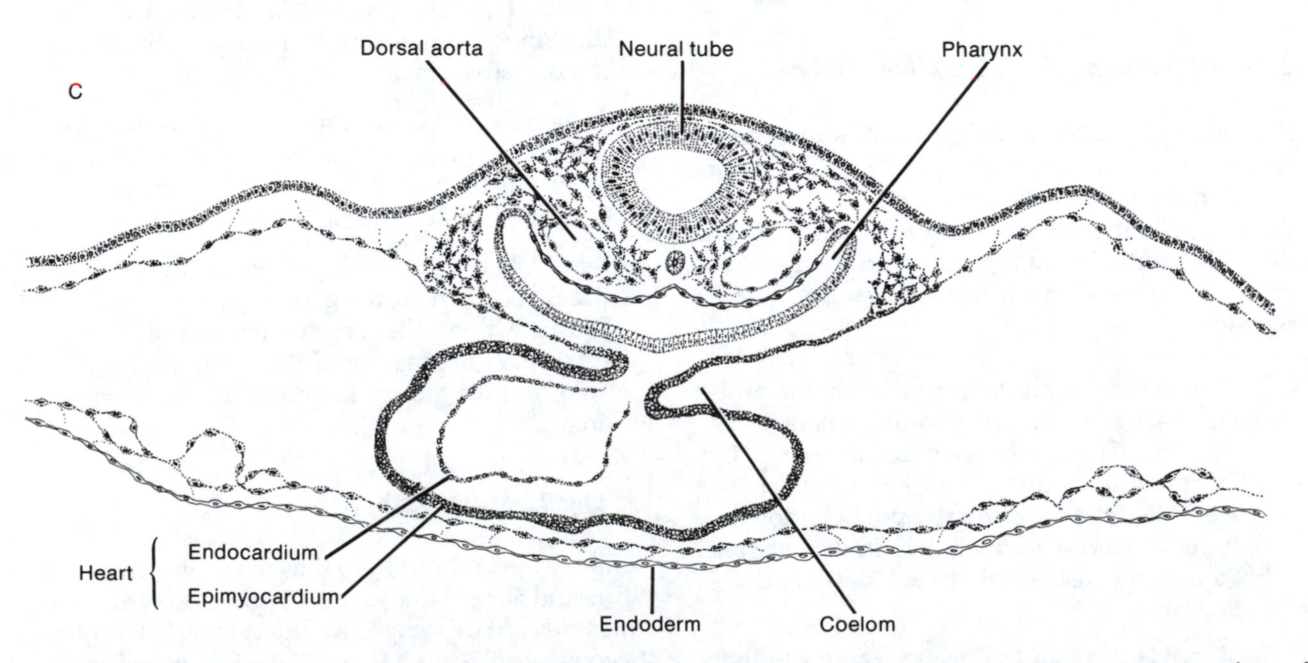

FIGURE 38-9 Serial sections through selected regions of a 33-hour (12–15 somite) chick embryo. The locations of these cross sections (A–H) on the embryo are shown in Fig. 38-7.

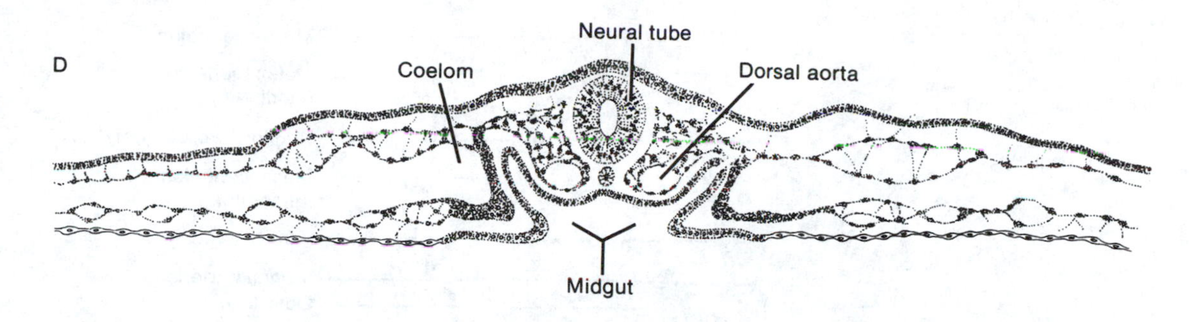

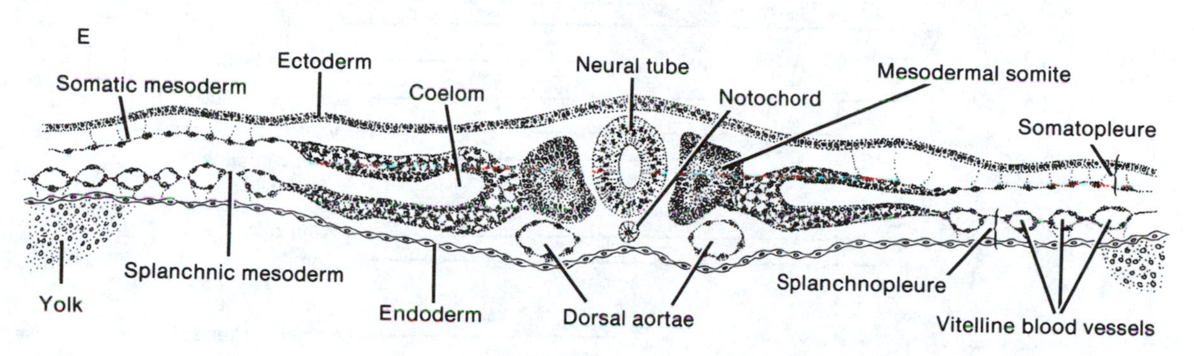

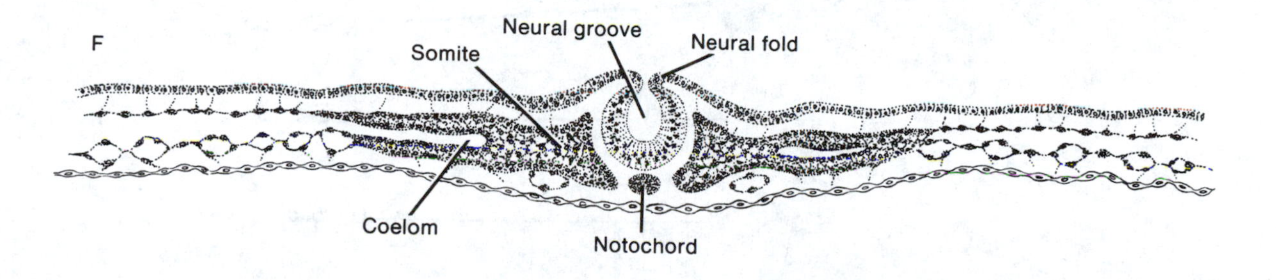

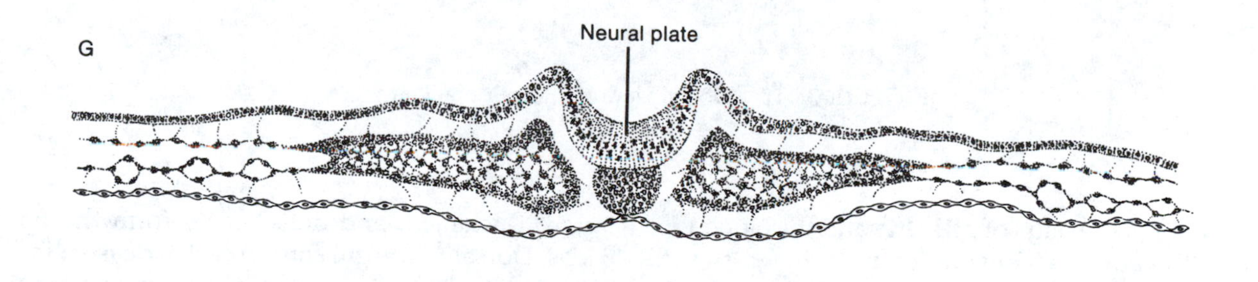

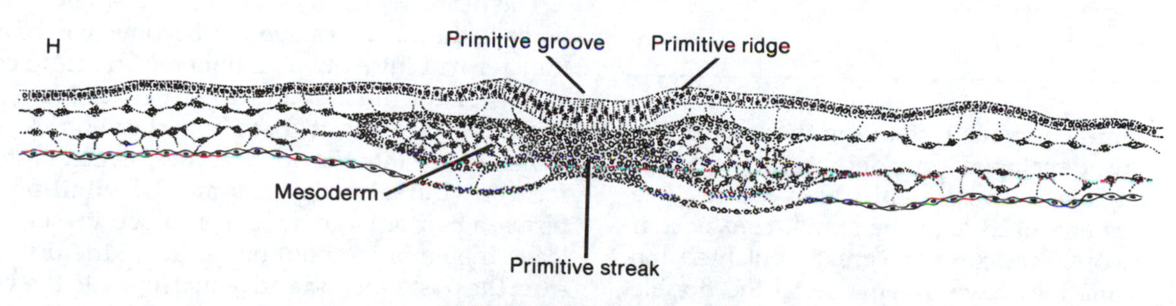

FIGURE 38-9 (*Continued*)

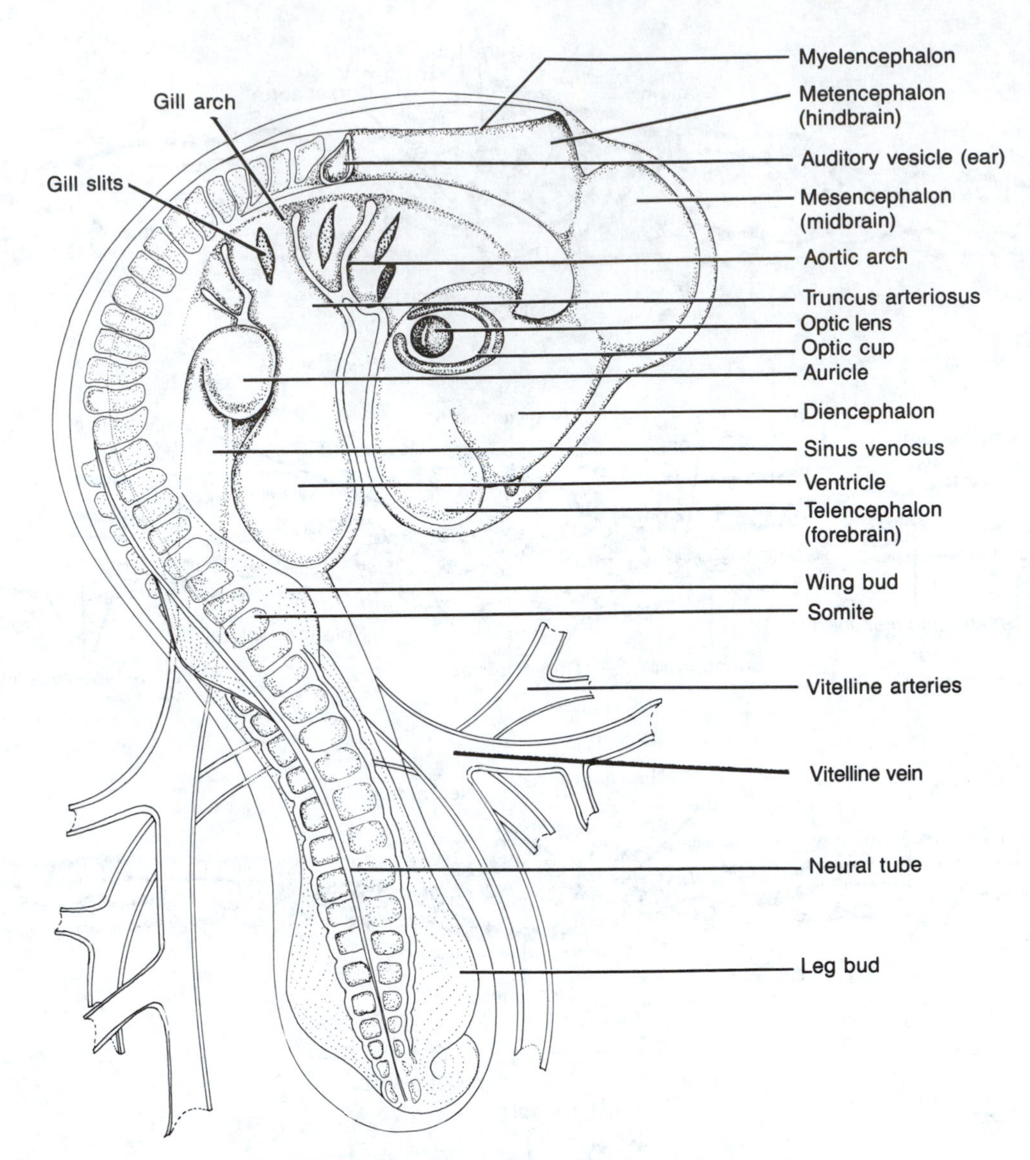

FIGURE 38-10 Dorsal view of 72-hour chick embryo

Note the three pairs of **gill slits** and the blood vessels (the **aortic arches**) that pass through the gill arches. What is the evolutionary significance of the appearance of these structures in the embryos of higher vertebrates?

____

The central nervous system and the sense organs have now developed considerably beyond their 33-hour stage. The neural tube has become a five-part brain anteriorly and the spinal cord posteriorly. The optic vesicles have greatly enlarged and, by invaginating, have given rise to the double-walled **optic cups.** The surface ectoderm of the head immediately over the optic cups has invaginated, thickened, and detached to form the optic lens. Dorsal to the gill slits on each side is an invaginated thickening of the ectoderm forming the **auditory vesicle** (ear).

The heart, which was tubular and single-chambered in the 33-hour stage, has become twisted and transformed into a two-chambered structure consisting of an **auricle** and a **ventricle.** Note the **truncus arteriosus;** it arises from the ventricle and branches out into the aortic arches, which unite to form the **dorsal aortae.** Locate the **vitelline arteries,** a pair of large transverse blood vessels that leave the embryo about one-third of the distance from the posterior end and branch out into the **yolk sac,** and the **vitelline veins,** which enter the auricle.

E. EXPERIMENTAL SURGERY ON CHICK EMBRYOS

Surgical procedures have been carried out on young chick embryos to determine the effects on subsequent development of altering the structure of various parts or of completely removing structures. The following study is an introduction to the fundamental techniques used in such studies.

1. Surgical Removal (Extirpation) of Leg Limb Bud

a. Preliminary Procedures

Thoroughly wipe the work area with a disinfectant and assemble the following equipment and instruments.

Several 72- to 84-hour eggs

Stereoscopic microscope with spotlight illuminator

Egg candler

Styrofoam or cotton egg holder

Sharp hacksaw blade

Single-edge razor blade

Clean Pasteur pipets

Beaker containing 70% ethyl alcohol for sterilizing instruments

Watchmaker's forceps

Microsurgical needle

Absorbent cotton

Sterile chick Ringer's solution

Alcohol lamp

Curad Tape (flesh color)

Note: When you use an instrument, remove it from the beaker of 70% ethanol and flame off the alcohol by lighting it with an alcohol lamp. Flaming completes the sterilization of the instrument. The microsurgical needle, however, should not be kept in alcohol. To sterilize this needle, merely pass the tip through the alcohol flame.

1. Place an egg that has been incubated for 72–84 hours on the egg candler to determine that the egg contains an embryo. Your instructor will demonstrate the procedure for candling eggs.

2. Place the egg on its side in a styrofoam or cotton holder. Wipe the upper surface of the egg with cotton moistened with 70% alcohol. After the surface has dried, use a hacksaw blade to make three cuts in the shell near the blunt end, as shown in Fig. 38-11A. Place your fingertip against the edge of the blade to prevent it from sliding around while sawing. Stop cutting when the blade "catches" rather than cuts smoothly.

3. Using a razor blade that has been flamed (passed through the flame of an alcohol lamp or Bunsen burner), lift up the shell flap as shown in Fig. 38-11B. If you cannot lift the flap, make the cuts a little deeper with the hacksaw blade.

4. Remove the shell flap to expose the shell membrane (Fig. 38-11B). Pipet a small volume of sterile chick Ringer's solution onto the membrane to wash away the debris left from sawing. Gently roll the egg back and forth to break any adhesions of the shell membrane to the membranes covering the embryo. Then, with your forceps held at an oblique angle (to prevent damaging the embryo), lightly puncture and grasp the shell membrane and strip it away as shown in Fig. 38-11C. This procedure makes a "window" that exposes the embryo (Fig. 38-11D).

5. Adjust the spotlight illuminator of the stereoscopic microscope to provide maximum illumination of the embryo. In doing this you may notice a reflection from the vitelline membrane, which surrounds the embryo, but do not be concerned if you do not see it. This membrane is easily ruptured and its absence will pose no problems during the surgical procedures.

6. After you have completed these preliminary procedures, cover the window with Curad tape to prevent drying. Return the egg in its holder to the incubator until you are ready to use it.

b. Removal of the Leg Limb Bud

Remove the Curad tape from the egg to expose the embryo. Using a stereoscopic microscope, locate the leg limb bud and, with a microsurgical needle or watchmaker's forceps, cut away a substantial part of it (Fig. 38-12). It may be necessary to use both instruments to hold the limb bud steady. Mark the date and procedure on the egg with a pencil. Cover the window with tape and replace the egg in the incubator. Record the results of your procedure using drawings and written observations. Eggs should not be opened until about 2 weeks after removal of the limb bud. At that time,

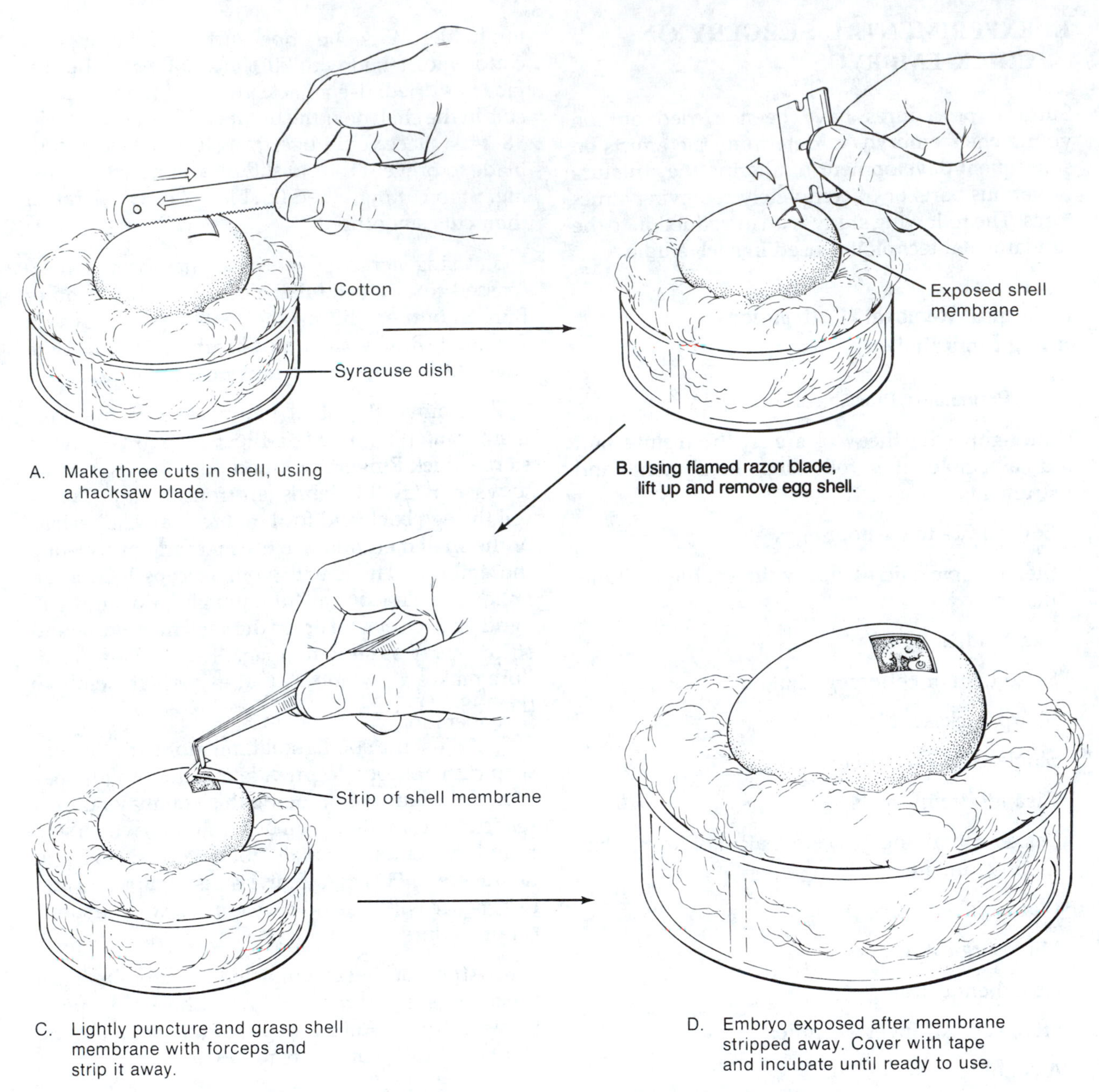

FIGURE 38-11 Procedure for cutting window in egg shell and exposing embryo

check for any deficiencies in growth or abnormal development.

After you have become proficient at this procedure, you might want to determine the effects of other such operations.

- Removing various amounts of limb tissue
- Removing a limb and replacing it upside down
- Removing a limb and implanting it on a different part of the embryo
- Placing carbon particles on various areas of the limb and—by observing the movement of the particles for a period of time—determining the fate of that particular area (e.g., Does it develop into the digits, the radius and ulna of the lower arm, the humerus of the upper arm, etc.?)

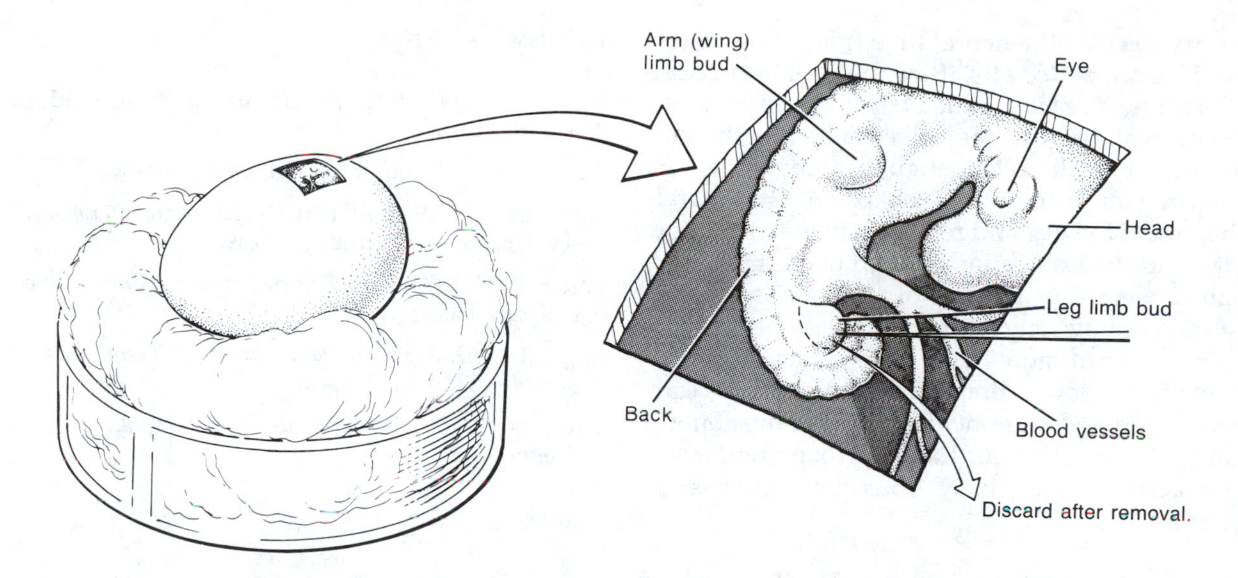

FIGURE 38-12 Removal of right limb bud from 72-hour embryo

2. Disruption of the Neural Tube

Normal development of the spinal cord depends on the neural tube remaining intact and closed. In a normal 72-hour chick embryo, the tube is closed. If it fails to close, a condition known as **spina bifida** results. The severity of this condition ranges from minor vertebral malformations to a severe deformation in which the unprotected spinal cord protrudes from the surface of the back. This is a common birth defect in human beings.

In this study, you will attempt to induce spina bifida experimentally by surgically cutting open the posterior region of the neural tube. In a 72-hour chick embryo, such an opening does not repair itself, and normal structural relationships are permanently disrupted. Be especially careful during this operation because a major blood vessel, the dorsal aorta, lies a short distance under the neural tube. Puncturing this blood vessel kills the embryo. Some damage to smaller vessels can be tolerated, but if you notice large amounts of spurting blood, discard the egg.

Using a microsurgical needle with the tip bent at an angle, cut open a small section of the neural tube at the level of the leg limb buds (Fig. 38-13). To do

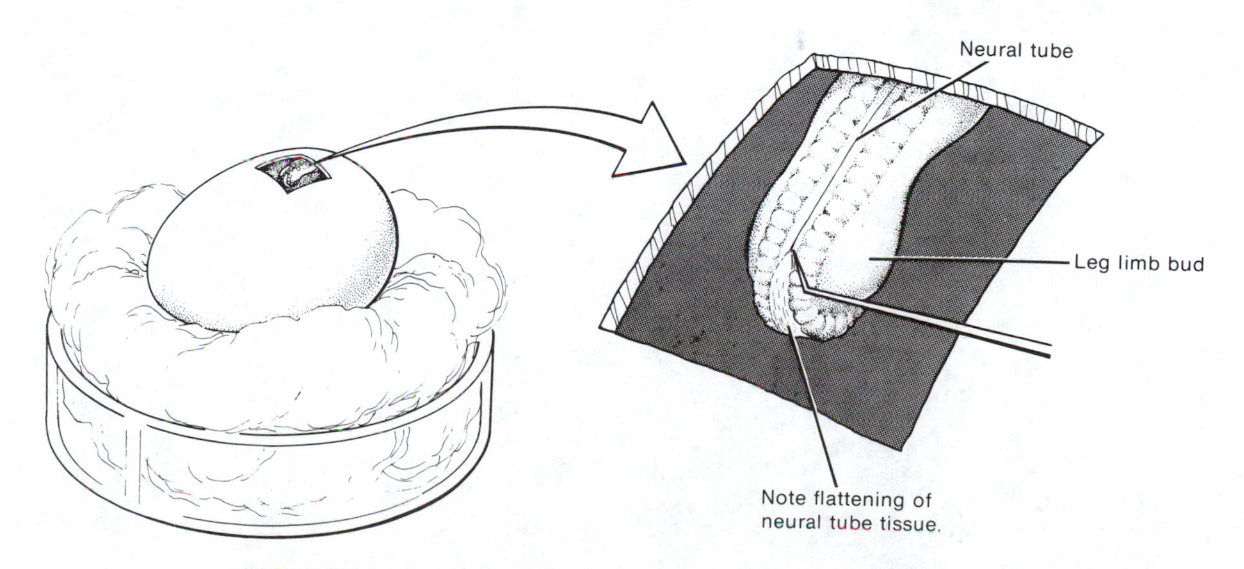

FIGURE 38-13 Disruption of neural tube in 72-hour embryo

this, try to enter the neural tube from its posterior end. This can be done by lifting the tip of the needle as it enters the neural canal. If you are successful, the side walls of the tube will collapse and flatten, and the tube will no longer appear tubular.

When you observe the collapse of the neural tube, reseal the egg and replace it in the incubator in its holder. Observations should not be made for about 2 weeks. At the end of this time, open the eggs and remove any feathers that may be present so that you can more easily observe the extent of the malformation. If time permits, the embryo can be dissected to determine the nature of the abnormality of the spinal cord and surrounding areas. Keep accurate records of your procedures and observations.

REFERENCES

Browder, L. W. 1991. *Developmental Biology.* 3d ed. Saunders.

Dale, B. 1983. *Fertilization in Animals.* Arnold.

Hamburger, V. 1960. *A Manual of Experimental Embryology.* University of Chicago Press.

Patten, B. M. 1971. *Early Embryology of the Chick.* 5th ed. McGraw-Hill.

Rugh, R. 1962. *Experimental Embryology: Techniques and Procedures.* 3d ed. Burgess.

Trinkaus, J. P. 1984. *Cells into Organs.* 2d ed. Prentice-Hall.

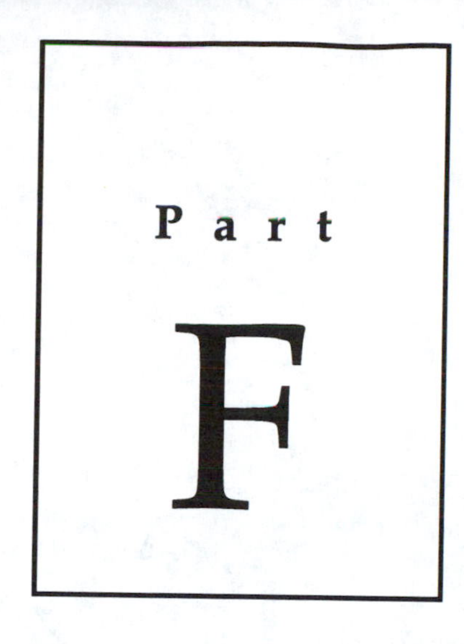

Environmental Biology

By 1993 the world's population reached 5.5 billion human beings, with 94 million new births that year alone. Although the human population growth rate has been rather slow for most of history, it has dramatically increased in the past 200 years. Whereas there were some 1 billion people on earth in 1850, almost 8 billion are projected for the year 2050. How did we ever get into such a predicament?

Humans, by virtue of their large, complex brains, have been able to expand into new environments. Their ability to learn and adapt has resulted in changes in characteristic existence: people have moved from being largely vegetarians to hunter–gatherers to communities who learned to build fires, assemble shelters, create clothing, and devise tools. These abilities ultimately led to the industrial and scientific revolutions, which brought about the dramatic population expansion.

As people became more educated and knowledgeable, they began to side-step factors that placed natural limitations on growth. For example, up to about 300 years ago, poor hygiene, lack of proper sewage disposal, malnutrition, and contagious diseases (spread by fleas and rats) kept the death rate at a high enough level to balance the birth rate. Then plumbing and sewage treatment facilities were developed, antitoxins, vaccines, and antibiotics were discovered, and medical research greatly extended human life. The result was a dramatic increase in life span and drop in death rate. Births now exceeded deaths and the population began growing at a rapid rate.

Modern technology, as represented by the following laboratory studies, has played an important part in analyzing the environment for various factors detrimental to our health and well-being. In a way, we are perpetuating our own population problems.

Exercise

39

Analysis of Surface Water Pollution by Microorganisms

In the clearing of land and the development of the United States, particularly the eastern areas, changes in ground cover and surface soils have greatly altered surface water runoff and ground seepage. Soil seepage, surface runoff and erosion have been further impaired by deforestation, fires, overgrazing, and improper drainage. As a result, during the past 50 years the water table in the eastern half of the country has been lowered some 60 feet. The lowered water table, coupled with an increased use of ground water, has created severe water shortages.

As population and industrial demands increase and ground water supplies become more and more inadequate, communities are turning to surface waters, such as rivers, lakes, streams, or reservoirs, for their water supplies. This change to using surface water has created a host of problems for those engaged in procuring and treating water for domestic use. While ground water is generally free of bacteria and other nuisance organisms such as algae, surface water frequently contains numerous microorganisms that interfere with its domestic use. Many of these nuisance organisms are algae that affect the odor and taste of water, clog filters, grow in water pipes, cooling towers, and reservoir walls, produce luxuriant mats or "blooms" that interfere with other aquatic life, or produce toxic substances. In addition, especially in heavily populated areas, the pollution of surface waters by sewage and animal feces is an ever-present hazard. Several diseases have been traced to polluted drinking water, among them typhoid fever and a group of intestinal disorders generally referred to as dysentery.

The following procedures should help you determine the kinds of algae present and measure the degree of algae pollution in water samples.

A. BACTERIAL CONTAMINATION OF WATER SUPPLIES

Health authorities routinely check for the presence of certain **coliform bacteria** that act as pollution "indicators." Coliform bacteria are **enteric organisms:** they are normally found in the intestines of humans and other animals. Even healthy people normally have some coliform bacteria in their intestines, which may be introduced into "raw" sewage water in excreted feces. Not only are coliform bacteria always present in sewage, but they are always found in the presence of disease-producing bacteria such as *Salmonella.*

Another reason that coliform bacteria are used as indicators of pollution is that they are very resistant organisms and are harder to kill than most disease-producing bacteria. Thus, if these organisms are absent from the water "test sample," one can be reasonably sure that no disease-producing bacteria are present, either.

Properly treated tap water should be free of coliform bacteria, while raw or untreated water probably would contain an appreciable number. In this part of the exercise, you will isolate coliform bacteria from a selected water sample by trapping them on the surface of a membrane filter that you then culture on a selective growth medium. This procedure has been developed by the Millipore Corporation.

In addition to promoting the growth of coliform bacteria, the Millipore MF-Endo medium used is also selective: it discourages the growth of most other species of bacteria. This is a significant help in this test because "raw" water sources contain hundreds of species of microorganisms that have nothing to do with pollution and would otherwise completely overgrow the test filter and mask the presence of any coliform colonies.

How does MF-Endo medium work? Coliform bacteria have the ability to break down a complex "sugar" called lactose, forming a number of simpler substances, among which are a group of chemicals known as aldehydes. Endo-medium contains lactose, other nutrients, and a stain called basic fuchsin, which reacts with aldehyde molecules to form a complex that appears as a shiny green coating on the colony of coliform bacteria. Since few microorganisms (other than members of the coliform family) make aldehydes out of lactose, the "green sheen" colonies are quickly identified as coliform bacteria.

The mere presence of coliform bacteria does not mean that the water is "polluted," however. Keep in mind that all surface water supplies are subject to animal excretion as well as seepage from soil. What is important in determining the pollution level is the number of coliform bacteria found in the water sample. When health officials find that the number of coliform bacteria exceeds the standards set for particular areas and types of water, they then assume that disease-producing bacteria are also present. Additional tests must then be performed to isolate and culture the suspected pathogens.

1. Determining Number of Coliform Bacteria in a Water Sample

The following procedure is similar to the coliform test performed in health department laboratories.

1. Sterilize a Millipore Sterifil apparatus (Fig. 39-1) in an autoclave or by immersing it in boiling water for 3 minutes. If this is not practical, such as in a field experiment, dip the Sterifil apparatus in 70% alcohol for a few minutes, shake it off thoroughly, and let it air dry.

2. Using sterilized forceps, center a type HA (0.45 mμ) membrane filter on the filter support. Then screw the filter funnel onto the support so that the O ring seals the filter in place (Figs. 39-2 and 39-3).

(Caution: *Do NOT overly tighten the filter or it will become wrinkled, leading to erroneous results.*)

3. To the Sterifil funnel, add about 20 ml of room-temperature sterile water or tap water that has been boiled for several minutes. The exact volume added is not critical. Its only purpose is to disperse the bacteria present evenly in the measured sample.

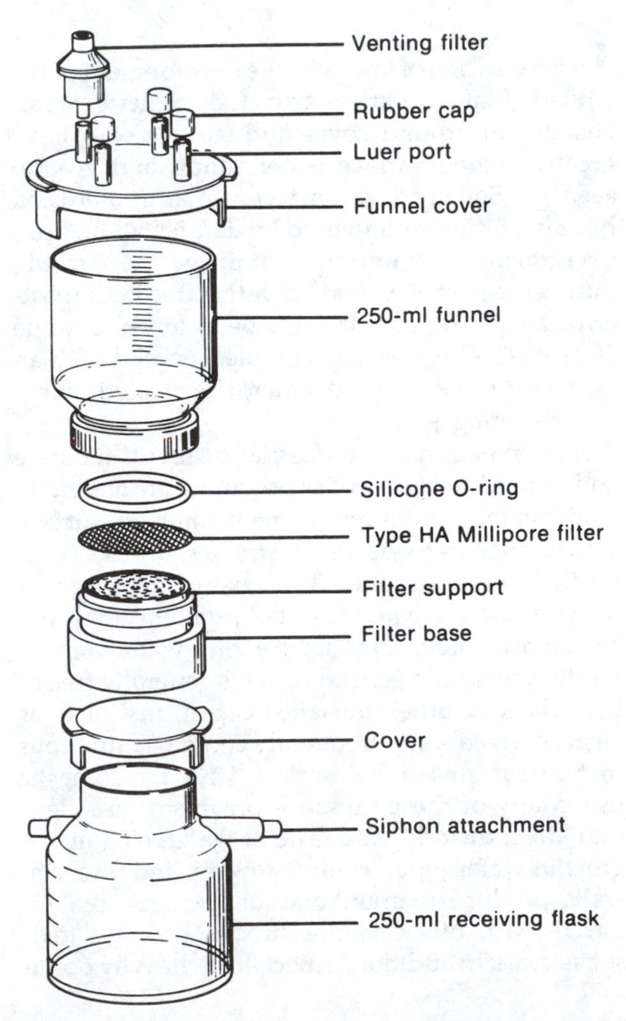

FIGURE 39-1 Components of Millipore Sterifil apparatus

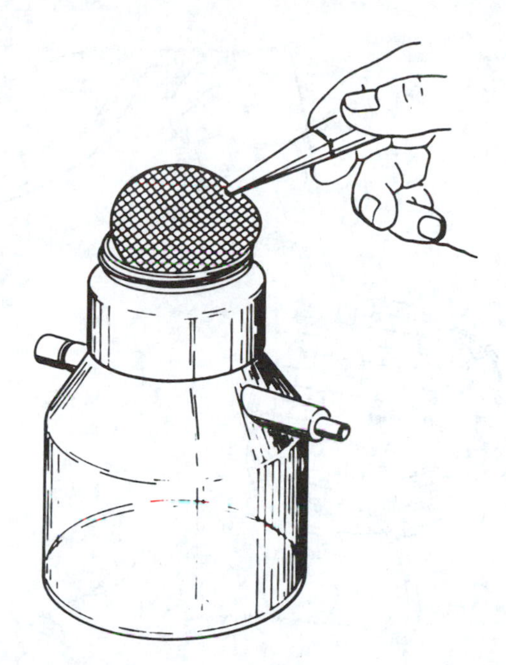

FIGURE 39-2 Placement of Millipore filter on filter support

4. Pipet a known volume of the sample water from a pond or stream into the Sterifil funnel. Attach the cover and then gently swirl the funnel to mix the sample with the sterile dilution water. The size of the sample will vary with the contamination level of the water being tested.

To determine proper sample size, start with 1 ml. After the first cultures have grown, you will be able to determine whether a larger or smaller sample size should be used. You should adjust the sample size so that you have no more than 20 to 80 coliform colonies growing on the surface of the filter (Fig. 39-4). The total of all colonies, including coliform, should not exceed 200.

Alternately, you can run the first experiment using 0.1, 0.5, 1, and 2 ml simultaneously to determine the optimum sample size, or you can prepare a serial dilution of the undiluted sample (i.e., 1×10^{-1} to 1×10^{-6}) and test each dilution at the same time.

5. Using a hand vacuum pump or some other suitable vacuum source (such as a water aspirator), slowly apply vacuum to the receiver flask (Fig. 39-5). This will result in the water in the top funnel being pulled through the filter, leaving the bacteria trapped on the surface of the filter.

6. While waiting for filtration to be completed, place a sterile absorbent pad into a 47 mm petri dish. Then pour contents of ampule (containing endo-medium) onto pad (Fig. 39-6) and replace cover on dish.

7. After filtration, release the vacuum by first removing the vacuum pump tubing from the side arm of the receiver flask. Then unscrew the funnel and, using alcohol-dipped forceps, remove the filter and place it, grid side up, on the endo-medium-saturated absorbent pad in the petri dish (Fig. 39-7). Carefully line up the filter with one edge of the petri dish and gently lower it over the pad so that it is evenly centered. This will prevent

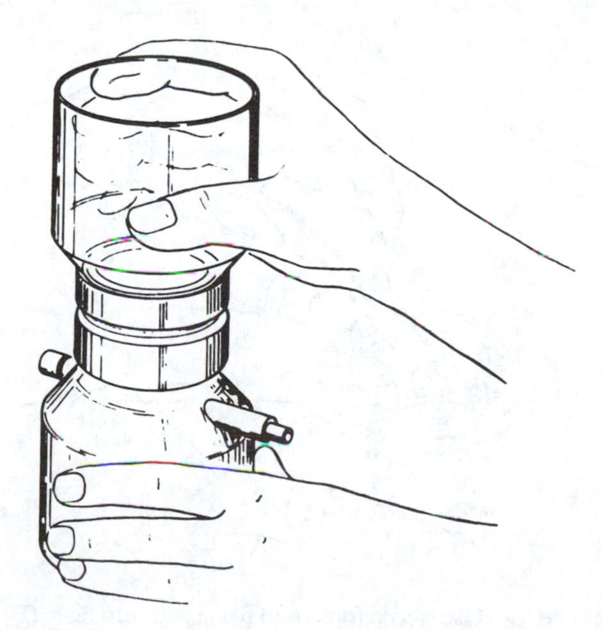

FIGURE 39-3 Assembling the filter apparatus

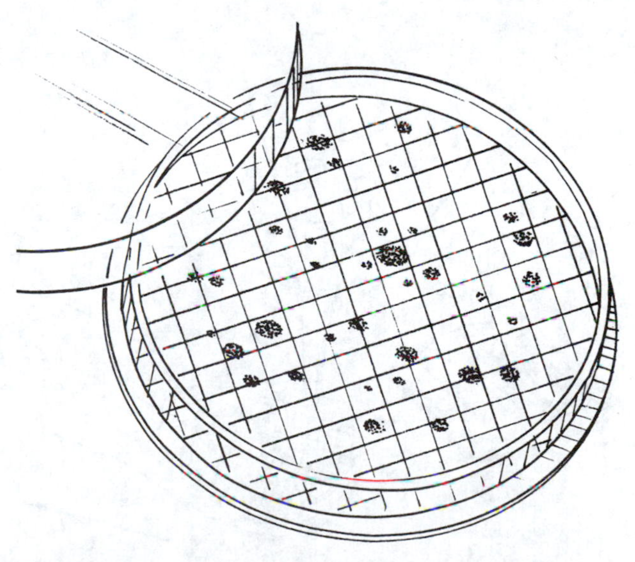

FIGURE 39-4 Appearance of coliform colonies on Millipore filter following incubation. Original sample size should be adjusted to obtain between 20 and 80 colonies.

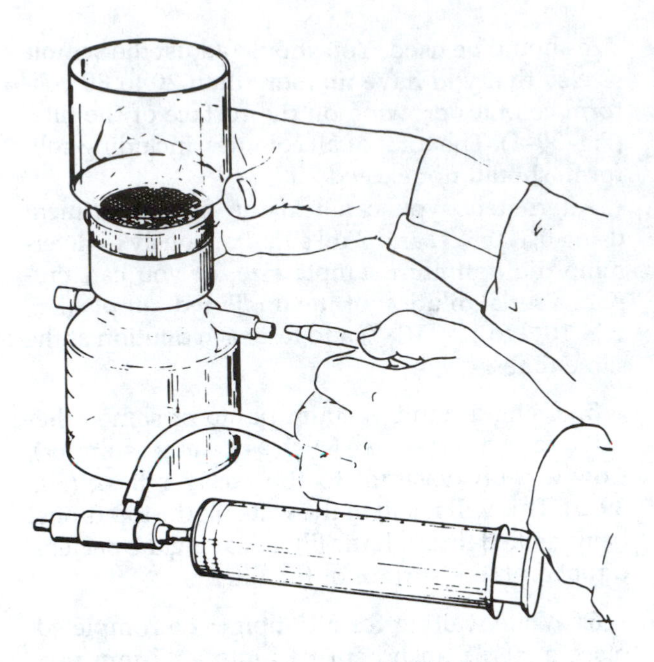

FIGURE 39-5 Method of evacuating filter apparatus using hand vacuum (25–50 ml syringe)

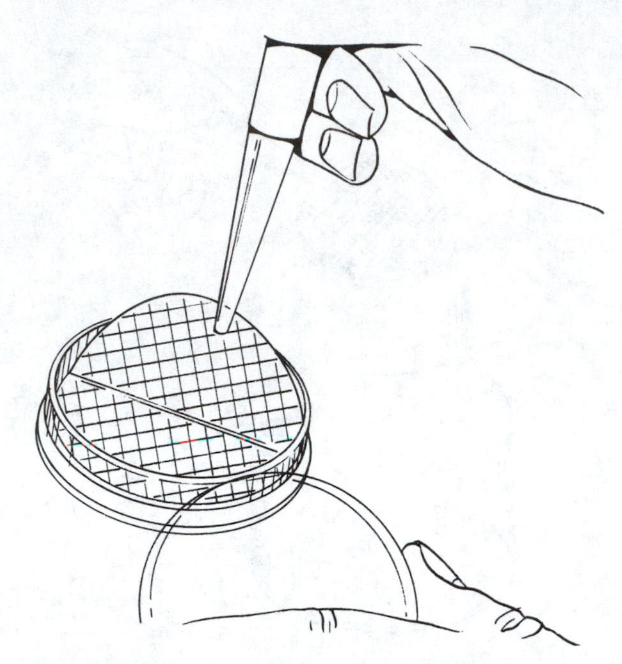

FIGURE 39-7 Placement of filter onto surface of pad saturated with endo-medium

air from being trapped under the filter. In terms of the validity of the results, why is it important not to trap any air bubbles under the filter?

____

____

____

8. Replace the cover, invert the petri dish, and incubate for 48 hours at room temperature, or 24 hours at 37°C in an incubator.

If the filter has been placed on the pad properly, the medium, as it passes up through the filter, will supply all areas of the filter with nutrients necessary for growth of the microorganisms on its surface. It is important to invert the petri dish because moisture often condenses on the inside surface of

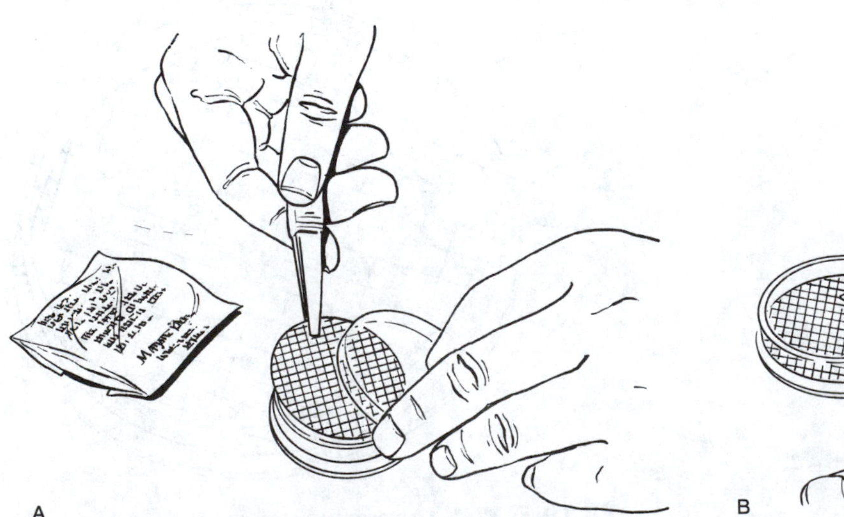

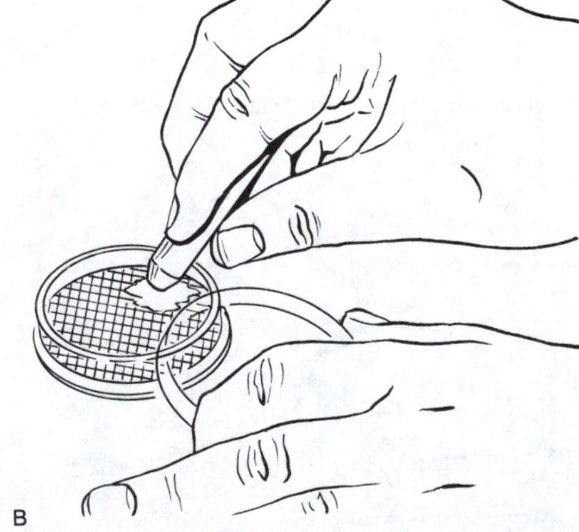

FIGURE 39-6 Preparation of filter dish with sterile absorbent pad. (A) Use sterile forceps to place pad into dish. (B) Pour contents of ampule (containing endo-medium) onto pad.

the cover during incubation. If the dish were set right side up, droplets could form that might fall onto the surface of the filter, causing the colonies to run together.

After colonies develop (Fig. 39-4), remove the filter from the petri dish and allow it to dry on a clean blotter or paper towel for 30 minutes.

9. With a hand magnifier or stereoscopic microscope, examine the surface of the filter for colonies having a shiny, greenish surface. Count the total number of "green sheen" colonies on the filter.

2. Calculations

To determine the number of coliform bacteria present in the water sample, the following formula may be used.

$$\frac{\text{no. of coliform colonies counted} \times 100}{\text{milliliters of sample added to funnel}} = \text{no. of coliform per 100 ml}$$

For example, a 1-ml sample was added to the filter funnel containing approximately 20 ml of sterile water. After incubation, 60 "green sheen" colonies were counted on the filter. Therefore, $60 \times 100/1 = 6000$ coliform bacteria per 100 ml.

As indicated earlier, coliform standards vary from one community to another. Check with your local health department to determine the standards for your area for different types of water (i.e., untreated water, well water, drinking water, and public swimming areas).

B. ALGAE IN WATER SUPPLIES

1. Determining Number of Algae in Surface Water

It has been shown that algal populations numbering over 1000 organisms per milliliter indicate that the water supply is "overenriched" from sewage or other nutrients. Use the following procedure to quantify the number of algae in a water sample.

a. Collect a sample of water (500 ml to 1 liter) from a pond, pool, or body of free-standing surface water. Alternately, collect some pool water and add some scrapings from the side of the pool.

b. Using a 47-mm Millipore filter in the filter apparatus, set up the apparatus as shown in Figures 39-1, 2, and 3 and add 250 ml of the water sample to the upper funnel. Replace the cover on the funnel.

c. Apply a vacuum to the receiver flask using a hand vacuum pump, syringe, or water aspirator (Fig. 39-5). This will cause the fluid in the upper funnel to be pulled through the filter, leaving the algae trapped on the surface of the Millipore filter.

d. After filtration, release the vacuum in the system by removing the vacuum pump tubing from the side arm of the flask. Then unscrew the top funnel and, using forceps, remove the filter and place it on a clean paper towel to air dry for 15–20 minutes. Alternately, the filter can be quickly dried in a 45°C oven for a few minutes. (*Note:* If you observe that the filter is heavily coated, you may have to reduce the sample size from 250 ml to 10, 25, or 50 ml.)

2. Counting Number of Algae in Sample

a. Float the dried filter on about 5 ml of room temperature immersion oil in a 47- or 60-mm petri dish (Fig. 39-8). Place the filter into the immersion oil to prevent air bubbles from becoming trapped under the filter. If the filter is thoroughly dry, it will become transparent as it absorbs the oil. If the filter remains opaque, place the petri dish and filter onto a warm surface, or into a 37°C incubator, until it clears.

b. Remove the filter from the dish by drawing it across the edge of the petri dish. This will remove excess oil from the surface of the filter.

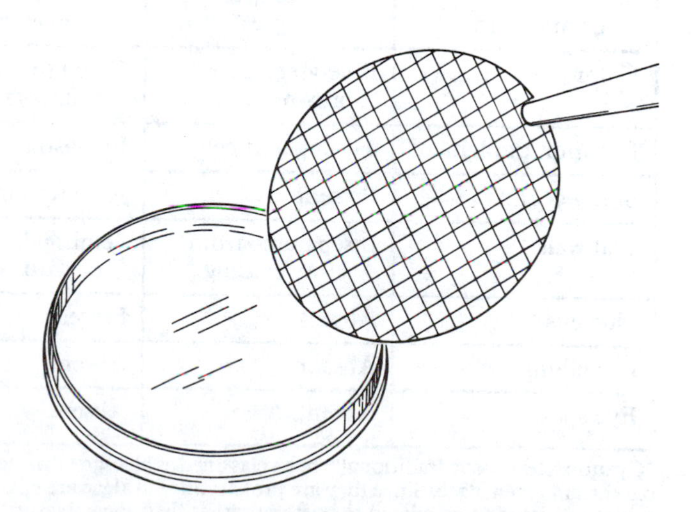

FIGURE 39-8 Placement of filter into petri dish containing immersion oil

c. Place the filter on a 2-inch by 3-inch microscope slide. Using low magnification (10×), count the number of algae in each of ten randomly selected fields of view and calculate the number of algae in a water sample using the following formulas.

$$\text{(1)}\quad \frac{1380\ \text{mm}^2}{\text{area of field (mm}^2\text{)*}} \times \text{number of fields counted} = \text{"factor"}$$

$$\text{(2)}\quad \text{Total number of algae counted} \times \text{"factor"} = \text{number of algae in sample}$$

or calculate the number of algae per milliliter of sample as follows.

$$\frac{\text{number of algae in sample}}{\text{volume of sample filtered (ml)}} = \text{number of algae per milliliter}$$

*The area of the field can be calculated by using a stage micrometer or, more simply, by laying a transparent metric rule on the stage and measuring the diameter of the field. The area can be calculated using the formula πr^2. The value of 1380 mm^2 represents the total area of a 47-mm Millipore filter used in the Millipore Sterifil apparatus.

3. Identification of Common Nuisance Algae

In addition to the number of algae present in a water sample, another indication of water quality is the kinds of algae present. It is known, for example, that when water becomes polluted there is a sharp decline in green algae and diatoms and an increase in the numbers of cyanobacteria (blue-green algae) and flagellates.

Obtain samples of various waters you suspect of having algal pollution. Examine the samples microscopically. Use Table 39-1 and Figures 39-9 to 39-12 to help you identify various algae. Once you have the name of an alga, use Table 39-2 to determine the group to which it belongs [e.g., green algae, cyanobacteria (blue-green algae), diatoms, or flagellates]. You can then use this information to help you determine whether the water is polluted, recalling that in polluted water there is an increase in various algae.

TABLE 39-1 Comparison of the Four Major Groups of Algae in Water Supplies

	Algal groups			
Characteristics	Cyanobacteria (blue-green algae)*	Green algae	Diatoms	Pigmented flagellates
Color	Blue-green to brown	Green to yellow-green	Brown to light green	Green or brown
Location of pigment	Throughout cell	In plastids	In plastids	In plastids
Slimy coating	Present	Absent in most	Absent in most	Absent in most
Cell wall	Inseparable from slimy coating	Semirigid, smooth, or with spines	Very rigid, with regular marking	Thin, thick, or absent
Nucleus	Absent	Present	Present	Present
Flagellum	Absent	Absent	Absent	Present
Eye spot	Absent	Absent	Absent	Present

*Cyanobacteria have traditionally been classified as blue-green algae because they perform photosynthesis and they generally resemble certain green algae. Since they are prokaryotic (all algae are eukaryotic) and have certain other bacterial characteristics, they are now considered to be related to bacteria rather than algae. Nevertheless, they play a significant role in surface water pollution.

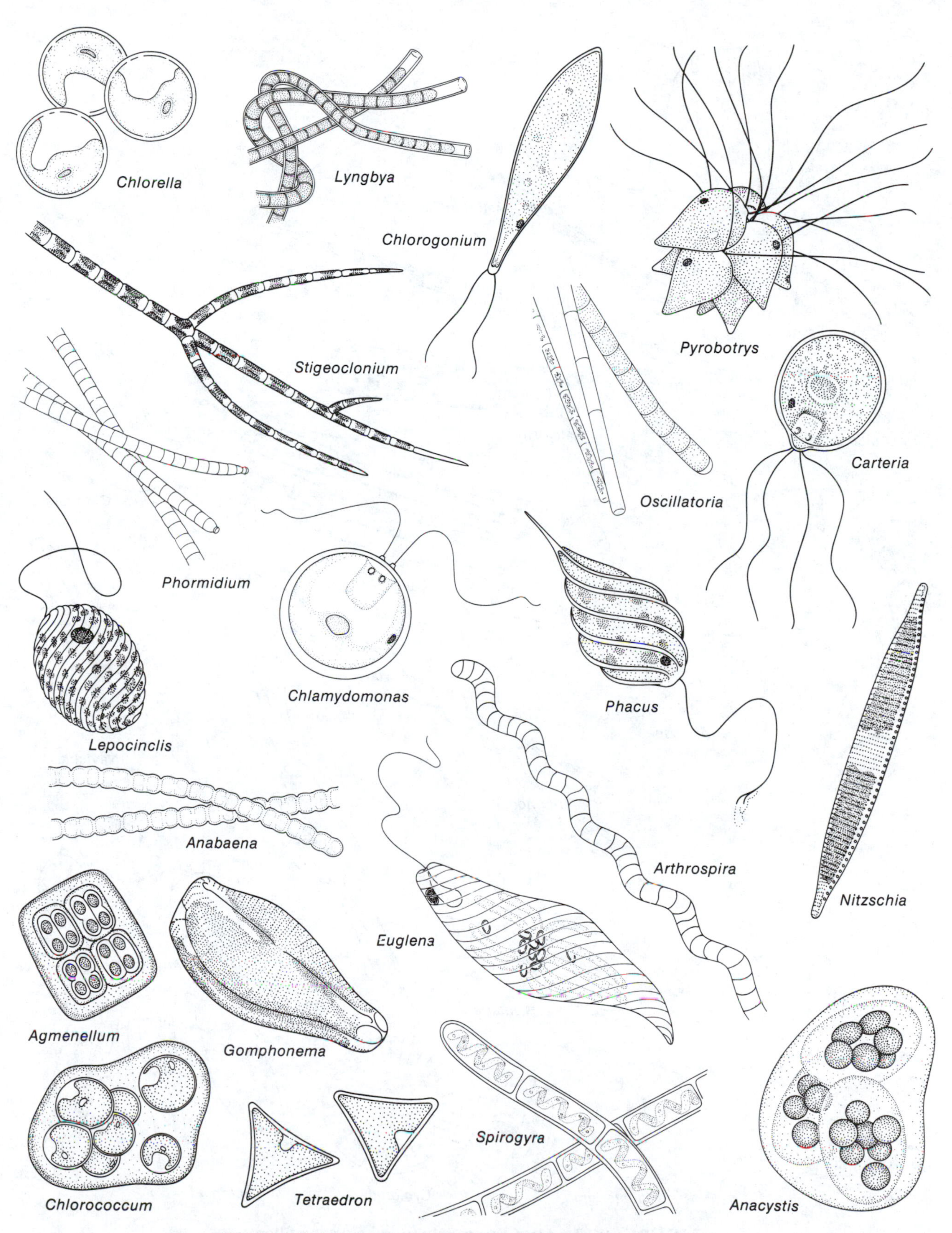

FIGURE 39-9 Common algae found in polluted surface waters

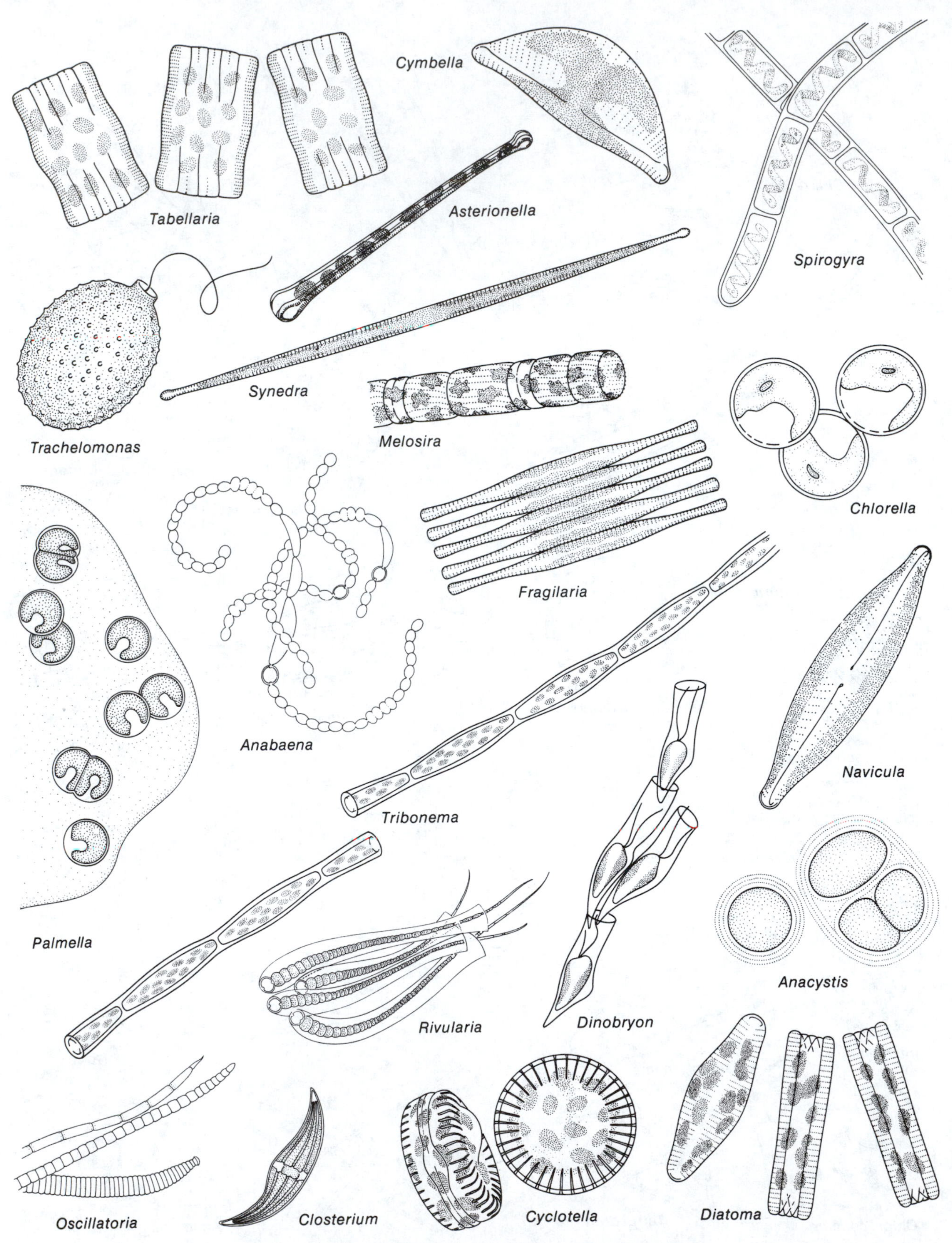

FIGURE 39-10 Algae commonly found clogging filters

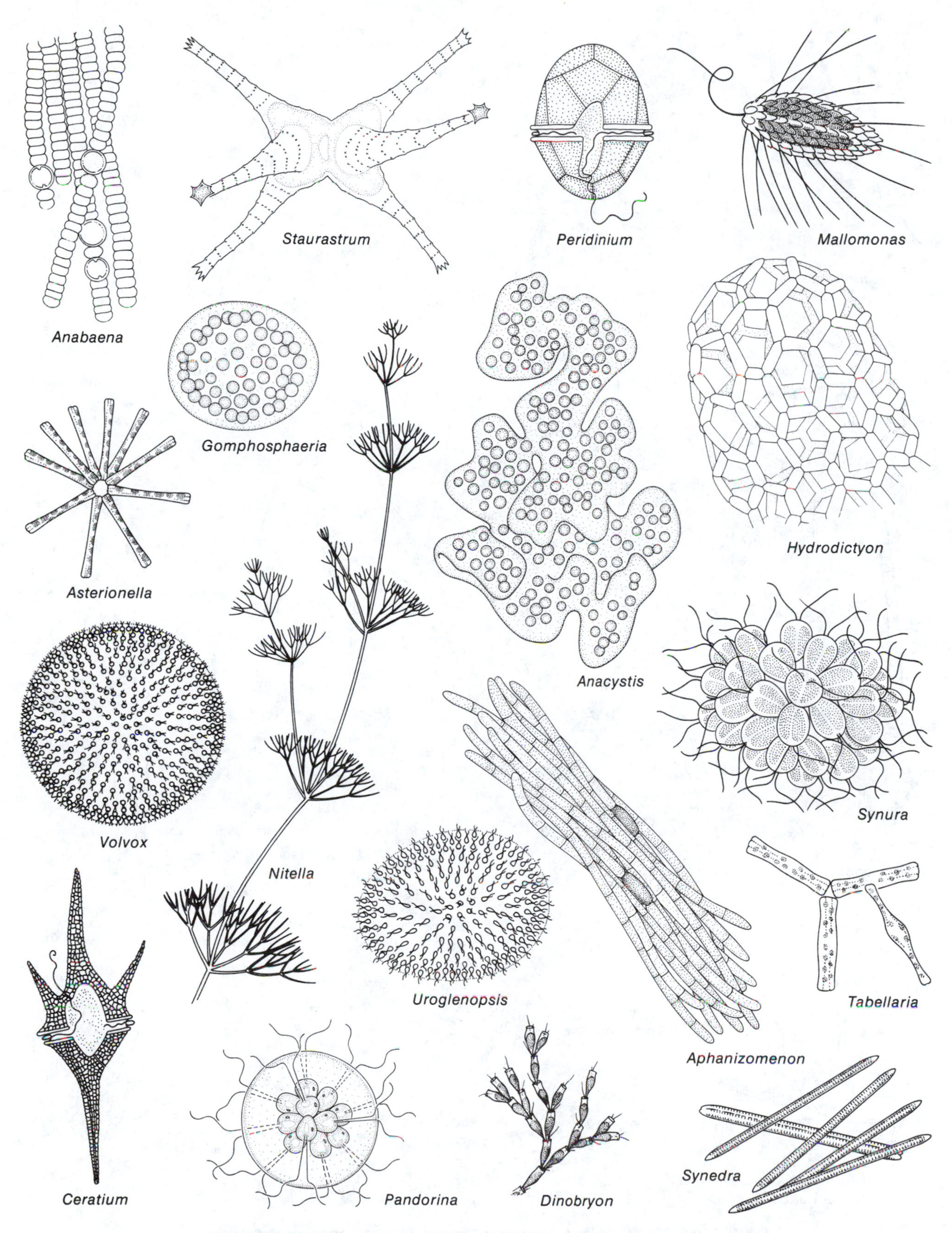

FIGURE 39-11 Algae contributing to taste and odor of water supplies

FIGURE 39-12 Algae commonly found growing on reservoir walls

TABLE 39-2 The More Important Algae Contributing to Surface Water Pollution

Key to columns:
1. Algae name
2. Group: D, diatoms; G, green; BG, cyanobacteria (blue-green); R, red; Fl, flagellates; De, desmids; YG, yellow-green.
3. Characteristics: A, attached; S, surface; F, filter; C, clean; P, polluted; T, taste and odor.

1	2	3	1	2	3
Achnanthes	D	A	*Mallomonas*	F1	C,T
Agmenellum	BG	C	*Melosira*	D	S,F,P
Anabaena	BG	T	*Microspora*	G	A
Anacystis	BG	T,F,P,S	*Navicula*	D	P,C,F,S
Aphanizomenon	BG	T	*Nitella*	G	A,T
Arthrospira	BG	P	*Oedogonium*	G	A,S
Asterionella	D	F,T	*Oscillatoria*	BG	F,P
Audouinella	R	A	*Palmella*	G	A,F
Batrachospermum	R	A	*Pandorina*	F1	P,T
Bulbochaete	G	A,C	*Pediastrum*	G	S,T
Carteria	F1	P	*Peridinium*	F1	T,F
Chaetophor	G	A	*Phacus*	F1	C
Chara	G	A,T	*Phormidium*	BG	P
Chlamydomonas	F1	T,P	*Phytoconis*	G	A
Chlorella	G	F,P	*Pyrobotrys*	F1	P
Chlorococcum	G	P	*Rivularia*	BG	F,T
Chlorogonium	F1	P	*Spriogyra*	G	P
Cladophora	G	F,A,S,C,T	*Staurastrum*	De	T,S,C
Closterium	De	S,F	*Stigeoclonium*	G	A,S,P
Compsopogon	R	A	*Synedra*	D	F,C
Cyclotella	D	C,T,S,F	*Synura*	F1	T
Cybella	D	C,A,F	*Tabellaria*	D	F,T
Diatoma	D	F,T	*Tetraedron*	G	P
Dinobryon	F1	T,F,S,C	*Tetraspora*	G	A
Draparnaldia	G	A,C	*Tolypothrix*	BG	A
Euglena	F1	P,S	*Trachelomonas*	F1	F
Fragilaria	D	S,T,F	*Tribonema*	YG	F
Gomphonema	D	A,P	*Ulothrix*	G	C,S,F,A
Gomphosphaeria	BG	S,T	*Uroglenopsis*	F1	T
Lepocinclis	F1	P	*Vaucheria*	G	A,C
Lyngbya	BG	P,A,S	*Volvox*	F1	T

REFERENCES

Biological Analysis of Water and Waste Water. 1973. AM-302. Bedford, MA: Millipore Corp.

Blum, J. L. 1956. The ecology of river algae. *Bot. Rev.* 22:291–341.

Clark, H. F., E. E. Geldreich, H. L. Jeter, and P. W. Kobler. 1951. The member filter in sanitary bacteriology. *Pub. Hlth. Rep.* 66:951–977.

Garney, P. L., and T. H. Lord. 1952. *Microbiology of Water and Sewage.* Englewood Cliffs, NJ: Prentice-Hall.

Ingram, W. M., and G. W. Prescott. 1954. Toxic freshwater algae. *Am. Midland Nat.* 52:75–87.

Lackey, J. B. 1941. Two groups of flagellated algae serving as indicators of clean water. *J. Am. Water Works Assoc.* 33:1099–1110.

Laubusch, E. J., E. E. Geldreich, and H. L. Jeter. 1953. Membrane filter procedure applied in the field. *Pub. Hlth. Rep.* 68:1118–1122.

Microbiological Analysis of Water. 1961. AR-81. Bedford, MA: Millipore Corp.

Palmer, C. M. 1959. *Algae in Water Supplies.* Public Health Service Publication No. 657, U.S. Government Printing Office, Washington, DC.

Standard Methods for the Analysis of Water and Waste Water. 1971. New York: American Public Health Association.

Exercise

40

Analysis of Solids in Water

Suspended solids found in water may come from a variety of sources. Some types of silt may come from soil erosion following forest fires, dredging, or landfills located near the water. Effluents from industrial operations also add sediment to water. For example, underground water contains large amounts of ferrous salts (e.g., ferrous hydroxide). During mining operations these salts are brought to the surface and then, through surface water runoff, find their way into streams and rivers where the ferrous salts react with oxygen dissolved in the water to form an insoluble suspension of ferric hydroxide. A high concentration of such suspended solids reduces the amount of sunlight reaching the underwater vegetation, thus reducing photosynthesis and the oxygen produced in this process.

In some instances the water may become so turbid that fish suffocate when their gills are clogged by the suspended particles.

A. PATCH TESTING FOR SUSPENDED SOLIDS

In this study, water collected from various sources is filtered through membrane filters. The amount of solids trapped on the surface of different filters can be compared as to relative amounts of suspended particles in each of the samples.

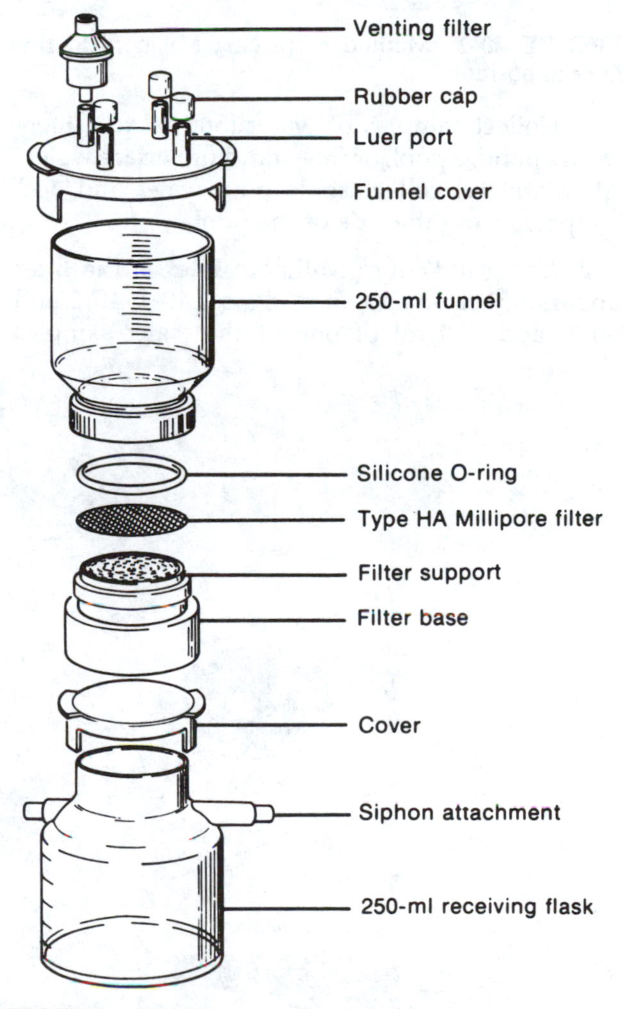

FIGURE 40-1 Components of Millipore Sterifil apparatus

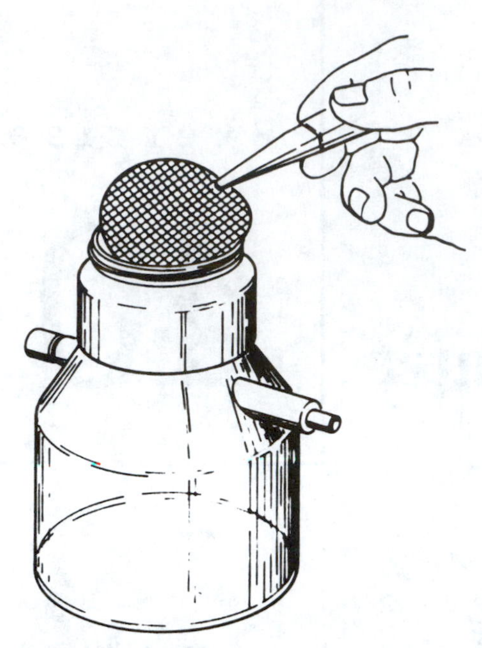

FIGURE 40-2 Method of placing Millipore Sterifil filter in position

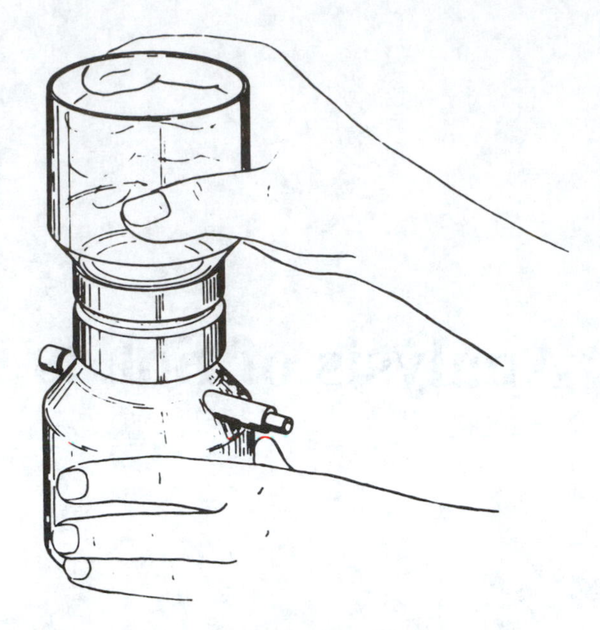

FIGURE 40-3 Assembling Sterifil filter apparatus

1. Collect samples of water (500 ml to a liter) from a pond, a pool, or free-standing surface water. Alternatively, collect some pool water and add scrapings from the side of the pool.

2. Using a 47-mm Millipore filter in the filter apparatus set-up shown in Figures 40-1, 40-2 and 40-3, add 250 ml of one of the water samples to the upper funnel. Replace the cover on the funnel.

3. Apply a vacuum to the lower receiver flask using a hand vacuum pump, syringe, or water aspirator. This will cause the fluid in the upper funnel to be pulled through the filter, leaving the solids trapped on the surface of the filter.

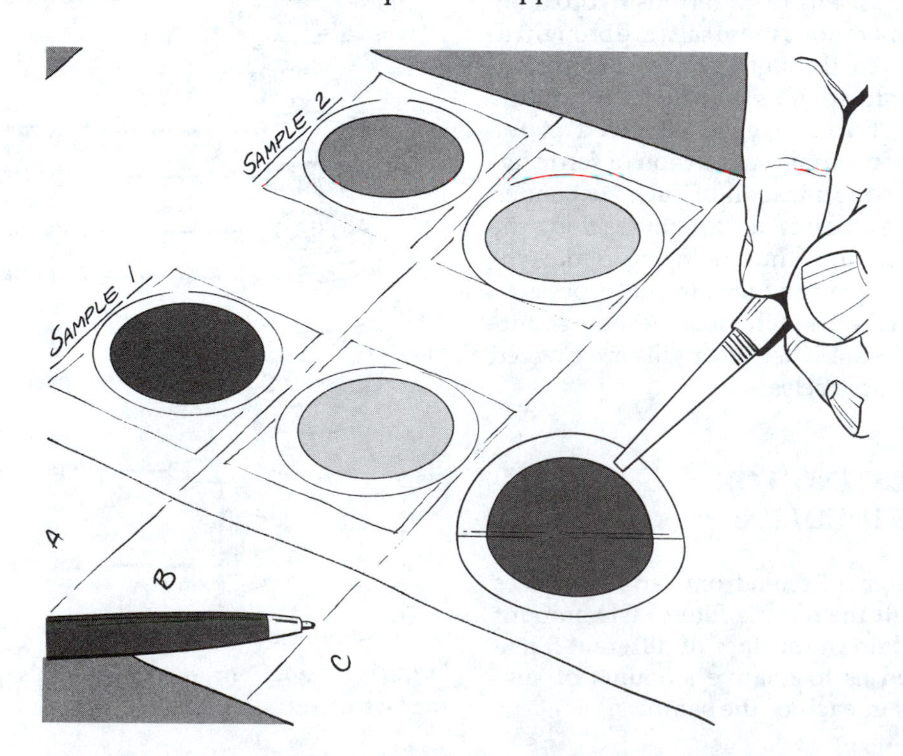

FIGURE 40-4 The difference in color or density of the various filters indicates the relative levels of collected samples. The amount of suspended solids from different samples can thus be compared.

4. After filtration, release the vacuum in the system by removing the vacuum pump tubing from the side arm of the flask. Then unscrew the top funnel and, using forceps, remove the filter and place it on a clean paper towel to air dry for 15–20 minutes. Alternatively, the filter can be quickly dried in a 45°C oven.

5. When dry, the filter can be mounted along with samples from other sources for comparison (Fig. 40-4).

6. Repeat steps 1–5 for each water sample.

B. GRAVIMETRIC ANALYSIS OF SUSPENDED SOLIDS

Gravimetric analysis essentially means to analyze by weight. Thus, if you have an accurate balance, you can determine the weight of the solids trapped on the surface of each filter as follows.

1. Weigh the dry filter before it is placed in the filter apparatus and again after the sample has been filtered and thoroughly dried.

$$\text{wt. of dry filter after test (mg)} - \text{wt. of dry filter before test (mg)} = \text{wt. of solids (mg)}$$

2. Determine the concentration of solids per liter of water sampled using the following formula.

$$\frac{\text{wt. of solids collected (mg)}}{\text{sample size in ml}} \times 1000 = \text{mg of suspended solids in 1000 ml}$$

REFERENCES

Engelbrecht, R. S., and R. E. McKinney. 1956. Membrane Filter Method Applied to Suspended Solids Determinations. *Sewage Works,* vol. 28, no. 11.

Sohn, B. I. 1969. Membrane Microfiltration for the Science Lab. *Sci. Teach.* 36(8):48–51.

Weatherford, R. L., and T. E. Larson. 1959. Preparation of Suspended Solids Samples for Radioactivity Counting. *Anal. Chem.* 31:11.

Part G

Appendices

Appendix

A

Use of the Milton Roy Spectronic 20® Colorimeter

The Milton Roy Spectronic 20® colorimeter is an extremely versatile instrument that is useful for spectrophotometric or colorimetric determinations of solutions.

The optical system found within this instrument is diagrammed in Fig. A-1. White light is focused by a field lens onto an entrance slit, where it is collected by a second (objective) lens and refocused on the exit slit after being reflected and dispersed by a diffraction grating. Rotation of this grating by a wavelength cam enables one to select various wavelengths of light in the range 340–600 nm. Addition of a filter can extend the usable wavelength to 950 nm. After the light passes through the exit slit, it goes through the sample being measured and is picked up by a measuring phototube. A mirrored meter indicates the amount of light absorbed by the sample.

A. COLORIMETRY

Directions for colormetric use are as follows (Fig. A–2).

1. Rotate the wavelength control until the desired wavelength appears on the wavelength dial. The wavelength for a given substance can be found by referring to the literature or by determining it experimentally.

2. Turn on the instrument by rotating the power switch/"0" control in a clockwise direction. Allow 5 minutes for the instrument to warm up.

3. Adjust the "0" control with the cover of the sample holder closed until the needle is at 0 on the transmittance scale.

4. Place a colorimeter tube containing water or another solvent in the sample holder and close the cover.

5. Rotate the light control so that the needle is at 100 on the transmittance scale (0.0 absorbance). This control regulates the amount of light passing through the exit slit to the phototube.

6. The unknown sample can then be placed in the tube holder and the percent transmittance or the absorbance read. The needle should always return to 0 when the tube is removed. Check the 0% and 100% transmittance occasionally with the solvent tube in the sample holder to make certain the unit is calibrated.

Note: Always check the wavelength scale to be certain that the desired wavelength is being used.

The colorimetric measurements made with this apparatus employ standard matched tubes. They are selected so that variation in light transmitted through the tubes due to slight differences in diameter and wall thickness is minimal. You will be is-

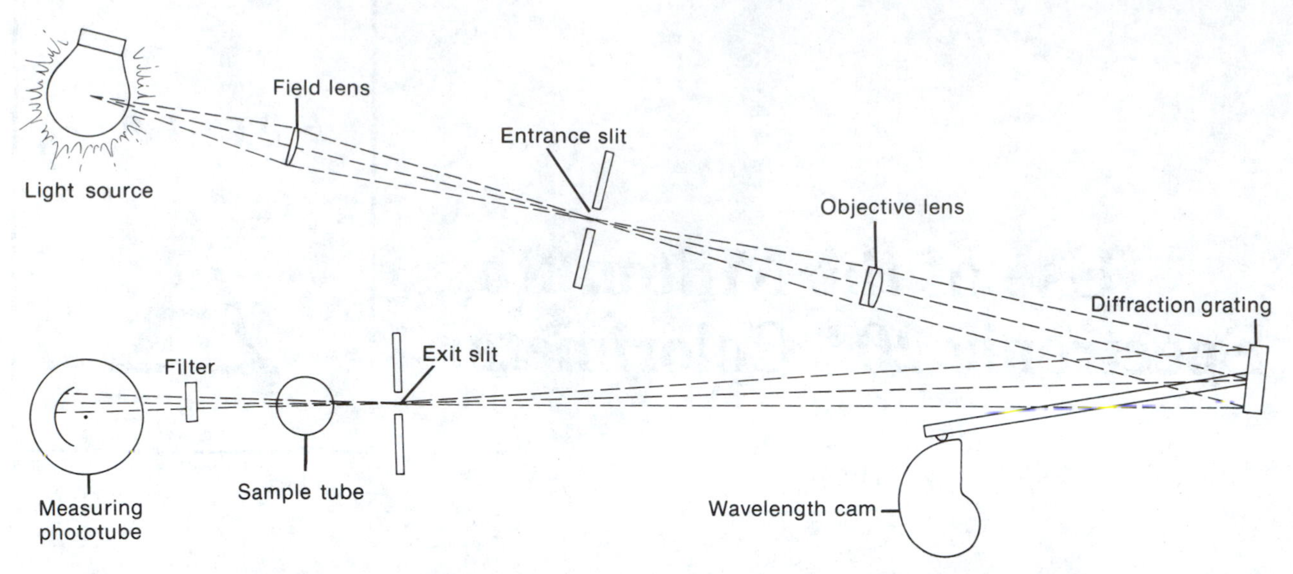

FIGURE A-1 Diagram of the optical system of the Spectronic 20® colorimeter

sued a set of matched tubes. *They are to be used only for colorimetry.* The matched tubes must be handled carefully so as not too etch or scratch the surfaces exposed to the light beam. Obviously, the tubes will no longer be "matched" if scratched or etched because such defects absorb and scatter light.

B. SPECTROPHOTOMETRY

The method used for spectrophotometry is essentially the same as that for colorimetry. The main difference is that the wavelength is reset for each reading, and thus a blank, or solvent, control must be readjusted to "0" and 100% (or "0" absorbance) at each new wavelength setting.

This procedure can be used when no information is available to determine the proper operating wavelength for an unknown substance. To do this, plot an absorption curve (absorbance versus wavelengths) of the unknown substance (Fig. A-3). An operating wavelength can then be chosen according to the following criteria. The details of this procedure are given in Appendix B.

- Choose the wavelength at which the substance maximally absorbs the light (the minimum transmittance) because the greatest sensitivity is obtained at this wavelength.
- Do not choose wavelengths on the slope of the curve because a small error in wavelength then causes a large error in reading.

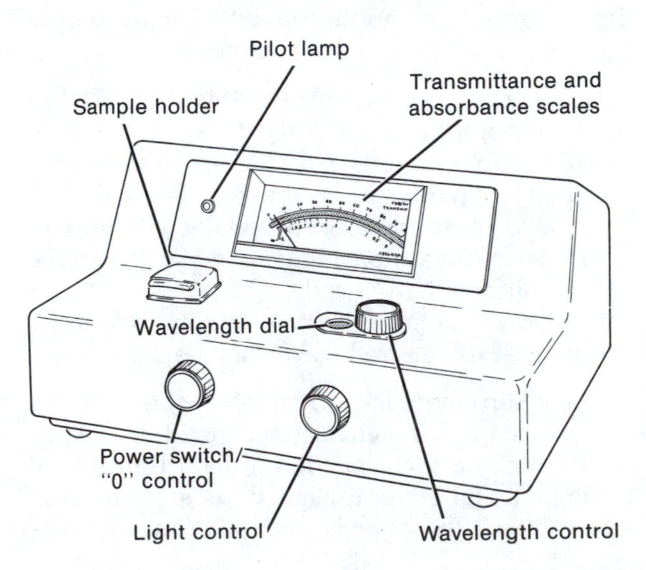

FIGURE A-2 Controls of the Spectronic 20® colorimeter

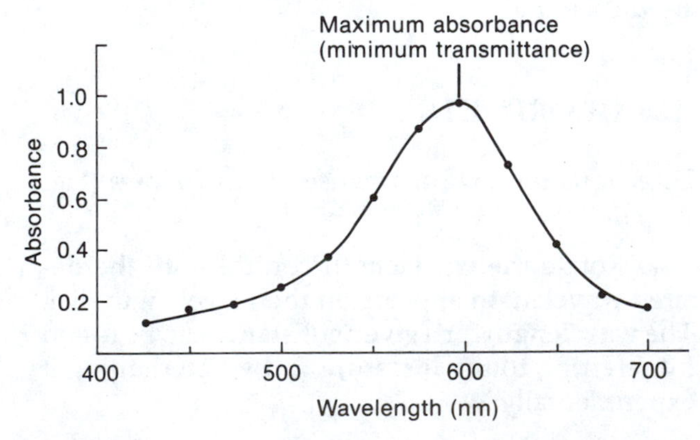

FIGURE A-3 Absorption spectrum of an unknown substance

Appendix B

Spectrophotometry

Many kinds of molecules interact with or absorb specific types of radiant energy in a predictable fashion. For example, when white light illuminates an object, the object absorbs one or more of the wavelengths (colors) that make up the white light. The remaining wavelength(s) are reflected (or transmitted) as a specific color, the color the eye perceives. Thus an object that appears red has absorbed the blue or green colors of light (or both) but not the red.

The perception of color as just described is qualitative. It indicates what is happening but says nothing about the extent to which the event is taking place; the eye is not a quantitative instrument. However, instruments called **spectrophotometers** electronically quantify the amount and kinds of light that are absorbed by molecules in solution. In its simplest form, a spectrophotometer has a source of white light (for visible spectrophotometry) that is focused on a prism or diffraction grating to separate the white light into its individual bands of radiant energy (Fig. B-1). Each wavelength (color), in turn, is focused through a narrow slit. The width of this slit is important to the precision of the measurement; the narrower the slit, the more closely the measured absorption is related to a specific wavelength of light. The broader the slit, the more wavelengths can pass through simultaneously, which results in less precision. This selected—hopefully monochromatic (single wavelength)—beam of light, called the **incident beam I_0,** then passes through the sample being measured. The sample, usually dissolved in a suitable solvent, is contained in a **cuvette,** which is standardized to have a light path 1 cm across. However, for special purposes, variations are available that are larger or smaller than the standard 1-cm size.

After passing through the sample, the selected wavelength of light (now referred to as the **transmitted beam I**) strikes a photoelectric tube. If the substance in the cuvette absorbs any of the incident beam, the transmitted light is reduced in total energy content. If the substance in the sample container does not absorb any of the incident beam, the radiant energy of the transmitted beam will be about the same as that of the incident beam. When the transmitted beam strikes the photoelectric tube, it generates an electric current proportional to the intensity of the light energy striking it. By connecting the photoelectric tube to a device that measures electric current (a galvanometer), one can directly measure the intensity of the transmitted beam. In the Milton Roy Spectronic 20® spectrophotometer, the galvanometer has two scales: one indicates the **percent transmittance (% T)**, and the other, a logarithmic scale with unequal divisions graduated from 0.0 to 2.0, indicates the **absorbance A**.

If the biological molecules being studied are dis-

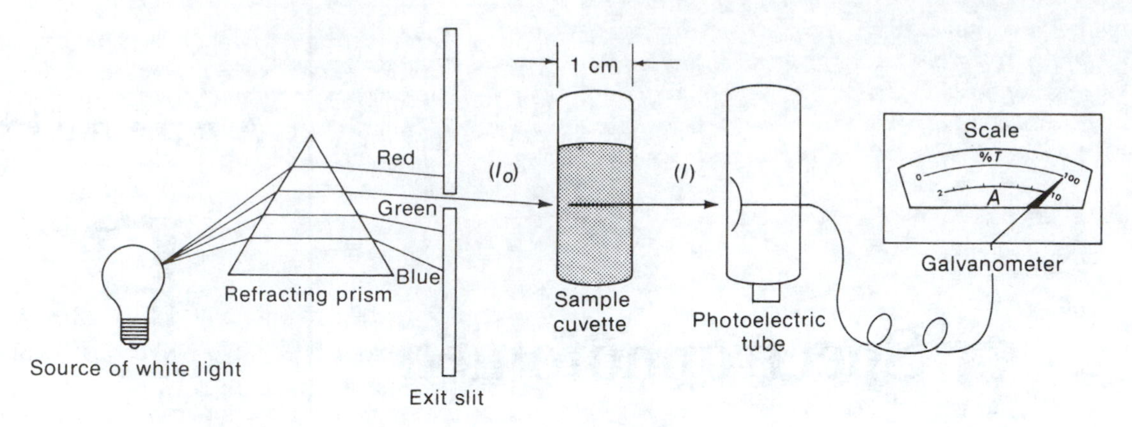

FIGURE B-1 Components of a typical photoelectric spectrophotometer

solved in a solvent before measurement, the absorption of light by the solvent can be a source of error. To assure that the spectrophotometric measurement reflects only the light absorption of the molecules being studied, a mechanism for "subtracting" the absorbance of the solvent is necessary. To achieve this the instrument is zero'd without anything in the tube holder. Then a blank (the solvent) is inserted into the instrument and the scale is set to read 100% transmittance (or 0.0 absorbance) for the solvent. Then the sample, containing the solute plus the solvent, is inserted into the instrument. Any reading on the scale that is less than 100% T (or greater than 0.0 A) is considered to be *due to absorbance by the solute only.* Other instruments are available that continuously give the desired ratio between sample and blank, both visually and on a strip-chart recorder. This kind of spectrophotometer "reads" the cuvettes for the blank and the sample simultaneously. Thus, the reading that appears on the absorbance scale is the ratio.

Spectrophotometers are not limited to detecting absorption of only visible light. Some also have a source of ultraviolet light (usually a hydrogen or mercury lamp) with wavelengths of about 180–400 nm. Ultraviolet wavelengths of 180–350 nm are particularly useful in studying such biological molecules as amino acids, proteins, and nucleic acids because each of these compounds has characteristic absorbances at various ultraviolet wavelengths. Other spectrophotometers have sources and suitable detectors of infrared radiation (780–25,000 nm). Numerous biological molecules can be effectively studied by means of their infrared absorption spectra.

A. UNITS OF MEASUREMENT

The following terminology is common in spectrophotometry.

- **Transmittance T:** Ratio of I, the light transmitted by the sample, to I_0, the light incident on the sample:

$$T = \frac{I}{I_0}$$

This value is multiplied by 100 to derive the %T. For example,

$$T = \frac{75}{100} = 0.75$$

and

$$\%T = \frac{75}{100} \times 100 = 75\%$$

- **Absorbance A:** Logarithm to the base 10 of the reciprocal of the transmittance (see part B for a discussion of logarithms):

$$A = \log_{10}\frac{1}{T}$$

For example, suppose a %T of 50 was recorded (equivalent to $T = 0.50$). Then

$$A = \log_{10}\frac{1}{0.50} = \log_{10}2.0$$

Thus $\log_{10}2.0 = 0.301$ (A equivalent to a $\%T$ of 50). Similarly, a $\%T$ of $25 = 0.602$ A; a $\%T$ of $75 = 0.125$ A; and so forth.

The absorbance scale is usually present on spectrophotometers, along with the transmittance scale. The chief usefulness of absorbance lies in the fact that it is a logarithmic rather than arithmetic function, allowing use of the Lambert–Beer law, which states that, for a given concentration range, the concentration of solute molecules is directly proportional to absorbance. The Lambert–Beer law can be expressed as

$$\log_{10}\frac{I_0}{I} = A$$

in which I_0 is the intensity of the incident light and I the intensity of the transmitted light.

The usefulness of absorbance can be seen in the graphs in Fig. B-2: one showing the percent transmittance plotted against concentration (Fig. B-2A) and the other showing absorbance plotted against concentration (Fig. B-2B). Using the Lambert–Beer relationship, you need to plot only three or four points to obtain the straight-line relationship shown in Fig. B-2B. However, certain conditions must prevail for the Lambert–Beer relationship to hold.

- Monochromatic light is used.
- A_{max} is used (i.e., the wavelength maximally absorbed by the substance being analyzed).
- The quantitative relationship between absorbance and concentration can be established.

The first condition can be met by using a prism or diffraction grating or other device that can disperse visible light into its spectra.

The second condition can be met by determining the absorption spectrum of the compound. This is done by graphing the absorbance of the substance at a number of different wavelengths. The wavelength at which absorbance is greatest, called A_{max} (or λ_{max}), is the most satisfactory wavelength to use, because on the slope the absorbance changes rapidly with slight variations in wavelength, whereas at the maximum absorbance the absorbance changes little with variations in wavelength. Figure B-3 is a graph of an absorption spectrum of a hypothetical substance. The A_{max} is approximately 650 nm. Some compounds, however, have several absorption peaks in the visible spectrum and in the ultraviolet range. An example of this is shown in Fig. B-4, the absorption spectrum for riboflavin.

To establish the quantitative relationship between absorbance and concentration of a substance, one must prepare a series of graded concentrations of the substance analyzed. Because absorbance is directly proportional to concentration, a plot of absorbance versus concentration of the standard yields a straight line. Such a plot is called a **concentration curve** or **standard curve** (Fig. B-5). After several points have been plotted, the intervening points can be extrapolated by connecting the known points with a straight line. It is

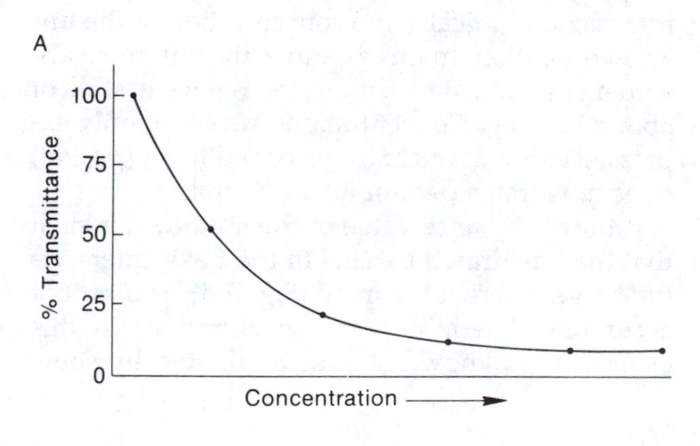

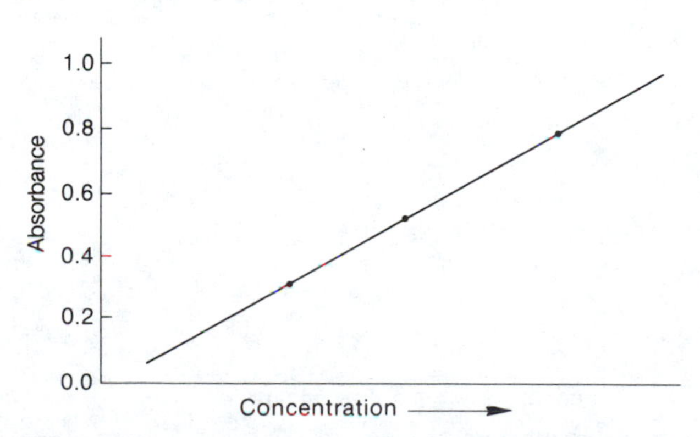

FIGURE B-2 (A) Percent transmittance versus concentration. (B) Absorbance versus concentration.

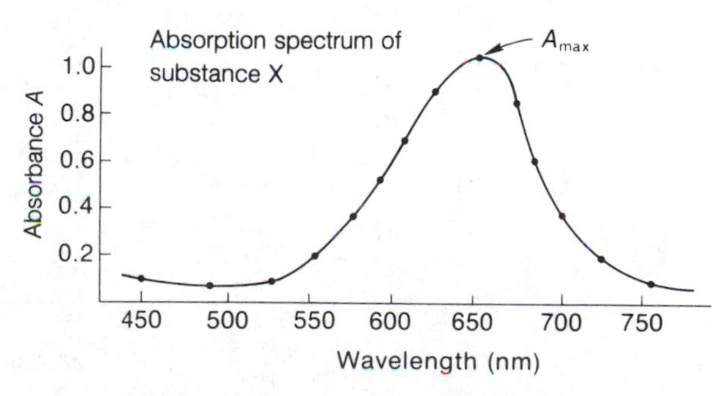

FIGURE B-3 Determination of A_{max}

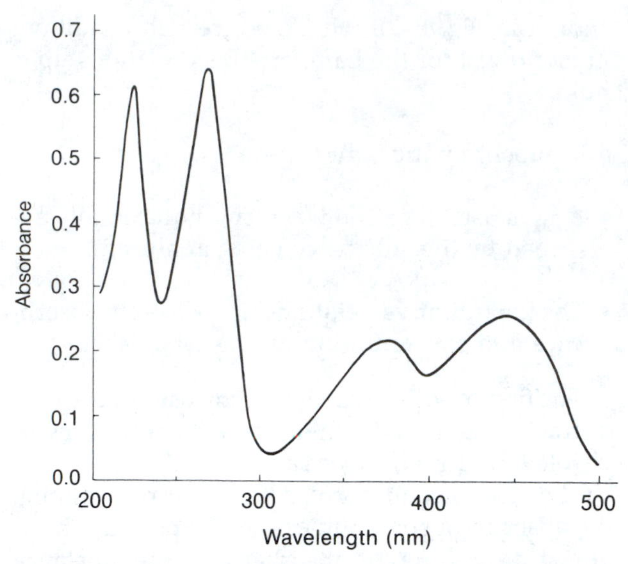

FIGURE B-4 Absorption spectrum of riboflavin. (From *Experimental Biochemistry,* 2d ed., edited by John M. Clark, Jr., and Robert L. Switzer. W. H. Freeman and Company © 1977.)

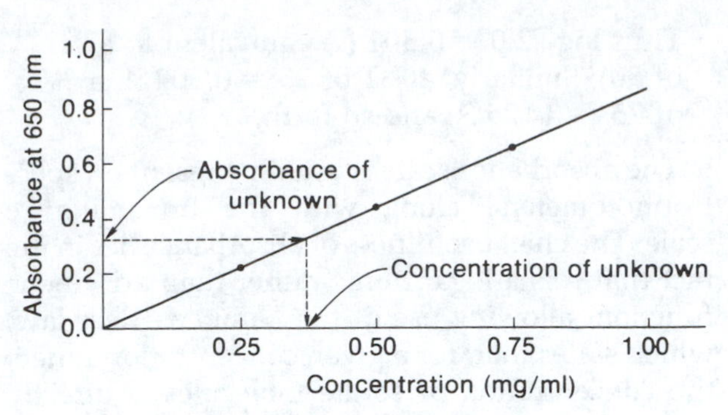

FIGURE B-6 Determining concentration of "unknown"

not necessary to use dotted lines to indicate extrapolation on graphs; a dotted line is used in the illustration to indicate the parts of the line for which points were not determined but were presumed. When the Lambert–Beer law is valid, this is an acceptable and time-saving assumption; otherwise, points must be plotted throughout the entire line. In general, your graph should extend from a minimum of about 0.025 A to a maximum of about 1.0 A, (94–10%T) this being the range of "readable" and reproducible values of absorbance. However, recall that the Lambert–Beer law operates at only certain concentrations. This is apparent in Fig. B-5, in which, at concentrations greater than 1.0 mg/ml, the curve slopes, indicating the loss of the linear concentration–absorbance relationship.

After a concentration curve for a given substance has been established, it is relatively easy to determine the quantity of that substance in a solution of unknown concentration by determining the absorbance of the unknown and locating it on the y axis, or ordinate (Fig. B-6). A straight line is then drawn parallel to the x axis, or abscissa, until it intersects with the curve. A perpendicular line is then dropped to the x axis, and the value at the point of intersection indicates the concentration of the unknown solution. In this example, the unknown absorbance is 0.32, which indicates a concentration of about 0.37 mg. Concentrations are commonly expressed either as micrograms per milliliter (μg/ml) or as milligrams per milliliter (mg/ml).

If the absorbance value of the unknown is such that the line drawn parallel to the x axis intersects the curve where it is curved (Fig. B-5), you cannot accurately determine the concentration. In this event, the unknown should be diluted by some

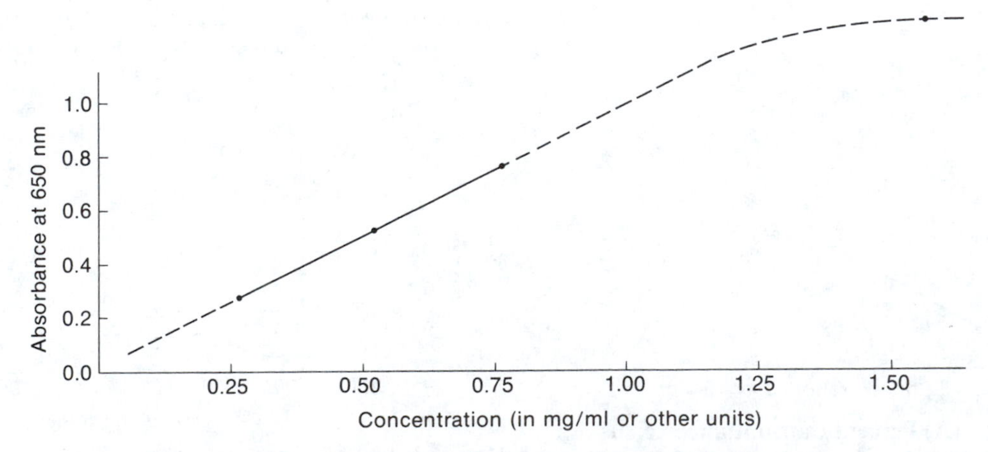

FIGURE B-5 Preparation of concentration (standard) curve

factor until the absorbance readings intersect with the straight-line part of the graph, where concentration is proportional to absorbance. You can then determine the unknown concentration and multiply the value by the reciprocal of the dilution. For example, if a 1/10 dilution was made of the unknown and this concentration was determined to be 55 mg/ml, then the concentration of the undiluted unknown would be 550 mg/ml (55 mg/ml × 10).

B. LOGARITHMS

If $a^x = y$, then the logarithmic relation is defined as $x = \log_a y$. Thus, the logarithm of a number to a given base is defined as the exponent of the power to which the base must be raised to give the number; that is, x is the logarithm of y, and y is the antilogarithm of x.

The fundamental theorems of logarithms are

$$\log(M \times N) = \log M + \log N$$

$$\log (M/N) = \log M - \log N$$

$$\log M^n = n \log M$$

$$\log{}^n\sqrt{M} = \log (M/n)$$

In Table B-1 the logarithms are given in the body of the table and the numbers from 1.0 to 9.9 are given in the left-hand column and the top row. To locate the logarithm of 4.7, for example, read down the left-hand column to 4 and across the columns to the .7 column to find 0.672 (the zero and decimal point are omitted in the table for convenience). To find an antilogarithm, reverse the procedure; for example, 0.532 is located at the intersection of the 3 row and the .4 column, and therefore its antilogarithm is 3.4.

Numbers less than 1.0 or greater than 10.0 are conveniently expressed as the product of a number between 1.0 and 10.0 and a power of 10. Thus, $4700 = 4.7 \times 10^3$, and because the logarithm of a product is equal to the sum of the individual logarithms,

$$\log 4700 = \log 4.7 + \log 10^3 = 0.672 + 3 = 3.672$$

Similarly, if a number is less than 1.0,

$$\log 0.000047 = \log 4.7 + \log 10^{-5} = 0.672 - 5 = -4.328$$

To find an antilogarithm, or the number represented by a given logarithm, the reverse procedure is employed, and the logarithm is converted into a sum consisting of a positive term between 0.0 and 1.0 and an integer. The antilogarithm of this sum equals the product of the individual antilogarithms. Thus,

$$\text{antilog } 2.532 = \text{antilog } 0.532 \times \text{antilog } 2 = 3.4 \times 10^2 = 340$$

$$\text{antilog } -7.468 = \text{antilog } 0.532 - 8 = 3.4 \times 10^8 = 0.000000034$$

Finally, the following relationships should be remembered:

$$\log 1 = \log 10^0 = 0$$

$$\log 10 = \log 10^1 = 1$$

$$\log 100 = \log 10^2 = 2$$

TABLE B-1 Three-place logarithms

	0	.1	.2	.3	.4	.5	.6	.7	.8	.9
0	—	000	301	477	602	699	778	845	903	954
1	000	041	079	114	146	176	204	230	255	279
2	301	322	342	362	380	400	415	431	447	462
3	477	491	505	519	532	544	556	568	580	591
4	602	613	623	634	644	653	663	672	681	690
5	699	708	716	724	732	740	748	756	763	771
6	778	785	792	799	806	813	820	826	833	839
7	845	851	857	863	869	875	880	886	892	898
8	903	909	914	919	924	929	935	940	945	949
9	954	959	964	969	973	978	982	987	991	996

$$\log 1000 = \log 10^3 = 3$$
$$\log 10000 = \log 10^4 = 4$$
$$\log 0.1 = \log 10^{-1} = -1$$
$$\log 0.01 = \log 10^{-2} = -2$$
$$\log 0.001 = \log 10^{-3} = -3$$
$$\log 0.0001 = \log 10^{-4} = -4$$

Although this is a mathematical definition of logarithms, and the one that must be employed in carrying out problems, a qualitative description may be useful. At first glance, the logarithm appears to be an arbitrary and artificial way of expressing any measurement compared with the simpler arithmetic expression. However, we are often more concerned, particularly in biology, with *relative* measurements or relations than with absolute magnitudes. The logarithmic expression has exactly this property: it gives equal weight to equal relative (rather than absolute) changes. Thus, a twofold change represents 0.3 logarithmic units anywhere on the scale, although this may represent a milligram, a kilogram, or a ton on an absolute scale. Further, measurements are usually made with the same *relative* accuracy—to 1%, or to 1 part in 10,000—rather than to a given *absolute* accuracy. In a logarithmic plot this given relative accuracy is expressed by the same distance; for example, 10% accuracy represents 0.04 log units (=log 1.10).

That this term is an equally "natural" means of expressing many relations is further indicated by the fact that many relations or physical "laws" are expressed in logarithmic form; that is, the activity or action is proportional to the logarithm of some concentration or ratio rather than to its absolute amount. Outstanding examples are the relations between pH and salt concentrations in buffer solution, between oxidation-reduction potential and the concentrations of the reduced and oxidized substance, and between electromotive force and the concentration of substances in concentration cells.

Logarithms are particularly useful in graphing relations that extend over a wide range of values because they have the property of giving equal relative weight to all parts of the scale. This is valuable in "spreading out" the values that would otherwise be concentrated at the lower end of the scale.

REFERENCES

Schleif, R. F., and P. C. Wensink. 1982. *Practical Methods in Molecular Biology.* Springer-Verlag.

Straughton, B. P., and S. Walker, eds. 1976. *Spectroscopy.* Chapman and Hall.

Appendix C

Chromatography

Chromatographic techniques are among the most useful methods for separating complex mixtures of solutes. Separation results from selective adsorption of solutes as they pass over adsorbents such as charcoal, starch, cellulose powders, ion-exchange resins, and filter paper. The most widely used chromatographic techniques are column chromatography on ion-exchange resins, paper-partition chromatography on filter paper, and thin-layer methods using an adsorbent bound to a substrate such as glass. These techniques are frequently employed to isolate proteins, enzymes, lipids, hormones, plant growth substances, pigments, and other naturally occurring organic materials.

A. PAPER CHROMATOGRAPHY

Paper chromatography has revolutionized the art of detecting and identifying small amounts of organic and inorganic substances. It enables us to separate mixtures on a very small scale, which no other simple method affords. The technique has found widespread application in many areas of biology and has been used to separate and identify amino acids, carbohydrates, fatty acids, antibiotics, and many other natural substances.

To employ this technique, you first place a small amount (spot) of the mixture to be chromatographed near one end of a length of filter paper. Then you immerse this end of the paper in a solvent system that is usually composed of two or more miscible substances. In **descending chromatography,** the solvent is contained in a trough near the top of the chromatographic chamber and is allowed to irrigate the paper in a downward flow by means of capillary action and gravity (Fig. C-1). In **ascending chromatography,** the solvent is placed at the bottom of the chamber and is allowed to irrigate the paper in an upward flow by means of capillary action.

Usually ascending chromatographic separation is used because the apparatus is quite simple. However, descending chromatography is faster (the solvent flow is made faster by gravity) and makes it possible to collect the solvent (and solute) as it runs off the paper. This permits additional analysis of solute components by other procedures.

Locating or visualizing the substances that have to be separated can be done in a variety of ways. Some substances have color and are thus easily seen on the chromatogram. For those compounds that are colorless, various procedures are employed. For example, spots of amino acids become pink or purple when sprayed with ninhydrin; spots of reducing sugars become gray-black when treated with analine phthalate, and many organic

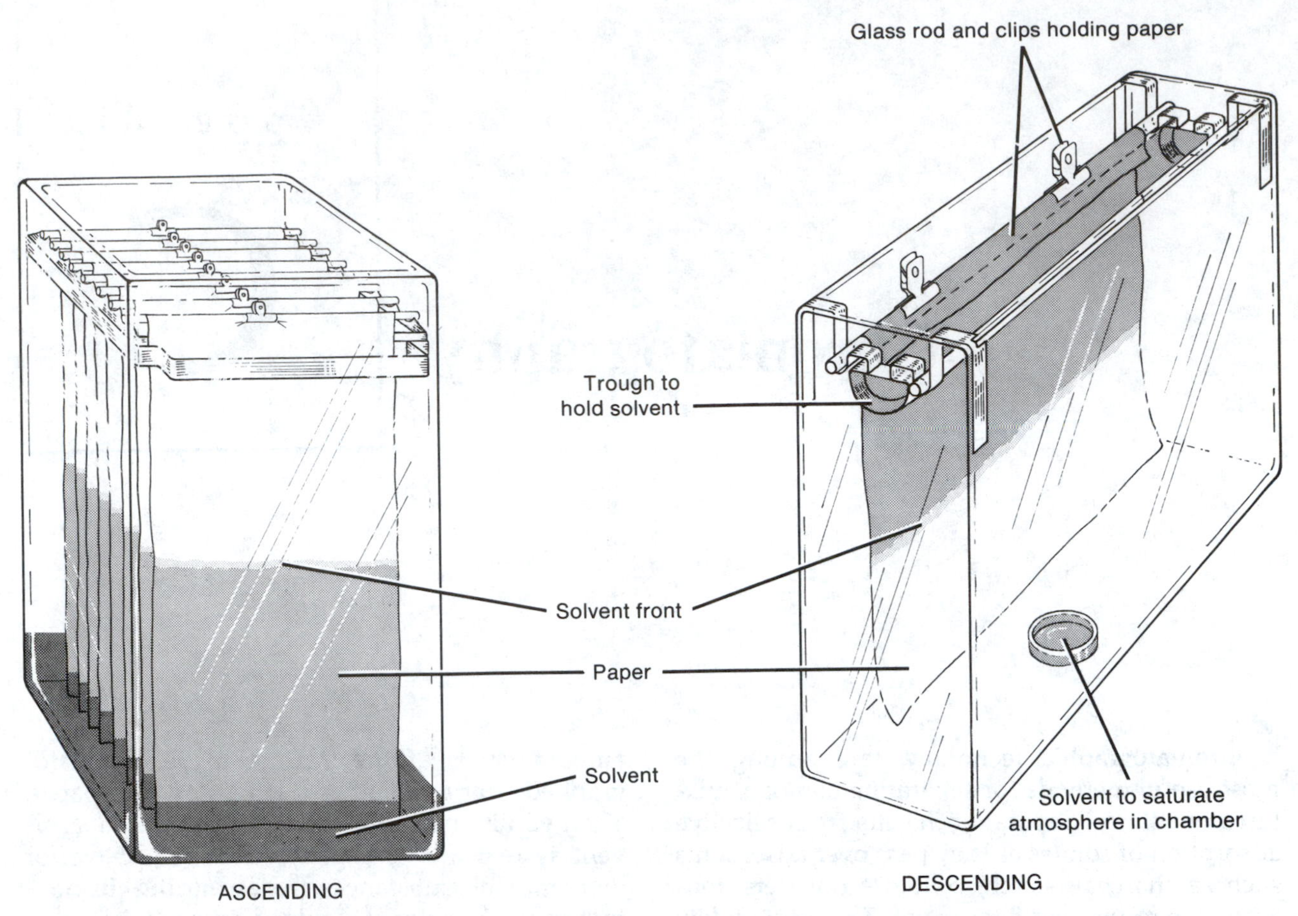

FIGURE C-1 Apparatus for ascending and descending chromatography

substances become brown spots on a yellow background when exposed to iodine vapors.

As long as conditions such as temperature and purity of solvent are carefully regulated, one can not only separate components of a mixture but also identify them. Under standardized conditions, a substance moves at a characteristic rate in a particular solvent system. Thus, one can calculate R_f values by comparing the distance the solvent travels with the distance the substance travels:

$$R_f = \frac{\text{distance of center of spot from origin}}{\text{distance of solvent front from origin}}$$

Tables of R_f values for various substances in various solvents permit preliminary identification of a substance by computing its R_f value. However, for more precise identification, one should chromatograph known substances on the same sheet so that R_f values of the known and suspected substances can be determined under identical conditions.

B. THIN-LAYER CHROMATOGRAPHY

In thin-layer chromatography, one applies an adsorbent to a substrate, frequently glass or aluminum, in a very thin layer. A binding agent is generally used to make the adsorbent adhere to the substrate.

In one of the more common procedures, a slurry of silica gel adsorbent, with calcium sulfate as the binder, is spread as a thin layer on 20-cm^2 glass plates. The plates are dried and then handled in much the same way as paper in ascending chromatography. After the separation, one can scrape the spots from the glass for detailed analysis, or one can elute and rechromatograph the spots.

Separation is accomplished very quickly on thin-layer plates. With paper chromatographic methods, the separation of some mixtures may take 24 hours. With thin-layer procedures, the same separation may take less than an hour.

Appendix D

Mendel's Laws of Inheritance

Gregor Mendel, an Austrian monk trained in mathematics, began his studies of heredity in 1857 using the common garden pea. He chose pea plants for his studies because they are easy to obtain and grow rapidly. Furthermore, different varieties have distinctly different traits that "breed" true and reappear year after year. Also, the pistil and stamens in the flower are entirely enclosed by the petals; thus, the flower typically self-pollinates (and thus self-fertilizes). Accidental cross-pollination could not occur and confuse the experimental results. In Mendel's own words (a truism that holds even today), "The value and utility of any experiment are determined by the fitness of the material to the purpose for which it is used."

Mendel chose two variants for each of seven different traits (form of the seed, color of the seed, position of the flower, flower color, form of the pod, color of the pod, and stem length). For each trait, Mendel noted the apparent loss of one alternative characteristic in the next generation. For example, when he crossed short plants with tall plants, he found only tall plants in the next generation (called the Filial 1 or F_1 generation). When he allowed these F_1 progeny to self-pollinate, both parental types reappeared in the succeeding, or F_2, generation. Furthermore, Mendel noted a consistent 3 : 1 ratio of tall plants to short plants. This 3 : 1 ratio is frequently called the **monohybrid** ratio; *hybrid* refers to the heterozygosity of the particular trait in the offspring or parents. Thus, in this example (and for all others he tested), the trait was not "lost" in the F_1 generation but was "masked" in some fashion such that it did not express itself. The trait that is expressed is **dominant** and that which is hidden, or not expressed, is **recessive.**

Mendel's research suggested to him that the "factors" (now called **genes**) that determined various traits were transmitted from generation to generation in a predictable manner. From his findings, Mendel suggested that each gene for a given trait can exist in two alternative forms (now called **alleles**) that are responsible for the phenotype expressed. Thus, plants or animals that express only one phenotype (either dominant or recessive) through several generations are considered to be **homozygous** (or pure breeding) for that trait. Organisms that produce two phenotypes are **heterozygous** (or hybrid).

Today we know that the gametes are the physical link between two generations and that homozygous individuals produce only one type of gamete, whereas heterozygous organisms produce two types of gametes and in equal numbers (Fig. D-1).

From his experiments involving the transmission of single characteristics, Mendel formulated his first law of inheritance — the **law of segregation.**

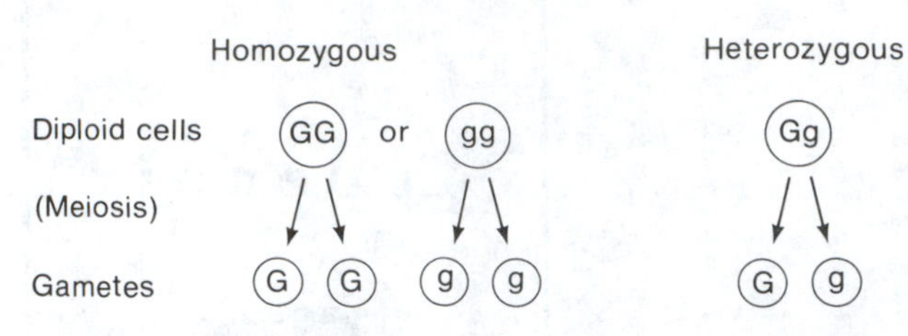

FIGURE D-1 Types of gametes produced in homozygous and heterozygous crosses

This law says that, in sexually reproducing organisms there are factors (genes) that come together at fertilization and segregate at meiosis. The alleles (or genes) involved do this without losing their identity; that is, they show no changes in the offspring, which thus rules out any "blending" theory of inheritance.

Mendel's second law, the **law of independent assortment,** evolved from his experiments with crosses between individuals differing in two gene pairs. For example, if you cross a homozygous plant having smooth (S), yellow (Y) seeds with a homozygous plant with wrinkled (s), green (y) seeds, the F_1 generation results in plants having smooth, yellow seeds. In this cross, what is the genotype for each of the parents (of the F_1 generation) for each of the traits considered?

____

____

Show the cross, gametes, and phenotype and genotype of the F_1 plants.

Cross of F_1 plants ________________

Gametes of F_1 plants ________________

Phenotype of F_1 plants ________________

Genotype of F_1 plants ________________

When the F_1 plants are self-fertilized, they give rise to an F_2 generation divided into four phenotypic classes as follows.

$9/16$ of the population are smooth and yellow.

$3/16$ of the population are wrinkled and yellow.

$3/16$ of the population are smooth and green.

$1/16$ of the population are wrinkled and green.

This ratio is referred to as a **dihybrid** ratio, where *dihybrid* is defined as being heterozygous in terms of two pairs of factors, or alleles. Show the phenotype and genotype of the F_1 plants, the gametes produced by the F_1 plants, the phenotype and genotype of the F_2 plants, and the dihybrid ratio.

Phenotype of F_1 plants ________________

Genotype of F_1 plants ________________

Gametes of F_1 plants ________________

Phenotype of F_2 plants ________________

Genotype of F_2 plants ________________

Dihybrid ratio ________________

You can still demonstrate the validity of Mendel's first law of segregation if you consider the data for only one pair of genes at a time. For example, if you combine all the smooth seeds ($9/16 + 3/16 = 12/16$) and all the wrinkled seeds ($3/16 + 1/16 = 4/16$), you obtain a $12/16 : 4/16$ or 3 : 1 monohybrid ratio. Thus, you again demonstrate that gene pairs act independently of one another and are not changed in their transmission from one generation to another. This indeed was the basis for Mendel's second law of inheritance.

Although Mendel's career as a "geneticist" was cut short by a beetle that devastated his pea crops and by being appointed head of the monastery, he left us with several important concepts, including his two laws of inheritance and the fact that results of various crosses are mathematically predictable.

REFERENCES

Avers, C. 1984. *Genetics.* 2d ed. W. Grant.

Gardner, E. J., and D. P. Snustad. 1984. *Principles of Genetics.* 7th ed. Wiley.

A. J. F. Griffiths, et al. 1993. *An Introduction to Genetic Analysis.* 5th ed. W. H. Freeman and Company.

Appendix E

Elementary Statistical Analysis

The inherent variability of organisms, coupled with the error normally encountered in most measuring systems, demands that you use some form of data evaluation before stating conclusions or drawing inferences. Particularly important for students of the biological sciences are the statistical methods of handling data. These methods fall into two groups: statistics that define the nature and distribution of the data and procedures that compare two or more sets of data.

This introduction to statistical methods is not to be regarded as complete, nor is it expected to substitute for formal training in this area. Wherever possible, the student should supplement the information given here by using the reference given at the end of this appendix.

A. DEFINING THE DATA

To define a sample of data, one must have some knowledge of the central tendencies and degree of dispersion of the data. The statistics usually used for this purpose are the arithmetic mean, the standard deviation, and the confidence interval of the mean.

1. Arithmetic Mean

The mean $\overline{X}$ is computed by summing (Σ) the individual sample measurements x_i and dividing by the total number of measurements N:

$$\overline{X} = \frac{\Sigma x_i}{N}$$

2. Standard Deviation

The mean of a group of data gives little information concerning the distribution of the data about the mean. Obviously, different numerical values can give the same mean value. For example,

Set 1: 32, 32, 36, 40, 40 $\overline{X} = 36$

Set 2: 2, 18, 24, 36, 100 $\overline{X} = 36$

The mean for both sets of figures is the same. Yet the variation from the mean in Set 2 is so great as to make the average meaningless. The standard deviation is a measure of data dispersion about the mean. The range of values covered by *mean plus or minus* ($\pm$) *one standard deviation* includes about 68% of the data for which the mean was calculated. The range covered by *mean* $\pm$ *two standard deviations* includes approximately 95% of the data. The calculation of the standard deviation is summarized as follows.

1. Compute the arithmetical mean $\overline{X}$ by summing the individual measurements x_i and dividing by the total number of measurements N: $\overline{X} = \Sigma x_i / N$.

2. Calculate the deviation from the mean for each measurement: $(x_i - \overline{X})$.

3. Square each of the individual deviations from the mean: $(x_i - \overline{X})^2$. This allows one to deal with positive values.

4. Determine the sum of the squared deviations: $\Sigma(x_i - \overline{X})^2$.

5. Calculate the standard deviation(s) of the sample using the formula

$$s = \sqrt{\frac{\Sigma(x_i - \overline{X})^2}{N - 1}}$$

A sample set of standard deviation measurements is shown in Table E-1, where

number of individual measurements $N = 10$

arithmetical mean $\overline{X} = 220/10 = 22$

degrees of freedom $(N - 1) = 9$

$$\text{standard deviation } s = \sqrt{\frac{\Sigma(x_i - \overline{X})^2}{N - 1}}$$

$$= \sqrt{\frac{42}{9}}$$

$$= \sqrt{4.66}$$

$$= 2.16.$$

3. Confidence Interval for a Sample Mean

The confidence interval C for a sample mean equals the standard deviation s of the sample divided by the square root of the number N in the sample and multiplied by a factor t that is determined by the probability level desired and the value of the sample number:

$$C = \pm t\left(\frac{s}{\sqrt{N}}\right)$$

It is highly improbable that a sample mean based on a relatively small series of data will correspond exactly to the true mean calculated from an infinitely large sample of the population. One must, therefore, define a range within which the true mean might be expected to lie. To define this range, the standard error of the mean (S.E. $\overline{X}$) must be known. The standard error equals the standard deviation divided by the square root of the number in the sample:

$$\text{S.E. } \overline{X} = \frac{s}{\sqrt{N}}$$

The confidence interval is then calculated by multiplying the S.E. $\overline{X}$ by t, whose value depends on the number in the sample N and the level of probability selected. Normally, a probability or significance level of 0.05 is accepted in biological studies. This implies that in only 5% of the samples taken separately from a given population would the parameters defined by the sample *fail* to have significance.

The t values can be obtained from the t table (Table E-2). These values are listed in columns for 0.10, 0.05, and 0.01 probability levels. Note that

TABLE E-1 Sample calculation of the standard deviation

Observation	Individual measurement (x_i) of stem length (mm)	Deviation from mean ($x_i - \bar{X}$)	$(x_i - \bar{X})^2$
x_1	20	−2	4
x_2	24	+2	4
x_3	22	0	0
x_4	19	−3	9
x_5	26	+4	16
x_6	22	0	0
x_7	24	+2	4
x_8	20	−2	4
x_9	22	0	0
x_{10}	21	−1	1
	TOTAL = 220 = Σx_i	0	$42 = \Sigma(x_i - \bar{X})^2$

TABLE E-2 Significance limits of Student's *t* distribution

(n) degrees of freedom	Confidence levels		
	0.10	0.05	0.01
1	6.314	12.706	63.657
2	2.920	4.303	9.925
3	2.353	3.182	5.841
4	2.132	2.776	4.604
5	2.015	2.571	4.032
6	1.943	2.447	3.707
7	1.895	2.365	3.499
8	1.860	2.306	3.355
9	1.833	2.262	3.250
10	1.812	2.228	3.169
11	1.796	2.201	3.106
12	1.782	2.179	3.055
13	1.771	2.160	3.012
14	1.761	2.145	2.977
15	1.753	2.131	2.947
16	1.746	2.120	2.921
17	1.740	2.110	2.898
18	1.734	2.101	2.878
19	1.729	2.093	2.861
20	1.725	2.086	2.845
21	1.721	2.080	2.831
22	1.717	2.074	2.819
23	1.714	2.069	2.807
24	1.711	2.064	2.797
25	1.708	2.060	2.797
26	1.706	2.056	2.779
27	1.703	2.052	2.771
28	1.701	2.048	2.763
29	1.699	2.045	2.756
30	1.697	2.042	2.750
40	1.684	2.021	2.704
60	1.671	2.000	2.660
120	1.658	1.980	2.617
∝	1.645	1.960	2.576

Source: Adapted from Table III of Fisher and Yates, *Statistical Tables for Biological, Agricultural, and Medical Research,* Oliver & Boyd, Edinburgh.

one must have a value for the number of degrees of freedom (D.F.) of the sample. In this case, the D.F. for the sample equals the number N in the sample minus one, or $(N - 1)$.

B. COMPARISON OF DATA

1. Standard Error of the Difference of Means

The standard error of the difference of means is computed by using the formula

$$\text{S.E. } \bar{X}_1 - \bar{X}_2 = \sqrt{\frac{s_1}{N_1} + \frac{s_2}{N_2}}$$

where s_1 and s_2 represent the standard deviations of two different groups, N_1 and N_2 represent the number of individuals in each group (preferably at least 20), and $\bar{X}_1$ and $\bar{X}_2$ are the respective means. Using this formula, if the difference between the two means is *larger than two times* the standard error of the difference, one can conclude that the difference between the groups is not due to chance alone but to the treatment given. One can also conclude that similar plants or animals under similar treatment could be expected to respond in a similar manner.

2. Student's *t* Test

Student's *t* test is used to determine whether, within a selected degree of probability, two groups of data represent samples taken from the same or from different populations of data. In other words, it is used to determine if two groups of data are significantly different. This test uses both the means and standard deviations of the two samples. It is calculated as

$$t = \frac{(\bar{X}_1 - \bar{X}_2)\left(\sqrt{\dfrac{N_1N_2}{N_1 + N_2}}\right)}{\sqrt{\dfrac{(N_1 - 1)(s_1^2) + (N_2 - 1)(s_2^2)}{N_1 + N_2 - 2}}}$$

where s_1 and s_2 represent standard deviations of two different groups, N_1 and N_2 represent the number of individuals in each group, and $\bar{X}_1$ and $\bar{X}_2$ are the respective means.

The calculated t value is then compared with the value in the t table (Table E-2) at the probability level chosen (usually 0.05) and at the combined degrees of freedom of the two samples $(N_1 + N_2 - 2)$. If the value for t is *less* than that found in the table, the two groups of data are not considered significantly different at the chosen level of probability. If the t value *exceeds* that in the table, the two groups of data are considered significantly different.

3. Wilcoxon Test

Although this test is mechanically easier to carry out than Student's t test, it should be used only on small amounts of data—such as in a preliminary study—to help determine if a particular experiment is worth pursuing.

Basically, this test employs a ranking system. A simple example is to compare the pulse rate (PR) of normal males and normal females to test for differences due to sex.

The hypothesis to be tested (H_0, the null hypothesis) is

$$PR♀ = PR♂$$

that is, there is no difference in heart rate (in terms of pulse beat) between males and females.

a. Raw Data

Sex	*Pulse Rate*					
♂	74	77	78	75	72	71
♀	80	83	73	84	82	79

b. Calculation of Means

$\Sigma♂ = 447 \qquad \bar{X} = 74.5$

$\Sigma♀ = 481 \qquad \bar{X} = 81.7$

c. Ranking of Data

1. Rank the data in order of increasing magnitude. Underline all values from the group having the smaller mean. In this example, it is the males.

2. Write the rank under the data values, beginning with 1 for the smallest value and proceeding to the largest value. Then underline the ranks corresponding to the male values.

$\underline{71}\ \underline{72}\ 73\ \underline{74}\ \underline{75}\ \underline{77}\ \underline{78}\ 79\ 80\ 82\ 83\ 84$

$\underline{1}\ \underline{2}\ 3\ \underline{4}\ \underline{5}\ \underline{6}\ \underline{7}\ 8\ 9\ 10\ 11\ 12$

3. Group the ranks by sex, and add them. If there is no real difference in pulse rate between sexes, then the sums of ranks should be equal.

$♂\ \underline{1} + \underline{2} + \underline{4} + \underline{5} + \underline{6} + \underline{7} = 25$

$♀\ 3 + 8 + 9 + 10 + 11 + 12 = 53$

d. Computing *U*

In this case, the sums of the ranks are different; but is there a significant difference between males and females? To determine this, the statistical value U must be determined:

$$U = W_1 - \tfrac{1}{2}n_1(n_1 + 1)$$

where W_1 = the total of the ranks belonging to the group with the smallest mean (i.e., the males: $W_1 = 25$)

n_1 = the number of individuals (or measurements) in the group having the smallest mean (i.e., the males: $n_1 = 6$)

Thus,

$$U = 25 - \tfrac{1}{2} \times 6(6 + 1) = 25 - 21 = 4$$

e. Determining Significance

1. Refer to Table E-3 (Wilcoxon table for unpaired data). Note that it has four columns labeled n_1, n_2, $C_{n_1n_2}$, and *Values of U.*

2. The values of n_1 and n_2 in the data are 6 and 6. Refer to this combination in the first two columns of the table. The corresponding figure in the $C_{n_1n_2}$ column is 924.

3. Run along the U columns to $U = 4$ and then down the $U = 4$ column to the row corresponding to $C_{n_1n_2} = 924$. The value at this intersection is 12. This value is the number of possible rank totals that are less than or equal to (≤) 25.

4. Suppose we decide that if an event occurs less than 5 times out of 100 (5/100 or 50/1000, probability $P = 0.05$), it is occurring not as a result of chance, but as a result of the treatment. In this case, it is occurring due to the difference in sex.

The value of 12 in the U column represents an occurrence of 12 times out of 1000 (12/100, $P = 0.012$). This probability is considerably lower than the 5/100 that we decided to accept as indicating that the difference in pulse rate is based on the difference in sex. Therefore, we reject the hypothesis that

$$PR♂ = PR♀$$

and say that the difference in pulse rates is significant and results from the difference in sex.

C. CHI-SQUARE ANALYSIS

The statistical test most frequently used to determine whether data obtained experimentally provide a good fit, or approximation, to the expected or theoretical data, is relatively simple to carry out. Basically, this test can be used to determine if deviations from the *expected values* are due to chance alone or to some factor or circumstance.

The formula for chi square (χ^2) is

$$\chi^2 = \Sigma\left[\frac{(O - E)^2}{E}\right]$$

TABLE E-3 Wilcoxon distribution (with no pairing). The numbers given in this table are the number of cases for which the sum of the ranks of the sample of size n_1, is less than or equal to W_1

n_1	n_2	$C_{n_1 n_2}$	Values of U, where $W_1 - \frac{1}{2}n_1(n_1+1)$																				
			0	1	2	3	4	5	6	7	8	9	10	11	12	13	14	15	16	17	18	19	20
3	3	20	1	2	4	7	10	13	16	18	19	20											
3	4	35	1	2	4	7	11	15	20	24	28	31	33	34	35								
4	4	70	1	2	4	7	12	17	24	31	39	46	53	58	63	66	68	69	70				
3	5	56	1	2	4	7	11	16	22	28	34	40	45	49	52	54	55	56					
4	5	126	1	2	4	7	12	18	26	35	46	57	69	80	91	100	108	114	119	122	124	125	126
5	5	252	1	2	4	7	12	19	28	39	53	69	87	106	126	146	165	183	199	213	224	233	240
3	6	84	1	2	4	7	11	16	23	30	38	46	54	61	68	73	77	80	82	83	84		
4	6	210	1	2	4	7	12	18	27	37	50	64	80	96	114	130	146	160	173	183	192	198	203
5	6	462	1	2	4	7	12	19	29	41	57	76	99	124	153	183	215	247	279	309	338	363	386
6	6	924	1	2	4	7	12	19	30	43	61	83	111	143	182	224	272	323	378	433	491	546	601
3	7	120	1	2	4	7	11	16	23	31	40	50	60	70	80	89	97	104	109	113	116	118	119
4	7	330	1	2	4	7	12	18	27	38	52	68	87	107	130	153	177	200	223	243	262	278	292
5	7	792	1	2	4	7	12	19	29	42	59	80	106	136	171	210	253	299	347	396	445	493	539
6	7	1716	1	2	4	7	12	19	30	44	63	87	118	155	201	253	314	382	458	539	627	717	811
7	7	3432	1	2	4	7	12	19	30	45	65	91	125	167	220	283	358	445	545	657	782	918	1064
3	8	165	1	2	4	7	11	16	23	31	41	52	64	76	89	101	113	124	134	142	149	154	158
4	8	495	1	2	4	7	12	18	27	38	53	70	91	114	141	169	200	231	264	295	326	354	381
5	8	1287	1	2	4	7	12	19	29	42	60	82	110	143	183	228	280	337	400	466	536	607	680
6	8	3003	1	2	4	7	12	19	30	44	64	89	122	162	213	272	343	424	518	621	737	860	994
7	8	6435	1	2	4	7	12	19	30	45	66	93	129	174	232	302	388	489	609	746	904	1080	1277
8	8	12870	1	2	4	7	12	19	30	45	67	95	133	181	244	321	418	534	675	839	1033	1254	1509

Source: Reproduced from Table H of Hodges and Lehmann, *Basic Concepts of Probability and Statistics,* published by Holden-Day, San Francisco, by permission of the authors and publisher.

where O = **observed** number of individuals
E = **expected** number of individuals
Σ = sum of all values of $(O - E)^2/E$ for the various categories of phenotypes

The following example shows how this type of analysis can be applied to genetics data. It can be used in many other analyses as long as numerical data are used and not percentages or ratios.

1. Example of Chi-Square Analysis

In a cross of tall maize (corn) plants to dwarf plants, the F_1 generation consisted entirely of tall plants. The F_2 generation consisted of 84 tall and 26 dwarf plants. The question we want to answer is whether the F_2 data fit the expected 3:1 monohybrid ratio. Using the data given in Table E-4, we calculate chi square:

$$\chi^2 = \Sigma\left[\frac{(O - E)^2}{E}\right]$$

$$= \Sigma[0.027 + 0.082]$$

$$= 0.109$$

What does this chi-square value of 0.109 mean? If the observed values were exactly equal to the expected values ($O = E$), we would have a perfect fit, and χ^2 would equal 0. Thus, a small value of χ^2 would indicate a close agreement of the observed and expected ratios, whereas a large value of χ^2 would indicate marked deviation from the expected ratios. However, deviations from the expected values are always bound to occur due to chance alone. The question is "Are the observed deviations within the limits expected by chance?"

Statisticians have generally agreed, for these types of studies, on the arbitrary limits of 1 chance in 20 (probability = 0.05 = 5%) for making the distinction between acceptance or rejection of the hypothesis that the data fit the expected ratio.

The chi-square value for a two-term ratio (e.g., 3:1) that corresponds to a 0.05, or 1 in 20, probability is 3.841. That is, you would expect to obtain this value *due to chance deviations* in only 5% of similar trials if the hypothesis is true. If you obtain a χ^2 for this two-term ratio that is larger than 3.841, then the probability that the variation is due to chance alone is less than 5%, or 1 in 20. You would therefore *reject* the hypothesis that the observed and expected ratios are in close agreement.

In our example, χ^2 was 0.109, which is considerably less than 3.841. Thus, we can say that the variation between the observed and expected values *was* due to chance alone, and we accept the data as fitting the 3:1 ratio.

Where did we obtain the value 3.841? Mathematicians have developed a variety of statistical tables for various applications. Table E-5 is an example of a chi-square table. This table is set up so that probability (P) values extend across the top and degrees-of-freedom ($df = N - 1$) values are down the left margin. The number of degrees of freedom in tests of genetic ratios is generally *one less than the number of terms* in the ratio being analyzed. Thus, tests of such ratios as 1:1 or 3:1 have one degree of freedom, a test of a 1:2:1 ratio has two degrees of freedom, and a test of a 1:2:1:2:4:2:1:2:1 has eight degrees of freedom. The idea of degrees of freedom can be exemplified by the situation met by a small boy putting on his shoes. He has two shoes but only one degree of freedom; that is, once one shoe is filled by a foot, correctly or incorrectly, the other shoe is automatically committed to being correct or incorrect too. Similarly, in a two-term table, one value can be filled arbitrarily, but the other is then fixed by the fact that the total must add up to the precise number of observations made in the experiment, and the deviations in the two terms must compensate for each other. When there are four terms any three are usually free, but the fourth is fixed. Thus, when there are four terms, there are usually three degrees of freedom.

In our example, we have two terms in the ratio (3:1) and therefore have one degree of freedom when we interpret the chi-square table. Look at the one-degree-of-freedom row under the .05 probability column and you will find the value 3.841. This number represents the **maximum value** of chi square that you should be willing to accept and still consider the deviations observed as due to chance

TABLE E-4 Summary of the calculations of chi squares for the hypothetical cross given

Phenotype	Genotype	O	E	$(O-E)$	$(O-E)^2$	$(O-E)^2/E$
Tall	T ___	84	82.5	1.5	2.25	0.027
Dwarf	tt	26	27.5	−1.5	2.25	0.082
Total		110	110.0	0		0.109

TABLE E-5 Distribution of X^2

	Probability (P)													
n	.99	.98	.95	.90	.80	.70	.50	.30	.20	.10	.05	.02	.01	.001
1	.00016	.00063	.00393	.0158	.0642	.148	.455	1.074	1.642	2.706	3.841	5.412	6.635	10.827
2	.0201	.0404	.103	.211	.446	.713	1.386	2.408	3.219	4.605	5.991	7.824	9.210	13.815
3	.115	.185	.352	.584	1.005	1.424	2.366	3.665	4.642	6.251	7.815	9.837	11.345	16.268
4	.297	.429	.711	1.064	1.649	2.195	3.357	4.878	5.989	7.779	9.488	11.668	13.277	18.465
5	.554	.752	1.145	1.610	2.343	3.000	4.351	6.064	7.289	9.236	11.070	13.388	15.086	20.517
6	.872	1.134	1.635	2.204	3.070	3.828	5.348	7.231	8.558	10.645	12.592	15.033	16.812	22.457
7	1.239	1.564	2.167	2.833	3.822	4.671	6.346	8.383	9.803	12.017	14.067	16.622	18.475	24.322
8	1.646	2.032	2.733	3.490	4.594	5.527	7.344	9.524	11.030	13.362	15.507	18.168	20.090	26.125
9	2.088	2.532	3.325	4.168	5.380	6.393	8.343	10.656	12.242	14.684	16.919	19.679	21.666	27.877
10	2.558	3.059	3.940	4.865	6.179	7.267	9.342	11.781	13.442	15.987	18.307	21.161	23.209	29.588

Source: Table IV of Fisher and Yates, *Statistical Tables for Biological, Agricultural and Medical Research,* published by Longman Group, Ltd., London (previously published by Oliver & Boyd, Edinburgh), by permission of the authors and publisher.

alone. If you were willing to accept a *P* value representing 1 chance in 10, what value of chi square would you accept as maximum?

In our example, chi square was calculated to be 0.109. Looking across the one-degree-of-freedom line in Table E-5, we find that this value falls between the .70 ($\chi^2 = .148$) and the .80 ($\chi^2 = .0642$) columns. This means that the probability that the deviations we obtained from the expected values could be attributed to chance alone is 70–80%. That is, if we were to repeat the study 100 times, we would obtain deviations as large as those observed about 70% of the time (7 out of every 10 experiments). We can thus reasonably regard this deviation as simply a sampling, or chance, error.

REFERENCE

Sokal, R. R., and F. J. Rohlf. 1994. *Biometry.* 3d ed. W. H. Freeman.

Appendix F

Culturing *Drosophila*

A. LIFE CYCLE OF *DROSOPHILA*

The life cycle of the fruit fly has four distinct stages (Fig. F-1): egg, larva, pupa, and adult. At 25°C, a fresh culture of *Drosophila* will produce new adults in 9 or 10 days: about 5 days for the egg and larval stages and 4 days for the pupal stage. The adult flies may live for several weeks. *Drosophila* cultures should be exposed neither to high temperatures (30°C or more), because such exposure results in sterilization or death of the flies, nor to low temperatures (10°C or less), because it prolongs the life cycle (perhaps 57 days) and reduces viability.

1. Egg

The adult *Drosophila* female starts to deposit eggs on the second day after emergence from the pupa. Each egg is about 0.5 mm long, ovoid, and white. Extending from its anterior end are two thin stalks that expand into flattened, spoonlike terminal parts. Embryonic development of the egg takes about one day at 25°C.

2. Larva

The larva is white, segmented, and wormlike. It has black mouth parts (jaw hooks) in a narrowed head region but no eyes, so it is completely blind. The larva also lacks appendages and must literally eat and push its way through its environment. It breathes by trachea and has a pair of conspicuous spiracles (air pores) at both the anterior end and the posterior end of the body.

The larval stage in the *Drosophila* life cycle is one of rapid eating and growing. It consists of three subdivisions called **instars.** The first and second instars terminate in molts. Each molt consists of a complete shedding of the skin and mouth parts of the larva and is the mechanism by which the animal grows. The third instar terminates in pupation. Just before pupation, the animal ceases to feed, crawls to some relatively dry surface, and everts its anterior spiracles. At 25°C the larval stage takes about 4 days, at which time the third instar is about 4.5 mm long.

3. Pupa

The pupal stage is a reorganizational stage of the fly's life cycle in which most larval structures are destroyed and adult structures are developed from embryonic tissues called **anlagen** (imaginal discs). These embryonic tissues have been lying dormant in the animal since their differentiation in the egg stage. The animal pupates within the last larval

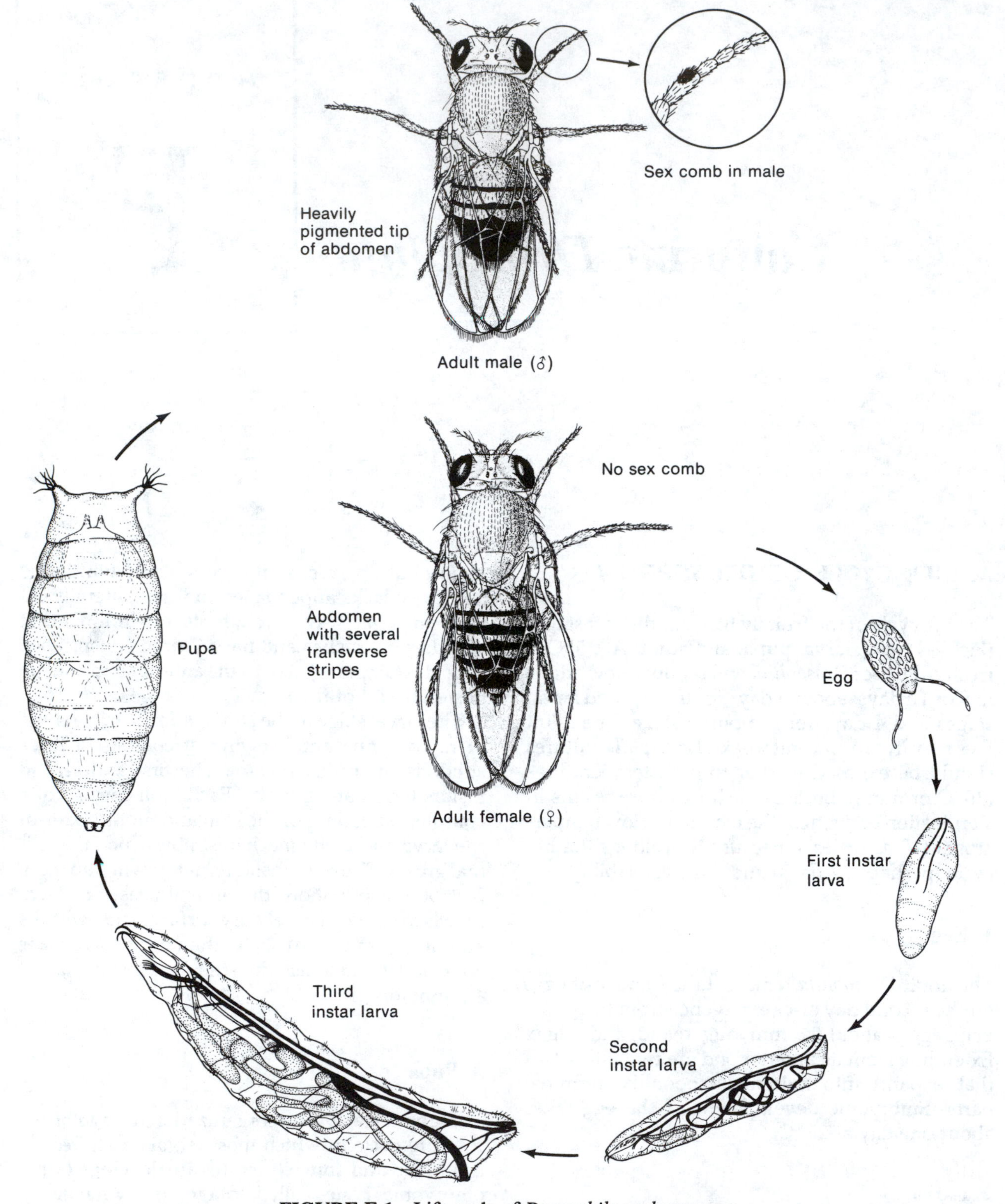

FIGURE F-1 Life cycle of *Drosophila melanogaster*

skin, which is at first soft and white but slowly hardens and darkens in color. The transformation that takes place in the pupa results in the development of the body form and structures of the adult **(imago).** At 25°C the pupal stage takes about 4 days, at which time the adult fly emerges from the puparium (pupal case).

4. Adult

The adult stage is the reproductive stage of the life cycle. The fly emerges from the puparium by forcing its way through the anterior end. At first the adult fly is greatly elongated, with its wings unexpanded. Within an hour, however, the wings expand, and the body gradually attains the more rounded form typical of an adult. The adult is also light in color when it first emerges, but within the first few hours it darkens to its characteristic adult color patterns.

Adult *Drosophila* mate about 6 hours after emerging from the puparium. The sperm are stored in the spermathecae and ventral receptacles of the female and are released gradually into the oviduct as eggs are produced and passed through the oviduct into the vagina. The female begins to deposit eggs about 2 days after it has emerged. It can deposit as many as 50 to 75 eggs per day for the first few days. Thereafter, its egg production decreases with time. The average lifespan of adult flies is 37 days at 25°C.

B. CULTURE MEDIUM

A number of media have been developed for the culture of *Drosophila.* However, the easiest to use are the "instant media" available from many biological supply houses. These concentrated media require only the addition of water to be immediately usable. No cooking is necessary.

Pour the medium into clean bottles. These bottles can be any size, but 120-ml wide-mouthed bottles or half-pint bottles are satisfactory. (It is best to transfer the mixture to a beaker for pouring. None of the medium should come into contact with the neck of the bottle.) Fill the bottles with medium to a depth of about 2.5 cm. Then place a piece of nonadsorbent paper such as brown wrapping paper in each bottle so that it extends down into the medium (a double piece of paper is more satisfactory). The paper should be about 2.5 cm wide and should extend upward to a point about 12 mm below the neck of the bottle. The paper provides a dry place on which the larvae can pupate. Stopper the bottles with cotton plugs (such as those used in bacteriology) or with disposable foam plugs, which are more convenient. Sterilize the bottles in an autoclave at 20 pounds pressure for 20 minutes. (At the same time, sterilize other materials that you may want to use in handling flies.) Before placing the flies in the cooled bottles of medium, shake a small amount of dry yeast on the medium. The yeast will grow and serve as food for the developing fly larvae.

C. ETHERIZATION OF FLIES

It is necessary to etherize flies to keep them inactive while they are being examined and when they are being transferred into culture bottles for matings. The etherizing bottle can be any small container that has an opening the same size as that of the culture bottle and a tight-fitting cork. Into the bottom of the cork, tack a pad of cotton. Just before the following procedure (Fig. F-2), douse the cotton with a few dropperfuls of ether.

1. Shake the flies down into the bottom of the culture bottle by tapping the bottle on a rubber pad on a desk.

2. Remove the cotton plug from the culture bottle and, in its place, *quickly* insert the empty etherizing bottle. Make certain you have blown any residual fumes from the etherizing bottle before you invert it over the culture bottle.

3. Reverse the position of the two bottles so that the etherizer is now on the bottom.

4. Hold the two bottles firmly together and shake the flies from the culture bottle into the etherizer. This is best done by tapping the sides of the culture bottle horizontally so that the flies are dislodged and fall into the etherizer. Make certain you do not shake the food loose in the culture bottle.

5. After you have shaken the flies into the etherizer, remove the culture bottle and quickly replug it. Simultaneously plug the etherizer with its cork. The cotton on the cork should feel *moist* with ether but not wet.

6. Observe the flies in the etherizer. Thirty seconds after they have stopped walking pour them out onto your counting plate. If the flies have been overetherized (killed), their wings will be extended at right angles to their bodies.

7. Normally, the flies will remain etherized for 5

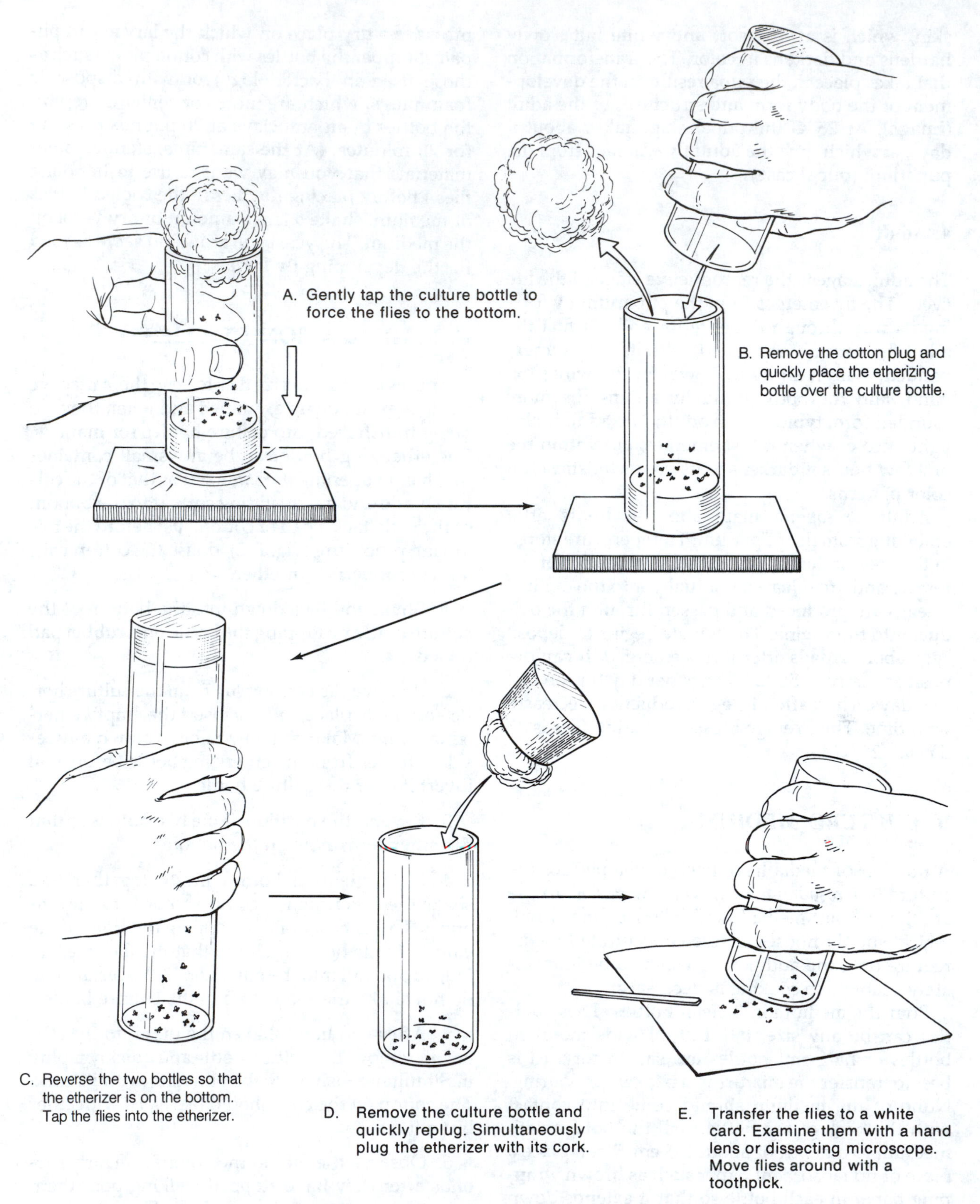

FIGURE F-2 Procedure for etherizing flies

to 10 minutes. If they begin to awaken on the plate, they can be reetherized as follows: on the inner surface of a petri dish, tape a piece of paper toweling. Moisten the paper toweling with ether, and place the petri dish over the flies to form an etherizing chamber.

D. DISTINGUISHING THE SEX

Examination of the external genitalia under magnification is the best means of distinguishing the sex of flies. Only male flies exhibit darkly colored external genitalia, which are visible on the ventral side of the tip of the abdomen. The following characteristics may also be helpful in distinguishing males from females.

1. Size: Females are usually larger than males.

2. Shape: In a dorsal view, the abdomen of the male is round and blunt, whereas that of the female is sharp and protruding. The abdomen of the male is relatively narrow and cylindrical, whereas that of the female is distended and appears spherical or ovate.

3. Color: Black pigment is more extensive on the abdomen of the male than on that of the female. On the male, the markings extend completely around the abdomen and meet on the ventral side. On the female, the pigment is present only in the dorsal region.

4. Sex Combs: Only males have a small tuft of black bristles, called a **sex comb,** on the anterior margin at the basal tarsal joint of each front leg. Magnification is necessary to see the sex combs.

E. ISOLATING VIRGINS

Females of *D. melanogaster* can store and use sperm from one insemination for a large part of their reproductive lives. As a result, only virgin females should be used in making crosses. Females of this species can mate 6 hours after they have emerged from the puparium. Therefore, a procedure must be followed that will ensure the collection of females that are no more than 6 hours old. A number of possible procedures are available to you, depending on your preference and the accessibility of the laboratory.

If the laboratory is usually available to students:

1. From the culture bottle, shake out and discard into the morgue (a bottle containing ethanol, oil, or detergent) all adult flies (10 or more days after the introduction of the parental flies). This step should be done early in the morning (8:00 – 10:00 A.M.) for most efficient results.

2. Return to the laboratory within 4 – 6 hours, and examine the newly hatched flies. The females in this group may be presumed to be virgin and can be used in experimental matings.

If the laboratory is not usually available:

1. Collect darkened pupae from the paper or sides of the bottle with a fine camel-hair brush.

2. Turn each pupa so that its legs are visible through the pupal case. Examine under high power the uppermost or proximal joint of the tarsi of the front legs. The presence of sex combs indicates a male.

3. Collect the pupae that are *without* sex combs, hence female, and place them individually in fresh culture bottles. After the flies have hatched from their pupal cases, examine them to be certain that no male pupae have accidentally been placed in a culture. All the flies that are female can be used in experimental matings.

REFERENCES

Demerec, M., and B. P. Kaufman. 1969. *Drosophila Guide: Introduction to the Genetics and Cytology of Drosophila.* 8th ed. Carnegie Institution.

Flagg, R. O. 1971. *Drosophila Manual.* Carolina Biological Supply Co.

Appendix

G

Aseptic Techniques

The term **aseptic technique** refers to a procedure that enables a microbiologist to keep the organisms used in an experiment separate from the millions of other microorganisms in the environment. After the working materials are sterilized, aseptic techniques become a matter of simple technical procedures designed to prevent contamination. These procedures consist of ways of transferring organisms from test tube to test tube, from test tube to flask or petri dish, and from petri dish to petri dish or flask.

A. PROCEDURES FOR TRANSFERRING BACTERIA

When a nutrient medium is being prepared in a test tube or flask, the mouth of the container should be plugged with long-fiber cotton, preformed plastic plugs, or plastic or metal caps before sterilization. To inoculate the tube or flask with the desired organism, use the following procedure (refer to Fig. G-1). Use an inoculating loop for liquid cultures and a loop or needle for cultures growing on solid media.

1. Hold the test tube (or flask) containing the organisms to be transferred and the tube to which they are to be transferred as shown in Fig. G-1A.

2. Sterilize the inoculating loop (or needle), holding it like a pencil, by heating its wire in the flame of an alcohol lamp or bunsen burner until it glows red.

3. Remove the cotton plugs or other closures by grasping them between the fingers and pulling, and flame the mouths of the test tubes briefly to eliminate loose cotton fibers and dust.

4. Cool the loop by touching it to the inside of the culture test tube (or to the agar). Pick up the organisms to be transferred by dipping the loop into the broth (or touching it lightly to the growth).

5. Transfer the culture to the new test tube by dipping the loop gently into the broth (or drawing it across the surface of the agar).

6. Pass the mouths of the containers rapidly through the flame again, reinsert the cotton plugs to prevent the entrance of other microbes, and kill the organisms remaining on the loop (or needle) by heating the wire to redness.

Caution: *To avoid potentially dangerous spatter, heat the wire gradually from the holder toward the tip.*

To transfer bacteria from one petri dish to another, use the following procedure (Fig. G-2).

1. Sterilize the transfer needle or loop.

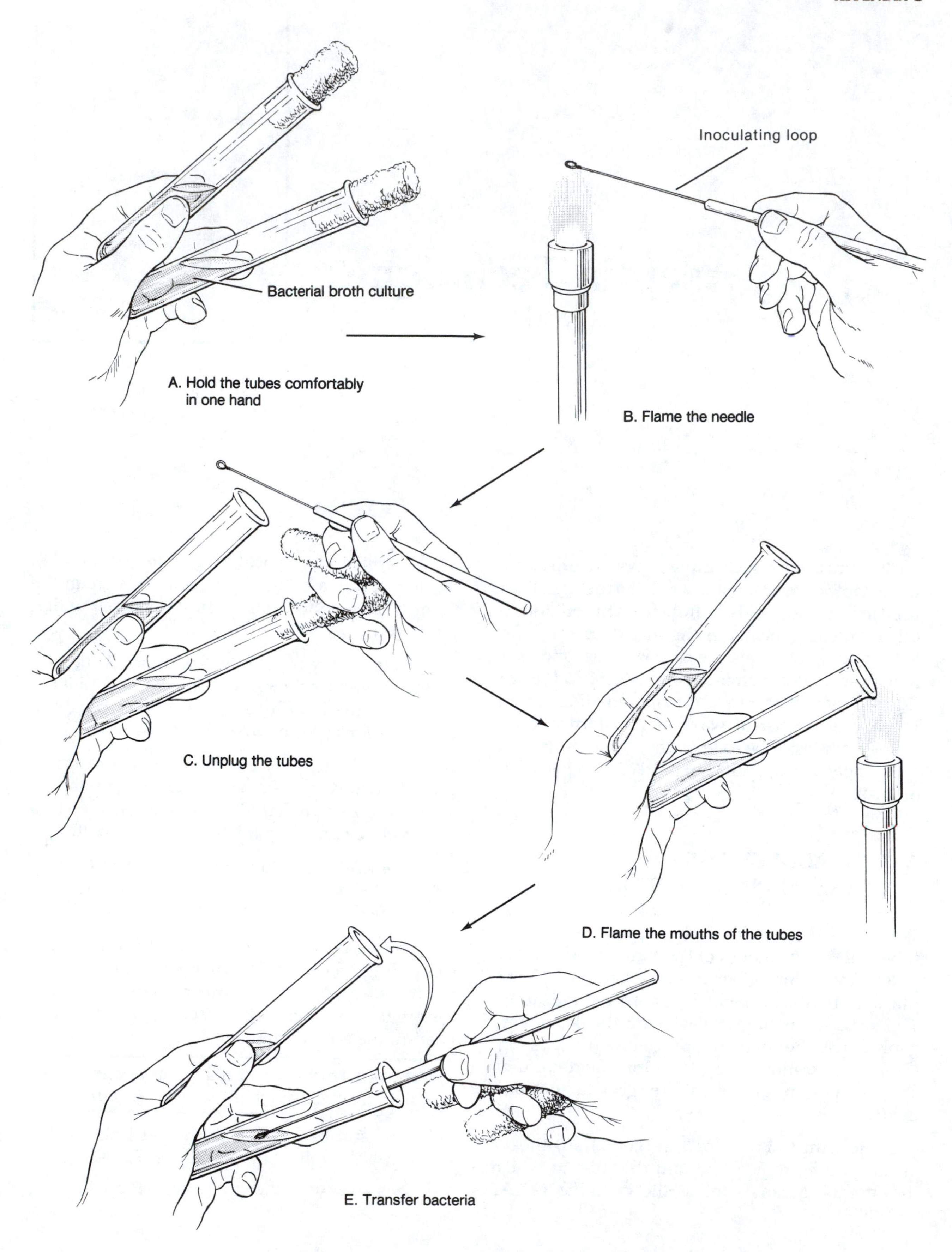

FIGURE G-1 Procedure for transferring bacteria

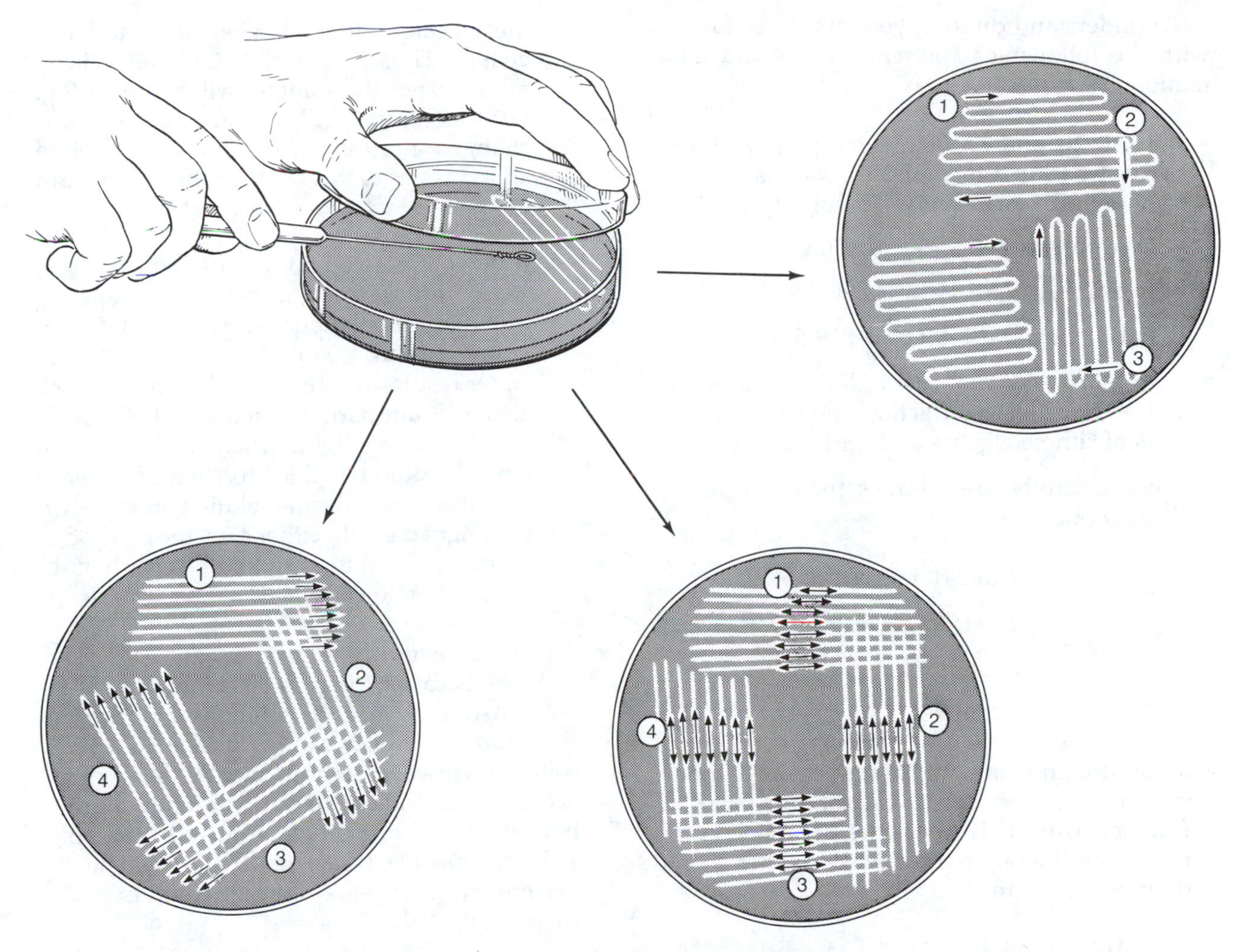

FIGURE G-2 Procedure for plating bacteria

2. Slightly lift the lid of the culture-containing dish.

3. Touch the hot needle or loop to the agar to cool it and then touch the bacterial growth on the agar.

4. Close the lid of the culture dish and slightly open the lid of the petri dish to which you are transferring a culture.

5. Streak the needle gently across the surface of the agar, using one of the streaking patterns shown in Fig. G-2. The objective of streaking is to isolate individual colonies at the completion of the third and fourth streak on the plate.

6. Close the lid and sterilize the needle.

B. DILUTION TECHNIQUES AND CALCULATIONS

Dilution techniques are among the more useful laboratory procedures. They provide simple and accurate methods for (1) changing the concentration of a solution, (2) indirectly "weighing" a solute whose weight is well below the usual limits of analytical balances, and (3) determining the quantity of bacteria in a culture. For example, suppose you want to prepare a solution having a solute concentration of 0.001 mg/ml. Because 0.001 mg is 1/100 of 0.1 mg, dissolving 0.1 mg in 100 ml would give the required concentration. By extrapolation it is easy to see how any amount of solute, regardless how small, could indirectly be weighed in this manner.

To understand dilution you should be familiar with the following basic terminology and information.

- 1:10, 1:20, and 1:100 mean 1 part in a total of 10, 1 part in a total of 20, and 1 part in a total of 100, respectively. Therefore,

  1:10 means 1 + 9 or 0.1 + 0.9

  1:20 means 1 + 19 or 0.1 + 1.9

  1:100 means 1 + 99 or 0.1 + 9.9

- $1:10 = 1/10 = 1:1 \times 10^1$; $1:100 = 1/100 = 1:1 \times 10^2$. This is a fraction and obeys the laws of simple algebra and arithmetic.
- Any unit can be used if the same unit is used throughout:

  1 ml:10 ml

  1 gal:10 gal

  0.1 ml + 0.9 ml

  1 g + 9 g

- In dilution, the amount present in the original suspension is reduced by the dilution factor or fraction. Thus, if 100 particles per milliliter are present in the original suspension, after a 1:10 dilution there are

$$\frac{100 \text{ particles}}{\text{ml}} \times \frac{1}{10} = \frac{10 \text{ particles}}{\text{ml}}$$

- There are three methods of preparing a 1:10 dilution.

  a. In the **weight-to-weight** (w:w) method, 1.0 g of solute is dissolved in 9.0 g of solvent, giving 10 total parts by weight, one part of which is solute.

  b. In the **weight-to-volume** (w:v) method, enough solvent is added to 1.0 g of solute to make a total volume of 10 ml. If a graduated cylinder is used, first place the dry solute in the cylinder, then fill the cylinder to exactly 10.0 ml. In this method, one part (by weight) is dispersed in 10 total parts (by volume).

  Most biological solutions used in the laboratory are very dilute. Therefore, the accuracy of most work is not affected if a previously weighed solute is dissolved in the desired volume of solvent because most dilute solutions do not appreciably change in volume after adding small quantities of solute. Of course, the weight of the solution will change. Thus, if 1 g of NaCl is dissolved in 10 ml of water, the solution will *weigh* 11.0 g, but the volume will remain essentially unchanged at 10 ml. If the amount of solute added is large, this does not hold true, and the volume increases measurably.

  c. If the solute is a liquid, a 1:10 solution is prepared **volume to volume** (v:v). In preparing a 1:10 dilution of pure ethyl alcohol, for example, adding 1.0 ml of ethanol to 9.0 ml of water results in a 10-part solution, of which alcohol is one part. Alternatively, because alcohol is a little lighter than water, 1.0 g of ethanol could be added to 9.0 g of water to obtain almost the same solution on a weight-to-weight basis. In either case, the method of preparing a solution should be clearly indicated to avoid confusion.

- *Percent concentration.* Percent means *parts per hundred.* Because 1 part in 10 is the same as 10 parts in 100, each of the solutions described in the dilution methods can be considered a 10% solution. However, each is slightly different in actual concentration, so clarity is assured only if solutions are properly labeled—for example, 10% (w:w), 10% (w:v), or 10% (v:v). The percent-by-weight-to-volume method is frequently used.
- The concentration of a solution is given as the amount of solute per volume, for example,

  particles/ml

  g/liter

  mg/ml

- Some calculations:

  a. To determine the total dilution of the following three consecutive dilutions of an initial solution

  1:10 (1 + 9)
  1:10 (1 + 9)
  1:5 (1 + 4)

  simply multiply each factor:

$$\frac{1}{10} \times \frac{1}{10} \times \frac{1}{5}$$

  Thus, the total dilution is $1:500 = 5 \times 10^{-2} = 1:5 \times 10^2$.

b. To obtain a dilution of 1:2000, factor the denominator (2000):

$$10 \times 10 \times 10 \times 2 = 2000$$

or

$$10 \times 10 \times 20 = 2000$$

Then combine the following dilutions:

$$1{:}10,\ 1{:}10,\ 1{:}10,\ 1{:}2 = 1{:}2000$$

or

$$1{:}10,\ 1{:}10,\ 1{:}20 = 1{:}2000$$

Note: Dilutions can be made of any multiple of the factors of the denominator of the total dilution fraction. This denominator is called the **dilution factor (df)** and equals the reciprocal of dilution: $1/10$ = d.f. of 10.

c. To calculate the original amount after determining the amount in dilution, multiply the amount in dilution by the dilution factor. For example, suppose that, after plating out 1 ml of a 1:2000 dilution of an original culture containing x cells per milliliter, you find that 300 colonies develop. How many bacteria were originally in the culture?

$$\frac{300 \text{ bacteria}}{1 \text{ ml plated}} \times 2000 = 600{,}000 = 6.0 \times 10^5 \text{ bacteria/ml}$$

in the original suspension.

d. Suppose you find 300 colonies on a plate. The volume you plated was 0.2 ml (0.2 ml = $1/5$ ml). The dilution was 1:5000. Determine the amount of bacteria in the original suspension.

$$300 \text{ bacteria} \times 5000 \times 5 = 7{,}500{,}000 \text{ bacteria/ml} = 7.5 \times 10^6 \text{ bacteria/ml}$$

in the original suspension, or

$$\frac{300 \text{ bacteria}}{0.2 \text{ ml}} \times 5000 = 7.5 \times 10^6 \text{ bacteria/ml}$$

REFERENCES

Atlas, R. M., and A. E. Brown. 1984. *Experimental Microbiology.* Macmillan.

Benson, H. J. 1985. *Microbiological Applications.* 4th ed. Wm C Brown.

Collins, C. H., and P. M. Lyne. 1984. *Microbiological Methods.* 5th ed. Butterworth.

Appendix

H

Glossary of Common Anatomical Terms

The following glossary consists of terms commonly used in descriptions of vertebrate anatomy.

acetabulum (Latin *acetabulum,* vinegar cup; from *acetum,* vinegar): Cup-shaped socket in the pelvic girdle that receives the head of the femur.

adrenal gland: Endocrine gland located cranial to the kidney (mammals) or on the ventral surface of the kidney (frogs). Its hormones help the body adjust to stress and help regulate sexual development and the metabolism of salts, minerals, carbohydrates, and proteins. Often called the suprarenal gland.

allantois (Greek *allas,* sausage): Extraembryonic membrane in reptiles, birds, and mammals that develops as an outgrowth of the urinary bladder. It accumulates waste products and functions as a gas exchange organ in reptiles and birds; in mammals, it is part of the fetal placenta.

alveoli (Latin *alveolus,* small cavity): Small, thin-walled vascular sacs at the ends of the mammalian respiratory tree. Gas exchange occurs in these structures.

anus (Latin *anus,* anus): The terminal opening of the digestive tract in vertebrates.

aorta (Greek *aorte,* to lift up): The major artery that carries blood from the heart to the various parts of the body. It is also called the *dorsal aorta* to distinguish it from the ventral aorta of fish, which carries blood to the gills.

aortic valve: Valve, consisting of three semilunar-shaped folds at the base of the aorta, that prevents blood from flowing back into the left ventricle.

arrector pili (Latin *arrectus,* upright; *pilus,* hair): Small muscles in the skin at the base of hair follicles that raise the hairs; the resulting small bumps on the skin surface are known as "goose flesh."

atlas (Greek mythology *Atlas*): First cervical vertebra, which supports the skull at the top of the vertebral column.

atrioventricular valve (Latin *atrium,* hall; *ventriculus,* little belly): Valves between the atria and ventricles of the heart, which prevent the backflow of blood. In mammals, the right valve has three flaps (thus, it is also called the tricuspid valve); the left one has two flaps (and is sometimes called the bicuspid or mitral valve).

atria (Latin *atrium,* hall): Two chambers of the heart that receive blood from the body (right atrium) and the lungs (left atrium).

axis (Latin *axis,* hub, axle): Second cervical vertebra. Rotation of the head occurs between the atlas and the axis.

Bowman's capsule: Thin-walled, expanded

proximal end of a kidney tubule, which surrounds the glomerulus.

bronchus (Greek *bronkhos,* windpipe): Major branch of the trachea, which leads to a lung.

caecum or **cecum** (Latin *caecus,* blind): First part of the large intestine, which forms a dilated pouch into which open the ileum, colon, and appendix. In many herbivores, including the rat, it is very long and often contains bacteria that digest cellulose.

canaliculi (Latin *canaliculus,* little canal): Microscopic canals in the bone matrix that enable the cell processes of bone cells to communicate with each other.

central canal: Cavity in the middle of the spinal cord. It connects with the fourth ventricle of the medulla oblongata and is filled with cerebrospinal fluid.

cervical (Latin *cervix,* neck): Pertaining to the neck, e.g., cervical vertebrae.

cervix: Neck of an organ, such as the cervix of the uterus.

chromatophore (Greek *chroma,* color; *phoros,* to bear): Cell in the skin of vertebrates that contains pigment granules.

cloaca (Latin *cloaca,* sewer): Exit chamber of the digestive system in lower vertebrates, which may also serve as the exit for the reproductive and urinary systems.

coelom (Greek *koilos,* hollow): Body cavity in vertebrates and many invertebrates that is completely lined by a simple, squamous epithelium of mesodermal origin.

collagen (Greek *kolla,* glue; Latin *genere,* to beget): Fibrous protein that forms most of the intercellular material in cartilage, tendons, and other connective tissues.

conus arteriosus: Chamber of the heart of lower vertebrates into which the ventricle empties.

coronary vessels (Latin *corona,* garland, crown): Blood vessels that encircle the heart and supply and drain the cardiac muscles.

corpus luteum (Latin *corpus,* body; *luteus,* yellow): Yellowish endocrine gland in the ovary that develops from the follicle after ovulation. It secretes estrogens and progesterone, which maintain the uterus during pregnancy.

cutaneous (Latin *cutis,* skin): Pertaining to the skin, e.g., cutaneous nerve.

dendrite (Greek *dendron,* tree): Filamentous process, usually branched, that carries nerve impulses toward the nerve cell body of a neuron.

diaphragm (Greek *dia,* across; *phragma,* partition): Complex of muscles and tendons that forms the partition between the thoracic and abdominal cavities in mammals.

ductus arteriosus: Fetal blood vessel connecting the pulmonary artery directly to the descending aorta, which permits much of the blood to bypass the lungs. It normally closes and atrophies after birth.

epidermis (Greek *epi-,* on or over; *derma,* skin): Outermost layers of cells in plants and animals.

epithelium (Greek *epi-; thele,* nipple): Tissue composed of cells that covers all body surfaces and lines all cavities including the lumen of blood vessels and ducts. Secretory cells of glands originate from epithelial layers during embryonic development.

foramen (Latin *foramen,* opening): Small opening in the skull or other organs.

foramen magnum (Latin *magnus,* great, large): Opening in the base of the skull through which the spinal cord passes.

foramen ovale (Latin *ovalis,* oval): Opening in the septum between the atria of a fetal mammalian heart. It permits much of the blood in the right atrium to enter the left atrium and thus bypass the lung. It closes at birth and becomes the fossa ovalis in the adult

fossa ovalis (Latin *fossa,* trench): Oval depression in the median wall of the right atrium of an adult mammal. It is a vestige of the foramen ovale found in the fetus.

glomerulus (Latin *glomeris,* ball): Ball-like cluster of capillaries found in the Bowman's capsule at the head of a nephron in vertebrate kidneys.

hepatic (Greek *hepar,* liver): Pertaining to the liver.

hepatic duct: The duct that carries bile from the liver. Hepatic ducts usually join the cystic duct to form the common bile duct.

hypophysis (Greek *hypo-,* under; *physis,* growth): Gland, often called the pituitary, that is attached to the underside of the hypothalamus. It produces a variety of hormones regulating growth, metabolism, sexual activity, and water balance.

hypothalamus (Greek *hypo-; thalamos,* inner chamber): Small region of the brain that lies just below the thalamus. It is an important center for the control of visceral activity and the regulation of the hypophysis.

insertion (Latin *in,* into; *serere,* to join): Attachment of a muscle to a bone or other structure that moves the most when the muscle contracts.

intestine (Latin *intestinus,* internal): Primary digestive and absorptive parts of the digestive tract. Located between the stomach and the cloaca or anus, it is divided into the small and large intestines.

islets of Langerhans: Patches of endocrine tissue in the pancreas that secrete hormones (insulin and glucagon) essential for regulation of blood glucose levels.

jejunoileum (Latin *jejunus,* empty; *ileum,* groin, flank): That part of the small intestine beyond the jejunum.

jejunum (Latin *jejunus,* empty): That part of the small intestine after the duodenum.

kidney (Middle English, *kidnenei,* kidney): Organ that removes nitrogenous waste products of metabolism from the blood and produces urine.

lacunae (Latin *lacuna,* plural *lacunae,* pool): Small cavities in the bone matrix that contain the bone cells (osteocytes).

ligament (Latin *ligamentum,* bone): Band of fibrous connective tissue that connects bones or cartilage. It supports and strengthens joints.

liver (Old English *lifer,* liver): Large gland located in the upper part of the abdominal cavity that secretes bile and metabolizes carbohydrates, proteins, and fats. It also synthesizes many plasma proteins, degrades toxins, and removes damaged red blood cells.

lymph nodes: Small oval structures associated with the lymphatic vessels of higher vertebrates, in which lymphocytes are produced, foreign particles phagocytosed, and some immune responses initiated.

mediastinum (Latin *mediastinus,* median): Mass of tissues and organs separating the lungs from other organs in the thoracic cavity. It contains the aorta, esophagus, pericardial cavity and heart, thymus, and vena cava.

melanin (Greek *melas,* black): Black or brown pigment in the skin. It is contained in the melanocytes in frogs and other lower vertebrates.

mesenteries (Latin *mesos,* middle; Greek *enteron,* gut): Double layers of mesoderm that suspend the digestive tract and other internal organs within the coelom.

mucosa (Latin *mucus,* mucus): Lining of the digestive and respiratory tracts that contains cells that secrete mucus.

muscle fiber (Latin *musculus,* little mouse, because of the mouselike shape of some muscles): Elongated muscle cell.

myofibrils (Greek *myos,* mouse; *fibrilla,* small fiber): Microscopic, fine longitudinal fibrils within a muscle fiber that act as the contractile elements. In striated muscle, they bear cross striations.

myofilaments (Latin *filum,* thread): Ultramicroscopic filaments of actin and myosin that are components of the myofibrils.

naris (Latin *naris,* nostril), pl. **nares:** External nostril.

nasal (Latin *nasus,* nose): Pertaining to the nose, e.g., nasal bone, nasal cavity.

nephron (Greek *nephros,* kidney): Functional unit of a vertebrate kidney, which includes the Bowman's capsule, glomerulus, and proximal and distal tubules.

nictitating membrane (Latin *nictare,* to wink): Membrane in the median corner of the eye of many terrestrial vertebrates that slides across the surface of the eyeball. In human beings, it consists of only a vestigial semilunar fold.

ocular (Latin *oculus,* eye): Pertaining to the eye, e.g., extrinsic ocular muscles.

omentum (Latin *omentum,* membrane): One of the two mesenteries that attach to the stomach. The greater omentum is a saclike fold between the body wall and the stomach; the lesser omentum extends from the liver to the stomach and duodenum.

optic (Greek *optikos,* sight): Pertaining to the eye.

oral cavity (Latin *os,* mouth): Mouth cavity; also known as the buccal (cheek) cavity.

origin of a muscle (Latin *origin,* beginning): Attachment of a muscle to a bone or other structure that moves the least when the muscle contracts.

ostium (Latin *ostium,* opening): Opening into a tubular organ (e.g., ostium of the uterine tube) or between two distinct cavities of the body.

ovarian follicle (Latin *ovum,* egg; *folliculus,* little bag): Group of cells in the ovary that envelop the developing egg. It is also an endocrine gland that produces estrogen.

ovulation (Latin *ovulum,* little egg): Release of egg(s) from the ovarian follicle(s) and ovary into the coelom, from which they enter the oviduct or uterine tube.

pancreas (Greek *pan,* all; *kreas,* flesh): Elongated gland attached to the duodenum that secretes enzymes and precursors of enzymes that act on all categories of food. It also contains the islets of Langerhans that produce insulin and glucagon.

parathyroid gland (Greek *para,* beside): One of several endocrine glands embedded on the surface of the thyroid gland. It secretes a hormone essential for calcium and phosphorus metabolism.

peritoneum (Greek *peri-,* around; *tonus,* stretched over): Membrane that lines the body cavity and forms the external covering of the visceral organs.

pharynx (Greek *pharynx,* pharynx): Part of the digestive tract that lies between the mouth cavity and the esophagus; the throat.

pituitary gland: See **hypophysis.**

placenta (Greek *plakoeis,* a flat object): Mammalian organ that connects a mother and her fetus and through which food, gases, and waste products are passed.

pleura (Greek *pleura,* side, rib): Epithelial membranes that cover the lungs (visceral pleura) and line the pleural cavities (parietal pleura).

polar body (Latin *polaris,* pole): Small cell located near the animal pole of a developing egg cell. Polar bodies result from an unequal division of the cytoplasm during the first and second meiotic divisions.

portal vein (Latin *portare,* to carry): Vein that carries blood from one organ to another rather than to the heart, e.g., hepatic portal vein.

pulmonary (Latin *pulmo,* lung): Pertaining to the lung.

pylorus (Greek *pylorus,* gate; *ourus,* guard): Distal opening of the stomach, which is surrounded by a strong band of tissue that closes the opening between the stomach and the duodenum.

rectum (Latin *rectus,* straight): The caudal part of the large intestine.

renal (Latin *ren,* kidney): Pertaining to the kidney, e.g., renal artery.

saliva (Latin *saliva,* saliva): Mucous secretions of several large glands that discharge into the mouth cavity. In mammals, it contains salivary amylase (ptyalin), which initiates the chemical breakdown of starch.

sarcolemma (Greek *sarkos,* flesh; *lemma,* peel): Thin covering of a muscle fiber.

sarcoplasm (Greek *sarkos; plasma,* form): Cytoplasm of a muscle fiber.

serosa (Latin *serum,* watery fluid): Epithelial and connective tissue membranes that line body cavities and cover visceral organs (e.g., peritoneum, pericardium, and pleura).

somatic (Greek *soma,* body): Pertaining to the body wall rather than the internal organs, e.g., somatic muscles.

sternum (Greek *sternon,* chest): Bone on the midventral surface of the chest; breastbone. Costal cartilages attach it to the ribs in mammals.

submucosa (Latin *sub-,* under; *mucus,* mucus): Layer of vascular connective tissue in the wall of the digestive or respiratory tract that lies beneath the mucosa.

tendon (Latin *tendere,* to stretch): Band of connective tissue that attaches muscles to bones or other muscles.

tendon of Achilles (Achilles, the hero of Homer's *Iliad,* was said to be invulnerable except for this tendon): Tendon that extends from the large muscle mass on the caudal surface of the leg to the calcaneus (heel) bone of the foot.

thalamus (Greek *thalamos,* inner chamber): One of two masses of gray matter located at the sides of the third ventricle of the brain. It is an important center for the sensory impulses traveling to the cerebrum.

thymus (Greek *thymos,* thymus): Lymphoid organ located in the ventral part of the thoracic cavity. It is well developed in the fetus, in which it participates in the maturation of bone marrow stem cells into immunologically competent T lymphocytes.

thyroid gland (Greek *thyreos,* shield): Bilobed endocrine gland located near the cranial end of the trachea and over the thyroid cartilage of the larynx in human beings (hence its name). It is the source of the hormone thyroxine, which regulates the general level of metabolism.

truncus arteriosus: One of two arterial trunks in a frog that lead from the front of the heart to arterial arches supplying the lungs and skin, the head, and the main part of the body.

umbilical (Latin *umbilicus,* navel): Pertaining to the navel, e.g., umbilical cord, umbilical artery.

urostyle (Greek *oura,* tail; *stylos,* pillar): spike-like caudal part of the frog's vertebral column. It consists of two fused caudal vertebrae and serves as the origin for certain muscles used in jumping.

vena cava (Latin *vena,* vein; *cavea,* hollow): One or more large veins in vertebrates that return blood to the heart from the body. It enters the sinus venosus (frogs) or the right atrium (mammals) of the heart.

ventricle (Latin *ventriculus,* stomach): (1) Muscular chamber of the heart that receives blood from an atrium and pumps blood out of the heart to either the lungs or the body tissues. Frogs have a single ventricle. (2) One of the large chambers within the brain.

villi (Latin *villus,* shaggy hair): Microscopic projections of the mucosa of the small intestine that increase its surface area.

white matter: Whitish material of the brain and

spinal cord. It is composed of the myelinated processes of neurons.

zygomatic arch (Greek *zygon,* yolk): Bony arch beneath the orbit (eye) of the mammalian skull, which joins the facial and cranial regions of the skull; the cheekbone.

REFERENCE

Nomina Anatomica. 1983. 5th ed. Williams and Wilkins.

Appendix I

Use of Live Animals in the Laboratory

A. GENERAL PROCEDURES

If you are using living vertebrate animals in your class, you must keep two principles in mind.

- There must be a genuine reason for using live animals. Be sure you understand why they are needed.
- Live animals must always be treated humanely: never cause them unnecessary irritation or injury. Accordingly, if organ or tissue damage can result from legitimate experimentation, you must first put the animal under an anesthetic or treat the nervous system so that the animal will feel no pain.

Special techniques are also required.

- Avoid injuring the animal's tissues or making the animal bleed; such damage makes the animal less capable of "normal" reactions.
- When handling organs or tissues that have been exposed or removed, use a glass hook or a small camel-hair brush moistened with Ringer's saline solution. Never handle living tissues or organs with your fingers.
- When you need to lay an excised organ or part down, never place it on the table or on dry paper; place it in a watch glass or other dish and keep it moistened. Apply Ringer's solution as often as necessary to keep the tissue from drying out.

B. DECEREBRATION PROCEDURE

A **decerebrate frog** is one in which the forebrain is no longer functional.

1. Select a large, active frog, and grasp it firmly in the left hand, pinioning the forelimbs and hindlimbs.

2. Position the blades of a pair of sharp scissors, as shown in Fig. I-1, just behind the posterior margins of the eyes. Cut quickly and cleanly to avoid excessive nerve shock.

3. Stop any blood flow with cotton or gauze pads. Lay the animal ventral side down on a damp paper towel. Be sure to moisten the skin from time to time, because the bulk of the respiration of such animals takes place through the skin.

C. PITHING PROCEDURE

A **spinal frog** (one whose entire brain has been destroyed) is prepared by a procedure called **pithing.**

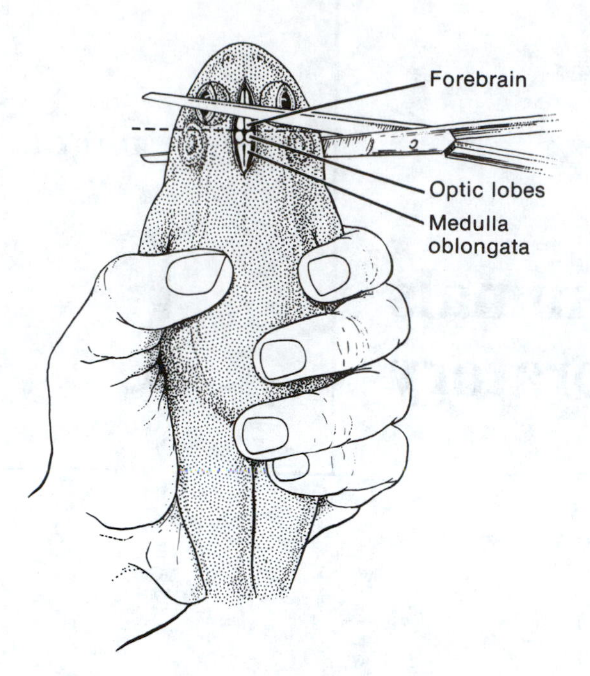

FIGURE I-1 Decerebrating a frog

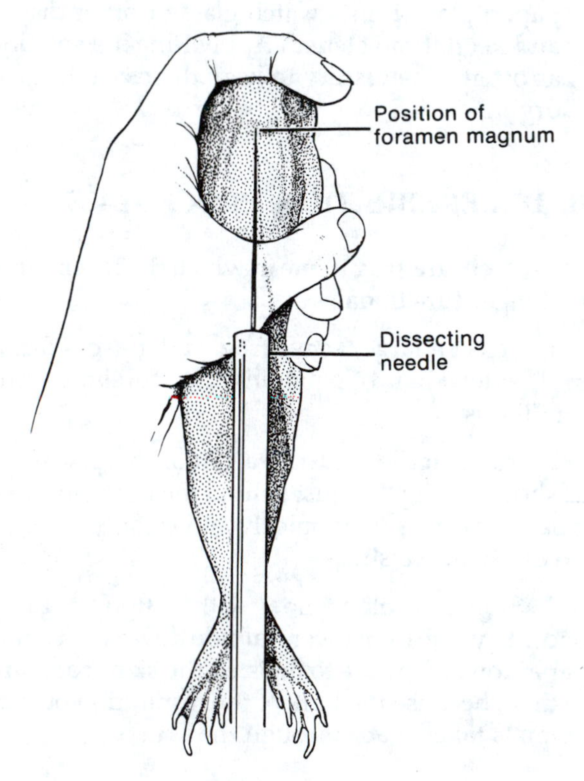

FIGURE I-2 Pithing a frog

1. Grasp the animal as indicated in Fig. I-2, using the thumb and fingers to secure the limbs. With the index finger, depress the snout so that the head is at a sharp angle to the body.

2. Run the tip of the dissecting needle down the midline of the head. At a point 2–3 mm behind the posterior border of the eardrums, the needle should dip into a small groove. This groove marks the location of the **foramen magnum,** a large opening in the skull through which the spinal cord emerges from the **cranium.**

3. Place the point of a dissecting needle in this groove, and with a sharp movement force it through the skin and foramen magnum into the brain. Twist and turn the needle to destroy the brain (Fig. 37-3 A–D).

4. Halt any bleeding, and treat the pithed frog in the same way as the decerebrated animal.